Geometrical and Physical OPTICS

Geometrical and Physical OPTICS

Dr P K Chakrabarti

Dept of Physics, St. Xavier's College, Kolkata

New Central Book Agency (P) Ltd

LONDON

HYDERABAD ERNAKULAM BHUBANESWAR

NEW DELHI KOLKATA PUNE GUWAHATI

NCBA

REGD OFFICE
8/1 Chintamoni Das Lane, Kolkata 700 009, India
email: ncbapvtltd@eth.net

OVERSEAS
NCBA (UK) Ltd, 149 Park Avenue North, Northampton, NN3 2HY, UK
email: ncbauk@yahoo.com

EXPORT
NCBA Exports Pvt. Ltd, 212 Shahpur Jat, New Delhi 110 049
email: ncbaexports@ncbapvtltd.com

BRANCHES

2/27 Ansari Road, 1st Floor
Daryaganj, **New Delhi** 110 002
email: ncbadel@ncbapvtltd.com

Shop Nos. 3 & 4, Vinayak Towers
681/B Budhwar Peth, Appa Balwant Chowk
Pune 411 002
email: ncbapune@ncbapvtltd.com

Shop No. 15, Bharati Towers, Ground Floor
Forest Park
Bhubaneswar 751 001
email: ncbabub@ncbapvtltd.com

House No. 3-1-315, 1st Floor
Nimboliadda, Kachiguda, **Hyderabad** 500 027
email: ncbahydb@ncbapvtltd.com

GSS Shopping Complex, 1st Floor
Opposite College Post Office, Convent Road
Ernakulam 682 035
0484-4021054 / 53

Radhamayee Complex [Second Floor]
Opp. Axis Bank, Panbazar, Jaswanta Road
Guwahati 781 001
email: ncbaguwahati@ncbapvtltd.com

GEOMETRICAL AND PHYSICAL OPTICS ● Dr P K Chakrabarti

First Published: 1992
Reprinted: 1997, 2000, 2001
Second Edition: 2004
Third Edition: 2005
Reprinted: 2007, 2009, 2012, 2015
Thoroughly Revised Fourth Edition: 2019

PUBLISHER AND TYPESETTER
New Central Book Agency (P) Ltd
8/1 Chintamoni Das Lane, Kolkata 700 009

PRINTER
New Central Book Agency (P) Ltd
Web-Offset Division, Dhulagarh, Sankrail, Howrah

PROJECT SUPERVISORS
Abhijit Sen and Soumen Paul

PROJECT TEAM
Subir Pal and Prabhat Jas

ISBN: 978 81 7381 748 9

Price ₹ 585.00

Preface to the Fourth Edition

The book *Geometrical and Physical Optics* has undergone several reprints and editions in the last few years which clearly indicates the usefulness and success of the title.

As I was revising the text for the fourth edition of the book, I myself felt satisfied to see its meticulous planning, carried out at a time when there was virtually no other book, even remotely comparable to this one, in the Indian market.

This edition had so many changes to be incorporated—inclusion of author index, question bank—all of which were aimed to make the book more comprehensive. It would be my pleasure to state that although all the previous editions and reprints contained solved numerical problems and exercise on theoretical questions, the addition of Question Bank in the present edition has nevertheless added to its usefulness. I would consider myself amply rewarded if the present edition of the book, like all its previous ones, caters to the need of those for whom additional pains have been generously put in.

56 RSC Mullick Road
Ekth Heights, Flat 3A3
Kolkata 700 032

Dr P. K. Chakrabarti

Preface to the Third Edition

Soon after its first publication the book on **'Geometrical and Physical Optics'** was warmly accepted by teachers and students all over India as a text-book on the subject. After three successive reprints in 1997, 2000 and 2001, the second revised edition came out in 2004. The revised edition was soon exhausted so that the third edition had to be published in 2005. It bespeaks of the popularity and usefulness of the book in the academic world.

Of course, there is many a reason behind this success. Just as a properly cut diamond sparkles more brightly than any other else, this book also shades out all other books under the same title. The physical clarity of this book is unparalleled, its mathematical exposition is other's envy. A few topics have been newly added in this edition and some logical re-arrangement made here and there.

However, everything was done under short time restriction and so there will be scope for further improvement and I would welcome suggestions from teachers and students alike so that the book really becomes a quality product.

March 2005
Selimpur Lane, Dhakuria
Kolkata 700031

P. K. Chakrabarti

Preface to the Second Edition

'Geometrical and Physical Optics' has been thoroughly revised and enlarged wherever necessary.

The beauty of this book that makes it completely different from others is that the whole geometrical optics portion has been written in matrix method formalism and that too is developed from the root. Use of Jone's vector and Jones's matrix to deal with the phenomenon of polarisation is a further beautification of the arena.

I acknowledge receipt of numerous letters of praise for the book both from teachers and students. This has encouraged me a lot and their suggestions have given me a new dimension.

I hope, this revised edition will also be cordially received by those for whom the book has been written.

Dhakuria
10th December 2003

P. K. Chakrabarti

Preface to the First Edition

This book in **Optics** is covered in two parts, **Part I** containing **Geometrical Optics** and **Part II** containing **Physical Optics** and recent developments in Optics.

There is no doubt that with the introduction of Laser and Fibre Optics, the subject matter of Optics has gained a new dimension. Consequently, there is a trend to revise the syllabus and a new formalism to be followed. In tune with the new trend, this volume on Optics has been written breaking away from the conventional approach.

In Part I (Geometrical Optics), I have followed Matrix Method Formalism. Since the Paraxial Ray Theory entails point to point correspondence between object space and image space, the linear transformation theory can be conveniently utilized to extract information out of the transformation processes. In this regard, Part I of the book is distinctly different from other books on Geometrical Optics available in the market. This new formalism, meeting today's computer demands, may at first-reading appear to be slightly difficult to a conventional reader but once he breaks through, he will find that Geometrical Optics, which was till now thought to be a subject as arid as a desert, would appear to blossom.

In Part II (Physical Optics), I have tried to give the physical explanation in a very lucid style and then follow it by mathematical interpretation. The mathematical exposition of the subject of physical optics is made different from other textbooks available in the market. Particularly the chapter on Polarization has been written in a completely new style; the formalism developed in this chapter is through Jones' vector and Jones' matrices. If the reader gets used to it, he will find real pleasure in going through the chapter. To understand the mathematics used in this book, the reader may find it useful to consult 'A Textbook of Mathematical Physics' by the author.

The great development that has taken place in Optics starting from classical phenomena to the modern discoveries is very difficult to be presented within the limit of a textbook meant for a certain course, but I have tried to do it as far as possible. For further details of some of the modern optics like Holography and Fibre communication, the reader may feel interested to consult advanced books on the subjects.

In this new journey there are reasons more than one for the author to take wrong steps and I would thankfully receive the comments from my erudite readers to give me a chance to improve the book further in subsequent editions. I hope, this book on Geometrical and Physical Optics will enable the reader to get a better understanding of the subject of Optics. In this connection I would frankly confess that I have freely consulted many of the standard treatises, some of which have been given in the reference list. As an author, I will feel encouraged if the reader offers constructive criticisms for further improvement of the book.

1st January, 1992

Department of Physics
St. Xavier's College, Calcutta 700 016

P. K. Chakrabarti

Contents

II. PHYSICAL OPTICS

Part I

Geometrical Optics

Chapter 1

A Brief Review of the Theory of Light

1.1 Introduction

Optics is that branch of Physics where we study the properties of light, its nature, its propagation, and the different phenomena exhibited by it. The whole branch is divided into—(1) Geometrical Optics involving geometrical consideration of image-formation based on the rectilinear propagation and (2) Physical Optics involving the consideration of the physical nature of light giving rise to various optical phenomena. Very recently two more branches of Optics have developed—(i) Laser Optics and (ii) Fibre Optics which are of current interest.

The Geometrical Optics deals with properties of light so far as its propagation, image-formation is concerned. It is based on the assumption of rectilinear propagation of light, a fact that is, of course, an approximate truth. So we sometimes call *Geometrical Optics* an approximate study. Here, we do not enquire into what the nature of light is and if the rectilinear propagation be granted, we can study the whole branch through geometrical construction. But there is an optical phenomenon known as *Diffraction*, where this basic principle of rectilinear propagation is at stake. The phenomenon of diffraction which shows that light bends round sharp obstacle which is appreciable when the irregularities of the diffracting surface are comparable to the wavelength of the wave. Sound wave can be diffracted by a brickwall but not the light wave, because the irregularities of the brickwall are comparable to the wavelength of sound, but the light wave whose wavelength is of the order of 10^{-8} cm has no comparison with the dimension of irregularities of the brickwall. The irregularities of the sharp edge of a blade are microscopic and comparable to the wavelength of light, hence when a blade is illuminated from one side, diffraction pattern is observed on a screen placed on the other side. No diffraction effect will be observed if $\lambda \to 0$, so Geometrical Optics is valid in case of short wavelength.

In physical optics we first consider the nature of light and then explain the phenomena exhibited by light in terms of suitable theories.

The 'LASER' (Light Amplification by Stimulated Emission of Radiation) optics started with the discovery of Laser beam which consists of perfectly parallel beam of highly coherent radiation, because of which Laser finds important applications as in holography, communications, drilling, etc.

The Fibre Optics is also a recent development where we study the transmission of Laser through fibres, called *optical transmitters*. The study of this branch has enabled us to make lens of continuously variable refractive index thus offering an auto-focussing system to our eyes.

In the present text we will be primarily concerned with the first two aspects of optics, but we will also give an introductory account of the two modern aspects of optics later.

1.2 The Different Theories about Light

Though in Geometrical Optics we are not at all concerned with the nature of light, we would give here a brief account of the different theories of light as a prelude to a better understanding of the principle of optics Here we will not expose ourselves to the details of the physical theories which we will gradually develop as we proceed on with different optical phenomena. From the time of Newton the nature of light was a pertinent question and different people proposed different ideas in the form of theories.

Corpuscular Theory (by Newton)

The first theory about the nature of light is the Corpuscular Theory proposed by Newton. According to Newton, light consists of *corpuscles* which are very tiny material particles having practically no mass but having high speed (3×10^8 m/s). Really corpuscles, according to Newton, are 'tiny yet terrific!' So according to this proposition when a source is switched on, the corpuscles are shot out and when they fall on the retina, they excite sensation of vision. If the corpuscles are granted having these properties, though somewhat contradictory, we can very simply explain some of the existing optical phenomena.

Since the corpuscles are supposed to be travelling with tremendously high speed, they travel in straight path. Also if we consider the collision between a corpuscle and a material particle of a reflector as elastic, it is not very difficult to explain the phenomenon of reflection.

Difficulty, however, arises when we apply the corpuscular theory to explain the phenomenon of refraction. We see that when the corpuscles (which are material particles) enter into a denser medium richer in material particles from a rarer medium with lesser number of material particles, each corpuscle will experience a greater pull in the denser medium urging it to move faster. So corpuscular idea leads to a conclusion in opposition to the experimental finding. The experimental fact is that the velocity of light in air is greater than that in any medium. Corpuscular theory predicts just the opposite.

Wave Theory (Young, Huygens and Fresnel, also known as Mechanical Wave Theory)

Corpuscular theory having failed miserably in the quantitative explanation of the phenomenon of refraction, a group of young scientists headed by Young, Fresnel and Huygens brought in a new concept in the name of *Wave Theory.*

According to them light does not consist of material particles like corpuscles but it is an energy which is carried from the source to some other point in the form of wave, this transference of energy from one point to the other takes place through the vibration of the particles of the medium.

By introducing the concept of wave, the laws of reflection and refraction could be nicely established. Not only that, the series of phenomena discovered subsequently in optics in the name of Interference, Diffraction and Polarisation (the details being explored in appropriate chapters) could also be successfully explained. The mere existence even of these phenomena could not be interpreted in terms of the corpuscular theory not to speak of the explanation of these phenomena.

No doubt the wave theory had enormous success in explaining facts of optics but at one point it fell flat. It could not explain why and how sunlight could reach the surface of earth passing through the vast void which is devoid of any material medium. The wave theorists tried to get rid of the trouble by proposing the existence of an all-pervasive medium which they called *aether*. Experiments had been carried out to detect the presence of such medium but with no success. So the idea of an aether medium had to be abandoned.

Electromagnetic Wave (Maxwell)

At this point of development Maxwell proposed that light is an electromagnetic disturbance and not a mechanical disturbance as conceived by earlier wave theorists. A mechanical wave (as in case of sound wave) requires a medium for its propagation, but an electromagnetic wave does not require such a medium as an absolute need. When a source of light is switched on, say a Na-vapour lamp is lit, the electrons in the Na-atom are excited, giving rise to a fluctuating electric field which in its train brings in fluctuating magnetic field also. Thus an electromagnetic disturbance consisting of an electric and magnetic wave propagates. Maxwell has also proved that these two disturbances are mutually at right angles and also at right angles to the direction of propagation. Hence light is a transverse wave. Since electromagnetic wave is also a wave, all the phenomena can easily be explained as in the earlier wave theory. In addition, since both electric field and magnetic field exist in vacuum, there is every reason to believe that their fluctuations producing electromagnetic wave can propagate through vacuum, the details being explored in a separate chapter.

Photon Theory

Up to this point everything was fine and every phenomenon of light could be explained. But trouble with wave theory again began with the advent of a new phenomenon called *photoelectric effect.* In this phenomenon, electrons are emitted from metal surfaces under the influence of suitable frequency light being made to fall on the metal surface. Thus, there is a certain frequency of light below which photoelectric emission is not possible. This frequency is called *threshold frequency.*

The characteristics of photoelectric effect, the existence of threshold frequency, etc. cannot be explained by any wave theory. So again the corpuscular idea about light had to be revived and it was revived in a new form, called *Photon Theory.* Here light is said to consist of photons which are energy particles, the minimum content of energy being $h\nu, h =$ Planck's constant, $\nu =$ frequency of radiation. Applying this concept of particle nature of light with minimum quantum of energy as mentioned above, the photoelectric effect could be explained quite satisfactorily.

Dual Theory

Thus we see that in some phenomena the wave theory could explain things with success, in some phenomena again the photon theory is the only explanation. We say, therefore, that light manifests itself sometimes in particle aspect, sometimes in wave aspect. Thus it is said to have dual characteristics, particle and wave.

In physical optics, these different theories are considered and applied to explain diverse phenomena in optics; but in geometrical optics, no idea about the nature of light is a pre-requisite; only if rectilinear propagation is granted to us, the whole branch of geometrical optics can be studied.

In the present text we will consider geometrical optics first, followed by physical optics consideration and then Laser optics and finally conclude our discussion with an introductory exposure to Fiber optics.

1.3 A Short Review of a Few Terms and Ideas in Geometrical Optics

Ray of Light

A ray in geometrical optics is represented by a straight line, with an arrow indicating the direction of flow of light energy.

Beam of Rays

A beam of rays consists of a number of rays forming a group of adjacent rays and it might be a parallel beam, converging beam or a diverging beam.

Parallel Beam

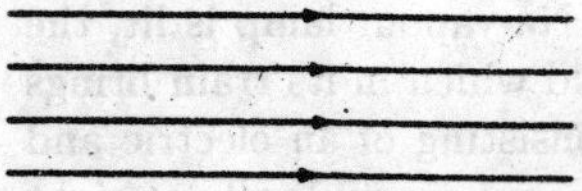

Fig 1.1

It consists of a number of parallel rays. When the source of light is at a very large distance, the light coming from the source is represented by a parallel beam as shown in fig.(1.1). It is possible to get a strictly parallel beam in case of Laser source.

Convergent Beam

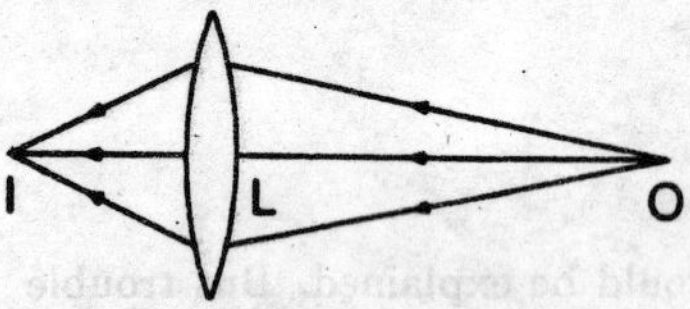

Fig 1.2

This consists of rays tending to converge to a point. Thus in case of a lens, rays from the source O after emergence from the lens converge to the point I, so the emergent rays form a converging beam.

Divergent Beam

If the constituent rays of the beam diverge from a point as in case of rays from O in fig.(1.2), it is said to form a diverging beam.

Rectilinear Propagation

In a homogeneous medium light travels in a straight path. The formation of the image in a pinhole camera, the formation of eclipses are taken as the evidences. But it is only an approximate truth. In case of the phenomenon of diffraction as we shall see later, light is found to bend round sharp obstacle, hence the principle of rectilinear propagation does not hold in this case. But in geometrical optics, this principle is assumed to hold.

Reversibility of Light

It is also a basic principle in optics. When light travels from one point to another in a certain medium, it will follow the same path back when the direction is reversed. It is because of this principle we find that when a lens forms an image of a source at some point, the image is formed at the previous source point when the source is placed at the previous image point. Thus we say that the object and image points are conjugate points.

Optical Path in a Medium

By optical path l in a medium we mean the path or distance travelled by light in air in the time it takes to cover the length l in that medium. Thus,

let the velocity of light in air $= c$,

let the velocity of light in the medium $= v$.

Therefore, time taken to cover the distance is given by, $t = l/v$. In this time, light travels a distance s in air given as

$$s = ct = \frac{c}{v} \cdot l = nl \tag{1.3.1}$$

where n is the r.i. of the medium $= \frac{c}{v}$.

Fermat's Principle

It is also a very important principle in optics, similar to the principle of least action in mechanics and a consequence of the principle of the reversibility of light. It states that —

The optical path between two points of an optical medium connected by actual path followed by light is an extremum, i.e., either a minimum or a maximum.

In practically all cases it is minimum but in some special cases it is maximum. This principle follows immediately from the reversibility of light as it is evident from the consideration of the fig.(1.3).

Let AOB be the path followed by light in going from A to B by reflection at O. The path from A to B is the least according to Fermat. So, if light path is reversed, it must follow the path back as otherwise there would be two minimal paths. Elaborate discussion on this principle is given in the next chapter.

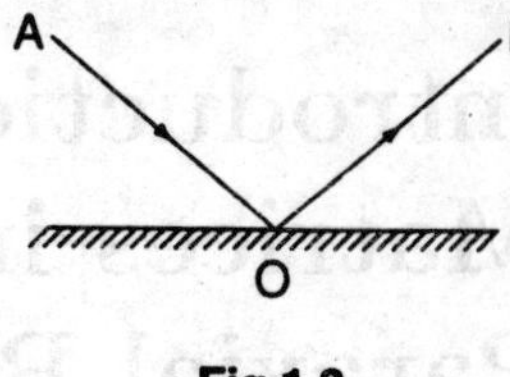

Fig 1.3

Sign Convention

In geometrical optics we too often have to measure distances, angles, etc., so some sign convention is to be followed. We will follow the same convention as that of Cartesian System. The source of light is to be considered on the right of the origin; then—any distance measured against the direction of light is considered positive, and in the direction of light, negative.

For measurement of angle, the angle measured in anti-clockwise sense is taken as positive, in clockwise sense, negative. Real translation will be considered positive while virtual translation, negative.

Domain of Validity of Geometrical Optics

Geometrical optics is said to be an approximate study, because it is based on rectilinear propagation. The rectilinear propagation is an approximate truth in view of the phenomena of diffraction. When light travels across a sharp obstacle, there is bending of light beam and there is pronounced bending effect when the dimension of the irregularities of the diffracting surface is comparable to the wavelength. Geometrical optics is valid only in short wave length limit $\lambda \to 0$. The wavelength of light is of the order of 10^{-8} cm which is pretty small compared to the size of the optical instruments that we ordinarily meet with. So the principle of rectilinear propagation may be granted as a good approximation in most of the ordinary cases. With the decrease in wavelength, the diffraction effect decreases and it disappears in the wavelength limit $\lambda \to 0$ when the domain of validity of the Geometrical Optics finds its ground.

Q1.1 'The limit (wavelength) 0 is treated to be the ray optics.'—Explain.

Chapter 2

Introduction to the Theory of Matrices in the Formulation of Paraxial Ray Optics

The whole branch of geometrical optics may be studied in two ways—(1) by geometrical consideration, drawing the ray-diagram based on rectilinear propagation, and (2) purely analytically through the matrix method formalism.

The first method of approach we have studied in the elementary course, now we will try to develop the matrix formalism for the study of the problems of geometrical optics in the present advanced discourse.

The basis of the matrix method formulation of geometrical optics is that here we are to establish relation between object and image formed due to reflection or refraction. As there is point to point correspondence between object space (the medium of incident rays) and image space (the medium into which light goes after some optical process) for paraxial nature of rays, every point (x, y, z) in object space admits of being linearly transformed into (x', y', z') in the image space. The transformation can, therefore, be carried through matrix operation. The advantage of matrix formalism is that we can avoid cumbersome calculation in the process of image-formation by a number of reflecting or refracting surfaces in this formalism if we can develop suitable matrices representing the effect of reflection or refraction. Moreover, matrix method formalism is amenable to computer usage.

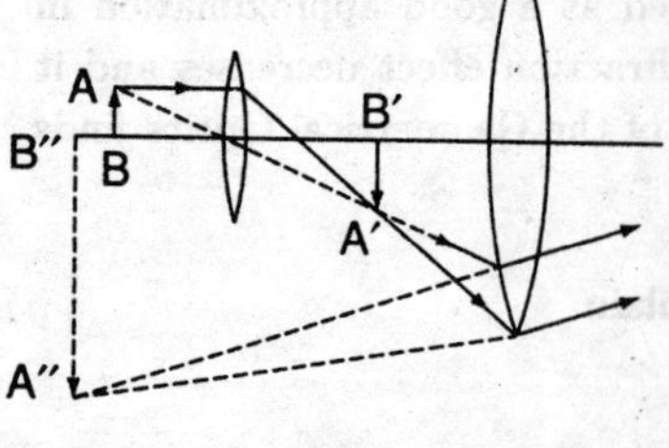

Fig 2.1

Actually in geometrical optics a final image is formed through a number of optical processes like refraction through a lens, translation of the ray in a certain medium through a certain distance, again there may be refraction through another lens and so on. As shown in fig.(2.1), the ray from the object AB passes in air (translation), then passes through the lens (refraction), again translates in air and again refracted by the second lens, finally the image is formed as $A''B''$. The whole process of image-formation can be carried through a number of matrix operation transforming object points into image-points. So a number of matrices representing successive optical operations need be developed. In general, we would require to develop the reflection matrix $[R_L]$, refraction matrix $[R]$ for both plane surface and curved surface and lastly the translation matrix $[T]$. In order to understand the contents of the text better, a fair knowledge of matrix operation is to be gathered.[1]

[1]For this requirement the reader is advised to consult 'A Textbook of Mathematical Physics'—by P.K. Chakrabarti and S.N. Kundu.

Here in this chapter we will develop the bare, basic necessities of matrix operations in the form of development of suitable matrices called *Reflection matrix, Translational matrix* and *Refraction matrix* which will be freely applied according to need as we pass gradually to different aspects of geometrical optics.

Essence of the Matrix Method Representation of the Object Ray and the Image Ray

Two parameters are used for representation of a ray either coming from the object or of a ray going to form the image. In fig. 2.1(a) we have shown a ray from the tip of the object. This ray is defined by referring to the position of the tip x_1, w.r.t. the principal axis. This forms one parameter. The other parameter is the angle θ_1. To get it, draw a line through the tip and parallel to the principal axis. From the parallel line rotate through shorter angle so as to grab the ray. This angle is θ_1, this is another parameter. Thus, the two elements x_1, θ_1 forming a column matrix $\begin{bmatrix} x_1 \\ \theta_1 \end{bmatrix}$ is the object-ray defining matrix. Similarly, the matrix $\begin{bmatrix} x_2 \\ \theta_2 \end{bmatrix}$, where x_2, θ_2 are the elements of the image-defining ray is called the image matrix. If rotation w.r.t. the horizontal line through the tip is anti-clockwise, it is reckoned as positive; if clockwise, it is negative. In fig. 2.1(a), θ_1 is negative.

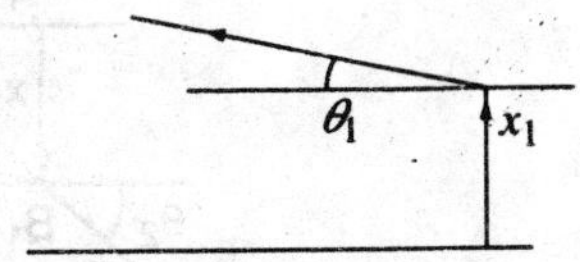

Fig 2.1(a)

2.1 Refraction and Reflection Matrix : Curved Surface and Plane Surface

It is to note that the theory of reflection at curved surface may be applied to plane surface reflection if we treat the plane surface as a curved surface with the radius of curvature infinity. So the general reflection matrix for spherical mirrors will be developed and this matrix with the modification of $r = \infty$, where r is the radius of curvature, may be applied for reflection at plane surface. Again, reflection law can be deduced from the laws of refraction only if we put $n_1 = -n_2$, where n_1 is the r.i. of the first medium and n_2 is the r.i. of the second medium.

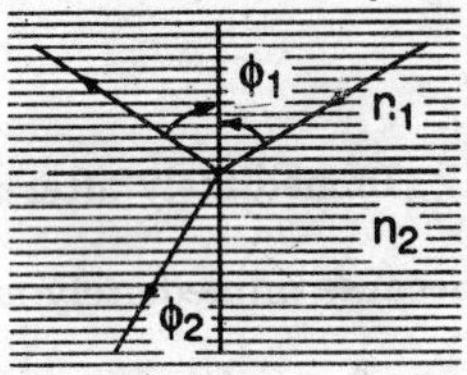

Fig 2.2

Thus, by Snell's law we get,

$$n_1 \sin\phi_1 = n_2 \sin\phi_2,$$

if we put $n_1 = -n_2$, we get

$$-\sin\phi_1 = \sin\phi_2 \quad \Rightarrow \quad \phi_1 = -\phi_2,$$

showing the phenomenon of reflection. So in general, we will develop the refraction matrix for a curved surface whence those for reflection matrix for plane and curved surface reflection may be deduced.

2.1.1 Scheme of the Method Explained

What is needed in this formalism is to first develop three matrices:

- Translation matrix $[T]$.
- Reflection matrix for curved reflector $[R_l]$.
- Refraction matrix for curved refractor.

With the help of these matrices, the image forming process can be represented in case of a reflector or a refractor.

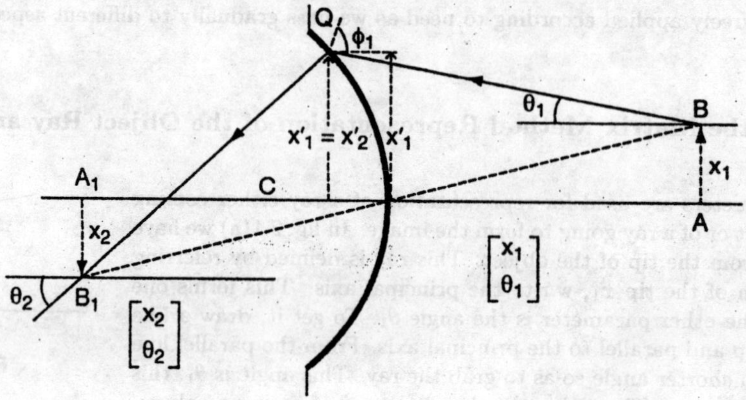

Fig 2.3

Consider fig. 2.3, where we have shown a curved refractor on which a ray defined by $\begin{bmatrix} x_1 \\ \theta_1 \end{bmatrix}$ matrix falls and suffers refraction. The translation matrix $[T_1]$ transforms $\begin{bmatrix} x_1 \\ \theta_1 \end{bmatrix}$ matrix to $\begin{bmatrix} x'_1 \\ \theta_1 \end{bmatrix}$ which defines the ray on incidence on the refractor. Thus,

$$\begin{bmatrix} x'_1 \\ \theta_1 \end{bmatrix} = [T_1] \begin{bmatrix} x_1 \\ \theta_1 \end{bmatrix}. \tag{2.1.1}$$

Immediately after refraction at the curved surface, the ray is defined by $\begin{bmatrix} x'_2 \\ \theta_2 \end{bmatrix}$ where $x'_1 = x'_2$, because immediately after refraction, the refracted ray starts from the same point, thus $x'_1 = x'_2$. This transformation from $\begin{bmatrix} x'_1 \\ \theta_1 \end{bmatrix}$ to $\begin{bmatrix} x'_2 \\ \theta_2 \end{bmatrix}$ is done by the refraction matrix for the curved refractor, thus,

$$\begin{bmatrix} x'_2 \\ \theta_2 \end{bmatrix} = [R] \begin{bmatrix} x'_1 \\ \theta_1 \end{bmatrix} = [R][T_1] \begin{bmatrix} x_1 \\ \theta_1 \end{bmatrix}. \tag{2.1.2}$$

The final image A_1B_1 defined by (x_2, θ_2) is obtained by translation again which is effected through another translation matrix $[T_2]$, thus

$$\begin{bmatrix} x_2 \\ \theta_2 \end{bmatrix} = [T_2] \begin{bmatrix} x'_2 \\ \theta_2 \end{bmatrix}.$$

So the whole image processing is represented by a series of matrix operation as shown in the above steps and so,

$$\begin{bmatrix} x_2 \\ \theta_2 \end{bmatrix} = [T_2] \begin{bmatrix} x'_2 \\ \theta_2 \end{bmatrix} = [T_2][R][T_1] \begin{bmatrix} x_1 \\ \theta_1 \end{bmatrix}. \tag{2.1.3}$$

2.2 Reflection Matrix for a Curved Reflector and hence a Plane Reflector

Reflection laws can be deduced from the laws of refraction only if we substitute $n_1 = -n_2$, thus we can construct reflection matrix for a curved reflector from the corresponding refraction matrix only replacing n_2 by $-n_1$.

But the reflection matrix for curved surface can also be independently developed w.r.t. fig.(2.4).

The column vector $\begin{bmatrix} x_1 \\ \theta_1 \end{bmatrix}$ represents the object just before reflection and the column vector $\begin{bmatrix} x_2 \\ \theta_2 \end{bmatrix}$ represents the image after reflection.

$$\therefore \quad \begin{bmatrix} x_2 \\ \theta_2 \end{bmatrix} = [R_l] \begin{bmatrix} x_1 \\ \theta_1 \end{bmatrix}.$$

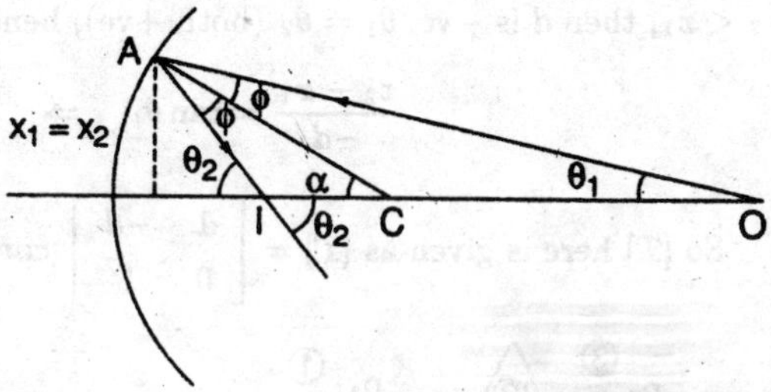

Fig 2.4

From fig.(2.4), we get, $\theta_2 = \phi + \alpha, \alpha = \phi + \theta_1$

$$\therefore \quad \theta_2 = (\alpha - \theta_1) + \alpha = 2\alpha - \theta_1$$

θ_2 is clockwise, hence negative; so also are α and θ_1.

$$\therefore \quad -\theta_2 = -2\alpha + \theta_1$$

$\Rightarrow \quad \theta_2 = 2\alpha - \theta_1$ (with sign convention) $= 2\frac{x_1}{r} - \theta_1$.

Thus, for reflection at curved reflector (concave)

$$x_2 = x_1$$

$$\theta_2 = \frac{2}{r}x_1 - \theta_1$$

$$\therefore \quad \begin{bmatrix} x_2 \\ \theta_2 \end{bmatrix} = \begin{bmatrix} 1 & 0 \\ 2/r & -1 \end{bmatrix} \begin{bmatrix} x_1 \\ \theta_1 \end{bmatrix}$$

$$\Rightarrow [R_l] = \begin{bmatrix} 1 & 0 \\ 2/r & -1 \end{bmatrix} \text{ for concave reflector.} \qquad (2.2.1)$$

For convex reflector, $[R_l] = \begin{bmatrix} 1 & 0 \\ -2/r & -1 \end{bmatrix}$.

For a plane surface reflection

$$[R_l'] = \begin{bmatrix} 1 & 0 \\ 0 & -1 \end{bmatrix}, \text{ since } r = \infty. \qquad (2.2.2)$$

2.3 Translational Matrix (T)

When a ray passes from A, specified by a column vector $\begin{bmatrix} x_1 \\ \theta_1 \end{bmatrix}$ to another point B in air specified by the column vector $\begin{bmatrix} x_2 \\ \theta_2 \end{bmatrix}$ where $\theta_1 = \theta_2$, since the

translation of the ray is in the direction of the ray, d is $-$ve; θ_1 is also negative. Then,

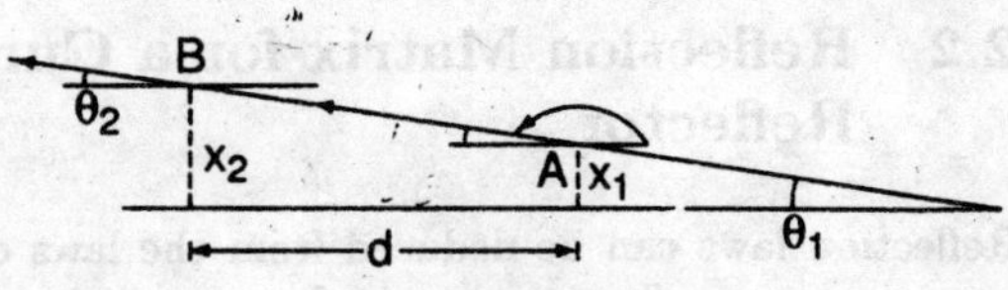

Fig 2.5

$$\frac{x_2 - x_1}{-d} = \tan(-\theta_1) \Rightarrow x_2 = (x_1 + d\tan\theta_1)$$

$$\therefore \quad x_2 = x_1 + \theta_1 d.$$

Also, $-\theta_2 = -\theta_1 \Rightarrow \theta_2 = \theta_1$

$$\therefore \begin{bmatrix} x_2 \\ \theta_2 \end{bmatrix} = \begin{bmatrix} 1 & d \\ 0 & 1 \end{bmatrix} \begin{bmatrix} x_1 \\ \theta_1 \end{bmatrix} \Rightarrow [T] = \begin{bmatrix} 1 & d \\ 0 & 1 \end{bmatrix} \text{ diverging translation matrix} \tag{2.3.1}$$

If on the other hand, translation of the ray takes place between two planes such that $x_2 < x_1$, then d is $-$ve, $\theta_1 = \theta_2$ (both +ve), hence

$$\frac{x_2 - x_1}{-d} = \tan\theta_1 \Rightarrow x_2 = x_1 - d\tan\theta_1 = x_1 - \theta_1 d. \tag{2.3.2}$$

So $[T]$ here is given as $[T] = \begin{bmatrix} 1 & -d \\ 0 & 1 \end{bmatrix}$ converging translational matrix.

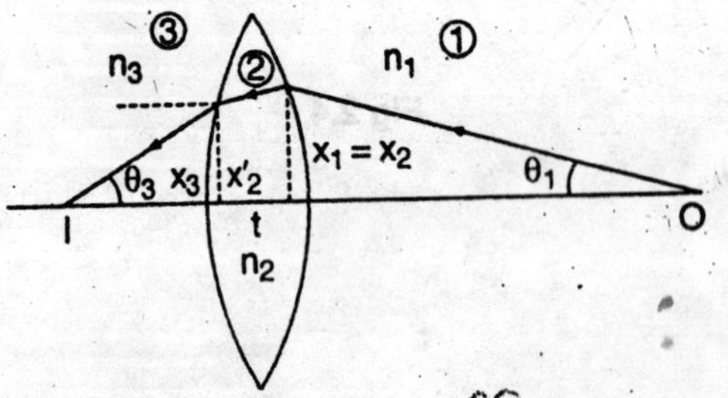

Fig 2.6

Any process of image-formation may be interpreted in terms of translation and refraction or reflection, hence the whole branch of geometrical optics dealing with the process of image-formation can be formulated in terms of translation matrix $[T]$ and refraction matrix $[R]$ or reflection matrix $[R_l]$. Let us clarify the point with reference to the following example:

The object space is of r.i. n_1 and is referred to as medium 1, the bounded medium forming the lens is of r.i. n_2 and it is referred to as medium 2, the emerging medium is the medium 3 of r.i. n_3. Then final image specified by the column vector $\begin{bmatrix} x_3 \\ \theta_3 \end{bmatrix}$ is obtained by refraction from second medium to the third medium, it is preceded by translation in the lens medium by t (thickness of the lens), preceded by refraction from medium 1 to medium 2 across the curved surface of radius of curvature r_1.

So the final column vector representing the image formed by the optical system can be written as

$$[I] = \begin{bmatrix} x_3 \\ \theta_3 \end{bmatrix} = [T_2][R_{23}][T][R_{12}][T_1] \begin{bmatrix} x_1 \\ \theta_1 \end{bmatrix} \tag{2.3.3}$$

Now, $[R_{23}] = \begin{bmatrix} 1 & 0 \\ -\frac{1}{r_2}\frac{(n_3 - n_2)}{n_3} & \frac{n_2}{n_3} \end{bmatrix}$; This refraction matrix is developed in Chapter 5.

$$[T] = \begin{bmatrix} 1 & -t \\ 0 & 1 \end{bmatrix}$$

$$[R_{12}] = \begin{bmatrix} 1 & 0 \\ -\frac{1}{r_1}\frac{(n_2 - n_1)}{n_2} & \frac{n_1}{n_2} \end{bmatrix} \tag{2.3.4}$$

This refraction matrix is also developed in Chapter 5.

In case of a thin lens, $t = 0$, hence the translational matrix $[T]$ becomes $[T] = \begin{bmatrix} 1 & 0 \\ 0 & 1 \end{bmatrix}$, a unit matrix, so $[R_{23}][T] = [T][R_{23}] = [R_{23}]$.

Eqn. (2.3) can be written as

$$\begin{bmatrix} x_3 \\ \theta_3 \end{bmatrix} = [R_{23}][R_{12}] \begin{bmatrix} x_1 \\ \theta_1 \end{bmatrix}$$

$$= \begin{bmatrix} 1 & 0 \\ -\frac{1}{r_2}\frac{(n_3 - n_2)}{n_3} & \frac{n_2}{n_3} \end{bmatrix} \begin{bmatrix} 1 & 0 \\ -\frac{1}{r_1}\frac{(n_2 - n_1)}{n_2} & \frac{n_1}{n_2} \end{bmatrix} \begin{bmatrix} x_1 \\ \theta_1 \end{bmatrix}$$

$$= \begin{bmatrix} 1 & 0 \\ P_{23} & \frac{n_2}{n_3} \end{bmatrix} \begin{bmatrix} 1 & 0 \\ P_{12} & \frac{n_1}{n_2} \end{bmatrix} \begin{bmatrix} x_1 \\ \theta_1 \end{bmatrix}$$

where, $P_{23} = -\frac{1}{r_2}\cdot\frac{n_3 - n_2}{n_3}$ and $P_{12} = -\frac{1}{r_1}\cdot\frac{n_2 - n_1}{n_2}$ respectively called the *refracting power* of the two surfaces. Performing the matrix product, we can write,

$$\begin{bmatrix} x_3 \\ \theta_3 \end{bmatrix} = \begin{bmatrix} 1 & 0 \\ \left\{P_{23} + P_{12}\cdot\left(\frac{n_2}{n_3}\right)\right\} & \frac{n_1}{n_3} \end{bmatrix} \begin{bmatrix} x_1 \\ \theta_1 \end{bmatrix} \Rightarrow x_3 = x_1$$

$$\text{and} \quad \theta_3 = \left(P_{23} + P_{12}\frac{n_2}{n_3}\right) x_1 + \frac{n_1}{n_3}\theta_1$$

$$= -\left(\frac{n_3 - n_2}{n_3 r_2} + \frac{n_2 - n_1}{n_2 r_1}\frac{n_2}{n_3}\right) x_1 + \frac{n_1}{n_3}\theta_1$$

$$= -\left(\frac{n_3 - n_2}{n_3 r_2} + \frac{n_2 - n_1}{n_3 r_1}\right) x_1 + \frac{n_1}{n_3}\theta_1$$

Put

$$\frac{1}{g} = \frac{n_3 - n_2}{n_3 r_2} + \frac{n_2 - n_1}{n_3 r_1} \tag{2.3.5}$$

Then,

$$\theta_3 = -\frac{x_1}{g} + \frac{n_1}{n_3}\theta_1.$$

Now from fig.(2.3.4) we get,

$$\frac{x_1}{u} = \tan(-\theta_1) = -\tan\theta_1 = -\theta_1$$

Similarly,

$$\frac{x_3}{-v} = \tan\theta_3 \simeq \theta_3$$

$$\therefore \quad -\frac{x_3}{v} = -\frac{x_1}{g} + \frac{n_1}{n_3}\theta_1 = -\frac{x_1}{g} - \frac{n_1}{n_3}\cdot\frac{x_1}{u}$$

$$\Rightarrow \quad \frac{x_3}{v} - \frac{n_1}{n_3}\cdot\frac{x_1}{u} = \frac{x_1}{g}$$

$$\Rightarrow \quad \frac{1}{v} - \frac{n_1}{n_3}\cdot\frac{1}{u} = \frac{1}{g} = \frac{n_3 - n_2}{n_3 r_2} + \frac{n_2 - n_1}{n_3 r_1}, \quad \because x_3 = x_1$$

$$\Rightarrow \quad \frac{n_3}{v} - \frac{n_1}{u} = \frac{n_3 - n_2}{r_2} + \frac{n_2 - n_1}{r_1}.$$

Now, when $u = \infty, v = f$, called *focal length* of the optical system.

$$\frac{n_3}{f} = \frac{n_3 - n_2}{r_2} + \frac{n_2 - n_1}{r_1} \Rightarrow \frac{1}{f} = \frac{n_3 - n_2}{n_3 r_2} + \frac{n_2 - n_1}{n_3 r_1} = \frac{1}{g}$$

Thus g as put in equation (2.3.5) is identified with the focal length of the optical system under the specified condition. Thus the thin lens formula is derived. The matrix $[R_{23}][T][R_{12}]$ is called the *system matrix* $[S]$. The system matrix $[S]$ for a thin lens is given as $[S] = [R_{23}][R_{12}]$, since here $[T]$ is a unit matrix. The system matrix $[S]$ in case of a thin lens can be written as

$$[S] = \begin{bmatrix} 1 & 0 \\ -(1/f) & n_1/n_3 \end{bmatrix}.$$

If $n_1 = n_3$, for a lens in air

$$\frac{1}{f} = \frac{n_1 - n_2}{n_1 r_2} + \frac{n_2 - n_1}{n_1 r_1} = \frac{n_1 - n_2}{n_1}\left(\frac{1}{r_2} - \frac{1}{r_1}\right) = \frac{n_2 - n_1}{n_1}\left(\frac{1}{r_1} - \frac{1}{r_2}\right).$$

Thus applying the matrix operation, the image-formation problem may be solved with more ease and comfort than the conventional approach through complicated geometrical consideration.

At first attempt the reader may find this formalism a bit difficult because he may not have any exposure to the method of matrix, but once he overcomes this difficulty and gets used to it, he will soon find that this formalism is a great relief to him and probably derive pleasure out of reading this branch of optics.

References

1. **A Textbook of Mathematical Physics**—*P.K. Chakrabarti* and *S.N. Kundu*, New Central Book Agency (P) Ltd., Kolkata.
2. **Mathematical Physics**—*C. Harper*, Prentice-Hall.
3. **Mathematics for Physicists and Engineers**—Melvin Pipes.
4. **Mathematical Physics**—Erfkan.

Q2.1 Using matrix method, build up the lens formula for a thin lens.

Chapter 3

Fermat's Principle and Its Applications

Fermat's principle in optics is a fundamental one giving rise to the phenomenon of reversibility and also the laws of reflection and refraction obeyed by light. It is the principle related to the nature of path followed by light. It can be looked upon as the optical analog of the principle of least action in mechanics.

This principle states that *of all the paths connecting two points, the path actually followed by light, is stationary, i.e., either maximum or minimum.*

Now, in case of reflection or refraction at plane surfaces, the path followed by light is minimum but in case of refraction at spherical surface, it may be a maximum; hence the term 'stationary' is used in the statement of the principle.

3.1 Fermat's Principle in Case of Reflection at Plane Surface

The two points A, B may be connected by actual ray via reflection at O of the plane reflector. It is represented by $(AO + OB)$. We can consider another path in the neighbourhood from A via a point C of the plane reflector to the point B. Its length is $(AC + CB)$. Drop a $\perp$ from A on the reflector and produce it to meet the extension of BO at the point A'. Since AOB is the path actually followed by light obeying law of reflection, it is easy to prove that $AP = PA'$; hence, PQ reflector is the perpendicular bisector of the line joining A and A'. Thus, $AO = OA'$ and also $AC = CA'$.

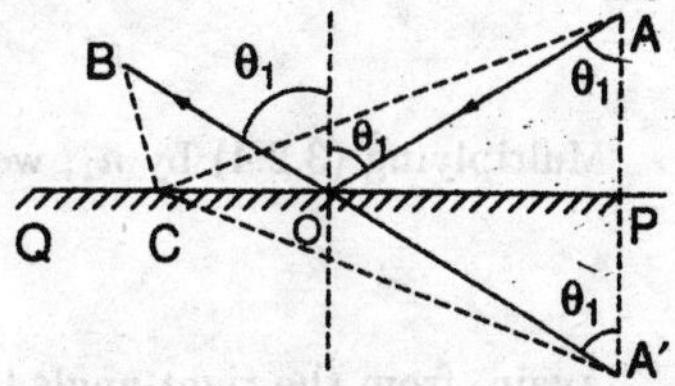

Fig 3.1

Now, actual optical path $L_{op} = AO + OB = OA' + OB = A'B$. Any other path in the neighbourhood of this reflected ray path is given by

$$L_o = AC + CB = CA' + CB.$$

Now, $CA' + CB > A'B$, because the sum of two sides of a triangle is always greater than the third side. Hence the Fermat's principle. Thus in case of reflection from a plane reflector, the actual path of light reaching two points by reflection on the surface is the least.

3.2 Refraction at Plane Surface

Consider now the case of refraction at a plane surface separating two media of r.i. values n_1 and n_2 as shown in fig.(3.2).

A ray passes from A to B via refraction at O on the plane surface separating the two media of r.i. values n_1 and n_2. Consider another path in its neighbourhood connecting the two points A, B via a point O_1 on the plane surface. Then the optical path actually followed by light is given by,

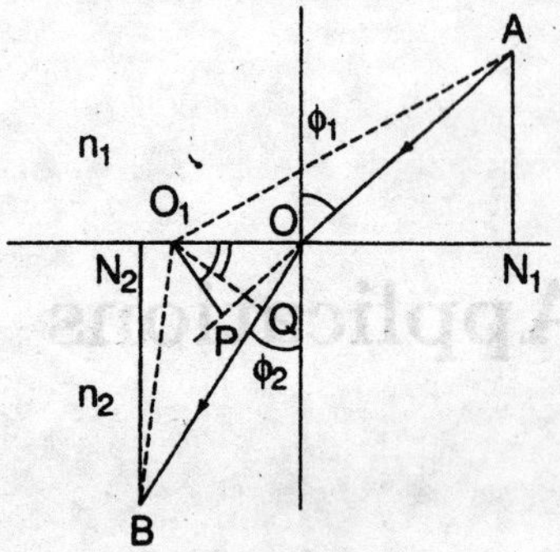

Fig 3.2

$$L_{op} = n_1 AO + n_2 OB \tag{3.2.1}$$

The other path in the neighbourhood has optical path

$$L_o = n_1 AO_1 + n_2 O_1 B. \tag{3.2.2}$$

Drop perpendiculars O_1P and O_1Q from O_1 on to AO produced at P and on OB at Q· $\angle OO_1P$ is the angle included between two lines O_1O and O_1P and it is equal to the angle included between normals to the lines O_1O and O_1P, thus,

$$\angle OO_1P = \phi_1 = \text{angle of incidence}$$

Similarly,

$$\angle OO_1Q = \phi_2 = \text{angle of refraction}$$

Now, since AOB is the actual path of light obeying laws of refraction, we must have,

$$n_1 \sin\phi_1 = n_2 \sin\phi_2 \quad \text{(Snell's law)}$$

$$\Rightarrow \quad n_1 \frac{OP}{OO_1} = n_2 \frac{OQ}{OO_1} \quad \Rightarrow \quad n_1 OP = n_2 OQ. \tag{3.2.3}$$

Now, in the right-angled triangle AO_1P, we have,

$$O_1A > PA \Rightarrow (O_1A - AO) > (PA - AO)$$
$$\Rightarrow (O_1A - AO) > PO \tag{3.2.4}$$

Multiplying (3.2.4) by n_1, we have,

$$(n_1 O_1 A - n_1 AO) > n_1 PO \tag{3.2.5}$$

Again, from the right-angled triangle O_1QB, we have,

$$O_1B > BQ \quad \Rightarrow \quad BQ < O_1B \quad \Rightarrow \quad (BO - OQ) < O_1B \tag{3.2.6}$$

Multiplying (3.2.6) by n_2

$$(n_2 BO - n_2 OQ) < n_2 O_1 B \quad \Rightarrow \quad n_2 O_1 B > (n_2 BO - n_2 OQ) \tag{3.2.7}$$

Adding (3.2.5) and (3.2.7), we get,

$$\{(n_1 O_1 A - n_1 AO) + n_2 O_1 B\} > \{n_1 PO + (n_2 BO - n_2 OQ)\}$$
$$\Rightarrow \{(n_1 O_1 A + n_2 O_1 B) - n_1 AO\} > n_2 BO, \quad \text{using relation (3.2.3)}$$
$$\Rightarrow (n_1 O_1 A + n_2 O_1 B) > (n_1 AO + n_2 BO) \quad \Rightarrow \quad L_o > L_{op} \tag{3.2.8}$$

Thus, the actual path followed by light in going from A to B through refraction is the least.

In both the cases we see that the optical path connecting two points due to reflection or refraction at plane surface is less than any path in the neighbourhood connecting the same two points. But if we consider the process of refraction at curved surface, what conclusion we arrive at?

3.3 Refraction at Curved Surface

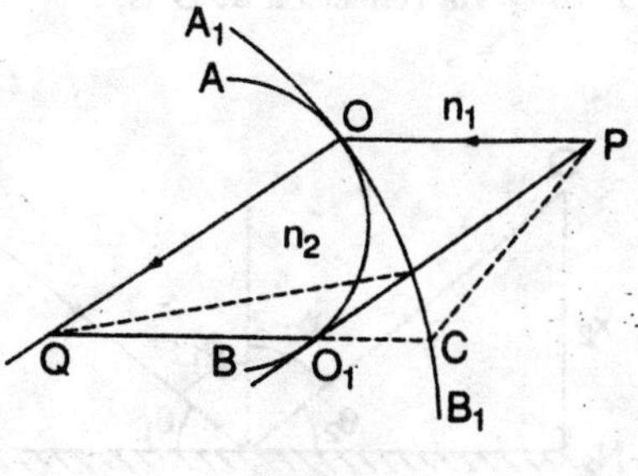

Fig 3.3

Let us consider a ray from P going to a point Q through refraction at O on the spherical surface AOB separating two media of r.i. values n_1 and $n_2 (n_1 < n_2)$.

The actual path followed by light from P through refraction at O to the point Q is POQ; consider another path connecting the two points P, Q through another point O_1 on the surface. Then the optical path for the actual traverse is $L_{op} = n_1 PO + n_2 OQ$.

The other path between the same two points $L_o = n_1 PO_1 + n_2 O_1 Q$.

There is a surface, called *aplanatic surface* and there are two points w.r.t. this surface such that if an object be placed at one point, the image is always formed at the second.

So, if A_1OB_1 be the trace of the aplanatic surface w.r.t. the two points P and Q as aplanatic points and touching the curved surface at O, the other path is PO_1Q, and QO_1 produced cuts the aplanatic surface at C, we would get by virtue of the property of aplanatic surface that

$$\text{optical path } PCQ = \text{optical path } POQ, \text{ i.e., } n_1 PC + n_2 CQ = L_{op}$$

$$\begin{aligned}\therefore \quad L_{op} - L_o &= (n_1 PO + n_2 OQ) - (n_1 PO_1 + n_2 O_1 Q) \\ &= (n_1 PC + n_2 CQ) - (n_1 PO_1 + n_2 O_1 Q) \\ &= n_2 (CQ - O_1 Q) - n_1 (PO_1 - PC) \\ &= n_2 \times CO_1 - n_1 (PO_1 - PC) \\ &= n_2 \times CO_1 - n_1 (PO_1 - PC) \end{aligned} \quad (3.3.1)$$

In the triangle PO_1C, we have,

$$PO_1 < PC + CO_1 \quad \Rightarrow \quad PO_1 - PC < O_1 C.$$

Since $n_1 < n_2$, we would get,

$$L_{op} - L_o = +\text{ve} \quad \Rightarrow \quad L_{op} > L_o.$$

Thus, the actual optical path followed by light is greater than any other path conceived.

Thus, the Fermat's principle can be stated as:

> ***The total optical path between any two points accessible to light is stationary, i.e., either a maximum or a minimum.***

Mathematically stated this principle stands as:

$L_{op} = \int_{P_1}^{P_2} n ds$ is stationary, i.e., either a maximum or a minimum. Again, $n = c/v$ where c is the velocity of light in air (constant) and v is the velocity in the medium. So, $ds = v dt$, hence, $L_{op} = c \int_{P_1}^{P_2} dt$, so it may be taken as the principle of stationary time also.

3.4 Laws of Reflection and Refraction on the Basis of Fermat's Principle

The laws of reflection and refraction can be derived if Fermat's principle is granted to us. We have seen how light rays obeying the laws of reflection and refraction follow Fermat's principle. Now we will see how Fermat's principle alone can give the laws of reflection as well as refraction.

Reflection from Fermat's Principle

The optical path $POQ = PO + OQ = \sqrt{x^2 + x_1^2} + \sqrt{x_2^2 + (l-x)^2}$. The time taken for light to go from P to Q via reflection at O is,

$$t = \frac{\sqrt{x^2 + x_1^2}}{v} + \frac{\sqrt{x_2^2 + (l-x)^2}}{v} \tag{3.4.1}$$

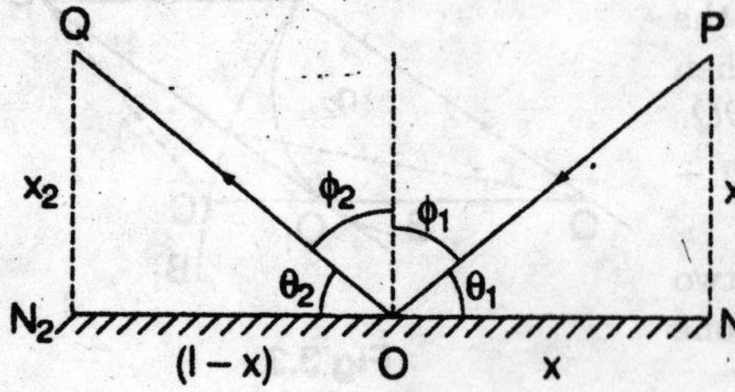

Fig 3.4

For t to be minimum, $dt/dx = 0$

$$\therefore \frac{1}{v}\left[\frac{1}{2}\cdot\frac{2x}{\sqrt{x^2+x_1^2}} - \frac{1}{2}\cdot\frac{2(l-x)}{\sqrt{x_2^2+(l-x)^2}}\right] = 0$$

$$\Rightarrow \frac{x}{\sqrt{x^2+x_1^2}} = \frac{(l-x)}{\sqrt{x_2^2+(l-x)^2}}$$

$$\Rightarrow \sin\phi_1 = \sin\phi_2$$

$$\Rightarrow \phi_1 = \phi_2 \quad \text{law of reflection} \tag{3.4.2}$$

Refraction from Fermat's Principle

Let a ray pass from P to Q through refraction at O on the surface separating two media of r.i. values n_1 and n_2. The time taken by light to travel from P to Q is given by,

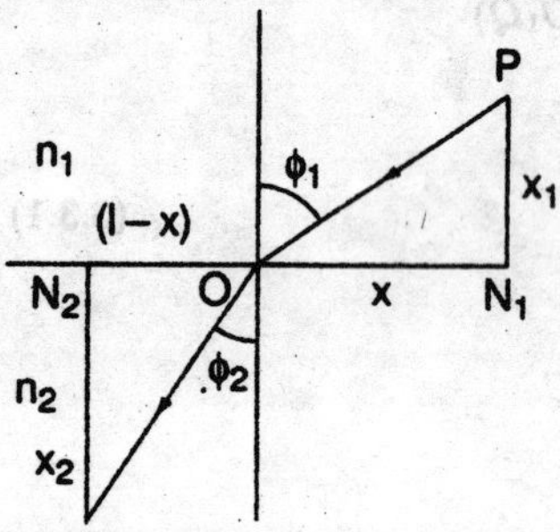

Fig 3.5

$$t = \frac{\sqrt{x_1^2 + x^2}}{v_1} + \frac{\sqrt{x_2^2 + (l-x)^2}}{v_2}.$$

In order to make t minimum, we must have,

$$\frac{dt}{dx} = 0 = \frac{1}{2v_1}\frac{2x}{\sqrt{x_1^2+x^2}} - \frac{1}{2v_2}\frac{2(l-x)}{\sqrt{x_2^2+(l-x)^2}}$$

$$\Rightarrow \frac{1}{v_1}\sin\phi_1 = \frac{1}{v_2}\sin\phi_2 \quad \Rightarrow \quad \frac{c}{v_1}\sin\phi_1 = \frac{c}{v_2}\sin\phi_2$$

$$\Rightarrow n_1\sin\phi_1 = n_2\sin\phi_2 \quad \text{(Snell's law)} \tag{3.4.3}$$

Reflection at Curved Surface : From Fermat's Principle

The object A is placed in air in front of the concave reflector, O is its pole. Let $OA = x(+\text{ve}), OR = r(+\text{ve}), OI = y$.

The angle θ is as indicated and is supposed to be small. Then, the optical path light follows is L_{op} given as,

$$L_{op} = AB + BI.$$

Now,

$$\begin{aligned} AB^2 &= r^2 + (x-r)^2 - 2r(x-r)\cos(180-\theta) \\ &= r^2 + (x-r)^2 + 2r(x-r)\cos\theta \\ &= r^2 + (x-r)^2 + 2r(x-r)(1-\theta^2/2) \end{aligned}$$

$$\therefore AB = [r^2 + (x-r)^2 + 2r(x-r)(1-\theta^2/2)]^{1/2}$$

$$\simeq x\left[1 - \frac{r\theta^2(x-r)}{2x^2}\right] = x - \frac{r\theta^2(x-r)}{2x}$$

Similarly,

$$BI = y + \frac{r\theta^2(r-y)}{2y}$$

$$\therefore \quad L_{op} = x + y - \frac{r\theta^2(x-r)}{2x} + \frac{r\theta^2(r-y)}{2y}$$

$$\frac{dL_{op}}{d\theta} = \left[-\frac{r(x-r)}{x} + \frac{r(r-y)}{y}\right]\theta.$$

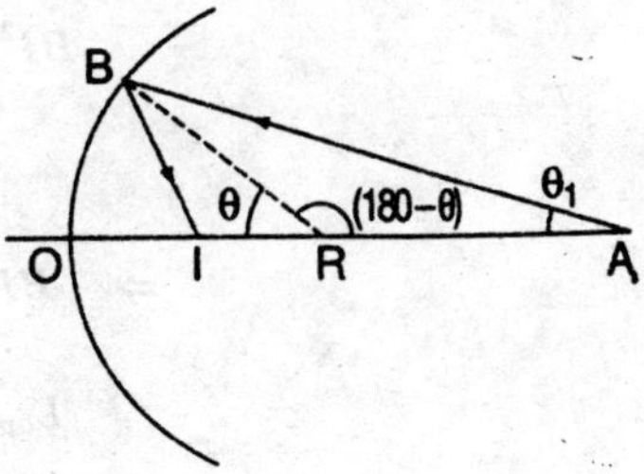

Fig 3.6

Since L_{op} is an extremum (minimum),

$$\frac{dL_{op}}{d\theta} = 0 \Rightarrow r^2\left(\frac{1}{x} + \frac{1}{y}\right) = 2r$$

$$\Rightarrow \frac{1}{x} + \frac{1}{y} = \frac{2}{r} \quad \text{(Reflection formula for curved reflector)} \tag{3.4.4}$$

Refraction at Curved Surface : From Fermat's Principle

Let PQ be the curved surface of radius of curvature r separating two media of r.i. values n_1 and $n_2 (n_2 > n_1)$. Thus the optical path L_{op} is given by,

$$L_{op} = n_1 AB + n_2 BI.$$

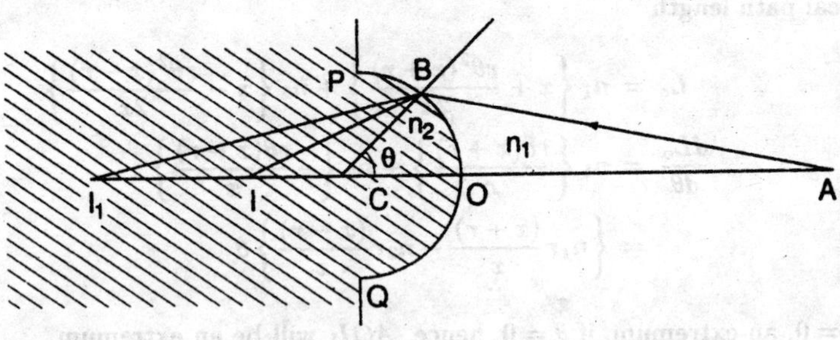

Fig 3.7

Now,

$$OA = x(+\text{ve}), OI = y(-\text{ve}), OC = r(-\text{ve})$$

$$\begin{aligned} \text{and} \quad AB^2 &= BC^2 + AC^2 - 2BCAC\cos\theta \\ &= r^2 + (x+r)^2 - 2r(x+r)\cos\theta \\ &= 2r^2 + x^2 + 2xr - 2r(x+r)(1-\theta^2/2) \\ &\simeq x^2 + r\theta^2(x+r) \end{aligned}$$

$$\Rightarrow \quad AB \simeq x + \frac{r\theta^2(x+r)}{2x} \tag{3.4.5}$$

Again,

$$\begin{aligned} BI^2 &= (y-r)^2 + r^2 - 2(y-r)r\cos(180-\theta) \\ &= y^2 + 2r^2 - 2yr + 2(y-r)r(1-\theta^2/2) \\ &= y^2 - yr\theta^2 + r^2\theta^2 = y^2 - r\theta^2(y-r) \end{aligned}$$

$$\Rightarrow \quad BI = y - \frac{r\theta^2(y-r)}{2y} \tag{3.4.6}$$

$$\therefore \quad L_{op} = n_1\left\{x + \frac{r\theta^2(x+r)}{2x}\right\} + n_2\left\{y - \frac{r\theta^2(y-r)}{2y}\right\}$$

$$\therefore \quad \frac{dL_{op}}{d\theta} = \left[\frac{n_1 r(x+r)}{x} - \frac{n_2 r(y-r)}{y}\right]\theta \tag{3.4.7}$$

So, for L_{op} to be an extremum, we have, $dL_{op}/d\theta = 0$

$$\therefore \quad \frac{n_1(x+r)}{x} - \frac{n_2(y-r)}{y} = 0$$

$$\Rightarrow \frac{n_2}{y} + \frac{n_1}{x} = \frac{n_2 - n_1}{r} \quad \text{(refraction formula for curved refractor)} \tag{3.4.8}$$

For convex surface refraction,

$$y = -\text{ve}, \quad r = -\text{ve}$$

$$\therefore \quad \frac{n_2}{-y} + \frac{n_1}{x} = \frac{n_2 - n_1}{-r} \quad \Rightarrow \quad \frac{n_2}{y} - \frac{n_1}{x} = \frac{n_2 - n_1}{r} \tag{3.4.9}$$

Again, when $\theta = 0, dL_{op}/d\theta = 0$, meaning that the path AOI is an extremum. Now, if I_1 is a point such that

$$OI_1 = x = OA, \quad \text{then} \quad BI_1 = x - \frac{r\theta^2(x-r)}{2x}$$

$\therefore$ the optical path length

$$L_o = n_1\left\{x + \frac{r\theta^2(x+r)}{2x}\right\} + n_2\left\{x - \frac{r\theta^2(x-r)}{2x}\right\}$$

$$\frac{dL_o}{d\theta} = n_1\left\{\frac{r\theta(x+r)}{x}\right\} + n_2\left\{-\frac{r\theta(x-r)}{x}\right\}$$

$$= \left\{n_1 r\frac{(x+r)}{x} - n_2 r\frac{(x-r)}{x}\right\}\theta \tag{3.4.10}$$

Thus $\frac{dL_o}{d\theta} = 0$, an extremum, if $\theta = 0$, hence, AOI_1 will be an extremum.
Differentiating (3.4.10) w.r.t. θ again,

$$\begin{aligned} \frac{d^2L_o}{d\theta^2} &= n_1 r\frac{(x+r)}{x} - \frac{n_2 r(x-r)}{x} \\ &= \frac{r}{x}\{n_1(x+r) - n_2(x-r)\} \\ &= n_1\frac{r}{x}\{(x+r) - \frac{n_2}{n_1}(x-r)\} \\ &= n_1\frac{r}{x}\{(x+r) - n(x-r)\}, \quad \text{putting } \tfrac{n_2}{n_1} = n \\ &= \frac{n_1 r}{x}\{(1-n)x + (n+1)r\} \\ &= \frac{n_1 r}{x}\{(n+1)r - (n-1)x\}. \end{aligned}$$

Thus,

$$\frac{d^2 L_o}{d\theta^2} = +\text{ve, hence } L_o \text{ is minimum if } x < \frac{(n+1)}{(n-1)} r < \frac{n_2 + n_1}{n_2 - n_1} r$$

$$= -\text{ve, hence } L_o \text{ is maximum if } x > \frac{(n+1)r}{(n-1)}$$

$$= 0, \text{ hence } L_o \text{ is stationary if } x = \frac{(n+1)r}{(n-1)}$$

3.5 The Path of a Ray in a Medium having r.i. Continuously Varying

If the medium be of constant refractive index, the ray path is a straight line, but if the r.i. of the medium varies from point to point, i.e., the r.i. is a function of the position, we would denote the refractive index function as $n(\bar{r})$.

In such a case the path will be a continuous curved path, say AP_1P_2. The optical path between two points P_1, P_2 is given by

$$\delta L_{P_1P_2} = n\delta s.$$

Now, let $\psi(\bar{r})$ = constant C_1 represent a wavefront, then another neighbouring wavefront is represented by $\psi(\bar{r} + d\bar{r}) = C_2$

$$\therefore \quad n\delta s = \psi(\bar{r} + \delta\bar{r}) - \psi(\bar{r}) = \bar{\nabla}\psi \cdot \delta\bar{r}$$

$$\Rightarrow \quad \bar{\nabla}\psi \cdot \lim_{\delta s \to 0} \frac{\delta\bar{r}}{\delta s} = n \quad \Rightarrow \quad \bar{\nabla}\psi \cdot \frac{d\bar{r}}{ds} = n$$

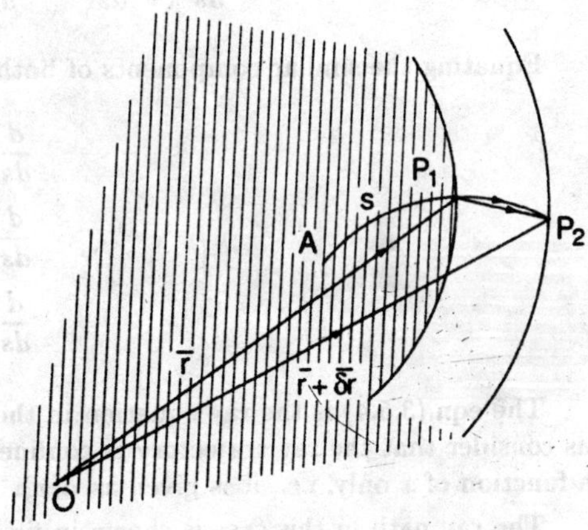

Fig 3.8

Now, $d\bar{r}/ds$ is a unit vector tangential to the path of the ray and normal to the wavefront surface $\psi(\bar{r}) = C_1$

$$\therefore \quad \frac{d\bar{r}}{ds} = \hat{s}.$$

So, $\bar{\nabla}\psi \cdot \frac{d\bar{r}}{ds} = n(r) \quad \Rightarrow \quad |\bar{\nabla}\psi|\hat{s} \cdot \hat{s} = n \quad \Rightarrow \quad |\bar{\nabla}\psi| = n$. Hence,

$$n\frac{d\bar{r}}{ds} = n\hat{s} = |\bar{\nabla}\psi|\hat{s} = \bar{\nabla}\psi. \tag{3.5.1}$$

The vector $n\hat{s}$ is referred to as ray vector.

Again,

$$n\frac{d\bar{r}}{ds} \cdot n\frac{d\bar{r}}{ds} = \bar{\nabla}\psi \cdot \bar{\nabla}\psi \quad \Rightarrow \quad n^2 = \left(\frac{\partial\psi}{\partial x}\right)^2 + \left(\frac{\partial\psi}{\partial y}\right)^2 + \left(\frac{\partial\psi}{\partial z}\right)^2. \tag{3.5.2}$$

The eqn.(3.5.1) is called *eikonal equation*.

Differentiating eqn.(3.5.1) w.r.t. s, we get,

$$\frac{d}{ds}\left(n\frac{d\bar{r}}{ds}\right) = \frac{d}{ds}(\bar{\nabla}\psi) = \bar{\nabla}\left(\frac{d\psi}{ds}\right).$$

Now, $d\psi = \psi(\bar{r} + d\bar{r}) - \psi(\bar{r}) = \bar{\nabla}\psi \cdot d\bar{r} \quad \Rightarrow \quad \frac{d\psi}{ds} = \bar{\nabla}\psi \cdot \frac{d\bar{r}}{ds}$

$$\therefore \quad \frac{d}{ds}\left(n\frac{d\bar{r}}{ds}\right) = \bar{\nabla}\left(\bar{\nabla}\psi \cdot \frac{d\bar{r}}{ds}\right) = \bar{\nabla}n. \tag{3.5.3}$$

The eqn.(3.5.3) is the vector equation of the light rays. If n is a constant, then eqn.(3.5.3) reduces to

$$n\frac{d^2\bar{r}}{ds^2} = 0 \quad \Rightarrow \quad \frac{d\bar{r}}{ds} = \hat{s} \quad \Rightarrow \quad \bar{r} = s\hat{s} + \bar{b},\ \bar{b} \text{ is a constant vector.}$$

This represents a straight line, hence if the r.i. is a constant, the path of a ray is a straight line.

From eqn.(3.5.3) we again get,

$$\frac{d}{ds}(n\hat{s}) = \bar{\nabla}n.$$

Now,

$$\hat{s} = i\frac{dx}{ds} + j\frac{dy}{ds} + k\frac{dz}{ds}$$

$$\therefore \quad \frac{d}{ds}\left(in\frac{dx}{ds} + jn\frac{dy}{ds} + kn\frac{dz}{ds}\right) = i\frac{dn}{dx} + j\frac{dn}{dy} + k\frac{dn}{dz}.$$

Equating the similar components of both sides, we get the three-component eqns. as,

$$\begin{aligned} \frac{d}{ds}\left(n\frac{dx}{ds}\right) &= \frac{dn}{dx} \\ \frac{d}{ds}\left(n\frac{dy}{ds}\right) &= \frac{dn}{dy} \\ \frac{d}{ds}\left(n\frac{dz}{ds}\right) &= \frac{dn}{dz} \end{aligned} \tag{3.5.4}$$

The eqn.(3.5.4) is the ray equation in the component form. As an application of this equation, let us consider that the ray trajectory is confined to the x-z plane and the r.i. function of the medium is a function of x only, i.e., it is given as $n(x)$.

The ray path in this case is shown in fig.(3.9).

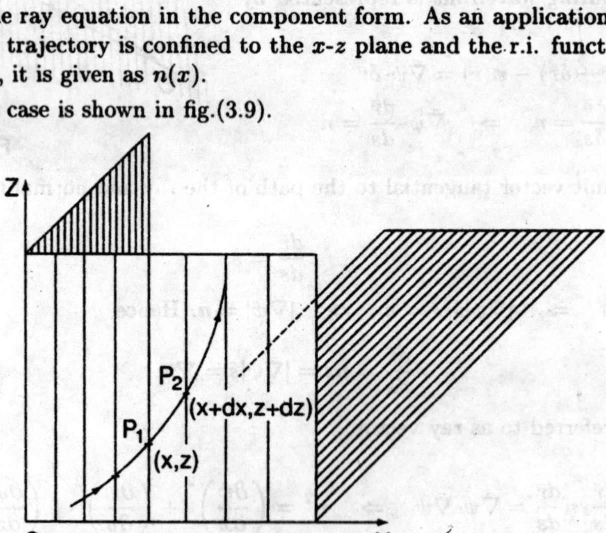

Fig 3.9

Since s, the arc length of the ray trajectory, lies in the x-z plane, it is a function of (x, z), so from eqn.(3.5.4) we get, also remembering that $n = n(x)$ only,

$$\frac{d}{ds}\left(n\frac{dx}{ds}\right) = \frac{dn}{dx} \tag{3.5.5}$$

$$\frac{d}{ds}\left(n\frac{dz}{ds}\right) = 0. \tag{3.5.6}$$

Solving (3.5.6) we get,

$$n\frac{dz}{ds} = \text{constant}.$$

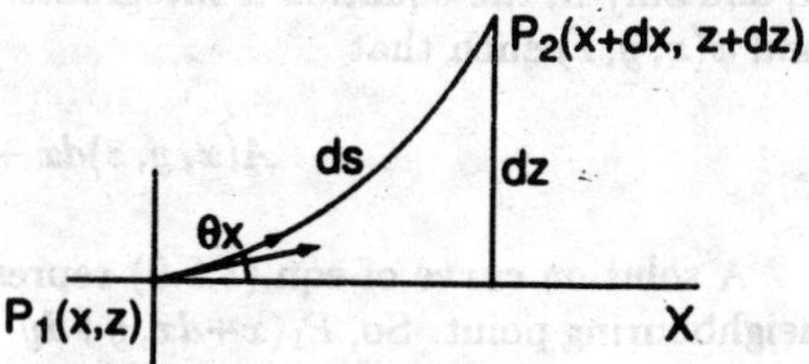

Fig 3.10

Now $dz/ds = \sin\theta_x$ (cf. fig.3.10), where θ_x is the angle made by the ray vector with the direction of X at the point $P_1(x, z)$.

$$\therefore \quad n(x)\sin\theta_x = \text{constant} = n_1 \sin\theta_1$$

Again,

$$ds^2 = dx^2 + dz^2 \quad \Rightarrow \quad \left(\frac{ds}{dz}\right)^2 = 1 + \left(\frac{dx}{dz}\right)^2$$

$$\Rightarrow \quad \left(\frac{dx}{dz}\right)^2 = \left(\frac{ds}{dz}\right)^2 - 1 = \frac{1}{\sin^2\theta_x} - 1$$

$$\therefore \quad \left(\frac{dx}{dz}\right)^2 = \frac{n^2}{n_1^2\sin^2\theta_1} - 1 \tag{3.5.7}$$

The solution of the eqn.(3.5.7) gives us relation between x and z determining the trajectory. Differentiating eqn.(3.5.7) w.r.t. z, we get,

$$2\frac{dx}{dz}\cdot\frac{d^2x}{dz^2} = \frac{1}{n_1^2\sin^2\theta_1}\frac{d}{dx}(n^2)\cdot\frac{dx}{dz} \quad \Rightarrow \quad \frac{d^2x}{dz^2} = \frac{1}{2n_1^2\sin^2\theta_1}\frac{d}{dx}(n^2). \tag{3.5.8}$$

The eqn.(3.5.8) is one-dimensional ray equation and its solution would connect x and z giving the ray trajectory.

Application

If n is a constant, then eqn.(3.5.8) reduces to

$$\frac{d^2x}{dz^2} = 0 \quad \Rightarrow \quad x = z + c,$$

representing a straight line passing through $x = c$ on the x-axis.

Many types of optical problems may be solved by solving eqn. (3.5.7) or eqn. (3.5.8) subject to the different forms of refractive index function $n(x)$. If the refractive index function is given by, $n^2(x) = n_0^2 - n_1 x^2$, the ray trajectory becomes sinusoidal but the period of the sinusoidal function depends on the angle of launching. Such a variation of r.i. finds its use in optical communication.

If the variation of the r.i. function be represented as $n^2(x) = n_0^2 + n_1^2 e^{-\alpha x}$, it represents the optical condition of the atmospheric air above the cold sea-water. In such a case an object appears to be above its actual position giving rise to the phenomenon of inferior mirage, also called *looming*.

Caratheodory's Principle and Fermat's Principle

According to Fermat's principle, there are points in an optical medium inaccessible to light ray by actual optical process. This reminds us of a mathematical theorem due to Caratheodory which states that

In the neighbourhood of any arbitrary point $P_0(x, y, z)$, there are points which are not accessible from P_0 along solution curves of the equation

$$A(x, y, z)dx + B(x, y, z)dy + C(x, y, z)dz = 0 \tag{3.5.9}$$

if, and only if, the equation is integrable. The eqn.(3.5.9)is integrable if we have two functions $\lambda(x,y,z)$ and $\phi(x,y,z)$ such that

$$A(x,y,z)dx + B(x,y,z)dy + C(x,y,z)dz = \lambda d\phi. \tag{3.5.10}$$

A solution curve of eqn.(3.5.9) represents a connecting link between P_0, the initial point and P_1, a neighbouring point. So, $P_1(x+dx, y+dy, z+dz)$ is not accessible from $P_0(x,y,z)$ by all curves connecting the two points, provided eqn.(3.5.9) is integrable. Let us try to use this principle in case of an optical system in which a ray passes from $P_0(x,y,z)$ a source point of light to a point $P_1(x+dx, y+dy, z+dz)$, the image point by some optical process.

This point P_1 is not accessible to light ray along solution curves of the eqn.(3.5.9) if it is integrable in the present case. Any solution curve represents a connecting link between the two points.

According to Fermat's principle, points like P_1 (image point) are not accessible from P_0 along any and every path connecting the two points; there is one and one connecting optical bridge between the two points. So the eqn.(3.5.9) in this case is to be integrable according to Caratheodory. So there are two functions λ and ϕ, such that eqn.(3.5.10) is satisfied. Now, in optical system, we can introduce n, the refractive index function, and s (connecting length in space) such that

$$Adx + Bdy + Cdz = nds. \tag{3.5.11}$$

Thus Caratheodory's principle initiates two functions $n(x,y,z)$, called *refractive index function* (RIF) and $s(x,y,z)$, called *path function.* Equation (3.5.9) implies that we can write, $nds = Adx + Bdy + Cdz = \bar{\nabla}\psi\cdot d\bar{r} = d\psi$ where $\psi(x,y,z)$is defined by the relation,

$$\frac{\partial\psi}{\partial x} = A,\ \frac{\partial\psi}{\partial y} = B,\ \frac{\partial\psi}{\partial z} = C.$$

From above, we get,

$$n = \bar{\nabla}\psi\cdot\frac{d\bar{r}}{ds}$$

$\bar{\nabla}\psi$ is a vector normal to the surface $\psi(x,y,z) = C_1$, also $\frac{d\bar{r}}{ds}$ is a vector tangential to the ray-path. If $\psi(x,y,z) = C_1$ be taken as representing the wavefront, then

$$\bar{\nabla}\psi = \hat{s}|\bar{\nabla}\psi|,\quad \frac{d\bar{r}}{ds} = \hat{s}$$

where $\hat{s}$ is a unit vector $\perp$ to the wavefront.

$$\therefore\ n = |\bar{\nabla}\psi|\hat{s}\cdot\hat{s} = |\bar{\nabla}\psi| \tag{3.5.12}$$

$$\therefore\ n\frac{d\bar{r}}{ds}\cdot n\frac{d\bar{r}}{ds} = |\bar{\nabla}\psi|\hat{s}\cdot|\bar{\nabla}\psi|\hat{s} = \bar{\nabla}\psi\cdot\bar{\nabla}\psi$$

$$\Rightarrow n^2 = \left(\frac{\partial\psi}{\partial x}\right)^2 + \left(\frac{\partial\psi}{\partial y}\right)^2 + \left(\frac{\partial\psi}{\partial z}\right)^2$$

This equation is called *eikonal equation* in optics. So just like Fermat's principle can be used to derive equations of optical process, Caratheodory's principle can also be used to derive optical equations representing optical processes.

Solved Numerical Problems

1. O is the centre of a spherical mirror of radius R. P_1 and P_2 are two points on a diameter and at the same distance w.r.t. the centre O (fig. 3.11).

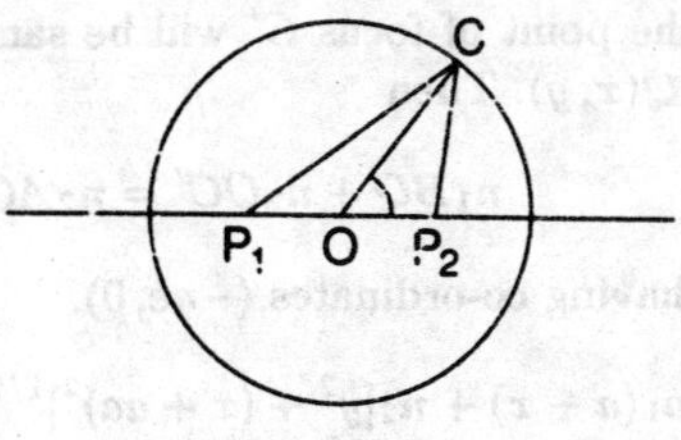

Fig 3.11

(a) find the optical path length $P_1C + P_2C$ as a function of θ.

(b) find the values of θ for which P_1CP_2 is a ray-path due to reflection at the sphere.

Solution:

$$P_1C^2 = P_1O^2 + OC^2 - 2P_1O \cdot OC\cos(180° - \theta)$$
$$= a^2 + R^2 + 2aR\cos\theta$$
$$\Rightarrow \quad P_1C = (a^2 + R^2 + 2aR\cos\theta)^{1/2}.$$

Similarly,

$$P_2C^2 = (a^2 + R^2 - 2aR\cos\theta)^{1/2}.$$

$$\therefore \quad L_{OP} = P_1C + P_2C = (a^2 + R^2 - 2aR\cos\theta)^{1/2} + (a^2 + R^2 - 2aR\cos\theta)^{1/2}.$$

For actual ray-path,

$$\frac{dL_{OP}}{d\theta} = 0 \Rightarrow \left[\frac{1}{2}\cdot\frac{-2aR\sin\theta}{\sqrt{a^2 + R^2 + 2aR\cos\theta}} + \frac{1}{2}\cdot\frac{2aR\sin\theta}{\sqrt{a^2 + R^2 - 2a\cos\theta}}\right] = 0$$

$$\Rightarrow \sin\theta\left[\frac{1}{\sqrt{a^2 + R^2 - 2aR\cos\theta}} - \frac{1}{\sqrt{a^2 + R^2 + 2aR\cos\theta}}\right] = 0.$$

So, $\sin\theta = 0 \Rightarrow \theta = 0$.

$$a^2 + R^2 + 2aR\cos\theta = a^2 + R^2 - 2aR\cos\theta \Rightarrow \cos\theta = 0 \Rightarrow \theta = \frac{\pi}{2}.$$

2. **Rotating an ellipse about its major axis, we get what is known as an ellipsoid of revolution. Show by using Fermat's principle that all rays parallel to the major axis of the ellipse will focus to one of the focal points of the ellipse (fig. 3.12) provided the eccentricity of ellipse equals (n_1/n_2).**

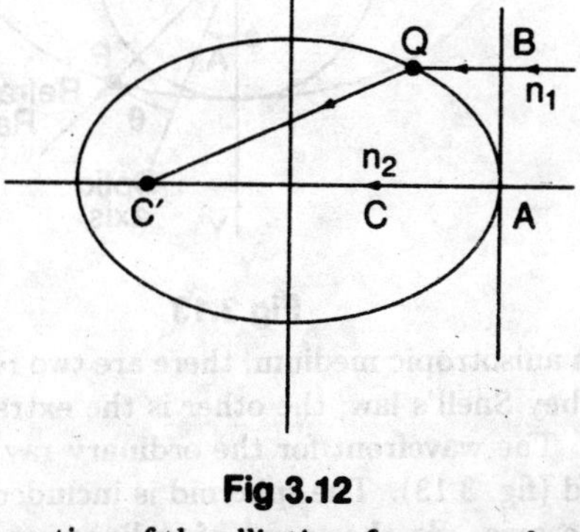

Fig 3.12

Solution. Fig. (3.12) shows a section of the elliptic refractor and a ray is incident on the refractor parallel to the major axis at the point Q of the refractor. AB is the incident wavefront to the refractor and so, according to Fermat's principle, the optical path actually travelled by light from the incident wavefront to the point of focus C' will be same. Let the ray parallel to the major axis strike the refractor at $Q(x, y)$. Then

$$n_1 BQ + n_2 QC' = n_2 AC'$$

C' is a focus of the ellipse having co-ordinates $(-ae, 0)$.

$$\therefore \quad n_1(a-x) + n_2[y^2 + (x+ae)^2]^{1/2} = n_2(a+ae).$$

If $e = \frac{n_1}{n_2}$, we would get,

$$n_2\left[y^2 + \left(x + a \times \frac{n_1}{n_2}\right)^2\right]^{1/2} = n_2\left(a + a\cdot\frac{n_1}{n_2}\right) - n_1(a-x)$$

$$= n_2 a + n_1 a - n_1 a + n_1 x$$

$$\Rightarrow \quad [n_2^2 y^2 + (n_2 x + n_1 a)^2]^{1/2} = n_2 a + n_1 x.$$

Squaring both sides, we get,

$$n_2^2 y^2 + (n_2 x + n_1 a)^2 = (n_2 a + n_1 x)^2$$

$$\Rightarrow \quad n_2^2 y^2 + n_2^2 x^2 + 2n_1 n_2 ax + n_1^2 a^2 = n_2^2 a^2 + 2n_1 n_2 ax + n_1^2 x^2$$

$$\Rightarrow \quad (n_2^2 - n_1^2)x^2 + n_2^2 y^2 = (n_2^2 - n_1^2)a^2$$

$$\Rightarrow \quad \frac{x^2}{a^2} + \frac{y^2}{(n_2^2 - n_1^2)/a^2} = 1.$$

So the point Q lies on the surface of the ellipse. So all such parallel rays after refraction would converge at C' which is the focus of the ellipse.

3. Application of Fermat's Principle in case of refraction into an anisotropic medium from an isotropic medium.

Solution. In case of refraction in between two isotropic medium, the refractive index remains contestant in all directions, but in case of an anisotropic medium (i.e., in case of a crystal), the refractive index is dependent on direction of propagation.

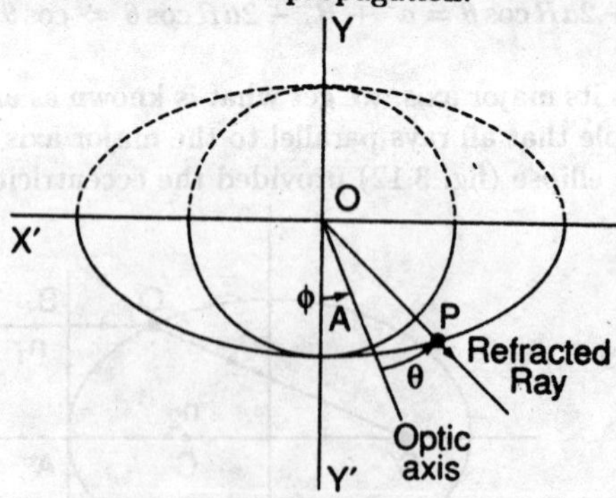

Fig 3.13

When light travels, into an anisotropic medium, there are two refracted rays—one is the ordinary refracted ray, because it obey Snell's law; the other is the extra-ordinary ray (E-ray), because it does not obey Snell's law. The wavefront for the ordinary ray is a spheroid and the wave-front for the E-ray is an ellipsoid (fig. 3.13). The spheroid is included within the ellipsoid and touches each other along the optic axis. So the speed of ordinary ray and the extraordinary is same along the optic axis. The speed of ordinary ray is same in all directions of the refracting medium but the speed of extraordinary ray depends on the direction. Let OP be a general direction

of refracted extraordinary ray, it makes an angle θ with the optic axis. The speed of the ray along this direction will be proportional to the length OP and can be written as $v(\theta)$ given by $v(\theta) = v_0 \frac{OP}{OA}$. OP is a general direction and OA is the direction of the optic axis along which speed is maximum and equal to v_0.
Now, $OP = r = \sqrt{x^2 + y^2}$ where $P(x, y)$ lies on the surface of the ellipsoid.

$$\therefore \quad v(\theta) = v_0 \times \frac{\sqrt{x^2 + y^2}}{a}.$$

Now, the ellipse can be represented parametrically as,

$$x = r\cos\theta, \quad y = r\sin\theta.$$

So, we get

$$\frac{x^2}{a^2} + \frac{y^2}{b^2} = 1 \Rightarrow r^2\left(\frac{\cos^2\theta}{a^2} + \frac{\sin^2\theta}{b^2}\right) = 1$$

$$\Rightarrow r = \frac{ab}{(b^2\cos^2\theta + a^2\sin^2\theta)^{1/2}}. \tag{3.5.13}$$

$$\therefore \quad v(\theta) = v_0 \times \frac{ab}{a(b^2\cos^2\theta + a^2\sin^2\theta)^{1/2}}$$

$$= \frac{v_0 b}{(b^2\cos^2\theta + a^2\sin^2\theta)^{1/2}}.$$

So, the r.i. along the general direction θ is given by

$$n(\theta) = \frac{c}{v(\theta)} = \frac{c}{v_0 b}(b^2\cos^2\theta + a^2\sin^2\theta)^{1/2}.$$

$$n^2(\theta) = n_{11}^2\cos^2\theta + n_\perp^2\sin^2\theta = n_e^2\cos^2\theta + n_0^2\sin^2\theta$$

$$\Rightarrow \quad n_{11} = \text{r.i. along the optic axis}$$

$$= \frac{c}{v_{op}} = \text{r.i. for ordinary ray.}$$

$$n_\perp = \frac{c}{v_{op}} \cdot \frac{a}{b} = \text{r.i. } \perp \text{ to the optic axis.}$$

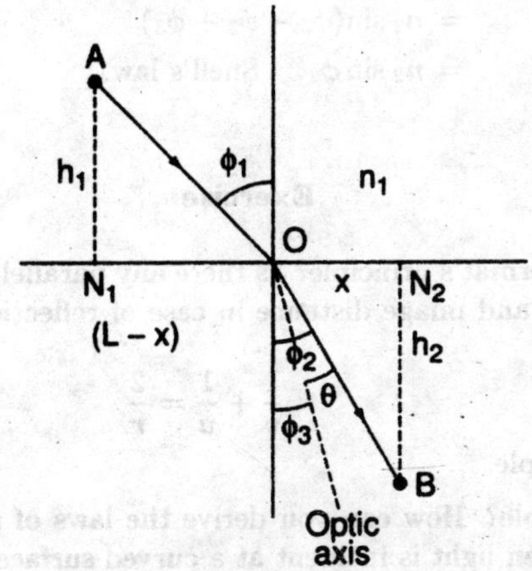

Fig 3.14

Consider fig. (3.14), O is the point of incidence in the first isotropic medium of r.i. n_1 and OB is the refracted ray, angle of incidence being ϕ_1, the angle of refraction ϕ_2; the angle made by the optic axis with the normal is ϕ_3.
The optical path from A to B is

$$L_{op} = n_1\sqrt{h_1^2 + (L-x)^2} + n(\theta)\sqrt{h_2^2 + x^2}.$$

In fig. (3.14) we have $\phi_3 = (\phi_2 - \theta)$

$$\begin{aligned} n^2(\theta) &= n_{11}^2 \cos^2\theta + n_\perp^2 \sin^2\theta \\ \Rightarrow \quad n(\theta) &= [n_{11}^2 \cos^2(\phi_2 - \phi_3) + n_\perp^2 \sin^2(\phi_2 - \phi_3)]^{1/2} \\ &= n_{11}^2[n_{11}^2(\cos\phi_2\cos\phi_3 + \sin\phi_2\sin\phi_3)^2 \\ &\qquad + n_\perp^2(\sin\phi_2\cos\phi_3 - \cos\phi_2\sin\phi_3)^2]^{1/2} \\ &= \left[n_{11}^2\left(\frac{h_2}{\sqrt{h_2^2+x^2}}\cos\phi_3 + \frac{x}{\sqrt{h_2^2+x^2}}\sin\phi_3\right)^2\right. \\ &\qquad \left. + n_\perp^2\left(\frac{x}{\sqrt{h_2^2+x^2}}\cos\phi_3 - \frac{h_2}{\sqrt{h_2^2+x^2}}\sin\phi_3\right)^2\right]^{1/2} \\ \Rightarrow \quad n(\theta) &= \frac{1}{\sqrt{h_2^2+x^2}}\left[n_{11}^2(h_2\cos\phi_3 + x\sin\phi_3)^2 + n_\perp^2(x\cos\phi_3 - h_2\sin\phi_3)^2\right]^{1/2}. \end{aligned}$$

$$\therefore \quad L_{op} = n_1\sqrt{n_1^2 + (L-x)^2} + [n_{11}^2(n_2\cos\phi_3 + x\sin\phi_3)^2 + n_\perp^2(x\cos\phi_3) - h_2\sin\phi_3)^2]^{1/2}.$$

If AOB is the actual ray-path, then

$$\frac{dL_{op}}{dx} = 0.$$

So, we get

$$n_1 \sin\phi_1 = \frac{n_{11}^2\cos\theta\sin\phi_3 + n_\perp^2\sin\theta\cos\phi_3}{[n_{11}^2\cos^2\theta + n_\perp^2\sin^2\theta]^{1/2}}.$$

From the relation above, if $n_{11} = n_\perp = n_2$, which is true for an isotropic medium

$$\begin{aligned} n_1\sin\phi_1 &= n_2[\cos(\phi_2 - \phi_3)\sin\phi_3 + \sin(\phi_2 - \phi_3)\cos\phi_3] \\ &= n_2\sin(\phi_2 - \phi_3 + \phi_3) \\ &= n_2\sin\phi_2, \quad \text{Snell's law.} \end{aligned}$$

Exercise

1. What do you mean by Fermat's principle? Is there any parallel principle? Show that the relation between object distance and image distance in case of reflection by a concave reflector is given by

$$\frac{1}{v} + \frac{1}{u} = \frac{2}{r}$$

applying Fermat's principle.

2. What is Fermat's principle? How can you derive the laws of reflection and refraction with the help of this principle when light is incident at a curved surface?

3. State Fermat's principle and deduce from it the law of refraction at the plane surface separating two media.

4. In general, when a ray passes from one point to another through different refracting media by obeying the laws of refraction, the total optical path may be a minimum or a maximum—justify the statement.
5. Explain 'Fermat's principle of stationary time', illustrating your account by using the principle to solve some problems in geometrical optics.
6. Show that in case of refraction at a curved refracting surface, the optical path may be a maximum or a minimum. Find the condition when it is maximum and when it is minimum.
7. Derive the eikonal equation and also the vector equation of the light rays. If n is a constant, show that the ray path is a straight line.
8. Solve for the ray-trajectory if the refractive index (r.i.) function is defined as $n^2(x) = n_0^2 - n_1 x^2$, where n_0 is the space independent part of the r.i. function.
9. What do you mean by 'looming'? Theoretically explain the formation of this phenomenon.
10. State Fermat's principle. Show how the laws of refraction can be obtained from Fermat's principle. **[CU 1990]**
11. Consider a refracting surface of radius of curvature R, separating two media of refractive indices n_1 and n_2. A straight line passing through the centre of curvature C intersects the refracting surface at O. P and Q are two points on the straight line on opposite sides of the refracting surface such that $PO = OQ = x$. Show that the straight line path POQ corresponds to an extremum. (cf. pp. 19–21) **[CU 1990]**
12. Show that the optical path connecting two points through the phenomenon of refraction may be a maximum.
13. In problem 11, show that the extremum will be a minimum when $x < R \times \left(\frac{n+1}{n-1}\right)$, maximum when $x > R \times \left(\frac{n+1}{n-1}\right)$ and stationary when $x = R\left(\frac{n+1}{n-1}\right)$ where $n = \left(\frac{n_2}{n_1}\right)$. **[BU 1992]**
14. Define optical path length between two points. State Fermat's principle of stationary optical path. How is reversibility incorporated in this principle? **[BU 1994]**
15. Consider an elliptic reflector whose focii are at the points S_1 and S_2. Show that the rays from S_1 will pass through S_2 after reflection. Comment on the nature of the path. **[BU 1992]**
[**Hints.** See pp. 106-107]
16. Define Fermat's principle of least time. Show that the optical path of a light ray during refraction between two plane media of r.i. n_1 and n_2 is least in comparison to any neighbouring geometric optical path.
17. Show how Fermat's principle leads to establish the conjugate focii for a thin lens.
18. Using Fermat's principle, prove that for a thin lens immersed in a medium of r.i. n_m, the following relation

$$\frac{1}{f} = \frac{n_1 - n_m}{n_m}\left(\frac{1}{R_1} - \frac{1}{R_2}\right),\ n_1 = \text{r.i. of the incident medium.}$$

References

1. *Principles of Optics*—Born and Wolf, Pergamon Press.
2. *Optics*—A. Ghatak, Tata McGraw-Hill Co., New Delhi.
3. *A Textbook of Light*—K.G. Majumder, Modern Book Agency, Calcutta.
4. *American Journal of Physics*—H.A. Buchdahl, 1949.
5. *Heat and Thermodynamics*—M.W. Zemansky, McGraw-Hill Book Company.

Chapter 4

Refraction at Plane Surface

4.1 Refraction at a Plane Surface

A ray of light of a particular colour passing through a certain medium of r.i. value n_1 when enters into a second medium of r.i. value n_2 at the point O, the incident ray and refracted ray are related by certain laws of optics, called *Snell's law*, which states

Law 1. *The incident ray, the refracted ray and the normal at the point of incidence are coplanar, i.e., all lie in one plane.*

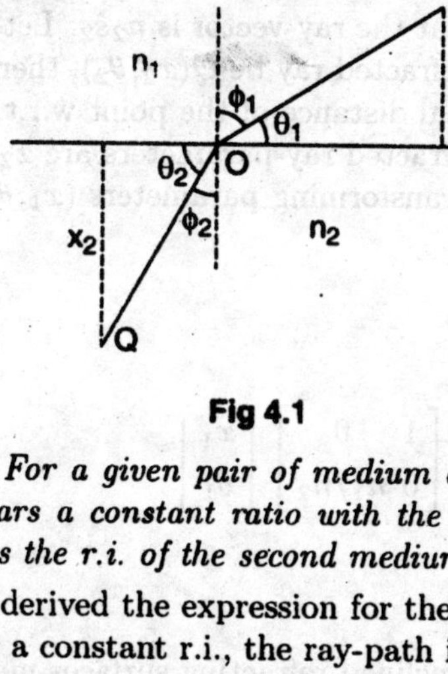

Fig 4.1

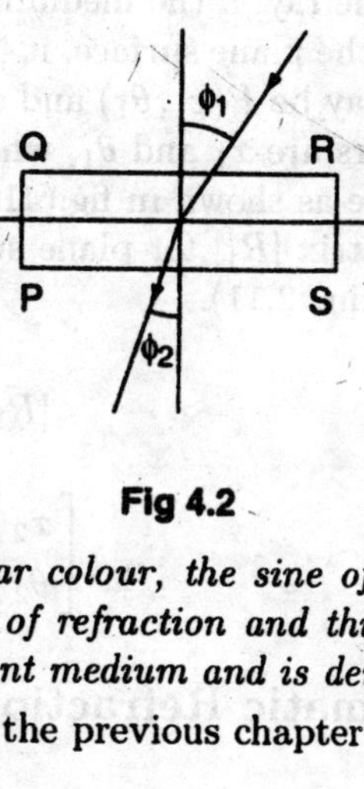

Fig 4.2

Law 2. *For a given pair of medium and for a particular colour, the sine of the angle of incidence bears a constant ratio with the sine of the angle of refraction and this constant is referred to as the r.i. of the second medium w.r.t. the incident medium and is denoted as* $^1n_2$.

We have derived the expression for the ray trajectory in the previous chapter and we have seen that for a constant r.i., the ray-path is a straight line.

The incident ray-vector is $n_1\bar{s}_1$, where n_1 is the r.i. of the incident medium, the refracted ray vector is $n_2\bar{s}_2$. Now, from eqn.(3.5.1) we get,

$$n\hat{s} = \vec{\nabla}\psi, \quad \therefore \quad \text{curl}\, n\hat{s} = \vec{\nabla} \times (\vec{\nabla}\psi) = 0. \tag{4.1.1}$$

If we consider a closed contour $PQRS$ enclosing certain area such that PS and QR are parallel to the surface of separation and PQ and $SR \perp$ to it, then in view of eqn.(4.1.1), we can write,

$$\int_A \text{curl}\,(n\hat{s})\cdot \vec{dA} = \oint (n\bar{s})\cdot \vec{dl} = 0. \tag{4.1.2}$$

Applying eqn.(4.1.2) to the contour of fig.(4.2), we get,

$$\oint (n\hat{s})\cdot d\vec{l} = \int_{PQ} n\hat{s}\cdot d\vec{l} + \int_{QR} n\hat{s}\cdot d\vec{l} + \int_{RS} n\hat{s}\cdot d\vec{l} + \int_{SP} n\hat{s}\cdot d\vec{l} = 0.$$

Since PQ and RS are very small lengths, line integral of $(n\hat{s})$ over these paths is negligibly small and hence,

$$\int_{QR} n\hat{s}\cdot d\vec{l} + \int_{SP} n\hat{s}\cdot d\vec{l} = 0 \quad \Rightarrow \quad \int_{L_1=QR} n_1\hat{s}_1\cdot d\vec{l} + \int_{L_2=SP} n_2\hat{s}_2\cdot d\vec{l} = 0 \tag{4.1.3}$$

Now,

$$\hat{s}_1\cdot d\vec{l} = \cos\left(\frac{\pi}{2}+\phi_1\right) dl = -\sin\theta_1 dl$$

$$\hat{s}_2\cdot d\vec{l} = \cos\left(\frac{\pi}{2}-\phi_2\right) dl = \sin\theta_2 dl$$

$\therefore$ Eqn.(4.1.3) gives

$$-n_1\sin\theta_1 L_1 + n_2\sin\theta_2 L_2 = 0 \quad \Rightarrow \quad n_1\sin\theta_1 = n_2\sin\theta_2$$

since $L_1 = L_2$, which is nothing but Snell's law.

Again, the incident ray is determined by the ray vector $n_1\hat{s}_1$, and this ray vector is in the direction of the ray; if the medium be homogeneous then the ray-path is straight; as a result of refraction at the plane surface, it is refracted such that the ray-vector is $n_2\hat{s}_2$. Let any point on the incident ray be $P(x_1,\theta_1)$ and any point on the refracted ray be $Q(x_2,\theta_2)$, then the incident ray-parameters are x_1 and θ_1, where x_1 is the normal distance of the point w.r.t. the surface, θ_1 is the angle as shown in fig.(41); similarly, the refracted ray-parameters are x_2 and θ_2. The refraction matrix $[R_1]$ for plane surface refraction transforming parameters (x_1,θ_1) to (x_2,θ_2) is given as (eqn. 2.11).

$$[R_1] = \begin{bmatrix} 1 & 0 \\ 0 & n_1/n_2 \end{bmatrix}$$

$$\therefore \quad \begin{bmatrix} x_2 \\ \theta_2 \end{bmatrix} = [R_1]\begin{bmatrix} x_1 \\ \theta_1 \end{bmatrix} = \begin{bmatrix} 1 & 0 \\ 0 & n_1/n_2 \end{bmatrix}\begin{bmatrix} x_1 \\ \theta_1 \end{bmatrix} \tag{4.1.4}$$

4.2 Prismatic Refraction

A prism is a transparent medium bounded by two inclined refracting surfaces meeting along a line, called *edge of the prism.* The angle between the inclined surfaces is called the *refracting angle* or *simply the angle of the prism.* A section of the medium cut at right-angles to the edge is called a *principal section of the prism,* shown in the diagram as ABC. It is easy to see that

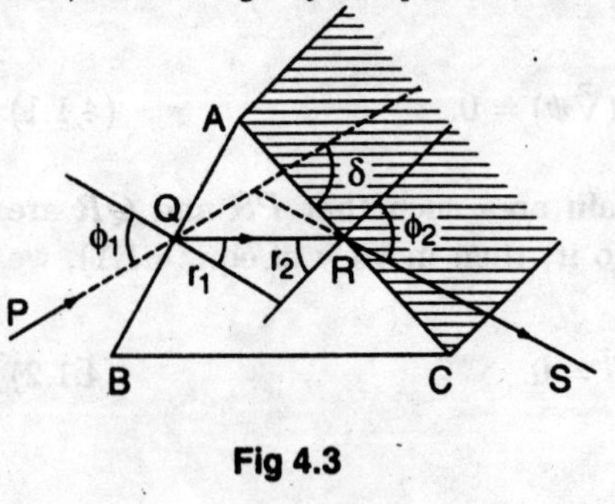

Fig 4.3

$$\delta = \text{total deviation by the prism} = \phi_1 + \phi_2 - (r_1 + r_2).$$

Again, $r_1 + r_2 = A$ (angle of the prism)

$$\therefore \quad \delta = \phi_1 + \phi_2 - A.$$

Also, $\sin\phi_1 = n\sin r_1$ and $\sin\phi_2 = n\sin r_2$.

Thus, we have a set of equations in case of refraction through a prism

$$\sin\phi_1 = n\sin r_1,\ \sin\phi_2 = n\sin r_2,\ \delta = \phi_1+\phi_2-A,\ A=(r_1+r_2). \quad (4.2.1)$$

From the third equation we easily get,

$$(\delta + A) = \phi_1 + \phi_2. \quad (4.2.2)$$

Now, A is a fixed angle; therefore, the condition for the left-hand side of eqn.(4.2.2) to be minimum corresponds to the condition of the deviation to be minimum.

Differentiating eqn.(4.2.2) w.r.t. ϕ_1 and setting it equal to zero for minimum, we get,

$$\frac{d\delta}{d\phi_1} = 0 = 1 + \frac{d\phi_2}{d\phi_1}. \quad (4.2.3)$$

Again, $\sin\phi_1 = n\sin r_1$ and $\sin\phi_2 = n\sin r_2$. Differentiating w.r.t. ϕ_1, we get,

$$\cos\phi_1 = n\cos r_1\frac{dr_1}{d\phi_1} \quad \text{and} \quad \cos\phi_2\frac{d\phi_2}{d\phi_1} = n\cos r_2\frac{dr_2}{d\phi_1}$$

$$\therefore \frac{d\phi_1}{d\phi_2}\cdot\frac{\cos\phi_1}{\cos\phi_2} = \frac{\cos r_1}{\cos r_2}\cdot\frac{dr_1}{d\phi_1}\cdot\frac{d\phi_1}{dr_2}$$

$$A = r_1 + r_2 \quad \Rightarrow \quad \frac{dA}{d\phi_1} = 0 = \frac{dr_1}{d\phi_1} + \frac{dr_2}{d\phi_1}$$

$$\therefore \frac{\cos\phi_1}{\cos\phi_2} = \frac{\cos r_1}{\cos r_2} \quad \Rightarrow \quad \frac{\cos^2\phi_1}{\cos^2\phi_2} = \frac{\cos^2 r_1}{\cos^2 r_2} = \frac{1-\sin^2 r_1}{1-\sin^2 r_2}$$

$$\Rightarrow \frac{1-\sin^2\phi_1}{1-\sin^2\phi_2} = \frac{1-\frac{\sin^2\phi_1}{n^2}}{1-\frac{\sin^2\phi_2}{n^2}} = \frac{n^2-\sin^2\phi_1}{n^2-\sin^2\phi_2}$$

$$= \frac{(1-n^2)}{(1-n^2)} = 1 \left(\text{cf. } \frac{a}{b} = \frac{c}{d} = \frac{a-c}{b-d}\right)$$

$$\Rightarrow \sin^2\phi_1 = \sin^2\phi_2 \quad \Rightarrow \quad \phi_1 = \phi_2.$$

Since,

$$n = \frac{\sin\phi_1}{\sin r_1} = \frac{\sin\phi_2}{\sin r_2}, \quad \therefore \quad r_1 = r_2.$$

The condition for minimum deviation implies $r_1 = r_2$ which in turn leads to the symmetric passage of light through the prism as shown in fig.(4.4) where the ray $PQRS$ passes symmetrically through the prism such that $AQ = AR \Rightarrow \angle AQR = \alpha = \angle ARQ$.

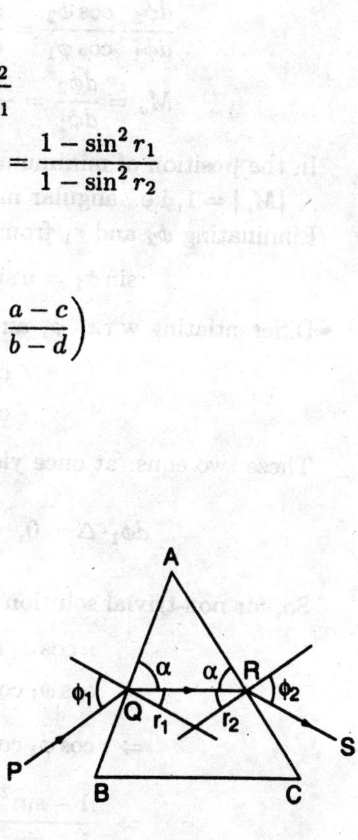

Fig 4.4

$$\therefore \quad \alpha + r_1 = \frac{\pi}{2} = \alpha + r_2 \quad \Rightarrow \quad r_1 = r_2,$$

a necessary consequence when light passing through the prism has minimum deviation.

4.3 Angular Magnification due to a Prism (M_a)

For two rays coming from two sources P and P_1 having an angular separation $d\phi_1$, w.r.t. the point Q, the ray-paths in and out of the prism are shown. After emergence through the prism the angles of emergence are respectively given by ϕ_2 and $(\phi_2 + d\phi_2)$ respectively. Thus the angular separation of the emergent rays is given by $\phi_2 - (\phi_2 + d\phi_2) = -d\phi_2$, which in, other words, means if the angle of incidence increases, the angle of emergence decreases. The angular magnification is given by, $M_a = \dfrac{d\phi_2}{d\phi_1}$ and it would come with a negative sign.

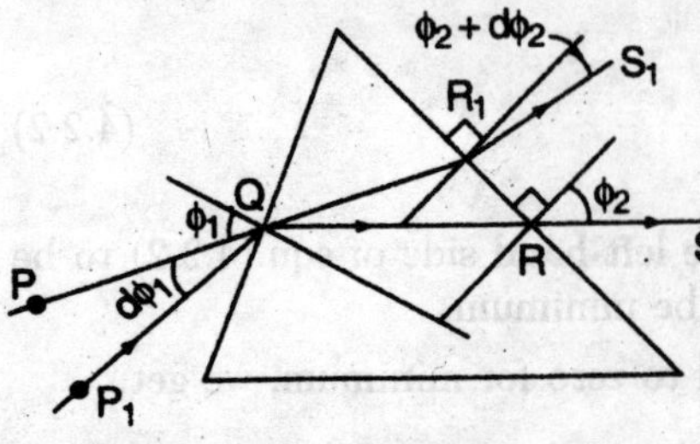

Fig 4.5

In case of prism,

$$\sin\phi_1 = n\sin r_1 \quad \text{and} \quad \sin\phi_2 = n\sin r_2.$$

Differentiating w.r.t. ϕ_1 and ϕ_2, we get,

$$\cos\phi_1 d\phi_1 = n\cos r_1 dr_1 \quad \text{and} \quad \cos\phi_2 d\phi_2 = n\cos r_2 dr_2.$$

Again, $r_1 + r_2 = A \quad \Rightarrow \quad dr_1 = -dr_2$

$$\therefore \frac{d\phi_2}{d\phi_1}\cdot\frac{\cos\phi_2}{\cos\phi_1} = \frac{\cos r_2}{\cos r_1}\cdot\frac{dr_2}{dr_1} = -\frac{\cos r_2}{\cos r_1} \Rightarrow \frac{d\phi_2}{d\phi_1} = -\frac{\cos\phi_1\cos r_2}{\cos\phi_2\cos r_1}$$

$$\therefore M_a = \frac{d\phi_2}{d\phi_1} = -\frac{\cos\phi_1\cos r_2}{\cos\phi_2\cos r_1} \tag{4.3.1}$$

In the position of minimum deviation, $\phi_1 = \phi_2, r_1 = r_2$.

$\therefore |M_a| = 1$, i.e., angular magnification is numerically equal to unity.

Eliminating ϕ_2 and r_1 from eqn.(4.2.1), we at once get,

$$\sin\phi_1 = n\sin(A - r_2) \quad \text{and} \quad \sin(\delta + A - \phi_1) = n\sin r_2.$$

Differentiating w.r.t. ϕ_1 and r_2, we get

$$\cos\phi_1 d\phi_1 + n\cos(A - r_2)dr_2 = 0$$
$$\cos(\delta + A - \phi_1)d\phi_1 + n\cos r_2 dr_2 = 0$$

These two eqns. at once yield

$$d\phi_1\cdot\Delta = 0, \quad \text{where} \quad \Delta = \begin{vmatrix} \cos\phi_1 & n\cos(A - r_2) \\ \cos(A + \delta - \phi_1) & n\cos r_2 \end{vmatrix}$$

So, for non-trivial solution we must have $\Delta = 0$

$$\therefore n\cos\phi_1\cos r_2 - n\cos(A - r_2)\cos(A + \delta - \phi_1) = 0$$
$$\Rightarrow \cos\phi_1\cos r_2 = \cos(A - r_2).\cos(A + \delta - \phi_1)$$
$$\Rightarrow \cos\phi_1\cos r_2 = \cos r_1\cos\phi_2 \quad \Rightarrow \quad \frac{\cos\phi_1}{\cos\phi_2} = \frac{\cos r_1}{\cos r_2}$$
$$\Rightarrow \frac{1 - \sin^2\phi_1}{1 - \sin^2\phi_2} = \frac{1 - \sin^2 r_1}{1 - \sin^2 r_2} = \frac{n^2 - \sin^2\phi_1}{n^2 - \sin^2\phi_2} = \frac{1 - n^2}{1 - n^2} = 1$$
$$\Rightarrow 1 - \sin^2\phi_1 = 1 - \sin^2\phi_2 \quad \Rightarrow \quad \phi_1 = \phi_2$$

When light of composite nature passes through a prism, the different wavelengths undergo differential amount of deviation as they pass through the prism and a spectrum of colour will be received on a screen with the red colour at the top and violet at the bottom. This type of separation of colours where red is least deviated and violet is most deviated is referred to as *normal dispersion*. But an atmosphere of Na-vapour with concentration gradient downwards behaves as a prism where the violet is least deviated and red is most deviated, thus we obtain anomalous dispersion, the detail being given later (cf. chapter on Dispersion, Vol.II).

4.4 Image Formed due to Refraction by a Prism

A prism is a very useful component of many an optical instrument. In any experiment on spectrometer, the source is virtually to be shifted off to infinity, called *Schuster's arrangement* and that is done by a prism. Let P be the source of light at a $\perp$ distance PN w.r.t. the face AB of the prism ABC. Two rays PQ and PQ_1 are incident on the surface at angles of incidence ϕ_1 and $\phi_1 + d\phi_1$; the corresponding refracted rays pass along QR and Q_1R_1 making angles of refraction r_1 and $r_1 + dr_1$. The refracted rays QR and Q_1R_1 produced back intersect at I_1 giving image due to refraction at the first surface. Let $PQ = u$, $QI_1 = v_1$ and let QN_1 and QN_2 be the normal sections of the incident and refracted cones of rays. For the second surface refraction, the rays are supposed to come from I_1 and get refracted at the second surface; the emergent rays as if come from I_2, called *final image* due to refraction by the prism.

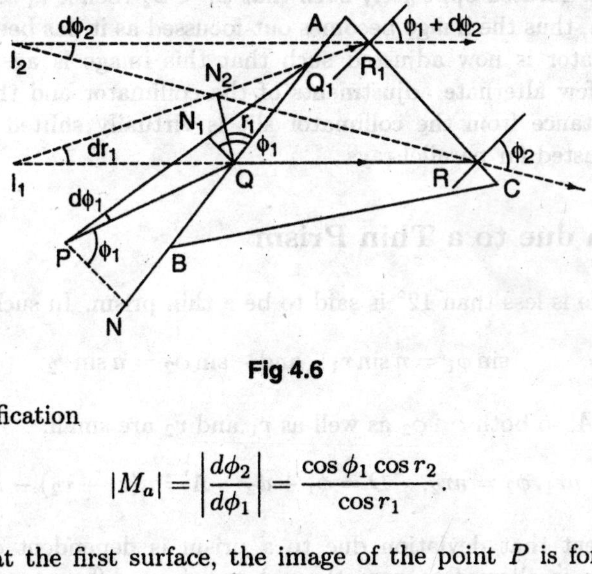

Fig 4.6

The angular magnification

$$|M_a| = \left|\frac{d\phi_2}{d\phi_1}\right| = \frac{\cos\phi_1 \cos r_2}{\cos r_1}.$$

Due to refraction at the first surface, the image of the point P is formed at I_1; QN_1 is dropped $\perp$ to PQ_1; QN_2 is $\perp$ to I_1Q_1 such that $\angle N_1QQ_1 = \phi_1$, $\angle N_2QQ_1 = r_1$. Let $PQ = u$ (*object distance*), $QI_1 = v_1$ (*first image distance*).

We have, $u\, d\phi_1 = QN_1 = QQ_1 \cos\phi_1$ and $v_1\, dr_1 = QN_2 = QQ_1 \cos r_1$

$$\therefore \quad \frac{u}{v_1} \cdot \frac{d\phi_1}{dr_1} = \frac{\cos\phi_1}{\cos r_1}.$$

Due to refraction at the second surface, we can easily get,

$$\frac{v_1}{v} \cdot \frac{dr_1}{d\phi_2} = \frac{\cos r_2}{\cos\phi_2} \Rightarrow \frac{u}{v} \cdot \frac{d\phi_1}{d\phi_2} = \frac{\cos r_2}{\cos\phi_2} \cdot \frac{\cos\phi_1}{\cos r_1}$$

$$\Rightarrow v = u\frac{\cos\phi_2\cos r_1}{\cos r_2\cos\phi_1}\cdot\frac{d\phi_1}{d\phi_2} = u\frac{\cos\phi_2\cos r_1}{\cos r_2\cos\phi_1}\cdot\frac{\cos\phi_2\cos r_1}{\cos\phi_1\cos r_2}$$

$$\Rightarrow v = u\left[\frac{\cos^2\phi_2}{\cos^2 r_2}\div\frac{\cos^2\phi_1}{\cos^2 r_1}\right]. \quad (4.4.1)$$

In case of min. deviation, since $\phi_2 = \phi_1$ and hence, $r_2 = r_1$, we get from eqn.(4.4.1), $v = u$. If $\phi_1 > \phi_2$,

$$r_1 > r_2 \quad\Rightarrow\quad \frac{\cos^2\phi_1}{\cos^2 r_1} < \frac{\cos^2\phi_2}{\cos^2 r_2} \quad\Rightarrow\quad v > u.$$

If $\phi_1 < \phi_2$,

$$r_1 < r_2 \quad\Rightarrow\quad \frac{\cos^2\phi_1}{\cos^2 r_1} > \frac{\cos^2\phi_2}{\cos^2 r_2} \quad\Rightarrow\quad v < u.$$

The eqn.(4.4.1) forms the basis of the Schuster's adjustment to be realised in all spectrometric experiments.

In the position of minimum deviation of the prism, when $\phi_1 = \phi_2, r_1 = r_2$, we get, $v = u$.

If the prism is so rotated from the minimum deviation position such that $\phi_1 > \phi_2$ (hence, $r_1 > r_2$), $v > u$, this position is referred to as *slanting position.* The telescope now is focussed for this image at a longer distance than the object.

The prism is now rotated oppositely such that $\phi_1 < \phi_2$ (hence $r_1 < r_2$ also), called *normal position* when $v < u$, thus the image becomes out-focussed as it has been focussed for a distant image. The collimator is now adjusted such that this image is again sharply seen by the telescope. Thus a few alternate adjustments of the collimator and the telescope, the object placed at finite distance from the collimator slit is virtually shifted off to infinity and the spectrometer is adjusted for parallel rays.

4.5 Deviation due to a Thin Prism

A prism whose angle is less than 12° is said to be a thin prism. In such a case since

$$\sin\phi_1 = n\sin r_1 \quad\text{and}\quad \sin\phi_2 = n\sin r_2.$$

Also, $r_1 + r_2 = A$, so both ϕ_1, ϕ_2 as well as r_1 and r_2 are small.

$$\therefore\quad \phi_1 = nr_1, \phi_2 = nr_2, \quad D = \phi_1 + \phi_2 - A = n(r_1 + r_2) - A = (n-1)A$$

Thus it is evident that deviation due to a prism is dependent on colour. So, if light of composite nature is allowed to pass through a prism, different colours will deviate to a different extent, hence after emergence, they will separate out giving rise to the phenomenon of dispersion. If a medium behaves such that its r.i. value decreases with increase of wavelength, the medium is referred to as normal dispersive medium, whereas if the r.i. increases with wavelength, the medium is referred to as anomalous dispersive medium.

Now, from the equation representing the deviation of a thin prism,

$$D = (n-1)A$$

$$\therefore\ D_v = (n_v - 1)A, D_r = (n_r - 1)A$$

$$\therefore\ D_v - D_r = \text{dispersion between red and violet}$$

$D =$ deviation of the mean colour

$$\therefore \omega = \frac{D_v - D_r}{D}, \text{ is called } \textit{dispersive power} \text{ of the medium}$$

$$\therefore \omega = \frac{D_v - D_r}{D} = \frac{(n_v - 1)A - (n_r - 1)A}{(n - 1)A} = \frac{(n_v - n_r)}{(n_1)} = \frac{\delta n}{(n - 1)} \quad (4.5.1)$$

4.6 Angle of Incidence on a Prism for No Emergence at the Second Surface

Let the angle of incidence be ϕ_1 and the angle of refraction at the first surface r_1; the angle of incidence at the second surface is r_2, and for no emergence, the emergent ray just grazes along the surface of separation.

So we have, $\sin\phi_1 = n\sin r_1$ and $n\sin r_2 = 1 \times \sin 90° = 1$

$$\begin{aligned}\therefore \quad \sin\phi_1 &= n\sin r_1 = n\sin(A - r_2)\\ &= n\sin A\cos r_2 - n\cos A\sin r_2\\ &= n\sin A\sqrt{1 - \sin^2 r_2} - \cos A\\ &= n\sin A\sqrt{\frac{n_2 - 1}{n^2}} - \cos A = \left\{\sin A\sqrt{n^2 - 1} - \cos A\right\}\end{aligned}$$

$$\Rightarrow \quad \phi_1 = \sin^{-1}\left\{\sin A\sqrt{n^2 - 1} - \cos A\right\}.$$

Solved Numerical Problems

1. The refractive index of a prism of refracting angle $A = 60°$ is $n = \sqrt{7/3}$. What is the limiting angle of incidence for which there will be no emergence?

Solution.

$$\begin{aligned}\sin\phi_1 &= \sin A\sqrt{n^2 - 1} - \cos A = \sin 60 \times \sqrt{\frac{7}{3} - 1} - \cos 60°\\ &= \frac{\sqrt{3}}{2} \times \frac{2}{\sqrt{3}} - \frac{1}{2} = \frac{1}{2} \quad \Rightarrow \quad \phi_1 = 30°\end{aligned}$$

2. Calculate the smallest r.i. of a right-angled prism with equal sides for which a ray entering normally at one of the equal sides, will be totally reflected.

Solution. The ray entering normally on one of the equal sides will be incident at an angle of 45° on the third face. As this ray is just totally reflected, critical angle = 45°.

$$\therefore \quad \mu = \frac{1}{\sin C} = \frac{1}{\sin 45°} = \sqrt{2}.$$

3. A prism of certain material has a refracting angle A. For two angles of incidence 40°6′ and 82°42′, it is found to produce a deviation angle of 62°48′. Find the angle A.

Solution. Here, $i_1 = 40°6'$, $i_2 = 82°42'$

$$\therefore \quad D = i_1 + i_2 - A \quad \Rightarrow \quad A = i_1 + i_2 - D = 122°48' - 62°48' = 60°.$$

Theory Questions

1. A prism has a r.i. $\sqrt{3/2}$ and refracting angle 90°. Calculate the minimum deviation produced by the prism and the corresponding angle of incidence. Show that the minimum value of the angle of incidence for which the emergent ray exists is 45°.
2. A prism in the form of an equilateral triangle is placed in the path of ray such that the refracted ray is internally reflected from the second face and emerges out of the prism so as to include an angle of 72° between the incident and emergent rays. Find the r.i. of the material of the prism. **($n = 1.45$)**
3. An equilateral prism is made of glass. A ray is to be totally internally reflected from the base. Find the maximum angle of incidence ($n = 3/2$). **($i = 28°54'$)**
4. Three right-angled prisms of r.i.'s n_1, n_2 and n_3 are fitted together so that the faces of the middle prism are in contact each with one of the faces of the outside prism. If a ray passes through the composite block without any deviation, find the relation connecting n_1, n_2 and n_3. **($n_1^2 + n_3^2 - n_2^2 = 1$)**
5. A ray of light makes an angle of incidence of 60° on one of the faces of a prism and suffers a total deviation of 30° on emergence from the other face. If the angle of the prism is 30°, find the angle of emergence. **($\theta = 90°$)**

Chapter 5

Refraction at Spherical Surface

5.1 Refraction through Spherical Surface—Refraction Matrix for Spherical Surfaces

There is relation between the object distance and image distance due to refraction at curved spherical surface separating two media. Longitudinal and angular magnifications also satisfy certain suitable relation, called Lagrange-Helmholtz relation. This formula for refraction at spherical interface may be extended to formulate the lens formula. So let us try to evolve an expression for the refraction matrix $[R]$ for spherical interface.

5.2 Concave Spherical Interface

Incident Medium Rarer, Refracted Medium Denser

The concave refractor shown in fig.(5.1) forms the virtual image at I of the object at O, the angles of incidence and refraction at the curved surface are respectively ϕ_1 and ϕ_2 as shown. The spherical interface separates two media of r.i. values n_1 and $n_2 (n_2 > n_1)$.

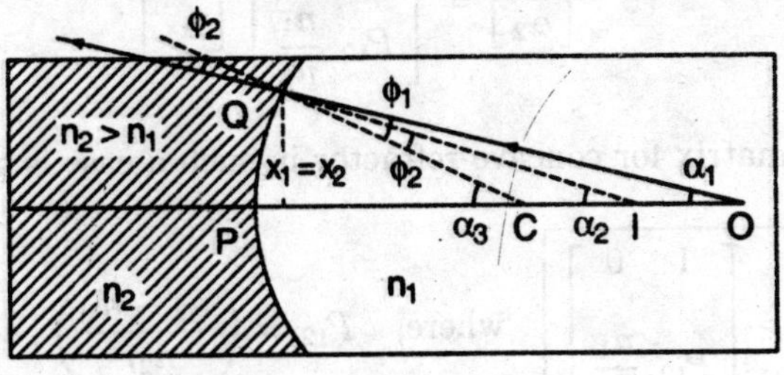

Fig 5.1

The incident ray on the concave refractor just before refraction is represented by the column vector $\begin{bmatrix} x_1 \\ \alpha_1 \end{bmatrix}$, and after refraction, the refracted ray starts from the same point Q and the angle made by the refracted ray with the horizontal is α_2 (fig. 5.1). One characteristic feature of such refraction at the concave refractor is $x_2 = x_1$.

What is the relation between α_2 and α_1?

From fig.(5.1), we get,

$$\begin{aligned} \alpha_3 &= \phi_1 + \alpha_1, \\ \Rightarrow \quad \phi_1 &= (\alpha_3 - \alpha_1), \end{aligned} \quad \Bigg| \quad \begin{aligned} \alpha_3 &= \phi_2 + \alpha_2 \\ \Rightarrow \quad \phi_2 &= (\alpha_3 - \alpha_2) \end{aligned}$$

By Snell's law, $n_1\phi_1 = n_2\phi_2$ (ϕ_1, ϕ_2 being small)

$$\Rightarrow \quad n_1(\alpha_3 - \alpha_1) = n_2(\alpha_3 - \alpha_2).$$

All of these angles α_1, α_2 and α_3 are −ve, so considering the sign convention also we get,

$$n_1(\alpha_3 - \alpha_1) = n_2(\alpha_3 - \alpha_2)$$

$$\Rightarrow n_2\alpha_2 - n_1\alpha_1 = (n_2 - n_1)\alpha_3$$

$$\Rightarrow \alpha_2 = \left(\frac{n_2 - n_1}{n_2}\right)\alpha_3 + \frac{n_1}{n_2}\alpha_1$$

$$\alpha_3 = \tan\alpha_3 = \frac{x_2}{r} = \frac{x_1}{r}$$

$$\therefore \alpha_2 = \frac{n_2 - n_1}{n_2 r}x_1 + \frac{n_1}{n_2}\alpha_1, \; r \text{ is +ve.}$$

So the two characteristic relations representing refraction at the concave refractor are given by

$$x_2 = x_1, \quad \alpha_2 = \frac{n_2 - n_1}{n_2 r}x_1 + \frac{n_1}{n_2}\alpha_1 = P_{12}x_1 + \frac{n_1}{n_2}\alpha_1 \tag{5.2.1}$$

where $P_{12} = \left(\frac{n_2 - n_1}{n_2 r}\right)$.

$$\therefore \begin{bmatrix} x_2 \\ \alpha_2 \end{bmatrix} = \begin{bmatrix} 1 & 0 \\ P_{12} & \frac{n_1}{n_2} \end{bmatrix} \begin{bmatrix} x_1 \\ \alpha_1 \end{bmatrix}. \tag{5.2.2}$$

Thus the refraction matrix for concave refractor in such a case is given by,

$$\boxed{[R_{12}] = \begin{bmatrix} 1 & 0 \\ P_{12} & \frac{n_1}{n_2} \end{bmatrix}} \quad \text{where,} \quad P_{12} = \left(\frac{n_2 - n_1}{n_2 r}\right), n_2 > n_1. \tag{5.2.3}$$

Incident Medium Denser and Refracted Medium Rarer

Fig.(5.2) shows refraction in this case. The image in this case is real and formed at I. The incident ray is represented by the parameters (x_1, α_1), the refracted ray by (x_2, α_2). The angles of incidence and refraction are respectively ϕ_1 and ϕ_2 and from fig.(5.2) we get,

$$\alpha_2 = \phi_1 + \alpha_1 \quad \text{and} \quad \phi_2 = \alpha_2 + \alpha_3.$$

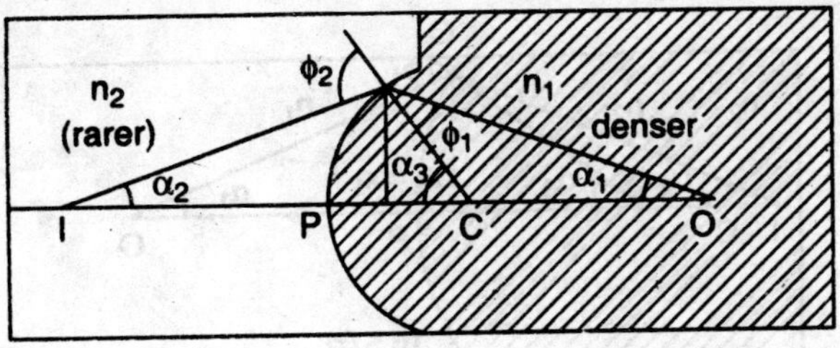

Fig 5.2

By Snell's law, we get, $n_1\phi_1 = n_2\phi_2 \quad \Rightarrow \quad \phi_2 = \dfrac{n_1}{n_2}\phi_1$

$$\therefore \quad \alpha_3 + \alpha_2 = \frac{n_1}{n_2}\phi_1 = \frac{n_1}{n_2}(\alpha_3 - \alpha_1)$$

$$\Rightarrow \quad \alpha_2 = \left(\frac{n_1}{n_2} - 1\right)\alpha_3 - \frac{n_1}{n_2}\alpha_1 = \frac{n_1 - n_2}{n_2}\alpha_3 - \frac{n_1}{n_2}\alpha_1.$$

Here, $\alpha_1 = -$ve, $\alpha_3 = -$ve, but $\alpha_2 = +$ve. Also, $\dfrac{x_1}{r} = \tan\alpha_3 \simeq \alpha_3$, $r$is +ve.

$$\therefore \quad \alpha_2 = \frac{n_1}{n_2}\alpha_1 - \left(\frac{n_1 - n_2}{n_2 r}\right)x_1 = \frac{n_1}{n_2}\alpha_1 + \left(\frac{n_2 - n_1}{n_2 r}\right)x_1.$$

Thus, $x_2 = x_1$

$$\alpha_2 = \frac{n_1}{n_2}\alpha_1 + \left(\frac{n_2 - n_1}{n_2 r}\right)x_1 = \frac{n_1}{n_2}\alpha_1 + P'_{12}x_1$$

$$\therefore \quad \begin{bmatrix} x_2 \\ \alpha_2 \end{bmatrix} = \begin{bmatrix} 1 & 0 \\ P'_{12} & \dfrac{n_1}{n_2} \end{bmatrix} \begin{bmatrix} x_1 \\ \alpha_1 \end{bmatrix} = [R'_{12}] \begin{bmatrix} x_1 \\ \alpha_1 \end{bmatrix},$$

$$\boxed{[R'_{12}] = \begin{bmatrix} 1 & 0 \\ P'_{12} & \dfrac{n_1}{n_2} \end{bmatrix}}, \quad \text{where} \quad P'_{12} = \left(\frac{n_2 - n_1}{n_2 r}\right), n_1 > n_2. \qquad (5.2.4)$$

5.3 Convex Spherical Surface

Incident Medium Rarer, Refracted Medium Denser

Here a convex refractor separates two media of r.i. values $n_1, n_2 (n_2 > n_1)$, the angles of incidence and refraction are respectively ϕ_1 and ϕ_2 as shown. The incident ray and hence, the object is represented by the parameters (x_1, α_1); the refracted ray and hence, the image by the parameters (x_2, α_2).

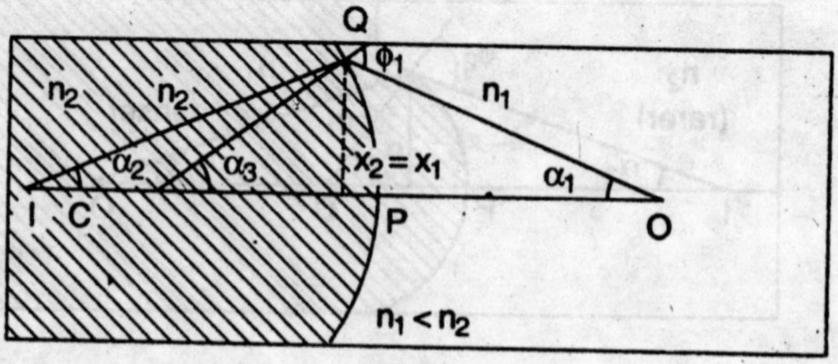

Fig 5.3

From fig.(5.3) we get,

$$\phi_1 = \alpha_3 + \alpha_1 \quad \text{and} \quad \alpha_3 = \phi_2 + \alpha_2.$$

By Snell's law we have,

$$n_1\phi_1 = n_2\phi_2 \quad \Rightarrow \quad \phi_2 = \frac{n_1}{n_2}\phi_1 .$$

$$\therefore \ \alpha_3 = \frac{n_1}{n_2}\phi_1 + \alpha_2 = \frac{n_1}{n_2}\alpha_3 + \frac{n_1}{n_2}\alpha_1 + \alpha_2$$

$$\Rightarrow \ \alpha_2 = -\frac{n_1}{n_2}\alpha_1 + \frac{n_2 - n_1}{n_2}\alpha_3$$

Also, $\dfrac{x_1}{-r} = \alpha_3$

$$\therefore \quad \alpha_2 = -\frac{n_1}{n_2}\alpha_1 - \frac{(n_2 - n_1)}{n_2 r}x_1 = \frac{n_1}{n_2}\alpha_1 + M_{12}x_1$$

(here, α_1 is $-$ve, but α_2, α_3 both $+$ve)

where, $M_{12} = -\dfrac{n_2 - n_1}{n_2 r}$.

Thus, $x_2 = x_1$ and $\alpha_2 = \dfrac{n_1}{n_2}\alpha_1 + M_{12}x_1$

$$\therefore \quad \begin{bmatrix} x_2 \\ \alpha_2 \end{bmatrix} = \begin{bmatrix} 1 & 0 \\ M_{12} & \dfrac{n_1}{n_2} \end{bmatrix} \begin{bmatrix} x_1 \\ \alpha_1 \end{bmatrix} \tag{5.3.1}$$

Thus, the refraction matrix in this case is

$$\boxed{[R_2] = \begin{bmatrix} 1 & 0 \\ M_{12} & \dfrac{n_1}{n_2} \end{bmatrix}}, \quad \text{where} \quad M_{12} = -\left(\frac{n_2 - n_1}{n_2 r}\right), \quad n_2 > n_1.$$

Incident Medium Denser, Refracted Medium Rarer

In this case, angle of refraction ϕ_2 is greater than the angle of incidence ϕ_1 as shown in fig.(5.4) and the image formed is virtual at I. From the figure it is evident that

$$\phi_1 = \alpha_3 + \alpha_1 \quad \text{and} \quad \phi_2 = \alpha_3 + \alpha_2.$$

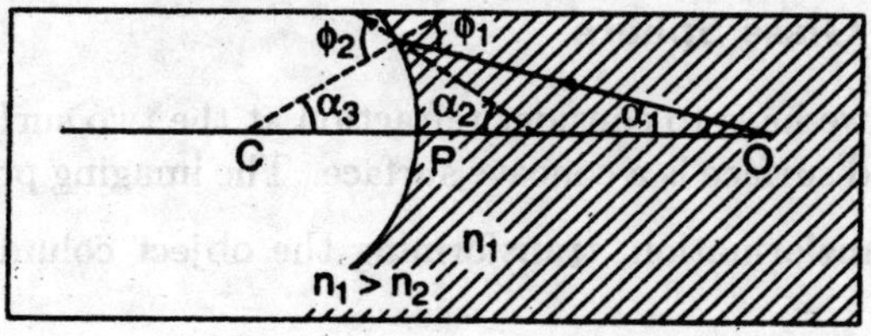

Fig 5.4

By Snell's law,

$$n_1\phi_1 = n_2\phi_2 \quad \Rightarrow \quad \phi_2 = \frac{n_1}{n_2}\phi_1$$

$$\therefore \quad \alpha_3 + \alpha_2 = \frac{n_1}{n_2}\phi_1 = \frac{n_1}{n_2}(\alpha_3 + \alpha_1)$$

$$\Rightarrow \quad \alpha_2 = \frac{n_1 - n_2}{n_2}\alpha_3 + \frac{n_1}{n_2}\alpha_1 = \frac{n_1}{n_2}\alpha_1 + \frac{n_1 - n_2}{n_2}\alpha_3$$

(here, both α_1 and α_2 −ve but α_3 +ve and $\frac{x_1}{-r} = \alpha_3$)

$$\therefore \quad \alpha_2 = \frac{n_1}{n_2}\alpha_1 + \frac{(n_1 - n_2)}{n_2 r}x_1 = \frac{n_1}{n_2}\alpha_1 + M'_{12}x_1.$$

Thus, $x_2 = x_1$ and $\alpha_2 = \frac{n_1}{n_2}\alpha_1 + M'_{12}x_1$

$$\therefore \quad \begin{bmatrix} x_2 \\ \alpha_2 \end{bmatrix} = \begin{bmatrix} 1 & 0 \\ M'_{12} & \frac{n_1}{n_2} \end{bmatrix} \begin{bmatrix} x_1 \\ \alpha_1 \end{bmatrix} \tag{5.3.2}$$

where, $M'_{12} = -\frac{(n_2 - n_1)}{n_2 r}, n_1 > n_2.$

Standard Formula for Refraction at Curved Surface

From eqn.(5.2.2),

$$\begin{bmatrix} x_2 \\ \alpha_2 \end{bmatrix} = \begin{bmatrix} 1 & 0 \\ P_{12} & \frac{n_1}{n_2} \end{bmatrix} \begin{bmatrix} x_1 \\ \alpha_1 \end{bmatrix} \text{ we get, } \alpha_2 = P_{12}x_1 + \frac{n_1}{n_2}\alpha_1.$$

If the image distance be v (+ve), object distance u (+ve), α_1, α_2 and α_3 are all negative, then

$$\frac{x_2}{v} = P_{12}x_1 + \frac{n_1}{n_2} \cdot \frac{x_1}{u}$$

$x_1 = x_2$, and we get,

$$\frac{n_2}{v} - \frac{n_1}{u} = n_2 P_{12} = \frac{n_2 - n_1}{r}.$$

Refraction through a Glass Sphere

In this case of refraction we have to consider refraction at the two surfaces, the first surface is a convex surface, the second surface is a concave surface. The imaging process may be represented by a series of matrix transformation, transforming the object column vector $\begin{bmatrix} x_1 \\ \alpha_1 \end{bmatrix}$ to $\begin{bmatrix} x_1' \\ \alpha_1 \end{bmatrix}$ by refraction matrix of the convex surface, then a translational matrix $[T]$ transforms $\begin{bmatrix} x_1' \\ \alpha_1 \end{bmatrix}$ to $\begin{bmatrix} x_2' \\ \alpha_2 \end{bmatrix}$, further transformation by the refraction matrix at the second surface transforms $\begin{bmatrix} x_2' \\ \alpha_2 \end{bmatrix}$ to $\begin{bmatrix} x_2 \\ \alpha_2 \end{bmatrix}$ which is the image column vector.

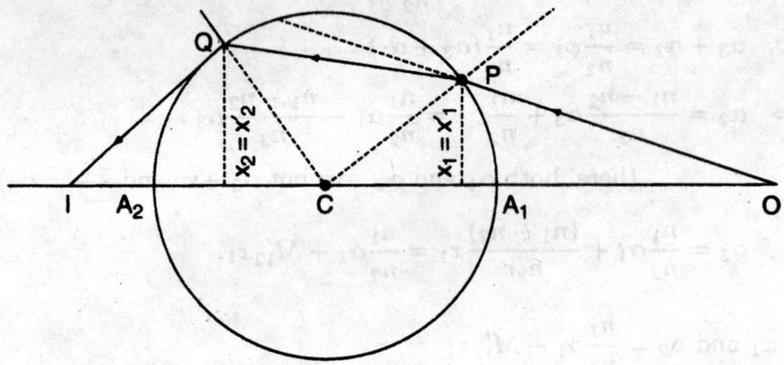

Fig 5.5

Thus,

$$\begin{bmatrix} x_2 \\ \alpha_2 \end{bmatrix} = [R_{23}'][T][R_{12}]\begin{bmatrix} x_1 \\ \alpha_1 \end{bmatrix} = \begin{bmatrix} 1 & 0 \\ P_{23}' & \dfrac{n_2}{n_3} \end{bmatrix}\begin{bmatrix} 1 & 2r \\ 0 & 1 \end{bmatrix}\begin{bmatrix} 1 & 0 \\ M_{12} & \dfrac{n_1}{n_2} \end{bmatrix}\begin{bmatrix} x_1 \\ \alpha_1 \end{bmatrix}$$

$$= \begin{bmatrix} 1 & 2r \\ P_{23}' & \left\{2rP_{23}' + \dfrac{n_2}{n_3}\right\} \end{bmatrix}\begin{bmatrix} 1 & 0 \\ M_{12} & \dfrac{n_1}{n_2} \end{bmatrix}\begin{bmatrix} x_1 \\ \alpha_1 \end{bmatrix}$$

$$= \begin{bmatrix} (1 + 2rM_{12}) & 2r\dfrac{n_1}{n_2} \\ P_{23}' + M_{12}\left(2rP_{23}' + \dfrac{n_2}{n_3}\right) & \dfrac{n_1}{n_2}\left(2rP_{23}' + \dfrac{n_2}{n_3}\right) \end{bmatrix}\begin{bmatrix} x_1 \\ \alpha_1 \end{bmatrix}$$

So, the overall system matrix for the given optical system can be written as

$$[N] = \begin{bmatrix} N_{11} & N_{12} \\ N_{21} & N_{22} \end{bmatrix}$$

where,

$$N_{11} = 1 + 2rM_{12}, N_{21} = P_{23}' + M_{12}\left(2rP_{23}' + \frac{n_2}{n_3}\right)$$

$$N_{12} = 2r\cdot\frac{n_2}{n_3}, N_{22} = \frac{n_1}{n_2}\left(2rP'_{23} + \frac{n_1}{n_2}\right)$$

$$\therefore \begin{bmatrix} x_2 \\ \alpha_2 \end{bmatrix} = \begin{bmatrix} N_{11} & N_{12} \\ N_{21} & N_{22} \end{bmatrix}\begin{bmatrix} x_1 \\ \alpha_1 \end{bmatrix} \Rightarrow \begin{cases} x_2 = N_{11}x_1 + N_{12}\alpha_1 \\ \alpha_2 = N_{21}x_1 + N_{22}\alpha_1 \end{cases} \quad (5.3.3)$$

5.4 Calculation of Lateral Magnification and Angular Magnification (For an Object of Finite Size)

For CONCAVE REFRACTOR

In fig.(5.6) a concave refractor forms an image of the object PQ at $P'Q'$, the object distance and image distance w.r.t the pole O being denoted as u and v respectively. The object space is having a refractive index n_1, that of the image space $n_2 (n_2 > n_1)$. The object is defined by the parameter (x_1, α_1), the ray strikes the refractor at A defined by (x'_1, α_1) in the object space, refracts into the image space and the refracted ray at A is defined by (x'_2, α_2) and finally the image is formed as $P'Q'$ defined by (x_2, α_2).

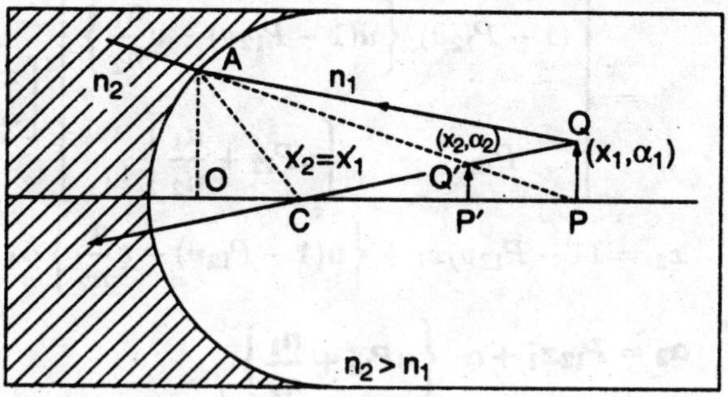

Fig 5.6

The whole process of image formation may be treated through a series of transformation followed in sequence (i) by translation through u transforming (x_1, α_1) into (x'_1, α_1), (ii) by refraction at the concave surface transforming (x'_1, α_1) in n_1 to (x'_2, α_2) in n_2, (iii) finally by translation through v to (x_2, α_2).

This sequence of transformation is carried through translational matrix $[T_1]$, refraction matrix (R_{12}) and again translational matrix $[T_2]$. Thus we have,

For $(x_1, \alpha_1) \to (x'_1, \alpha_1)$,

$$\begin{bmatrix} x_1 \\ \alpha_1 \end{bmatrix} = [T_1]\begin{bmatrix} x_1 \\ \alpha_1 \end{bmatrix} = \begin{bmatrix} 1 & u \\ 0 & 1 \end{bmatrix}\begin{bmatrix} x_1 \\ \alpha_1 \end{bmatrix}$$

For $(x'_1, \alpha_1) \to (x'_2, \alpha_2)$,

$$\begin{bmatrix} x'_2 \\ \alpha_2 \end{bmatrix} = [R_{12}]\begin{bmatrix} x'_1 \\ \alpha_1 \end{bmatrix} = \begin{bmatrix} 1 & 0 \\ P_{12} & \frac{n_1}{n_2} \end{bmatrix}\begin{bmatrix} x'_1 \\ \alpha_1 \end{bmatrix}$$

where, $P_{12} = \dfrac{n_2 - n_1}{n_2 r}$.

For $(x_2', \alpha_2) \to (x_2, \alpha_2)$,

$$\begin{bmatrix} x_2 \\ \alpha_2 \end{bmatrix} = [T_2] \begin{bmatrix} x_2' \\ \alpha_2 \end{bmatrix} = \begin{bmatrix} 1 & -v \\ 0 & 1 \end{bmatrix} \begin{bmatrix} x_2' \\ \alpha_2 \end{bmatrix}.$$

Thus, from above relation we at once get,

$$\begin{bmatrix} x_2 \\ \alpha_2 \end{bmatrix} = \begin{bmatrix} 1 & -v^* \\ 0 & 1 \end{bmatrix} \begin{bmatrix} 1 & 0 \\ P_{12} & \dfrac{n_1}{n_2} \end{bmatrix} \begin{bmatrix} 1 & u \\ 0 & 1 \end{bmatrix} \begin{bmatrix} x_1 \\ \alpha_1 \end{bmatrix} \tag{5.4.1}$$

$$\begin{bmatrix} x_2 \\ \alpha_2 \end{bmatrix} = \begin{bmatrix} (1 - P_{12}v) & -v\dfrac{n_1}{n_2} \\ P_{12} & \dfrac{n_1}{n_2} \end{bmatrix} \begin{bmatrix} 1 & u \\ 0 & 1 \end{bmatrix} \begin{bmatrix} x_1 \\ \alpha_1 \end{bmatrix}$$

$$= \begin{bmatrix} (1 - P_{12}v) & \left\{ u(1 - P_{12}v) - v\dfrac{n_1}{n_2} \right\} \\ P_{12} & \left\{ uP_{12} + \dfrac{n_1}{n_2} \right\} \end{bmatrix} \begin{bmatrix} x_1 \\ \alpha_1 \end{bmatrix}$$

$$\Rightarrow \quad x_2 = (1 - P_{12}v)x_1 + \left\{ u(1 - P_{12}v) - v\frac{n_1}{n_2} \right\} \alpha_1 \tag{5.4.2}$$

$$\alpha_2 = P_{12}x_1 + \alpha_1 \left\{ uP_{12} + \frac{n_1}{n_2} \right\} \tag{5.4.3}$$

The parameters (x_1, x_2) represent the size (or height) of the object and image respectively; α_1 and α_2 are the angles made by the object and image respectively with the axis of the optical system.

Now, for an axial object we must have $x_1 = 0 \Rightarrow x_2 = 0$ (also) but $\alpha_1 \neq 0$ and $\alpha_2 \neq 0$, physically meaning that an object point on the axis produces image point also on the axis.

$\therefore$ From eqn.(5.4.2), we get, therefore,

$$u(1 - P_{12}v) - v\frac{n_1}{n_2} = 0 \quad \Rightarrow \quad v\frac{n_1}{n_2} = +u(1 - P_{12}v). \tag{5.4.4}$$

So we get, $x_2 = (1 - P_{12}v)x_1$

$$\Rightarrow \quad \frac{x_2}{x_1} = \frac{\text{Size of the image}}{\text{Size of the object}} = \text{Lateral magnification } (m) = (1 - P_{12}v) \tag{5.4.5}$$

---

*The sign of translation is decided with reference to real or diverging translation (+ve), virtual or converging translation (−ve).

From (5.4.4) and (5.4.5), we get at once,

$$\frac{x_2}{x_1} = m = +\frac{v}{u}\cdot\frac{n_1}{n_2} \tag{5.4.6}$$

$$\therefore \quad m = \frac{v}{u}\cdot\frac{n_1}{n_2} = \frac{v/n_2}{u/n_1} \tag{5.4.7}$$

Angular Magnification

$M = \dfrac{\alpha_2}{\alpha_1}$, so for an axial point object, the lateral magnification is not meaningful but still angular magnification has significance. From eqn.(5.4.3), we have,

$$\frac{\alpha_2}{\alpha_1} = \left(uP_{12} + \frac{n_1}{n_2}\right) \tag{5.4.8}$$

From eqn.(5.4.4),

$$\frac{n_1}{n_2} = +\frac{u}{v} - uP_{12} \quad \Rightarrow \quad \left(uP_{12} + \frac{n_1}{n_2}\right) = +\frac{u}{v} \tag{5.4.9}$$

$$\therefore \quad \frac{\alpha_2}{\alpha_1} = +\frac{u}{v}, \quad \Rightarrow \quad \frac{\alpha_2}{\alpha_1} = \frac{u}{v} = M = \frac{1}{m}\frac{n_1}{n_2} \tag{5.4.10}$$

Axial Magnification

When the object of certain length stands on the axis, all points making up the object are at the same distance w.r.t. the refracting surface, so also true for the image points. Thus, when we say that the object distance is u, it means that it is the distance of all the object points. But if the object lies along the axis, the different points of the object are at different distances, so in such a case, object points at u and $u + du$ transform into image points at v and $v + dv$ respectively. Thus the axial magnification A is given by,

$$A = \frac{dv}{du}.$$

From eqn. (5.4.4), $\dfrac{n_1}{n_2}v = +u(1 - P_{12}v)$. Differentiating it w.r.t. u, we have,

$$\frac{n_1}{n_2}\cdot\frac{dv}{du} = +(1 - P_{12}v) - uP_{12}\frac{dv}{du} \quad \Rightarrow \quad \left(\frac{n_1}{n_2} + uP_{12}\right)\frac{dv}{du} = +(1 - P_{12}v)$$

$$\Rightarrow \quad +\frac{u}{v}\cdot A = +(1 - P_{12}v) \quad \Rightarrow \quad A = \frac{v}{u}\cdot m = m^2\frac{n_2}{n_1}. \tag{5.4.11}$$

For CONVEX REFRACTOR

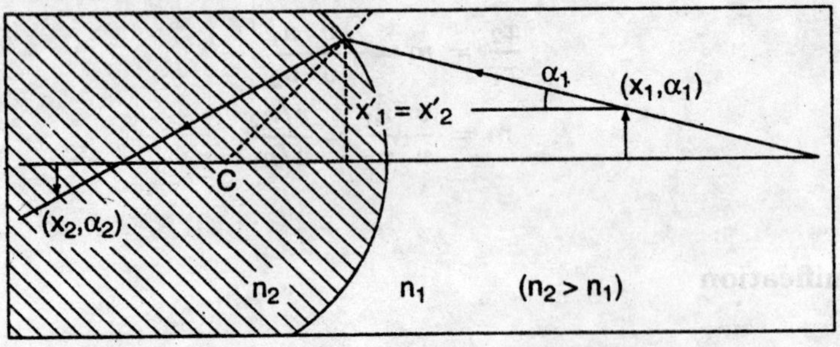

Fig 5.7

Refer to fig.(5.7), we can write here,

$$\begin{bmatrix} x_2 \\ \alpha_2 \end{bmatrix} = \begin{bmatrix} 1 & -v \\ 0 & 1 \end{bmatrix} \begin{bmatrix} 1 & 0 \\ M_{12} & \dfrac{n_1}{n_2} \end{bmatrix} \begin{bmatrix} 1 & u \\ 0 & 1 \end{bmatrix} \begin{bmatrix} x_1 \\ \alpha_1 \end{bmatrix}$$

$$= \begin{bmatrix} (1 - M_{12}v) & -v\dfrac{n_1}{n_2} \\ M_{12} & \dfrac{n_1}{n_2} \end{bmatrix} \begin{bmatrix} 1 & u \\ 0 & 1 \end{bmatrix} \begin{bmatrix} x_1 \\ \alpha_1 \end{bmatrix}$$

$$= \begin{bmatrix} (1 - M_{12}v) & \left\{ u(1 - M_{12}v) - v\dfrac{n_1}{n_2} \right\} \\ M_{12} & \left(uM_{12} + \dfrac{n_1}{n_2} \right) \end{bmatrix} \begin{bmatrix} x_1 \\ \alpha_1 \end{bmatrix} \tag{5.4.12}$$

Solving eqn.(5.4.12), we have,

$$x_2 = (1 - M_{12}v)x_1 + \alpha_1 \left\{ u(1 - M_{12}v) - v\frac{n_1}{n_2} \right\} \tag{5.4.13}$$

$$\alpha_2 = M_{12}x_1 + \left(uM_{12} + \frac{n_1}{n_2} \right) \alpha_1 \tag{5.4.14}$$

For an axial point, $x_1 = x_2 = 0$, but $\alpha_1 \neq 0$ and $\alpha_2 \neq 0$.

So from eqn.(5.4.13), we get,

$$u(1 - M_{12}v) - v\frac{n_1}{n_2} = 0 \quad \Rightarrow \quad \frac{v}{u} \cdot \frac{n_1}{n_2} = (1 - M_{12}v) \tag{5.4.15}$$

$$\therefore\ m = \frac{x_2}{x_1} = 1 - M_{12}v.$$

Now, $M_{12} = -\dfrac{n_2 - n_1}{n_2 r}$; hence, $M_{12} = -$ve, so $m = \dfrac{x_2}{x_1} = 1 + \left(\dfrac{n_2 - n_1}{n_2 r} \right) v.$

But, here x_2 is −ve and x_1 is +ve.

$$\therefore \quad \frac{x_2}{x_1} = m = -\left[1 + \frac{n_2 - n_1}{n_2 r} v\right]$$

the negative sign only indicates that x_2 and x_1 are opposite.

Now, $m = 1 - M_{12}v = \frac{v}{u} \cdot \frac{n_1}{n_2}$

$$\Rightarrow \quad \frac{n_1}{n_2} = \frac{u}{v}(1 - M_{12}v) = \frac{u}{v} - uM_{12} \quad \Rightarrow \quad uM_{12} + \frac{n_1}{n_2} = \frac{u}{v}.$$

From eqn.(5.4.14) we get,

$$\frac{\alpha_2}{\alpha_1} = \left(uM_{12} + \frac{n_1}{n_2}\right) = \frac{u}{v} = \frac{1}{m} \cdot \frac{n_1}{n_2}.$$

Actually, α_2 is +ve, but α_1 is −ve, hence,

$$\frac{\alpha_2}{\alpha_1} = -\left(uM_{12} + \frac{n_1}{n_2}\right). \tag{5.4.16}$$

5.5 Lagrange's Law and Helmholtz Relation

It is a relation connecting the linear and angular magnification of an optical system of refractors. This relationship was first established by Lagrange for a single refracting surface and Helmholtz extended the relation to a system of refractors. The generalised relation is now known as Lagrange-Helmholtz relation. The outcome of this relation is that for a particular position of the object defined by certain angle subtended by the object with the axis, there is one and only one position of the final image defined by certain angle subtended by it with the axis of the system.

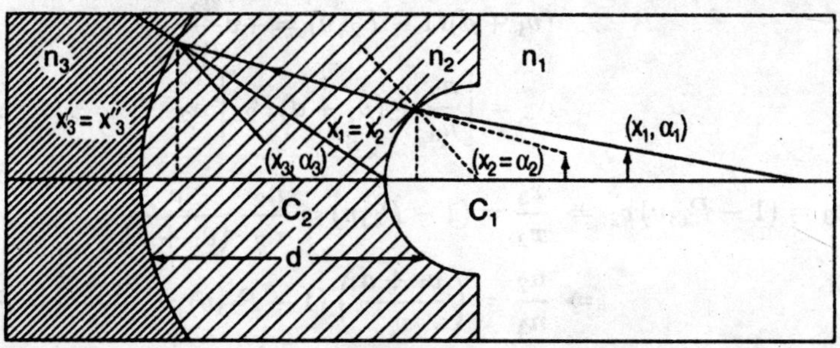

Fig 5.8

A composite optical system consisting of two concave refracting spherical surfaces of radii r_1 and r_2, and separating three media of r.i. values n_1, n_2, n_3 such that $n_3 > n_2 > n_1$ is shown in fig.(5.8).

Now, from the previous discussion, we can write,

$$\begin{bmatrix} x_2 \\ \alpha_2 \end{bmatrix} = \begin{bmatrix} 1 & -v_1 \\ 0 & 1 \end{bmatrix} \begin{bmatrix} 1 & 0 \\ P_{12} & \dfrac{n_1}{n_2} \end{bmatrix} \begin{bmatrix} 1 & u \\ 0 & 1 \end{bmatrix} \begin{bmatrix} x_1 \\ \alpha_1 \end{bmatrix}$$

$$= \begin{bmatrix} (1 - P_{12}v_1) & \left\{-v_1\dfrac{n_1}{n_2} + u(1 - P_{12}v_1)\right\} \\ P_{12} & \left(\dfrac{n_1}{n_2} + P_{12}u\right) \end{bmatrix} \begin{bmatrix} x_1 \\ \alpha_1 \end{bmatrix}$$

Again,

$$\begin{bmatrix} x_3 \\ \alpha_3 \end{bmatrix} = \begin{bmatrix} 1 & -v \\ 0 & 1 \end{bmatrix} \begin{bmatrix} 1 & 0 \\ P_{23} & \dfrac{n_2}{n_3} \end{bmatrix} \begin{bmatrix} 1 & v_1 + d \\ 0 & 1 \end{bmatrix} \begin{bmatrix} x_2 \\ \alpha_2 \end{bmatrix}$$

$$= \begin{bmatrix} (1 - P_{23}v) & -v\dfrac{n_2}{n_3} \\ P_{23} & \dfrac{n_2}{n_3} \end{bmatrix} \begin{bmatrix} 1 & v_1 + d \\ 0 & 1 \end{bmatrix} \begin{bmatrix} x_2 \\ \alpha_2 \end{bmatrix}$$

$$= \begin{bmatrix} (1 - P_{23}v) & \left\{(v_1 + d)(1 - P_{23}v) - v\dfrac{n_2}{n_3}\right\} \\ P_{23} & (v_1 + d)P_{23} + \dfrac{n_2}{n_3} \end{bmatrix} \begin{bmatrix} x_2 \\ \alpha_2 \end{bmatrix}$$

$$\therefore \quad x_3 = (1 - P_{23}v)x_2 + \alpha_2\left\{(v_1 + d)(1 - P_{23}v) - v\frac{n_2}{n_3}\right\} \tag{5.5.1}$$

$$\alpha_3 = P_{23}x_2 + \alpha_2\left\{(v_1 + d)P_{23} + \frac{n_2}{n_3}\right\} \tag{5.5.2}$$

Now, $x_3 = 0 \quad \Rightarrow \quad x_2 = 0$

$$\therefore \quad (v_1 + d)(1 - P_{23}v) = +v\frac{n_2}{n_3} \tag{5.5.3}$$

$$\alpha_3 = \left[\frac{n_2}{n_3} + (v_1 + d)P_{23}\right]\alpha_2 \tag{5.5.4}$$

$$\therefore \quad x_3 = (1 - P_{23}v)x_2 \Rightarrow \frac{x_3}{x_2} = (1 - P_{23}v) = \frac{n_2}{n_3}\cdot\frac{v}{(v_1 + d)} \tag{5.5.5}$$

$$\Rightarrow \frac{n_2}{n_3} = \left(\frac{v_1 + d}{v}\right)(1 - P_{23}v) = \frac{v_1 + d}{v} - P_{23}(v_1 + d)$$

$$\Rightarrow \frac{n_2}{n_3} + P_{23}(v_1 + d) = \frac{v_1 + d}{v}. \tag{5.5.6}$$

Now, eqn.(5.5.4) gives,

$$\frac{\alpha_3}{\alpha_2} = \frac{n_2}{n_3} + (v_1 + d)P_{23} = \frac{v_1 + d}{v}$$

$$\therefore \quad \frac{x_3}{x_2} = \frac{n_2}{n_3}\cdot\left(\frac{\alpha_2}{\alpha_3}\right) = \frac{n_2\alpha_2}{n_3\alpha_3} \quad \Rightarrow \quad x_3n_3\alpha_3 = n_2x_2\alpha_2$$

From eqns.(5.4.2) and (5.4.4), we get,

$$x_2n_2\alpha_2 = x_1n_1\alpha_1.$$

Therefore, for two refracting surfaces,

$$x_3\alpha_3n_3 = x_2n_2\alpha_2 = x_1n_1\alpha_1.$$

This can be generalised for $(k-1)$ refracting surfaces and we would get,

$$x_k\alpha_kn_k = x_{k-1}\alpha_{k-1}n_{k-1} = \cdots = x_1\alpha_1n_1. \tag{5.5.7}$$

Thus, for a particular object defined by the parameters (x_k, α_k), there is always one and only one image defined by the parameters x_n, α_n satisfying the relation given in (5.5.7).

5.6 Principal Foci and Conjugate Relation in Terms of the Principal Foci

The principal focal points of a curved refractor are (i) first principal focus (F_1) and (ii) second principal focus (F_2). The planes passing through these points and normal to the principal axis of the refractor are called the principal focal planes.

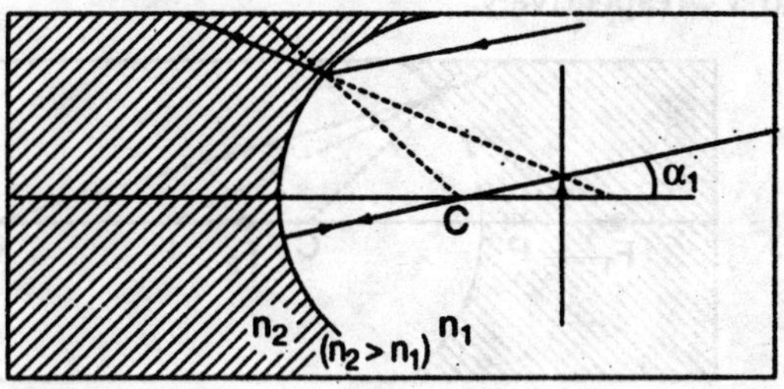

Fig 5.9

From eqns.(5.4.2) and (5.4.3) we have,

$$x_2 = (1 - P_{12}v)x_1 + \left\{u(1 - P_{12}v) - v\frac{n_1}{n_2}\right\}\alpha_1$$

$$\alpha_2 = P_{12}x_1 + \alpha_1\left\{uP_{12} + \frac{n_1}{n_2}\right\}$$

For axial point, $x_2 = x_1 = 0$

$$\therefore \quad v\frac{n_1}{n_2} = u(1 - P_{12}v).$$

If, $u = \infty, v = f_2$.

$$\therefore \quad \frac{v}{u}\cdot\frac{n_1}{n_2} = 1 - P_{12}v \quad \Rightarrow \quad \left(\frac{v}{u}\cdot\frac{n_1}{n_2}\right)_{u=\infty} = 1 - P_{12}f_2$$

$$\Rightarrow \quad f_2 = \frac{1}{P_{12}} = \frac{n_2r}{n_2 - n_1} \tag{5.6.1}$$

which shows that the second principal focus is on the same side as that of the object.

The first principal focus is a point on the principal axis such that rays diverging from it after refraction are rendered parallel. So here, $u = f_1$, $v = \infty$.

$$\therefore \quad \frac{n_1}{n_2} = \frac{u}{v} - P_{12}u$$
$$\Rightarrow \quad \frac{n_1}{n_2} = \left(\frac{u}{v}\right)_{v=\infty} - P_{12}f_1$$
$$\Rightarrow \quad f_1 = -\frac{n_1}{n_2} \cdot \frac{1}{P_{12}} = -\frac{n_1 r}{n_2 - n_1} \tag{5.6.2}$$

where the $-$ve sign means, the first principal focus is on the other side.

We can thus easily build up another important equation involving the two focal lengths as

$$\frac{f_1}{n_1} + \frac{f_2}{n_2} = 0 \tag{5.6.3}$$

Conjugate Relation in terms of f_1 and f_2

The two focal points F_2 and F_1 are defined by two eqns.(5.6.1) and (5.6.2) respectively, F_2 is to the right and F_1 to the left of the refracting surface. Refer to fig.(5.10) and suppose measurements are done w.r.t. the focal points. Then the object distance and image distance are represented as U and V, respectively.

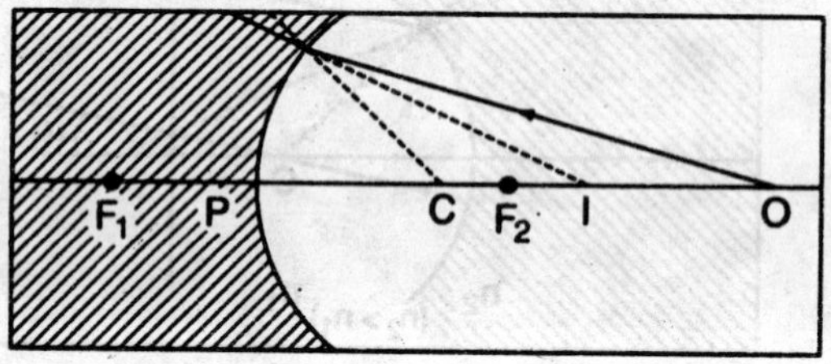

Fig 5.10

Now,

$$PF_1 = f_1 = -\frac{n_1 r}{n_2 - n_1} \quad \text{and} \quad PF_2 = f_2 = \frac{n_2 r}{n_2 - n_1}.$$

Then,

$$F_1P = f_1' = -f_1 = \frac{n_1 r}{n_2 - n_1} \quad \text{and} \quad F_2P = f_2' = -f_2 = -\frac{n_2 r}{n_2 - n_1}.$$

With reference to fig.(5.10) we can write,

$$U = F_1P + PO = -f_1 + u = \frac{n_1 r}{n_2 - n_1} + u$$
$$v = PI = PF_2 + F_2I = +f_2 + V = +\frac{n_2 r}{n_2 - n_1} + V.$$

Now, from eqn.(5.4.4) we have,

$$\frac{n_2}{v} - \frac{n_1}{u} = P_{12}n_2 = \frac{n_2 - n_1}{r}.$$

Multiplying throughout by $\left(\frac{r}{n_2 - n_1}\right)$ and putting the values of u, v we get,

$$\frac{n_2 r}{(n_2 - n_1)} \cdot \frac{1}{\left(V + \frac{n_2 r}{n_2 - n_1}\right)} - \frac{n_1 r}{(n_2 - n_1)} \cdot \frac{1}{\left(U - \frac{n_1 r}{n_2 - n_1}\right)} = 1$$

$$\Rightarrow \quad f_2 \cdot \frac{1}{V + f_2} + f_1 \cdot \frac{1}{U + f_1} = 1$$

$$\Rightarrow \quad \frac{f_2}{V + f_2} = 1 - \frac{f_1}{U + f_1} = \frac{U}{U + f_1}$$

$$\Rightarrow \quad UV + Uf_2 = Uf_2 + f_1 f_2 \quad \Rightarrow \quad UV = f_2 f_2 \qquad (5.6.4)$$

This eqn.(5.6.4) giving conjugate relation measured w.r.t. the focal points is known as Newton's relation.

Conjugate Relation w.r.t. Centre of Curvature

Referring again to fig.(5.10), we can write,

$$u = PO = PC + CO = U - r \text{ and } u = PI = PC + CI = (V - r)$$
$$PC = r, \qquad \text{but } CP = -r.$$

$$\therefore \quad \frac{n_2}{(V - r)} - \frac{n_1}{(U - r)} = \frac{n_2 - n_1}{-r} = \frac{n_1 - n_2}{r}$$

$$\Rightarrow \quad \frac{n_2}{V - r} = \frac{n_1 - n_2}{r} + \frac{n_1}{(U - r)} = \frac{(n_1 - n_2)U + n_2 r}{r(U - r)}$$

$$\Rightarrow \quad \frac{n_2 r}{V - r} = \frac{(n_1 - n_2)U + n_2 r}{(U - r)} = \frac{(n_1 - n_2)U}{(U - r) - (V - r)} = \frac{(n_1 - n_2)U}{(U - V)} \quad \left(\because \frac{a}{b} = \frac{c}{d} = \frac{c - a}{d - b}\right)$$

$$\Rightarrow \quad \frac{V - r}{n_2 r} = \frac{(U - V)}{(n_1 - n_2)U}$$

$$\Rightarrow \quad \frac{(n_1 - n_2)V}{n_2 r} - \frac{n_1 - n_2}{n_2} = 1 - \frac{V}{U}$$

$$\Rightarrow \quad \frac{n_1}{n_2} = \frac{(n_1 - n_2)V}{n_2 r} + \frac{V}{U}$$

$$\Rightarrow \quad \frac{n_1}{V} - \frac{n_2}{U} = \frac{n_1 - n_2}{r} \quad \Rightarrow \quad \frac{n_2}{V} - \frac{n_1}{U} = \frac{n_2 - n_1}{r}. \qquad (5.6.5)$$

5.7 Cardinal Points of an Optical System

By cardinal points we mean the three pairs of fundamental points of an optical system such that a knowledge of the positions of these points will enable us to calculate all the conjugate points of the optical system. These characteristic points define the characteristics of the optical system.

The cardinal points are—

(1) The two principal focal points, first and second (F_1 and F_2). The planes passing through these points and perpendicular to the optic axis of the system are called *focal planes*.

(2) The two principal or unit points H_1 and H_2. The planes passing through these points and perpendicular to the optic axis are called *principal* or *unit planes.*

(3) The two nodal points N_1 and N_2. The planes passing through these points and perpendicular to the axis are called *nodal planes.*

Principal Foci and Focal Planes

The principal foci of an optical system (curved refracting surface, thin or thick lens either single or a combination) refer to the two points F_1 and F_2 respectively called the first and second principal foci of the optical system. The planes passing through them and normal to the axis are called *focal planes.*

Principal or Unit Points and Principal Planes

They refer to the pair of points of an optical system such that an object placed at one will have image at the other of the same size and on the same side of the principal axis as the object. Thus principal planes are conjugate planes of *unit positive lateral magnification.* These planes will be situated depending on the nature of the optical system.

Nodal Points and Nodal Planes

They refer to the conjugate points having unit angular magnification. A ray of light proceeding from or towards one of them, the refracted ray proceeds or apparently proceeds from the second point in a direction parallel to the incident ray-direction.

General Formulation for the Location of Cardinal Points

Any optical system forming image of an object, the imaging process may be looked upon as a combination of translation in object space represented by $[T_1]$, refraction in the optical medium represented by a refraction matrix and then again translation in the image space represented by $[T_2]$, so we can always analytically simulate the imaging process by a series of matrix operation as

$$\begin{bmatrix} x_i \\ \alpha_i \end{bmatrix} = [T_2][R][T_1] \begin{bmatrix} x_o \\ \alpha_o \end{bmatrix} \tag{5.7.1}$$

where the column vector $\begin{bmatrix} x_i \\ \alpha_i \end{bmatrix}$ characterises the image and the column vector $\begin{bmatrix} x_o \\ \alpha_o \end{bmatrix}$ characterises the object. The refraction matrix, in general, can be written as

$$[R] = \begin{bmatrix} s_{11} & s_{12} \\ s_{21} & s_{22} \end{bmatrix}$$

This is also called the **optical element matrix.**

The matrix defined by $[T_2][R][T_1]$ is called the overall system matrix $[N]$ or simply the system matrix which transforms each point of the object from the object space to each point of the image in the image space. Thus any system matrix may be put as,

$$[N] = \begin{bmatrix} N_{11} & N_{12} \\ N_{21} & N_{22} \end{bmatrix} \tag{5.7.2}$$

So,

$$\begin{bmatrix} x_i \\ \alpha_i \end{bmatrix} = [T_2][R][T_1]\begin{bmatrix} x_o \\ \alpha_o \end{bmatrix} = \begin{bmatrix} 1 & -l_i \\ 0 & 1 \end{bmatrix}\begin{bmatrix} s_{11} & s_{12} \\ s_{21} & s_{22} \end{bmatrix}\begin{bmatrix} 1 & l_o \\ 0 & 1 \end{bmatrix}\begin{bmatrix} x_o \\ \alpha_o \end{bmatrix}$$

where l_o is the object distance and l_i is the image distance respectively measured against and in the direction of light.

Carrying out the matrix multiplication we can write,

$$\begin{aligned} \begin{bmatrix} x_i \\ \alpha_i \end{bmatrix} &= \begin{bmatrix} (s_{11} - s_{21}l_i) & (s_{12} - s_{22}l_i) \\ s_{21} & s_{22} \end{bmatrix}\begin{bmatrix} 1 & l_o \\ 0 & 1 \end{bmatrix}\begin{bmatrix} x_o \\ \alpha_o \end{bmatrix} \\ &= \begin{bmatrix} (s_{11} - s_{21}l_i) & l_o(s_{11} - s_{21}l_i) + (s_{12} - s_{22}l_i) \\ s_{21} & s_{21}l_o + s_{22} \end{bmatrix}\begin{bmatrix} x_o \\ \alpha_o \end{bmatrix} \\ &= \begin{bmatrix} N_{11} & N_{12} \\ N_{21} & N_{22} \end{bmatrix}\begin{bmatrix} x_o \\ \alpha_o \end{bmatrix} \end{aligned} \quad (5.7.3)$$

where $[N] = \begin{bmatrix} N_{11} & N_{12} \\ N_{21} & N_{22} \end{bmatrix}$ is the overall system matrix. The elements of the overall system matrix are defined as

$$\begin{aligned} N_{11} &= s_{11} - s_{21}l_i \text{ and } N_{12} = l_o(s_{11} - s_{21}l_i) + (s_{12} - s_{22}l_i) \\ N_{21} &= s_{21} \qquad\quad\ \text{and } N_{22} = s_{21}l_o + s_{22} \end{aligned}$$

From eqn.(5.7.3)

$$x_i = N_{11}x_o + N_{12}\alpha_o \quad (5.7.4)$$

$$\alpha_i = N_{21}x_o + N_{22}\alpha_o \quad (5.7.5)$$

The four elements of the overall system matrix N_{11}, N_{12}, N_{21} and N_{22} atually are used to locate the positions of the cardinal points of the system.

Principal Points and Principal Planes

If $N_{12} = 0, x_i = N_{11}x_o$, it means all the rays from a plane H_1 containing x_o will pass through a point in the plane H_2 containing the image point x_i, thus the planes H_1 and H_2 are conjugate planes. If in addition, $x_i/x_o = m = +1$, these conjugate planes become the principal planes or planes of *unit positive lateral magnification*.

Focal Points F_1, F_2 and Focal Planes

If $N_{22} = 0$, eqn.(5.7.5) gives us

$$\alpha_i = N_{21}x_o,$$

it means that rays from a point defined by x_o in the plane F_1 emerge in the same direction, i.e., as a parallel beam, hence F_1 plane is the *first principal focal plane*.

If $N_{11} = 0$, we get from eqn.(5.7.4)

$$x_i = N_{12}\alpha_o,$$

so rays incident as a parallel beam pass through a point x_i, thus it defines the second *principal focal plane*.

Nodal Points and Nodal Planes

If $N_{21} = 0$, $\alpha_i = N_{22}\alpha_o$, it means that emergent ray-direction is determined by incident ray-direction. Thus, if a collimated beam enters the system, the emergent beam also will be collimated. The system is then said to be collimated. If further $\alpha_i/\alpha_o = N_{22} = 1$, the conjugate points are called nodal points and the planes passing through them are the nodal planes.

5.8 Cardinal Points in Case of a Single Concave Refractor

We get from eqn.(5.4.1)

$$\begin{bmatrix} x_2 \\ \alpha_2 \end{bmatrix} = \begin{bmatrix} 1 & -v \\ 0 & 1 \end{bmatrix} \begin{bmatrix} 1 & 0 \\ P_{12} & \dfrac{n_1}{n_2} \end{bmatrix} \begin{bmatrix} 1 & u \\ 0 & 1 \end{bmatrix} \begin{bmatrix} x_1 \\ \alpha_1 \end{bmatrix}$$

$$= \begin{bmatrix} 1 & -v \\ 0 & 1 \end{bmatrix} \begin{bmatrix} s_{11} & s_{12} \\ s_{21} & s_{22} \end{bmatrix} \begin{bmatrix} 1 & u \\ 0 & 1 \end{bmatrix} \begin{bmatrix} x_1 \\ \alpha_1 \end{bmatrix}$$

$$= \begin{bmatrix} (s_{11} - vs_{21}) & (s_{12} - vs_{22}) \\ s_{21} & s_{22} \end{bmatrix} \begin{bmatrix} 1 & u \\ 0 & 1 \end{bmatrix} \begin{bmatrix} x_1 \\ \alpha_1 \end{bmatrix}$$

$$= \begin{bmatrix} (s_{11} - vs_{21}) & \{u(s_{11} - vs_{21}) + (s_{12} - vs_{22})\} \\ s_{21} & (us_{21} + s_{22}) \end{bmatrix} \begin{bmatrix} x_1 \\ \alpha_1 \end{bmatrix}$$

$$= \begin{bmatrix} N_{11} & N_{12} \\ N_{21} & N_{22} \end{bmatrix} \begin{bmatrix} x_1 \\ \alpha_1 \end{bmatrix}$$

Here,

$$N_{11} = s_{11} - vs_{21}, \quad N_{12} = u(s_{11} - vs_{21}) + (s_{12} - vs_{22})$$
$$N_{21} = s_{21}, \quad N_{22} = us_{21} + s_{22}$$

and
$$s_{11} = 1, \quad s_{12} = 0$$
$$s_{21} = P_{12}, \quad s_{22} = \frac{n_1}{n_2}$$

Principal Planes

Here, $N_{12} = 0 \quad \Rightarrow \quad u(s_{11} - vs_{21}) + (s_{12} - vs_{22}) = 0.$

For principal planes,

$$N_{11} = 1 = s_{11} - vs_{21} \quad \Rightarrow \quad v = \frac{(s_{11} - 1)}{s_{21}} = 0$$

$$\therefore \; u(s_{11}) + s_{12} = 0 \quad \Rightarrow \quad u = -\frac{s_{12}}{s_{11}} = 0.$$

So we conclude that the principal planes for a single refracting surface coincide with the plane passing through the poles and the unit points coincide with the poles.

Nodal Planes and Nodal Points

For nodal planes,

$$N_{22} = 1 \quad \Rightarrow \quad u = \frac{1 - s_{22}}{s_{21}} = \frac{1 - (n_1/n_2)}{P_{12}} = r.$$

So the nodal planes pass through the centre of curvature.

Focal Points and Focal Planes

Here,

$$N_{22} = 0 \quad \Rightarrow \quad u = -\frac{s_{22}}{s_{21}} = -\frac{n_1/n_2}{P_{12}} = -\frac{n_1}{n_2} \cdot \frac{n_2 r}{n_2 - n_1} = -\frac{n_1 r}{n_2 - n_1} = f_1.$$

The overall system matrix becomes

$$\begin{bmatrix} N_{11} & N_{12} \\ N_{21} & 0 \end{bmatrix} \quad \Rightarrow \quad \det N = -N_{12}N_{21} = \frac{n_1}{n_2} \quad \Rightarrow \quad v = \infty.$$

This is the property of matrix. Thus we get first principal focal plane.

Second Principal Focal Plane

This is defined by

$$N_{11} = 0 \quad \Rightarrow \quad s_{11} - v s_{21} = 0 \quad \Rightarrow \quad v = \frac{s_{11}}{s_{21}} = \frac{1}{P_{12}}.$$

So,

$$f_2 = v = \frac{1}{P_{12}} = \frac{n_2 r}{n_2 - n_1}.$$

Also, the overall system matrix becomes

$$[N] = \begin{bmatrix} 0 & N_{12} \\ N_{21} & N_{22} \end{bmatrix} \quad \Rightarrow \quad \det |N| = -N_{12} \cdot N_{21} = \frac{n_1}{n_2} \quad \therefore \quad u = \infty.$$

Thus we get the second principal focus of the surface. Here we have used the property that the determinant of the product matrix is equal to the products of the determinants of the matrices.

Refraction through a Composite Refracting Surface

Consider refraction through composite refracting surfaces separating three media of r.i. values n_1, n_2, n_3 ($n_1 < n_2 < n_3$). The imaging process consists of a series of operations which can be represented by suitable matrix transformation, as discussed in art. 5.3. From that consideration we can write,

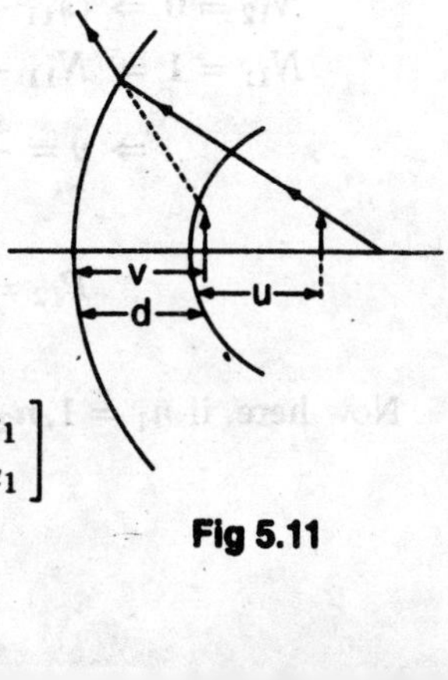

Fig 5.11

$$\begin{bmatrix} x_3 \\ \alpha_3 \end{bmatrix} = \begin{bmatrix} 1 & -v \\ 0 & 1 \end{bmatrix} \underbrace{\begin{bmatrix} 1 & 0 \\ P_{23} & \frac{n_2}{n_3} \end{bmatrix} \begin{bmatrix} 1 & v_1 + d \\ 0 & 1 \end{bmatrix} \begin{bmatrix} 1 & -v_1 \\ 0 & 1 \end{bmatrix} \begin{bmatrix} 1 & 0 \\ P_{12} & \frac{n_1}{n_2} \end{bmatrix}}_{s\text{-matrix}} \begin{bmatrix} 1 & u \\ 0 & 1 \end{bmatrix} \begin{bmatrix} x_1 \\ \alpha_1 \end{bmatrix}$$

The refraction matrix (also called the optical element matrix) in this case is

$$[R] = \begin{bmatrix} 1 & 0 \\ P_{23} & \frac{n_2}{n_3} \end{bmatrix} \begin{bmatrix} 1 & v_1 + d \\ 0 & 1 \end{bmatrix} \begin{bmatrix} 1 & -v_1 \\ 0 & 1 \end{bmatrix} \begin{bmatrix} 1 & 0 \\ P_{12} & \frac{n_1}{n_2} \end{bmatrix}$$

$$= \begin{bmatrix} (1 + P_{12}d) & d\frac{n_1}{n_2} \\ P_{23} + P_{12}\left(P_{23}d + \frac{n_2}{n_3}\right) & \frac{n_1}{n_2}\left(P_{23}d + \frac{n_2}{n_3}\right) \end{bmatrix} = \begin{bmatrix} s_{11} & s_{12} \\ s_{21} & s_{22} \end{bmatrix} \tag{5.8.1}$$

$$\therefore \quad \begin{bmatrix} x_3 \\ \alpha_3 \end{bmatrix} = \begin{bmatrix} 1 & -v \\ 0 & 1 \end{bmatrix} \begin{bmatrix} s_{11} & s_{12} \\ s_{21} & s_{22} \end{bmatrix} \begin{bmatrix} 1 & u \\ 0 & 1 \end{bmatrix} \begin{bmatrix} x_1 \\ \alpha_1 \end{bmatrix}$$

$$= \begin{bmatrix} (s_{11} - vs_{21}) & (s_{12} - vs_{22}) \\ s_{21} & s_{22} \end{bmatrix} \begin{bmatrix} 1 & u \\ 0 & 1 \end{bmatrix} \begin{bmatrix} x_1 \\ \alpha_1 \end{bmatrix}$$

$$= \begin{bmatrix} (s_{11} - vs_{21}) & \{(s_{11} - vs_{21})u + (s_{12} - vs_{22})\} \\ s_{21} & (us_{21} + s_{22}) \end{bmatrix} \begin{bmatrix} x_1 \\ \alpha_1 \end{bmatrix}$$

$$= \begin{bmatrix} N_{11} & N_{12} \\ N_{21} & N_{22} \end{bmatrix} \begin{bmatrix} x_1 \\ \alpha_1 \end{bmatrix}$$

Thus, the overall system matrix in this case is $[N] = \begin{bmatrix} N_{11} & N_{12} \\ N_{21} & N_{22} \end{bmatrix}$, where the elements of the matrix are defined as

$$N_{11} = s_{11} - vs_{21} = (1 + P_{12}d) - v\left\{P_{23} + P_{12}\left(P_{23}d + \frac{n_2}{n_3}\right)\right\}$$

$$N_{12} = (s_{11} - vs_{21})u + (s_{12} - vs_{22}); \; N_{21} = s_{21}, \; N_{22} = us_{21} + s_{22}.$$

Principal Planes of the Optical System

The principal planes are defined by $N_{12} = 0$ and also $N_{11} = 1$

$$N_{12} = 0 \Rightarrow (s_{11} - vs_{21})u + (s_{12} - vs_{22}) = 0$$

$$N_{11} = 1 \Rightarrow N_{11} - vs_{21} = 1$$

$$\Rightarrow v = \frac{s_{11} - 1}{s_{21}} = \frac{1 + P_{12}d - 1}{P_{23} + P_{12}\left(P_{23}d + \frac{n_2}{n_3}\right)} = \frac{P_{12}d}{P_{23} + P_{12}\left(P_{23}d + \frac{n_2}{n_3}\right)}$$

$$P_{12} = +\frac{n_2 - n_1}{n_2 r_1}, P_{23} = +\frac{n_3 - n_2}{n_3 r_2}.$$

Now here, if $n_1 = 1, n_2 = n, n_3 = 1$ and $r_1 = r_2 = r$.

$$\therefore \quad P_{12} = \frac{(n-1)}{nr}, P_{23} = \frac{1-n}{r}.$$

So,

$$v = \frac{\frac{(n-1)}{nr}\cdot d}{\frac{1-n}{r} + \frac{n-1}{nr}\left\{n + \frac{(1-n)d}{r}\right\}} = -\frac{r}{(n-1)}$$

giving the position of one principal plane.

Corresponding

$$u = -\frac{s_{12} - vs_{22}}{s_{11} - vs_{21}} = vs_{22} - s_{12}. \tag{5.8.2}$$

Now,

$$s_{22} = \frac{n_1}{n_2}\left(P_{23}d + \frac{n_2}{n_3}\right) = \frac{1}{n}\left\{-\frac{(n-1)d}{r} + n\right\}, s_{12} = \frac{dn_1}{n_2} = \frac{d}{n}.$$

∴ Eqn.(5.8.2) gives,

$$u = -\frac{r}{(n-1)}\left\{-\frac{(n-1)d}{nr} + 1\right\} - \frac{d}{n} = \frac{d}{n} - \frac{r}{(n-1)} - \frac{d}{n} = -\frac{r}{n-1}.$$

So under such a situation, the two principal planes coincide in position.

Refraction Through a Glass-Sphere

The final image by refraction through a glass-sphere is formed at I; the first refraction takes place at the convex surface, then a translation by $2r$ follows and then again refraction occurs at the second concave surface, then the final image is formed at I.

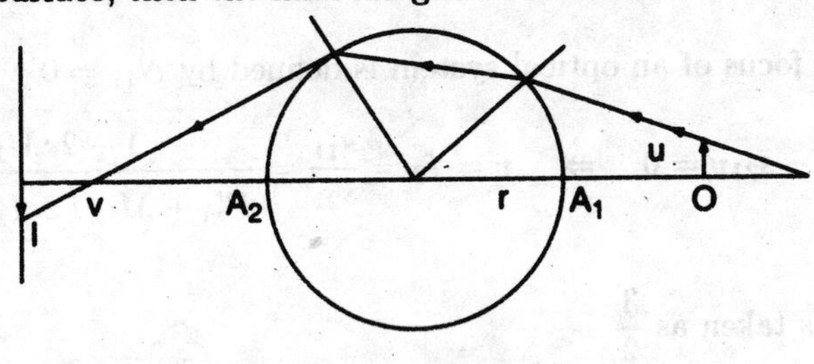

Fig 5.12

So the image parameters (x_2, α_2) are given by

$$\begin{bmatrix} x_2 \\ \alpha_2 \end{bmatrix} = \begin{bmatrix} 1 & -v \\ 0 & 1 \end{bmatrix} [R_2] \begin{bmatrix} 1 & 2r \\ 0 & 1 \end{bmatrix} [R_1] \begin{bmatrix} 1 & u \\ 0 & 1 \end{bmatrix} \begin{bmatrix} x_1 \\ \alpha_1 \end{bmatrix}$$

where, $[R_1]$ = refraction matrix at the first convex refractor, light going from air to glass

$$= \begin{bmatrix} 1 & 0 \\ M_{12} & \frac{n_1}{n_2} \end{bmatrix}, n_1 = 1, n_2 = n \quad \text{and} \quad M_{12} = -\frac{n_2 - n_1}{n_2 r} = -\frac{(n-1)}{nr}$$

$[R_2]$ = refraction matrix for light refracting at the concave refractor from glass to air

$$= \begin{bmatrix} 1 & 0 \\ P'_{23} & \frac{n_2}{n_3} \end{bmatrix}, P'_{23} = -\frac{n_2 - n_3}{n_3 r} = \frac{1-n}{r}.$$

So the refraction matrix for the glass-sphere is

$$[R] = [R_2]\begin{bmatrix} 1 & 2r \\ 0 & 1 \end{bmatrix} [R_1] = \begin{bmatrix} 1 & 0 \\ P_{23} & \frac{n_2}{n_3} \end{bmatrix}\begin{bmatrix} 1 & 2r \\ 0 & 1 \end{bmatrix}\begin{bmatrix} 1 & 0 \\ M_{12} & \frac{n_1}{n_2} \end{bmatrix}$$

$$\Rightarrow \quad [R] = \begin{bmatrix} 1 + 2rM_{12} & 2r\cdot\frac{n_1}{n_2} \\ P'_{23} + M_{12}\left(2rP'_{23} + \frac{n_2}{n_3}\right) & \frac{n_1}{n_2}\left(P_{23}\cdot 2r + \frac{n_2}{n_3}\right) \end{bmatrix} = \begin{bmatrix} s_{11} & s_{12} \\ s_{21} & s_{22} \end{bmatrix}$$

$$\therefore \quad \begin{bmatrix} x_2 \\ \alpha_2 \end{bmatrix} = \begin{bmatrix} 1 & -v \\ 0 & 1 \end{bmatrix}\begin{bmatrix} s_{11} & s_{12} \\ s_{21} & s_{22} \end{bmatrix}\begin{bmatrix} 1 & u \\ 0 & 1 \end{bmatrix}\begin{bmatrix} x_1 \\ \alpha_1 \end{bmatrix}$$

$$= \begin{bmatrix} (s_{11} - s_{21}v) & \{u(s_{11} - s_{21}v) + (s_{12} - s_{22}v)\} \\ s_{21} & us_{21} + s_{22} \end{bmatrix}\begin{bmatrix} x_1 \\ \alpha_1 \end{bmatrix}$$

$$= \begin{bmatrix} N_{11} & N_{12} \\ N_{21} & N_{22} \end{bmatrix}\begin{bmatrix} x_1 \\ \alpha_1 \end{bmatrix}$$

where, $[N] = \begin{bmatrix} N_{11} & N_{12} \\ N_{21} & N_{22} \end{bmatrix}$ is called the overall system matrix representing the behaviour of the system under the prescribed condition. As discussed in art.(5.5) we have—

Focal Points and Focal Planes

The second principal focus of an optical system is defined by $N_{11} = 0$

$$\therefore \quad s_{11} - s_{21}v = 0 \quad \Rightarrow \quad v = f_2 = \frac{s_{11}}{s_{21}} = \frac{1 + 2rM_{12}}{P'_{23} + M_{12}\left(2rP'_{23} + \frac{n_2}{n_3}\right)}$$

In this case, if n is taken as $\frac{3}{2}$

$$M_{12} = -\frac{n-1}{nr} = -\frac{1}{3r} \quad \text{and} \quad P'_{23} = \frac{1-n}{r} = -\frac{1}{2r}$$

$$\therefore \quad s_{11} = 1 + 2rM_{12} = \left(1 - \frac{2}{3}\right) = \frac{1}{3}$$

$$s_{21} = P'_{23} + M_{12}\left(2rP_{23} + \frac{n_2}{n_3}\right) = -\frac{1}{2r} - \frac{1}{3r}(-1+n) = -\frac{1}{2r} - \frac{1}{6r} = -\frac{2}{3r}$$

$\therefore \quad f_2 = \dfrac{1/3}{-(2/3r)} = -\dfrac{r}{2}$, so the second principal focus is at a distace $\dfrac{r}{2}$ behind the second surface.

The first principal focus is defined by $N_{22} = 0$

$$\therefore \quad us_{21} + s_{22} = 0 \quad \Rightarrow \quad u = f_1 = -\frac{s_{22}}{s_{21}}$$

Now,

$$s_{22} = \frac{n_1}{n_2}\left(P'_{23}.2r + \frac{n_2}{n_3}\right) = \frac{1}{n}(-1+n) = \frac{n-1}{n} = \frac{1}{3}$$

and $s_{21} = -\frac{2}{3r}$. $\therefore$ $f_1 = \frac{1/3}{-(2/3r)} = \frac{r}{2}$ (to the right of the first surface).

Principal Planes

The principal planes are defined by $N_{12} = 0$ and $N_{11} = 1$.

$$\therefore\ u(s_{11} - s_{21}v) + (s_{12} - s_{22}v) = 0 \tag{5.8.3}$$

$$\text{and}\ \ (s_{11} - s_{21}v) = 1 \tag{5.8.4}$$

$$\Rightarrow v = A_2H_2 = \frac{s_{11}-1}{s_{21}} = \frac{\frac{1}{3}-1}{-\frac{2}{3r}} = r$$

thus the second principal plane is to the right of the second surface at a distance of r, hence at the centre. From (5.8.3) we get,

$$u = A_1H_1 = -(s_{12} - s_{22}v)$$

$$s_{12} = 2r\cdot\frac{n_1}{n_2} = \frac{2r}{n} = \frac{4r}{3} \quad \text{and} \quad s_{22} = \frac{1}{3}$$

$$\therefore \quad u = A_1H_1 = -\left(\frac{4r}{3} - \frac{r}{3}\right) = -r \quad \text{(to the left of the first surface).}$$

Thus, we see that the two principal or unit planes coincide with the centre of the glass-sphere.

Nodal Planes

Here, $N_{21} = 0$ and $N_{22} = 1$.

Now, $N_{22} = 1$

$$\Rightarrow \quad us_{21} + s_{22} = 1 \quad \Rightarrow \quad u = A_1N_1 = \frac{1-s_{22}}{s_{21}} = \frac{1-\frac{1}{3}}{-\frac{2}{3r}} = -r.$$

The overall system matrix $[N]$ now becomes

$$[N] = \begin{bmatrix} N_{11} & N_{12} \\ 0 & N_{22} \end{bmatrix}$$

Since, $\det N = 1$,

$$\therefore \quad N_{11}\cdot N_{22} = 1 \Rightarrow N_{11} = 1$$

$$\Rightarrow v = A_2N_2 = \frac{s_{11}-1}{s_{21}} = r.$$

Thus the nodal planes also coincide at the centre. Thus the cardinal points of the optical system of a glass-sphere are as shown in fig. (5.13), where H_1 and H_2 indicate the principal planes; N_1 and N_2, the nodal planes.

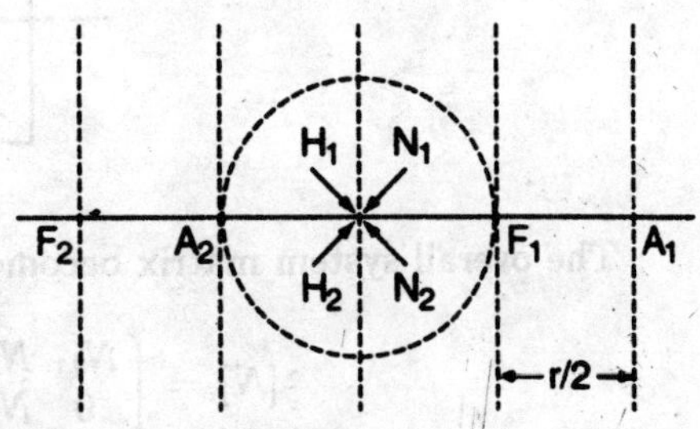

Fig 5.13

Short Questions with Answers

1. **What do you mean by conjugate planes? How many conjugate planes are there for an optical system?**

   Ans. The conjugate planes are those pair of planes such that if an object be placed at one, the image is formed at the other. The conjugate planes of an optical system are principal planes (H_1, H_2) and the nodal planes (N_1, N_2).

2. **Define principal planes. Where do these principal points lie in case of a glass-sphere?**

   Ans. Principal planes, also called unit planes are those pair of planes such that object in one plane produces image in the other with unit positive lateral magnification. The principal planes of a glass-sphere pass through the centre of curvature of the sphere.

3. **Define nodal planes. Where do the nodal planes lie in case of a glass hemisphere?**

   Ans. Nodal planes are pair of conjugate planes such that a ray directed towards one nodal point leaves the optical system from the other nodal point parallel to the incident ray, thus nodal planes are planes of unit angular magnification. One nodal point is defined by

$$N_{22} = 1 \quad \Rightarrow \quad us_{21} + s_{22} = 1 \quad \Rightarrow \quad u = A_1N_1 = \frac{1 - s_{22}}{s_{21}}$$

where $s_{22} = \dfrac{n_1}{n_2}\left(P_{23}\cdot r + \dfrac{n_2}{n_3}\right)$ and where translation through the hemisphere is by $r, n_1 = 1 = n_3$, and $n_2 = n$.

$$\therefore \quad P_{23} = \frac{n_3 - n_2}{\infty} = 0, \quad M_{12} = -\frac{n_2 - n_1}{n_2 r} = -\frac{n-1}{nr} = -\frac{1}{3r}.$$

Hence,

$$s_{22} = \frac{1}{n}\cdot n = 1$$

$$s_{21} = P_{23} + M_{12}\left(rP_{23} + \frac{n_2}{n_3}\right) = 0 - \frac{1}{3r}(0 + n) = -\frac{n}{3r}$$

$$\therefore \quad u = \frac{1 - s_{22}}{s_{21}} = 0.$$

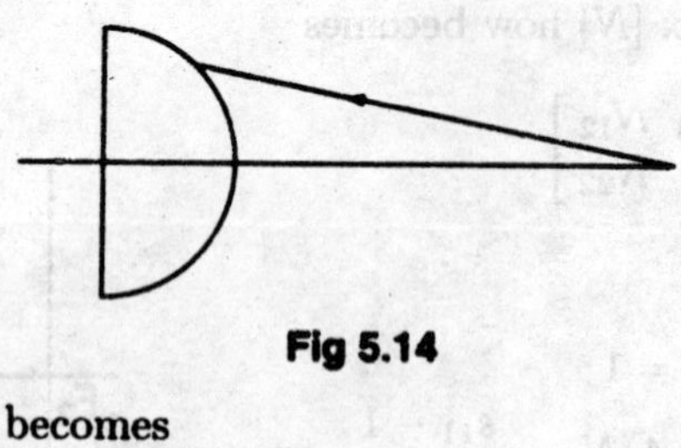

Fig 5.14

The overall system matrix becomes

$$[N] = \begin{bmatrix} N_{11} & N_{12} \\ 0 & N_{22} \end{bmatrix}$$

$$\therefore \quad |N| = N_{11}\cdot N_{22} = 1 \quad \Rightarrow \quad N_{11} = 1 \quad \Rightarrow \quad v = \frac{s_{11} - 1}{s_{21}}$$

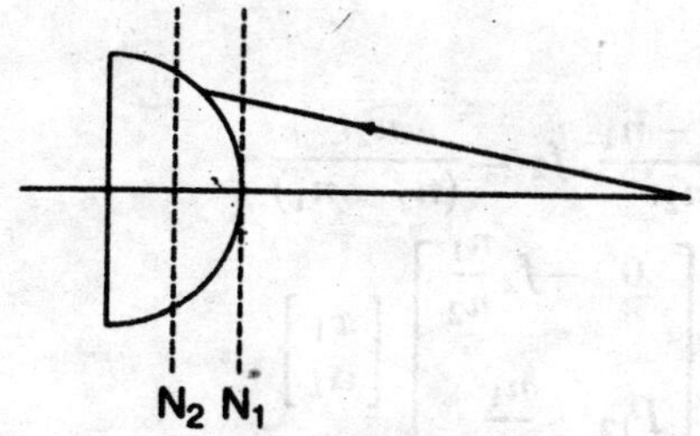

Fig 5.15

Now, $s_{11} = 1 + r \cdot M_{12}$

$$\Rightarrow \quad s_{11} - 1 = -r \cdot \frac{1}{3r} = -\frac{1}{3}$$

$$s_{21} = -\frac{n}{3r}$$

$$\therefore \quad v = \frac{-1/3}{-n/3r} = \frac{r}{n} = \frac{2}{3}r$$

So, the nodal planes for a hemisphere are shown in fig.(5.15).

4. **What is Lagrange-Helmholtz equation? Write down the expression for a number of refracting surfaces. How is the expression modified in case the object is at an infinite distance?**

Ans. It is a relation connecting the linear and angular magnification by a system of refracting surfaces. For $(k-1)$ refracting surfaces separating a number of media having refractive indices n_1, n_2, n_3, etc., this eqn. can be written as

$$x_1 n_1 \alpha_1 = x_2 n_2 \alpha_2 = \cdots = x_k \alpha_k n_k.$$

Consider rays coming from infinity and the marginal ray strikes the refracting surface at a height x_1. Then,

$$\begin{bmatrix} x_2 \\ \alpha_2 \end{bmatrix} = \begin{bmatrix} 1 & -f_2 \\ 0 & 1 \end{bmatrix} \begin{bmatrix} 1 & 0 \\ P_{12} & \frac{n_1}{n_2} \end{bmatrix} \begin{bmatrix} x_1 \\ \alpha_1 \end{bmatrix} \begin{bmatrix} (1 - P_{12}f_2) & -f_2\frac{n_1}{n_2} \\ P_{12} & \frac{n_1}{n_2} \end{bmatrix} \begin{bmatrix} x_1 \\ \alpha_1 \end{bmatrix}$$

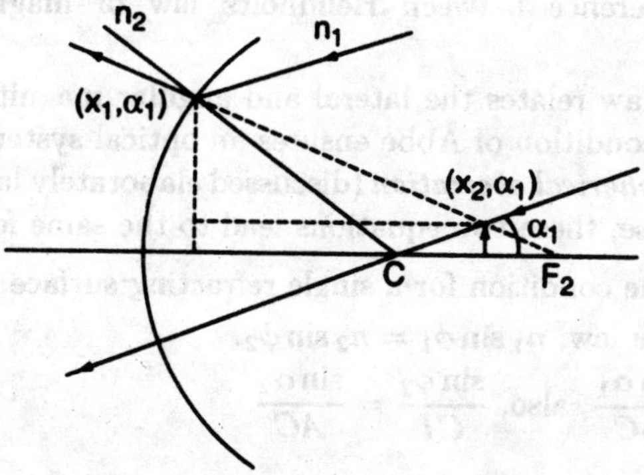

Fig 5.16

Now,

$$P_{12} = \frac{n_2 - n_1}{n_2 r}, f_2 = \frac{n_2 r}{(n_2 - n_1)}$$

$$\therefore \begin{bmatrix} x_2 \\ \alpha_2 \end{bmatrix} = \begin{bmatrix} 0 & -f_2\dfrac{n_1}{n_2} \\ P_{12} & \dfrac{n_1}{n_2} \end{bmatrix} \begin{bmatrix} x_1 \\ \alpha_1 \end{bmatrix}$$

$$\Rightarrow x_2 = -f_2\frac{n_1}{n_2}\alpha_1, \alpha_2 = P_{12}x_1 + \frac{n_1}{n_2}\alpha_1$$

$$\therefore \frac{n_2\alpha_2}{n_1\alpha_1} = \frac{n_2}{n_1}P_{12}\frac{x_1}{\alpha_1} + 1 = -\frac{n_2}{n_1}P_{12}\frac{x_1}{x_2}\cdot\frac{n_1 f_2}{n_2} + 1 = 1 - \frac{x_1}{x_2} = \frac{x_2 - x_1}{x_2}.$$

Here, $\alpha_1 = +\text{ve}$, $\alpha_2 = -\text{ve}$, hence,

$$\frac{n_2\alpha_2}{n_1\alpha_1} = \frac{x_1 - x_2}{x_2} \quad \Rightarrow \quad (x_1 - x_2)n_1\alpha_1 = n_2 x_2 \alpha_2.$$

Since, $x_2 \ll x_1$, we get,

$$x_1 n_1 \alpha_1 = x_2 n_2 \alpha_2. \tag{5.8.5}$$

Eqn.(5.8.5) is the form of Helmholtz-Lagrange relation for an object at infinity.

5. What do you mean by a telescopic system? What will be the form of Lagrange-Helmholtz relation in case of a telescopic system consisting of a number of refracting surfaces?

Ans. In case of a telescopic system of a number of refractors, the angular magnification and linear magnification are reciprocally related.

Consider a system of $(k-1)$ refracting surfaces, then Lagrange-Helmholtz relation gives us,

$$x_1 n_1 \tan\alpha_1 = x_2 n_2 \tan\alpha_2 = \cdots = x_k n_k \tan\alpha_k.$$

If $n_1 = n_k$, then we get,

$$x_1 \tan\alpha_1 = x_k \tan\alpha_k \quad \Rightarrow \quad \frac{x_k}{x_1} = m = \frac{\tan\alpha_1}{\tan\alpha_k} = \frac{1}{M}.$$

6. What is the difference between Helmholtz law of magnifications and Abbe's sine condition?

Ans. Helmholtz law relates the lateral and angular magnification by an optical system whereas the sine condition of Abbe ensures an optical system to be free from the defects called *coma* and *spherical aberration* (discussed elaborately later). Only in case of paraxial condition, of course, these two equations lead to the same form.

7. Obtain Abbe's sine condition for a single refracting surface.

Ans. From Snell's law, $n_1 \sin\phi_1 = n_2 \sin\phi_2$.

Now, $\dfrac{\sin\phi_1}{CO} = \dfrac{\sin\alpha_1}{AC}$; also, $\dfrac{\sin\phi_2}{CI} = \dfrac{\sin\alpha_2}{AC}$

$$\therefore \frac{\sin\phi_1}{\sin\phi_2} = \frac{CO}{CI}\cdot\frac{\sin\alpha_1}{\sin\alpha_2} = \frac{n_2}{n_1} \quad \Rightarrow \quad \frac{n_1\sin\alpha_1}{n_2\sin\alpha_2} = \frac{CI}{CO} = \frac{v-r}{u-r}$$

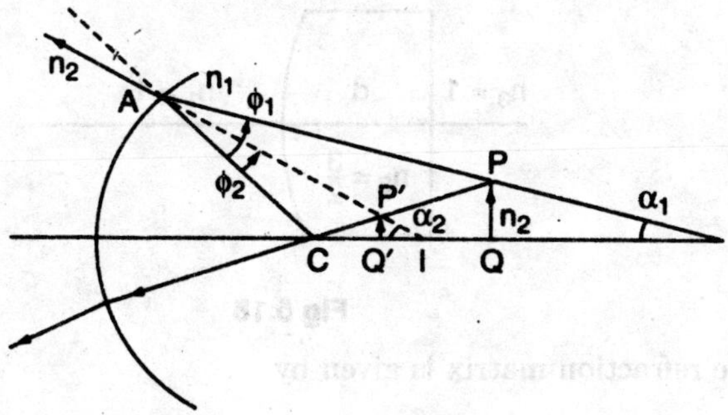

Fig 5.17

Again,

$$\frac{x_2}{x_1} = \frac{P'Q'}{PQ} = \frac{CQ'}{CQ} = \frac{v-r}{u-r}$$

$$\therefore \quad \frac{n_1 \sin\alpha_1}{n_2 \sin\alpha_2} = \frac{x_2}{x_1} \quad \Rightarrow \quad x_1 n_1 \sin\alpha_1 = x_2 n_2 \sin\alpha_2.$$

Solved Numerical Problems

1. Consider a sphere of radius 20 cm and r.i. 3/2. Find the position of the paraxial focal point.

Solution. From the problem of refraction through glass-sphere,

$$f_2 = \frac{1 + 2rM_{12}}{P_{23} + M_{12}\left(2rP_{23} + \frac{n_2}{n_3}\right)}.$$

Here,

$$M_{12} = -\frac{n-1}{nr} = -\frac{1}{3r} \quad \text{and} \quad P_{23} = \frac{1-n}{r} = \frac{1}{-2r}$$

$$\therefore \quad 1 + 2rM_{12} = 1 - \frac{2}{3} = \frac{1}{3} \quad \text{and} \quad P_{23} + M_{12}\left(2rP_{23} + \frac{n_2}{n_3}\right)$$

$$= -\frac{1}{2r} - \frac{1}{3r}(-1+n) = -\frac{1}{2r} - \frac{1}{6r} = -\frac{4}{6r} = -\frac{2}{3r}$$

$$\therefore \quad f_2 = \frac{1/3}{-(2/3r)} = -\frac{r}{2}, \quad \text{i.e., to the left of the second surface.}$$

So its distance from the centre $= \frac{3}{2}r = 30$ cm.

2. Find the refraction matrix for the transparent medium bounded by a curved surface and a plane surface as shown in fig.(5.18).

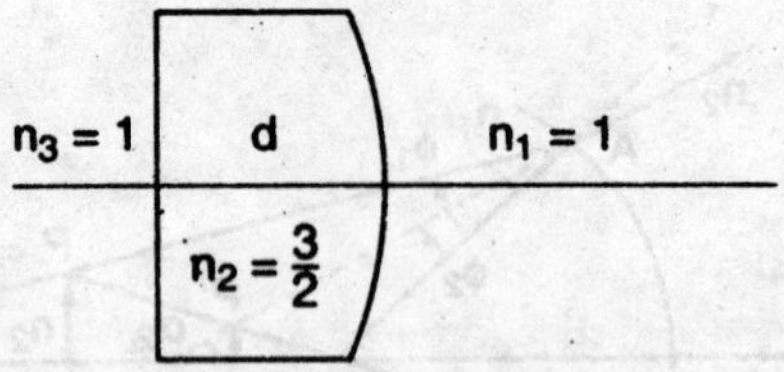

Fig 5.18

Solution. The refraction matrix is given by

$$[R] = \begin{bmatrix} 1 & 0 \\ P_{23} & \dfrac{n_2}{n_3} \end{bmatrix} \begin{bmatrix} 1 & d \\ 0 & 1 \end{bmatrix} \begin{bmatrix} 1 & 0 \\ M_{12} & \dfrac{n_1}{n_2} \end{bmatrix}$$

Here, $P_{23} = \dfrac{n_3 - n_2}{n_3 r}$, since the surface of separation is plane $r = \infty$, hence, $P_{23} = 0$.

$$M_{12} = -\frac{n_2 - n_1}{n_2 r} = -\frac{3/2 - 1}{(3/2)r} = -\frac{1}{3r}$$

$n_1 = n_3 = 1, n_2 = 3/2$

$$\therefore \quad [R] = \begin{bmatrix} 1 & 0 \\ 0 & \dfrac{3}{2} \end{bmatrix} \begin{bmatrix} 1 & d \\ 0 & 1 \end{bmatrix} \begin{bmatrix} 1 & 0 \\ -\dfrac{1}{3r} & \dfrac{2}{3} \end{bmatrix} = \begin{bmatrix} \left(1 - \dfrac{d}{3r}\right) & \dfrac{2d}{3} \\ -\dfrac{1}{2r} & 1 \end{bmatrix}$$

3. Find the refraction matrix in case of refraction as shown in fig.(5.19).
Solution. Here again, the refraction matrix of the optical system is given by,

$$[R] = \begin{bmatrix} 1 & 0 \\ M'_{23} & \dfrac{n_2}{n_3} \end{bmatrix} \begin{bmatrix} 1 & d \\ 0 & 1 \end{bmatrix} \begin{bmatrix} 1 & 0 \\ P_{12} & \dfrac{n_1}{n_2} \end{bmatrix}$$

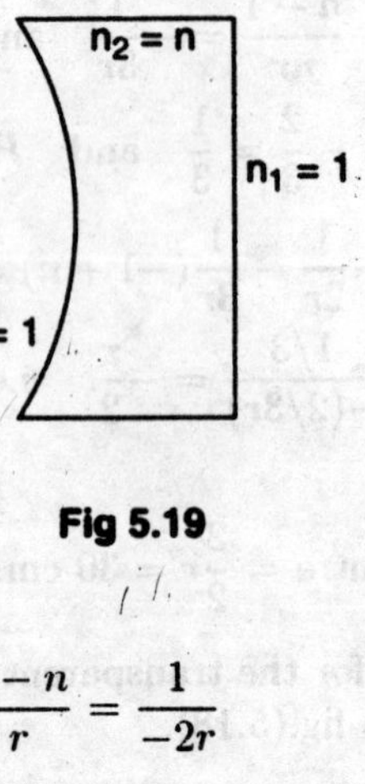

Fig 5.19

Here, $P_{12} = 0$,

$$M'_{23} = +\frac{n_3 - n_2}{n_3 r} = \frac{1 - n}{r} = \frac{1}{-2r}$$

$$\therefore \quad [R] = \begin{bmatrix} 1 & 0 \\ -\frac{1}{2r} & \frac{3}{2} \end{bmatrix} \begin{bmatrix} 1 & d \\ 0 & 1 \end{bmatrix} \begin{bmatrix} 1 & 0 \\ 0 & \frac{2}{3} \end{bmatrix}$$

$$= \begin{bmatrix} 1 & d \\ -\frac{1}{2r} & \left(-\frac{d}{2r}+\frac{3}{2}\right) \end{bmatrix} \begin{bmatrix} 1 & 0 \\ 0 & \frac{2}{3} \end{bmatrix} = \begin{bmatrix} 1 & \frac{2}{3}d \\ -\frac{1}{2r} & \left\{1-\left(\frac{d}{3r}\right)\right\} \end{bmatrix}$$

4. **A hemisphere of glass ($n = 3/2$) of 3 inch radius serves as an optical system. Find the position of the image when an object is placed 4 inch in front of the curved surface.**

Solution. Here the refraction matrix is given as in problem 2 by

$$[R] = \begin{bmatrix} \left(1-\frac{d}{3r}\right) & \frac{2d}{3} \\ -\frac{1}{2r} & 1 \end{bmatrix} = \begin{bmatrix} \left(1-\frac{r}{3r}\right) & \frac{2r}{3} \\ -\frac{1}{2r} & 1 \end{bmatrix}.$$

$$\begin{bmatrix} x_2 \\ \alpha_2 \end{bmatrix} = \begin{bmatrix} 1 & -v \\ 0 & 1 \end{bmatrix} [R] \begin{bmatrix} 1 & u \\ 0 & 1 \end{bmatrix} \begin{bmatrix} x_1 \\ \alpha_1 \end{bmatrix} = \begin{bmatrix} 1 & -v \\ 0 & 1 \end{bmatrix} \begin{bmatrix} \left(1-\frac{1}{3}\right) & 2 \\ -\frac{1}{6} & 1 \end{bmatrix} \begin{bmatrix} 1 & u \\ 0 & 1 \end{bmatrix} \begin{bmatrix} x_1 \\ \alpha_1 \end{bmatrix}$$

$$= \begin{bmatrix} \left(\frac{2}{3}+\frac{v}{6}\right) & (2-v) \\ -\frac{1}{6} & 1 \end{bmatrix} \begin{bmatrix} 1 & u \\ 0 & 1 \end{bmatrix} \begin{bmatrix} x_1 \\ \alpha_1 \end{bmatrix}$$

$$= \begin{bmatrix} \left(\frac{2}{3}+\frac{v}{6}\right) & \left\{u\left(\frac{2}{3}+\frac{v}{6}\right)+(2-v)\right\} \\ -\frac{1}{6} & -\frac{1}{6}u+1 \end{bmatrix} \begin{bmatrix} x_1 \\ \alpha_1 \end{bmatrix}$$

$$\therefore \quad x_2 = \left(\frac{2}{3}+\frac{v}{6}\right)x_1 + \alpha_1\left\{u\left(\frac{2}{3}+\frac{v}{6}\right)+(2-v)\right\} \quad \text{and} \quad \alpha_2 = -\frac{1}{6}x_1 + \alpha_1\left(1-\frac{1}{6}u\right)$$

For an axial point, $x_2 = 0 = x_1$ and $\alpha_1 \neq 0$

$$\therefore \quad u\left(\frac{2}{3}+\frac{v}{6}\right)+(2-v)=0 \quad \Rightarrow \quad 4\left(\frac{2}{3}+\frac{v}{6}\right)+(2-v)=0$$

$$\Rightarrow \quad \frac{8}{3}+\frac{2}{3}v+2-v=0 \quad \Rightarrow \quad \frac{14}{3}-\frac{1}{3}v=0$$

$\Rightarrow$ **$v = 14$ inch from plane surface towards object side.**

5. **A thin transparent medium bounded by two surfaces of same radius of 3 cm is situated just below the surface of water in a tank which is 40 cm deep. A point object is at the**

bottom of the tank. Find the position of the image. The refractive indices of the medium and water are respectively 3/2 and 4/3.

Solution. The image parameter (x_2, α_2) and object parameter (x_1, α_1) are given by,

$$\begin{bmatrix} x_2 \\ \alpha_2 \end{bmatrix} = \begin{bmatrix} 1 & -v \\ 0 & 1 \end{bmatrix} [R] \begin{bmatrix} 1 & u \\ 0 & 1 \end{bmatrix} \begin{bmatrix} x_1 \\ \alpha_1 \end{bmatrix}$$

$$[R] = \begin{bmatrix} 1 & 0 \\ 0 & \dfrac{n_1}{n_3} \end{bmatrix} \begin{bmatrix} 1 & 0 \\ P'_{21} & \dfrac{n_2}{n_1} \end{bmatrix} \begin{bmatrix} 1 & 0 \\ M_{12} & \dfrac{n_1}{n_2} \end{bmatrix}$$

$$= \begin{bmatrix} 1 & 0 \\ 0 & \dfrac{4}{3} \end{bmatrix} \begin{bmatrix} 1 & 0 \\ -\dfrac{1}{24}, & \dfrac{9}{8} \end{bmatrix} \begin{bmatrix} 1 & 0 \\ -\dfrac{1}{27} & \dfrac{8}{9} \end{bmatrix} = \begin{bmatrix} 1 & 0 \\ -\dfrac{1}{9} & \dfrac{4}{3} \end{bmatrix}$$

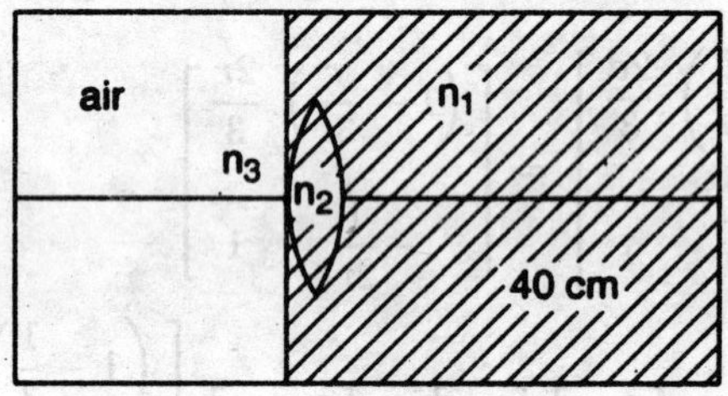

Fig 5.20

$$\therefore \begin{bmatrix} x_2 \\ \alpha_2 \end{bmatrix} = \begin{bmatrix} 1 & -v \\ 0 & 1 \end{bmatrix} \begin{bmatrix} 1 & 0 \\ -\dfrac{1}{9} & \dfrac{4}{3} \end{bmatrix} \begin{bmatrix} 1 & u \\ 0 & 1 \end{bmatrix} \begin{bmatrix} x_1 \\ \alpha_1 \end{bmatrix}$$

$$= \begin{bmatrix} \left(1 + \dfrac{v}{9}\right) & \left\{u\left(1 + \dfrac{v}{9}\right) - \dfrac{4}{3}v\right\} \\ -\dfrac{1}{9} & \left(-\dfrac{u}{9} + \dfrac{4}{3}\right) \end{bmatrix} \begin{bmatrix} x_1 \\ \alpha_1 \end{bmatrix}$$

$$\Rightarrow \quad x_2 = \left(1 + \frac{v}{9}\right) x_1 + \alpha_1 \left\{u\left(1 + \frac{v}{9}\right) - \frac{4}{3}v\right\}$$

For an axial point object $x_2 = x_1 = 0$ but $\alpha_1 \neq 0$

$$\therefore \quad u\left(1 + \frac{v}{9}\right) - \frac{4}{3}v = 0 \quad \Rightarrow \quad 40\left(1 + \frac{v}{9}\right) - \frac{4}{3}v = 0 \quad \Rightarrow v = -12\frac{6}{7} \text{ cm.}$$

6. A ray of light emerging from an axial point object, subtends an angle $\alpha_1 = +6°$ with the optic axis. At a height $h = 2$ cm, it intersects a convex spherical surface of radius 5 cm, the r.i. $n = 1.62$. Find the angle the ray subtends with the optic axis after refraction.

Solution.

$$\begin{bmatrix} x_2 \\ \alpha_2 \end{bmatrix} = \begin{bmatrix} 1 & 0 \\ M_{12} & \dfrac{n_1}{n_2} \end{bmatrix} \begin{bmatrix} x_1 \\ \alpha_1 \end{bmatrix}$$

$$\alpha_1 = +6° = +0.1047 \text{ rad} \quad \text{+ve sign indicates anti-clockwise}$$

$$M_{12} = -\frac{n_2 - n_1}{n_2 r} = -\frac{1.62 - 1}{1.62 \times 5} = -.0765 \quad \text{and} \quad \frac{n_1}{n_2} = \frac{1}{1.62} = .617$$

$$\therefore \quad x_2 = x_1 \quad \text{and} \quad \alpha_2 = M_{12}x_1 + \frac{n_1}{n_2}\alpha_1$$

$$= -.0765 \times 2 + .617 \times .1047 = -.088 \text{ rad} = -5° \text{ (app.)}$$

−ve sign means clockwise.

7. A small object is located 0.1 m in front of a convex surface of radius 0.03 m radius and r.i. 1.66 ground on a glass block. Find the position of the image
   (a) if object and image are in air;
   (b) if both are immersed in water ($n = 4/3$).

Solution. (a)

$$\begin{bmatrix} x_2 \\ \alpha_2 \end{bmatrix} = \begin{bmatrix} 1 & -v \\ 0 & 1 \end{bmatrix} \begin{bmatrix} 1 & 0 \\ P_{12} & \frac{n_1}{n_2} \end{bmatrix} \begin{bmatrix} 1 & u \\ 0 & 1 \end{bmatrix} \begin{bmatrix} x_1 \\ \alpha_1 \end{bmatrix}$$

$$= \begin{bmatrix} (1 - vP_{12}) & -v\frac{n_1}{n_2} \\ P_{12} & \frac{n_1}{n_2} \end{bmatrix} \begin{bmatrix} 1 & u \\ 0 & 1 \end{bmatrix} \begin{bmatrix} x_1 \\ \alpha_1 \end{bmatrix}$$

$$= \begin{bmatrix} (1 - vP_{12}) & \left\{ u(1 - vP_{12}) - v\frac{n_1}{n_2} \right\} \\ P_{12} & \left(uP_{12} + \frac{n_1}{n_2} \right) \end{bmatrix} \begin{bmatrix} x_1 \\ \alpha_1 \end{bmatrix}$$

$$\Rightarrow \quad x_2 = (1 - vP_{12})x_1 + \alpha_1 \left\{ u(1 - vP_{12}) - v\frac{n_1}{n_2} \right\}$$

For an axial point $x_2 = x_1 = 0$, but $a_1 \neq 0$.

$$\therefore \quad u(1 - vP_{12}) - v\frac{n_1}{n_2} = 0$$

$$\Rightarrow \frac{1}{v}(1 - vP_{12}) - \frac{1}{u}\frac{n_1}{n_2} = 0$$

$$\Rightarrow \frac{n_2}{v} - \frac{n_2}{u} = n_2 P_{12} = -\frac{n_2 - n_1}{r}$$

$$\therefore \quad \frac{1.66}{v} = \frac{1}{0.1} - \frac{1.66 - 1}{0.03}$$

$$\Rightarrow v = -\frac{1.66}{12} = -13.8 \text{ cm.} \quad \text{(towards the right of the curved surface.)}$$

$$\text{(b)} \quad \frac{1.66}{v} = \frac{4/3}{0.1} - \frac{1.66 - 4/3}{.03} \quad \Rightarrow v = 0.679 \text{ m} = 67.9 \text{ (towards self).}$$

8. Using matrix multiplication, determine the new co-ordinates (x', y') in terms of initial co-ordinates (x_o, y_o)

$$\begin{bmatrix} 3 & 1 \\ 2 & 0 \end{bmatrix} \begin{bmatrix} 2 & 0 \\ -1 & 3 \end{bmatrix} \begin{bmatrix} x_o \\ y_o \end{bmatrix} = \begin{bmatrix} x' \\ y' \end{bmatrix}$$

Solution.

$$\begin{bmatrix} x' \\ y' \end{bmatrix} = \begin{bmatrix} 3 & 1 \\ 2 & 0 \end{bmatrix}\begin{bmatrix} 2 & 0 \\ -1 & 3 \end{bmatrix}\begin{bmatrix} x_\circ \\ y_\circ \end{bmatrix} = \begin{bmatrix} 5 & 3 \\ 4 & 0 \end{bmatrix}\begin{bmatrix} x_\circ \\ y_\circ \end{bmatrix}$$

$$\Rightarrow \quad x' = x_\circ + 3y_\circ \quad \text{and} \quad y' = 4x_\circ.$$

Theory Questions

1. **Why and under what condition can we use the matrix method to solve geometrical optics problems? Develop the translation and refraction matrices. [BU 1996]**

2. **Obtain the system matrix for a glass-sphere of radius r placed in air.**

3. **Find the refraction matrix of a plano-concave refractor.**

4. **Using matrix method to build up Lagrange-Helmholtz law in case of a number of refracting surfaces.**

5. **A refracting surface separates two media, one called incident medium having r.i. n_1, the other refracting medium having r.i. n_2. If f_1 and f_2 are image side focal length and object side focal length respectively, then prove, $(f_1/n_1) + (f_2/n_2) = 0$.**

6. **Find the principal focal length of a glass-sphere of radius 2.5 cm w.r.t. the second surface. [1.25 cm]**

7. **What do you mean by a telescopic system? What will be the form of Lagrange-Helmholtz relation in case of a telescopic system of a number of refracting surfaces?**

8. **What are cardinal points of an optical system? Find their positions in case of a glass-sphere of radius r placed in air.**

9. **Derive the Helmholtz-Lagrange relation between linear and angular magnification of an optical system and discuss the modification for a telescopic system. [CU 1995]**

10. **(a) What is an aplanatic surface? Can a concave mirror act as an aplanatic surface?**
    (b) When and under what condition can we use matrix method to solve geometrical optics problems?

11. **A lens is in the form of a spherical ball of radius R. The r.i. is n. Show that the principal planes of the lens coincide and pass through the centre of the sphere. Determine the focal length of the lens.**

12. **Show that in case of a single spherical surface, the principal points coincide with the pole of the surface, while the nodal points coincide with the centre of the sphere.**

Chapter 6

Thin Lens and the Lens System

Lens is an essential component in many of the optical instruments. So the formation of image by a lens is an important consideration.

A lens is usually defined as an optically transparent medium bounded by surfaces. These surfaces may be both curved convex outwards giving bi-convex lens, both may be concave giving bi-concave lens; one curved, the other plane giving plano-concave or plano-convex lens, etc. So optical behaviour of a lens may be deduced from a consideration of refraction at the two surfaces. In this chapter, we will develop the optics of a thin lens and also a thin lens system.

6.1 Refraction Matrix for a Thin Convex Lens $[R_L]$

The object is defined by the column vector $\begin{bmatrix} x_1 \\ \alpha_1 \end{bmatrix}$. The ultimate image is formed through a number of optical processes—

(1) A translation through a distance u bringing the object on to the first curved surface (convex), represented by a matrix $[T_1]$. If the transformed object is defined by a column vector $\begin{bmatrix} x_1' \\ \alpha_1 \end{bmatrix}$, we can write,

$$\begin{bmatrix} x_1' \\ \alpha_1 \end{bmatrix} = [T_1] \begin{bmatrix} x_1 \\ \alpha_1 \end{bmatrix} = \begin{bmatrix} 1 & u \\ 0 & 1 \end{bmatrix} \begin{bmatrix} x_1 \\ \alpha_1 \end{bmatrix}.$$

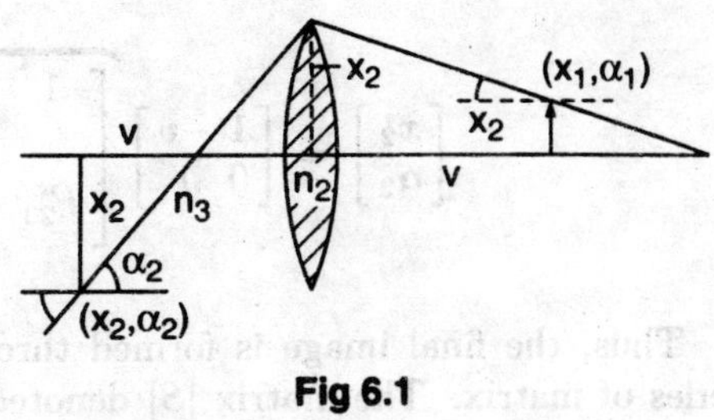

Fig 6.1

(2) Refraction through the convex surface from the medium of r.i. n_1 to the medium of r.i. n_2.

Let the refraction matrix $[R_1]$ transform the column vector $\begin{bmatrix} x_1' \\ \alpha_1 \end{bmatrix}$ into another column vector $\begin{bmatrix} x_2' \\ \alpha_1' \end{bmatrix}$ representing the image due to refraction, thus,

$$\begin{bmatrix} x_2' \\ \alpha_1' \end{bmatrix} = [R_1] \begin{bmatrix} x_1' \\ \alpha_1 \end{bmatrix} = \begin{bmatrix} 1 & 0 \\ M_{12} & \dfrac{n_1}{n_2} \end{bmatrix} \begin{bmatrix} x_1' \\ \alpha_1 \end{bmatrix} = \begin{bmatrix} 1 & 0 \\ M_{12} & \dfrac{n_1}{n_2} \end{bmatrix} \begin{bmatrix} 1 & u \\ 0 & 1 \end{bmatrix} \begin{bmatrix} x_1 \\ \alpha_1 \end{bmatrix}$$

(3) Since the lens is thin, translation of $\begin{bmatrix} x'_2 \\ \alpha'_1 \end{bmatrix}$ through zero distance is represented by a unit matrix, transforming $\begin{bmatrix} x'_2 \\ \alpha'_1 \end{bmatrix}$ into $\begin{bmatrix} x''_2 \\ \alpha''_1 \end{bmatrix}$ at the second curved surface within the lens material medium, thus,

$$\begin{bmatrix} x''_2 \\ \alpha''_1 \end{bmatrix} = \begin{bmatrix} 1 & 0 \\ 0 & 1 \end{bmatrix} \begin{bmatrix} x'_2 \\ \alpha'_1 \end{bmatrix}$$

(4) Second refraction at the surface separating n_2 and n_3.

The refraction matrix $[R_2]$ transforms $\begin{bmatrix} x''_2 \\ \alpha''_1 \end{bmatrix}$ within the lens medium to $\begin{bmatrix} x'''_2 \\ \alpha'''_1 \end{bmatrix}$ in medium of r.i. n_3 just outside the concave boundary, thus,

$$\begin{bmatrix} x'''_2 \\ \alpha'''_1 \end{bmatrix} = [R_2] \begin{bmatrix} x''_2 \\ \alpha''_1 \end{bmatrix} = \begin{bmatrix} 1 & 0 \\ P'_{23} & \dfrac{n_2}{n_3} \end{bmatrix} \begin{bmatrix} x''_2 \\ \alpha''_1 \end{bmatrix} = \begin{bmatrix} 1 & 0 \\ P'_{23} & \dfrac{n_2}{n_3} \end{bmatrix} \begin{bmatrix} 1 & 0 \\ 0 & 1 \end{bmatrix} \begin{bmatrix} x'_2 \\ \alpha'_1 \end{bmatrix}$$

(5) Finally, by translation of the ray characterised by $\begin{bmatrix} x'''_2 \\ \alpha'''_1 \end{bmatrix}$ through a distance v in the last homogeneous medium, we get the final image defined by $\begin{bmatrix} x_2 \\ \alpha_2 \end{bmatrix}$, thus for real image

$$\begin{bmatrix} x_2 \\ \alpha_2 \end{bmatrix} = [T_2] \begin{bmatrix} x'''_2 \\ \alpha'''_1 \end{bmatrix} = \begin{bmatrix} 1 & -v \\ 0 & 1 \end{bmatrix} \begin{bmatrix} 1 & 0 \\ P'_{23} & \dfrac{n_2}{n_3} \end{bmatrix} \begin{bmatrix} 1 & 0 \\ 0 & 1 \end{bmatrix} \begin{bmatrix} x'_2 \\ \alpha'_1 \end{bmatrix}$$

$$\begin{bmatrix} x_2 \\ \alpha_2 \end{bmatrix} = \begin{bmatrix} 1 & -v \\ 0 & 1 \end{bmatrix} \overbrace{\begin{bmatrix} 1 & 0 \\ P'_{23} & \dfrac{n_2}{n_3} \end{bmatrix} \begin{bmatrix} 1 & 0 \\ 0 & 1 \end{bmatrix} \begin{bmatrix} 1 & 0 \\ M_{12} & \dfrac{n_1}{n_2} \end{bmatrix}} \begin{bmatrix} 1 & u \\ 0 & 1 \end{bmatrix} \begin{bmatrix} x_1 \\ \alpha_1 \end{bmatrix}$$

Thus, the final image is formed through a number of optical processes represented by a series of matrix. The matrix $[S]$ denoted as S above representing the optical function of the optical element only is called *optical element matrix* and is given by

$$[S] = \begin{bmatrix} 1 & 0 \\ P'_{23} & \dfrac{n_2}{n_3} \end{bmatrix} \begin{bmatrix} 1 & 0 \\ 0 & 1 \end{bmatrix} \begin{bmatrix} 1 & 0 \\ M_{12} & \dfrac{n_1}{n_2} \end{bmatrix} = \begin{bmatrix} 1 & 0 \\ P'_{23} & \dfrac{n_2}{n_3} \end{bmatrix} \begin{bmatrix} 1 & 0 \\ M_{12} & \dfrac{n_1}{n_2} \end{bmatrix}$$

$$= \begin{bmatrix} 1 & 0 \\ \left(P'_{23} + \dfrac{n_2}{n_3} M_{12} \right) & \dfrac{n_1}{n_3} \end{bmatrix} = \begin{bmatrix} S_{11} & S_{12} \\ S_{21} & S_{22} \end{bmatrix}$$

Thus,

$$\begin{aligned}\begin{bmatrix} x_2 \\ \alpha_2 \end{bmatrix} &= \begin{bmatrix} 1 & -v \\ 0 & 1 \end{bmatrix}\begin{bmatrix} S_{11} & S_{12} \\ S_{21} & S_{22} \end{bmatrix}\begin{bmatrix} 1 & u \\ 0 & 1 \end{bmatrix}\begin{bmatrix} x_1 \\ \alpha_1 \end{bmatrix} \\ &= \begin{bmatrix} 1 & -v \\ 0 & 1 \end{bmatrix}\begin{bmatrix} S_{11} & (uS_{11}+S_{12}) \\ S_{21} & (uS_{21}+S_{22}) \end{bmatrix}\begin{bmatrix} x_1 \\ \alpha_1 \end{bmatrix} \\ &= \begin{bmatrix} (S_{11}-vS_{21}) & \{(uS_{11}+S_{12})-v(uS_{21}+S_{22})\} \\ S_{21} & (uS_{21}+S_{22}) \end{bmatrix}\begin{bmatrix} x_1 \\ \alpha_1 \end{bmatrix} \\ &= \begin{bmatrix} N_{11} & N_{12} \\ N_{21} & N_{22} \end{bmatrix}\begin{bmatrix} x_1 \\ \alpha_1 \end{bmatrix} \end{aligned} \tag{6.1.1}$$

The matrix $[N]$ may be called the ***transformation matrix*** or ***overall system matrix***.

$$\therefore \quad x_2 = N_{11}x_1 + N_{12}\alpha_1 \quad \text{and} \quad \alpha_2 = N_{21}x_1 + N_{22}\alpha_1.$$

If we set $N_{12} = 0$, then $x_2 = N_{11}x_1$, which means, for any value of α_1 for a particular value of x_1, the rays pass through a point x_2, so the conjugate relations between object plane and image plane are given by,

$$N_{12} = 0 \quad \Rightarrow \quad (uS_{11}+S_{12}) - v(uS_{21}+S_{22}) = 0 \quad \Rightarrow \quad v = \frac{uS_{11}+S_{12}}{uS_{21}+S_{22}}$$

If it gives $-$ve value, the image is on the left, if it is +ve, it is to the right of the lens. Suppose $u = \infty$, then $v = f_2$ (second focal length or image side focal length).

$$\therefore \quad f_2 = \lim_{u\to\infty} \frac{S_{11}+S_{12}/u}{S_{21}+S_{22}/u} = \frac{S_{11}}{S_{21}} = \frac{1}{P'_{23} + (n_2/n_3)M_{12}}.$$

Now,

$$M_{12} = -\frac{n_2-n_1}{n_2r_1} \quad \text{and} \quad P'_{23} = \frac{n_3-n_2}{n_3r_2} \quad \Rightarrow \quad P'_{23} + \frac{n_2}{n_3}M_{12} = \frac{n_3-n_2}{n_3r_2} - \frac{n_2-n_1}{n_3r_1}$$

$$\therefore \quad \frac{1}{f_2} = -\left(\frac{n_2-n_1}{n_3r_1} - \frac{n_3-n_2}{n_3r_2}\right) \quad \Rightarrow \quad -\frac{1}{f_2} = \frac{n_2-n_1}{n_3r_1} - \frac{n_3-n_2}{n_3r_2}$$

$\therefore$ the system matrix for a convex lens giving real image is given by,

$$[R_L] = [S] = \begin{bmatrix} 1 & 0 \\ -\frac{1}{f_2} & \frac{n_1}{n_3} \end{bmatrix}, \quad \text{where} \quad -\frac{1}{f_2} = \left(\frac{n_2-n_1}{n_3r_1} - \frac{n_3-n_2}{n_3r_2}\right) \tag{6.1.2}$$

But in case the convex lens produces virtual image, the system matrix or refraction matrix for the lens is written as

$$[R_L] = \begin{bmatrix} 1 & 0 \\ \frac{1}{f_2} & \frac{n_1}{n_3} \end{bmatrix}$$

If the lens is placed in air,

$$[R'_L] = \begin{bmatrix} 1 & 0 \\ -\frac{1}{f} & 1 \end{bmatrix}, \quad -\frac{1}{f} = \left(\frac{n_2 - 1}{r_1} + \frac{n_2 - 1}{r_2}\right) = (n_2 - 1)\left(\frac{1}{r_1} + \frac{1}{r_2}\right).$$

Usually focal length means second focal length, so the convex lens matrix is $[R_L] = \begin{bmatrix} 1 & 0 \\ -\frac{1}{f} & 1 \end{bmatrix}$ when in air.

6.2 Case of a Thin Diverging Lens

In this case, the sequence of matrix transformation giving rise to the final image formation can be written as

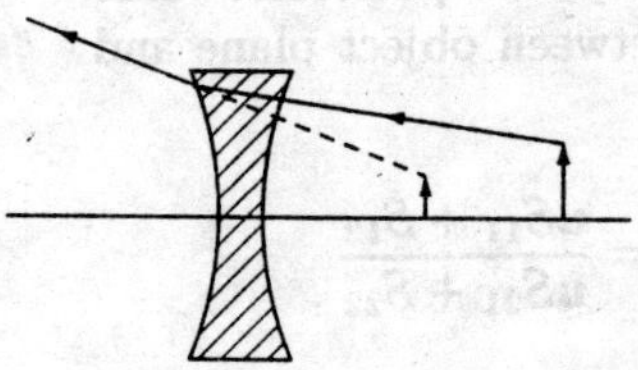

Fig 6.2

$$\begin{bmatrix} x_2 \\ \alpha_2 \end{bmatrix} = \begin{bmatrix} 1 & -v \\ 0 & 1 \end{bmatrix} \begin{bmatrix} S_{11} & S_{12} \\ S_{21} & S_{22} \end{bmatrix} \begin{bmatrix} 1 & u \\ 0 & 1 \end{bmatrix} \begin{bmatrix} x_1 \\ \alpha_1 \end{bmatrix}$$

The optical element matrix $[S]$ can be written as

$$[S] = \begin{bmatrix} 1 & 0 \\ M'_{23} & \frac{n_2}{n_3} \end{bmatrix} \begin{bmatrix} 1 & 0 \\ 0 & 1 \end{bmatrix} \begin{bmatrix} 1 & 0 \\ P_{12} & \frac{n_1}{n_2} \end{bmatrix} = \begin{bmatrix} 1 & 0 \\ M'_{23} & \frac{n_2}{n_3} \end{bmatrix} \begin{bmatrix} 1 & 0 \\ P_{12} & \frac{n_1}{n_2} \end{bmatrix}$$

$$= \begin{bmatrix} 1 & 0 \\ \left(M'_{23} + P_{12}\frac{n_2}{n_3}\right) & \frac{n_1}{n_3} \end{bmatrix} = \begin{bmatrix} S_{11} & S_{12} \\ S_{21} & S_{22} \end{bmatrix}$$

Here,

$$P_{12} = \frac{n_2 - n_1}{n_2 r_1} \quad \text{and} \quad M'_{23} = -\frac{n_3 - n_2}{n_3 r_2}$$

$$\therefore \begin{bmatrix} x_2 \\ \alpha_2 \end{bmatrix} = \begin{bmatrix} 1 & -v \\ 0 & 1 \end{bmatrix} \begin{bmatrix} S_{11} & S_{12} \\ S_{21} & S_{22} \end{bmatrix} \begin{bmatrix} 1 & u \\ 0 & 1 \end{bmatrix} \begin{bmatrix} x_1 \\ \alpha_1 \end{bmatrix} = \begin{bmatrix} 1 & v \\ 0 & 1 \end{bmatrix} \begin{bmatrix} S_{11} & (uS_{11} + S_{12}) \\ S_{21} & (uS_{21} + S_{22}) \end{bmatrix} \begin{bmatrix} x_1 \\ \alpha_1 \end{bmatrix}$$

$$= \begin{bmatrix} (S_{11} + vS_{21}) & \{(uS_{11} + S_{12}) - v(uS_{21} + S_{22})\} \\ S_{21} & (uS_{21} + S_{22}) \end{bmatrix} \begin{bmatrix} x_1 \\ \alpha_1 \end{bmatrix}$$

$$= \begin{bmatrix} N_{11} & N_{12} \\ N_{21} & N_{22} \end{bmatrix} \begin{bmatrix} x_1 \\ \alpha_1 \end{bmatrix}$$

The conjugate plane relation is given by $S_{12} = 0$.

$$\therefore \quad uS_{11} + S_{12} = v(uS_{21} + S_{22}) \quad \Rightarrow \quad v = \frac{uS_{11} + S_{12}}{uS_{21} + S_{22}}.$$

The second principal focus is defined mathematically as, $v = f_2'$, when $u = \infty$.

$$\therefore \quad f_2' = \frac{S_{11}}{S_{21}} = \frac{1}{S_{21}} \quad \Rightarrow \quad \frac{1}{f_2'} = S_{21}$$

$$\Rightarrow \quad S_{21} = M_{23}' + P_{12}\frac{n_2}{n_3} = -\frac{n_3 - n_2}{n_3 r_2} + \frac{n_2 - n_1}{n_3 r_1} = \left(\frac{n_2 - n_1}{n_2 r_1} - \frac{n_3 - n_2}{n_3 r_2}\right)$$

$$\therefore \quad \frac{1}{f_2'} = \left(\frac{n_2 - n_1}{n_3 r_1} - \frac{n_3 - n_2}{n_3 r_2}\right). \tag{6.2.1}$$

So the system matrix for the concave lens becomes

$$[R_L'] = \begin{bmatrix} 1 & 0 \\ \frac{1}{f_2'} & \frac{n_1}{n_3} \end{bmatrix}. \tag{6.2.2}$$

In case it is placed in air on both sides,

$$[R_L'] = \begin{bmatrix} 1 & 0 \\ \frac{1}{f_2'} & 1 \end{bmatrix} \quad \text{where} \quad \frac{1}{f_2'}\left(\frac{n_2 - 1}{r_1} + \frac{n_2 - 1}{r_2}\right).$$

6.3 Relation between the Object, Image Distances and the Object Side and the Image Side Focal Lengths

To show that $f_1/u + f_2/v = 1$ for a thin convex lens of r.i. n and radii of curvatures R_1 and R_2, the r.i. of the media on two sides of the lens being n_1 and n_2.

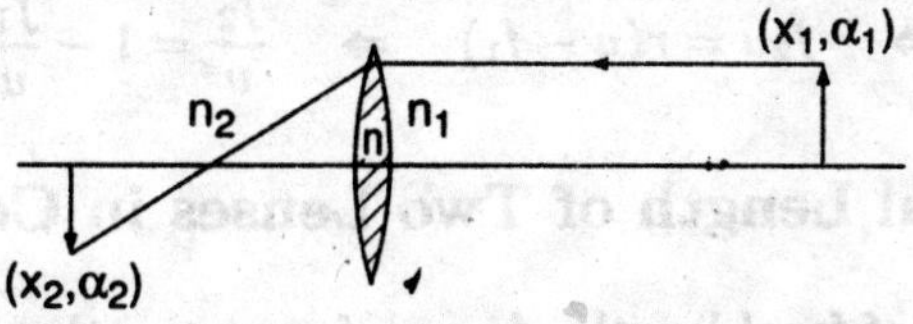

Fig 6.3

Consider fig.(6.3) where the object column vector is (x_1, α_1), the image column vector (x_2, α_2); the refraction matrix for a thin lens is

$$[R]\begin{bmatrix} 1 & 0 \\ \left(\frac{n_3 - n_2}{n_3 r_2} - \frac{n_2 - n_1}{n_3 r_1}\right) & \frac{n_1}{n_3} \end{bmatrix} = \begin{bmatrix} 1 & 0 \\ \left(\frac{n_2 - n}{n_2 R_2} - \frac{n - n_1}{n_2 R_1}\right) & \frac{n_1}{n_2} \end{bmatrix}$$

as given in this case.

Thus,

$$[R] = \begin{bmatrix} s_{11} & s_{12} \\ s_{21} & s_{22} \end{bmatrix}, \quad \text{is called the } \textit{system matrix}$$

$$\begin{bmatrix} x_2 \\ \alpha_2 \end{bmatrix} = \begin{bmatrix} 1 & -v \\ 0 & 1 \end{bmatrix} \begin{bmatrix} s_{11} & s_{12} \\ s_{21} & s_{22} \end{bmatrix} \begin{bmatrix} 1 & u \\ 0 & 1 \end{bmatrix} \begin{bmatrix} x_1 \\ \alpha_1 \end{bmatrix}$$

$$= \begin{bmatrix} (s_{11} + vs_{21}) & \{(us_{11} + s_{12}) - v(us_{21} + s_{22})\} \\ s_{21} & (us_{21} + s_{22}) \end{bmatrix} \begin{bmatrix} x_1 \\ \alpha_1 \end{bmatrix}$$

$$= \begin{bmatrix} N_{11} & N_{12} \\ N_{21} & N_{22} \end{bmatrix} \begin{bmatrix} x_1 \\ \alpha_1 \end{bmatrix}$$

where $[S]$ is the overall system matrix.

The conjugate plane relation is obtained by setting

$$N_{12} = 0 \quad \Rightarrow \quad (us_{11} + s_{12}) = v(us_{21} + s_{22}) \quad \Rightarrow \quad (u + 0) = v(us_{21} + s_{22}).$$

Now, the second principal focus f_2 is obtained by setting $N_{11} = 0$.

$$\therefore \quad f_2 = \frac{s_{11}}{s_{21}} = \frac{1}{s_{21}} = -\frac{1}{\left(\dfrac{n-n_1}{n_2R_1} - \dfrac{n_2-n}{n_2R_2}\right)} = -\frac{n_2}{\left(\dfrac{n-n_1}{R_1} - \dfrac{n_2-n}{R_2}\right)}.$$

The object side focal length is obtained by setting $N_{22} = 0$.

$$\therefore \quad f_1 = -\frac{s_{22}}{s_{21}} = +\frac{\dfrac{n_1}{n_2}}{\left(\dfrac{n-n_1}{n_2R_1} - \dfrac{n_2-n}{n_2R_2}\right)} = \frac{n_1}{\left(\dfrac{n-n_1}{R_1} - \dfrac{n_2-n}{R_2}\right)}.$$

So,

First relation: $\dfrac{f_1}{n_1} + \dfrac{f_2}{n_2} = 0.$

Another relation: $u = r\left(u \cdot \dfrac{1}{f_2} - s_{21}f_1\right) = r\left(\dfrac{u}{f_2} - \dfrac{f_1}{f_2}\right)$

$$\Rightarrow \quad f_2u = r(u - f_1) \quad \Rightarrow \quad \frac{f_2}{v} = 1 - \frac{f_1}{u} \quad \Rightarrow \quad \frac{f_1}{u} + \frac{f_2}{v} = 1.$$

6.4 Equivalent Focal Length of Two Lenses in Contact in Air

Consider two convex lenses of focal lengths f_1 and f_2 put together. The refraction matrix for this optical system is

$$[R_C] = \begin{bmatrix} 1 & 0 \\ -\dfrac{1}{f_1} & \dfrac{n_1}{n_3} \end{bmatrix} \begin{bmatrix} 1 & 0 \\ -\dfrac{1}{f_2} & \dfrac{n_1}{n_3} \end{bmatrix}$$

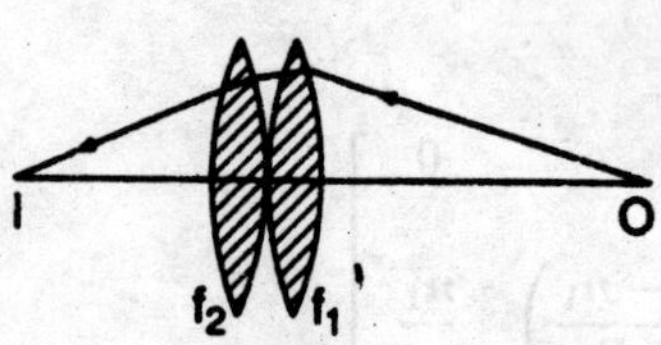

Fig 6.4

If $n_1 = 1, n_3 = 1,$

$$[R_C] = \begin{bmatrix} 1 & 0 \\ -\dfrac{1}{f_1} & 1 \end{bmatrix} \begin{bmatrix} 1 & 0 \\ -\dfrac{1}{f_2} & 1 \end{bmatrix}$$

$$\therefore \begin{bmatrix} x_2 \\ \alpha_2 \end{bmatrix} = \begin{bmatrix} 1 & 0 \\ -\dfrac{1}{f_1} & 1 \end{bmatrix} \begin{bmatrix} 1 & 0 \\ -\dfrac{1}{f_2} & 1 \end{bmatrix} \begin{bmatrix} x_1 \\ \alpha_1 \end{bmatrix} = \begin{bmatrix} 1 & 0 \\ \left(-\dfrac{1}{f_1} - \dfrac{1}{f_2}\right) & 1 \end{bmatrix} \begin{bmatrix} x_1 \\ \alpha_1 \end{bmatrix}$$

$$\Rightarrow \quad x_2 = x_1 \quad \text{and} \quad \alpha_2 = -\left(\frac{1}{f_1} + \frac{1}{f_2}\right) x_1 + \alpha_1.$$

When the object is at infinity $\alpha_1 = 0$ and $x_2 = F\alpha_2$, where F is the focal length of the combined lens

$$\therefore \quad -\frac{x_2}{F} = -\left(\frac{1}{f_1} + \frac{1}{f_2}\right) x_1 \quad \Rightarrow \quad \frac{1}{F} = \frac{1}{f_1} + \frac{1}{f_2}. \tag{6.4.1}$$

6.5 Cardinal Points in case of Two Thin Lenses Separated by a Distance 'd' in Air (Convex Lens System)

In fig.(6.5) we have shown two convex lenses of focal lengths f_1 and f_2 placed in air, separated by a distance d. The refraction matrix for the first lens is $[R_1]$, followed by translational matrix $[T]$ then a refraction matrix $[R_2]$ ultimately transforms the object on the first lens to the image at the second lens.

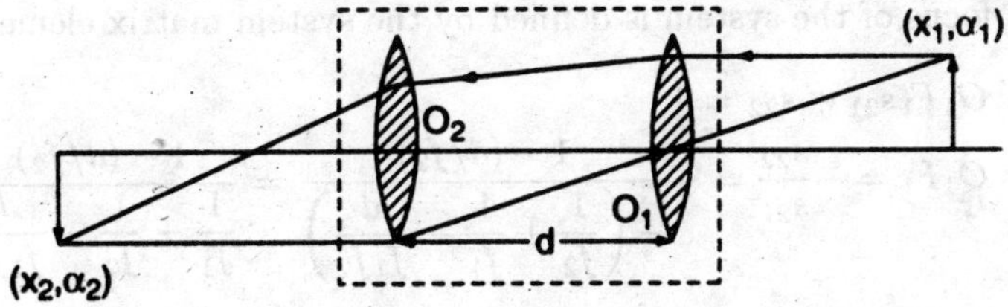

Fig 6.5

Following the matrix method formalism, we can write,

$$\begin{bmatrix} x_2 \\ \alpha_2 \end{bmatrix} = \begin{bmatrix} 1 & -v \\ 0 & 1 \end{bmatrix} [R_2][T][R_1] \begin{bmatrix} 1 & u \\ 0 & 1 \end{bmatrix} \begin{bmatrix} x_1 \\ \alpha_1 \end{bmatrix}.$$

The first matrix $\begin{bmatrix} 1 & u \\ 0 & 1 \end{bmatrix}$ is to transform the object (x_1, α_1) to the first lens by translation, the last matrix transforms the image out of the system to the image (x_2, α_2) by translation. Thus,

$$\begin{bmatrix} x_2 \\ \alpha_2 \end{bmatrix} = \begin{bmatrix} 1 & -v \\ 0 & 1 \end{bmatrix} \begin{bmatrix} 1 & 0 \\ -(1/f_2) & 1 \end{bmatrix} \begin{bmatrix} 1 & d \\ 0 & 1 \end{bmatrix} \begin{bmatrix} 1 & 0 \\ -(1/f_1) & 1 \end{bmatrix} \begin{bmatrix} 1 & u \\ 0 & 1 \end{bmatrix} \begin{bmatrix} x_1 \\ \alpha_1 \end{bmatrix} \tag{6.5.1}$$

Refraction matrix for the lens system is given by,

$$[R] = \begin{bmatrix} 1 & 0 \\ -(1/f_2) & 1 \end{bmatrix} \begin{bmatrix} 1 & d \\ 0 & 1 \end{bmatrix} \begin{bmatrix} 1 & 0 \\ -(1/f_1) & 1 \end{bmatrix}$$

$$= \begin{bmatrix} 1 & d \\ -(1/f_2) & \{-(d/f_2) + 1\} \end{bmatrix} \begin{bmatrix} 1 & 0 \\ -(1/f_1) & 1 \end{bmatrix}$$

$$= \begin{bmatrix} \{1 - (d/f_1)\} & d \\ -(1/f_2) - (1/f_1)\{1 - (d/f_2)\} & \{1 - (d/f_2)\} \end{bmatrix} = \begin{bmatrix} s_{11} & s_{12} \\ s_{21} & s_{22} \end{bmatrix} \tag{6.5.2}$$

$$\therefore \begin{bmatrix} x_2 \\ \alpha_2 \end{bmatrix} = \begin{bmatrix} 1 & -v \\ 0 & 1 \end{bmatrix} \begin{bmatrix} s_{11} & s_{12} \\ s_{21} & s_{22} \end{bmatrix} \begin{bmatrix} 1 & u \\ 0 & 1 \end{bmatrix} \begin{bmatrix} x_1 \\ \alpha_1 \end{bmatrix}$$

$$= \begin{bmatrix} (s_{11} - vs_{21}) & (s_{12} - vs_{22}) \\ s_{21} & s_{22} \end{bmatrix} \begin{bmatrix} 1 & u \\ 0 & 1 \end{bmatrix} \begin{bmatrix} x_1 \\ \alpha_1 \end{bmatrix}$$

$$= \begin{bmatrix} (s_{11} - vs_{21}) & \{u(s_{11} - vs_{21}) + (s_{12} - vs_{22})\} \\ s_{21} & us_{21} + s_{22} \end{bmatrix} \begin{bmatrix} x_1 \\ \alpha_1 \end{bmatrix}$$

$$= \begin{bmatrix} N_{11} & N_{12} \\ N_{21} & N_{22} \end{bmatrix} \begin{bmatrix} x_1 \\ \alpha_1 \end{bmatrix} \tag{6.5.3}$$

where the overall system matrix is $[S] = \begin{bmatrix} N_{11} & N_{12} \\ N_{21} & N_{22} \end{bmatrix}$, the elements of matrix being

$$N_{11} = s_{11} - vs_{21},\ N_{12} = u(s_{11} - vs_{21}) + (s_{12} - vs_{22})$$
$$N_{21} = s_{21}, \qquad N_{22} = us_{21} + s_{22}.$$

According to the discussion made in art. 5.5, the cardinal points of the optical system can be derived.

Focal Points and Focal Planes

The first principal focus of the system is defined by the system matrix element $N_{22} = 0$.

$$\therefore\ O_1F_1 s_{21} + s_{22} = 0$$

$$\Rightarrow\ O_1F_1 = -\frac{s_{22}}{s_{21}} = -\frac{1 - (d/f_2)}{-\left(\dfrac{1}{f_2} + \dfrac{1}{f_1} - \dfrac{d}{f_1 f_2}\right)} = \frac{1 - (d/f_2)}{\dfrac{1}{f_1} + \dfrac{1}{f_2} - \dfrac{d}{f_1 f_2}} \tag{6.5.4}$$

This focal length is measured w.r.t. the first lens optical centre. The second focal point of the optical system is defined by $N_{11} = 0$.

$$\therefore\ s_{21} \times O_2F_2 - s_{11} = 0 \quad \Rightarrow \quad O_2F_2 = \frac{s_{11}}{s_{21}} = \frac{(1 - d/f_1)}{-\left(\dfrac{1}{f_1} + \dfrac{1}{f_2} - \dfrac{d}{f_1 f_2}\right)} \tag{6.5.5}$$

measured from the centre of the second lens.

Principal Planes of the Optical System

The principal planes of the optical system are defined by $N_{12} = 0$ and $N_{11} = 1$

$$N_{12} = 0 \Rightarrow u(s_{11} - vs_{21}) + (s_{12} - vs_{22}) = 0$$

and $N_{11} = 1 \Rightarrow s_{11} - vs_{21} = 1 \quad \Rightarrow \quad v = \dfrac{s_{11} - 1}{s_{21}}.$

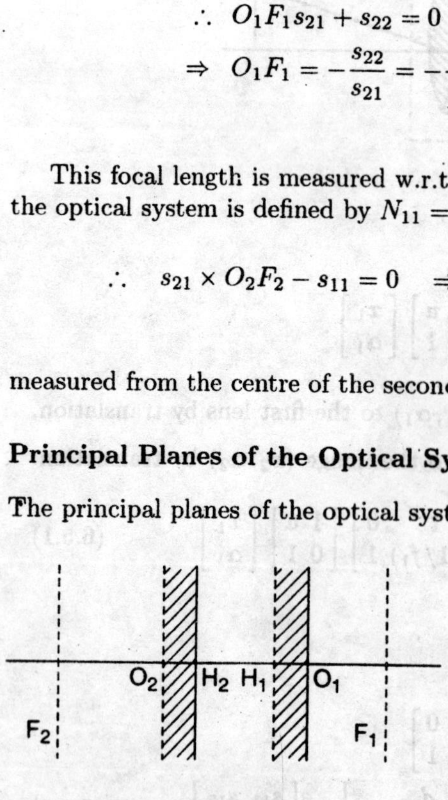

Fig 6.6

Thus,

$$O_2H_2 = v = \frac{s_{11} - 1}{s_{21}} = \frac{d/f_1}{\dfrac{1}{f_1} + \dfrac{1}{f_2} - \dfrac{d}{f_1 f_2}} \tag{6.5.6}$$

to the right of the optical centre of the second lens. The corresponding u defining the position of the first principal plane is given by,

$$u = O_1H_1 = -(s_{12} - vs_{22}) = -\frac{d/f_2}{\frac{1}{f_1} + \frac{1}{f_2} - \frac{d}{f_1 f_2}}. \qquad (6.5.7)$$

The two focal planes and the two principal planes of the present optical system are shown in fig.(6.6).

The significance of the principal planes will be understood if we remember that when a ray is incident on the first principal plane, its corresponding emergent ray leaves the second principal plane at the same height. Another important point to remember is that in case of two convex lenses separated by a small distance apart [$d < (f_1 + f_2)$], the two H-planes are as shown in fig.(6.6). But if $d > (f_1 + f_2)$, $u = +$ve, $v = -$ve, so the two H-planes are on opposite sides of the two lenses as shown in fig.(6.7). Whichever may be the case, we can consider rays coming from infinity and getting refracted by the two lenses and the emergent rays passing through the second principal focus of the optical system (see fig.6.8). The incident ray strikes the first principal plane at P, the final emergent ray must leave the second principal plane from Q at the same height as P. Thus, H_2F_2 may be termed as the focal length of the optical system. The equivalent single lens may be placed at the principal plane H_2 having the focal length H_2F_2 forming the image at the same position as done by the optical system.

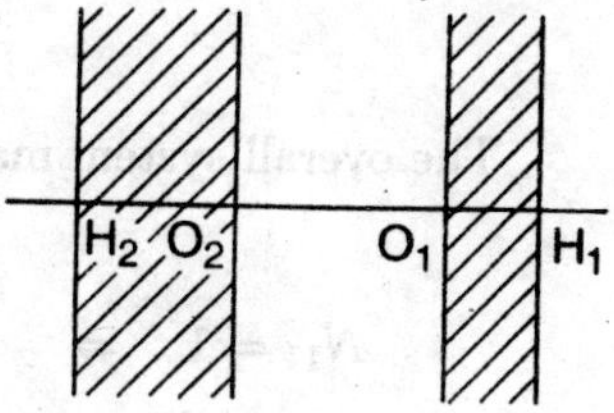

Fig 6.7

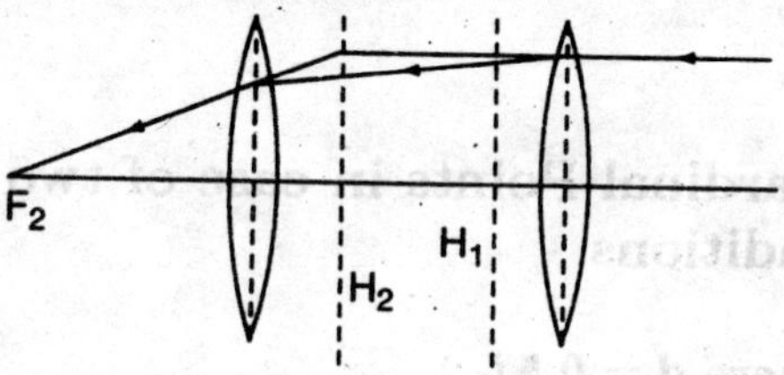

Fig 6.8

Thus the equivalent focal length or the focal length of the optical system is given by,

$$H_2F_2 = O_2F_2 - O_2H_2 = \frac{(d/f_1 - 1)}{\left(\frac{1}{f_1} + \frac{1}{f_2} - \frac{d}{f_1 f_2}\right)} - \frac{d/f_1}{\left(\frac{1}{f_1} + \frac{1}{f_2} - \frac{d}{f_1 f_2}\right)} = -\frac{1}{\left(\frac{1}{f_1} + \frac{1}{f_2} - \frac{d}{f_1 f_2}\right)}$$

$$\therefore \quad \frac{1}{F_e} = -\left(\frac{1}{f_1} + \frac{1}{f_2} - \frac{d}{f_1 f_2}\right) \qquad (6.5.8)$$

assuming H-planes on opposite sides of the two lenses.

The $-$ve indicates that it is beyond the second lens. The position of the equivalent lens is at H_2, such that its distance w.r.t. the optical centre O_2 is given by,

$$X_o = O_2H_2 = \frac{d/f_1}{\frac{1}{f_1} + \frac{1}{f_2} - \frac{d}{f_1 f_2}} = \frac{d/f_1}{1/F_e} = \frac{dF_e}{f_1} \qquad (6.5.9)$$

Again, $H_1F_1 = O_1F_1 - O_1H_1 = \dfrac{1}{(1/f_1 + 1/f_2 - d/f_1 f_2)} = -H_2F_2.$

Nodal Planes and Nodal Points (N_1, N_2)

The nodal points of an optical system are defined by $N_{21} = 0$ and $N_{22} = 1$.

$N_{22} = 1 \Rightarrow us_{21} + s_{22} = 1$

$$\Rightarrow O_1N_1 = \frac{1-s_{22}}{s_{21}} = \frac{1-(1-d/f_2)}{-\left(\frac{1}{f_1}+\frac{1}{f_2}-\frac{d}{f_1f_2}\right)} = -\frac{d/f_2}{\left(\frac{1}{f_1}+\frac{1}{f_2}-\frac{d}{f_1f_2}\right)} = O_1H_1 \quad (6.5.10)$$

The overall system matrix in this case becomes $[N] = \begin{bmatrix} N_{11} & N_{12} \\ 0 & 1 \end{bmatrix}$; $|N| = 1$, so we get,

$$N_{11} = 1 \quad \Rightarrow \quad s_{11} - vs_{21} = 1 \quad \Rightarrow \quad v = O_2N_2 = \frac{s_{11}-1}{s_{21}} = O_2H_2. \quad (6.5.11)$$

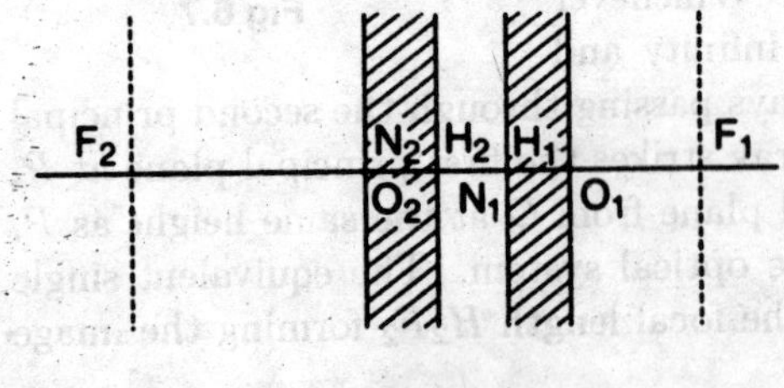

Fig 6.9

Thus from eqns.(6.5.10) and (6.5.11) we conclude that when the lens system is surrounded by the same medium, the nodal points and principal points are coincident. So now the complete representation of the cardinal points and planes are possible and they are shown in fig.(6.9) where the first principal point and nodal point are respectively denoted by (H_1, N_1), the corresponding quantities for the second by (H_2, N_2).

6.6 Position of Cardinal Points in case of two Lenses in Air under Specified Conditions

1. Two thin lenses where $d = 0.5f_1$:

Here, $O_1F_1 = \dfrac{f_1(2f_2 - f_1)}{2f_2 + f_1}$

If $f_1 = f_2 = f$, then $O_1F_1 = \dfrac{f}{3}$ and $O_2F_2 = -\dfrac{f_1f_2}{2f_2 + f_1}$.

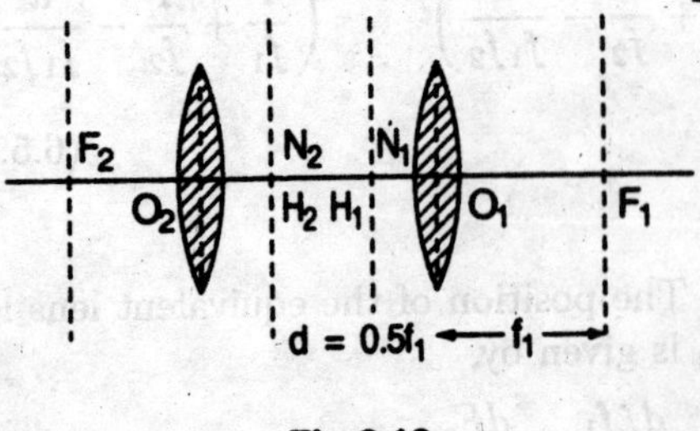

Fig 6.10

For $f_1 = f_2$,

$$O_2F_2 = -\frac{f}{3},$$

$$O_1H_1 = -\frac{d/f_2}{\frac{1}{f_1}+\frac{1}{f_2}-\frac{d}{f_1f_2}} = -\frac{f_1^2}{2f_2+f_1}$$

and $O_2H_2 = \dfrac{f_2}{(2f_2+f_1)}$.

For $f_1 = f_2 = f$,

$$O_1H_1 = -f/3 \quad \text{and} \quad O_2H_2 = f/3.$$

2. Two thin lenses where $d = f_1$:

Here, $O_1F_1 = \frac{f_1}{f_2}(f_2 - f_1)$.

If $f_1 = f_2 = f$,

$$O_1F_1 = 0 \quad \text{and} \quad O_2F_2 = 0,$$

the focal points coincide with the optical centres.

$$O_1H_1 = -\frac{d/f_2}{\frac{1}{f_1}+\frac{1}{f_2}-\frac{d}{f_1f_2}} = -\frac{f_1/f_2}{\frac{1}{f_1}+\frac{1}{f_2}-\frac{1}{f_2}} = \frac{-f_1^2}{f_2}$$

$$O_2H_2 = \frac{d/f_1}{\frac{1}{f_1}+\frac{1}{f_2}-\frac{d}{f_1f_2}} = \frac{1}{\frac{1}{f_1}} = f_1.$$

So, if $f_1 = f_2 = f, O_1H_1 = -f_1, O_2H_2 = f_1$.

The focal length of the system is given by,

$$H_2F_2 = F_e = -\frac{1}{\frac{1}{f_1}+\frac{1}{f_2}-\frac{d}{f_1f_2}} = -f_1.$$

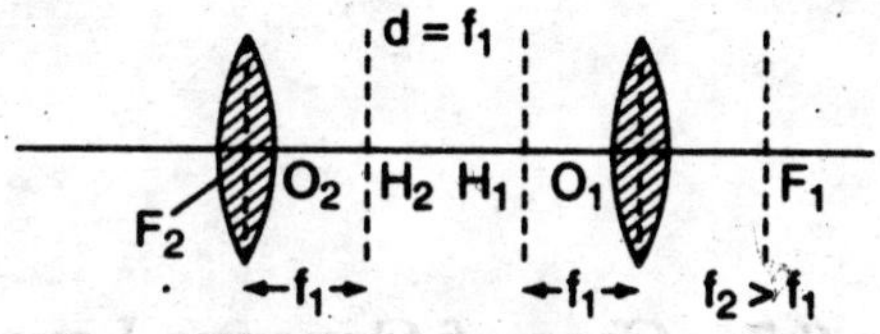

Fig 6.11

3. Huygens Eyepiece:

It consists of two plano-convex lenses, objective lens O and eyelens E having focal lengths $f_1 = 3f$, $f_2 = f$ and the distance of separation $d = 2f$.

The cardinal points in this case are,

$$O_1F_1 = \frac{1 - d/f_2}{\frac{1}{f_1}+\frac{1}{f_2}-\frac{d}{f_1f_2}}.$$

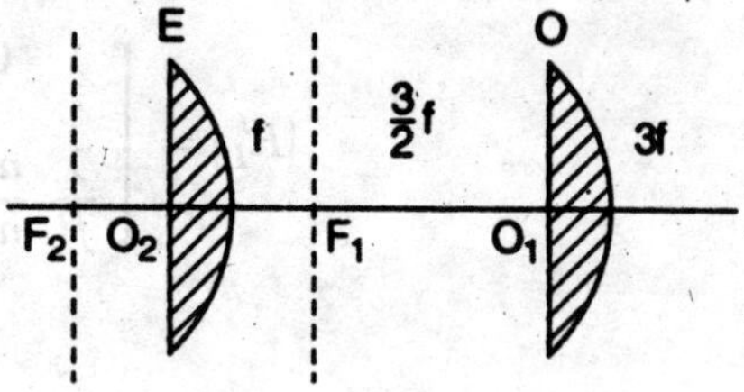

Fig 6.12

Now,

$$\frac{1}{f_1}+\frac{1}{f_2}-\frac{d}{f_1f_2} = \frac{1}{3f}+\frac{1}{f}-\frac{2f}{3f^2} = \frac{2}{3f}$$

$$\therefore \quad O_1F_1 = \frac{1-2}{(2/3f)} = -\frac{3}{2}f$$

$$O_2F_2 = \frac{1 - d/f_1}{-\left(\frac{1}{f_1}+\frac{1}{f_2}-\frac{d}{f_1f_2}\right)} = -\frac{1-2/3}{\frac{2}{3f}} = -\frac{f}{2}$$

$$O_1H_1 = -\frac{d/f_1}{\left(\frac{1}{f_1}+\frac{1}{f_2}-\frac{d}{f_1f_2}\right)} = -\frac{2}{\frac{2}{3f}} = -3f$$

$$O_2H_2 = \frac{d/f_1}{\left(\frac{1}{f_1}+\frac{1}{f_2}-\frac{d}{f_1f_2}\right)} = \frac{2/3}{\frac{2}{3f}} = f$$

H_2F_2 = Second focal length of the system

$$= F_e = -\frac{1}{\left(\frac{1}{f_1}+\frac{1}{f_2}-\frac{d}{f_1f_2}\right)} = -\frac{1}{2/3f} = -\frac{3}{2}f.$$

Thus the cardinal points and planes of Huygens eyepiece (more details being given later) are as shown in the figure. So it is evident that

$$H_1F_1 = \frac{3}{2}f \quad \text{and} \quad H_2F_2 = -\frac{3}{2}f \quad \text{(in the direction of light).}$$

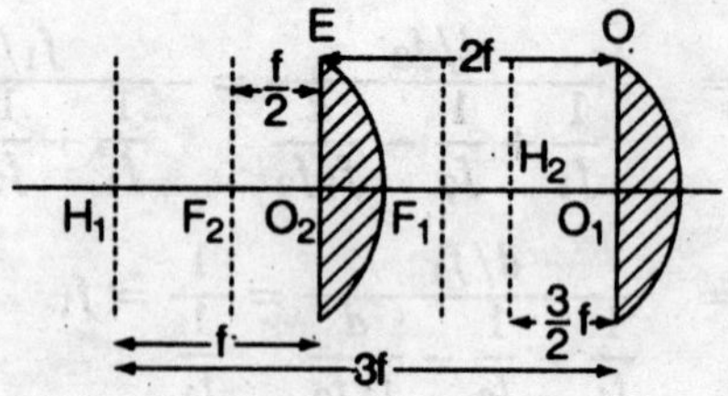

Fig 6.13

6.7 Case of Concave Lens System

In this case we evaluate the system matrix as we have done in case of convex lens system remembering that the refraction matrix for a concave lens is given by,

$$[R'_L] = \begin{bmatrix} 1 & 0 \\ \frac{1}{f'} & \frac{n_1}{n_3} \end{bmatrix}, \quad \text{where} \quad \frac{1}{f'} = \frac{n_2 - n_1}{n_3 r_1} - \frac{n_2 - n_3}{n_3 r_2}.$$

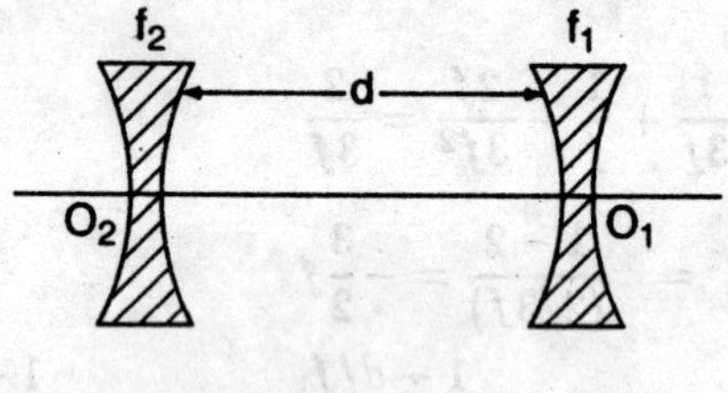

Fig 6.14

If $n_1 = n_3 = 1, n_2 = n$, then

$$\frac{1}{f'} = \frac{n-1}{r_1} - \frac{n-1}{r_2} = (n-1)\left(\frac{1}{r_1} - \frac{1}{r_2}\right).$$

$$\therefore \quad \begin{bmatrix} x_2 \\ \alpha_2 \end{bmatrix} = \begin{bmatrix} 1 & -v \\ 0 & 1 \end{bmatrix} \begin{bmatrix} 1 & 0 \\ +\frac{1}{f'_2} & 1 \end{bmatrix} \begin{bmatrix} 1 & d \\ 0 & 1 \end{bmatrix} \begin{bmatrix} 1 & 0 \\ \frac{1}{f'_1} & 1 \end{bmatrix} \begin{bmatrix} 1 & u \\ 0 & 1 \end{bmatrix} \begin{bmatrix} x_1 \\ \alpha_1 \end{bmatrix}$$

Proceeding exactly as in convex lens system the whole sequence of discussion can be repeated and the cardinal points may be determined.

Short Theory Questions with Answers

1. What do you mean by a thin lens? In the matrix formalism what advantage do we get if we consider the lens to be thin?

   Ans. A thin lens is a transparent medium bounded by two surfaces where the thickness of the medium is supposed to be very small. If the lens is a thin one, the translational matrix transforming the object parameter from one surface of the lens to the other through the lens need not be considered and the poles of the curved surfaces may be taken to be coincident with the optical centre of the lens.

2. A lens is usually defined as a trânsparent medium bounded by surfaces. Can a glass-slab be treated as a lens?

   Ans. A glass-slab can be treated as a lens whose focal length is infinity, hence no practical purpose of a lens is served. But theoretically at least such a medium can be treated as a lens.

3. In case of a lens system consisting of two thin lenses separated by a certain distance in air, why do we define the equivalent focal length of the system as the length of the point of focus by the system corresponding to parallel incident rays?

   Ans. Because the equivalent lens producing the image at the same position as formed by the lens system may be supposed to be placed in the second principal plane.

4. Is it possible to have both the principal planes in between two lenses forming the lens system? To have one in the space between the lenses, the other beyond?

   Ans. Yes, if two convex lenses are used and the distance of separation is small, the two principal planes will lie in space between the two lenses. But it is also possible to have the principal planes one inside, the other outside as in case in Huygens eyepiece.

Solved Numerical Problems

1. A telephoto lens consists of a convergent lens of focal length 12 cm facing the object and a divergent lens of focal length 5 cm placed 8 cm behind the former. Find the position where a plate is to be placed to photograph a distant object.

   Solution. The lens system here consists of two lenses, one objective lens which is convex of focal length $f_1 = 12$ cm, the other a diverging lens of focal length $f_2 = 5$ cm, $d = 8$ cm. The focal point F_2 is the point of convergence of rays coming from a distant object. Now,

$$O_2F_2 = \frac{1-(d/f_1)}{-\left[\dfrac{1}{f_1}-\dfrac{1}{f_2}-\dfrac{d}{f_1(-f_2)}\right]} = \frac{1-(8/12)}{-\left[\dfrac{1}{12}-\dfrac{1}{5}+\dfrac{8}{60}\right]} = -20 \text{ cm},$$

   so the plate is to be placed 20 cm, behind the second lens.

2. The convex lenses $f_1 = 12$ cm, $f_2 = 4$ cm are kept separated by a distance $d = 8$ cm in air. Find the positions of the cardinal points.

   Solution.

$$O_1F_1 = \frac{1-d/f_2}{\dfrac{1}{f_1}+\dfrac{1}{f_2}-\dfrac{d}{f_1f_2}}$$

Now,

$$\frac{1}{f_1}+\frac{1}{f_2}-\frac{d}{f_1 f_2}=\frac{1}{12}+\frac{1}{4}-\frac{8}{12\times 4}=\frac{1}{6}$$

$$\therefore\quad O_1F_1=\frac{1-8/4}{1/6}=-6 \text{ cm}$$

$$O_2F_2=-\frac{1-d/f_1}{\left(\frac{1}{f_1}+\frac{1}{f_2}-\frac{d}{f_1 f_2}\right)}=-\frac{1-(8/12)}{1/6}=-2 \text{ cm}$$

$$O_1H_1=-\frac{d/f_2}{\frac{1}{f_2}+\frac{1}{f_2}-\frac{d}{f_1 f_2}}=-\frac{8/4}{1/6}=-12 \text{ cm}$$

$$O_2H_2=\frac{d/f_1}{\frac{1}{f_1}+\frac{1}{f_2}-\frac{d}{f_1 f_2}}=\frac{8/12}{\frac{1}{6}}=4 \text{ cm}$$

$$H_2F_2=F_e=-\frac{1}{\frac{1}{f_1}+\frac{1}{f_2}-\frac{d}{f_1 f_2}}=-\frac{1}{1/6}=-6 \text{ cm}$$

$$H_1F_1=-H_2F_2=6 \text{ cm.}$$

3. **Find the optical separation between two convex lenses of focal lengths 30 cm and 10 cm functioning as a telescope of effective focal length 20 cm.**

Solution.

$$\frac{1}{F_e}=-\left(\frac{1}{f_1}+\frac{1}{f_2}-\frac{d}{f_1 f_2}\right)\quad\Rightarrow\quad d=55 \text{ cm.}$$

4. **A system of two thin lenses, one is the convex lens $f_1 = 20$ cm, and the other concave $f_1' = 10$ cm has separation between the lenses equal to 8 cm. An object 1 cm high is placed at a distance of 40 cm from the convex lens. Calculate the size and position of the image.**

Solution.

$$\begin{bmatrix} x_2 \\ \alpha_2 \end{bmatrix}=\begin{bmatrix} 1 & -v \\ 0 & 1 \end{bmatrix}\begin{bmatrix} 1 & 0 \\ \frac{1}{f_1'} & 1 \end{bmatrix}\begin{bmatrix} 1 & d \\ 0 & 1 \end{bmatrix}\begin{bmatrix} 1 & 0 \\ -\frac{1}{f_1} & 1 \end{bmatrix}\begin{bmatrix} 1 & u \\ 0 & 1 \end{bmatrix}\begin{bmatrix} x_1 \\ \alpha_1 \end{bmatrix}$$

$$=\begin{bmatrix} 1 & -v \\ 0 & 1 \end{bmatrix}\begin{bmatrix} 1 & 0 \\ \frac{1}{10} & 1 \end{bmatrix}\begin{bmatrix} 1 & 8 \\ 0 & 1 \end{bmatrix}\begin{bmatrix} 1 & 0 \\ -\frac{1}{20} & 1 \end{bmatrix}\begin{bmatrix} 1 & 40 \\ 0 & 1 \end{bmatrix}\begin{bmatrix} x_1 \\ \alpha_1 \end{bmatrix}$$

$$=\begin{bmatrix} 1 & -v \\ 0 & 1 \end{bmatrix}\begin{bmatrix} 0.6 & 32 \\ 0.01 & 2.2 \end{bmatrix}\begin{bmatrix} x_1 \\ \alpha_1 \end{bmatrix}=\begin{bmatrix} (.6-.01v) & (32-2.2v) \\ .01 & 2.2 \end{bmatrix}\begin{bmatrix} x_1 \\ \alpha_1 \end{bmatrix}$$

$$\Rightarrow\quad x_2=x_1(.6-.01v)+(32-2.2v)\alpha_1 \quad\text{and}\quad \alpha_2=.01x_1+2.2\alpha_1$$

Now, for an axial point $x_1 = 0 = x_2$, but $\alpha_1 \neq 0$

$$\therefore \quad 32 - 2.2v = 0 \quad \Rightarrow \quad v = 14.5 \text{ cm},$$

which is translation against the direction of light from the second lens.

$$\therefore \quad \frac{x_2}{x_1} = .6 - .01v = .6 - .01 \times 14.5 = .455$$

$$\therefore \quad x_2 = .455 \times 1 = 0.455 \text{ cm}.$$

5. A convex lens and a convave lens each of 10 cm focal length are held co-axially at a distance of 3 cm apart. Find the position of the image of an object when it is held at a distance of 15 cm beyond the (a) convex lens, (b) the concave lens.

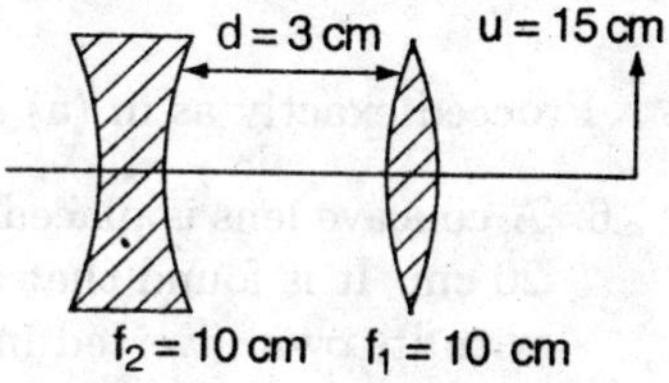

Fig 6.15

Solution. (a) From eqn.(6.5.1), f_2 is for a concave lens,

$$\begin{bmatrix} x_2 \\ \alpha_2 \end{bmatrix} = \begin{bmatrix} 1 & -v \\ 0 & 1 \end{bmatrix} [R_2][T][R_1] \begin{bmatrix} 1 & u \\ 0 & 1 \end{bmatrix}, \begin{bmatrix} x_1 \\ \alpha_1 \end{bmatrix}$$

The refraction matrix

$$[R] = \begin{bmatrix} 1 & 0 \\ \frac{1}{f_2} & 1 \end{bmatrix} \begin{bmatrix} 1 & d \\ 0 & 1 \end{bmatrix} \begin{bmatrix} 1 & 0 \\ -\frac{1}{f_1} & 1 \end{bmatrix} = \begin{bmatrix} 1 & 0 \\ \frac{1}{10} & 1 \end{bmatrix} \begin{bmatrix} 1 & 3 \\ 0 & 1 \end{bmatrix} \begin{bmatrix} 1 & 0 \\ -\frac{1}{10} & 1 \end{bmatrix}$$

$$= \begin{bmatrix} 1 & 3 \\ \frac{1}{10} & \frac{13}{10} \end{bmatrix} \begin{bmatrix} 1 & 0 \\ -\frac{1}{10} & 1 \end{bmatrix} = \begin{bmatrix} \frac{1}{10} & 3 \\ \left(\frac{1}{10} - \frac{13}{100}\right) & \frac{13}{10} \end{bmatrix}$$

$$\therefore \begin{bmatrix} x_2 \\ \alpha_2 \end{bmatrix} = \begin{bmatrix} 1 & -v \\ 0 & 1 \end{bmatrix} \begin{bmatrix} \frac{7}{10} & 3 \\ -\frac{3}{100} & \frac{13}{10} \end{bmatrix} \begin{bmatrix} 1 & 15 \\ 0 & 1 \end{bmatrix} \begin{bmatrix} x_1 \\ \alpha_1 \end{bmatrix}$$

$$= \begin{bmatrix} 1 & -v \\ 0 & 1 \end{bmatrix} \begin{bmatrix} \frac{7}{10} & \frac{27}{2} \\ -\frac{3}{100} & \left(-\frac{9}{20} + \frac{13}{10}\right) \end{bmatrix} \begin{bmatrix} x_1 \\ \alpha_1 \end{bmatrix}$$

$$= \begin{bmatrix} \left(\frac{7}{10} + \frac{3v}{100}\right) & \frac{27}{2} - v\left(\frac{13}{10} - \frac{9}{20}\right) \\ -\frac{3}{100} & \left(\frac{13}{10} - \frac{9}{20}\right) \end{bmatrix} \begin{bmatrix} x_1 \\ \alpha_1 \end{bmatrix}$$

$$\Rightarrow \quad x_2 = \left(\frac{7}{10} + \frac{3v}{100}\right) x_1 + \alpha_1 \left\{\frac{27}{2} \cdot v\left(\frac{13}{10} - \frac{9}{20}\right)\right\}$$

For an axial point $x_2 = x_1 = 0$, but $\alpha_1 \neq 0$

$$\therefore \quad \frac{27}{2} = v\left(\frac{13}{10} - \frac{9}{20}\right) = v \times \frac{17}{20}$$

$$\Rightarrow \quad v = \frac{270}{17} = 15\frac{15}{17}\ \text{cm} \quad \text{(to the right of the second lens).}$$

(b) Here concave lens faces the object first. So in this case

$$\begin{bmatrix} x_2 \\ \alpha_2 \end{bmatrix} = \begin{bmatrix} 1 & -v \\ 0 & 1 \end{bmatrix} \begin{bmatrix} 1 & 0 \\ -\frac{1}{f_1} & 1 \end{bmatrix} \begin{bmatrix} 1 & d \\ 0 & 1 \end{bmatrix} \begin{bmatrix} 1 & 0 \\ \frac{1}{f_2} & 1 \end{bmatrix} \begin{bmatrix} 1 & u \\ 0 & 1 \end{bmatrix} \begin{bmatrix} x_1 \\ \alpha_1 \end{bmatrix}$$

Proceed exactly as in (a) and find the image distance. Here, $v = 90$ cm.

6. A concave lens is placed at a distance of 25 cm in front of a concave mirror of focal length 20 cm. It is found that a pin placed at a distance of 68.6 cm in front of the lens coincides with its own inverted image formed by the lens and mirror. Find the focal length of the lens.

Solution. The object and image coincide in position, such a system is termed as self-adjusted conjugate system. So the concave lens forms the image at the centre of curvature of the concave mirror at a distance = 40 – 25 = 15 cm from the lens.

$$\therefore \quad \begin{bmatrix} x_2 \\ \alpha_2 \end{bmatrix} = \begin{bmatrix} 1 & -15 \\ 0 & 1 \end{bmatrix} \begin{bmatrix} 1 & 0 \\ \frac{1}{f} & 1 \end{bmatrix} \begin{bmatrix} 1 & 68.6 \\ 0 & 1 \end{bmatrix} \begin{bmatrix} x_1 \\ \alpha_1 \end{bmatrix}$$

$$= \begin{bmatrix} 1 & -15 \\ 0 & 1 \end{bmatrix} \begin{bmatrix} 1 & 68.6 \\ \frac{1}{f} & \left(\frac{68.6}{f} + 1\right) \end{bmatrix} \begin{bmatrix} x_1 \\ \alpha_1 \end{bmatrix}$$

$$= \begin{bmatrix} \left(1 - \frac{15}{f}\right) & \left\{68.6 - 15\left(\frac{68.6}{f} + 1\right)\right\} \\ \frac{1}{f} & \left(\frac{68.6}{f} + 1\right) \end{bmatrix} \begin{bmatrix} x_1 \\ \alpha_1 \end{bmatrix}$$

$$\Rightarrow \quad x_2 = \left(1 - \frac{15}{f}\right)x_1 + \left\{68.6 - 15\left(\frac{68.6}{f} + 1\right)\right\}\alpha_1.$$

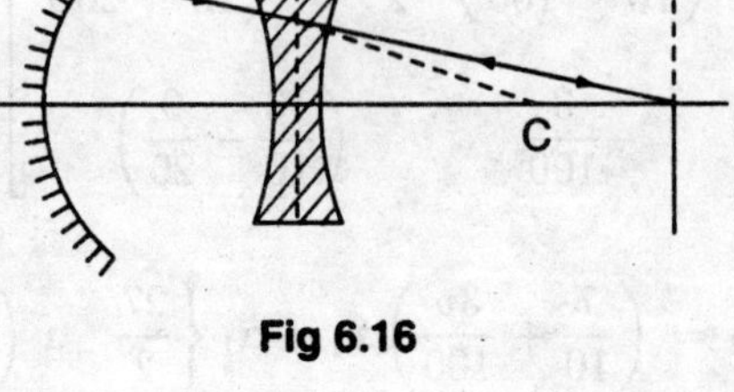

Fig 6.16

For an axial point $x_2 = x_1 = 0$, but $\alpha_1 \neq 0$.

$$\therefore\ 68.6 = 15\left(\frac{68.6}{f}+1\right) \Rightarrow \frac{68.6}{f} = \frac{68.6}{15} - 1 = \frac{53.6}{15} \Rightarrow f = \frac{15\times 68.6}{53.6} = 19.2 \text{ cm.}$$

7. A camera lens system consists of a convergent lens of focal length 12 cm facing the object and a divergent lens of focal length 5 cm is placed 8 cm behind the former. Find the position where the photographic plate should be placed to photograph a distant object.

Solution. We are to find the focal length of the combination; the image of a distant object will form the image in the focal plane.

The system matrix $[S]$ can be written as

$$[S] = \begin{bmatrix} 1 & 0 \\ \frac{1}{f_2} & 1 \end{bmatrix}\begin{bmatrix} 1 & 8 \\ 0 & 1 \end{bmatrix}\begin{bmatrix} 1 & 0 \\ -\frac{1}{f_1} & 1 \end{bmatrix} = \begin{bmatrix} 1 & 0 \\ \frac{1}{5} & 1 \end{bmatrix}\cdot\begin{bmatrix} 1 & 8 \\ 0 & 1 \end{bmatrix}\begin{bmatrix} 1 & 0 \\ -\frac{1}{12} & 1 \end{bmatrix}$$

$$= \begin{bmatrix} 1 & 0 \\ \frac{1}{5} & 1 \end{bmatrix}\begin{bmatrix} \left(1-\frac{2}{3}\right) & 8 \\ -\frac{1}{12} & 1 \end{bmatrix} = \begin{bmatrix} 1 & 0 \\ \frac{1}{5} & 1 \end{bmatrix}\begin{bmatrix} \frac{1}{3} & +8 \\ -\frac{1}{12} & 1 \end{bmatrix}$$

$$= \begin{bmatrix} \frac{1}{3} & 8 \\ \left(\frac{1}{15}-\frac{1}{12}\right) & \left(1+\frac{8}{5}\right) \end{bmatrix} = \begin{bmatrix} \frac{1}{3}, & 8 \\ -\frac{1}{60} & \frac{13}{5} \end{bmatrix}$$

$$\therefore\ \begin{bmatrix} x_2 \\ \alpha_2 \end{bmatrix} = \begin{bmatrix} 1 & -v \\ 0 & 1 \end{bmatrix}[S]\begin{bmatrix} x_1 \\ \alpha_1 \end{bmatrix} = \begin{bmatrix} 1 & -v \\ 0 & 1 \end{bmatrix}\begin{bmatrix} \frac{1}{3} & 8 \\ -\frac{1}{60} & \frac{13}{5} \end{bmatrix}\begin{bmatrix} x_1 \\ \alpha_1 \end{bmatrix}$$

$$\begin{bmatrix} x_2 \\ \alpha_2 \end{bmatrix} = \begin{bmatrix} \left(\frac{1}{3}+\frac{v}{60}\right) & \left(+8-\frac{13}{5}v\right) \\ -\frac{1}{60} & +\frac{13}{5} \end{bmatrix}\begin{bmatrix} x_1 \\ \alpha_1 \end{bmatrix}.$$

$$\Rightarrow\ x_2 = \left(\frac{1}{3}+\frac{v}{60}\right)x_1 + \alpha_1\left(+8-\frac{13}{5}v\right) \quad\text{and}\quad \alpha_2 = -\frac{1}{60}x_1 + \alpha_1\cdot\frac{13}{5}. \qquad (1)$$

For rays incident parallel to the axis, $\alpha_1 = 0$, for focal point $x_2 = 0$

$$\therefore\ x_1\left(\frac{1}{3}+\frac{v}{60}\right) = 0.$$

Since, $x_1 \neq 0$, $\frac{1}{3} = \frac{-v}{60} \Rightarrow v = f = -20$ cm. So the system behaves as a convex lens.

8. A hemisphere of glass ($\mu = 3/2$) of 3 inch radius serves as a lens. Find the positions of the image when an object is placed 4 inch in front of the (a) plane surface, (b) curved surface.

Solution. (a) The system matrix can be written as

$$[S] = \begin{bmatrix} 1 & 0 \\ P_{23} & \frac{n_2}{n_3} \end{bmatrix} \begin{bmatrix} 1 & r \\ 0 & 1 \end{bmatrix} \begin{bmatrix} 1 & 0 \\ P_{12} & \frac{n_1}{n_3} \end{bmatrix}$$

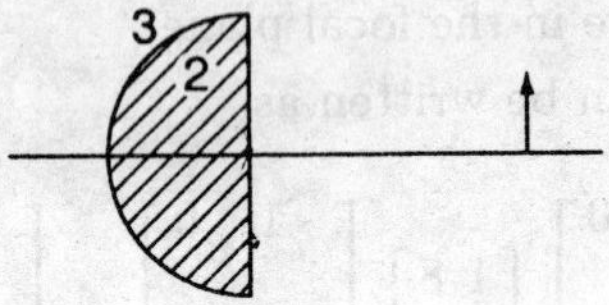

Fig 6.17

Here, $P_{12} = 0, P_{23} = \dfrac{1-n}{r}, n_1 = n_3 = 1, n_2 = n$

$$\therefore \quad [S] = \begin{bmatrix} 1 & 0 \\ P_{23} & n \end{bmatrix} \begin{bmatrix} 1 & r \\ 0 & 1 \end{bmatrix} \begin{bmatrix} 1 & 0 \\ 0 & \frac{1}{n} \end{bmatrix} = \begin{bmatrix} 1 & 0 \\ P_{23} & \frac{1}{n} \end{bmatrix} \begin{bmatrix} 1 & \frac{r}{n} \\ 0 & \frac{1}{n} \end{bmatrix} = \begin{bmatrix} 1 & \frac{r}{n} \\ P_{23} & \left(P_{23}\frac{r}{n} + 1\right) \end{bmatrix}$$

$$\begin{bmatrix} x_2 \\ \alpha_2 \end{bmatrix} = \begin{bmatrix} 1 & -v \\ 0 & 1 \end{bmatrix} \begin{bmatrix} 1 & \frac{r}{n} \\ P_{23} & \left(P_{23}\frac{r}{n} + 1\right) \end{bmatrix} \begin{bmatrix} 1 & u \\ 0 & 1 \end{bmatrix} \begin{bmatrix} x_1 \\ \alpha_1 \end{bmatrix}$$

$$= \begin{bmatrix} 1 & -v \\ 0 & 1 \end{bmatrix} \begin{bmatrix} 1 & \left(u + \frac{r}{n}\right) \\ P_{23} & \left\{uP_{23} + P_{23}\frac{r}{n} + 1\right\} \end{bmatrix} \begin{bmatrix} x_1 \\ \alpha_1 \end{bmatrix}$$

$$= \begin{bmatrix} (1 - vP_{23}) & \left(u + \frac{r}{n}\right) - v\left\{u + P_{23} + P_{23}\frac{r}{n} + 1\right\} \\ P_{23} & uP_{23} + P_{23}\frac{r}{n} + 1 \end{bmatrix} \begin{bmatrix} x_1 \\ \alpha_1 \end{bmatrix} = \begin{bmatrix} N_{11} & N_{12} \\ N_{21} & N_{22} \end{bmatrix} \begin{bmatrix} x_1 \\ \alpha_1 \end{bmatrix}$$

$$\Rightarrow \quad x_2 = N_{11}x_1 + N_{12}\alpha_1 \quad \text{and} \quad \alpha_2 = N_{21}x_1 + N_{22}\alpha_1$$

The conjugate positions are given by $N_{12} = 0$

$$\therefore \; u + \frac{r}{n} = v\left\{uP_{23} + P_{23}\frac{r}{n} + 1\right\} \Rightarrow v = \frac{u + \frac{r}{n}}{P_{23}\left(u + \frac{r}{n}\right) + 1}$$

$$P_{23} = \frac{1-n}{r} = \frac{1 - 3/2}{3} = -\frac{1}{6} \quad \text{and} \quad \frac{1}{6} = \frac{3}{3/2} + 2$$

$$\therefore \; v = \frac{4+2}{-\frac{1}{6} \times 6 + 1} = \infty; \quad \text{the image is formed at } \infty.$$

(b) The system matrix in this case

$$[S] = \begin{bmatrix} 1 & 0 \\ P_{23} & \frac{n_2}{n_3} \end{bmatrix} \begin{bmatrix} 1 & r \\ 0 & 1 \end{bmatrix} \begin{bmatrix} 1 & 0 \\ P_{12} & \frac{n_1}{n_2} \end{bmatrix}$$

$$P_{12} = -\frac{n_2 - n_1}{n_2 r} = -\frac{n-1}{nr} = -\frac{1}{9}$$

$$n_1 = n_3 = 1, n_2 = n \quad \text{and} \quad P_{23} = 0.$$

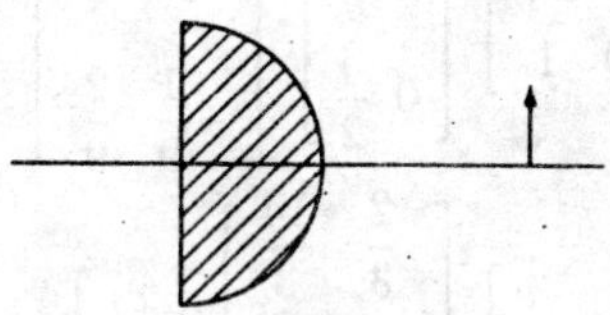

Fig 6.18

Here, the convex surface faces the direction of light.

$$\therefore \begin{bmatrix} x_2 \\ \alpha_2 \end{bmatrix} = \underset{[T_3]}{\begin{bmatrix} 1 & -v \\ 0 & 1 \end{bmatrix}} \underset{[R_2]}{\begin{bmatrix} 1 & 0 \\ P_{23} & \frac{n_2}{n_3} \end{bmatrix}} \underset{[T_2]}{\begin{bmatrix} 1 & r \\ 0 & 1 \end{bmatrix}} \underset{[R_1]}{\begin{bmatrix} 1 & 0 \\ P_{12} & \frac{n_1}{n_2} \end{bmatrix}} \underset{[T_1]}{\begin{bmatrix} 1 & u \\ 0 & 1 \end{bmatrix}} \begin{bmatrix} x_1 \\ \alpha_1 \end{bmatrix}$$

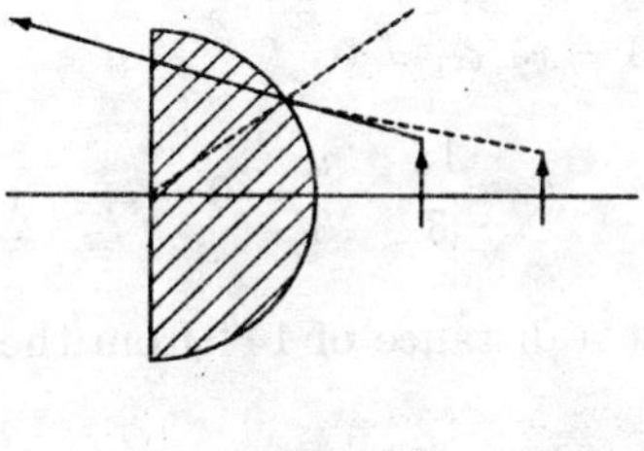

Fig 6.19

$[T]$ = Translation matrix, brings the object to the curved surface
$[R_1]$ = Refraction matrix at the curved surface transforming the object to glass medium
$[T_2]$ = Translation matrix bringing the image from the first surface to the second surface
$[R_2]$ = Refraction matrix transforming the image from glass to air across the plane surface
$[T_3]$ = Translation matrix forming the final image

Also,

$$P_{23} = -\frac{n_2 - n_3}{n_3 r} = -\frac{n_2 - n_3}{n_3 \infty} = 0$$

$$n_1 = n_3 = 1, n_2 = n = \frac{3}{2} \quad \text{and} \quad P_{12} = -\frac{n_2 - n_1}{n_2 r} = -\frac{1}{9}$$

$$\therefore \quad \begin{bmatrix} x_2 \\ \alpha_2 \end{bmatrix} = \begin{bmatrix} 1 & -v \\ 0 & 1 \end{bmatrix} \begin{bmatrix} 1 & 0 \\ 0 & \dfrac{3}{2} \end{bmatrix} \begin{bmatrix} 1 & 3 \\ 0 & 1 \end{bmatrix} \begin{bmatrix} 1 & 0 \\ -\dfrac{1}{9} & \dfrac{2}{3} \end{bmatrix} \begin{bmatrix} 1 & 4 \\ 0 & 1 \end{bmatrix} \begin{bmatrix} x_1 \\ \alpha_1 \end{bmatrix}$$

$$= \begin{bmatrix} 1 & -v \\ 0 & 1 \end{bmatrix} \begin{bmatrix} 1 & 0 \\ 0 & \dfrac{3}{2} \end{bmatrix} \begin{bmatrix} 1 & 3 \\ 0 & 1 \end{bmatrix} \begin{bmatrix} 1 & 4 \\ -\dfrac{1}{9} & \dfrac{2}{9} \end{bmatrix} \begin{bmatrix} x_1 \\ \alpha_1 \end{bmatrix}$$

$$= \begin{bmatrix} 1 & -v \\ 0 & 1 \end{bmatrix} \begin{bmatrix} 1 & 0 \\ 0 & \dfrac{3}{2} \end{bmatrix} \begin{bmatrix} \dfrac{2}{3} & \dfrac{14}{3} \\ -\dfrac{1}{9} & \dfrac{2}{9} \end{bmatrix} \begin{bmatrix} x_1 \\ \alpha_1 \end{bmatrix}$$

$$= \begin{bmatrix} 1 & -v \\ 0 & 1 \end{bmatrix} \begin{bmatrix} \dfrac{2}{3} & \dfrac{14}{3} \\ -\dfrac{1}{6} & \dfrac{1}{3} \end{bmatrix} \begin{bmatrix} x_1 \\ \alpha_1 \end{bmatrix}$$

$$= \begin{bmatrix} \left(\dfrac{2}{3} + \dfrac{v}{6}\right) & \left(\dfrac{14}{3} - \dfrac{v}{3}\right) \\ -\dfrac{1}{6} & \dfrac{1}{3} \end{bmatrix} \begin{bmatrix} x_1 \\ \alpha_1 \end{bmatrix}$$

$$\Rightarrow \quad x_2 = \left(\frac{2}{3} + \frac{v}{6}\right) x_1 + \left(\frac{14}{9} - \frac{v}{3}\right) \alpha_1 \quad \text{and} \quad \alpha_2 = -\frac{1}{6} x_1 + \frac{1}{3} \alpha_1$$

For an axial point, $x_1 = 0 = x_2, \alpha_1 \neq 0$

$$\therefore \quad \frac{14}{3} - \frac{v}{3} = 0 \quad \Rightarrow \quad v = 14.$$

So the image is formed at a distance of 14″ from the plane surface on the same side as the object.

The magnification m is given by,

$$\frac{x_2}{x_1} = m = \left(\frac{2}{3} + \frac{14}{6}\right) = 3.$$

Theory Questions

1. What is a lens? What do you mean by a thin lens? Develop the refraction matrix for a thin convex lens. Show that for a thin convex lens

$$\frac{f_1}{n_1} + \frac{f_2}{n_2} = 0$$

where f_1 and f_2 are the object side focal length and image side focal length respectively, and n_1 and n_2 are the r.i. on the object side and image side medium.

2. Find the equivalent focal length of two thin lenses of focal lengths f_1 and f_2; separated by a distance a in air. Find also the position of the equivalent lens.

3. Find the cardinal points in case of two thin lenses separated by a distance a if (a) both the lenses are convex, (b) both the lenses are concave, (c) one concave, the other convex.

4. Find the cardinal points of Huygens eyepiece. Represent their positions in a diagram.

5. A thin lens of r.i. n and radii of curvature R_1 and R_2 is placed at the interface between two media with r.i.'s n_1 and n_2. If u and v are the object and image distances respectively and f_1 and f_2 are the object and image side focal lengths, prove that $f_1/u + f_2/v = 1$.

6. Find the cardinal points of a Ramsden's eyepiece. Also represent their positions in a diagram.

7. A thin symmetrical converging lens of focal length 12.5 cm and r.i. 1.60 is mounted such that one of its faces is in air and the other in a transparent liquid of r.i. 1.40. Find the position of the image of a small object which lies in the liquid at a distance of 15 cm from the lens.

[Hints. Proceeding as in art 6.3, you will get

$$\begin{bmatrix} x_2 \\ \alpha_2 \end{bmatrix} = \begin{bmatrix} 1 & -v \\ 0 & 1 \end{bmatrix} \overbrace{\begin{bmatrix} s_{11} & s_{12} \\ s_{21} & s_{22} \end{bmatrix}}^{S\text{-matrix}} \begin{bmatrix} 1 & u \\ 0 & 1 \end{bmatrix} \begin{bmatrix} x_1 \\ \alpha_1 \end{bmatrix}$$

$$= \begin{bmatrix} (s_{11} - vs_{21}) & (vs_{11} + s_{12}) - v(us_{21} + s_{22}) \\ s_{21} & (us_{21} + s_{22}) \end{bmatrix} \begin{bmatrix} x_1 \\ \alpha_1 \end{bmatrix} = \begin{bmatrix} N_{11} & N_{12} \\ N_{21} & N_{22} \end{bmatrix} \begin{bmatrix} x_1 \\ \alpha_1 \end{bmatrix}$$

The conjugate relation is obtained through

$$N_{12} = 0 \quad \Rightarrow \quad v = \frac{us_{11} + s_{12}}{(us_{21} + s_{12})}.$$

Here,

$$s_{12} = 0, \qquad s_{11} = 1$$

$$s_{21} = \frac{1-n}{R_2} - \frac{n - n_1}{R_1}, \quad s_{22} = \frac{n_1}{1} = n_1$$

$$\therefore \quad v = \frac{u}{(us_{21} + s_{22})}.$$

Now,

$$s_{21} = \frac{1-n}{R_1} - \frac{n - 1.4}{R_1} = \frac{1}{R_1}(1 - 2n + 1.4) = \frac{1}{R_1}(2.4 - 3.2) = -\frac{.8}{R_1}$$

$$\therefore \quad s_{22} = 1.40.$$

To find R_1, refer to the given focal length 12.5 cm, v = 25 cm (in the liquid)]

8. A thin biconvex lens of r.i. 1.6 and a thin diverging lens of r.i. 1.5 are cemented together to form a doublet. Two of the surfaces have a radius of 50.0 cm and the other two of 40.0 cm; the cemented surfaces have curvatures of the same value. Find the focal length of the doublet if (a) it is in air, (b) the source, screen, lens are immersed in water ($n = 4/3$). (−40.8 cm, −119 cm).

9. A thin glass lens of focal length f and r.i. 3/2 is placed under a liquid of r.i. 4/3. Find the geometrical focus of a pencil of parallel rays directly incident upon the lens.

10. Explain the term 'refraction matrix' and 'system matrix'. Construct the system matrix for refraction through a thick lens. Reduce it to that of a thin lens. Hence prove that for a thin convex lens, $\dfrac{1}{f} = (n-1)\left(\dfrac{1}{r_1} + \dfrac{1}{r_2}\right)$. [CU 1999]

11. Derive the system matrix for a plano-convex lens of a material of r.i. $\frac{3}{2}$. The convex surface has a radius of curvature 2.5 cm and is facing the incident light. The centre thickness of the lens is 6 mm. **[CU 2001]**
12. What is the function of an eyepiece? Why does an eyepiece system consist of two lenses instead of one? Find the focal points and principal points of a Huygens eyepiece. **[CU 2001]**
13. Obtain the translation matrix and refraction matrix. Hence calculate the system matrix for a thin lens. **[BU 1994]**
14. Derive expression for the positions of the principal points and focal points of a system of two thin lenses placed co-axially in air and separated by a distance. **[BU 1995]**
15. (a) Describe Huygens and Ramsden's eyepieces, and discuss their relative merits.
    (b) The lenses in Huygens eyepiece have focal lengths of 2 cm and 4 cm. Find the distance of separation of the lenses and locate the positions of principal points and principal focus.
16. Consider a combination of a pair of co-axial thin lenses of the same material—one convex and the other a concave lens of lengths 20 cm and 10 cm respectively. They are separated by a distance of 8 cm. Find the principal points and also the nodal points of the system. **[CU Similar 1991, 93]**
17. Why does an eyepiece system consist of two lenses instead of one? Show that the Huygens' eyepiece satisfies the condition of minimum spherical aberration and achromatism. **[CU 1992]**
18. Describe be construction of Ramsden's eyepiece and find its cardinal points.
19. A plano-convex lens ($\mu = 3/2$) is of thickness 3 cm and radius of curvature 8 cm. Calculate the minimum distance between an object and its real image. **[CU 1996]**
20. Find by matrix method the conjugate relationship in case of refraction at a curved surface. **[CU 1997]**
21. When a system of lenses is surrounded on both sides by the same medium (air), prove that principal points and nodal points coincide.
22. A lens is in the form of a spherical ball of radius R. The r.i. is n. Show that the principal planes of the lens coincide and pass through the centre of the sphere. Determine the focal lengths of the lens.
23. Find the distances of principal points and focal points of a glass hemisphere of radius r and r.i. n, when surrounded by air.

Chapter 7

Thick Lens

7.1 Introduction

A thick lens is a transparent medium bounded by surfaces such that the thickness of the medium is of considerable value. So in this case, the conjugate relationship, the positions of the cardinal points or planes are different when referred to the optical centre or the poles. The thick lens formulation leads to thin lens case if only the thickness is considered to be of negligible value. In case of a thick lens, both the principal planes may be inside the lens or one may be inside, the other outside depending on the nature of curvature and thickness of the lens. A glass sphere, a glass hemisphere, etc., may be treated as a thick lens.

In case of a thick lens, there are six cardinal points defining every detail about the characteristic feature of the lens, so far as image formation and magnification by the lens is concerned.

7.2 Introducing the Cardinal Points of a Thick Lens

Principal Points

It refers to a pair of conjugate points on the principal axis such that if an object be placed at one point, an image of the same size as that of the object is formed at the other point and on the same side of the principal axis as the object, thus these conjugate planes are planes of unit positive magnification (lateral).

Focal Points

These refer to a pair of points on the principal axis. The first focal point is a point on the principal axis such that rays diverging from this point emerge parallel to the principal axis.

The second focal point (F_2) is a point on the principal axis such that rays incident parallel to the principal axis pass through it after emergence.

Nodal Points (N_1 and N_2)

These also refer to a pair of conjugate points such that if a ray be allowed to pass through or travel towards one point its corresponding emergent ray from the lens will be parallel to the incident ray and will appear to diverge from, or pass through the second point. Thus, the nodal planes may be taken as planes of unit angular magnification.

With the above idea in the background we now set to develop the matrix formulation of a thick lens and hence find out mathematical expression for the cardinal point distances.

Let us consider a thick lens bounded by two curved surfaces having poles A_1 and A_2 respectively, the thickness of the lens being d. An object ray defined by the column vector $\begin{bmatrix} x_1 \\ \alpha_1 \end{bmatrix}$ is transformed into the image ray defined by the column vector $\begin{bmatrix} x_2 \\ \alpha_2 \end{bmatrix}$ through a series of matrix transformation, the first through translation bringing the object P_1P on to the curved surface having pole A_1; this is represented by a translation matrix $[T_1]$; then refraction at the first curved surface, thus bringing the incident ray within the lens material of r.i. n_2 from outside (r.i. n_1). By another translation $[T_2]$, the ray refracting at the first curved surface is brought on to the second curved surface. Another refraction at the second surface transforms the ray from inside the lens into the emergent ray across the second curved surface, the process of transformation being represented by suitable refraction matrix. Finally, the image ray defined by $\begin{bmatrix} x_2 \\ \alpha_2 \end{bmatrix}$ is obtained through another translation through the image distance v. Thus, the whole process of image formation may be represented by a series of matrix operation representing the process of refraction, translation, etc. Thus,

$$\begin{bmatrix} x_2 \\ \alpha_2 \end{bmatrix} = \begin{bmatrix} 1 & -v \\ 0 & 1 \end{bmatrix} \underbrace{[R_2][T][R_1]}_{s\text{-matrix}} \begin{bmatrix} 1 & u \\ 0 & 1 \end{bmatrix} \begin{bmatrix} x_1 \\ \alpha_1 \end{bmatrix}$$

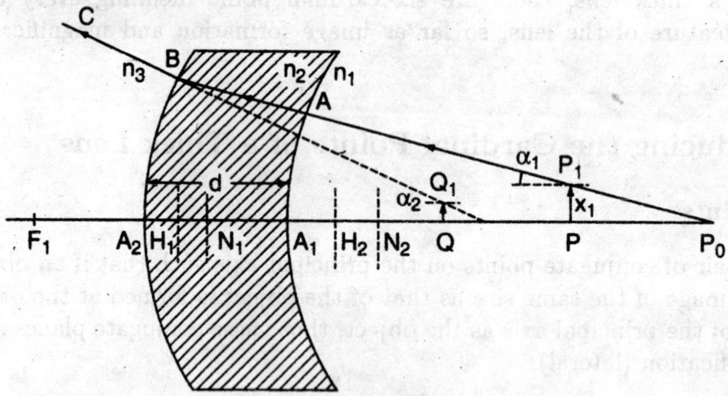

Fig 7.1

The system matrix can be written as,

$$[S] = [R_2][T][R_1] = \begin{bmatrix} 1 & 0 \\ P_{23} & n_2/n_3 \end{bmatrix} \begin{bmatrix} 1 & d \\ 0 & 1 \end{bmatrix} \begin{bmatrix} 1 & 0 \\ P_{12} & n_1/n_2 \end{bmatrix}$$

$$= \begin{bmatrix} 1 & 0 \\ P_{23} & n_2/n_3 \end{bmatrix} \begin{bmatrix} (1+P_{12}d) & n_1d/n_2 \\ P_{12} & n_1/n_2 \end{bmatrix}$$

$$= \begin{bmatrix} (1+P_{12}d) & n_1d/n_2 \\ \{P_{23}(1+P_{12}d)+(n_2/n_3)P_{12}\} & \{(n_1d/n_2)P_{23}+(n_1/n_3)\} \end{bmatrix} = \begin{bmatrix} S_{11} & S_{12} \\ S_{21} & S_{22} \end{bmatrix}$$

$$S_{11} = 1 + P_{12}d = 1 + \frac{(n_2 - n_1)}{n_2 r_1} = 1 + \frac{d}{f_1 n_2}, \text{ where } \frac{1}{f_1} = +\frac{n_2 - n_1}{r_1}; S_{12} = \frac{n_1 d}{n_2}$$

$$P_{23} = \frac{n_3 - n_2}{n_3 r_2} = \frac{1}{n_3 f_2}, \text{ where } \frac{1}{f_2} = \frac{n_3 - n_2}{r_2}$$

$$\therefore \quad \begin{bmatrix} x_2 \\ \alpha_2 \end{bmatrix} = \begin{bmatrix} 1 & -v \\ 0 & 1 \end{bmatrix} \begin{bmatrix} S_{11} & S_{12} \\ S_{21} & S_{22} \end{bmatrix} \begin{bmatrix} 1 & u \\ 0 & 1 \end{bmatrix} \begin{bmatrix} x_1 \\ \alpha_1 \end{bmatrix} = \begin{bmatrix} 1 & -v \\ 0 & 1 \end{bmatrix} \begin{bmatrix} S_{11} & (uS_{11} + S_{12}) \\ S_{21} & (uS_{21} + S_{22}) \end{bmatrix} \begin{bmatrix} x_1 \\ \alpha_1 \end{bmatrix}$$

$$= \begin{bmatrix} (S_{11} - vS_{21}) & \{(uS_{11} + S_{12}) - v(uS_{21} + S_{22})\} \\ S_{21} & (uS_{21} + S_{22}) \end{bmatrix} \begin{bmatrix} x_1 \\ \alpha_1 \end{bmatrix} = \begin{bmatrix} N_{11} & N_{12} \\ N_{21} & N_{22} \end{bmatrix} \begin{bmatrix} x_1 \\ \alpha_1 \end{bmatrix}$$

$$\therefore \quad x_2 = N_{11} x_1 + N_{12} \alpha_1 \quad \text{and} \quad \alpha_2 = N_{21} x_1 + N_{22} \alpha_1 \tag{7.2.1}$$

where $N_{11} = S_{11} - vS_{21}$, $N_{12} = (uS_{11} + S_{12}) - v(uS_{21} + S_{22})$, $N_{21} = S_{21}$ and $N_{22} = uS_{21} + S_{22}$.

7.3 Cardinal Points of Thick Lens

First Focus with respect to the Poles

The first principal focus is the value of u for which the conjugate point is at infinity. Thus, if in the second relation of eqn.(7.2.1) we set $N_{22} = 0$, we get,

$$\alpha_2 = N_{21} x_1$$

which means, that rays passing through the object point for any value of α_1 (i.e., rays diverging from the object) after emergence become rays having same value of α_2, i.e., are rendered parallel. The plane containing such object points is called *First Principal Focal Plane.* Thus the first principal focus is defined by the transformation matrix element

$$N_{22} = 0 \quad \Rightarrow \quad u_1 S_{21} + S_{22} = 0 \quad \Rightarrow \quad u_1 = A_1 F_1 = -\frac{S_{22}}{S_{21}}.$$

Now,

$$\begin{aligned} S_{21} &= P_{23}(1 + P_{12} d) + \frac{n_2}{n_3} P_{12} = P_{23} S_{11} + \frac{n_2}{n_3} P_{12} \\ &= \left(1 + \frac{d}{f_1 n_2}\right) \frac{1}{n_3 f_2} + \frac{n_2}{n_3} \cdot \frac{1}{f_1 n_2} = \frac{1}{n_3 f_2} + \frac{d}{n_2 n_3 f_1 f_2} + \frac{1}{n_3 f_1} \\ &= \frac{1}{n_3} \left(\frac{1}{f_1} + \frac{1}{f_2} + \frac{d}{n_2 f_1 f_2} \right) = \frac{1}{n_3} \cdot \frac{f_1 + f_2 + d/n_2}{(f_1 f_2)} = \frac{\Delta}{n_3 f_1 f_2} \\ S_{22} &= \left(\frac{n_1 d}{n_2} \right) P_{23} + \frac{n_1}{n_3} = \frac{n_1 d}{n_2} \cdot \frac{1}{n_3 f_2} + \frac{n_1}{n_3} = \frac{n_1}{n_3} \left(1 + \frac{d}{n_2 f_2} \right) \end{aligned}$$

$$\therefore \quad A_1 F_1 = \text{distance of the first focal point } F_1 \text{ w.r.t. the pole}$$

$$= -\frac{(n_1/n_3)(1 + d/n_2 f_2)}{\Delta/(n_3 f_1 f_2)} = -\frac{(f_2 + d/n_2) n_1 f_1}{\Delta}. \tag{7.3.1}$$

Second Focus F_2

The second principal focus F_2 is that point where the emergent rays converge or appear to come from, after refraction in the optical system. If we set $N_{11} = 0$ in eqn.(7.2.1) we get,

$x_2 = N_{12}\alpha_1$; it means, all the incident rays have same value of α_1, i.e., incident parallel rays pass through the image point x_2. Thus the second focus F_2 is defined by

$$N_{11} = 0 \quad \Rightarrow \quad S_{11} - v_2 S_{21} = 0$$

$$\therefore \quad v_2 = A_2F_2 = \frac{S_{11}}{S_{21}} = \frac{\left(1 + \frac{d}{f_1 n_2}\right)}{\Delta/(n_3 f_1 f_2)} = \frac{\left(f_1 + \frac{d}{n_2}\right) n_3 f_2}{\Delta}. \tag{7.3.2}$$

Conjugate Relations

If we set $N_{12} = 0$, we get, $x_2 = N_{11}x_1$; it means, for an object point x_1, there is a unique image point x_2, so the conjugate relation is defined by

$$N_{12} = 0 \quad \Rightarrow \quad (uS_{11} + S_{12}) - v(uS_{21} + S_{22}) = 0. \tag{7.3.3}$$

Conjugate Relation w.r.t. the Foci : Newton's Equation in case of Thick Lens

Eqn.(7.3.3) defines conjugate relationship w.r.t. the poles. If we want the relation w.r.t. the foci (F_1, F_2), we can transform eqn.(7.3.3) suitably. The positions of the foci of the thick lens defined by eqns.(7.3.1) and (7.3.2) are shown in fig.(7.2).

$$A_1F_1 = -\frac{S_{22}}{S_{21}} \quad \text{and} \quad A_2F_2 = \frac{S_{11}}{S_{21}}.$$

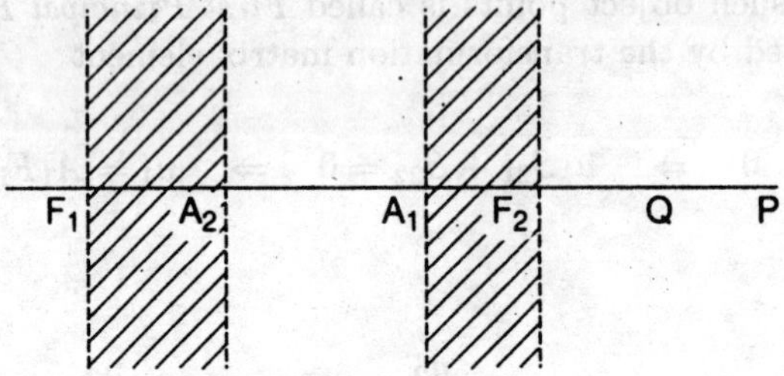

Fig 7.2

If measurement is taken with (F_1 and F_2) as the origin, we can write,

$$U = \text{object distance w.r.t. } F_1 = F_1P = F_1A_1 + A_1P = \left(u + \frac{S_{22}}{S_{21}}\right) \tag{7.3.4}$$

$$\text{and} \quad V = \text{image distance w.r.t. } F_2 = F_2Q = A_2Q - A_2F_2 = \left(v - \frac{S_{11}}{S_{21}}\right). \tag{7.3.5}$$

Putting the values of u and v given by eqns.(7.3.4) and (7.3.5), we get,

$$UV = -\frac{S_{11}S_{22}}{S_{21}^2} + \frac{S_{12}}{S_{21}}. \tag{7.3.6}$$

So the conjugate relation is given by eqn.(7.3.6) if object and image distances are measured w.r.t. foci. The expression on the right-hand side of eqn.(7.3.6) will be shown presently as related to the product of the two focal lengths of the thick lens.

Principal Points and Principal Planes

The principal planes are planes of unit +ve magnification. If we set $N_{12} = 0$, we get,

$$x_2 = N_{11}x_1 \quad \Rightarrow \quad \frac{x_2}{x_1} = N_{11}, \text{ so for unit magnification, } \frac{x_2}{x_1} = 1.$$

Therefore, $N_{11} = 1$ with $N_{12} = 0$ defines completely the principal planes. Now,

$$N_{11} = 1 \quad \Rightarrow \quad S_{11} - vS_{21} = 1 \quad \Rightarrow \quad v = A_2H_2 = \frac{S_{11} - 1}{S_{21}} \tag{7.3.7}$$

Putting this value in eqn.

$$N_{12} = 0 \Rightarrow uS_{11} + S_{12} = \frac{S_{11} - 1}{S_{21}}(uS_{21} + S_{22})$$

$$\Rightarrow u(S_{11} - S_{11} + 1) = \frac{(S_{11} - 1)S_{22}}{S_{21}} - S_{12}$$

$$\Rightarrow u = A_1H_1 = (S_{11} - 1)\frac{S_{22}}{S_{21}} - S_{12}. \tag{7.3.8}$$

Thus, the second principal plane is given by,

$$A_2H_2 = \frac{S_{11} - 1}{S_{21}} = \frac{\left(1 + \dfrac{d}{f_1 n_2}\right) - 1}{(\Delta / n_3 f_1 f_2)} = \frac{\left(\dfrac{d}{n_2}\right) n_3 f_2}{\Delta}. \tag{7.3.9}$$

The first principal plane is given by,

$$A_1H_1 = \frac{(S_{11} - 1)}{S_{21}} \cdot S_{22} - S_{12} = \frac{\left(\dfrac{d}{n_2}\right) n_3 f_2}{\Delta} \cdot \left(\frac{n_1 d}{n_2} P_{23} + \frac{n_1}{n_3}\right) - \frac{n_1 d}{n_2}$$

$$= \frac{1}{\Delta}\left(\frac{n_3 f_2 d}{n_2^2} \cdot \frac{n_1 d}{n_3 f_2} + \frac{n_3 f_2 d}{n_2} \cdot \frac{n_1}{n_3}\right) - \frac{n_1 d}{n_2} = \frac{1}{\Delta}\left(\frac{n_1 d^2}{n_2^2} + \frac{n_1}{n_2} f_2 d\right) - \frac{n_1 d}{n_2}$$

$$= -\frac{n_1 d}{n_2} \cdot \frac{\left(\Delta - \dfrac{d}{n_2} - f_2\right)}{\Delta} = -\frac{\left(\dfrac{n_1 d}{n_2}\right) f_1}{\Delta}. \tag{7.3.10}$$

Focal Lengths of the Thick Lens

The distances of the focal points F_1 and F_2 from the principal points are called the *focal lengths* (first and second) F' and F respectively.

$$\therefore \quad F' = H_1F_1 = A_1F_1 - A_1H_1$$

$$= -\frac{(f_2 + d/n_2) n_1 f_1}{\Delta} + \frac{(n_1 d / n_2) f_1}{\Delta} = -\frac{n_1 f_1 f_2}{\Delta} \tag{7.3.11}$$

Similarly,

$$F = A_2F_2 - A_2H_2 = H_2F_2$$

$$= \frac{(f_1 + d/n_2) n_3 f_2}{\Delta} - \frac{(d/n_2) n_3 f_2}{\Delta} = \frac{n_3 f_1 f_2}{\Delta}. \tag{7.3.12}$$

Conjugate Relation in terms of F' and F, and Newton's Equation

From eqn.(7.3.6), we write,

$$
\begin{aligned}
UV &= -\frac{S_{11}S_{22}}{S_{21}^2} + \frac{S_{12}}{S_{21}} = \frac{\left(\frac{n_1}{n_2}\right) d \times n_3 f_1 f_2}{\Delta} - \frac{\left(f_1 + \frac{d}{n_2}\right)\left(f_2 + \frac{d}{n_2}\right) n_1 n_3 f_1 f_2}{\Delta^2} \\
&= \frac{n_1 n_3 f_1 f_2}{\Delta^2}\left[\Delta \times \frac{d}{n_2} - f_1 f_2 - f_1 \frac{d}{n_2} - f_2 \frac{d}{n_2} - \frac{d^2}{n_2^2}\right] \\
&= \frac{n_1 n_3 f_1 f_2}{\Delta^2}\left[\left(f_1 + f_2 + \frac{d}{n_2}\right)\frac{d}{n_2} - f_1 f_2 - f_1 \frac{d}{n_2} - f_2 \frac{d}{n_2} - \frac{d^2}{n_2^2}\right] \\
&= -n_1 n_3 \left(\frac{f_1 f_2}{\Delta}\right)^2 = -n_1 \left(\frac{f_1 f_2}{\Delta}\right) \cdot n_3 \left(\frac{f_1 f_2}{\Delta}\right) = FF'.
\end{aligned}
\tag{7.3.13}
$$

Eqn.(7.3.13) is known as *Newton's equation.*

Nodal Planes

Nodal planes correspond to conjugate planes of unit angular magnification. So, if in eqn.(7.2.1) we set $N_{21} = 0$, we get,

$$\alpha_2 = N_{22}\alpha_1 \quad \Rightarrow \quad \frac{\alpha_2}{\alpha_1} = \text{angular magnification} = N_{22}.$$

So the nodal planes are defined by

$$N_{21} = 0 \quad \text{and} \quad N_{22} = 1.$$

Now, $N_{22} = 1 \quad \Rightarrow \quad uS_{21} + S_{22} = 1$

$$
\begin{aligned}
\therefore \quad u &= A_1N_1 = \frac{1 - S_{22}}{S_{21}} = \frac{1 - \frac{n_1}{n_2}\left(P_{23}d + \frac{n_2}{n_3}\right)}{\left(\frac{\Delta}{n_3 f_1 f_2}\right)} \\
&= \frac{1 - \frac{n_1}{n_2}\left(\frac{d}{n_3 f_2} + \frac{n_2}{n_3}\right)}{(\Delta / n_3 f_1 f_2)} = \frac{n_3 f_1 f_2}{\Delta} - \frac{n_3 f_1 f_2 \cdot n_1}{n_2 \Delta}\left(\frac{d}{n_3 f_2} + \frac{n_2}{n_3}\right) \\
&= -\frac{n_1 f_1 d}{n_2 \Delta} + \frac{(n_3 - n_1)}{\Delta} f_1 f_2 = -\frac{n_1 f_1 d}{n_2 \Delta} + (n_3 - n_1)P; \left[P = \frac{f_1 f_2}{\Delta}\right].
\end{aligned}
\tag{7.3.14}
$$

Using the conjugate relation expressed in equation (7.3.3), we get the second nodal points as

$$
\begin{aligned}
v &= A_2N_2 = \frac{uS_{11} + S_{12}}{(uS_{21} + S_{22})} = (A_1N_1)S_{11} + S_{12} \\
&= \left[-\frac{n_1 f_1 d}{n_2 \Delta} + (n_3 - n_1)P\right]\left(1 + \frac{d}{f_1 n_2}\right) + \frac{n_1 d}{n_2} = (n_3 - n_1)P + \frac{n_3 f_2 d}{\Delta n_2}.
\end{aligned}
\tag{7.3.15}
$$

7.4 The Principal and Nodal Points w.r.t. the Focal Points

Principal Points

$$F_1H_1 = F_1A_1 + A_1H_1 = \frac{\left(f_2 + \dfrac{d}{n_2}\right) n_1 f_1}{\Delta} - \frac{\left(\dfrac{n_1 d}{n_2}\right) f_1}{\Delta} = \frac{n_1 f_1 f_2}{\Delta} = n_1 P \qquad (7.4.1)$$

$$F_2H_2 = F_2A_2 + A_2H_2 = -\frac{\left(f_1 + \dfrac{d}{n_2}\right) n_3 f_2}{\Delta} + \frac{\left(\dfrac{d}{n_2}\right) n_3 f_2}{\Delta} = -n_3 \frac{f_1 f_2}{\Delta} = -n_3 P. \qquad (7.4.2)$$

Nodal Points

$$A_1N_1 = A_1F_1 + F_1N_1$$

$$\therefore \quad F_1N_1 = A_1N_1 - A_1F_1 = -\frac{n_1 f_1 d}{n_2 \Delta} + (n_3 - n_1)P + \frac{\left(f_2 + \dfrac{d}{n_2}\right) n_1 f_1}{\Delta}$$

$$= (n_3 - n_1)P + \frac{n_1 f_1 f_2}{\Delta} = n_3 P.$$

Again, $A_2N_2 = A_2F_2 + F_2N_2$

$$\therefore \quad F_2N_2 = A_2N_2 - A_2F_2 = (n_3 - n_1)P + \frac{n_3 f_2 d}{n_2 \Delta} - \frac{\left(\dfrac{f_1 + d}{n_2}\right) n_3 f_2}{\Delta}$$

$$= (n_3 - n_1)P - \frac{n_3 f_1 f_2}{\Delta} = -n_1 P. \qquad (7.4.3)$$

7.5 The Distances of the Focal Points and Nodal Points w.r.t. the Principal Points (H_1 and H_2)

The distance of the first principal focal point F_1 from H_1 is called the *first focal length* of the thick lens, and the distance of focal point F_2 from H_2 is called the *second focal length of the thick lens.*

From eqns.(7.3.11) and (7.3.12) we have,

$$F' = H_1F_1 = -n_1 P \text{ , and } \quad F = H_2F_2 = n_3 P.$$

Nodal Points

$$A_1N_1 = A_1H_1 + H_1N_1$$

$$\Rightarrow \quad -\frac{n_1 f_1 d}{n_2 \Delta} + (n_3 - n_1)P = -\frac{(n_1 d/n_2) f_1}{\Delta} + H_1N_1$$

$$\Rightarrow \quad H_1N_1 = (n_3 - n_1)P \qquad (7.5.1)$$

Again, $A_2N_2 = A_2H_2 + H_2N_2$

$$\Rightarrow \quad (n_3 - n_1)P + \frac{n_3 f_2 d}{\Delta n_2} = \frac{(d/n_2)n_3 f_2}{\Delta} + H_2N_2$$

$$\Rightarrow \quad H_2N_2 = (n_3 - n_1)P. \tag{7.5.2}$$

Thus, the nodal point N_1 w.r.t. the first principal plane, and the nodal point N_2 w.r.t. the second principal plane, are equidistant. Also, if $n_3 = n_1$; both $H_1N_1 = H_2N_2 = 0$, i.e., N_1 coincides with H_1 and N_2 coincides with H_2 when the lens is surrounded by the same medium. In such a case, $n_1 = n_3$, so we get, $F' = F$, i.e., the focal lengths are equal.

7.6 Cardinal Points in Special Cases

Equiconvex Lens : Surrounded by Air

$$\frac{1}{f_1} = -\frac{n_2 - n_1}{r_1} = -\frac{n-1}{r}$$

$$\frac{1}{f_2} = \frac{n_3 - n_2}{r_2} = \frac{1-n}{r} = -\frac{n-1}{r} \quad \Rightarrow \quad f_1 = f_2 = f.$$

$$\therefore \quad A_1H_1 = -\frac{n_1 f_1 d}{\Delta n_2} = -\frac{f_1 d}{(f_1 + f_2 + d/n)n} = -\frac{fd}{(2f + d/n)n} = -\frac{fd}{2fn + d}$$

$$= -\frac{d}{2n + d/f} = -\frac{d}{2n - d(n-1)/r} \tag{7.6.1}$$

$$\text{and} \quad A_2H_2 = \frac{d}{2n - (d/r)(n-1)} \tag{7.6.2}$$

The two principal points are equidistant w.r.t. the poles. If

$$d \ll r, \quad A_1H_1 = -d/2n, \quad A_2H_2 = d/2n.$$

In case of a Sphere of Glass

$$A_1H_1 = -\frac{d}{2n - (d/r)(n-1)} = -\frac{2r}{2n - 2(n-1)} = -r$$

$$A_2H_2 = \frac{d}{(2n - (d/r)(n-1)} = r$$

Thus, the two principal points H_1 and H_2 are coincident with the centre of the sphere.

Focal Distances

$$H_1F_1 = -n_1P = -n_1\frac{f_1 f_2}{\Delta}$$

$$\Rightarrow \quad \frac{1}{H_1F_1} = -\frac{1}{n_1}\cdot\frac{\Delta}{f_1 f_2} = -\frac{1}{n_1}\cdot\frac{f_1 + f_2 + d/n_2}{f_1 f_2} = -\frac{1}{n_1}\left(\frac{1}{f_1} + \frac{1}{f_2} + \frac{d}{n_2}\cdot\frac{1}{f_1 f_2}\right). \tag{7.6.3}$$

In case the surrounding medium is air

$$f_1 = f_2 = -\frac{n-1}{r}, n_1 = 1 = n_3, n_2 = n$$

$$\therefore \quad \frac{1}{H_1F_1} = -\left(\frac{1}{f_1} + \frac{1}{f_2} + \frac{d/n}{f_1 f_2}\right) = -\left[-\frac{2(n-1)}{r} + \frac{d}{n}\left(\frac{n-1}{r}\right)^2\right]$$

$$= +\left(2 - \frac{d}{n}\cdot\frac{n-1}{r}\right)\frac{(n-1)}{r} \qquad (7.6.4)$$

Similarly,

$$\frac{1}{H_2F_2} = -\left(2 - \frac{d}{n}\cdot\frac{n-1}{r}\right)\frac{n-1}{r}. \qquad (7.6.5)$$

In case of Sphere of Glass

Here, $d = 2r$

$$\therefore \quad \frac{1}{H_1F_1} = \left(2 - \frac{2r}{n}\cdot\frac{n-1}{r}\right)\left(\frac{n-1}{r}\right) = \left\{2 - \frac{2}{n}\cdot(n-1)\right\}\frac{n-1}{r} = \frac{2(n-1)}{nr} \qquad (7.6.6)$$

$$\text{and} \quad \frac{1}{H_2F_2} = -\frac{2(n-1)}{nr}. \qquad (7.6.7)$$

Plano-convex Lens : Surrounding Medium is Air

Here, $\frac{1}{f_1} = -\frac{n_2 - n_1}{r_1} = -\frac{n-1}{r}$ and $\frac{1}{f_2} = 0$

$$\therefore \quad A_1H_1 = -\frac{f_1 d}{\left(f_1 + f_2 + \frac{d}{n}\right)n} = -\frac{(f_1/f_2)d}{\left(\frac{f_1}{f_2} + 1 + \frac{d}{nf_2}\right)n} = 0$$

$$A_2H_2 = \frac{(d/n)f_2}{\left(f_1 + f_2 + \frac{d}{n}\right)} = \frac{d/n}{1} = \frac{d}{n}$$

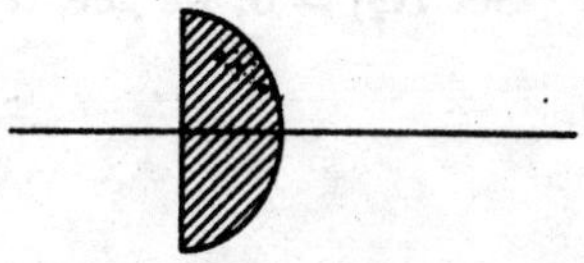

Fig 7.3

In case of a Hemisphere of Glass

Here, $d = r, A_1H_1 = 0, A_2H_2 = \frac{r}{n}$

$$\therefore \quad A_1N_1 = -\frac{n_1 f_1 d}{n_2 \Delta}, \quad \text{since} \quad n_1 = n_3 = -\frac{f_1 r}{n(f_1 + f_2 + d/n)} = 0$$

$$\text{and} \quad A_2N_2 = \frac{n_3 f_2 d}{n(f_1 + f_2 + d/n)} = \frac{r}{n}$$

Thus, the principal points and nodal points are coincident.

Focal Distances

$$\therefore \quad H_1F_1 = \frac{r}{(n-1)} \quad \text{and} \quad H_2F_2 = -\frac{r}{(n-1)}.$$

7.7 Helmholtz's Equation Connecting Transverse and Angular Magnification in case of a Thick Lens

From eqn.(7.3.4)

$$U = \text{Object distance w.r.t. } F_1 = u + \frac{S_{22}}{S_{21}} = \frac{uS_{21} + S_{22}}{S_{21}} = \frac{N_{22}}{S_{21}}$$

$$\therefore \quad N_{22} = S_{21}U, \quad V = v - \frac{S_{11}}{S_{21}} = \frac{vS_{21} - S_{11}}{S_{21}} = -\frac{N_{11}}{S_{21}}$$

$$\therefore \quad N_{11} = -VS_{21} = N'_{11}$$

Putting the value of u and v in the conjugate relation $\{(uS_{11} + S_{12}) - v(uS_{21} + S_{22})\}$ we get,

$$N_{12} = \left(S_{12} - \frac{S_{11}S_{22}}{S_{21}} - UVS_{21}\right) = N'_{12}$$

$$N_{21} = S_{21}, N'_{22} = US_{21}$$

$$\therefore \begin{bmatrix} x_2 \\ \alpha_2 \end{bmatrix} = \begin{bmatrix} -VS_{21} & \left(S_{12} - \frac{S_{11}S_{22}}{S_{21}} - UVS_{21}\right) \\ S_{21} & US_{21} \end{bmatrix} \begin{bmatrix} x_1 \\ \alpha_1 \end{bmatrix} = \begin{bmatrix} N'_{11} & N'_{12} \\ N_{21} & N'_{22} \end{bmatrix} \begin{bmatrix} x_1 \\ \alpha_1 \end{bmatrix}$$

$$\Rightarrow x_2 = N'_{11}x_1 + N'_{12}\alpha_1 \quad \text{and} \quad \alpha_2 = N_{21}x_1 + N'_{22}\alpha_1$$

Set $N'_{12} = 0$, we get, $x_2 = N'_{11}x_1 \quad \Rightarrow \quad \frac{x_2}{x_1} = N'_{11}$

Set $N_{21} = 0$, we get, $\alpha_2 = N'_{22}\alpha_1 \quad \Rightarrow \quad \frac{\alpha_2}{\alpha_1} = N'_{22}$

$$\therefore \quad \frac{x_2\alpha_2}{x_1\alpha_1} = N'_{11} \cdot N'_{22} = -VS_{21} \cdot S_{21}U$$

$$= -UVS_{21}^2 = -UV\left(\frac{\Delta}{n_3 f_1 f_2}\right)^2 = -\frac{UV}{F^2}$$

$$= -\frac{FF'}{F_2} = -\frac{F'}{F} = \frac{n_1}{n_3}$$

$$\Rightarrow \quad x_2\alpha_2 n_3 = n_1\alpha_1 x_1. \tag{7.7.1}$$

Eqn.(7.7.1) is the paraxial form of *Helmholtz's equation.*

Telescopic System

If the cardinal points of an optical system are at infinity, the optical system is said to be *telescopic system.* We have developed the expression for the distances of the cardinal points $(F_1, F_2), (H_1, H_2)$ w.r.t. the poles of the thick lens. We have,

$$A_1F_1 = -\frac{S_{22}}{S_{21}} = -\frac{(f_2 + d/n_2)n_1 f_1}{\Delta}$$

$$A_2F_2 = \frac{S_{11}}{S_{21}} = \frac{(f_1 + d/n_2)n_3 f_2}{\Delta}$$

$$A_1H_1 = (S_{11}-1)S_{22} - S_{12}S_{21} = -\frac{(n_1 d/n_2)f_1}{\Delta}$$

$$A_2H_2 = \frac{(d/n_2)n_3 f_2}{\Delta}$$

So if the cardinal point-distances are to be infinity,

$$\Delta = 0 \;\Rightarrow\; f_1 + f_2 + \frac{d}{n_2} = 0$$

$$\Rightarrow\; \frac{d}{n_2} = -(f_1+f_2) = -\left[\frac{r_1}{n_2-n_1} + \frac{r_2}{n_3-n_2}\right]$$

If $n_1 = n_3 = 1, n_2 = n$, then,

$$\frac{d}{n} = -\left[\frac{r_1}{(n-1)} + \frac{r_2}{1-n}\right] = -\left[\frac{r_1}{(n-1)} - \frac{r_2}{(n-1)}\right] = -\frac{1}{(n-1)}(r_1 - r_2).$$

So, if a thick lens is to behave as a telescopic system, d must be numerically equal to $\frac{n}{(n-1)}(r_1 - r_2)$. Thus we are now in a position to draw the ray-diagram in the formation of image by the thick lens when the incident and emergent medium is air.

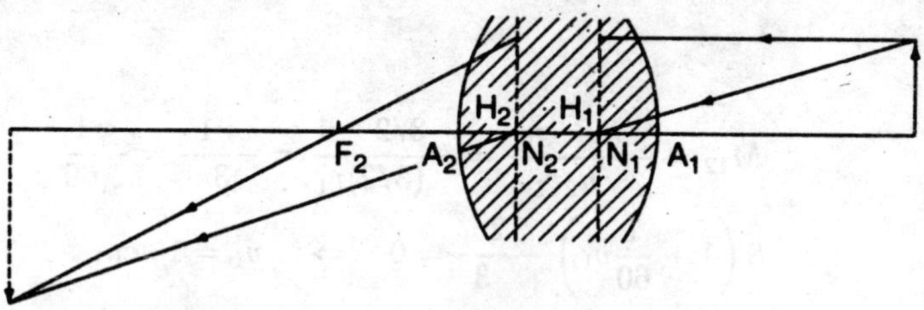

Fig 7.4

Example 1. *Consider a thick bi-convex lens in air with radii of curvature* $r_1 = 20$ *cm and* $r_2 = -10$ *cm, refractive index* $3/2$ *and thickness* 5 *cm. An object is located* 8 *cm to the right of the first surface (fig.* 7.5).

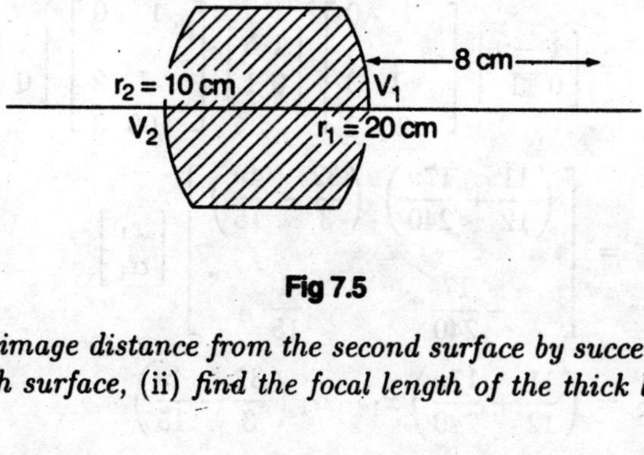

Fig 7.5

Then (i) *calculate the image distance from the second surface by successively using the single surface refraction for each surface,* (ii) *find the focal length of the thick lens.*

Solution.

$$\begin{bmatrix} x_2' \\ \alpha_2' \end{bmatrix} = \begin{bmatrix} 1 & -v_1 \\ 0 & 1 \end{bmatrix} \begin{bmatrix} 1 & 0 \\ M_{12} & \frac{n_1}{n_2} \end{bmatrix} \begin{bmatrix} 1 & u \\ 0 & 1 \end{bmatrix} \begin{bmatrix} x_1 \\ \alpha_1 \end{bmatrix}$$

$$= \begin{bmatrix} (1 - M_{12}v_1) & -v_1\frac{n_1}{n_2} \\ M_{12} & \frac{n_1}{n_2} \end{bmatrix} \begin{bmatrix} 1 & u \\ 0 & 1 \end{bmatrix} \begin{bmatrix} x_1 \\ \alpha_1 \end{bmatrix}$$

$$= \begin{bmatrix} (1 - M_{12}v_1) & u\left\{(1 - M_{12}v_1) - v_1\frac{n_1}{n_2}\right\} \\ M_{12} & uM_{12} + \frac{n_1}{n_2} \end{bmatrix} \begin{bmatrix} x_1 \\ \alpha_1 \end{bmatrix}$$

$$\therefore \quad x_1' = (1 - M_{12}v_1)x_1 + \alpha_1\left\{u(1 - M_{12}v_1) - v_1\frac{n_1}{n_2}\right\}$$

For an axial point $x_2' = x_1 = 0$, but $\alpha_1 \neq 0$

$$\therefore \quad u(1 - M_{12}v_1) - v_1\frac{n_1}{n_2} = 0.$$

Now,

$$M_{12} = -\frac{n_2 - n_1}{n_2 r_1} = -\frac{3/2 - 1}{(3/2)r_1} = -\frac{1}{3r_1} = -\frac{1}{60}$$

$$\therefore \quad 8\left(1 + \frac{1}{60}v_1\right) - \frac{2v_1}{3} = 0 \quad \Rightarrow \quad v_1 = 15 \text{ cm.}$$

The final image is defined by (x_2, α_2) given as,

$$\begin{bmatrix} x_2 \\ \alpha_2 \end{bmatrix} = \begin{bmatrix} 1 & -v \\ 0 & 1 \end{bmatrix} \begin{bmatrix} 1 & 0 \\ P_{23}' & \frac{n_2}{n_3} \end{bmatrix} \begin{bmatrix} 1 & d \\ 0 & 1 \end{bmatrix} \begin{bmatrix} 1 & 0 \\ M_{12} & \frac{2}{3} \end{bmatrix} \begin{bmatrix} 1 & u \\ 0 & 1 \end{bmatrix} \begin{bmatrix} x_1 \\ \alpha_1 \end{bmatrix}$$

$$= \begin{bmatrix} 1 & -v \\ 0 & 1 \end{bmatrix} \begin{bmatrix} 1 & 0 \\ -\frac{1}{20} & \frac{3}{2} \end{bmatrix} \begin{bmatrix} 1 & 5 \\ 0 & 1 \end{bmatrix} \begin{bmatrix} 1 & 0 \\ -\frac{1}{60} & \frac{2}{3} \end{bmatrix} \begin{bmatrix} 1 & 8 \\ 0 & 1 \end{bmatrix} \begin{bmatrix} x_1 \\ \alpha_1 \end{bmatrix}$$

$$= \begin{bmatrix} \left(\frac{11}{12} + \frac{17v}{240}\right) & \left(\frac{32}{3} - \frac{4v}{15}\right) \\ -\frac{17}{240} & \frac{4}{15} \end{bmatrix} \begin{bmatrix} x_1 \\ \alpha_1 \end{bmatrix}$$

$$\therefore \quad x_2 = \left(\frac{11}{12} + \frac{17v}{240}\right)x_1 + \alpha_1\left(\frac{32}{3} - \frac{4v}{15}\right)$$

For an axial point $x_1 = x_2 = 0$, but $\alpha_1 \neq 0$

$$\therefore \quad \frac{32}{3} - \frac{4v}{15} \quad \Rightarrow \quad v = 40 \text{ cm}.$$

The focal length of the thick lens is given by,

$$H_2F_2 = \frac{n_3 f_1 f_2}{\Delta},$$

$$f_1 = \frac{1}{n_2 M_{12}} = -\frac{1}{\frac{3}{2}\cdot\frac{1}{60}} = -40$$

$$f_2 = \frac{1}{n_3 P'_{23}} = \frac{1}{-(1/20)} = -20$$

$$\therefore \quad H_2F_2 = \frac{40 \times 20}{\Delta}$$

$$\Delta = f_1 + f_2 + \frac{d}{n_2} = -60 + \frac{5}{3/2} = -60 + \frac{10}{3} = -\frac{170}{3}$$

$$\therefore \quad H_2F_2 = -\frac{800}{(170/3)} = -\frac{800 \times 3}{170} = -\frac{240}{17} = -14.12 \text{ cm}$$

$$H_1F_1 = -\frac{n_1 f_1 f_2}{\Delta} = 14.12 \text{ cm}.$$

Example 2 *Considering a glass sphere as a thick lens, find the image formed due to refraction in a glass sphere of radius* 10 *cm, r.i.* 3/2 *of an object situated at a distance of* 120 *cm from the first refracting surface.*

Solution. The refraction matrix for a thick lens is

$$[R] = \begin{bmatrix} 1 & 0 \\ P'_{23} & \frac{n_2}{n_3} \end{bmatrix} \begin{bmatrix} 1 & 2r \\ 0 & 1 \end{bmatrix} \begin{bmatrix} 1 & 0 \\ M_{12} & \frac{n_1}{n_2} \end{bmatrix} = \begin{bmatrix} 1 & 0 \\ -\frac{1}{20} & \frac{3}{2} \end{bmatrix} \begin{bmatrix} 1 & 20 \\ 0 & 1 \end{bmatrix} \begin{bmatrix} 1 & 0 \\ -\frac{1}{30} & \frac{2}{3} \end{bmatrix}$$

$$= \begin{bmatrix} 1 & 0 \\ -\frac{1}{20} & \frac{3}{2} \end{bmatrix} \begin{bmatrix} \left(1 - \frac{2}{3}\right) & \frac{40}{3} \\ -\frac{1}{30} & \frac{2}{3} \end{bmatrix} = \begin{bmatrix} 1 & 0 \\ -\frac{1}{20} & \frac{3}{2} \end{bmatrix} \begin{bmatrix} \frac{1}{3} & \frac{40}{3} \\ -\frac{1}{30} & \frac{2}{3} \end{bmatrix}$$

$$= \begin{bmatrix} \frac{1}{3} & \frac{40}{3} \\ \left(-\frac{1}{60} - \frac{1}{20}\right) & \left(-\frac{2}{3} + 1\right) \end{bmatrix} = \begin{bmatrix} \frac{1}{3} & \frac{40}{3} \\ -\frac{1}{15} & \frac{1}{3} \end{bmatrix}$$

$$\therefore \quad \begin{bmatrix} x_2 \\ \alpha_2 \end{bmatrix} = \begin{bmatrix} 1 & -v \\ 0 & 1 \end{bmatrix} [R] \begin{bmatrix} 1 & u \\ 0 & 1 \end{bmatrix} \begin{bmatrix} x_1 \\ \alpha_1 \end{bmatrix} = \begin{bmatrix} 1 & -v \\ 0 & 1 \end{bmatrix} \begin{bmatrix} \frac{1}{3} & \frac{40}{3} \\ -\frac{1}{15} & \frac{1}{3} \end{bmatrix} \begin{bmatrix} 1 & 120 \\ 0 & 1 \end{bmatrix} \begin{bmatrix} x_1 \\ \alpha_1 \end{bmatrix}$$

$$= \begin{bmatrix} 1 & -v \\ 0 & 1 \end{bmatrix} \begin{bmatrix} \frac{1}{3} & \left(40 + \frac{40}{3}\right) \\ -\frac{1}{15} & -8 + \frac{1}{3} \end{bmatrix} \begin{bmatrix} x_1 \\ \alpha_1 \end{bmatrix} = \begin{bmatrix} 1 & -v \\ 0 & 1 \end{bmatrix} \begin{bmatrix} \frac{1}{3} & \frac{160}{3} \\ -\frac{1}{15} & -\frac{23}{3} \end{bmatrix} \begin{bmatrix} x_1 \\ \alpha_1 \end{bmatrix}$$

$$= \begin{bmatrix} \left(\frac{1}{3} + \frac{v}{15}\right) & \left(\frac{160}{3} + v \times \frac{23}{3}\right) \\ -\frac{1}{15} & -\frac{23}{3} \end{bmatrix} \begin{bmatrix} x_1 \\ \alpha_1 \end{bmatrix}$$

$$\Rightarrow \quad x_2 = \left(\frac{1}{3} + \frac{v}{15}\right) x_1 + \alpha_1 \left(\frac{160}{3} + \frac{v \times 23}{3}\right)$$

For an axial point $x_1 = x_2 = 0$, but $\alpha_1 \neq 0$

$$\therefore \quad v \times 23 = -160 \quad \Rightarrow \quad v = -\frac{160}{23}\text{ cm} \quad \text{(behind the second surface).}$$

Example 3 *A glass hemisphere of radius r and r.i. μ is treated as a lens. Rays passing through it are limited to those nearly coinciding with the axis. Find out the principal points of this lens. Prove that the focal length is given by* $f = \dfrac{r}{(\mu - 1)}$.

Solution. The whole process of transformation is represented by a series of matrix operation given as

$$\begin{bmatrix} x_2 \\ \alpha_2 \end{bmatrix} = \begin{bmatrix} 1 & -v \\ 0 & 1 \end{bmatrix} \underbrace{[R_2] \begin{bmatrix} 1 & r \\ 0 & 1 \end{bmatrix} [R_1]}_{\text{system matrix}} \begin{bmatrix} 1 & u \\ 0 & 1 \end{bmatrix} \begin{bmatrix} x_1 \\ \alpha_1 \end{bmatrix}$$

Here,

$$[R_2] = \begin{bmatrix} 1 & 0 \\ P_{23} & \frac{n_2}{n_3} \end{bmatrix} = \begin{bmatrix} 1 & 0 \\ 0 & \mu \end{bmatrix}, P_{23} = 0$$

$$[R_1] = \begin{bmatrix} 1 & 0 \\ P_{12} & \frac{n_1}{n_2} \end{bmatrix} = \begin{bmatrix} 1 & 0 \\ P_{12} & \frac{1}{\mu} \end{bmatrix}, P_{12} = -\frac{(\mu - 1)}{\mu r}$$

$$\therefore \quad \begin{bmatrix} x_2 \\ \alpha_2 \end{bmatrix} = \begin{bmatrix} 1 & -v \\ 0 & 1 \end{bmatrix} \overbrace{\begin{bmatrix} 1 & 0 \\ 0 & \mu \end{bmatrix} \begin{bmatrix} 1 & r \\ 0 & 1 \end{bmatrix} \begin{bmatrix} 1 & 0 \\ P_{12} & 1/\mu \end{bmatrix}}^{\text{system matrix}} \begin{bmatrix} 1 & u \\ 0 & 1 \end{bmatrix} \begin{bmatrix} x_1 \\ \alpha_1 \end{bmatrix}$$

$$[S] = \begin{bmatrix} 1 & 0 \\ 0 & \mu \end{bmatrix} \begin{bmatrix} 1 & r \\ 0 & 1 \end{bmatrix} \begin{bmatrix} 1 & 0 \\ P_{12} & 1/\mu \end{bmatrix} = \begin{bmatrix} 1 & 0 \\ 0 & \mu \end{bmatrix} \begin{bmatrix} (1 + rP_{12}) & r/\mu \\ P_{12} & 1/\mu \end{bmatrix}$$

$$= \begin{bmatrix} (1 + rP_{12}) & r/\mu \\ \mu P_{12} & 1 \end{bmatrix}$$

$$\therefore \quad [S] = \begin{bmatrix} S_{11} & S_{12} \\ S_{21} & S_{22} \end{bmatrix}$$

$$\therefore \quad \begin{bmatrix} x_2 \\ \alpha_2 \end{bmatrix} = \begin{bmatrix} 1 & -v \\ 0 & 1 \end{bmatrix} \begin{bmatrix} S_{11} & S_{12} \\ S_{21} & S_{22} \end{bmatrix} \begin{bmatrix} 1 & u \\ 0 & 1 \end{bmatrix} \begin{bmatrix} x_1 \\ \alpha_1 \end{bmatrix} = \begin{bmatrix} 1 & -v \\ 0 & 1 \end{bmatrix} \begin{bmatrix} S_{11} & uS_{11} + S_{12} \\ S_{21} & uS_{21} + S_{22} \end{bmatrix} \begin{bmatrix} x_1 \\ \alpha_1 \end{bmatrix}$$

$$= \begin{bmatrix} (S_{11} - vS_{21}) & (uS_{11} + S_{12}) - v(uS_{21} + S_{22}) \\ S_{21} & (uS_{21} + S_{22}) \end{bmatrix} \begin{bmatrix} x_1 \\ \alpha_1 \end{bmatrix} = \begin{bmatrix} N_{11} & N_{12} \\ N_{21} & N_{22} \end{bmatrix} \begin{bmatrix} x_1 \\ \alpha_1 \end{bmatrix}$$

7.8 The Cardinal Points of the Hemispherical Lens

Focal Points and Focal Planes

The first principal focus of the system is defined by the transformation matrix element $N_{22} = 0$

$$\therefore \quad uS_{21} + S_{22} = 0$$

$$\Rightarrow \quad u = A_1F_1 = -\frac{S_{22}}{S_{21}} = -\frac{1}{\mu P_{12}} = \frac{1}{\left[\mu \times \dfrac{(\mu-1)}{\mu r}\right]} = +\frac{r}{(\mu-1)}$$

The second principal focus is defined by $N_{11} = 0$

$$\therefore \quad S_{11} - vS_{21} = 0$$

$$\Rightarrow \quad v = A_2F_2 = \frac{S_{11}}{S_{21}} = \frac{1 + rP_{12}}{\mu P_{12}} = \frac{1 - (\mu-1)/\mu}{-(\mu-1)/r}$$

$$\Rightarrow \quad v = -\frac{1}{\mu} \times \frac{r}{(\mu-1)} = -\frac{r}{\mu(\mu-1)}$$

Principal Points and Principal Planes

The principal points are defined by,

$$N_{12} = 0 \quad \text{and} \quad N_{11} = 1.$$

Now, $N_{11} = S_{11} - vS_{21} = 1$

$$\Rightarrow \quad v = A_2H_2 = \frac{S_{11} - 1}{S_{21}} = \frac{1 + rP_{12} - 1}{\mu P_{12}}$$

$$= \frac{-\dfrac{(\mu-1)}{\mu}}{-\dfrac{(\mu-1)}{r}} = \frac{(\mu-1)}{\mu} \times \frac{r}{(\mu-1)} = \frac{r}{\mu} \quad \text{(second point)}.$$

The corresponding u value gives the first principal plane w.r.t. the curved surface pole A_1. Thus, from $N_{12} = 0$, we get,

$$uS_{11} + S_{12} = v(uS_{21} + S_{22}) \Rightarrow u(S_{11} - vS_{21}) = vS_{22} - S_{12}$$

$$\Rightarrow u = \frac{vS_{22} - S_{12}}{S_{11} - vS_{21}}$$

$$\therefore \quad u = A_1H_1 = \frac{(r/\mu)S_{22} - S_{12}}{S_{11} - (r/\mu)S_{21}} = \frac{(r/\mu) - (r/\mu)}{1 + rP_{12} - (r/\mu)\mu P_{12}} = 0$$

Thus, the first principal plane coincides with the pole A_1.

Focal Length of the Lens

The first principal focal length of the lens is the distance of the first principal focal point F_1 w.r.t. the first principal plane. Hence,

$$f_1 = A_1F_1 - A_1H_1 = \frac{r}{(\mu - 1)}.$$

Similarly, the second principal focal length

$$f_2 = A_2F_2 - A_2H_2 = -\frac{r}{\mu(\mu-1)} - \frac{r}{\mu} = -\frac{r}{\mu}\left(\frac{1}{\mu-1}+1\right) = -\frac{r}{(\mu-1)}.$$

Thus the two focal lengths are equal in magnitude.

Example 4 *What are the cardinal points of a thick lens? Find out the positions of these points mathematically with reference to the optic centre of the thick lens.* **[C.U. Hons. '75]**

Solution. We have found out mathematically the expressions for the cardinal points of a thick lens w.r.t. the principal planes or the poles. We follow the same scheme but here the cardinal points are to be measured w.r.t. the optical centre of the thick lens. What is the position of the optical centre? Consider the figure (7.6) as shown. Let C_1 and C_2 represent the centres of curvatures of the first and second surfaces with radii of curvature r_1 and r_2 respectively. A ray incident at some point P passes through the lens as PQ and emerges out along QR such that the incident ray is parallel to the emergent ray. The ray PQ extended back intersects the principal axis at O, called the *optical centre.* The tangent planes at P and Q being parallel, the radii C_1P and C_2Q are also parallel.

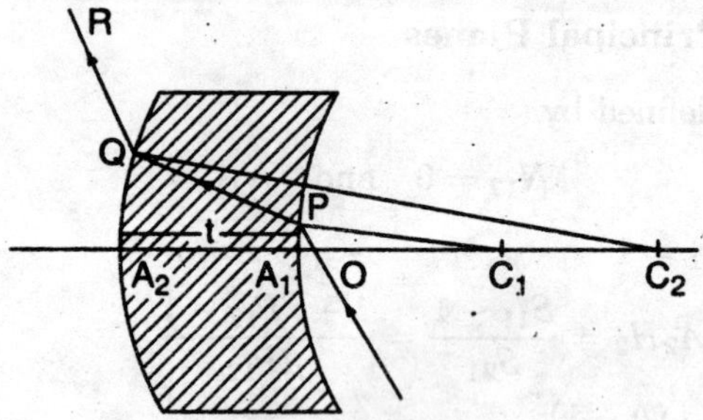

Fig 7.6

From similar Δs OPC_1 and OQC_2, we get,

$$\frac{PC_1}{QC_2} = \frac{OC_1}{OC_2}$$

$$\Rightarrow \frac{r_1}{r_2} = \frac{A_1C_1 - A_1O}{A_2C_2 - A_2O} = \frac{r_1 - A_1O}{r_2 - A_2O} = \frac{A_1O}{A_2O}$$

$$\Rightarrow 1 - \frac{r_1}{r_2} = 1 - \frac{A_1O}{A_2O} \quad \Rightarrow \quad \frac{r_2 - r_1}{r_2} = \frac{A_2O - A_1O}{A_2O} = \frac{d}{A_2O}$$

$$\Rightarrow A_2O = \frac{r_2 d}{(r_2 - r_1)}$$

Similarly,

$$A_1O = \frac{r_1 d}{(r_2 - r_1)}.$$

Thus, the distances of the optical centre w.r.t. the poles A_1 and A_2 are

$$t_1 = \frac{r_1 d}{(r_2 - r_1)} \quad \text{and} \quad t_2 = \frac{r_2 d}{(r_2 - r_1)}.$$

So, taking the expression derived for a thick lens w.r.t. the poles A_1 and A_2, we can express them w.r.t. the optical centre O. Thus,

$$A_1F_1 = \frac{\Delta}{n_3 f_1 f_2} \quad \text{defines the position of } F_1 \text{ w.r.t. the pole } A_1.$$

$\therefore$ The distance of F_1 from the optical centre

$$x_1 = A_1F_1 - A_1O = -\frac{\frac{n_1}{n_2}\left(\frac{d}{n_3 f_2} + \frac{n_2}{n_3}\right) n_3 f_1 f_2}{\Delta} - t_1 = -\frac{\left(f_2 + \frac{d}{n_2}\right) n_1 f_1}{\Delta} - t_1.$$

The distance of F_2 from the optical centre

$$x_2 = A_2F_2 - t_2 = \frac{\left(f_1 + \frac{d}{n_2}\right) n_3 f_2}{\Delta} - t_2, \text{ etc.}$$

Conjugate Relation w.r.t. the Principal Points (H_1, H_2)

We have already established conjugate relation, i.e., relation between object distance and image distance w.r.t. the focal points F_1 and F_2. The focal lengths, first and second, of the thick lens are given by,

$$F' = H_1F_1 = -n_1\frac{f_1 f_2}{\Delta} = -n_1 P, \quad F = H_2F_2 = \frac{n_3 f_1 f_2}{\Delta} = n_3 P$$

w.r.t. the focal points (F_1, F_2), the object distance U and image distance V are related as

$$UV = FF'.$$

Now, let U_1 and V_1 represent object distance and image distance w.r.t. H_1 and H_2.

$$\begin{aligned}
\therefore\ & U_1 = H_1P = H_1F_1 + F_1P = F' + U \\
& V_1 = H_2Q = H_2F_2 + F_2Q = F + V \\
\therefore\ & U = (U_1 - F'), V = (V_1 - F) \\
\therefore\ & UV = FF' = (U_1 - F')(V_1 - F) = U_1V_1 - U_1F - V_1F' + FF' \\
\Rightarrow\ & FF' = U_1V_1 - U_1F - V_1F' + FF' \\
\Rightarrow\ & U_1F + V_1F' = U_1V_1 \\
\Rightarrow\ & \frac{F'}{U_1} + \frac{F}{V_1} = 1 \quad \Rightarrow \quad \frac{-n_1P}{U_1} + \frac{n_3P}{V_1} = 1 \\
\Rightarrow\ & \frac{n_3}{V_1} - \frac{n_1}{U_1} = +\frac{1}{P} \quad \text{where,} \quad P = \frac{f_1 f_2}{\Delta}.
\end{aligned}$$

This shows that the formula for a thick lens is identical with a thin lens formula if measured with respect to the principal planes.

When the thick lens is surrounded by air, $n_1 = n_3 = 1$, we have,

$$\frac{1}{V_1} - \frac{1}{U_1} = \frac{1}{P}.$$

If real image is formed, $V_1 = -\text{ve}$, P is also $-\text{ve}$,

$$\therefore \quad \frac{1}{V_1} + \frac{1}{U_1} = \frac{1}{P}.$$

Since magnification is unity,

$$\frac{V_1}{U_1} = 1 \quad \Rightarrow \quad V_1 = U_1 \quad \therefore \quad U_1 = 2P, \quad V_1 = 2P.$$

The distance between the object and image is given by,

$$L = (U_1 + V_1) + H_2H_1 \quad \Rightarrow \quad L - H_2H_1 = (U_1 + V_1) = 4P.$$

Exercise

1. Find the expressions for the cardinal points of an optical system represented in fig. (7.7).

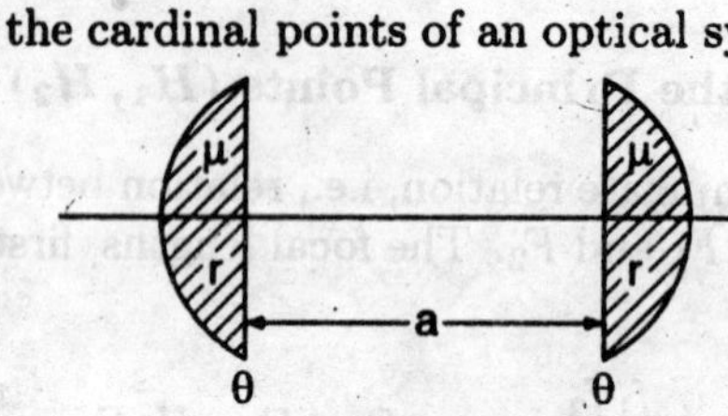

Fig 7.7

2. Explain what do you mean by cardinal points of a thick lens.
3. Show that in case of a transparent sphere surrounded by air, the principal points, the nodal points and optical centre are all coincident with the centre. Calculate also the focal lengths of the sphere.
4. On one side of a spherical glass lens of radius 1 cm and r.i. (3/2), the medium is air and on the other side it is water (r.i. 4/3). Find the distances of the cardinal points of the system. Calculate also the positions and magnification of the image when the object is situated in air (a) 4 cm and (b) 1.5 cm from the centre of the sphere.

   ($A_1H_1 = -0.4$ cm, $A_2H_2 = 1.6$ cm, $A_1F_1 = 14$ cm
   $A_2F_2 = -0.8$ cm, $A_1N_1 = -1$ cm, $A_2N_2 = 1$ cm)
5. Find the equivalent focal length of two lenses separated by a distance.
6. What do you mean by a telescopic system? What condition is to be satisfied so that an optical system consisting of two thin lenses of focal lengths f and f' separated by a distance a in air behaves as a telescopic system?

7. Define unit points and nodal points of an optical system. Show that when the system is surrounded by air, the principal points coincide with the nodal points. Indicate how this fact may be utilised to determine the focal length of a convergent system of two thin convex lenses separated by a distance in air.

8. Show that when the size of the real image produced by a thick lens surrounded by air is equal to the size of the object, the distance between the object and its image, minus the distance between the principal planes will be equal to four times the focal length of the lens.

9. A thin biconvex lens of r.i. 1.6 and a thin diverging lens of r.i. 1.5 are cemented together to form a doublet. Two of the surfaces have a radius of 50.0 cm and the other two of 40.0 cm; the cemented surfaces have curvatures of the same value. Find the focal length of the doublet if (a) it is in air (b) the source, screen, lens are immersed in water ($\mu = 4/3$).

(−40.8 cm, −119 cm)

10. A thin glass lens of focal length f and r.i. 3/2 is placed under a liquid of r.i. 4/3. Find the geometrical focus of a pencil of parallel rays directly incident upon the lens.

11. Obtain an expression for the focal length of a thick lens of thickness d, radii of curvature R_1 and R_2 of the curved surfaces, refractive index n placed in air. Find the principal planes from the respective vertices.

[BU 1993]

12. A double convex lens (thick) of thickness 1 cm, radii of the curved surfaces 20 cm and 40 cm, and r.i. 1.5 is placed in air. Find the image distance for the object 30 cm from one of the vertices.

[BU 1993]

13. Explain the meaning of cardinal points of a thick lens. Show that when a thick lens is placed in a homogeneous medium, the nodal points coincide with unit points of the lens.

[CU 1997]

14. (a) Obtain an expression for the focal length of a thick lens of thickness d, radii of curvature R_1 and R_2 of the curved surfaces, r.i. n, placed in air.

    (b) Find the principal planes from the respective vertices.

    (c) A double convex thick lens of thickness 1 cm, radii of the curved surfaces 30 cm and 40 cm, and r.i. 1.5 is placed in air. Find the image distance for an object 40 cm from one of the vertices.

15. (a) Obtain the translation and refraction matrices. Hence calculate the system matrix for a thick lens.

    (b) Show that the plane surface of a plano-convex or plano-concave lens does not contribute to the system matrix.

    (c) Show that for such a lens one of the principal planes will always be a tangent to the curved surface.

[BU 1994]

16. Obtain an expression for the focal length of a thick lens using translation as well as refraction matrices. Show that it reduces to that of a thin lens when the thickness $d \to 0$.

    Find the positions of the principal planes from the respective vertices.

17. What are the cardinal points of a thick lens? Show that in case of a thick lens in a homogeneous medium, the nodal points coincide with the unit points of the lens.

[CU 1997]

18. What is your idea about the cardinal points of a thick lens? Show that when a thick lens is surrounded by a homogeneous medium, the nodal points coincide with the unit points of the lens. Deduce the system matrix for a thick lens and hence derive the formula for the focal length of a thin converging lens under paraxial approximation.

Chapter 8

Aberrations

8.1 Introduction

All through the formulation of Geometrical Optics in the previous chapters we considered only paraxial rays for which there is point to point correspondence between object space and image space. We have evolved suitable transformation matrix effecting transformation between the two spaces as a result of optical processes. By the term 'paraxial rays' we mean those rays which make small angles with the *principal* axis of the optical system, such small angle that sine of the angle can be replaced by the angle itself or those rays which are incident on the reflecting or refracting surface very near to the pole. This consideration is sometimes referred to as *paraxial ray theory or first-order theory* because only the first-order term in the expansion of $\sin\theta$ is retained. We know that, $\sin\theta = \theta - \dfrac{\theta^3}{3!} + \dfrac{\theta^5}{5!} - \cdots$. So when we consider the paraxial rays, $\sin\theta$ is replaced by the first-order in θ in the expansion. The advantage of this consideration is that here we will get for a punctual object a unique punctual image and hence, there is no aberration or defect associated with the image.

But it is also easy to see that a paraxial beam which really consists of only a limited number of rays cannot produce a bright image. To improve the brightness of the optical image more rays are to be involved, this means that we have to consider non-paraxial rays also. As a result of this deviation from the first-order theory, i.e., as a result of consideration of more and more oblique and non-paraxial rays, the image loses its distinctive sharpness and the departures from the first-order theory, called *aberrations* can be calculated by considering the second, third-order term in the sine expansion. The inclusion of the more oblique rays lead to five correction terms leading to five *Seidal aberrations* for light of single colour or frequency. They are — (i) *spherical aberration,* (ii) *coma,* (iii) *astigmatism,* (iv) *distortion,* (v) *curvature.* These five members belong to monochromatic aberration family; of course all members are not equally important for consideration in optical instruments.

There is another family of aberration, called *chromatic aberration* which occurs when we use a polychromatic light. Even if an optical system be corrected for the five monochromatic aberrations due to departure from the first-order theory, chromatic defect will creep in due to variation of r.i. of the optical element like a lens for different frequencies. In the present chapter we will discuss the two families of aberrations, their qualitative as well as quantitative aspects with the remedial measure. It is to remember here that in the third-order theory point to point correspondence between the object space and image space fails and since the transformation is non-linear, the matrix formulation of the geometrical optics which we followed so far will not apply in the present problem.

Spherical Aberration

8.2 Introduction

If we consider a large aperture reflecting or refracting medium, then paraxial as well as non-paraxial rays are involved in the image formation. In fig.8.1(a) we have shown a large aperture reflecting surface AB and two marginal rays incident parallel to the axis at A and B denote the non-paraxial ray limit in the incident beam; similarly in fig.8.1(b), a refracting medium of large aperture, a lens, is shown where non-paraxial ray limit in the incident parallel beam is shown. The marginal non-partial rays come to a point F_m, called *marginal ray focus.* The paraxial rays, very few in number, incident in the immediate vicinity of the pole converge to a point of focus F_p, called *paraxial ray focus.* All the other rays beyond the paraxial region will be brought to different points of focus in between the marginal ray focus F_m and paraxial ray focus F_p. Thus we do not get a punctual image of a punctual object; we get, instead, every point in the image formed by marginal rays surrounded by a diffuse disc of light formed by the paraxial rays. Thus the image is blurred. In fig.8.1(a), if we place a screen perpendicular to the axis and passing through the paraxial ray focus F_p, we would receive a circular patch of light whose central region is brighter compared to its outskirts; if the screen is placed passing through the marginal ray focus F_m, the image received would be a circular patch whose outskirts are brighter compared to the central region. So, neither of these two images gives the faithful representation of the object in the image space.

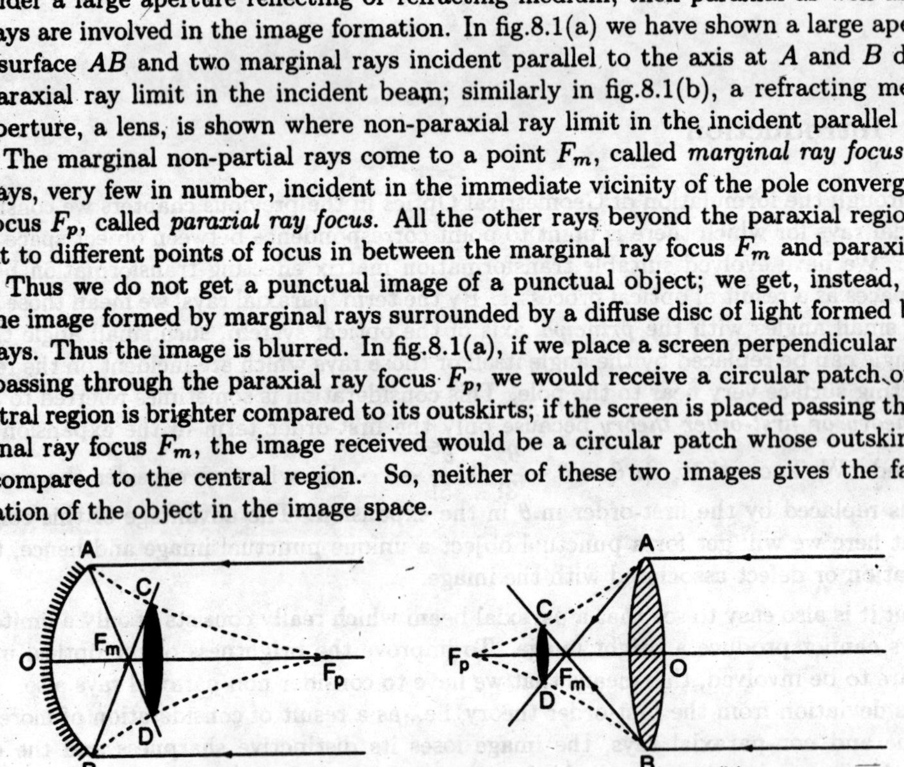

Fig 8.1

If two consecutive reflecting elements of a reflector (fig.8.1a) or refractor (fig.8.1b) be considered, the reflected or refracted rays from or through them intersect one another at a point before they reach the axis. Such points of intersections of the reflected or refracted rays when joined give rise to a continuous curve like ACF_pDB with its *cusp* at F_p. This curve is called *caustic curve.* Now, if we hold a screen perpendicular to the axis and passing through the points of intersection C and D of the reflected or refracted rays BC and AD with the caustic curve, we would receive a circular patch of almost uniform illumination all through, called the *circle of least confusion.*

The distance of separation between F_p and F_m measures the Longitudinal Spherical Aberration (LSA) while the radius (or diameter) measures the Transverse Spherical Aberration (TSA). In all optical instruments, therefore, arrangements should be incorporated so as to minimise the separation between F_p and F_m.

8.3 Longitudinal Spherical Aberration

A Spherical Reflector

Consider a reflector of linear aperture AB, semi-aperture of length h (see fig.8.2). The ray PA is incident at the marginal point of the aperture at an angle ϕ, gets reflected making the same angle ϕ with AC. The image corresponding to the marginal ray is formed at I_m; the paraxial ray image is formed at I_p. The difference between the I_p and I_m is referred to as the longitudinal spherical aberration.

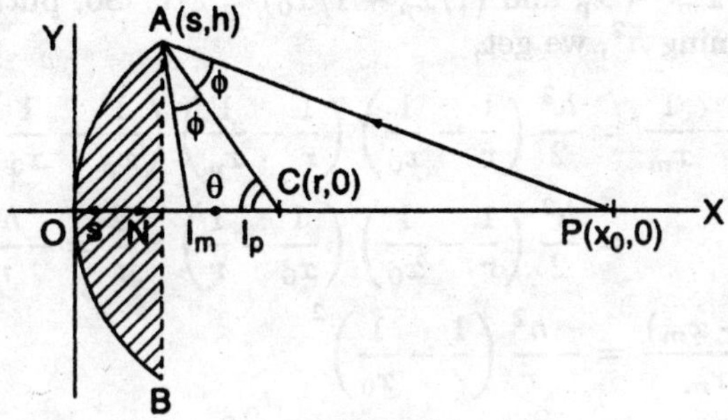

Fig 8.2

Set up the co-ordinate frame with X-axis along the principal axis and Y-axis perpendicular to it; O, the pole is at the origin.

$$AP^2 = (x_0 - s)^2 + h^2, NC = (r - s), AN = h$$

$$\therefore h^2 + (r-s)^2 = r^2 \quad \Rightarrow \quad (r-s)^2 = (r^2 - h^2) \quad \Rightarrow \quad r - s = \sqrt{(r^2 - h^2)}$$

$$\Rightarrow \quad s = r - \sqrt{(r^2 - h^2)} = r - \left(1 - \frac{1}{2}\cdot\frac{h^2}{r^2}\right) r = \frac{1}{2}\frac{h^2}{r}$$

$$\therefore \quad AP^2 = \left(x_0 - \frac{1}{2}\cdot\frac{h^2}{r}\right)^2 + h^2 = x_0^2\left(1 - \frac{1}{2}\cdot\frac{h^2}{x_0 r}\right)^2 + h^2 = x_0^2\left(1 - \frac{h^2}{x_0 r}\right) + h^2$$

$$= x_0^2\left(1 - \frac{h^2}{x_0 r} + \frac{h^2}{x_0^2}\right) = x_0^2\left[1 - \frac{h^2}{x_0}\left(\frac{1}{r} - \frac{1}{x_0}\right)\right]$$

$$\Rightarrow \quad AP = x_0\left[1 - \frac{1}{2}\cdot\frac{h^2}{x_0}\left(\frac{1}{r} - \frac{1}{x_0}\right)\right] \tag{8.3.1}$$

Similarly,

$$AI_m = x_m\left[1 - \frac{1}{2}\cdot\frac{h^2}{x_m}\left(\frac{1}{r} - \frac{1}{x_m}\right)\right] \tag{8.3.2}$$

Now, since AC is the internal bisector of the angle I_mAP, we get,

$$\frac{I_mA}{AP} = \frac{CI_m}{CP} = \frac{r - x_m}{(x_0 - r)}$$

$$\Rightarrow \quad (x_0 - r)x_m\left[1 - \frac{1}{2}\cdot\frac{h^2}{x_m}\left(\frac{1}{r} - \frac{1}{x_m}\right)\right] = (r - x_m)x_0\left[1 - \frac{1}{2}\cdot\frac{h^2}{x_0}\left(\frac{1}{r} - \frac{1}{x_0}\right)\right]$$

Dividing by $x_m x_0 r$, we get,

$$\left(\frac{1}{r}-\frac{1}{x_0}\right)\left[1-\frac{1}{2}\cdot\frac{h^2}{x_m}\left(\frac{1}{r}-\frac{1}{x_m}\right)\right]=\left(\frac{1}{x_m}-\frac{1}{r}\right)\left[1-\frac{1}{2}\cdot\frac{h^2}{x_0}\left(\frac{1}{r}-\frac{1}{x_0}\right)\right]$$

$$\Rightarrow \left(\frac{1}{r}-\frac{1}{x_0}\right)-\frac{1}{2}\cdot\frac{h^2}{x_m}\left(\frac{1}{r}-\frac{1}{x_m}\right)\left(\frac{1}{r}-\frac{1}{x_0}\right)=\left(\frac{1}{x_m}-\frac{1}{r}\right)-\frac{1}{2}\cdot\frac{h^2}{x_0}\left(\frac{1}{r}-\frac{1}{x_0}\right)\left(\frac{1}{x_m}-\frac{1}{r}\right)$$

$$\Rightarrow \frac{2}{r}-\left(\frac{1}{x_0}+\frac{1}{x_m}\right)=\frac{h^2}{2}\left(\frac{1}{r}-\frac{1}{x_m}\right)\left(\frac{1}{r}-\frac{1}{x_0}\right)\left(\frac{1}{x_m}+\frac{1}{x_0}\right)$$

For paraxial image $x_m \to x_p$ and $(1/x_p + 1/x_0) = 2/r$. So, putting $x_m = x_p$ in the right-hand expression containing h^2, we get,

$$\therefore \quad \frac{1}{x_p}-\frac{1}{x_m}=\frac{h^2}{2}\left(\frac{1}{r}-\frac{1}{x_0}\right)\left(\frac{1}{r}-\frac{1}{x_p}\right)\left(\frac{1}{x_p}+\frac{1}{x_0}\right)$$

$$=\frac{h^2}{2}\left(\frac{1}{r}-\frac{1}{x_0}\right)\left(\frac{1}{x_0}-\frac{1}{r}\right)\times\frac{2}{r}=-\frac{h^2}{r}\left(\frac{1}{r}-\frac{1}{x_0}\right)^2$$

$$\Rightarrow -\frac{(x_p-x_m)}{x_p x_m}=-\frac{h^2}{r}\left(\frac{1}{r}-\frac{1}{x_0}\right)^2$$

$$\Rightarrow (x_p-x_m)=\text{LSA}=\frac{h^2}{r}\left(\frac{1}{r}-\frac{1}{x_0}\right)^2 x_p x_m$$

$$=\frac{h^2}{r}x_p^2\left(\frac{1}{r}-\frac{1}{x_0}\right)^2=\frac{h^2}{r}\cdot\frac{\left(\frac{1}{r}-\frac{1}{x_0}\right)^2}{\left(\frac{2}{r}-\frac{1}{x_0}\right)^2} \qquad (8.3.3)$$

In case of parallel incident beam striking the reflector at the marginal point A, the marginal image distance $= F_m$, paraxial image distance $= F_p$.

$$\therefore \quad F_p - F_m = (h^2/r)\cdot(1/r^2)\cdot(r^2/4) = (h^2/4r), \text{ because here } x_0 = \infty.$$

Aberration due to Refraction at a Spherical Surface

Fig.(8.3) shows the case of refraction at the spherical surface separating the two media of r.i.'s μ_1 and $\mu_2 (\mu_2 > \mu_1)$. The marginal ray image is formed at $I_m, (x_m, 0)$ and I_p is the paraxial image at $(x_p, 0)$.

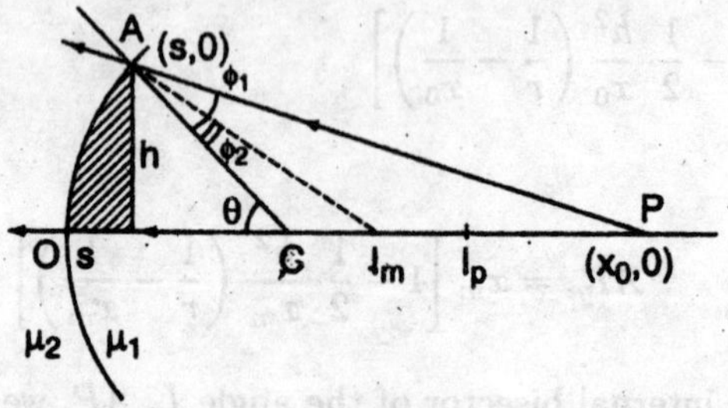

Fig 8.3

As usual as calculated earlier,

$$AP = x_0\left[1-\frac{1}{2}\cdot\frac{h^2}{x_0}\left(\frac{1}{r}-\frac{1}{x_0}\right)\right], \quad AI_m = x_m\left[1-\frac{1}{2}\cdot\frac{h^2}{x_m}\left(\frac{1}{r}-\frac{1}{x_m}\right)\right].$$

Considering refraction at A, we get, $\mu_1 \sin\phi_1 = \mu_2 \sin\phi_2$.

Now,

$$\frac{CP}{\sin\phi_1} = \frac{AP}{\sin(180-\theta)} = \frac{AP}{\sin\theta}, \quad \frac{CI_m}{\sin\phi_2} = \frac{AI_m}{\sin(180-\theta)} = \frac{AI_m}{\sin\theta}.$$

$$\therefore \quad \mu_1 \sin\phi_1 = \mu_2 \sin\phi_2 \Rightarrow \mu_1 \frac{CP}{AP} = \mu_2 \times \frac{CI_m}{AI_m}$$

$$\Rightarrow \mu_1 \frac{(x_0 - r)}{AP} = \mu_2 \frac{(x_m - r)}{AI_m}$$

$$\Rightarrow \mu_1 (x_o - r) AI_m = \mu_2 \times (x_m - r) AP$$

$$\Rightarrow \mu_1 (x_0 - r) x_m \left[1 - \frac{1}{2}\cdot\frac{h^2}{x_m}\left(\frac{1}{r} - \frac{1}{x_m}\right)\right]$$

$$= \mu_2 (x_m - r) x_0 \left[1 - \frac{1}{2}\cdot\frac{h^2}{x_0}\left(\frac{1}{r} - \frac{1}{x_0}\right)\right]$$

Dividing by $x_0 x_m r$, we get,

$$\mu_1\left(\frac{1}{r} - \frac{1}{x_0}\right)\left[1 - \frac{1}{2}\frac{h^2}{x_m}\left(\frac{1}{r} - \frac{1}{x_m}\right)\right] = \mu_2\left(\frac{1}{r} - \frac{1}{x_m}\right)\left[1 - \frac{1}{2}\frac{h^2}{x_0}\left(\frac{1}{r} - \frac{1}{x_0}\right)\right]$$

$$\Rightarrow \frac{\mu_2}{x_m} - \frac{\mu_1}{x_0} - \left(\frac{\mu_2 - \mu_1}{r}\right) = \mu_1\left(\frac{1}{r} - \frac{1}{x_0}\right)\cdot\frac{h^2}{2x_m}\left(\frac{1}{r} - \frac{1}{x_m}\right) - \mu_2\left(\frac{1}{r} - \frac{1}{x_m}\right)\cdot\frac{h^2}{2x_0}\left(\frac{1}{r} - \frac{1}{x_0}\right)$$

$$\Rightarrow \frac{\mu_2}{x_m} - \frac{\mu_1}{x_0} - \left(\frac{\mu_2 - \mu_1}{r}\right) = \frac{h^2}{2}\left(\frac{1}{r} - \frac{1}{x_0}\right)\left(\frac{1}{r} - \frac{1}{x_m}\right)\left(\frac{\mu_1}{x_m} - \frac{\mu_2}{x_0}\right).$$

For paraxial ray image,

$$\frac{\mu_2}{x_p} - \frac{\mu_1}{x_0} = \frac{(\mu_2 - \mu_1)}{r} \quad \text{and} \quad x_p \to x_m.$$

$$\therefore \quad \frac{\mu_2}{x_m} - \frac{\mu_2}{x_p} = \frac{h^2}{2}\left(\frac{1}{r} - \frac{1}{x_0}\right)\left(\frac{1}{r} - \frac{1}{x_p}\right)\left(\frac{\mu_1}{x_p} - \frac{\mu_2}{x_0}\right)$$

Dividing throughout by μ_2, we get, putting $\mu_2/\mu_1 = \mu$,

$$\frac{1}{x_m} - \frac{1}{x_p} = \frac{h^2}{2}\left(\frac{1}{r} - \frac{1}{x_0}\right)\left(\frac{1}{r} - \frac{1}{x_p}\right)\left(\frac{1}{\mu x_p} - \frac{1}{x_0}\right).$$

Again,

$$\frac{\mu_2}{x_p} - \frac{\mu_1}{x_0} = \frac{\mu_2 - \mu_1}{r} \quad \Rightarrow \quad \frac{1}{x_p} - \frac{1}{\mu x_0} = \frac{1 - \frac{1}{\mu}}{r} = \frac{\mu - 1}{\mu r}$$

$$\therefore \quad \frac{1}{x_m} - \frac{1}{x_p} = \frac{h^2}{2}\left(\frac{1}{r} - \frac{1}{x_0}\right)\left(-\frac{1}{\mu x_0} + \frac{1}{\mu r}\right)\left(\frac{1}{r} - \frac{\mu + 1}{\mu^2}\right)\frac{\mu - 1}{\mu^2}$$

$$= +\frac{h^2}{2}\left(\frac{1}{r} - \frac{1}{x_0}\right)\frac{1}{\mu}\left(\frac{1}{r} - \frac{1}{x_0}\right)\cdot\frac{\mu - 1}{\mu^2}\left(\frac{1}{r} - \frac{\mu + 1}{x_0}\right)$$

$$= \frac{h^2}{2}\left(\frac{1}{r} - \frac{1}{x_0}\right)^2\cdot\frac{\mu - 1}{\mu^3}\left(\frac{1}{r} - \frac{\mu + 1}{x_0}\right)$$

$$\Rightarrow \quad (x_p - x_m) = \text{LSA} = \left(\frac{1}{r} - \frac{1}{x_0}\right)^2 \frac{\mu - 1}{\mu^3}\left(\frac{1}{r} - \frac{\mu+1}{x_0}\right)\frac{h^2 x_p^2}{2}$$

$$= \frac{\mu - 1}{\mu} \cdot \frac{\left(\frac{1}{r} - \frac{1}{x_0}\right)^2 \left(\frac{1}{r} - \frac{\mu+1}{x_0}\right)\frac{h^2}{2}}{\left(\frac{\mu}{x_p}\right)^2}$$

$$= \frac{h^2}{2}\frac{(\mu - 1)}{\mu} \cdot \frac{\left(\frac{1}{r} - \frac{1}{x_0}\right)^2 \left(\frac{1}{r} - \frac{\mu+1}{x_0}\right)}{\left(\frac{\mu-1}{r} + \frac{1}{x_0}\right)^2} \tag{8.3.4}$$

If the incident beam be parallel to the principal axis, then $x_0 = \infty, x_p = F_p, x_m = F_m$.

$$\therefore \quad F_p - F_m = \frac{h^2}{2}\left(\frac{\mu - 1}{\mu}\right) \cdot \frac{1}{r^3} \times \frac{r^2}{(\mu - 1)^2} = \frac{h^2}{2\mu r(\mu - 1)}. \tag{8.3.5}$$

In case the aperture is small, $h \to 0, F_p - F_m = 0$, so the LSA disappears.

Longitudinal Spherical Aberration for a Thin Lens

A lens consists of two curved surfaces of radii of curvature r_1 and r_2. Due to refraction at first surface we can write,

$$\frac{\mu}{x'_m} - \frac{\mu_1}{x_0} = \frac{\mu - \mu_1}{r_1} + \mu_1\left(\frac{1}{r_1} - \frac{1}{x_0}\right) \cdot \frac{h^2}{2x'_m}\left(\frac{1}{r_1} - \frac{1}{x'_m}\right)$$
$$-\mu\left(\frac{1}{r_1} - \frac{1}{x'_m}\right)\frac{h^2}{2x_0}\left(\frac{1}{r_1} - \frac{1}{x_0}\right).$$

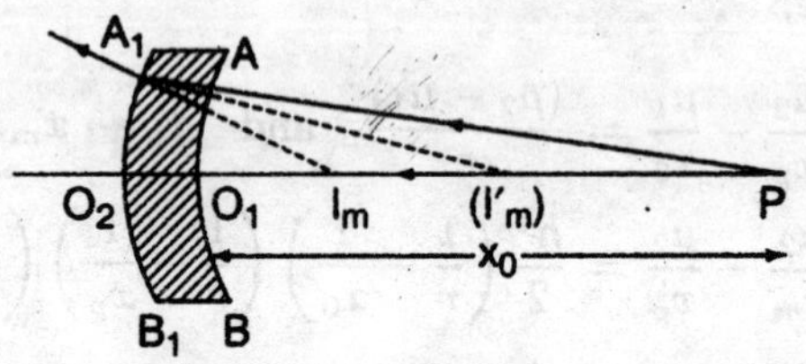

Fig 8.4

For the refraction at second surface,

$$\frac{\mu_2}{x_m} - \frac{\mu}{x'_m} = \frac{\mu_2 - \mu}{r_2} + \mu\left(\frac{1}{r_2} - \frac{1}{x'_m}\right)\frac{h^2}{2x_m}\left(\frac{1}{r_2} - \frac{1}{x_m}\right)$$
$$-\mu_2\left(\frac{1}{r_2} - \frac{1}{x_m}\right)\frac{h^2}{2x'_m}\left(\frac{1}{r_2} - \frac{1}{x'_m}\right).$$

Adding we get,

$$\frac{\mu_2}{x_m} - \frac{\mu_1}{x_0} = \left(\frac{\mu_2 - \mu}{r_2} + \frac{\mu - \mu_1}{r_1}\right) + \mu_1\left(\frac{1}{r_1} - \frac{1}{x_0}\right)\frac{h^2}{2x'_m}\left(\frac{1}{r_1} - \frac{1}{x'_m}\right)$$
$$-\mu\left(\frac{1}{r_1} - \frac{1}{x'_m}\right)\frac{h^2}{2x_0}\left(\frac{1}{r_1} - \frac{1}{x_0}\right) + \mu\left(\frac{1}{r_2} - \frac{1}{x'_m}\right)\frac{h^2}{2x_m}\left(\frac{1}{r_2} - \frac{1}{x_m}\right)$$
$$-\mu_2\left(\frac{1}{r_2} - \frac{1}{x_m}\right)\frac{h^2}{2x'_m}\left(\frac{1}{r_2} - \frac{1}{x'_m}\right).$$

If the lens is placed in air, $\mu_2 = \mu_1 = 1$, we get,

$$\frac{1}{x_m} - \frac{1}{x_0} = (\mu - 1)\left(\frac{1}{r_1} - \frac{1}{r_2}\right) + \left(\frac{1}{r_1} - \frac{1}{x_0}\right)\frac{h^2}{2x'_m}\left(\frac{1}{r_1} - \frac{1}{x'_m}\right)$$
$$-\mu\left(\frac{1}{r_1} - \frac{1}{x'_m}\right)\frac{h^2}{2x_0}\left(\frac{1}{r_1} - \frac{1}{x_0}\right) + \mu\left(\frac{1}{r_2} - \frac{1}{x'_m}\right)\frac{h^2}{2x_m}\left(\frac{1}{r_2} - \frac{1}{x_m}\right)$$
$$-\left(\frac{1}{r_2} - \frac{1}{x_m}\right)\frac{h^2}{2x'_m}\left(\frac{1}{r_2} - \frac{1}{x'_m}\right).$$

Now, $x'_m \simeq x'_p$ in the term containing h^2

$x'_m \simeq x'_p$ in the term containing h^2

Also,

$$\left.\begin{aligned}\frac{\mu}{x'_p} - \frac{1}{x_0} &= \frac{\mu - 1}{r_1} \\ \frac{1}{x_p} - \frac{\mu}{x'_p} &= \frac{1-\mu}{r_2}\end{aligned}\right\} \quad \begin{aligned}&\Rightarrow \frac{1}{x_p} - \frac{1}{x_0} = (\mu - 1)\left(\frac{1}{r_1} - \frac{1}{r_2}\right) \\ &\Rightarrow \frac{1}{x_p} = \frac{1}{x_0} + (\mu - 1)\left(\frac{1}{r_1} - \frac{1}{r_2}\right)\end{aligned}$$

$$\therefore \left(\frac{1}{r_1} - \frac{1}{x_0}\right)\frac{h^2}{2x'_m}\left(\frac{1}{r_1} - \frac{1}{x'_m}\right) - \mu\left(\frac{1}{r_1} - \frac{1}{x'_m}\right)\frac{h^2}{2x_0}\left(\frac{1}{r_1} - \frac{1}{x_0}\right)$$
$$= \frac{h^2}{2}\left(\frac{1}{r_1} - \frac{1}{x_0}\right)\left[\frac{1}{x'_p}\left(\frac{1}{r_1} - \frac{1}{x'_p}\right) - \frac{\mu}{x_0}\left(\frac{1}{r_1} - \frac{1}{x'_p}\right)\right]$$
$$= \frac{h^2}{2}\left(\frac{1}{r_1} - \frac{1}{x_0}\right)\left(\frac{1}{r_1} - \frac{1}{x'_p}\right)\left(\frac{1}{x'_p} - \frac{\mu}{x_0}\right)$$
$$= \frac{h^2}{2}\cdot\frac{1}{\mu}\left(\frac{1}{r_1} - \frac{1}{x_0}\right)^2\cdot\frac{(\mu - 1)}{\mu}\left(\frac{1}{r_1} - \frac{\mu + 1}{x_0}\right)$$
$$= \frac{h^2}{2}\cdot\frac{(\mu - 1)}{\mu^2}\left(\frac{1}{r_1} - \frac{1}{x_0}\right)^2\left(\frac{1}{r_1} - \frac{\mu + 1}{x_0}\right).$$

Similarly,

$$\frac{h^2}{2}\left(\frac{1}{r_2} - \frac{1}{x_m}\right)\left[\frac{\mu}{x_m}\left(\frac{1}{r_2} - \frac{1}{x'_m}\right) - \frac{1}{x'_m}\left(\frac{1}{r_2} - \frac{1}{x'_m}\right)\right]$$
$$= \frac{h^2}{2}\left(\frac{1}{r_2} - \frac{1}{x_m}\right)\left(\frac{1}{r_2} - \frac{1}{x'_m}\right)\left(\frac{\mu}{x_m} - \frac{1}{x'_m}\right)$$
$$= \frac{h^2}{2}\left(\frac{1}{r_2} - \frac{1}{x_p}\right)\left(\frac{1}{r_2} - \frac{1}{x'_p}\right)\left(\frac{\mu}{x_p} - \frac{1}{x'_p}\right).$$

Now, $\frac{1}{x_p} - \frac{\mu}{x'_p} = \frac{1-\mu}{r_2}$

$$\left(\frac{1}{x_p} - \frac{1-\mu}{r_2}\right) = \frac{\mu}{x'_p} \quad \Rightarrow \quad \frac{1}{x'_p} = \frac{1}{\mu}\left(\frac{1}{x_p} - \frac{1-\mu}{r_2}\right)$$

$$\therefore \frac{1}{r_2} - \frac{1}{x'_p} = \frac{1}{r_2} - \frac{1}{\mu x_p} + \frac{1-\mu}{\mu r_2} = \frac{1}{\mu r_2} - \frac{1}{\mu x_p} = \frac{1}{\mu}\left(\frac{1}{r_2} - \frac{1}{x_p}\right)$$

$$\frac{\mu}{x_p} - \frac{1}{x_p'} = \frac{\mu}{x_p} - \frac{1}{\mu}\left(\frac{1}{x_p} - \frac{1-\mu}{r_2}\right) = \frac{\mu}{x_p} - \frac{1}{\mu x_p} + \frac{1-\mu}{\mu r_2}$$

$$= \frac{\mu^2 r_2 - r_2 - x_p(\mu-1)}{\mu r_2 x_p} = \frac{(\mu^2-1)r_2 - x_p(\mu-1)}{\mu r_2 x_p} = \frac{(\mu-1)}{\mu}\left(\frac{\mu+1}{x_p} - \frac{1}{r_2}\right)$$

$$\therefore \frac{h^2}{2}\left(\frac{1}{r_2} - \frac{1}{x_m}\right)\left(\frac{1}{r_2} - \frac{1}{x_m'}\right)\left(\frac{\mu}{x_m} - \frac{1}{x_m'}\right) = \frac{h^2}{2}\left(\frac{1}{r_2} - \frac{1}{x_p}\right)^2 \cdot \frac{1}{\mu} \cdot \frac{\mu-1}{\mu}\left(\frac{\mu+1}{x_p} - \frac{1}{r_2}\right)$$

$$= -\frac{h^2}{2}\left(\frac{1}{r_2} - \frac{1}{x_p}\right)^2 \cdot \frac{\mu-1}{\mu^2}\left(\frac{1}{r_2} - \frac{\mu+1}{x_p}\right)$$

Thus, the longitudinal spherical aberration (LSA) is given by,

$$\frac{1}{x_m} - \frac{1}{x_0} = (\mu-1)\left(\frac{1}{r_1} - \frac{1}{r_2}\right) + \frac{h^2}{2} \cdot \frac{\mu-1}{\mu^2}$$
$$\left[\left(\frac{1}{r_1} - \frac{1}{x_0}\right)^2\left(\frac{1}{r_1} - \frac{\mu+1}{x_0}\right) - \left(\frac{1}{r_2} - \frac{\mu+1}{x_p}\right)\left(\frac{1}{r_2} - \frac{1}{x_p}\right)^2\right].$$

For paraxial image formed by the lens,

$$\frac{1}{x_p} - \frac{1}{x_0} = (\mu-1)\left(\frac{1}{r_1} - \frac{1}{r_2}\right)$$

$$\therefore \quad \frac{1}{x_m} - \frac{1}{x_p} = \frac{h^2}{2} \cdot \frac{(\mu-1)}{\mu^2}\left[\left(\frac{1}{r_1} - \frac{1}{x_0}\right)^2\left(\frac{1}{r_1} - \frac{\mu+1}{x_0}\right) - \left(\frac{1}{r_2} - \frac{1}{x_p}\right)^2\left(\frac{1}{r_2} - \frac{\mu+1}{x_p}\right)\right]$$

$$\Rightarrow \quad (x_p - x_m) = \text{LSA}$$

$$= \frac{h^2}{2}x_p^2 \cdot \frac{(\mu-1)}{\mu^2}\left[\left(\frac{1}{r_1} - \frac{1}{x_0}\right)^2\left(\frac{1}{r_1} - \frac{\mu+1}{x_0}\right) - \left(\frac{1}{r_2} - \frac{1}{x_p}\right)^2\left(\frac{1}{r_2} - \frac{\mu+1}{x_p}\right)\right] \tag{8.3.6}$$

From eqn.(8.3.6) we can get the eqn. in terms of x_0 in place of x_p, using the relation

$$\frac{1}{x_p} - \frac{1}{x_0} = (\mu-1)\left(\frac{1}{r_1} - \frac{1}{r_2}\right) \quad \Rightarrow \quad \frac{1}{x_p} = \frac{1}{x_0} + (\mu-1)\left(\frac{1}{r_1} - \frac{1}{r_2}\right)$$

$$\therefore \quad \left(\frac{1}{r_2} - \frac{1}{x_p}\right)^2 = \left[\frac{1}{r_2} - \frac{1}{x_0} - (\mu-1)\left(\frac{1}{r_1} - \frac{1}{r_2}\right)\right]^2 = \left(\frac{1}{x_0} + \frac{\mu-1}{r_1} - \frac{\mu}{r_2}\right)^2$$

$$\frac{1}{r_2} - \frac{\mu+1}{x_p} = \frac{1}{r_2} - \frac{\mu+1}{x_0} - (\mu^2-1)\left(\frac{1}{r_1} - \frac{1}{r_2}\right) = \left(\frac{\mu^2}{r_2} - \frac{\mu+1}{x_0} - \frac{\mu^2-1}{r_1}\right).$$

$\therefore$ Eqn.(8.3.6) gives us,

$$\frac{1}{x_m} - \frac{1}{x_p} = \frac{h^2}{2}\frac{\mu-1}{\mu^2}\left[\left(\frac{1}{r_1} - \frac{1}{x_0}\right)^2\left(\frac{1}{r_1} - \frac{\mu+1}{x_0}\right)\right.$$
$$\left. - \left(\frac{1}{x_0} + \frac{\mu-1}{r_1} - \frac{\mu}{r_2}\right)^2 \times \left(\frac{\mu^2}{r_2} - \frac{\mu+1}{x_0} - \frac{\mu^2-1}{r_1}\right)\right]$$

If we now introduce

$$\sigma_s = \frac{r_2 + r_1}{r_2 - r_1}, \quad \text{called } \textit{shape factor}$$

$$\text{and} \quad \sigma_p = \frac{x_p + x_0}{x_p - x_0}, \quad \text{called } \textit{position factor}$$

We can easily get

$$\frac{1}{r_2} = \frac{\sigma_s - 1}{2(\mu - 1)f}, \frac{1}{r_1} = \frac{\sigma_s + 1}{2(\mu - 1)f}$$

$$\frac{1}{x_0} = -\frac{\sigma_p + 1}{2f}, \frac{1}{x_p} = \frac{1 - \sigma_p}{2f}.$$

Introducing these parameters in the above equation, we would get

$$\frac{1}{x_m} - \frac{1}{x_p} = \frac{h^2}{f^3}(a\sigma_s^2 + b\sigma_s\sigma_p + c\sigma_p^2 + d) \tag{8.3.7}$$

where $a = \frac{\mu + 2}{8\mu(\mu - 1)^2}, b = \frac{\mu + 1}{2\mu(\mu - 1)}, c = \frac{3\mu + 2}{8\mu}$ and $d = \frac{\mu^2}{8(\mu - 1)^2}$.

So spherical aberration of a thin lens depends on both σ_s, the shape factor and σ_p, the position factor. We write

$$Y = a\sigma_s^2 + b\sigma_s\sigma_p + c\sigma_p^2 + d \tag{8.3.8}$$

If $Y = 0$, spherical aberration is eliminated. But it is evident from eqn.(8.3.8) that the graph of Y vs σ_s for a particular value of σ_p is a parabola, hence $Y \neq 0$ for any value of σ_s.

But Y attains its minimum value when

$$\frac{dY}{d\sigma_s} = 0 \quad \Rightarrow \quad 2a\sigma_s + b\sigma_p = 0 \quad \Rightarrow \quad \sigma_s = -\frac{b}{2a}\sigma_p = \sigma_m \tag{8.3.9}$$

Eqn.(8.3.9) implies minimum spherical aberration in case of a thin lens.

$$\therefore \quad Y_{\min} = A\sigma_m^2 - 2A\sigma_m + 4A^2 = a\sigma_m^2 - 2a\sigma_m^2 + \frac{4a^2c}{b^2}\sigma_m^2 + d = \left(\frac{4a^2c}{b^2} - a\right)\sigma_m^2 + d \tag{8.3.10}$$

So, from eqn.(8.3.10) it is also clear that $(Y)_{\min} \neq 0$ for any real value of σ_m. So we conclude that for a single thin lens spherical aberration cannot be eliminated. But it can be minimised only, for the condition in equation (8.3.9).

For parallel rays, $x_0 = \infty$, so from eqn.(8.3.9) we get

$$\sigma_m = \frac{r_2 + r_1}{r_2 - r_1} = -\frac{b}{2a}(\sigma_p)_{x_0 = \infty}.$$

Now,

$$(\sigma_p)_{x_0=\infty} = \frac{x_p + x_0}{x_p - x_0} = -1$$

$$\therefore \quad \frac{r_2 + r_1}{r_2 - r_1} = \frac{b}{2a} \quad \Rightarrow \quad \frac{r_1}{r_2} = \frac{b - 2a}{b + 2a} = \frac{2\mu^2 - \mu - 4}{\mu(2\mu + 1)}. \tag{8.3.11}$$

This is a very important formula and it has been applied to minimise spherical aberration in actual practice. Also,

$$\frac{r_2 + r_1}{r_2 - r_1} = \sigma_s$$

So, for minimum spherical aberration for parallel rays

$$\sigma_s = \frac{r_2 + r_1}{r_2 - r_1} = \frac{b}{2a} = \frac{\mu + 1}{2\mu(\mu - 1)} \cdot \frac{8\mu(\mu - 1)^2}{2(\mu + 2)} = \frac{2(\mu^2 - 1)}{(\mu + 2)}$$

For parallel rays $(\sigma_p)_{x_0=\infty} = -1$

$$\therefore \quad \sigma_s = -\frac{2(\mu^2 - 1)}{(\mu + 2)}(\sigma_p)_{x_0=\infty}.$$

Remedial Measures

In optical instruments because of large aperture reflecting or refracting surfaces being used, spherical aberration is inevitably associated with the image which becomes blurred. So optical devices should be so constructed that the spherical aberration is removed or minimised. The remedial measures discussed below, some have practical applications, some are of theoretical interest only. The methods are —

(a) By using a Crossed Lens

A crossed lens is a double convex lens whose radii of curvature bear a ratio of 1 : 6. The minimisation of spherical aberration by a lens is done with reference to the minimisation formula for the spherical aberration which we have derived earlier.

We have shown that for minimum spherical aberration for a lens of given focal length and for parallel incidence, the ratio of the two radii of curvature of the lens should satisfy

$$\frac{r_1}{r_2} = \frac{2\mu^2 - \mu - 4}{\mu(2\mu + 1)}.$$

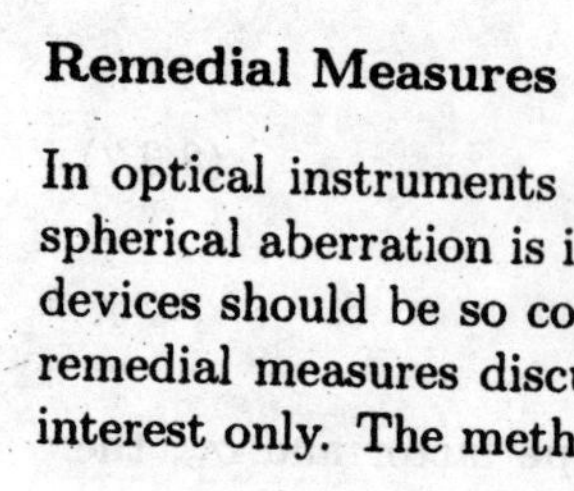

Fig 8.5

Now, if we consider a plano-convex lens with the curved surface facing the light, we have, $r_1 = -\text{ve}$, $r_2 = \infty$, hence,

$$\frac{2\mu^2 - \mu - 4}{\mu(2\mu + 1)} = 0 \quad \Rightarrow \quad 2\mu^2 - \mu - 4 = 0.$$

Solving this equation we get $\mu = 1.6861$. So a plano-convex lens can be used provided the r.i. of the material is given by the above value. It is difficult to make a material having the prescribed r.i. value but it is not impossible. The advantage is that the radius of curvature of the curved surface is of no concern. If, on the other hand, the flat surface is made to face the incident light, $r_1 = \infty$, $r_2 = +\text{ve}$, and so, for minimum spherical aberration,

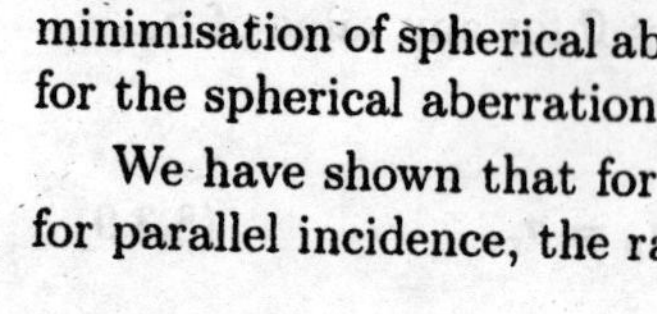

Fig 8.6

$$\frac{2\mu^2 - \mu - 4}{\mu(2\mu + 1)} = \infty \quad \Rightarrow \quad \mu(2\mu + 1) = 0 \quad \Rightarrow \quad \mu = 0, \mu = -\frac{1}{2}$$

both of which are impossible. So, by using a plano-convex lens of r.i. value 1.6861 with its curved surface facing the light, spherical aberration may be minimised. But here there lies the difficulty of getting suitable r.i. material, so taking ordinary glass (r.i. $\frac{3}{2}$) and using the minimisation formula, we get

$$\frac{r_1}{r_2} = \frac{2\mu^2 - \mu - 4}{\mu(2\mu + 1)} = -\frac{1}{6} \tag{8.3.12}$$

So, adjusting the radii of curvature to be in the ratio 1 : 6, a double convex lens will minimise spherical aberration when placed with its more curved face facing the direction of light as shown in fig.(8.6), such a lens is called a *crossed lens.*

(b) By Dividing the Total Deviation amongst a Number of Refracting Surfaces

Sometimes instead of using a single lens we have to use a pair of lenses as in the construction of eyepices; in this case the minimisation formula cannot be made to be satisfied by each of the doublet. In this case, the distance of separation of the two lenses is kept such that the deviation produced by each lens is same.

We know that each lens produces certain angular deviation and the deviation produced is dependent on the angle of incidence which in turn also controls the spherical aberration. So, if the total deviation of a composite system be equally divided amongst a number of refracting elements, deviation will be a minimum and hence the spherical aberration also will be minimum.

Thus, consider two convex lenses of focal lengths f_1 and f_2, kept separated by a distance equal to a as shown in fig.(8.7).

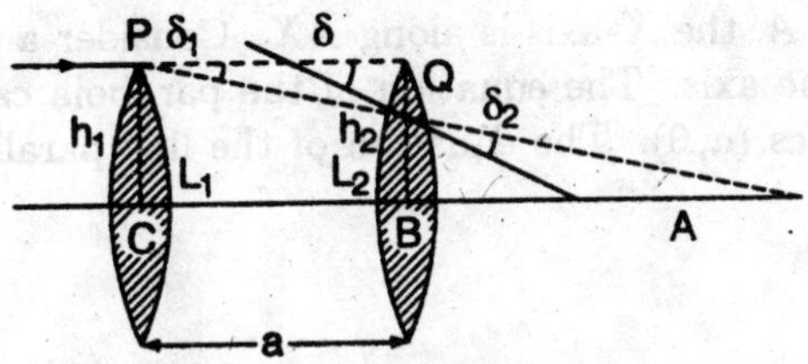

Fig 8.7

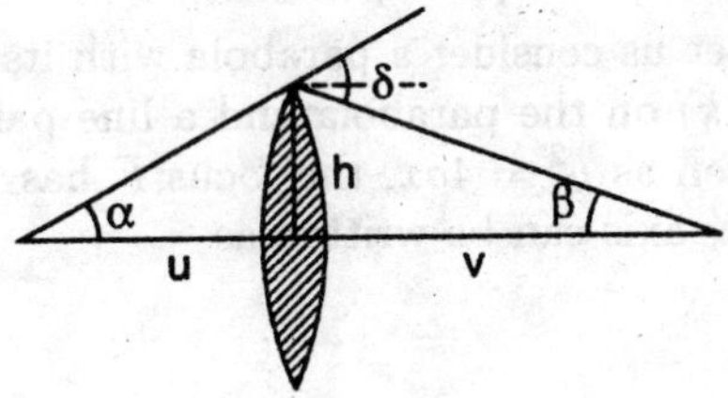

Fig 8.8

The total deviation is $\delta = \delta_1 + \delta_2$.

Now, the individual deviation δ_1 or δ_2 can be found w.r.t. fig. (8.8).

$$\delta = \alpha + \beta = \frac{h}{v} - \frac{h}{u} = h\left(\frac{1}{v} - \frac{1}{u}\right) = \frac{h}{f}. \quad \text{So, } \delta_1 = \frac{h_1}{f_1}, \delta_2 = \frac{h_2}{f_2}.$$

From similar Δs AQB and APC,

$$\frac{h_2}{h_1} = \frac{f_1 - a}{f_1} \quad \Rightarrow \quad h_2 = \frac{h_1(f_1 - a)}{f_1} \quad \therefore \quad \delta_2 = \frac{h_2}{f_2} = \frac{h_1}{f_1 f_2}(f_1 - a).$$

So, for minimum spherical aberration, $\delta_1 = \delta_2$

$$\therefore \quad \frac{h_1}{f_1} = \frac{h_1(f_1 - a)}{f_1 f_2} \quad \Rightarrow \quad a = (f_1 - f_2). \tag{8.3.13}$$

So, if the two lenses are kept separated by a distance $a = (f_1 - f_2)$, the deviation due to individual lens is equal, hence the spherical aberration will be minimum. This conclusion has been practically utilised in the construction of Huygens' eyepiece where the distance of separation is $a = 3f - f = 2f$, where f and $3f$ are the focal lengths of eye-lens and objective lens. We shall see later that this arrangement also removes the chromatic aberration.

(c) By using Aplanatic Surface

Aplanatic surfaces are those surfaces which behave w.r.t. two points in such a way that if an object is placed at one of the two points, the image is always formed at the other, irrespective of whether paraxial or non-paraxial rays are considered in the formation of the image. Such surfaces are termed as aplanatic surfaces and the two points with respect to which they behave as aplanatic surfaces are called *aplanatic points.*

A variety of surfaces may act as aplanatic surfaces belonging to both reflecting and refracting type, and we consider those which find practical use.

Paraboloidal Mirror

If a parabola be rotated about its axis, the surface generated is called ***paraboloidal surface.*** This surface has two points, one is the focus F of the parabola, the other is a point at infinity with respect to which the paraboloidal reflector would behave as aplanatic surface. The rays incident parallel to the principal axis will pass through F after reflection whatever may be the points of incidence or rays diverging from F after reflection from the surface will be rendered parallel to the principal axis.

Let us consider a parabola with its vertex at A, the X-axis is along AX. Consider a point $P(h, k)$ on the parabola and a line parallel to the axis. The equation of the parabola can be written as $y^2 = 4ax$, the focus F has co-ordinates $(a, 0)$. The equation of the line parallel to the X-axis can be written as

$$y = c_1.$$

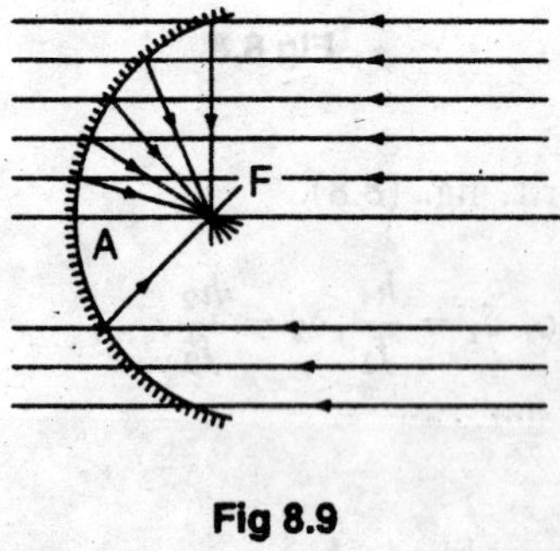

Fig 8.9

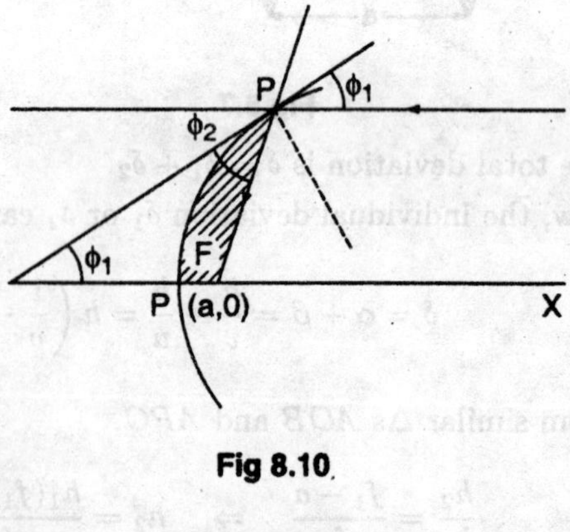

Fig 8.10

Since it passes through (h, k), we have $c_1 = k$.

$\therefore \quad y = k$ is the parallel line equation.

The point also lies on the parabola, hence, $k^2 = 4ah$.

$$\therefore \quad h = \frac{k^2}{4a}.$$

So the co-ordinates of $P\left(\frac{k^2}{4a}, k\right)$.

The eqn. of the tangent to the parabola at (h, k) is,

$$yk = 2a(x+h) \quad \Rightarrow \quad y = \frac{2a}{k}x + \frac{2ah}{k}, \text{ its slope } m_1 = \frac{2a}{k}.$$

The eqn. of the parallel line is given as $y = k, m_1' = 0$

$$\therefore \tan\phi_1 = \frac{m_1 - m_1'}{1+m_1m_1'} = \frac{2a/k}{1} = \frac{2a}{k}.$$

The eqn. to the line joining two points P and F, is,

$$\frac{y-k}{k-0} = \frac{x-h}{h-a} \quad \Rightarrow \quad y - k = \frac{k}{(h-a)}x - \frac{hk}{(h-a)} \quad \Rightarrow \quad y = \frac{k}{(h-a)}x - \frac{ka}{(h-a)}.$$

So its slope $m_2 = k/(h-a)$. So the angle between the tangent and the line joining P and F is given by,

$$\tan\phi_2 = \frac{m_2 - m_1}{1+m_1m_2} = \frac{\frac{k}{h-a} - \frac{2a}{k}}{1+\frac{2ak}{k(h-a)}} = \frac{2a}{k}, \text{ putting the values of } h.$$

So, $\phi_1 = \phi_2$, i.e., the glancing angle of incidence is equal to glancing angle of reflection. So we conclude that for a parabolic mirror, all rays incident parallel to the principal axis, pass through the focus F irrespective of the points of incidence. Hence paraxial rays and non-paraxial rays form the image at the same point and there is no spherical aberration. Thus a paraboloidal surface is an aplanatic surface with respect to two points, the focus of the parabola and a point at infinity.

Ellipsoidal Mirror

If we consider an ellipsoidal mirror, it will behave as an aplanatic surface with resppect to two points F_1 and F_2, called the *focal points of the ellipse* such that if the source be placed at F_1, the rays after reflection pass through F_2 and vice versa. The property of the ellipse demands that for any point P on the ellipse, the sum of the distances $F_1P + PF_2 =$ constant, i.e.,

$$r_1 + r_2 = \text{constant}.$$

Since light takes shortest route in going from F_1 to F_2 via reflection from any point of the reflector, so if for the path F_1 to P and then from P to F_2 is a minimum and any other path via reflection at the reflector will be equal to this minimum, we can say all rays starting from F_1 after reflection should pass through F_2. Hence, F_1 and F_2 are aplanatic points.

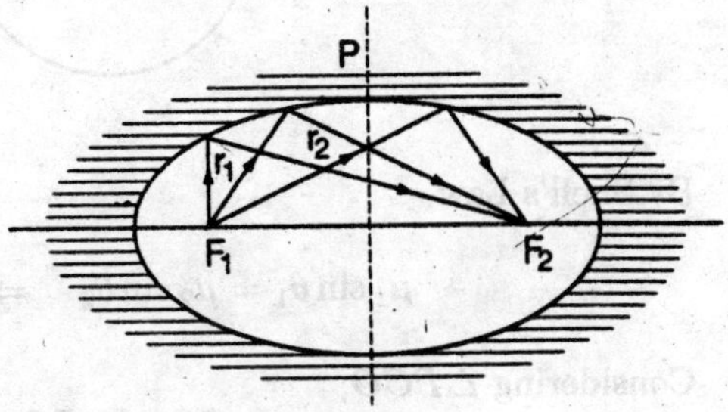

Fig 8.11

Mathematically, let $\vec{F_1P} = \bar{r}_1$, $\vec{F_2P} = \bar{r}_2$ be the vectors drawn respectively from foci F_1 and F_2 to the point P (fig.8.11a) on the ellipse and $\vec{T}$ is a unit vector tangent to the ellipse at P.

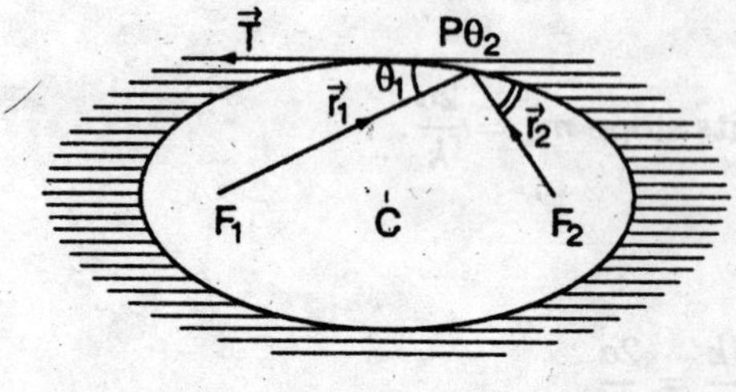

Fig 8.12

The equation of the ellipse is given by,

$$r_1 + r_2 = \rho \text{ (constant)}$$

$\vec{\nabla}(r_1+r_2)$ represents a vector normal to the ellipse.

So, $\vec{\nabla}(r_1+r_2)\cdot\vec{T} = 0$, since $\vec{T}$ is a tangential unit vector.

$$\Rightarrow \vec{\nabla}r_1\cdot\vec{T} = -\vec{\nabla}r_2\cdot\vec{T} = \vec{\nabla}r_2\cdot(-\vec{T}).$$

Now, the co-ordinates of F_1 are $(-ae, 0)$ and those of P are (x, y), hence

$$\vec{F_1P} = \bar{r}_1 = i(x+ae) + jy$$

$$r_1^2 = (x+ae)^2 + y^2 \quad\Rightarrow\quad r_1 = \sqrt{(x+ae)^2+y^2}$$

$$\therefore \quad \vec{\nabla}r_1 = \left(i\frac{\partial}{\partial x} + j\frac{\partial}{\partial y}\right)[(x+ae)^2+y^2]^{\frac{1}{2}}$$

$$= i\frac{(x+ae)}{\sqrt{(x+ae)^2+y^2}} + j\frac{y}{\sqrt{(x+ae)^2+y^2}} = \frac{\bar{r}_1}{r_1} = \text{unit vector along } \bar{r}_1.$$

So, from the above eqn. we get,

$$\cos\,(180-\theta_1) = \cos\,(180-\theta_2) \quad\Rightarrow\quad \theta_1 = \theta_2.$$

It proves that if P be a point of incidence of a ray from a source at F_1, it will pass through F_2 after reflection.

Aplanatic Points of a Glass-sphere

A glass-sphere can behave as an aplanatic surface w.r.t. two points, one inside the glass-sphere, say O, the other outside, say I, such that all rays from O after refraction by the glass-sphere appear to come from I, whatever may be the points of incidence of the rays.

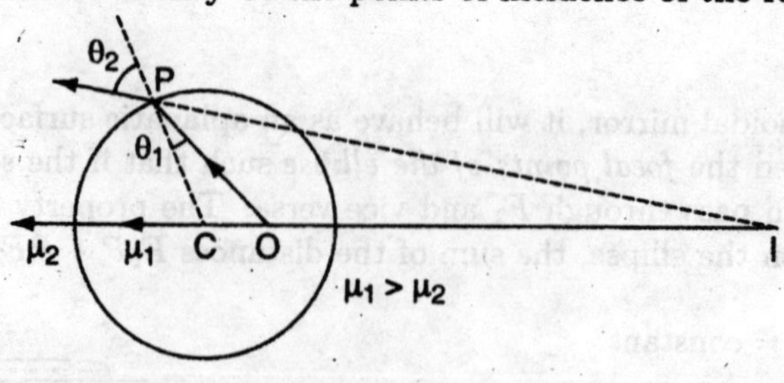

Fig 8.13

By Snell's Law,

$$\mu_1 \sin\theta_1 = \mu_2 \sin\theta_2 \quad\Rightarrow\quad \frac{\sin\theta_1}{\sin\theta_2} = \frac{\mu_2}{\mu_1} = \mu_2 = \mu \text{ (say)}$$

Considering ΔPCO,

$$\frac{CO}{\sin\theta_1} = \frac{PO}{\sin\phi} \quad\Rightarrow\quad \frac{u}{\sin\theta_1} = \frac{PO}{\sin\phi}$$

Now, $PO^2 = CO^2 + CP^2 - 2CO \cdot CP \cos\phi = u^2 + r^2 - 2ur\cos\phi$

$$\therefore \frac{u^2}{\sin^2\theta_1} = \frac{PO^2}{\sin^2\phi} = \frac{u^2 + r^2 - 2ur\cos\phi}{\sin^2\phi} \Rightarrow \frac{\sin^2\phi}{\sin^2\theta_1} = \frac{u^2 + r^2 - 2ur\cos\phi}{u^2}$$

Again, considering ΔCPI,

$$\frac{CI}{\sin\theta_2} = \frac{PI}{\sin\phi} \Rightarrow \frac{v}{\sin\theta_2} = \frac{PI}{\sin\phi} \Rightarrow \frac{v^2}{\sin^2\theta_2} = \frac{PI^2}{\sin^2\phi}$$

$$\Rightarrow \frac{\sin^2\phi}{\sin^2\theta_2} = \frac{PI^2}{v^2} = \frac{v^2 + r^2 - 2vr\cos\phi}{v^2}$$

$$\therefore \frac{\sin^2\theta_1}{\sin^2\theta_2} = \frac{u^2}{v^2} \cdot \frac{v^2 + r^2 - 2vr\cos\phi}{u^2 + r^2 - 2ur\cos\phi} = \mu^2$$

$$\Rightarrow \mu^2 \frac{(u^2 + r^2 - 2ur\cos\phi)}{u^2} = \frac{v^2 + r^2 - 2vr\cos\phi}{v^2}$$

$$\Rightarrow \mu^2\left(1 + \frac{r^2}{u^2} - 2\frac{r}{u}\cos\phi\right) = \left(1 + \frac{r^2}{v^2} - \frac{2r}{v}\cos\phi\right)$$

$$\Rightarrow \mu^2\left(1 + \frac{r^2}{u^2}\right) - \left(1 + \frac{r^2}{v^2}\right) = \left(2\mu^2\frac{r}{u} - \frac{2r}{v}\right)\cos\phi$$

In order that the relation is independent of ϕ, we must have,

$$\frac{2\mu^2 r}{u} - \frac{2r}{v} = 0 \Rightarrow \frac{\mu^2}{u} = \frac{1}{v} \Rightarrow \mu^2 v = v$$

$$\therefore (\mu^2 - 1) + \frac{\mu^2 r^2}{u^2} - \frac{r^2}{v^2} = 0 \Rightarrow (\mu^2 - 1) + \frac{\mu^2 r^2}{\mu^4 v^2} - \frac{r^2}{v^2} = 0$$

$$\Rightarrow (\mu^2 - 1) + \frac{r^2}{v^2}\left(\frac{1}{\mu^2} - 1\right) = 0 \Rightarrow (\mu^2 - 1) - \frac{(\mu^2 - 1)}{\mu^2}\frac{r^2}{v^2} = 0$$

$$\Rightarrow (\mu^2 - 1)\left(1 - \frac{r^2}{\mu^2 v^2}\right) = 0 \Rightarrow 1 - \frac{r^2}{\mu^2 v^2} = 0$$

$$\Rightarrow v^2 = \frac{r^2}{\mu^2} \Rightarrow v = \frac{r}{\mu} = \frac{r}{\mu_2/\mu_1} = \frac{\mu_1 r}{\mu_2}$$

The corresponding object distance is given by,

$$u = \mu^2 v = \mu^2 \cdot \frac{r}{\mu} = \mu r = \frac{\mu}{\mu_1} r \tag{8.3.14}$$

Now, since in the diagram drawn $\mu_1 > \mu_2, u < r, v > r$, so for an object at a point O at a distance $u = \frac{\mu_2}{\mu_1} r$, the image is formed at the other point I such that $CI = v = \frac{\mu_1}{\mu_2} r$, irrespective of the nature of rays.

This property of the glass-sphere has been used in the construction of high power objective in the oil-immersion type of microscope.

The construction of the objective is shown in fig.(8.13). A glass-hemispherical lens with its centre C is placed in contact with a drop of cedar oil (whose refractive index is same as that of glass) such that the glass-hemisphere in presence of cedar oil is equivalent to a complete sphere of glass and thus allows us to place an object at O, such that $u = \frac{\mu_2}{\mu_1} r = \frac{r}{\mu_1}$, the image is formed at $v = \mu_1 r / \mu_2 = \mu_1 r$ by all rays diverging from the object O. A concavo-convex lens L is so adjusted in position that the image I_1 is at the centre of curvature of the concave surface, so the rays from I_1 pass into the lens undeviated and fall on the convex surface. The radius of curvature of this convex surface is such that I_1 is one of the aplanatic points. Rays appearing to come from I_1 after refraction at the convex surface form the virtual image I_2 at the second aplanatic point at I_2, thus the spherical aberration is altogether removed.

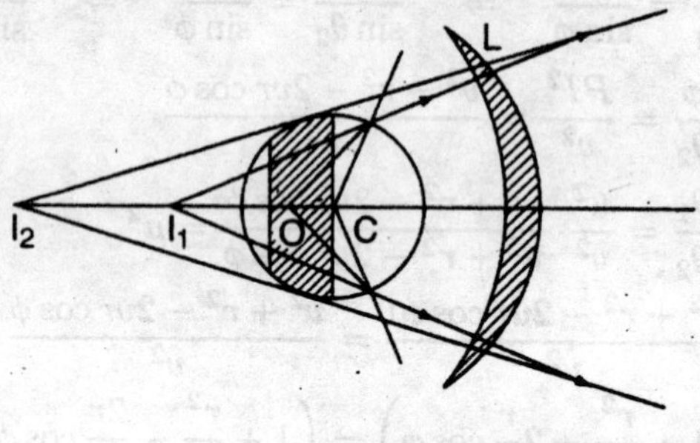

Fig 8.14

Linear Magnification

Let the object placed at O is of height H, the height of the first virtual image is h_1. Then,

$$m_1 = \frac{h_1}{H} = \frac{\text{Image distance}}{\text{Object distance}} = \frac{\left(\frac{\mu_1}{\mu_2}\right)}{\left(\frac{\mu_2}{\mu_1}\right)} = \left(\frac{\mu_1}{\mu_2}\right)^2, \text{ since } \mu_1 > \mu_2, m_1 > 1. \tag{8.3.15}$$

COMA

The second member of the monochromatic aberration family is named as COMA, most probably from the appearance of the image associated with this defect; the image is found to have a comet-like appearance. The reason for this defect is that the different zones of a lens have different amount of magnifications. This comatic defect can be very simply demonstrated by holding a magnifying glass in the path of sunlight and then tilting the glass, a comet-like flare of light appears.

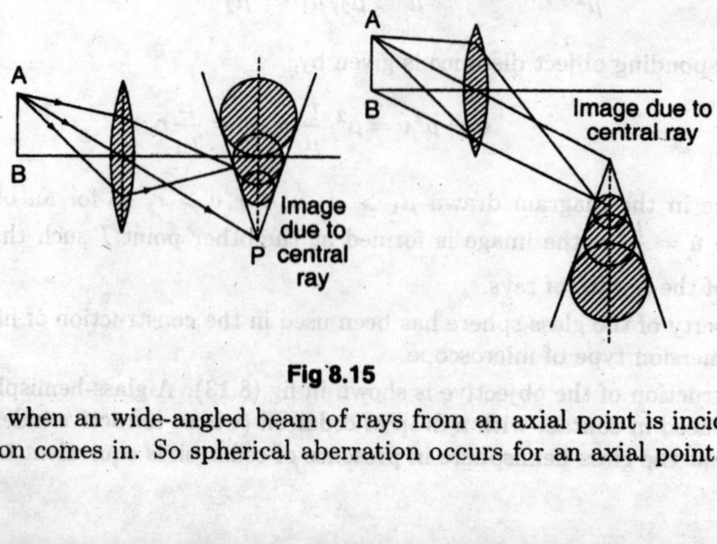

Fig 8.15

We know that when an wide-angled beam of rays from an axial point is incident on a lens, spherical aberration comes in. So spherical aberration occurs for an axial point. The off-axial object points lying close to the axis, give rise to the defect, called **Coma**. So we can say that for object points lying close to the axis we will have both spherical aberration and coma. If spherical aberration is minimised or removed, comatic aberration still persists.

For an off-axis point the image formed by different zones of the lens will be at different points in the image plane. So different zones have different magnifications.

If we consider the off-axis point A of the object AB, the central ray forms a point image at P; if we consider rays passing through different zones of the lens. We would get circular patch of increasing radius. The image point P is brightest and the other circular patches are diffuse. The diffuse tail of the flare can lie in a direction away from the axis, and this is called *positive coma*, if it lies in a direction towards the axis, it is called *negative coma*.

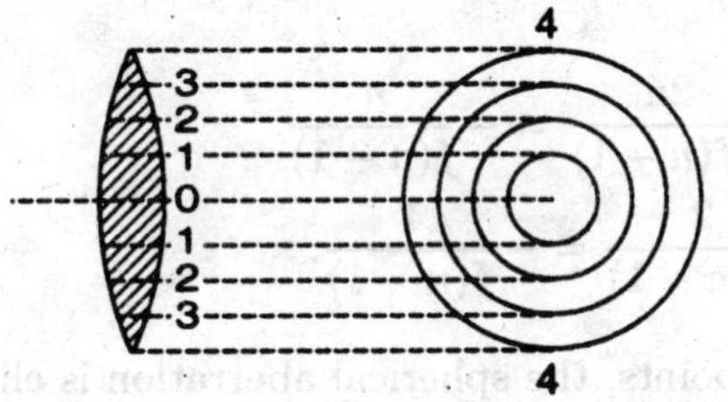

Fig 8.16

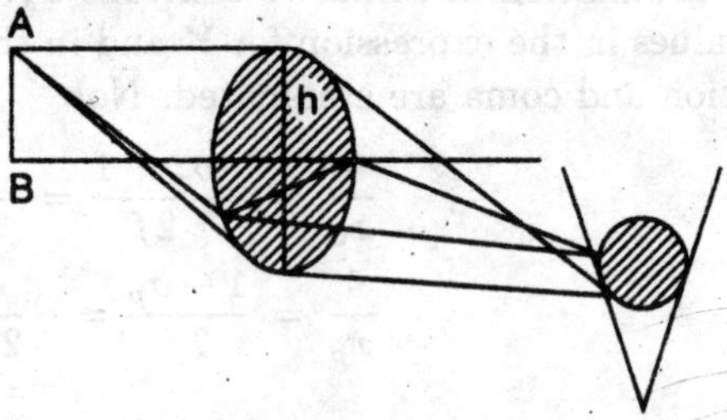

Fig 8.17

Consider a lens and let it be divided into a number of zones $(1,1)$, $(2,2)$, $(3,3)$, $(4,4)$ etc. As rays from the off-axis point A refract through different zones, the refracted rays meet at points which lie on a circle, called *comatic circle* (fig.8.16).

Remedy

The point P is the image of an oblique radiant A formed by the central ray and Q is the image formed by two rays refracting through the outer zones. The distance of separation between P, Q can be found to be given by

$$PQ = R = \frac{3\rho^2 \tan\beta}{3fn(n-1)}[(n+1)\sigma_s + (2n+1)(n-1)\sigma_p]\frac{1}{(1-\sigma_s)} \tag{8.3.16}$$

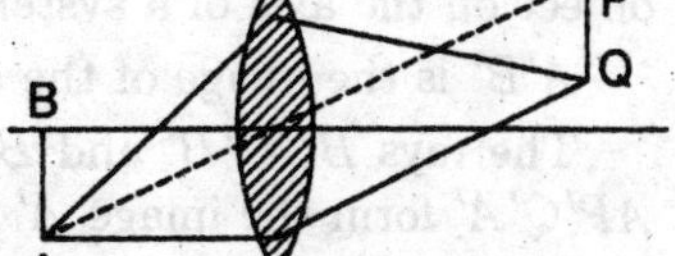

Fig 8.18

where,

ρ = distance of the zone from the axis

β = angle subtended by the chief ray with the axis

$\sigma_s = \dfrac{r_2 + r_1}{r_2 - r_1}$, also called *Coddington's shape factor*

r_1 and r_2 = radii of curvature

f = focal length of the lens

n = r.i. of the lens material

$\sigma_p = \dfrac{x_p + x_0}{x_p^2 x_0}$, x_0, x_p being paraxial conjugate distances, also called Coddington's position factor

So, for elimination of Coma, we must have,

$$O = \frac{3\rho^2 \tan\beta}{2fn(n-1)}[(n+1)\sigma_s + (2n+1)(n-1)\sigma_p]\frac{1}{(1-\sigma_s)}$$

$$\Rightarrow \quad \sigma_s = -(2n+1)\frac{(n-1)}{(n+1)}\sigma_p.$$

Aplanatic Lens

A lens free of spherical aberration and coma is called an *aplanatic lens.*

For elimination of coma we can take $\sigma_s = -(2n+1)$ and $\sigma_p = (n+1)/(n-1)$. Putting these values in the expression for Y and in the expression for coma, we see that both spherical aberration and coma are eliminated. Now

$$\frac{1}{x_0} = -\frac{\sigma_p + 1}{2f} = -\frac{2n}{2f(n-1)} = -\frac{n}{f(n-1)}$$

$$\frac{1}{x_p} = \frac{1-\sigma_p}{2f} = -\frac{2}{2f(n-1)} = -\frac{1}{f(n-1)}.$$

So, if $x_0 = r_1$, $x_p = nr_1$, and for these two points, the spherical aberration is eliminated.

We know that spherical aberration occurs for an axial point and coma occurs for a point slightly off-axial. So a system corrected for spherical aberration does not exhibit the defect called *Coma* if, in addition, the system satisfies the *sine condition* due to Abbe. A lens or a system which rectifies both spherical as well as comatic aberration is referred to as *aplanatic lens.*

Since coma is due to different lateral magnifications produced by different zones of a lens and since condition is a condition based on same lateral magnification of all zones, so comatic aberration will be removed if the system is made to satisfy Abbe's sine condition.

Abbe's Sine Condition

Abbe's sine condition is an equation expressing the fact that the lateral magnification of a small object on the axis of a system will be same for all zones of image-forming element.

$A'B'$ is the image of the object AB formed by a convex lens L.

The rays $BCC'B'$ and $BC_1C_1'B'$ form the image B' of the point B; the rays $APQA'$ and $AP'Q'A'$ form the image A' of the point A.

The ray BC is parallel to AP and the ray BC_1 is parallel to AP'.

Now, the optical path between corresponding points of the object and image is same, B' being the image of B, we get

$$\mu_1 BC + \mu CC' + \mu_2 C'B' = \mu_1 BC_1 + \mu C_1C_1' + \mu_2 C_1'B'. \quad (8.3.17)$$

A' being the image of A, we get

$$\mu_1 AP + \mu PQ + \mu_2 QA' = \mu_1 AP' + \mu P'Q' + \mu_2 Q'A' \quad (8.3.18)$$

F being the focus for the parallel rays BC and AP, we get

$$\mu_1 BC + \mu CC' + \mu_2 C'F = \mu_1 AP + \mu PQ + \mu_2 QF \quad (8.3.19)$$

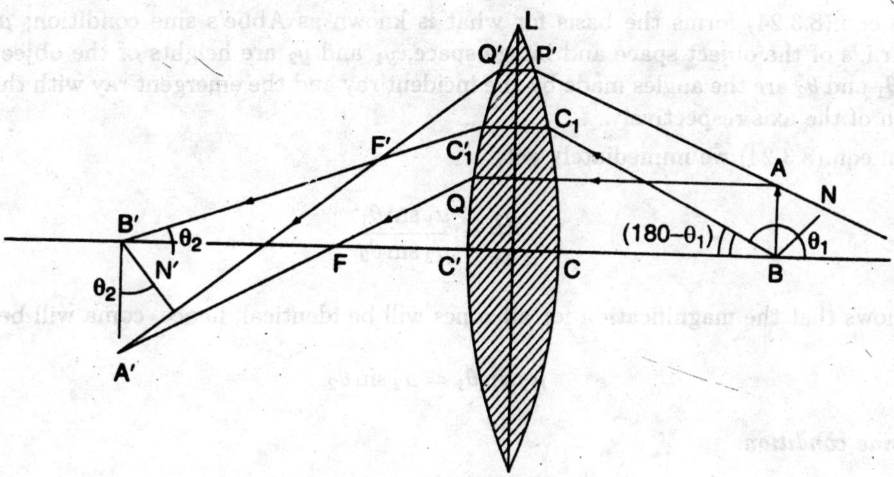

Fig 8.19

Similarly, F' being the focus for the parallel rays BC_1 and AP', we also get

$$\mu_1 BC_1 + \mu C_1C_1' + \mu_2 C_1'F' = \mu_1 NA' + \mu_1 AP' + \mu P'Q' + \mu_2 Q'F' \tag{8.3.20}$$

Now, object and image are supposed to be small, hence

$$\mu_2 FB' = \mu_2 FA' \tag{8.3.21}$$

Adding eqns.(8.19) and (8.21), we have,

$$\mu_1 BC + \mu CC' + \mu_2 C'B' = \mu_1 AP + \mu PQ + \mu_2 QA' \tag{8.3.22}$$

Also, the object and image being small,

$$\mu_2 F'B' = \mu_2 F'N' \tag{8.3.23}$$

Adding eqns.(8.3.23) and (8.3.20) we get,

$$\mu_1 BC_1 + \mu C_1C_1' + \mu_2 C_1'B' = \mu_1 NP' + \mu P'Q' + \mu_2 Q'N'$$

$$\Rightarrow \quad \mu_1 BC + \mu CC' + \mu_2 C'B' = \mu_1 NP' + \mu P'Q' + \mu_2 Q'N' \quad \text{(using relation 8.3.17)}$$

$$\Rightarrow \quad \mu_1 AP + \mu PQ + \mu_2 QA' = \mu_1 NP' + \mu P'Q' + \mu_2 Q'N' \quad \text{(using eqn. 22)}$$

$$\Rightarrow \quad \mu_1 AP' + \mu P'Q' + \mu_2 Q'A' = \mu_1 NP' + \mu P'Q' + \mu_2 Q'N'$$

$$= \mu_1 NA + \mu_1 AP' + \mu P'Q' + \mu_2 Q'N' \quad \text{(using eqn. 18)}$$

$$\Rightarrow \quad \mu_2 Q'A' = \mu_1 NA + \mu_2 Q'N'$$

$$\Rightarrow \quad \mu_2 (Q'A' - Q'N') = \mu_1 NA$$

$$\Rightarrow \quad \mu_2 A'N' = \mu_1 NA$$

$$\Rightarrow \quad \mu_2 y_2 \sin\theta_2 = \mu_1 y_1 \sin(180 - \theta_1)$$

$$\Rightarrow \quad \mu_1 y_1 \sin\theta_1 = \mu_2 y_2 \sin\theta_2 \tag{8.3.24}$$

This eqn.(8.3.24) forms the basis for what is known as Abbe's sine condition; μ_1 and μ_2 are the r.i.'s of the object space and image space, y_1 and y_2 are heights of the object and the image; θ_1 and θ_2 are the angles made by the incident ray and the emergent ray with the positive direction of the axis respectively.

From eqn.(8.3.24) we immediately get,

$$\frac{y_2}{y_1} = \frac{\mu_1 \sin\theta_1}{\mu_2 \sin\theta_2} \tag{8.3.25}$$

It shows that the magnification for all zones will be identical, hence, coma will be absent if

$$\mu_1 \sin\theta_1 = \mu_2 \sin\theta_2 \tag{8.3.26}$$

called *sine condition.*

Astigmatism

Like comatic aberration astigmatism is also an off-axis aberration. But astigmatism is a defect associated with object points widely off-axial. When coma is removed, rays further from the axis do not meet at one image off-axis point, but they form different image points depending on their path through the lens in the vertical and horizontal planes (cf. fig.8.19). This is known as *astigmatism* or astigmatic aberration.

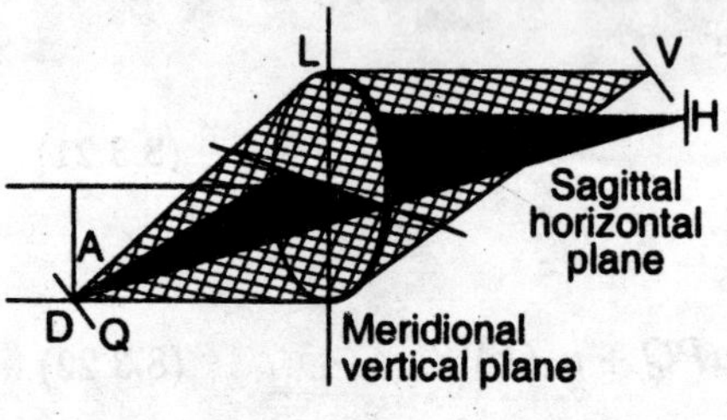

Fig 8.20

For the object point Q, the two image points corresponding to the rays in the vertical plane and horizontal plane are V and H respectively. For object points on the line A, the image points will lie on the line through H, parallel to A and the vertical plane; the image points are due to rays in the horizontal plane. All the rays in the vertical plane coming from object points on D meet at a line V, parallel to D and the horizontal plane.

If the object be a spoked wheel, the spokes being equivalent to lines like A, the rim is equivalent lines like D, then a screen placed passing through V will contain an image where the rim is sharp but spokes become fuzzy whereas in the plane through H, the image will contain sharp spokes but fuzzy rim. This defect of the image is referred to as astigmatism. There is a position in between V and H, where we will get an image of the spoked wheel where both spokes and rim are equally sharp. This is referred to as the image of *least confusion.*

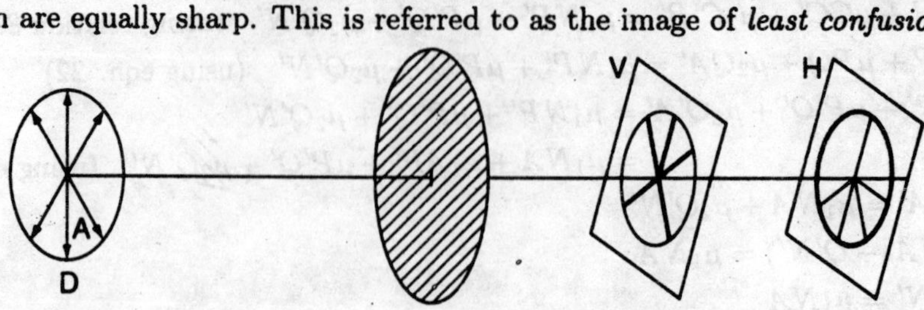

Fig 8.21

Remedy

The astigmatism cannot be eliminated for a single lens. But it can be reduced by employing a stop. However, lens systems may be designed where astigmatism is eliminated. Such a lens system is said to be an anastigmatic system.

Curvature Defect

It is a longitudinal defect, it causes the image of a two-dimensional flat object to become curved. It is because, the focal length of a lens for oblique centric pencil is less than that for direct rays. So the central part of the image is flat (it is formed by direct centric pencil) but the edges of the image formed by the oblique centric pencil are curved. This defect of the image is referred to as *curvature*. The radius of curvature of the image is +ve for convex lens and −ve for concave lens.

Remedy

Since the radius of curvature of the image due to convex lens is positive, while that for a concave lens is −ve, hence by using two lenses, one convex, the other concave, kept separated by a distance, the curvature may be eliminated.

Distortion

The defect distortion appears with the image formed by an optical system free of spherical aberration, coma and astigmatism. This defect is associated with the non-uniform lateral magnification of the optical system.

To demonstrate the effect we consider the fig.(8.21) where we have shown a stop before the lens.

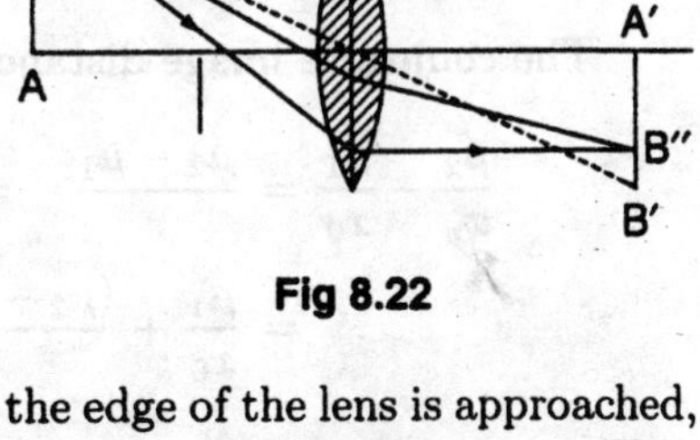

Fig 8.22

The image points A' and B' are the paraxial image points of A and B. But because of the presence of some stop as more and more obique rays are incident, the cone of light from the point B passes through only a small portion of the lens. This portion is not at the centre of the lens. The ray from B passing through the centre of the stop forms the image at B''. Since the object distance increases as the edge of the lens is approached, the image point occurs nearer to the lens than the corresponding paraxial image points. Thus a rectangular object appears to be barrel-shaped (fig.8.22).

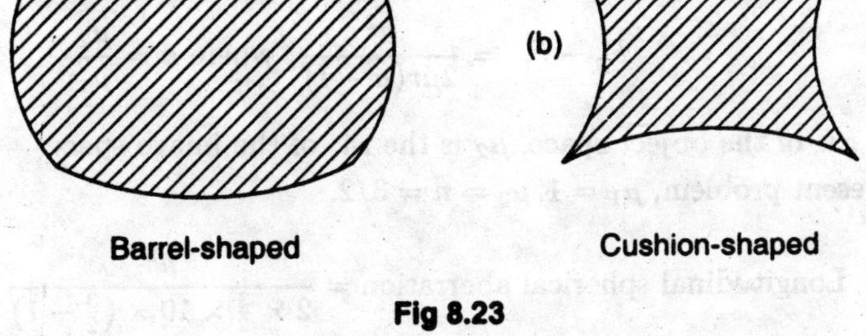

Fig 8.23

If the stop is placed after the lens, the rectangular object will appear to have a shape of a pin-cushion (fig.8.22b). The barrel-shaped distortion is also negative distortion and the pin-cushion distortion is called *+ve distortion.*

Remedy

This defect of the image may be remedied by placing a stop between two convex lenses, the first combination tends to produce pin-cushion distortion which is counter-acted by the second combination producing barrel-type distortion.

Solved Numerical Problems

1. A spherical refracting surface of radius r separates two media of r.i. values μ_1 and μ_2; μ_1 being the r.i. of the object space.

   Show that for an object at $x_0 = \dfrac{\mu_1+\mu_2}{\mu_1}r$, spherical aberration is eliminated.

   Solution. From eqn.(8.3.4), we get,

$$\text{LSA} = \frac{\frac{h^2}{2}\frac{(\mu-1)}{\mu}\left(\frac{1}{r}-\frac{1}{x_0}\right)^2\left(\frac{1}{r}-\frac{\mu+1}{x_0}\right)}{\left(\frac{\mu-1}{r}+\frac{1}{x_0}\right)^2}$$

   We at once conclude that LSA = 0 if

$$\frac{1}{r} = \frac{\mu+1}{x_0} \text{ where, } \mu = \frac{\mu_2}{\mu_1} \quad \Rightarrow \quad x_0 = (\mu+1)r = \frac{\mu_2+\mu_1}{\mu_1}r.$$

   The conjugate image distance is given by,

$$\frac{\mu_2}{x_p} - \frac{\mu_1}{x_0} = \frac{\mu_2-\mu_1}{r} \quad \Rightarrow \quad \frac{\mu_2}{x_p} = \frac{\mu_1}{x_0} + \frac{(\mu_2-\mu_1)}{r}$$

$$= \frac{\mu_1}{x_0} + \frac{(\mu_2-\mu_1)(\mu_2+\mu_1)}{\mu_1 x_0} = \frac{1}{x_0}\left[\frac{\mu_1^2+\mu_2^2-\mu_1^2}{\mu_1}\right] \quad \Rightarrow \quad x_p = x_0\frac{\mu_1}{\mu_2}.$$

2. A single convex spherical surface with $r = 10.0$ cm separates air and a medium of r.i. 3/2. Calculate the (*i*) longitudinal spherical aberration and (*ii*) lateral spherical aberration for parallel light through a zone at a height of 2 cm.

   Solution. From eqn.(8.3.5), we get,

$$F_p - F_m = \frac{h^2}{2\mu r(\mu-1)}, \text{ where } \mu = \frac{\mu_2}{\mu_1}$$

   μ_1 is the r.i. of the object space, μ_2 is the r.i. of the image space.
   In the present problem, $\mu_1 = 1, \mu_2 = n = 3/2$.

$$\therefore \text{Longitudinal spherical aberration} = \frac{h^2}{2\times\frac{3}{2}\times 10\times\left(\frac{3}{2}-1\right)} = \frac{2^2}{15} = \frac{4}{15}.$$

3. Find the radii of curvature for a lens of focal length $f = 10$ cm made of a material of r.i. value 1.5, which for parallel rays has minimum spherical aberration.

Solution. For minimum spherical aberration for parallel rays in case of a lens, the ratio r_1/r_2 is given by

$$\frac{r_1}{r_2} = \frac{2\mu^2 - \mu - 4}{\mu(2\mu + 1)}.$$

For a convex lens, $r_1 = -$ve, hence

$$\frac{r_1}{r_2} = \frac{1}{6}.$$

Again

$$\frac{1}{f} = (\mu - 1)\left(\frac{1}{r_1} + \frac{1}{r_2}\right) \quad \Rightarrow \quad \frac{1}{10} = \frac{1}{2} \cdot \frac{r_1 + r_2}{r_1 r_2} \quad \Rightarrow \quad \frac{r_1 r_2}{r_1 + r_2} = 5.$$

From above we get

$$\frac{r_1 + r_2}{r_2} = \frac{7}{6} \Rightarrow \frac{r_2}{r_1 + r_2} = \frac{6}{7}$$

$$\therefore \; r_1 \times \frac{6}{7} = 5 \Rightarrow r_1 = \frac{35}{6} \text{ cm}$$

$$r_2 = 6r_1 = 35 \text{ cm}.$$

Theory Questions

1. What do you mean by monochromatic aberration? What is spherical aberration? What is it due to? How can you minimise spherical aberration of a lens?
2. What is a crossed lens? What is its utility?
3. Find the aplanatic points of a glass refracting sphere. Show how you can construct microscope objective free from spherical aberration.
4. What is astigmatism? How can you eliminate astigmatic aberration?
5. What do you mean by chromatic error? Derive a suitable relation for an achromatic doublet of two lenses separated by a distance a in air.
6. What do you mean by shape factor and position factor? How are they to be related so as to minimise spherical aberration?
7. (a) What do you mean by the ray-optics limit? (b) Define optical path length between two points. (c) Consider an elliptical reflector whose focii are at the points S_1 and S_2, so that rays emanating from the point S_1 will pass through the point S_2 after undergoing a reflection. Comment on the nature of the path. **[BU 1994]**
8. Show that rays incident parallel to the axis of a paraboloidal reflector pass through the focus of the mirror.
9. Find the positions of the aplanatic points of a glass-sphere of r.i. n_1 surrounded by a medium of r.i. n_2. Where do you find the application of this to remove spherical aberration?

Chromatic Aberration

8.4 Introduction

Even when an optical system like a lens is corrected for all monochromatic defects, some colour defect is associated with the image due to the use of polychromatic light. This defect is referred to as *chromatic aberration.* Due to this defect we do not get a white image of a white object, but, instead, we get an image where some colour is in prominence while others seem to be out-focussed or blurred. This is due to differential deviation of different colours present in the polychromatic light by the dispersive optical element. This chromatic defect requires to be rectified in optical instruments like telescope, camera, etc.

8.5 What is Dispersion and Dispersive Power?

Dispersion is the phenomenon in which a composite light is separated out into its constitute colours by an optical medium, called *dispersive material.* The separation into the constituent colours is due to differential deviation suffered by the different colours, the different amount of deviation is again due to difference in the refractive index of the medium corresponding to different colours. If the refractive index of a medium decreases with wavelength, the medium is referred to as *normal dispersive medium* while if it increases, the medium is referred to as the *anomalous dispersive medium.* Thus, if the refractive index for violet colour denoted as μ_v is greater than that for red denoted as μ_r, the medium is normal dispersive medium and if it is less, it is an anomalous dispersive medium. Also deviation depends on the refractive index, hence deviation for violet colour is more than the deviation for red colour in case of normal dispersive medium and hence, in the spectrum through a normal dispersive medium the red colour will occupy the topmost position while the violet will come at the bottom. The details of the normal and the anomalous dispersion with their necessary theories have been discussed in the chapter on '*Dispersion*' given in Vol.II. However, it is to note here that $\frac{d\mu}{d\lambda}$ is known as chromatic dispersion; for $\frac{d\mu}{d\lambda} = -\text{ve}$, dispersion is said to be normal, and for $\frac{d\mu}{d\lambda} = +\text{ve}$, it is said to be anomalous.

8.6 Dispersion in a Glass Prism

In fig.(8.24), a composite light is made incident on the refracting face of a prism; the different colours present will get separated as they pass through the prism and the emergent red and violet colours are shown as R and V. When these two rays are produced back, they intersect at the point O; through O, a line is drawn parallel to the incident direction and the deviations of red and violet are shown as D_r and D_v. It is evident from the figure that $D_v > D_r$. The difference $(D_v - D_r)$ is referred to as dispersion between the two colours. Thus,

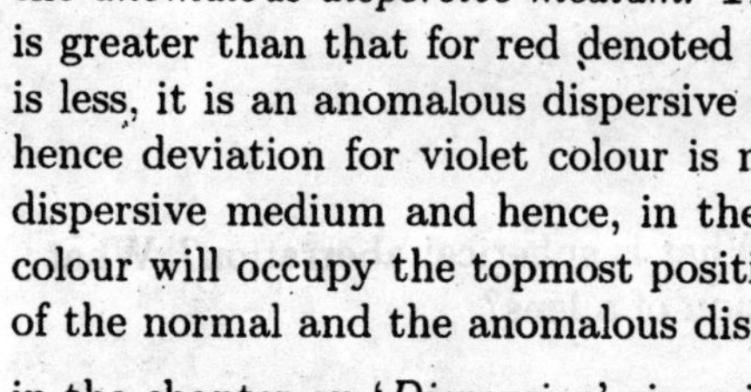

Fig 8.24

$$\text{Dispersion} = (D_v - D_r).$$

Let the deviation of the mean colour in the visible spectrum be denoted as D, then the dispersive power ω is defined as

$$\omega = \frac{D_v - D_r}{D}.$$

In case of a thin prism,

$D = (\mu - 1)A, D$ and μ represent the deviation and r.i. values for the mean colour (i.e., yellow)

$$D_v = (\mu_v - 1)A, \quad D_r = (\mu_r - 1)A$$

$$\therefore \quad \text{Dispersion} = (D_v - D_r) = (\mu_v - \mu_r)A.$$

Dispersive power

$$\omega = \frac{(\mu_v - \mu_r)A}{(\mu - 1)A} = \frac{(\mu_v - \mu_r)}{(\mu - 1)} = \frac{\delta\mu}{(\mu - 1)}$$

$\delta\mu$ represents the change in refractive index as wavelength changes from violet to red colour.

Dispersion in Case of a Lens

We know that in case of a lens, the deviation is given by,

$$D = \frac{h}{f} \qquad \therefore \quad D_v = \frac{h}{f_v}, \quad D_r = \frac{h}{f_r}.$$

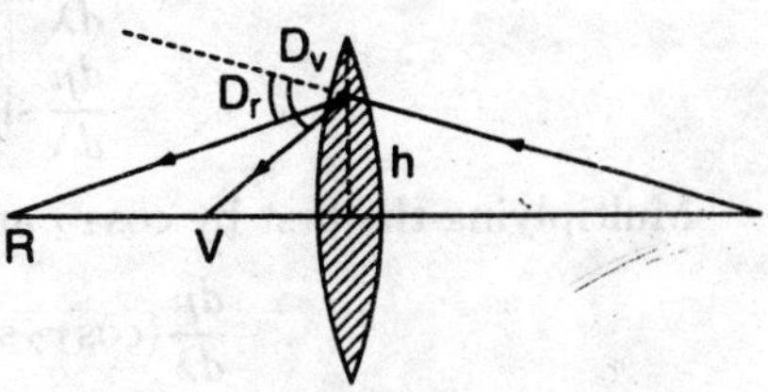

Fig 8.25

Now,

$$\frac{1}{f} = (\mu - 1)\left(\frac{1}{r_1} - \frac{1}{r_2}\right) = (\mu - 1)k; k = \left(\frac{1}{r_1} - \frac{1}{r_2}\right)$$

$$\therefore \quad \frac{1}{f_v} = (\mu_v - 1)k, \frac{1}{f_r} = (\mu_r - 1)k, \text{ since } \mu_v > \mu_r, \frac{1}{f_v} > \frac{1}{f_r} \quad \Rightarrow \quad f_v < f_r$$

$$\therefore \quad \text{Dispersion} = (D_v - D_r) = h\left(\frac{1}{f_v} - \frac{1}{f_r}\right)$$

$$\therefore \quad \omega = \frac{D_v - D_r}{D} = \frac{h(1/f_v - 1/f_r)}{(h/f)} = \frac{(\mu_v - 1)k - (\mu_r - 1)k}{(\mu - 1)k} = \frac{\mu_v - \mu_r}{(\mu - 1)} = \frac{\delta\mu}{(\mu - 1)}.$$

Dispersion in Case of a Thick Prism

Consider a thick prism refraction, the path of a certain colour is shown in fig.(8.25). The deviation can be written as $D = \theta_1 + \theta_2 - A$. Then the angular dispersion is given by $dD/d\lambda$, so

$$\frac{dD}{d\lambda} = \frac{dD}{d\mu} \cdot \frac{d\mu}{d\lambda} = G\frac{d\mu}{d\lambda}$$

where $G = dD/d\mu$ is called the *geometric factor*, and $d\mu/d\lambda$ is the chromatic dispersion, which is a factor dependent on the prism characteristic.

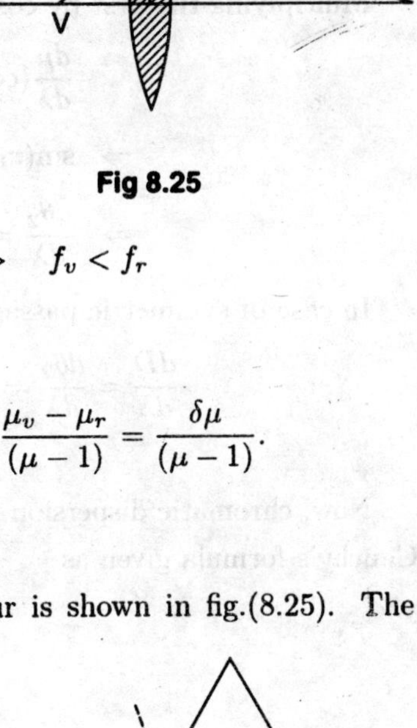

Fig 8.26

The factor G depends on the geometry of the prism, hence called *geometric factor*.

Now, if the incident beam of rays makes same angle of incidence, θ_1 is fixed for all colours.

$$\therefore \quad \frac{dD}{d\lambda} = \frac{d\theta_1}{d\lambda} + \frac{d\theta_2}{d\lambda} = \frac{d\theta_2}{d\lambda}, \text{ since } \frac{d\theta_1}{d\lambda} = 0$$

$$\sin\theta_1 = \mu\sin r_1, \sin\theta_2 = \mu\sin r_2, r_1 + r_2 = A$$

If we consider two wavelengths λ and $\lambda + d\lambda$, for which r.i. of the medium is μ and $\mu + d\mu$, then,

$$\frac{d\mu}{d\lambda}\sin r_1 + \mu\cos r_1\frac{dr_1}{d\lambda} = 0$$

$$\frac{d\mu}{d\lambda}\sin r_2 + \mu\cos r_2\frac{dr_2}{d\lambda} = \cos\theta_2\frac{d\theta_2}{d\lambda}$$

$$\frac{dr_1}{d\lambda} + \frac{dr_2}{d\lambda} = 0$$

The last of the triplet relation gives $dr_2/d\lambda = -dr_1/d\lambda$, with this relation, the first two eqns. just above now become,

$$\frac{d\mu}{d\lambda}\sin r_1 + \mu\cos r_1\frac{dr_1}{d\lambda} = 0$$

$$\frac{d\mu}{d\lambda}\sin r_2 - \mu\cos r_2\frac{dr_1}{d\lambda} = \cos\theta_2\frac{d\theta_2}{d\lambda}$$

Multiplying the first by $\cos r_2$ and second by $\cos r_1$, and adding we get,

$$\frac{d\mu}{d\lambda}(\cos r_2\sin r_1 + \sin r_2\cos r_1) = \cos r_1\cos\theta_2\frac{d\theta_2}{d\lambda}$$

$$\Rightarrow \quad \sin(r_1 + r_2)\frac{d\mu}{d\lambda} = \cos r_1\cos\theta_2\frac{d\theta_2}{d\lambda}$$

$$\Rightarrow \quad \frac{d\theta_2}{d\lambda} = \frac{\sin A}{\cos r_1\cos\theta_2}\frac{d\mu}{d\lambda}.$$

In case of symmetric passage through the prism, $\theta_1 = \theta_2, r_1 = r_2 = A/2$, so in this case,

$$\frac{dD}{d\lambda} = \frac{d\theta_2}{d\lambda} = \frac{\sin A}{\cos(A/2)\cos\theta_1}\cdot\frac{d\mu}{d\lambda} = \frac{2\sin(A/2)}{\sqrt{1 - \mu^2\sin^2(A/2)}}\cdot\frac{d\mu}{d\lambda}$$

Now, chromatic dispersion $\frac{d\mu}{d\lambda}$ can be calculated w.r.t the standard dispersion formula, say Cauchy's formula given as

$$\mu = A + \frac{B}{\lambda^2} \quad \Rightarrow \quad \frac{d\mu}{d\lambda} = -\frac{2B}{\lambda^3}$$

$$\therefore \quad \frac{dD}{d\lambda} = \frac{d\theta_2}{d\lambda} = -\frac{4B\sin(A/2)}{\lambda^3\sqrt{1 - \mu^2\sin^2(A/2)}}$$

The negative sign indicates that the deviation decreases with increase in wavelength.

$$\therefore \quad \left|\frac{dD}{d\lambda}\right| = \frac{4B\sin(A/2)}{\lambda^3\sqrt{1 - \mu^2\sin^2(A/2)}}$$

so it shows that the angular dispersion is inversely proportional to the cube of the wavelength.

Hence, dispersion or spread is greater towards the shorter wavelength side (i.e., towards the violet end) than towards the longer wavelength side (i.e., towards the red end). So, the prismatic spectrum is not normal for which $\left|\frac{dD}{d\lambda}\right|$ should be a constant.

8.7 Chromatic Aberration

We have just seen that the refractive index of a medium depends on wavelength or colour. So, if we use a lens and a polychromatic source is taken, then a point white object will not give rise to a white point image. We have also shown that $f_v < f_r$, hence the red rays will come to a point of focus at a further distance than the point of focus for the violet.

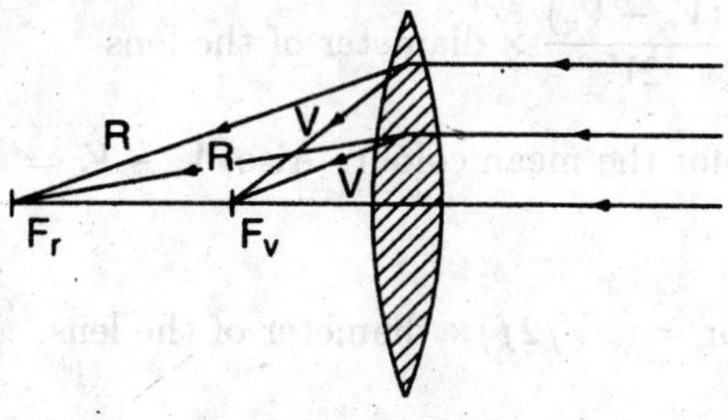

Fig 8.27

Consider a beam of parallel rays of composite nature to be incident on the lens; the violet component will form the image at F_v and the red component at F_r, called respectively the *violet-ray focus* and the *red-ray focus*. So, if a screen be held at F_v, we will get a circular patch of light where violet colour will be in prominence while other colours present will be out-focussed. Similarly, if the screen be held at F_r, the red will be in prominence while other colours will be out-focussed. This defect of the image is called *chromatic error* or *chromatic aberration*.

The difference between F_v and F_r represents the axial chromatic error for parallel rays. If the rays are non-parallel, then the axial chromatic error can be found by considering fig.(8.27).

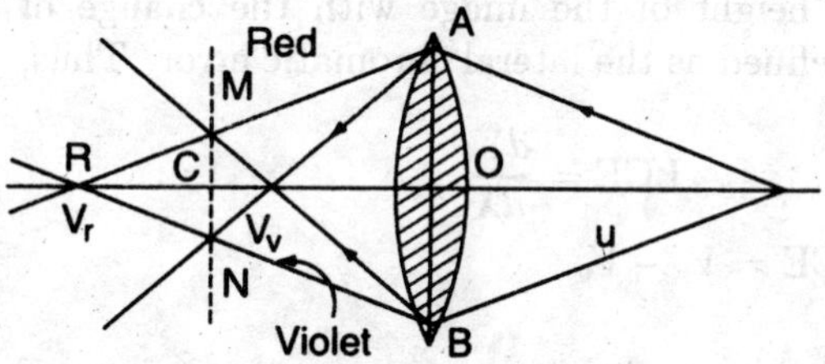

Fig 8.28

We have,

$$\frac{1}{V_v} - \frac{1}{u} = \frac{1}{f_v}, \quad \frac{1}{V_r} - \frac{1}{u} = \frac{1}{f_r}.$$

Taking the difference, we get

$$\frac{1}{V_v} - \frac{1}{V_r} = \frac{1}{f_v} - \frac{1}{f_r} \Rightarrow \frac{V_r - V_v}{V_r V_v} = \frac{1}{f_v} - \frac{1}{f_r}.$$

Now,

$$\frac{1}{f_v} = (\mu_v - 1)k, k = \frac{1}{r_1} - \frac{1}{r_2}, \frac{1}{f_r} = (\mu_r - 1)k$$

$$\therefore \quad \frac{1}{f_v} - \frac{1}{f_r} = (\mu_v - \mu_r)k = \frac{(\mu_v - \mu_r)}{(\mu - 1)f} = \frac{\omega}{f}$$

$$\therefore \quad V_r - V_v = \frac{\omega}{f} \cdot V_r V_b \simeq \frac{\omega}{f} V^2,$$

where V is the image distance for the mean colour image.

If the screen be placed at MN, we would get a circular patch where all colours appear in almost equal prominence. This circle is called *circle of least chromatic error*.

Diameter of the Circle of Least Chromatic Error

From the similar Δs RMN and ARB, we get,

$$\frac{MN}{AB} = \frac{CR}{RO} = \frac{CR}{V_r}.$$

Also from the similar Δs, MVN and VAB, we get,

$$\frac{MN}{AB} = \frac{CV}{OV} = \frac{CV}{V_v}.$$

$$\therefore \quad \frac{CR}{V_r} = \frac{CV}{V_v} \quad \Rightarrow \quad \frac{CR}{V_r} = \frac{CV}{V_v} = \frac{CR+CV}{V_r+V_v} = \frac{RV}{V_r+V_v} = \frac{V_r - V_v}{V_r+V_v}$$

$$\therefore \quad \frac{MN}{AB} = \frac{V_r - V_v}{V_r+V_v} \quad \Rightarrow \quad MN = \frac{(V_r - V_v)}{(V_r + V_v)} \cdot AB = \frac{(V_r - V_v)}{2V} \times \text{diameter of the lens}$$

where we have put $V_r + V_v = 2V$, V being image distance for the mean colour. Also, $V_r - V_v = \frac{\omega}{f}V^2$.

$\therefore \quad MN =$ diameter of the circle of least chromatic error $= (\omega V/2f) \times$ diameter of the lens.

Lateral or Transverse Chromatic Error

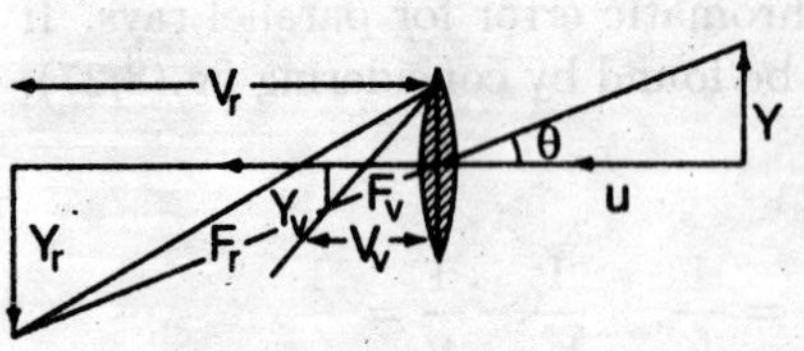

Fig 8.29

The object is of finite size, the sizes of the image corresponding to the violet and red are respectively Y_v and Y_r, so the heights of different colour-images are different. The change of height of the image with the change of wavelength is defined as the lateral chromatic error. Thus,

$$\text{LCE} = \frac{dY}{d\lambda}.$$

The axial chromatic error (ACE) is given by, ACE $= V_r - V_v$.

Now, size of the violet image $= Y_v$
size of the red image $= Y_r$
size of the object $= Y$

From the fig.(8.28), we get,

$$\frac{Y_r}{Y} = \frac{V_r}{U}, \quad \frac{Y_v}{Y} = \frac{V_v}{U}$$

$$\therefore \quad \frac{Y_r - Y_v}{Y} = \frac{V_r - V_v}{U} \quad \Rightarrow \quad (Y_r - Y_v) = \frac{Y}{U} \cdot (V_r - V_v) = \tan\theta \times \frac{\omega}{f}V^2$$

where V is the image distance corresponding to the mean colour.

Remedy from Chromatic Aberration

The optical element like a lens used in an optical instrument like telescope, microscope or a camera is to be rectified for chromatic error, as otherwise a white object, whether punctual or not, will not produce a white image or in other words there will be colour blurring. For a punctual axial source, there is no lateral chromatic error but there is axial chromatic error only. But for an object of finite size, there is both axial as well as lateral chromatic error.

A system or arrangement which removes chromatic aberration is called an *achromatic system* and the process of removal is referred to as *achromatisation.* It is to note here that perfect achromatic system which removes both axial as well as lateral chromatic aberration at the same time is almost impossible, what is possible is removal of either the axial or the lateral chromatic aberration and that too with respect to two colours and their mean. Such achromatisation is referred to as *partial achromatisation.* Which of the two chromatic errors is to be removed and which part of the spectrum is to be achromatised depend on the nature and purpose of the optical instrument under consideration. Thus in case of a telescope, the objective is to be achromatised for axial error and for the part of the spectrum very effective in producing illumination (i.e., the yellow part of the spectrum). This achromatisation is referred to as *visual achromatisation.* On the other hand, in a camera, the lens that is used must be corrected for the part of the spectrum to which the photographic film is very sensitive. It is the blue end of the spectrum and this achromatisation is referred to as *actinic achromatisation.*

Apochromatic and Semi-apochromatic System

Since an achromatic system is achromatised for two colours and their mean, chromatic spread for other colours present gives rise to *secondary spectrum.* To remove this secondary spectrum, achromatisation w.r.t. a number of colours is needed.

Again, when an wide-angled beam of composite light is made incident on a lens, spherical aberration comes into play. If the lens is corrected for spherical aberration for the marginal rays for a single colour, it will still predominantly persist for marginal rays of other colours. It will produce image coloured at the edge. This defect is called *chromatic effects of spherical aberration.*

A system where the secondary spectrum is removed but chromatic effects of spherical aberration still persists is called an *apochromatic system.* The system will be called *semi-apochromatic* if the chromatic effect of the spherical aberration is removed but not the secondary spectrum.

8.8 Mechanisms of Removal

By Two Lenses in Contact

We can make an achromatic system by placing two lenses in contact. The equivalent focal length of two lenses in contact is to be made independent of wavelength.

The equivalent focal length of two lenses of focal lengths f_1 and f_2 is given by

$$\frac{1}{F} = \frac{1}{f_1} + \frac{1}{f_2}$$

In order that this acts as an achromat, we must have

$$\frac{d}{d\lambda}\left(\frac{1}{F}\right) = 0 \quad \Rightarrow \quad \frac{d}{d\lambda}\left(\frac{1}{f_1}\right) + \frac{d}{d\lambda}\left(\frac{1}{f_2}\right) = 0.$$

Now,

$$\frac{1}{f} = (\mu - 1)k$$

or, $$\frac{d}{d\lambda}\left(\frac{1}{f}\right) = k\frac{d\mu}{d\lambda} \Rightarrow \frac{d}{d\lambda}\left(\frac{1}{f}\right) \times \delta\lambda = k\frac{d\mu}{d\lambda} \times \delta\lambda,$$

where $\delta\lambda$ represents the difference in wavelengths of two colours w.r.t. which achromatisation is to be made.

$$\therefore \quad \frac{d}{d\lambda}\left(\frac{1}{f}\right)\delta\lambda = k\delta\mu = \frac{\delta\mu}{f(\mu-1)} = \frac{\omega}{f}.$$

Thus, the condition of achromatisation of two lenses in contact can be written as

$$\frac{d}{d\lambda}\left(\frac{1}{f_1}\right)\delta\lambda + \frac{d}{d\lambda}\left(\frac{1}{f_2}\right)\delta\lambda = 0 \Rightarrow \frac{\omega_1}{f_1} + \frac{\omega_2}{f_2} = 0 \qquad (8.8.1)$$

where ω_1 and ω_2 are the dispersive powers of the materials of the two lenses.

From the above equation we at once get,

$$\frac{\omega_1}{f_1} = -\frac{\omega_2}{f_2}.$$

Now, since ω_1, ω_2 both are positive, this relation suggests that a convex lens is to be placed in contact with a concave lens so as to make an achromat. Also because this achromat is achromatised for focal lengths for two colours, it will act as an achromat for all positions of the object. If f_1 be the focal length of the convex lens and f_2, that of the concave lens, then the combination acting as a converging lens requires.

$$-\frac{1}{F} = -\frac{1}{|f_2|} + \frac{1}{f_2} \Rightarrow \frac{1}{f_1} > \frac{1}{f_2} \Rightarrow f_1 < f_2,$$

i.e., convex lens must be more powerful than the concave lens. So the condition of achromatism given by,

$$\frac{\omega_1}{|f_1|} = \frac{\omega_2}{|f_2|}$$

requires $\omega_1 < \omega_2$, meaning the convex lens must be made of less dispersive material than the concave lens. Usually the convex lens is made of crown glass and the concave lens, made of flint glass is a plano-concave lens (see fig.8.29). The achromat is to be placed such that the convex lens faces the incident light. Such an arrangement also minimises spherical aberration. The two lenses are joined together by a cementing material called *Canada Balsam* which has r.i. nearly equal to that of glass, thus the passage of light through the doublet is almost without any loss of light due to reflection. This type of achromat is referred to as old type achromat. New type achromats are made from more recently introduced glass containing a high proportion of barium called *new achromat*.

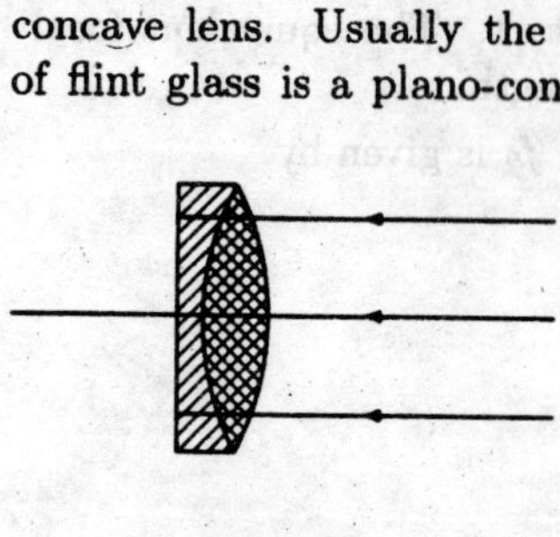

Fig 8.30

By Two Lenses kept separated by a Distance in Air

In some instruments like eyepieces we use two lenses kept separated by certain distance apart. To make this system an achromat, the distance of separation is related to the focal lengths of the two lenses and also the dispersive power of the lens materials.

Let us consider two lenses of focal lengths f_1 and f_2, and they are separated by a distance a in air. If both the lenses are convex lenses, the equivalent focal length is given by,

$$\frac{1}{F} = \frac{1}{f_1} + \frac{1}{f_2} - \frac{a}{f_1 f_2} \tag{8.8.2}$$

In order that this system of two lenses acts as an achromat w.r.t. two selected colours and their mean, we would get,

$$\frac{d}{d\lambda}\left(\frac{1}{F}\right) = 0$$

$$\Rightarrow \frac{d}{d\lambda}\left(\frac{1}{f_1}\right) + \frac{d}{d\lambda}\left(\frac{1}{f_2}\right) - \frac{a}{f_2}\frac{d}{d\lambda}\left(\frac{1}{f_1}\right) - \frac{a}{f_1}\frac{d}{d\lambda}\left(\frac{1}{f_2}\right) = 0$$

$$\Rightarrow \frac{d}{d\lambda}\left(\frac{1}{f_1}\right)\delta\lambda + \frac{d}{d\lambda}\left(\frac{1}{f_2}\right)\delta\lambda - \frac{a}{f_2}\frac{d}{d\lambda}\left(\frac{1}{f_1}\right)\delta\lambda - \frac{a}{f_1}\frac{d}{d\lambda}\left(\frac{1}{f_2}\right)\delta\lambda = 0$$

$$\Rightarrow \frac{\omega_1}{f_1} + \frac{\omega_2}{f_2} - \frac{a}{f_1 f_2}(\omega_1 + \omega_2) = 0 \tag{8.8.3}$$

Eqn.(8.8.3) is the required condition for the formation of an achromatic system. If $\omega_1 = \omega_2 = \omega$, we would get from the above eqn.,

$$\frac{1}{f_1} + \frac{1}{f_2} - \frac{2a}{f_1 f_2} = 0 \quad \Rightarrow \quad a = \frac{f_1 + f_2}{2} \tag{8.8.4}$$

Thus two convex lenses of the same material kept separated by a distance a equal to the average of two focal lengths may be used as an achromat.

Prisms in Combination

By a suitable combination of two prisms we can also make an achromat producing deviation but no dispersion. Also by combination of prisms we can produce dispersion but no deviation.

Two Prism Combination for Deviation without Dispersion

Two prisms of small refracting angles A and A' combined with their refracting angles in opposite directions can produce deviation without dispersion. The dispersion in one is counteracted by the dispersion in the other.

$\therefore$ Dispersion by the first prism = Dispersion by the second prism

or, $(\mu_v - \mu_r)A = (\mu'_v - \mu'_r)A'$

$$\Rightarrow \frac{(\mu_v - \mu_r)}{(\mu - 1)} \times (\mu - 1)A = \frac{\mu'_v - \mu'_r}{(\mu' - 1)} \times (\mu' - 1)A'$$

$$\Rightarrow \omega(\mu - 1)A = \omega'(\mu' - 1)A' \tag{8.8.5}$$

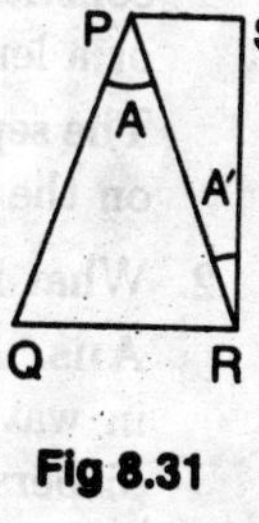

Fig 8.31

Eqn.(8.8.5) is the required condition for an achromatic combination of prisms.

The deviation D in the first prism is given by,

$$D = (\mu - 1)A$$

The deviation in the second in opposite direction is given by

$$D' = (\mu' - 1)A'.$$

$\therefore$ The resultant deviation by the prism-combination is given as,

$$D - D' = (\mu - 1)A - (\mu' - 1)A' = (\mu - 1)A - \frac{\omega}{\omega'}(\mu - 1)A = (\mu - 1)A\left[1 - \frac{\omega}{\omega'}\right] \quad (8.8.6)$$

So, if the two prisms are made of the same material such that $\omega = \omega'$ and also $\mu = \mu'$, achromatism of the combination requires $A = A'$. In this case, the resultant deviation $(D - D')$ reduces to zero also.

Dispersion but no Deviation

In this case, no deviation is to be produced. Since the prisms are placed in contact with their refracting angles in opposite direction, so both the dispersion and deviation would occur in opposite sense.

The resultant deviation $= D - D' = (\mu - 1)A - (\mu' - 1)A'$.

So, for no deviation, the required condition is,

$$(\mu - 1)A = (\mu' - 1)A'. \quad (8.8.7)$$

The resultant dispersion is given by,

$$(\mu_v - \mu_r)A - (\mu'_v - \mu'_r)A' = (\mu - 1)\left[\frac{\mu_v - \mu_r}{(\mu - 1)}A\right] - \left[\frac{\mu' - \mu'_r}{(\mu' - 1)}\right](\mu' - 1)A'$$
$$= (\mu - 1)A\omega - \omega'(\mu - 1)A = (\mu - 1)A[\omega - \omega'].$$

Short Theory Questions with Answers

1. What is dispersion? What it is due to?

   Ans. Dispersion is the phenomenon in which the composite light is broken up into its constituents when it is allowed to pass through a dispersing optical element like a prism or a lens.

   The separation of different colours is due to the fact that the refractive index is dependent on the wavelength or colour, the refractive index again determines the deviation.

2. What is the normal dispersive medium?

   Ans. A normal dispersive medium is one in which refractive index decreases with increase in wavelength. Thus, if $\mu_v > \mu_r$ in a medium, that medium is referred to as normal dispersive medium.

3. Does the focal length of a lens depend on colour of the light? In case of a glass lens, which is greater, f_v or f_r?

Ans. The focal length of a lens depends on the r.i. of the material of the lens, which again depends on the colour. Thus,

$$\frac{1}{f} = (\mu - 1)\left(\frac{1}{r_1} - \frac{1}{r_2}\right) \quad \therefore \quad \frac{1}{f_v} = (\mu_v - 1)k, \frac{1}{f_r} = (\mu_r - 1)k$$

Since glass is a normal dispersive material, $\mu_v > \mu_r$, hence $1/f_v > 1/f_r \Rightarrow f_v < f_r$.

4. What is a normal spectrum? Is prismatic spectrum normal?

Ans. The spectrum in which change in deviation is proportional to the change in wavelength is called a *normal spectrum.*

A prismatic spectrum is not normal because in this case $\left|\frac{dD}{d\lambda}\right|$ is inversely proportional to the cube of the wavelength.

5. What is chromatic aberration?

Ans. Chromatic aberration is the defect of the image due to use of composite light. If we use white light and image is formed by a lens, a white object will not give rise to a white image. We will get instead some colour in prominence while other colour will be out-focussed. This colour defect associated with the image is referred to as the chromatic aberration.

6. What is the meaning of an achromatic system? What is a perfect achromatic system?

Ans. A system or an optical device by which chromatic aberration is removed is called an *achromatic system.*

A perfect achromatic system is one in which both axial as well as lateral chromatic aberration are removed.

7. What is an apochromatic system? What is a semi-apochromatic system?

Ans. An achromatic system is achromatised for two colours and their mean. So for other colours present there is the formation of *secondary spectrum.* Also, when an wide-angled beam is used and spherical aberration is removed for one colour, there would occur spherical aberration for the other colours. The image as a result will be coloured at the edge. This defect is called *chromatic effects of spherical aberration.* A system which removes both the secondary spectrum and the chromatic effects of spherical aberration is called an *apochromatic system.*

In a semi-apochromatic system chromatic effects of spherical aberration are removed but not the secondary spectrum.

8. What is an achromatic doublet?

Ans. An achromatic doublet may mean two lenses in contact or separated by a certaindistance and this arrangement removes chromatic error.

9. In an achromatic doublet, what type of lenses is to be used?

Ans. To make an achromatic doublet in contact a convex lens of greater dispersive power is to be cemented to a (plano- or double) concave lens, whereas in case of two lenses separated by a distance, the two lenses may be both convex or concave.

10. What is the condition satisfied by an achromat?

Ans. An achromat satisfies an eqn. written as

$$\frac{\omega_1}{f_1} + \frac{\omega_2}{f_2} = 0.$$

11. Can a prism-combination serve as an achromat? What is the requirement for such combination?

Ans. Yes, two small angled prisms of angles A and A', placed in opposite directions and satisfying the relation $\omega(\mu - 1)A = \omega'(\mu' - 1)A'$, will act as an achromat.

Solved Numerical Problems

1. Design a crown-flint doublet of focal length 10 cm achromatic for blue and red. Given $\mu_b = 1.5293$, $\mu_r = 1.5204$, $\mu_{(mean)} = 1.5230$ for crown; $\mu_b = 1.7378$, $\mu_r = 1.7130$, $\mu_{(mean)} = 1.7200$ for flint glass.

Solution.

$$\frac{1}{F} = \frac{1}{f_1} + \frac{1}{f_2}, \frac{\omega_1}{f_1} + \frac{\omega_2}{f_2} = 0$$

$$\therefore \ \omega_1 = \frac{\mu_b - \mu_r}{(\mu - 1)} = \frac{1.5293 - 1.5204}{(1.5230 - 1)} = \frac{1}{58.764}$$

$$\therefore \ \omega_2 = \frac{\mu_b' - \mu_r'}{(\mu' - 1)} = \frac{1.7378 - 1.7130}{(1.7200 - 1)} = \frac{1}{29.032}$$

$$\frac{1}{f_1} + \frac{1}{f_2} = \frac{1}{F} = -\frac{1}{10} \quad \text{and} \quad \frac{1}{58.764} \times \frac{1}{f_1} + \frac{1}{29.032} \times \frac{1}{f_2} = 0$$

Solving these two eqns. we get, $f_1 = -5.06$ cm, $f_2 = 10.24$ cm

$$\frac{1}{f_1} = (\mu - 1) \times \frac{2}{R_1} = \frac{2 \times (1.523 - 1)}{R_1} = \frac{1}{5.06} \quad \Rightarrow \quad R_1 = 2 \times 5.06 \times .523 = 5.29 \text{ cm.}$$

The concave lens has first surface of radius of curvature 5.29 cm.

$$\therefore \ \frac{1}{f_2} = (\mu - 1)\left(\frac{1}{R_1'} - \frac{1}{R_2'}\right) = (1.720 - 1)\left(\frac{1}{5.29} - \frac{1}{R_2'}\right)$$

$$\Rightarrow \frac{1}{10.24} = 0.720\left(\frac{1}{5.29} - \frac{1}{R_2'}\right) \quad \Rightarrow \quad R_2' = 18.72 \text{ cm.}$$

2. An achromatic converging combination of focal length 50 cm is formed with a convex lens of crown glass and a concave lens of flint glass placed in contact with each other. Calculate their focal lengths if the dispersive powers of the crown glass and flint glass are respectively 0.03 and 0.05.

Solution.

$$\frac{1}{f_1} + \frac{1}{f_2} = -\frac{1}{F} \quad \text{and} \quad \frac{\omega_1}{f_1} + \frac{\omega_2}{f_2} = 0$$

$$\therefore \ \frac{1}{f_1} + \frac{1}{f_2} = -\frac{1}{50} \quad \text{and} \quad \frac{.03}{f_1} + \frac{.05}{f_2} = 0$$

$$\Rightarrow \frac{.05}{f_1} + \frac{.05}{f_2} = -\frac{.05}{50} \quad \text{and} \quad \frac{.03}{f_1} + \frac{.05}{f_2} = 0$$

Subtracting, we get

$$\frac{(.05-.03)}{f_1} = -\frac{.05}{50} \Rightarrow f_1 = -20 \text{ cm}$$

$$\therefore \quad f_2 = 33.3 \text{ cm}.$$

3. The r.i.'s for crown glass for C, D and F lines are respectively 1.514, 1.517, 1.523; while corresponding values for flint glass are 1.644, 1.650 and 1.664. Calculate the refracting angle of a flint glass prism which, when combined with a crown glass prism of refracting angle 5°, produces no deviation corresponding to D line. Find also the separation produced by the combination for C and F lines.

Solution. C, F, D are Fraunhofer designation of red, blue and yellow lines.

For no deviation, we require

$$(\mu-1)A = (\mu'-1)A' \text{ for } D \text{ line}$$

$$\Rightarrow (1.517-1)5^\circ = (1.650-1)A' \Rightarrow A' = 3°58.6'$$

Separation produced by the crown glass prism $= (\mu_b - \mu_r)A = (1.523 - 1.514)5^\circ = 0.045^\circ$.

Separation produced by the flint glass prism $= (\mu'_b - \mu'_r)A' = (1.664 - 1.644)3^\circ 58.6' = .0795$

$\therefore$ Resultant separation $= 0.0795 - 0.045 = .0345^\circ = 2.07'$.

4. A plano-convex lens of crown glass is cemented to a concavo-convex lens of flint glass so that the convex surface of the first and the concave surface of the second fit exactly. If the combination is achromatic and the combined focal length is −50 cm, determine the radii of curvature of the faces of the flint glass lens. The dispersive power of crown glass is one-half of that of flint glass and the r.i.'s for yellow light are respectively 1.500 and 1.625.

Solution.

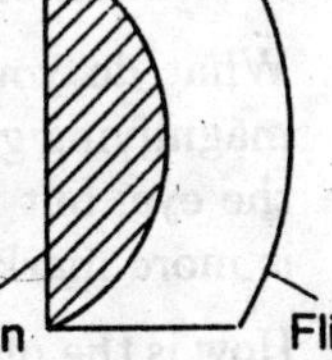

Fig 8.32

$$\frac{1}{f_1} + \frac{1}{f_2} = \frac{1}{F} = -\frac{1}{50} \quad \text{and} \quad \frac{\omega_1}{f_1} + \frac{\omega_2}{f_2} = 0$$

$$\Rightarrow \frac{1}{2}\frac{\omega_2}{f_1} + \frac{\omega_2}{f_2} = 0 \Rightarrow \frac{1}{2f_1} + \frac{1}{f_2} = 0$$

$$\therefore \frac{1}{f_1} - \frac{1}{2f_1} = -\frac{1}{50} \Rightarrow \frac{1}{2f_1} = -\frac{1}{50} \Rightarrow f_1 = -25 \text{ cm}$$

$$\therefore \frac{1}{f_2} = \frac{1}{50} \Rightarrow f_2 = 50 \text{ cm}$$

$$\frac{1}{f_1} = -\frac{1}{25} = (\mu-1)\left(+\frac{1}{\infty} - \frac{1}{r_2}\right) \Rightarrow \frac{\mu-1}{r_2} = \frac{1}{25}$$

$$\Rightarrow \frac{3/2-1}{r_2} = \frac{1}{25} \Rightarrow r_2 = 12.5 \text{ cm}$$

$$\frac{1}{50} = (\mu'-1)\left(\frac{1}{r_2} - \frac{1}{r_3}\right) = (1.625-1)\left(\frac{2}{25} - \frac{1}{r_3}\right) \Rightarrow r_3 = 20.8 \text{ cm}.$$

5. The mean r.i. of two specimens of glass are 1.52 and 1.66 respectively. The difference in the r.i.'s for the same two lines in the spectrum are 0.013 for the first and 0.022 for the

second. Find the focal length of the lens of the second glass which when combined with a convex lens of 50 cm focal length of the first will make an object-glass achromatic.

Solution.

$$\omega_1 = \frac{\mu_v - \mu_r}{(\mu - 1)} = \frac{0.013}{(1.52 - 1)} = .025$$

$$\omega_2 = \frac{\mu'_v - \mu'_r}{(\mu' - 1)} = \frac{0.022}{(1.66 - 1)} = .033$$

For achromatisation,

$$\frac{\omega_1}{f_1} + \frac{\omega_2}{f_2} = 0 \quad \Rightarrow \quad -\frac{.025}{50} + \frac{.033}{f_2} = 0 \quad \Rightarrow \quad f_2 = \frac{50 \times .033}{.025} = 66 \text{ cm}.$$

Theory Exercise

1. What do you mean by chromatic aberration? Explain axial and lateral chromatic aberration. What is an achromat? How can you make an achromat out of two lenses in contact?
2. Can you make a system achromatised w.r.t. all colours? If not, which part of the spectrum you should consider when you try to make achromatisation?
3. What is an apochromatic system? Find an expression for the diameter of the circle of least chromatic error.
4. Can you make an achromatic combination of two thin prisms? Derive the necessary condition.
5. Explain what do you mean by the dispersion of light, dispersive power and chromatic aberration of a lens. Derive the condition of a contact achromat.
6. What do you mean by chromatic aberration? When a single convex lens is used as a magnifying glass, the image almost appears free from colour when the lens is held close to the eye, but when such a lens is used as the object glass of a telescope, the colour effect is more marked. Explain.
7. How is the phenomenon of dispersion in a prism explained? How is it related to refraction? How can you produce deviation without dispersion and dispersion without deviation?
8. Explain what is meant by chromatic aberration. Find the condition for obtaining an achromatic combination of two thin lenses made of the same material.
9. Define dispersive power of a lens. Explain how you would combine conveniently a lens system to satisfy achromatic condition when they are (a) made of the same material, (b) made of different materials.

Numerical Problem Exercise

1. A system consisting of two crown glass lenses, is to be achromatic for the focal length, which is 40 cm. The focal length of one lens being 25 cm, find the focal length of the other and the required distance between the lenses.

2. A convex lens of focal length 35 cm, achromatic for C and F lines is to be made from two glasses, the data for which are as follows:
Crown glass : $\mu_d = 1.5176$, $\mu_f - \mu_c = 0.00856$
Flint glass : $\mu_d = 1.6264$, $\mu_f - \mu_c = 0.01722$
Calculate the focal lengths of the two component lenses. [$f_1 = -13.94$ cm, $f_2 = 22.7$ cm.]

3. A beam of white light is made incident on a thin bi-convex lens in a direction parallel to its axis. If the r.i.'s of the lens are 1.514 and 1.524 for red and violet lights respectively and if the radii of curvature of its faces are 30 cm and 20 cm, calculate the separation of its foci for red and violet lights. [$f_r - f_v = 0.4454$ cm.]

4. A combination of two small-angle prisms is to be achromatised w.r.t. C and H_γ lines of hydrogen spectrum. The glasses used have the following r.i. values:

$C(\lambda = 6563$ Å$)$	$D(\lambda = 5893$ Å$)$	$H_\gamma(\lambda = 4340$ Å$)$
1.5210	1.5243	1.5345
1.6421	1.6475	1.6730

For a deviation of D line to be 1.5°, find the angles of the prisms. [6° 13′, 2° 43′]

5. An achromatic converging combination of focal length 100 cm is to be made of two lenses of glass as shown below. Find focal length of each of the component lenses:

	Glass 1	Glass 2
$\mu_r =$	1.5155	1.641
$\mu_v =$	1.5245	1.659

[$f_1 = -37.5$ cm, $f_2 = 60$ cm]

6. A camera lens ($f = 20$ cm) is made of two components, one of crown glass ($\omega_1 = 0.2$) and the other of flint glass ($\omega_2 = 0.4$). Find the focal lengths of the component lenses. [$f_1 = -10$ cm, $f_2 = 20$ cm]

7. A convex lens of crown glass ($\omega = 0.21$ and $\mu = 1.5$) is combined with a plano-concave lens of flint glass ($\omega = 0.45$ and $\mu = 1.6$) to form an achromatic doublet of power $+2D$. Calculate the radii of curvature of the two surfaces of the convex lens. [$r_1 = -21.82$ cm, $r_2 = 34.3$ cm]

8. A combination consisting of two thin lenses, one of crown glass and the other of flint glass, has the same focal length for a given pair of wavelengths. Derive expression for the powers of the crown and flint lenses in terms of the power of the combination and dispersive power of the glasses. [$P_1 = \dfrac{\omega_2 P}{\omega_2 - \omega_1}, P_2 = \dfrac{\omega_1 P}{\omega_1 - \omega_2}$ cm]

9. The focal length of an achromatic combination of two lenses in contact is 150 cm. If the dispersive powers of the lenses are 0.018 and 0.027, calculate the focal lengths of the two lenses. [$f_1 = -50$ cm, $f_2 = 75$ cm]

10. A telescope objective of focal length 100 cm is to be made from a contact combination of flint and crown glass lenses, achromatism being desired between Fraunhofer C (red) and F (blue) lines. If the r.i.'s for C and F lines are 1.5079 and 1.5152 respectively for crown glass and 1.5744 and 1.5885 respectively for flint glass, calculate the focal length for the crown glass lens for the mean colour between C and F. [$f = -36.36$ cm]

11. A flint glass lens and a crown glass lens placed in contact form an achromatic doublet of focal length 100 cm. The dispersive powers of crown and flint glass are 0.015 and 0.030

respectively. Find the focal lengths of the lenses. If one of the surfaces of flint glass is plane, calculate the radii of curvature of the other surfaces of the two lenses. Refractive indices of crown and flint glass for the mean ray are 1.52 and 1.65 respectively.

[f (flint) = 100 cm, f (crown) = −50 cm, r_1 = 160.1 cm, r_2 = 154 meter.]

12. If μ_b and μ_r be the r.i. values for blue and red light in crown glass and μ_b' and μ_r', those in flint glass and an achromatic doublet is formed out of an equiconvex crown glass lens of radius r placed in contact with a double concave lens of radii r and s, prove that

$$\frac{(\mu_b - \mu_r)}{(\mu_b' - \mu_r')} = \frac{r+s}{2s}.$$

13. Two convex lenses of focal lengths 20 and 30 cm and of crown glass and flint glass respectively are placed at a certain distance apart for achromatism. The dispersive powers for the two materials are 0.01 and 0.02. Find the distance between them. [26.7 cm]

14. (a) In which type of optical element chromatic aberration does not occur? What is chromatic aberration? Obtain an expression for the chromatic aberration of a thin lens.

(b) Two thin lenses of focal lengths f_1 and f_2 are placed co-axially in air at a distance of d. Find an expression for the equivalent focal length of the combination. Hence obtain the condition for the combination to be achromatic. Can you make an achromatic doublet with the two lenses in contact?

Chapter 9

Optical Instruments

9.1 Introduction

There is a long list of optical instruments starting with our eyes and ending with telescope and microscope. These instruments are designed on the basis of the consideration of Geometrical Optics. In this chapter we will consider some of them which have significance of considerable magnitude and practical utility.

In most of the cases our eye is the ultimate receptor to see objects. The construction of the eýe and its defects are given in elementary text. Here we will consider only the other instruments like telescopes and microscopes.

9.2 Telescope

It is an optical instrument by means of which distant objects are made distinct and enlarged. The first telescope was made by Jan Lippershey in Holland in the year 1608. After one year, in 1609, Galileo made an improved variety of telescope which enabled him to observe the ring of the saturn and also the satellite of venus. The telescope meant for observing the celestial bodies is termed as Astronomical Telescope while a telescope used for observing distant terrestrial objects is called *Terrestrial Telescope.*

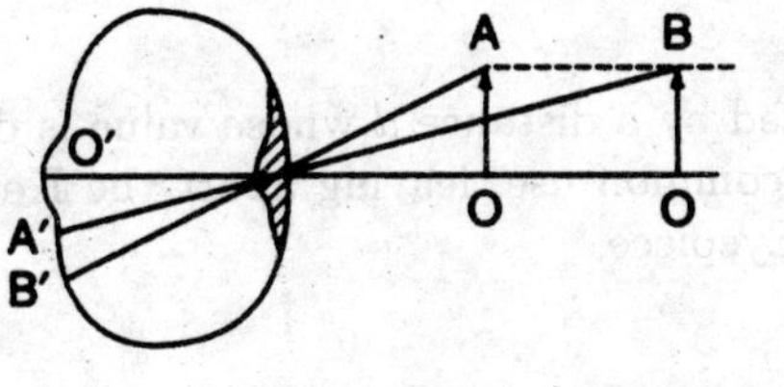

Fig 9.1(a)

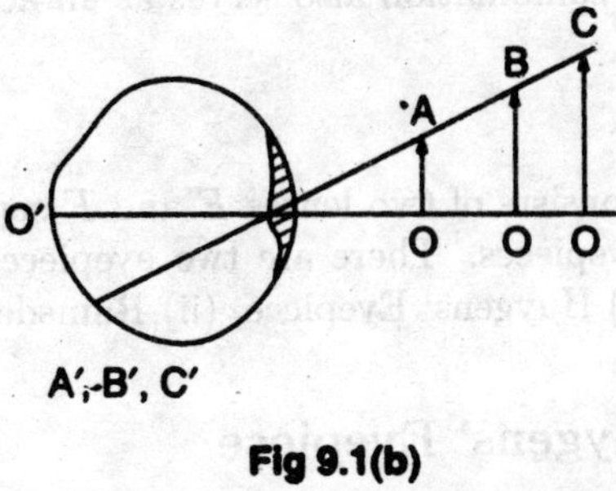

Fig 9.1(b)

The function of any optical instrtument is to increase the *visual angle.* When an object is placed before our eyes, depending on the size, it subtends a smaller or larger angle. In fig.9.1(a), the same object when shifted away and away from the eye makes small visual angle, hence forming an as image on the retina of smaller and smaller size.

In fig.9.1(b), objects of larger and larger size but placed at greater and greater distance make the same visual angle at the eye-lens and produce image of the same size. Thus it appears that greater the visual angle, greater becomes the size of the image. The aim of any optical device is to increase the visual angle.

Astronomical Telescope

The essential parts of an astronomical telescope are (i) objective lens and (ii) eyepiece.

Objective Lens

This is the lens system facing the object. Usually it is a combination of the double convex crown glass lens and a plano-concave flint glass lens. The combination behaves as a plano-convex lens of large focal length say, f_o.

In such a combination, when the convex surface faces the incoming rays, it behaves so as to minimise spherical aberration.

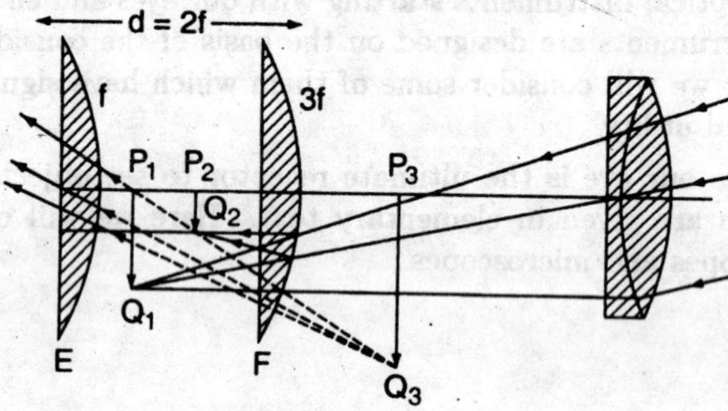

Fig 9.2

The dispersive powers and focal lengths are so chosen that they satisfy

$$\frac{\omega_1}{f_1} + \frac{\omega_2}{f_2} = 0 \quad (9.2.1)$$

So this combination also serves as an achromatic doublet.

Eyepiece

Eyepiece consists of two lenses E and F separated by a distance d whose value is different for different eyepieces. There are two eyepieces in common use, leaving apart the Kellner's one; they are (i) Huygens' Eyepiece, (ii) Ramsden's Eyepiece.

9.3 Huygens' Eyepiece

In fig.(9.2), the astronomical telescope is fitted with a Huygens' Eyepiece. This eyepiece consists of two plano-convex lenses E, called the *eye-lens* nearer to the eye of the observer and F, called the *field-lens* which is away from the eye of the observer. The eye-lens is of focal length f and the field-lens, of focal length $3f$. They are kept separated by a distance $d = 3f - f = 2f$ with the convex face of each towards the incident light.

The Optics of the Huygens' Eyepiece

The process of image-formation by this optical device consists of first translation, represented by the translation matrix $[T_1]$, then refraction through the field-lens of focal length f_1 (say) represented by a refraction matrix $[R_1]$, followed by translation in air represented by translation matrix $[T_2]$. Then again refraction through the eye-lens of focal length f_2 (say) represented by refraction matrix $[R_2]$ and finally, translation to the final image-position, represented by $[T_3]$. So the column vector representing the image can be expressed through a series of matrix product written as

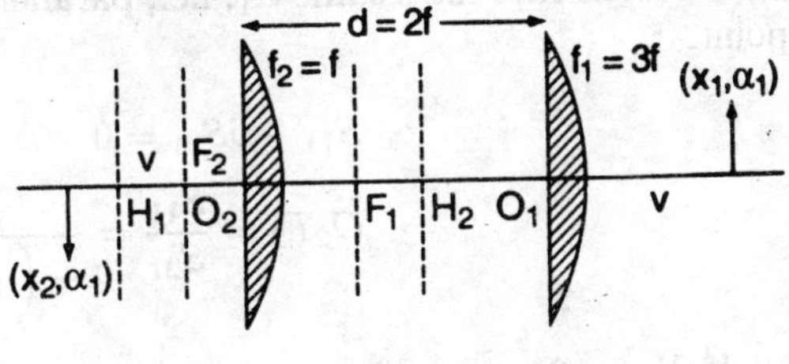

Fig 9.3

$$\begin{bmatrix} x_2 \\ \alpha_2 \end{bmatrix} = [T_3[R_2][T_2][R_1][T_1]\begin{bmatrix} x_1 \\ \alpha_1 \end{bmatrix}$$
$$= \begin{bmatrix} 1 & -v \\ 0 & 1 \end{bmatrix}\begin{bmatrix} 1 & 0 \\ -1/f_2 & 1 \end{bmatrix}\begin{bmatrix} 1 & +d \\ 0 & 1 \end{bmatrix}\begin{bmatrix} 1 & 0 \\ -1/f_1 & 1 \end{bmatrix}\begin{bmatrix} 1 & u \\ 0 & 1 \end{bmatrix}\begin{bmatrix} x_1 \\ \alpha_1 \end{bmatrix}$$
$$= \begin{bmatrix} 1 & -v \\ 0 & 1 \end{bmatrix}[S]\begin{bmatrix} 1 & u \\ 0 & 1 \end{bmatrix}\begin{bmatrix} x_1 \\ \alpha_1 \end{bmatrix} \qquad (9.3.1)$$

where $[S]$ is the system matrix defined as

$$[S] = [R_2][T_2][R_1]$$
$$\begin{bmatrix} 1 & 0 \\ -1/f_2 & 1 \end{bmatrix}\begin{bmatrix} 1 & +d \\ 0 & 1 \end{bmatrix}\begin{bmatrix} 1 & 0 \\ -1/f_1 & 1 \end{bmatrix} = \begin{bmatrix} 1 & 0 \\ -1/f_2 & 1 \end{bmatrix}\begin{bmatrix} (1-d/f_1) & d \\ -1/f_1 & 1 \end{bmatrix}$$
$$= \begin{bmatrix} (1-d/f_1) & d \\ \left\{-\frac{1}{f_2}\left(1-\frac{d}{f_1}\right)-\frac{1}{f_1}\right\} & \left(-\frac{d}{f_2}+1\right) \end{bmatrix} = \begin{bmatrix} S_{11} & S_{12} \\ S_{21} & S_{22} \end{bmatrix} \qquad (9.3.2)$$

$$\begin{bmatrix} x_2 \\ \alpha_2 \end{bmatrix} = \begin{bmatrix} 1 & -v \\ 0 & 1 \end{bmatrix}\begin{bmatrix} S_{11} & S_{12} \\ S_{21} & S_{22} \end{bmatrix}\begin{bmatrix} 1 & u \\ 0 & 1 \end{bmatrix}\begin{bmatrix} x_1 \\ \alpha_1 \end{bmatrix} = \begin{bmatrix} 1 & -v \\ 0 & 1 \end{bmatrix}\begin{bmatrix} S_{11} & (uS_{11}+S_{12}) \\ S_{21} & (uS_{21}+S_{22}) \end{bmatrix}\begin{bmatrix} x_1 \\ \alpha_1 \end{bmatrix}$$
$$= \begin{bmatrix} (S_{11}-vS_{21}) & \{(uS_{11}+S_{12})-(uS_{21}+S_{22})v\} \\ S_{21} & (uS_{21}+S_{22}) \end{bmatrix}\begin{bmatrix} x_1 \\ \alpha_1 \end{bmatrix}$$
$$= \begin{bmatrix} N_{11} & N_{12} \\ N_{21} & N_{22} \end{bmatrix}\begin{bmatrix} x_1 \\ \alpha_1 \end{bmatrix} \qquad (9.3.3)$$

$$\therefore \quad x_2 = N_{11}x_1 + N_{12}\alpha_1 \quad \text{and} \quad \alpha_2 = N_{21}x_1 + N_{22}\alpha_1.$$

So, if $N_{22} = 0$, we get, $\alpha_2 = N_{21}x_1$, which physically means that rays from the object point defined by x_1 for any value of α_1 after passing through the optical system are rendered parallel (same α_2 value). Hence, $N_{22} = 0$ defines the first principal focus of the system w.r.t. the first lens optical centre O_1.

$$\therefore \quad uS_{21} + S_{22} = 0 \quad \Rightarrow \quad O_1F_1S_{21} + S_{22} = 0$$
$$\Rightarrow \quad O_1F_1 = -\frac{S_{22}}{S_{21}} \quad \Rightarrow \quad \frac{1}{O_1F_1} = -\frac{S_{21}}{S_{22}}$$
$$\therefore \quad \frac{1}{O_1F_1} = -\frac{-\left(\frac{1}{f_1}+\frac{1}{f_2}-\frac{d}{f_1f_2}\right)}{1-\frac{d}{f_2}} = \frac{\frac{1}{3f}+\frac{1}{f}-\frac{2f}{3f^2}}{1-\frac{2f}{f}}$$

$$\Rightarrow \frac{1}{O_1F_1} = -\frac{(2/3f)}{-1} = \frac{2}{3f} \quad \Rightarrow O_1F_1 = -\frac{3f}{2}.$$

The second principal focus is defined by the eqn. $N_{11} = 0 \quad \Rightarrow \quad x_2 = N_{12}\alpha_1$, implies that the incident rays have same α_1, i.e., parallel incident rays after refraction pass through a single point.

$$\therefore \; S_{11} - vS_{21} = 0$$

$$\Rightarrow O_2F_2 = \frac{S_{11}}{S_{21}} = \frac{1 - \frac{d}{f_1}}{-\left(\frac{1}{f_1} + \frac{1}{f_2} - \frac{d}{f_1 f_2}\right)} = \frac{1 - \frac{2}{3}}{-\frac{2}{3f}} = -\frac{f}{2}$$

If $N_{12} = 0$, we get,

$$x_2 = N_{11}x_1 \quad \Rightarrow \quad N_{11} = \frac{x_2}{x_1}.$$

If $\frac{x_2}{x_1} = 1$, it defines the planes of unit magnification, hence defines the principal planes:

$$\therefore \; S_{11} - vS_{21} = 1$$

$$\Rightarrow O_2H_2 = \frac{(S_{11} - 1)}{S_{21}} = \frac{\left(1 - \frac{d}{f_1}\right) - 1}{-\left(\frac{1}{f_1} + \frac{1}{f_2} - \frac{d}{f_1 f_2}\right)} = \frac{-\frac{2}{3}}{-\frac{2}{3f}} = +f$$

The corresponding conjugate plane, called the *first principal plane* is given by,

$$\therefore \; uS_{11} + S_{12} - uvS_{21} - vS_{22} = 0$$

$$\Rightarrow u(S_{11} - vS_{21}) = -(S_{12} - vS_{22}) \quad \Rightarrow \quad u = -(S_{12} - vS_{22})$$

Now,

$$\frac{S_{11} - 1}{S_{21}} = v$$

$$\therefore \; S_{12} - \frac{(S_{11} - 1)}{S_{21}} S_{22} = d - \frac{\left(-\frac{d}{f_1}\right)\left(1 - \frac{d}{f_2}\right)}{-\left[\frac{1}{f_1} + \frac{1}{f_2} - \frac{d}{f_1 f_2}\right]} = d + \frac{\frac{d^2}{f_1 f_2} - \frac{d}{f_1}}{\frac{1}{f_1} + \frac{1}{f_2} - \frac{d}{f_1 f_2}}$$

$$\therefore \; u = O_1H_1 = -\frac{\frac{d}{f_2}}{\frac{1}{f_1} + \frac{1}{f_2} - \frac{d}{f_1 f_2}} = -3f$$

The equivalent focal length of the optical system is given by,

$$H_2F_2 = O_2F_2 - O_2H_2 = -\frac{f}{2} - f = -\frac{3}{2}f.$$

The position of the equivalent lens is given by,

$$X_o = O_2H_2 = f.$$

Thus, the Huygens' eyepiece is equivalent to a lens of focal length (second principal focal length) $-\frac{3}{2}f$ and to be considered at a distance of f to the right of the eye-lens E. The positions of the different cardinal points and planes are indicated in fig.9.3.

9.4 Ramsden's Eyepiece

This eyepiece also consists of two plano-convex lenses of equal focal lengths, but the field-lens is placed with its plane face towards the incident ray. So the system matrix in this case can be written as

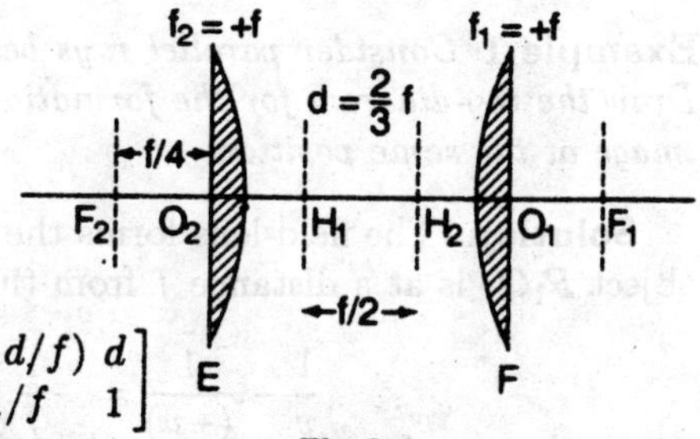

Fig 9.4

$$[S] = \begin{bmatrix} 1 & 0 \\ -1/f_2 & 1 \end{bmatrix}\begin{bmatrix} 1 & d \\ 0 & 1 \end{bmatrix}\begin{bmatrix} 1 & 0 \\ -1/f_1 & 1 \end{bmatrix}$$

$$= \begin{bmatrix} 1 & 0 \\ -1/f & 1 \end{bmatrix}\begin{bmatrix} 1 & d \\ 0 & 1 \end{bmatrix}\begin{bmatrix} 1 & 0 \\ -1/f & 1 \end{bmatrix} = \begin{bmatrix} 1 & 0 \\ -1/f & 1 \end{bmatrix}\begin{bmatrix} (1-d/f) & d \\ -1/f & 1 \end{bmatrix}$$

$$= \begin{bmatrix} (1-d/f) & d \\ -(1/f)(1-d/f) - 1/f & (1-d/f) \end{bmatrix} = \begin{bmatrix} S_{11} & S_{12} \\ S_{21} & S_{22} \end{bmatrix}$$

The discussion is similar to the case with a Huygens' eyepiece with only difference in the system matrix elements.

$$\therefore \quad O_1F_1 = -\frac{S_{22}}{S_{21}} = -\frac{1-d/f}{-\frac{1}{f}\left(1-\frac{d}{f}\right)-\frac{1}{f}} = -\frac{1-2/3}{-\frac{1}{f}\left(1-\frac{2}{3}\right)-\frac{1}{f}} = -\frac{1/3}{-(4/3f)} = \frac{1}{4}f$$

$$\text{and} \quad O_2F_2 = \frac{S_{11}}{S_{21}} = \frac{1-d/f}{-(4/3f)} = -\frac{1/3}{(4/3f)} = -\frac{1}{4}f.$$

$$\therefore \quad O_2H_2 = \frac{S_{11}-1}{S_{21}} = \frac{-d/f}{-(4/3f)} = +\frac{2}{3}\cdot\frac{3f}{4} = \frac{1}{2}f$$

$$\text{and} \quad O_1H_1 = -\left\{S_{12} - \frac{(S_{11}-1)S_{22}}{S_{21}}\right\} = -\left\{d - \frac{(-d/f)\cdot(1-d/f)}{-(4/3f)}\right\}$$

$$= -\left\{d - \frac{(-2/3)\cdot(1/3)}{-(4/3f)}\right\} = -\left(\frac{2}{3}f - \frac{f}{6}\right) = -\frac{1}{2}f.$$

The equivalent focal length is given by,

$$H_2F_2 = O_2F_2 - O_2H_2 = -\left(\frac{1}{4}+\frac{1}{2}\right)f = -\frac{3f}{4}.$$

The position of the equivalent lens is

$$X_o = O_2H_2 = +\frac{1}{2}f$$

in front of the eye-lens.

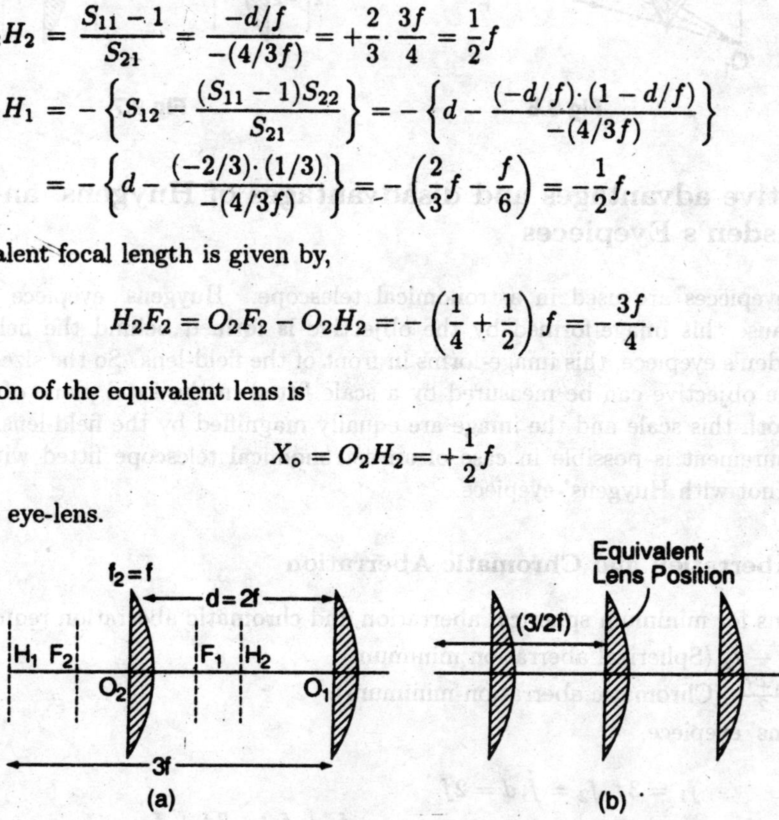

Fig 9.5

So the cardinal points of Huygens' and Ramsden's eyepieces are shown in the figs.9.5(a) and 9.5(b).

Example 1 *Consider parallel rays being incident on the field-lens of the Huygens' eyepiece. Draw the ray-diagram for the formation of image. Show that the equivalent lens also forms the image at the same position.*

Solution. The field-lens forms the image at the focal plane of the field-lens. So the virtual object P_1Q_1 is at a distance f from the eye-lens.

$$\therefore \quad \frac{1}{v} - \frac{1}{(-u)} = -\frac{1}{f} \; \Rightarrow \quad \frac{1}{v} = -\frac{1}{u} - \frac{1}{f} = -\frac{1}{f} - \frac{1}{f} = -\frac{2}{f},$$

$v = -f/2$ behind the eye-lens. The equivalent lens is considered at a distance f from the field-lens. If the parallel rays fall on the equivalent lens, the image will be formed at the focal plane of this lens, hence it is at a distance $(f + 3f/2)$, i.e., $(5/2)f$ from the field-lens. Hence the equivalence.

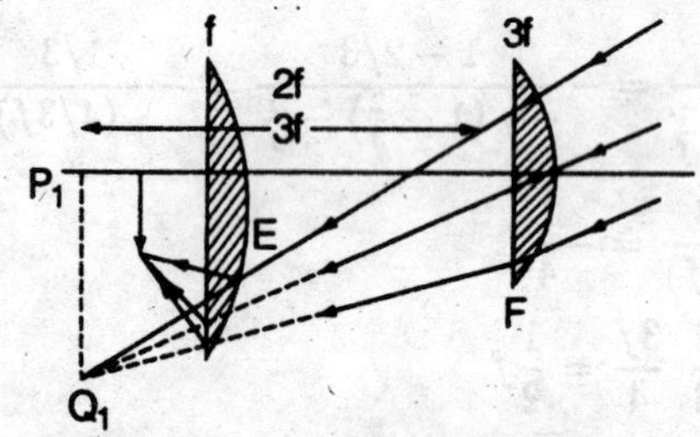

Fig 9.6

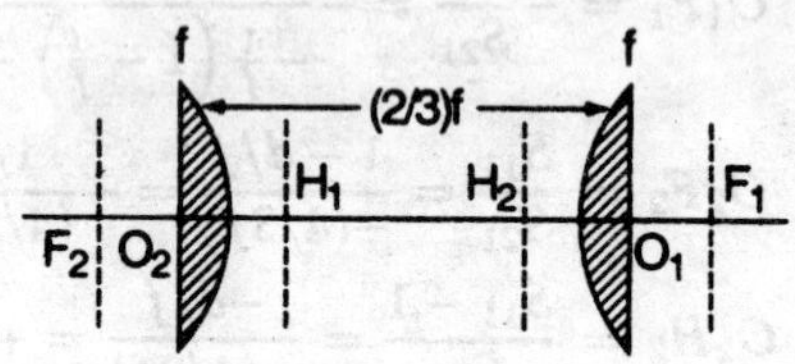

Fig 9.7

9.5 Relative advantages and disadvantages of Huygens' and Ramsden's Eyepieces

Both these eyepieces are used in astronomical telescope. Huygens' eyepiece is a negative eyepiece because this image-formed by the objective is formed behind the field-lens but in case of Ramsden's eyepiece, this image-forms in front of the field-lens. So the size of the image-formed by the objective can be measured by a scale fitted in the focal plane of the objective lens. Since both this scale and the image are equally magnified by the field-lens and eye-lens, reliable measurement is possible in case of an astronomical telescope fitted with Ramsden's eyepiece and not with Huygens' eyepiece.

Spherical Aberration and Chromatic Aberration

The conditions for minimum spherical aberration and chromatic aberration require

(i) $d = f_1 - f_2$ (Spherical aberration minimum)

(ii) $d = \frac{f_1+f_2}{2}$ (Chromatic aberration minimum).

In Huygens' eyepiece,

$$f_1 = 3f, f_2 = f, d = 2f$$

$$\therefore \; f_1 - f_2 = 3f - f = 2f = d; \quad \frac{f_1 + f_2}{2} = \frac{3f + f}{2} = 2f = d.$$

Thus in Huygens' eyepiece both the minimisation conditions are satisfied.

In Ramsden's eyepiece,

$$f_1 = f = f_2, \quad d = (3/2)f$$

thus the conditions of minimisation for neither of spherical aberration nor of chromatic aberration are satisfied.

Now we are in a position to consider astronomical telescope fitted with eyepieces.

The formation of the image in an astronomical telescope fitted with Huygens' eyepiece is shown in fig.9.2. The image by the objective was going to be formed as P_1Q_1 at the focal plane of the objective but the rays are intercepted by the field-lens at P_2Q_2. If P_2Q_2 is formed exactly at the focus of the eye-lens, the final image is formed at ∞, and the astronomical telescope is said to be in normal adjustment. But if P_2Q_2 is within the focus such that the final image is formed at the least distance of distinct vision ($D = 25$ cm).

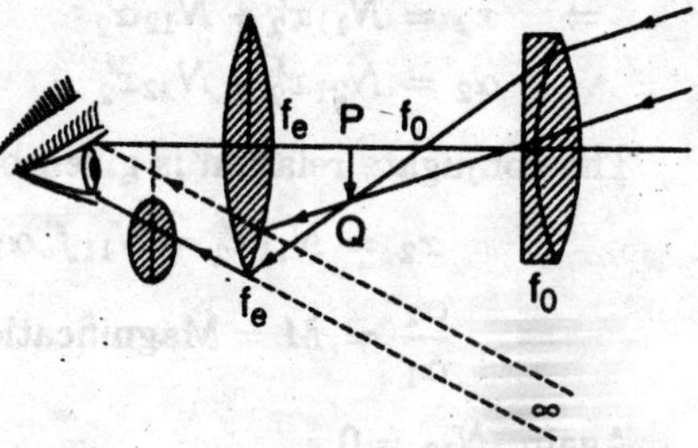

Fig 9.8

The fig.9.2 showing the image-formation in an astronomical telescope is re-drawn in fig.(9.8). where the Huygens' eyepiece is shown by its equivalent representation. The image PQ formed by the objective is formed in front of the equivalent lens, equivalent of the eyepiece. If it is at the focus of the equivalent lens of focal length f_e, the final image is formed at ∞, and the telescope is said to be in normal adjustment. But, if it is slightly within the focus such that the final image is at the least distance of distinct vision, then it is said to be in distinct vision, then it is said to be in distinct vision.

9.6 Astronomical Telescope in Distinct Vision

The eye is placed close to the equivalent eye-lens of focal length f_e. The first image is formed at PQ at the focal plane of the objective and it is within the focus of the equivalent eye-lens such that the final image P_1Q_1 is formed at the least distance of distinct vision.

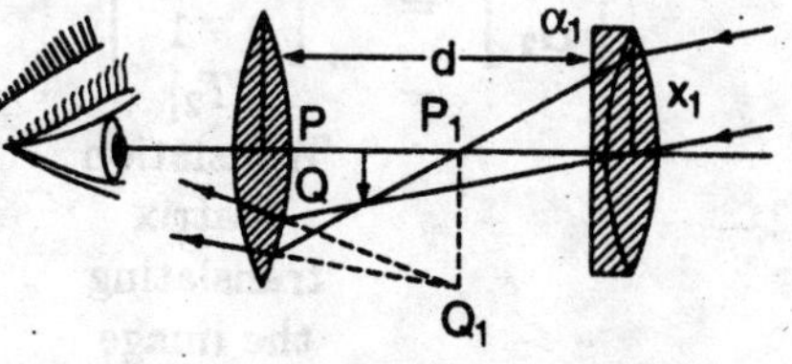

Fig 9.9

The parallel rays incident on the objective lens are represented by the column vector $\begin{bmatrix} x_1 \\ \alpha_1 \end{bmatrix}$. The formation of the image PQ represented by the column vector $\begin{bmatrix} x_2' \\ \alpha_2' \end{bmatrix}$ is given as

$$\begin{bmatrix} x_2' \\ \alpha_2' \end{bmatrix} = \begin{bmatrix} 1 & -v_1 \\ 0 & 1 \end{bmatrix} \begin{bmatrix} 1 & v \\ -1/f_\circ & 1 \end{bmatrix} \begin{bmatrix} x_1 \\ \alpha_1 \end{bmatrix} = \begin{bmatrix} 1 + v_1/f_\circ & -v_1 \\ -1/f_\circ & 1 \end{bmatrix} \begin{bmatrix} x_1 \\ \alpha_1 \end{bmatrix}$$

$$x_2' = (1 + v_1/f_\circ)x_1 - v_1\alpha_1 \quad (9.6.1)$$

$$\alpha_2' = -\frac{x}{f_\circ} + \alpha_1 \quad (9.6.2)$$

Putting $(1 + v_1/f_\circ) = 0$, in eqn.(9.6.1) we get $x_2' = -v_1\alpha_1$, which means parallel rays pass through a point in the image-space.

$\therefore \quad v_1 = -f_o$, meaning the image PQ is formed at a distance f_o behind the objective lens. Let this image be at a distance u in front of the eye-lens. Then the final image represented by the column vector $\begin{bmatrix} x_2 \\ \alpha_2 \end{bmatrix}$ formed at the least distance of distinct vision D is given by,

$$\begin{bmatrix} x_2 \\ \alpha_2 \end{bmatrix} = \begin{bmatrix} 1 & -D \\ 0 & 1 \end{bmatrix} \begin{bmatrix} 1 & 0 \\ -1/f_e & 1 \end{bmatrix} \begin{bmatrix} 1 & u \\ 0 & 1 \end{bmatrix} \begin{bmatrix} x_2' \\ \alpha_2' \end{bmatrix} = \begin{bmatrix} 1 & -D \\ 0 & 1 \end{bmatrix} \begin{bmatrix} 1 & u \\ -1/f_e & 1-u/f_e \end{bmatrix} \begin{bmatrix} x_2' \\ \alpha_2' \end{bmatrix}$$

$$= \begin{bmatrix} \left(1+\frac{D}{f_e}\right) & \left\{u - D\left(1-\frac{u}{f_e}\right)\right\} \\ -\frac{1}{f_e} & \left(1-\frac{u}{f_e}\right) \end{bmatrix} \begin{bmatrix} x_2' \\ \alpha_2' \end{bmatrix} = \begin{bmatrix} N_{11} & N_{12} \\ N_{21} & N_{22} \end{bmatrix} \begin{bmatrix} x_2' \\ \alpha_2' \end{bmatrix}$$

$$\Rightarrow \quad x_2 = N_{11}x_2' + N_{12}\alpha_2' \tag{9.6.3}$$

$$\alpha_2 = N_{21}x_2' + N_{22}x_2' \tag{9.6.4}$$

The conjugate relation is given by $N_{12} = 0$, using this we get,

$$x_2 = N_{11}x_2' = N_{11}f_o\alpha_1 \quad \Rightarrow \quad -D\alpha_2 = N_{11}f_o\alpha_1$$

$$\Rightarrow \quad \frac{\alpha_2}{\alpha_1} = M = \text{Magnification} = -\frac{N_{11}f_o}{D} = -\frac{f_o}{D}\left(1+\frac{D}{f_e}\right) = -\frac{f_o}{f_e}\left(1+\frac{f_e}{D}\right). \tag{9.6.5}$$

Again, $N_{12} = 0$

$$\Rightarrow \quad u - D\left(1-\frac{u}{f_e}\right) = 0 \quad \Rightarrow \quad u\left(1+\frac{D}{f_e}\right) = D \quad \Rightarrow \quad u = \frac{Df_e}{D+f_e} \tag{9.6.6}$$

which proves that the image PQ is formed in front of the eye-lens.

9.7 Telescope in Normal Adjustment

Refer to fig.9.9. The matrix equivalent representation of the process of image-formation is shown as,

$$\begin{bmatrix} x_2 \\ \alpha_2 \end{bmatrix} = \begin{bmatrix} 1 & -v \\ 0 & 1 \end{bmatrix} \begin{bmatrix} 1 & 0 \\ -f_e & 1 \end{bmatrix} \begin{bmatrix} 1 & d \\ 0 & 1 \end{bmatrix} \begin{bmatrix} 1 & 0 \\ -1/f_o & 1 \end{bmatrix} \begin{bmatrix} x_1 \\ \alpha_1 \end{bmatrix}$$

	$[T_2]$	$[R_e]$	$[T_1]$	$[R_o]$
=	Translation matrix translating the image of eye-lens	Eye-lens refraction matrix	Translation matrix between objective lens and eye-lens	Refraction matrix for objective

$$\begin{bmatrix} x_2 \\ \alpha_2 \end{bmatrix} = \begin{bmatrix} 1 & -v \\ 0 & 1 \end{bmatrix} \begin{bmatrix} 1 & 0 \\ -1/f_e & 1 \end{bmatrix} \begin{bmatrix} (1-d/f_o) & d \\ -1/f_o & 1 \end{bmatrix} \begin{bmatrix} x_1 \\ \alpha_1 \end{bmatrix}$$

$$= \begin{bmatrix} 1 & -v \\ 0 & 1 \end{bmatrix} \begin{bmatrix} (1-d/f_o) & d \\ \left\{-\frac{1}{f_e}(1-d/f_o) - \frac{1}{f_o}\right\} & (1-d/f_e) \end{bmatrix} \begin{bmatrix} x_1 \\ \alpha_1 \end{bmatrix} = \begin{bmatrix} 1 & -v \\ 0 & 1 \end{bmatrix} \begin{bmatrix} S_{11} & S_{12} \\ S_{21} & S_{22} \end{bmatrix} \begin{bmatrix} x_1 \\ \alpha_1 \end{bmatrix}$$

$$= \begin{bmatrix} (S_{11} - vS_{21}) & (S_{12} - vS_{22}) \\ S_{21} & S_{22} \end{bmatrix} \begin{bmatrix} x_1 \\ \alpha_1 \end{bmatrix} = \begin{bmatrix} N_{11} & N_{12} \\ N_{21} & N_{22} \end{bmatrix} \begin{bmatrix} x_1 \\ \alpha_1 \end{bmatrix}$$

$$x_2 = N_{11}x_1 + N_{12}\alpha_1 \tag{9.7.1}$$

$$\alpha_2 = N_{21}x_1 + N_{22}\alpha_1 \tag{9.7.2}$$

For an axial point, $x_1 = 0$, so we get,

$$\alpha_2 = N_{22}\alpha_1 \quad \Rightarrow \quad \frac{\alpha_2}{\alpha_1} = M = \text{Magnification} = N_{22} = \left(1 - \frac{d}{f_e}\right)$$

$$\therefore \quad M = \left(1 - \frac{d}{f_e}\right) \text{ but } d = f_o + f_e \quad \Rightarrow \quad M = -\frac{f_o}{f_e} \tag{9.7.3}$$

The expression for M in eqn.(9.7.3) can also be deduced from eqn.(9.6.5).

Relation between linear magnification (m) and angular magnification in case of normal adjustment of a telescope

Putting $N_{12} = 0$, we get,

$$m = \frac{x_2}{x_1} = N_{11} \quad \text{and} \quad M = \frac{\alpha_2}{\alpha_1} = N_{22}$$

$$\therefore \quad \frac{M}{(1/m)} = N_{22} \cdot N_{11} = 1$$

Remember, if $N_{21} = 0$, the system is said to be telescopic system.

9.8 Galilean Telescope

It produces an erect image of the object, hence suitable for use as a terrestrial telescope.

It consists of a plano-convex objective and a concave eye-lens. The ray diagram for the image-formation is shown in fig.9.10.

Parallel rays form image PQ due to refraction through the objective of focal length f_o. Due to refraction by the eye-lens, the final image is formed as P_1Q_1 at the least distance of distinct vision.

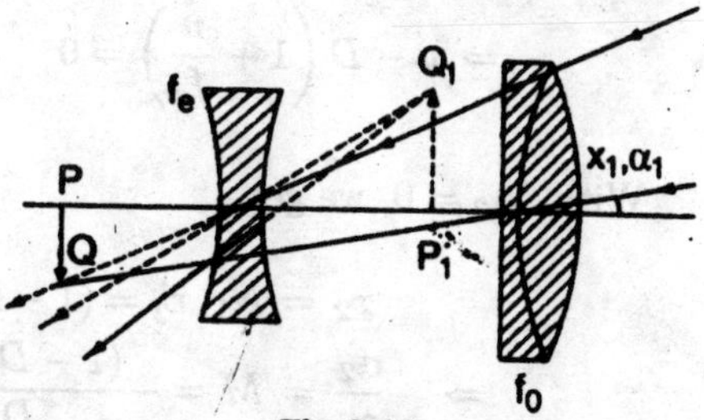

Fig 9.10

Matrix formulation of the image

$\begin{bmatrix} x_2' \\ \alpha_2' \end{bmatrix}$ **represents the column vector of the image** PQ which can be written as,

$$\begin{bmatrix} x_2' \\ \alpha_2' \end{bmatrix} = \underset{\text{Translation matrix}}{\begin{bmatrix} 1 & -v_1 \\ 0 & 1 \end{bmatrix}} \quad \underset{\text{Refraction matrix of the objective}}{\begin{bmatrix} 1 & 0 \\ -1/f_o & 1 \end{bmatrix}} \quad \begin{bmatrix} x_1 \\ \alpha_1 \end{bmatrix}$$

$$\Rightarrow \begin{bmatrix} x_2' \\ \alpha_2' \end{bmatrix} = \begin{bmatrix} (1 + v_1/f_o) & -v_1 \\ -1/f_o & 1 \end{bmatrix} \begin{bmatrix} x_1 \\ \alpha_1 \end{bmatrix}$$

$$\Rightarrow x_2' = \left(1 + \frac{v_1}{f_o}\right) x_1 - \alpha_1 v_1 \quad \text{and} \quad \alpha_2' = -\frac{x_1}{f_o} - \left(\frac{\alpha_1}{f_o}\right).$$

Setting $\left(1+\frac{v_1}{f_\circ}\right) = 0$, we get, $x_2' = -\alpha_1 v_1$ representing parallel rays (same α_1) passing through a point,

$$\therefore \quad v_1 = -f_2, \text{ is the image by the objective.}$$

The final image P_1Q_1 represented by the column vector $\begin{bmatrix} x_2 \\ \alpha_2 \end{bmatrix}$ is obtained as,

$$\begin{bmatrix} x_2 \\ \alpha_2 \end{bmatrix} = \begin{bmatrix} 1 & -D \\ 0 & 1 \end{bmatrix} \begin{bmatrix} 1 & 0 \\ (1/f_e) & 1 \end{bmatrix} \begin{bmatrix} 1 & u \\ 0 & 1 \end{bmatrix} \begin{bmatrix} x_2' \\ \alpha_2' \end{bmatrix}$$

Image	Translation matrix	Refraction matrix for the eye-lens	Translation matrix	Object

$$= \begin{bmatrix} 1 & -D \\ 0 & 1 \end{bmatrix} \begin{bmatrix} 1 & u \\ (1/f_e) & \{1+(u/f_e)\} \end{bmatrix} \begin{bmatrix} x_2' \\ \alpha_2' \end{bmatrix} = \begin{bmatrix} 1 & -D \\ 0 & 1 \end{bmatrix} \begin{bmatrix} S_{11} & S_{12} \\ S_{21} & S_{22} \end{bmatrix} \begin{bmatrix} x_2' \\ \alpha_2' \end{bmatrix}$$

$$= \begin{bmatrix} (1-DS_{21}) & (S_{12}-DS_{22}) \\ S_{21} & S_{22} \end{bmatrix} \begin{bmatrix} x_2' \\ \alpha_2' \end{bmatrix} = \begin{bmatrix} N_{11} & N_{12} \\ N_{21} & N_{22} \end{bmatrix} \begin{bmatrix} x_2' \\ \alpha_2' \end{bmatrix}$$

$$\Rightarrow x_2 = N_{11}x_2' + N_{12}\alpha_2' \quad \text{and} \quad \alpha_2 = N_{21}x_2' + N_{22}\alpha_2'.$$

$$N_{12} = 0 \quad \text{defines conjugate relation} \quad \Rightarrow \quad S_{12} - DS_{22} = 0$$

$$\Rightarrow u - D\left(1+\frac{u}{f_e}\right) = 0 \quad \Rightarrow \quad u\left(1-\frac{D}{f_e}\right) = D \Rightarrow \quad u = \frac{Df_e}{(f_e - D)}.$$

With $N_{12} = 0$, we get,

$$x_2 = N_{11}x_2' = (1-DS_{21})f_\circ\alpha_1 \quad \Rightarrow \quad -D\alpha_2 = (1-DS_{21})f_\circ\alpha_1$$

$$\Rightarrow \quad \frac{\alpha_2}{\alpha_1} = M = -\frac{(1-DS_{21})}{D}f_\circ = -\left(1-\frac{D}{f_e}\right)\frac{f_\circ}{D}$$

$$= -\frac{f_\circ}{f_e}\left(\frac{f_e}{D}-1\right) = \frac{f_\circ}{f_e}\left(1-\frac{f_e}{D}\right).$$

Advantage of this telescope is that we can get erect image in this telescope whereas in astronomical telescope, the image is inverted. The magnification M for an astronomical telescope is usually greater than that of a Galilean telescope. Measurement of the size of the image is not possible.

9.9 Microscope

It is an instrument which magnifies tiny objects placed near the instrument. Starting from a mere magnifying glass used as a simple microscope, the gradual developments have led to the invention of compound microscope whose resolving power (discussed later in Vol. II – Physical Optics) is limited by the diffraction effect, then to the electron microscope and then to the proton microscope of high resolution. We will consider here only the optical microscopes.

Magnifying Glass

An object PQ is placed within the focus of the convex lens of focal length $f_\circ$; the final image is formed as P_1Q_1 at the least distance of distinct vision (D).

Let the object be at a distance $u < f_\circ$, the object ray is defined by the column vector $\begin{bmatrix} x_1 \\ \alpha_1 \end{bmatrix}$.

The formulation of the image $\begin{bmatrix} x_2 \\ \alpha_2 \end{bmatrix}$ through matrix transformation is represented as

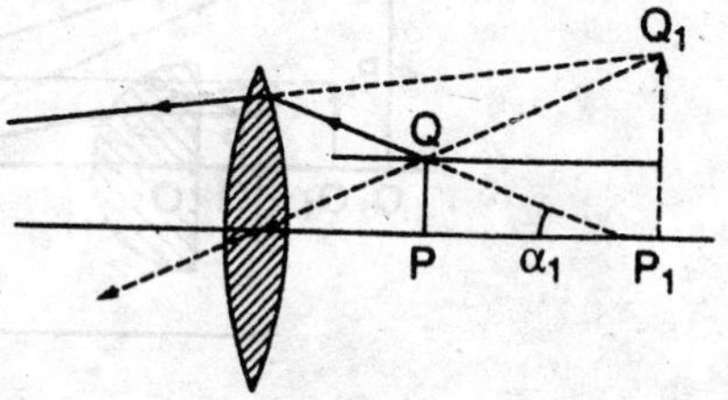

Fig 9.11

$$\begin{bmatrix} x_2 \\ \alpha_2 \end{bmatrix} = \begin{bmatrix} 1 & -D \\ 0 & 1 \end{bmatrix} \begin{bmatrix} 1 & 0 \\ -\frac{1}{f_\circ} & 1 \end{bmatrix} \begin{bmatrix} 1 & u \\ 0 & 1 \end{bmatrix} \begin{bmatrix} x_1 \\ \alpha_1 \end{bmatrix} = \begin{bmatrix} 1 & -D \\ 0 & 1 \end{bmatrix} \begin{bmatrix} 1 & u \\ -\frac{1}{f_\circ} & \left(1 - \frac{u}{f_\circ}\right) \end{bmatrix} \begin{bmatrix} x_1 \\ \alpha_1 \end{bmatrix}$$

$$= \begin{bmatrix} 1 & -D \\ 0 & 1 \end{bmatrix} \begin{bmatrix} S_{11} & S_{12} \\ S_{21} & S_{22} \end{bmatrix} \begin{bmatrix} x_1 \\ \alpha_1 \end{bmatrix} = \begin{bmatrix} (1 - DS_{21}) & (S_{12} - DS_{22}) \\ S_{21} & S_{22} \end{bmatrix} \begin{bmatrix} x_1 \\ \alpha_1 \end{bmatrix}$$

$$= \begin{bmatrix} N_{11} & N_{12} \\ N_{21} & N_{22} \end{bmatrix} \begin{bmatrix} x_1 \\ \alpha_1 \end{bmatrix}$$

$$\Rightarrow x_2 = N_{11}x_1 + \alpha_1 N_{12} \quad \text{and} \quad \alpha_2 = N_{21}x_1 + \alpha_1 N_{22}$$

$N_{12} = 0$ defines the conjugate rotation $\Rightarrow \quad S_{12} - DS_{22} = 0$

$$\Rightarrow u - D\left(1 - \frac{u}{f_\circ}\right) = 0 \Rightarrow u\left(1 + \frac{d}{f_\circ}\right) = D \quad \Rightarrow \quad u = \frac{Df_\circ}{D + f_\circ}$$

$$\therefore \quad D\alpha_2 = x_2 = N_{11}x_1$$

If the object is placed at a distance D, the object ray passing through the optical centre is defined by the column vector $\begin{bmatrix} x_1 \\ \alpha_2' \end{bmatrix}$ such that

$$x_1 = D\alpha_2'$$

$$\therefore \quad D\alpha_2 = N_{11} \cdot D\alpha_2' \quad \Rightarrow \quad \frac{\alpha_2}{\alpha_2'} = N_{11} = 1 - DS_{21}$$

$$= 1 - D\left(-\frac{1}{f_\circ}\right) = 1 + \frac{D}{f_\circ} = \text{angular magnification} = M_\alpha.$$

High Power Compound Microscope

An improved version of a simple microscope is a compound microscope having greater magnification and defect-free image. It consists mainly of the following components:

(i) Objective, (ii) Eyepiece.

Objective

It is a short-focus system of lenses. One component is incorporated to minimise spherical aberration consisting of a hemispherical lens immersed in an oil, then a concave-convex lens L_1 followed by an achromat lens L_2 (see fig.9.12).

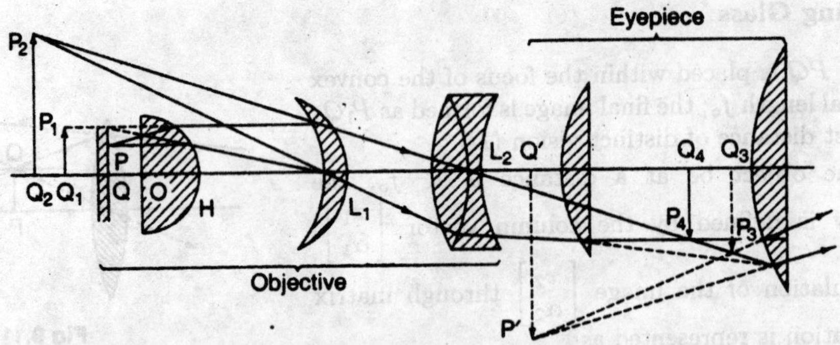

Fig 9.12

The object PQ is filled with cedar oil which also fills the space between the plane surface of the hemisphere and the thin slide; cedar oil has the same r.i. as that of glass. As discussed in the chapter on aberration, this arrangement serves to eliminate spherical aberration. The object PQ is placed at a distance $\frac{r}{\mu} = OQ$, where r is the radius and μ is the r.i. of the hemisphere H (same as that of the oil). If Q is the position of one aplanatic focus, hence PQ is at an aplanatic plane, the image by the semispherical lens will be formed at the other aplanatic plane as P_1Q_1 at a distance $OQ_1 = \mu r$ by all types of rays. The image P_1Q_1 is acting as the virtual source for the concavo-convex lens such that P_1Q_1 is at the centre of curvature of the concave surface of the lens L_1. So rays from Q_1 fall normally on the concave refractor and pass on to the convex surface undeviated. P_1Q_1 lies on one of the aplanatic plane and the image, P_2Q_2 is formed at the other aplanatic plane. The rays from this image-free from spherical aberration fall on the achromat L_2 which tends to converge the rays to form the real image P_3Q_3 but before. That the field-lens of the Huygens' eyepiece forms the image at P_4Q_4 and finally a magnified image is formed at $P'Q'$. This eyepiece is used because in this eyepiece both spherical aberration and chromatic aberration are minimised.

Magnification : Linear

The magnification of the entire instrument consisting of two magnifying units called *objective* and *eyepiece* is given by,

$$M = M_o \cdot M_e$$

where M_o is the magnifying power of the objective, M_e is the magnifying power of the eyepiece.

To find M_o

Let us consider simplified diagram as shown in fig.(9.13).

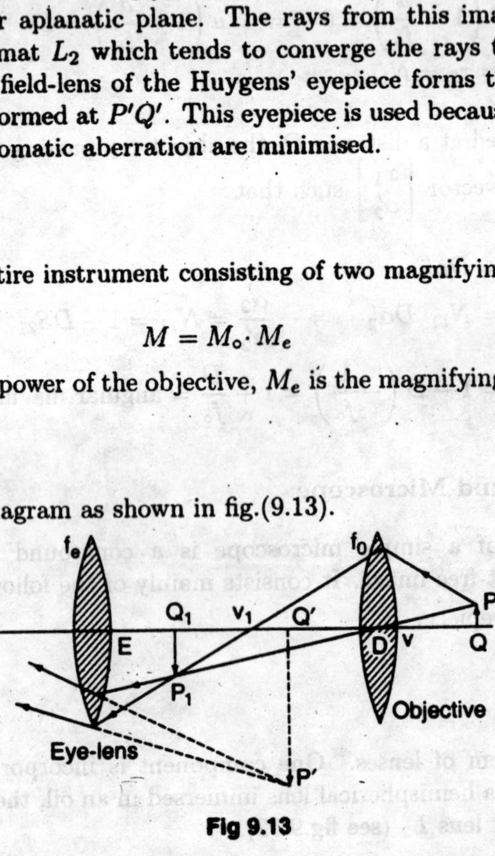

Fig 9.13

Let the objective arrangement be equivalent to a lens of focal length f_o and the eyepiece arrangement be equivalent to a convex lens of focal length f_e.

The image-formed by the objective lens at a distance v_1 from it is represented by the column vector $\begin{bmatrix} x_2' \\ \alpha_2' \end{bmatrix}$ given by,

$$\begin{bmatrix} x_2' \\ \alpha_2' \end{bmatrix} = \begin{bmatrix} 1 & -v_1 \\ 0 & 1 \end{bmatrix}\begin{bmatrix} 1 & 0 \\ -1/f_o & 1 \end{bmatrix}\begin{bmatrix} 1 & u \\ 0 & 1 \end{bmatrix}\begin{bmatrix} x_1 \\ \alpha_1 \end{bmatrix} = \begin{bmatrix} 1 & -v_1 \\ 0 & 1 \end{bmatrix}\begin{bmatrix} 1 & u \\ -1/f_o & (1-u/f_o) \end{bmatrix}\begin{bmatrix} x_1 \\ \alpha_1 \end{bmatrix}$$

$$= \begin{bmatrix} \left(1+\frac{v_1}{f_o}\right) & u - v_1\left(1-\frac{u}{f_o}\right) \\ -\frac{1}{f_o} & \left(1-\frac{u}{f_o}\right) \end{bmatrix}\begin{bmatrix} x_1 \\ \alpha_1 \end{bmatrix} = \begin{bmatrix} N_{11}' & N_{12}' \\ N_{21}' & N_{22}' \end{bmatrix}\begin{bmatrix} x_1 \\ \alpha_1 \end{bmatrix}$$

$$\Rightarrow x_2' = x_1 N_{11}' + N_{12}'\alpha_1 \quad \text{and} \quad \alpha_2' = x_1 N_{21}' + \alpha_1 N_{22}'.$$

If we put $N_{22}' = 0$, we get,

$$x_2' = N_{11}' x_1 \quad \Rightarrow \quad \frac{x_2'}{x_1} = M_o = N_{11}' = \left(1 + \frac{v_1}{f_o}\right).$$

Now, for a convex lens forming real image $v_1 = -\text{ve} = -v'$

$$\therefore \quad M_o = 1 - \frac{v'}{f_o} = \frac{f_o - v'}{f_o} = \frac{x}{f_o}.$$

For the second lens, let the object distance be u_1, then,

$$\begin{bmatrix} x_2 \\ \alpha_2 \end{bmatrix} = \begin{bmatrix} 1 & -D \\ 0 & 1 \end{bmatrix}\begin{bmatrix} 1 & 0 \\ -\frac{1}{f_e} & 1 \end{bmatrix}\begin{bmatrix} 1 & u_1 \\ 0 & 1 \end{bmatrix}\begin{bmatrix} x_2' \\ \alpha_2' \end{bmatrix} = \begin{bmatrix} 1 & -D \\ 0 & 1 \end{bmatrix}\begin{bmatrix} 1 & u_1 \\ -\frac{1}{f_e} & \left(1-\frac{u_1}{f_e}\right) \end{bmatrix}\begin{bmatrix} x_2' \\ \alpha_2' \end{bmatrix}$$

$$= \begin{bmatrix} \left(1+\frac{D}{f_e}\right) & \left\{u_1 - D\left(1-\frac{u_1}{f_e}\right)\right\} \\ -\frac{1}{f_e} & \left(1-\frac{u_1}{f_e}\right) \end{bmatrix}\begin{bmatrix} x_2' \\ \alpha_2' \end{bmatrix} = \begin{bmatrix} N_{11} & N_{12} \\ N_{21} & N_{22} \end{bmatrix}\begin{bmatrix} x_2' \\ \alpha_2' \end{bmatrix}$$

$N_{12} = 0$ defines conjugate relation, hence,

$$u_1 - D\left(1 - \frac{u_1}{f_e}\right) = 0 \quad \Rightarrow \quad u_1\left(1 + \frac{D}{f_e}\right) = D \quad \Rightarrow \quad u_1 = \frac{D}{1 + \frac{D}{f_e}}$$

$$\therefore \quad M_e = \left(1 + \frac{D}{f_e}\right)$$

Exercises

1. Give the advantages of using a compound eyepiece. Describe the eyepiece that is usually fitted into an astronomical telescope and explain its usefulness.
2. Describe (a) Huygens' eyepiece and (b) Ramsden's eyepiece. Compare their relative advantages and disadvantages.
3. A Ramsden's eyepiece is made of two thin lenses, focal length 22.5 mm each and separated by 16 mm apart. Find (a) its equivalent focal length, (b) magnifying power in normal and distinct vision adjustment. [f_E = 1.75 cm. M_d = 15.3, M_n = 14.3]

4. A compound microscope has an objective and eyepiece of focal lengths 1 cm and 5 cm respectively. An object is placed 11 mm from the objective and the final image is 25 cm from the eye. What is the magnification produced and what is the distance between the lenses?
[60; 15.16 cm]
5. A Galilean telescope is adjusted for normal vision. Calculate its magnifying power. What change in the adjustment is necessary for short-sighted person and what will be magnifying power when the distance of distinct vision is D?
6. Show that for a telescopic system, the reciprocal of the lateral magnification is equal to angular magnification.
7. Describe the different parts of a modern microscope. On what factors does its magnifying power depend?
8. Describe Huygens' and Ramsden's eyepieces, and discuss their relative merits. Find the distance of separation between the lenses in Huygens' eyepiece having focal lengths of 2 cm and 4 cm. Find also the locations of principal planes and principal foci of the eyepiece.
[B.U. 1993]

---

9. **What do you mean by telescopic system? Show that for a telescope system the lateral magnification is the reciprocal of angular magnification.**
10. **Explain the function of an eyepiece used with optical device.**
11. **What is the function of an eyepiece? Why does an eyepiece system consist of two lenses instead of one lens?**
12. **Using deviation method find the equivalent focal length of two thin lens separated by a distance in air.**
13. **Show that in case of a transparent sphere surrounded by air, the principal points, the nodal points and optical centre are all coincident with the centre of the sphere.**
14. **Establish Lagrange's law regarding linear and angular magnification of a single refracting surface. In what Helmholtz extended Lagrange's law in case of a series of coaxial refracting surfaces of p media? How is the relation modified in case of (a) telescopic system (b) object plane at infinity?**

Part II

Physical Optics

Chapter 1

Light as a Wave

1.1 Introduction

It has been already stated that light has dual character—wave and particle (cf. Geometrical Optics). The corpuscular theory of Newton failed to explain the very basic phenomena of light and so Young, Fresnel, Huygens and others brought in the concept of light as wave. it means light energy from the source is conveyed from one point to the other as a result of vibration of the particles of the medium. Later on, this concept of wave had to be revised to accommodate one basic property of radiation or light. The earlier concept of light as a wave due to Young and others is really related to mechanical wave character while later Maxwell changed the concept of light as an electromagnetic wave. We will first try to explain the different phenomena exhibited by light in terms of earlier wave theory, then we will consider Maxwell's electromagnetic theory. It is to be noted that mechanical wave of Young could explain all known facts about light but could not explain why light could travel through vacuum. Electromagnetic wave theory, since it also attributes wave character to light, could be used to explain all the phenomena which the mechanical wave theory could do, and, in addition, it could give answer very simply to why light can travel in vacuum.

1.2 A few Concepts about a Wave

What is a Wave?

A wave is a disturbance that carries energy from one place to the other through the vibration of the particles of the medium.

The creation of disturbance in a medium is due to vibration of a body at some place. If we trace the origin of a wave we go to a vibrating body. This vibrating body, the source, sends out energy in the surrounding, such that the particles of the medium do also vibrate. This vibration of the particles takes place about the mean position of rest, either along the direction of propagation of the disturbance or along the normal to the direction of propagation, depending on the nature of disturbance. These vibrations of the particles of the medium are supposed to be simple harmonic and if the medium be non-dissipative, the amplitude of vibration of all the particles will be same but the attainment of the amplitude of vibration will be at different times depending on the position of the particles w.r.t. the source.

So the origin of any disturbance is the simple harmonic vibration of a particle, call it the source particle, but the carriers of their disturbance are the particles of the medium and the transmission takes place through the vibration of the particles.

Simple Harmonic Vibration

If the force or acceleration acting on a particle is proportional to the first power of the displacement (i.e., linear), then the motion of the particle is referred to as Simple Harmonic Motion (SHM). In simple harmonic motion the displacement x is given as

$$x = \hat{A}\cos(\omega t + \phi_1) \tag{1.2.1}$$

where $\hat{A}$ represents the amplitude or maximum displacement. The instantaneous displacement can be obtained as the resolved part of a rotating vector of magnitude $\hat{A}$ and making an angle $(\omega t + \phi_1)$ with the direction of a fixed unit vector $\vec{\nu}$ at the time instant t. This concept gives rise to the essence of the Rotating Vector Method[1]. Thus the suitable rotating vector representing the vibration of the particle can be written as

$$\vec{X} = \hat{A}e^{j(\omega t + \phi_1)}\vec{\nu}. \tag{1.2.2}$$

The instantaneous displacement is given by the projection of the rotating vector on the fixed direction of $\vec{\nu}$ and j refers to the imaginary quantity $\sqrt{-1}$.

$$\therefore \quad x = \hat{A}e^{j(\omega t + \phi_1)}\vec{\nu}\cdot\vec{\nu} = \hat{A}\cos(\omega t + \phi_1). \tag{1.2.3}$$

The Operator $(a + bj)$

If we set

$$a = r\cos\theta \quad \text{and} \quad b = r\sin\theta$$

then, $a + bj = r(\cos\theta + j\sin\theta) = re^{j\theta}$, where $r = \sqrt{a^2+b^2}$, $\tan\theta = \frac{b}{a}$.

So, $(a + bj)\vec{\nu} = re^{j\theta}\vec{\nu}$ meaning $(a + bj)$ operates on unit vector $\vec{\nu}$ produces a vector whose magnitude is $\sqrt{a^2+b^2}$ times the magnitude of the vector operated upon and there is a rotation by θ given as $\tan\theta = \frac{b}{a}$. Thus,

$$(a + bj)\vec{X} = (a + bj)\hat{A}e^{j(\omega t + \phi_1)}\vec{\nu} = \sqrt{a^2+b^2}\hat{A}e^{j(\omega t + \phi_1 + \theta)}\vec{\nu}. \tag{1.2.4}$$

Thus the eqn.(1.2.3) can be geometrically represented as shown in fig.(1.1), its resolution representing the displacement of the particle is also shown.

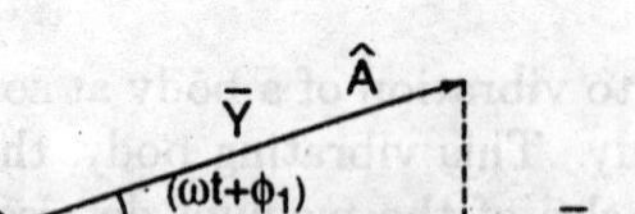

Fig 1.1

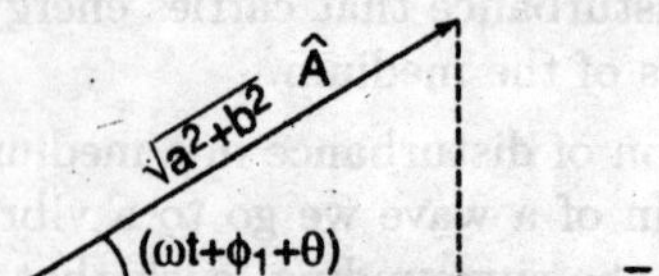

Fig 1.2

The eqn.(1.2.4) is represented geometrically as in fig.(1.2), where the magnitude is $\sqrt{a^2+b^2}$ times the magnitude of the vector and it makes an angle $(\omega t + \phi_1 + \theta)$ with the direction of $\vec{\nu}$.

Thus, here, $x = \sqrt{a^2+b^2}\hat{A}\cos(\omega t + \phi_1 + \theta)$.

---

[1] The details of this method have been discussed in 'Advanced Textbook on Electricity and Magnetism', Vol.II, by the author.

Composition of two or more Collinear SHMs of Equal Periods but Different Phases and Amplitudes

Consider two simple harmonic vibrations represented by suitable rotating vectors $\vec{X}_1$ and $\vec{X}_2$ as

$$\vec{X}_1 = \hat{A}_1 e^{j(\omega t+\phi_1)}\vec{\nu} \quad \text{and} \quad \vec{X}_2 = \hat{A}_2 e^{j(\omega t+\phi_2)}\vec{\nu}.$$

Geometrically, these two vectors are represented as in fig.(1.3). The resultant of the two vectors is given by,

$$\begin{aligned}\vec{X} = \vec{X}_1 + \vec{X}_2 &= \hat{A}_1 e^{j(\omega t+\phi_1)}\vec{\nu} + \hat{A}_2 e^{j(\omega t+\phi_1)}\vec{\nu} \\ &= e^{j\omega t}\left[\hat{A}_1 e^{j\phi_1}\vec{\nu} + \hat{A}_2 e^{j\phi_2}\vec{\nu}\right] \\ &= e^{j\omega t}\left[\hat{A}_1\cos\phi_1 + j\hat{A}_1\sin\phi_1 + \hat{A}_2\cos\phi_2 + j\hat{A}_2\sin\phi_2\right]\vec{\nu} \\ &= e^{j\omega t}\left[\left(\hat{A}_1\cos\phi_1 + \hat{A}_2\cos\phi_2\right) + j\left(\hat{A}_1\sin\phi_1 + \hat{A}_2\sin\phi_2\right)\right]\vec{\nu} \\ &= e^{j\omega t}(a+bj)\vec{\nu} = \sqrt{a^2+b^2}e^{j\omega t}e^{j\theta}\vec{\nu} = \hat{A}e^{j(\omega t+\theta)}\vec{\nu} \end{aligned} \qquad (1.2.5)$$

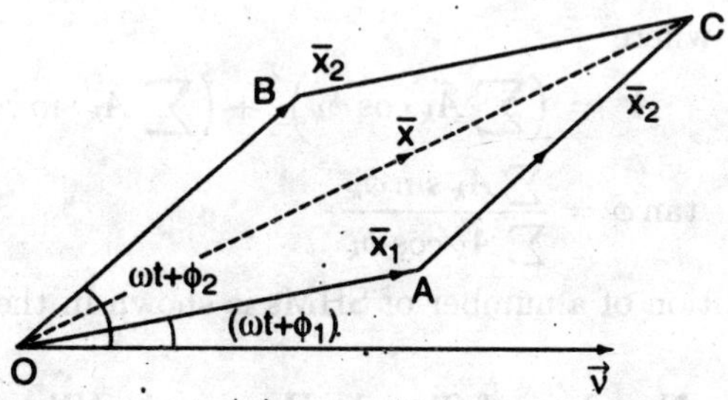

Fig 1.3

Thus the resultant vibration is represented by a rotating vector having magnitude $\hat{A}$ and at some instant it makes an angle $(\omega t+\theta)$ with the direction of $\vec{\nu}$. It is geometrically represented in fig.(1.3) by $\overline{OC}$.

$$\therefore \quad \hat{A}^2 = a^2 + b^2 = \hat{A}_1^2 + \hat{A}_2^2 + 2\hat{A}_1\hat{A}_2\cos(\phi_2-\phi_1) \qquad (1.2.6)$$

$$\tan\theta = \frac{b}{a} = \frac{\hat{A}_1\sin\phi_1 + \hat{A}_2\sin\phi_2}{\hat{A}_1\cos\phi_1 + \hat{A}_2\cos\phi_2}. \qquad (1.2.7)$$

The phase difference between two component SHMs is given by,

$$\delta = (\phi_2 - \phi_1).$$

So the resultant amplitude depends on the value of δ. If $\delta = 2n\pi, n \in I$ including 0, we have, $\cos\delta = 1$.

$$\therefore \quad \hat{A}^2 = (\hat{A}_1 + \hat{A}_2)^2 \quad \Rightarrow \quad \hat{A} = \hat{A}_1 + \hat{A}_2, \quad \text{maximum.}$$

So if the component vibrations are in phase, the resultant amplitude of vibration is maximum.

If $\delta = (2n+1)\pi, n \in I, \cos\delta = -1$, so we have,

$$\hat{A}^2 = (\hat{A}_1 \sim \hat{A}_2)^2 \quad \Rightarrow \quad \hat{A} = \hat{A}_1 \sim \hat{A}_2 \quad \text{minimum.}$$

Composition of a Number of SHMs of Different Phases and Amplitudes

Let us consider a number of SHMs given by,

$$\begin{aligned}
\vec{Y}_1 &= \hat{A}_1 e^{j(\omega t+\phi_1)}\vec{\nu} \\
\vec{Y}_2 &= \hat{A}_2 e^{j(\omega t+\phi_2)}\vec{\nu} \\
\vec{Y}_3 &= \hat{A}_3 e^{j(\omega t+\phi_3)}\vec{\nu} \\
\cdots &\cdots\cdots\cdots \\
\vec{Y}_n &= \hat{A}_n e^{j(\omega t+\phi_n)}\vec{\nu}
\end{aligned}$$

The resultant vibration is defined by

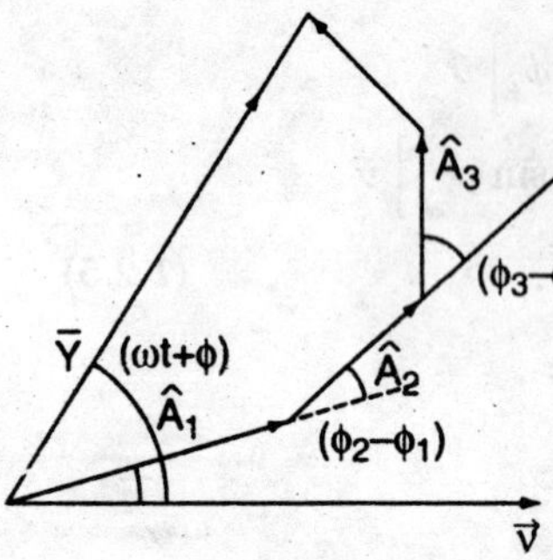

Fig 1.4

$$\begin{aligned}
\vec{Y} &= \vec{Y}_1 + \vec{Y}_2 + \vec{Y}_3 + \cdots + \vec{Y}_n \\
&= e^{j\omega t}\Big[(\hat{A}_1\cos\phi_1 + \hat{A}_2\cos\phi_2 + \hat{A}_3\cos\phi_3 + \cdots + \hat{A}_n\cos\phi_n) \\
&\qquad + j(\hat{A}_1\sin\phi_1 + \hat{A}_2\sin\phi_2 + \cdots + \hat{A}_n\sin\phi_n)\Big] \\
&= \hat{A}e^{j(\omega t+\phi)}\vec{\nu}
\end{aligned} \tag{1.2.8}$$

where

$$\hat{A}^2 = \left(\sum \hat{A}_k\cos\phi_k\right)^2 + \left(\sum \hat{A}_k\sin\phi_k\right)^2 \tag{1.2.9}$$

$$\tan\phi = \frac{\sum \hat{A}_k\sin\phi_k}{\sum \hat{A}_k\cos\phi_k} \tag{1.2.10}$$

Geometrically, such addition of a number of SHMs is shown in the fig.(1.4).

Superposition of a Large Number of Simple Harmonic Vibrations of Small but Equal Amplitudes and Constant but Small Phase Difference

This case is very important and it will be useful in the consideration of the single slit diffraction intensity.

In this case we have the rotating vector representing the resultant vibration as,

$$\begin{aligned}
\vec{X} &= \hat{A}e^{j\omega t}\left(e^{j\phi} + e^{j2\phi} + e^{j3\phi} + \cdots + e^{jn\phi}\right)\vec{\nu} \\
&= \hat{A}e^{j\omega t}e^{jn\bar{\phi}}\left(e^{-j(n-1)\phi} + e^{-j(n-2)\phi} + \cdots + e^{-j\phi} + 1\right)\vec{\nu} \\
&= \hat{A}e^{j(\omega t+n\phi)}\left(1 + e^{-j\phi} + e^{-j^2\phi} + \cdots + e^{-j(n-1)\phi}\right)\vec{\nu} \\
&= \hat{A}e^{j(\omega t+n\phi)}\vec{\nu}\times\frac{(1-e^{-jn\phi})}{(1-e^{-j\phi})} = \hat{A}\frac{(1-e^{-jn\phi})}{(1-e^{-j\phi})}e^{j(\omega t+n\phi)}\vec{\nu}
\end{aligned} \tag{1.2.11}$$

Thus eqn.(1.2.11) shows that the resultant displacement is represented by a rotating vector of magnitude $A_r = \hat{A}\frac{(1-e^{-jn\phi})}{(1-e^{-j\phi})}$ and it makes an angle $(\omega t + n\phi)$ with the direction of $\vec{\nu}$. Now,

$$A_r = \hat{A}\frac{(1-e^{-jn\phi})}{(1-e^{-j\phi})}. \tag{1.2.12}$$

The magnitude of any complex quantity, say $A = x + jy$, is given by,

$$|A| = \sqrt{x^2 + y^2}.$$

This is also available by multiplying A by its complex conjugate. Thus we get,

$$AA^* = (x+jy)(x-jy) = x^2+y^2 \quad \therefore \quad |A| = \sqrt{x^2+y^2} = \sqrt{AA^*}$$

So to get the magnitude of A_r, we can write,

$$|A_r|^2 = A_r A_r^* = \hat{A}^2 \frac{(1-e^{-jn\phi})(1-e^{jn\phi})}{(1-e^{-j\phi})(1-e^{j\phi})}$$

Now,

$$(1-e^{j\phi})(1-e^{-j\phi}) = 1-(e^{j\phi}+e^{-j\phi})+1 = 2(1-\cos\phi) = 4\sin^2\frac{\phi}{2}$$

$$\text{or,} \quad (1-e^{jn\phi})(1-e^{-jn\phi}) = 2(1-\cos n\phi) = 4\sin^2\frac{n\phi}{2}$$

$$\therefore \quad |A_r|^2 = \hat{A}^2 \frac{\sin^2\frac{n\phi}{2}}{\sin^2\left(\frac{\phi}{2}\right)} \tag{1.2.13}$$

If we set $n\phi = 2\theta$, then eqn.(1.2.13) gives,

$$A_r = \hat{A}\frac{\sin\frac{n\phi}{2}}{\sin\frac{\phi}{2}} = \hat{A}\frac{\sin\theta}{\sin\left(\frac{\theta}{n}\right)} = \hat{A}\frac{\sin\theta}{\frac{\theta}{n}}$$

since $\frac{\theta}{n}$ is small, n being a large number.

So we get, $A_r = n\hat{A}\frac{\sin\theta}{\theta}$

$$\therefore \quad \vec{X} = n\hat{A}\frac{\sin\theta}{\theta}e^{j(\omega t+n\phi)}\vec{\nu}. \tag{1.2.14}$$

Eqn.(1.2.14) is graphically shown in fig.(1.5), the displacement vectors lie along a circular arc, the resultant displacement vector is represented by $\overline{OR}$, the magnitude of which is given by,

$$A_r = n\hat{A}\frac{\sin\theta}{\theta}.$$

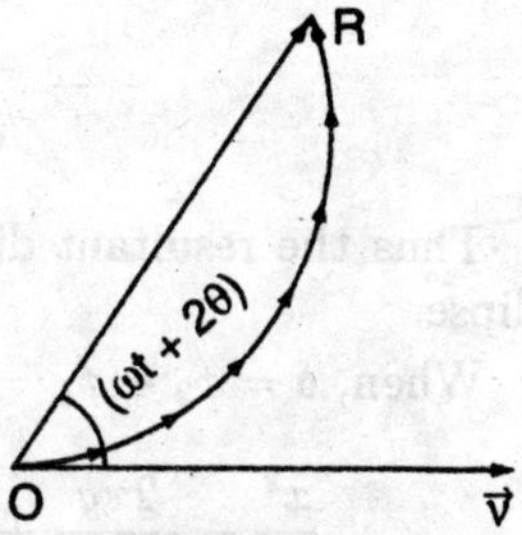

Fig 1.5

Now, ϕ representing the phase difference between consecutive vibrations, it will be zero when all the vibrations are in phase.

$$\therefore \quad (A_r)_{\max} = n\hat{A}\lim_{\theta\to 0}\frac{\sin\theta}{\theta} = n\hat{A} = A_0$$

$$\therefore \quad A_r = A_0\frac{\sin\theta}{\theta}, \tag{1.2.15}$$

Superposition of two Rectangular Simple Harmonic Motions of Equal Periods but Different Phase and Amplitude

Let us consider two fixed directions defined by two unit vectors $\vec{\nu}$ and $\vec{\nu}_1$, and they are at right angles to each other. Thus we can write,

$$\vec{\nu}_1 = j\vec{\nu}.$$

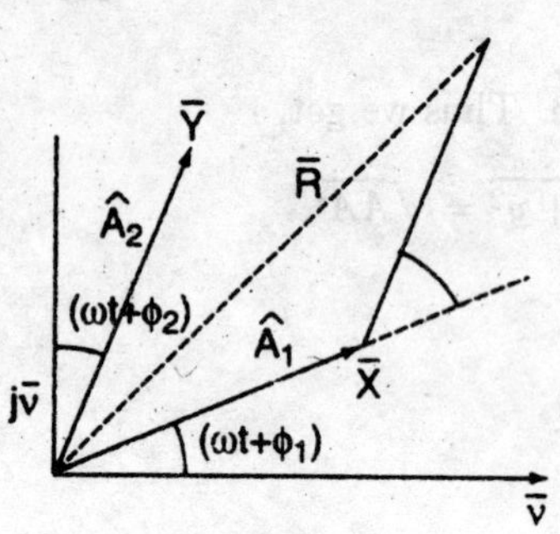

Fig 1.6

Let the first vibration be represented as

$$\vec{X} = \hat{A}_1 e^{j(\omega t+\phi_1)}\vec{\nu}, \quad \text{(cf fig.1.6)}$$

The second vibration along a perpendicular axis is given by,

$$\vec{Y} = \hat{A}_2 e^{j(\omega t+\phi_2)} j\vec{\nu}$$

$$x = \vec{X}\cdot\vec{\nu} = \hat{A}_1\cos(\omega t+\phi_1)$$

$$y = \vec{Y}\cdot j\vec{\nu} = \hat{A}_2\cos(\omega t+\phi_2)$$

$$\therefore \quad \frac{y}{\hat{A}_2} = \cos(\omega t+\phi_2) = \cos(\omega t+\phi_1+\overline{\phi_2-\phi_1}) = \cos(\omega t+\phi_1+\delta)$$

$$= \cos(\omega t+\phi_1)\cos\delta + \sin(\omega t+\phi_1)\sin\delta = \frac{x}{\hat{A}_1}\cos\delta + \sqrt{1-\cos^2(\omega t+\phi_1)}\sin\delta$$

$$= \frac{x}{\hat{A}_1}\cos\delta + \sqrt{1-\frac{x^2}{\hat{A}_1^2}}\sin\delta \quad\Rightarrow\quad \left(\frac{y}{\hat{A}_2} - \frac{x}{\hat{A}_1}\cos\delta\right) = \sqrt{1-\frac{x^2}{\hat{A}_1^2}}\sin\delta$$

Squaring both sides, we get,

$$\frac{y^2}{\hat{A}_2^2} + \frac{x^2}{\hat{A}_1^2}\cos^2\delta - \frac{2xy}{\hat{A}_1\hat{A}_2}\cos\delta = \left(1-\frac{x^2}{\hat{A}_1^2}\right)\sin^2\delta$$

$$\Rightarrow \frac{x^2}{\hat{A}_1^2} + \frac{y^2}{\hat{A}_2^2} - \frac{2xy}{\hat{A}_1\hat{A}_2}\cos\delta = \sin^2\delta \tag{1.2.16}$$

Thus the resultant displacement satisfies eqn.(1.2.16) which is the general equation of an ellipse.

When, $\delta = \phi_2 - \phi_1$ = phase difference $= 2n\pi$, then $\cos\delta = 1, \sin\delta = 0$ and we get,

$$\frac{x^2}{\hat{A}_1^2} - \frac{2xy}{\hat{A}_1\hat{A}_2} + \frac{y^2}{\hat{A}_2^2} = 0 \quad\Rightarrow\quad \left(\frac{x}{\hat{A}_1} - \frac{y}{\hat{A}_2}\right)^2 = 0 \quad\Rightarrow\quad y = \frac{\hat{A}_2}{\hat{A}_1}x \tag{1.2.17}$$

Thus, when two rectangular vibrations at right angles to each other and having a phase difference as even multiple of π superpose, we get the resultant vibration due to superposition to be linear along a direction making an angle θ given by $\tan\theta = \frac{\hat{A}_2}{\hat{A}_1}$ with the direction of $\vec{\nu}$.

When, $\delta = (2n+1)\pi$, hence, the phase diff. is an odd multiple of π, $\sin\delta = 0$, $\cos\delta = -1$.

$$\therefore \quad \frac{x^2}{\hat{A}_1^2} + \frac{2xy}{\hat{A}_1\hat{A}_2} + \frac{y^2}{\hat{A}_2^2} = 0 \quad\Rightarrow\quad \left(\frac{x}{\hat{A}_1} + \frac{y}{\hat{A}_2}\right)^2 = 0 \quad\Rightarrow\quad y = -\frac{\hat{A}_2}{\hat{A}_1}x \tag{1.2.18}$$

Thus, the particle on which superposition takes place vibrates linearly along a direction making an angle θ given by, $\tan\theta = -\frac{\hat{A}_2}{\hat{A}_1}$. If $\delta = \frac{\pi}{2}$, the two component vibrations are in quadrature, then eqn.(1.2.16) becomes,

$$\frac{x^2}{\hat{A}_1^2} + \frac{y^2}{\hat{A}_2^2} = 1 \tag{1.2.19}$$

The eqn.(1.2.16) is the general eqn. of an ellipse as shown in fig.(1.7). The elliptic vibration degenerates into linear vibrations as represented by eqns.(1.2.17) and (1.2.18), and shown diagrammatically as in fig.(1.8).

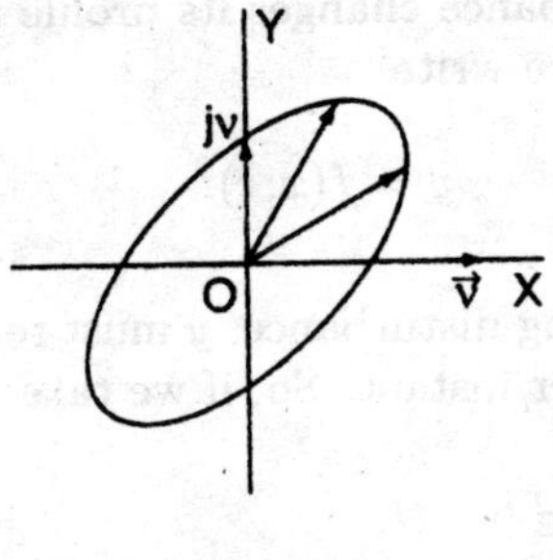

Fig 1.7

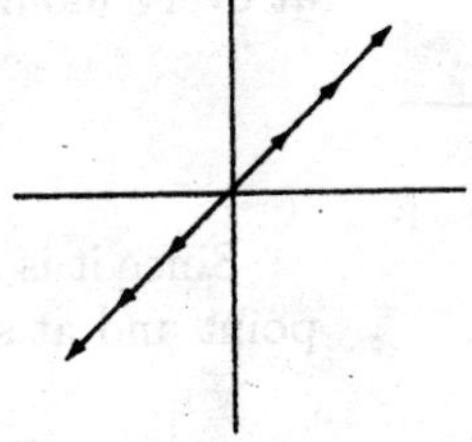

Fig 1.8

This consideration will be useful in explaining the elliptic and circularly polarised light, which we will see later.

Wave Motion

This is a basic concept of Wave Theory. A wave may be defined as the fluctuations of some physical quantity in space with respect to time. It is a form of disturbance that moves through a medium and is produced due to certain periodic motion in succession of the constituent particles of the medium. The simultaneous positions of the successive vibrating particles at any particular instant determine the shape of the wave or in other words, a wave profile may be looked upon as the position of the constituent particles of the medium frozen at some instant of time t. As the vibration of the particles continues, the envelope or the wave profile connecting the particles appears to move steadily in the forward direction and this motion of the curve constitutes what is known as *moving wave*.

Let us consider an example and that of water waves. When a stone is dropped on a calm surface of water in a pond, a circular pattern consisting of crest and trough spreads out from the point where the stone is dropped. This is called a wave. Water is pushed down at the point where the stone is dropped, thus creating a disturbance at that point. The water molecule at that point is pushed down, and because of elasticity of the water medium, this particle is set into oscillation. As this particle moves down, an elevation of the surrounding layer takes place giving rise to circular ridge, called crest. The depression due to downward movement of a layer is referred to as trough. The disturbance created at some point does not remain confined to the point of origin but spreads out in larger and larger circles through simultaneous up and down motion of water particles. Although the disturbance moves forward, particles of the medium do not move out but are merely set into up and down motion. In case of longitudinal wave, the particle of the medium do vibrate along the direction of propagation but in transverse wave, they vibrate at right angles.

What we have stated above is one way of looking into a wave where the displacement of the particles of the medium are considered corresponding to a certain instant t. Thus, in this picture the wave is represented by writing the displacement as a function of space, i.e., $y = f(x)$ for t fixed.

Another way of looking at it is to consider the displacement of a particle as a function of time as the disturbance passes across it. Thus we write here, $y = f(t)$ for x fixed in fig.(1.9).

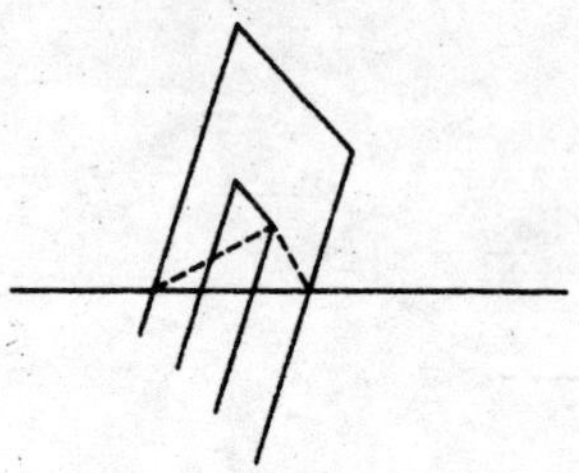

Fig 1.9

So a moving disturbance changes its profile at every point and at every moment. So we write

$$y = f(x, t).$$

Since it is a travelling disturbance, y must repeat at some other point and at some other instant. So, if we take

$$y = f(ct - x),$$

it repeats itself at $(x + c\delta t)$ and at the instant $(t + \delta t)$ because

$$y' = f[c(t + \delta t) - (x + c\delta t)] = f(ct - x) = y.$$

Thus, $f(ct - x)$ is a progressive disturbance with a velocity c. Thus the equation in general can be written as

$$y = f(ct - x).$$

A particular form of the functional is

$$\sin \frac{2\pi}{\lambda}(ct - x), \text{ hence } y = a \sin \frac{2\pi}{\lambda}(ct - x) = a \sin(\omega t - \phi_s)$$

represents a harmonic progressive wave.

Since y represents the displacement of a particle at x at time t, it can be represented by suitable rotating vector as

$$\vec{Y} = ae^{j\left(\frac{\pi}{2} + \omega t - \phi_s\right)}\vec{\nu}$$

where, ϕ_s, a function of x brings in the phase difference between the vibration of the source particle and the particle under consideration.

Since in a progressive wave displacements repeat, so the particles separated by a distance equal to λ from each other will be in the same state of vibration. So, if the source particle displacement at time instant t be represented as

$$\vec{Y}_0 = ae^{j\left(\frac{\pi}{2} + \omega t\right)}\vec{\nu}$$

the displacement of any other particle in the medium along which disturbance travels can be written as

$$\vec{Y} = ae^{j\left(\frac{\pi}{2} + \omega t - \phi_s\right)}\vec{\nu} \tag{1.2.20}$$

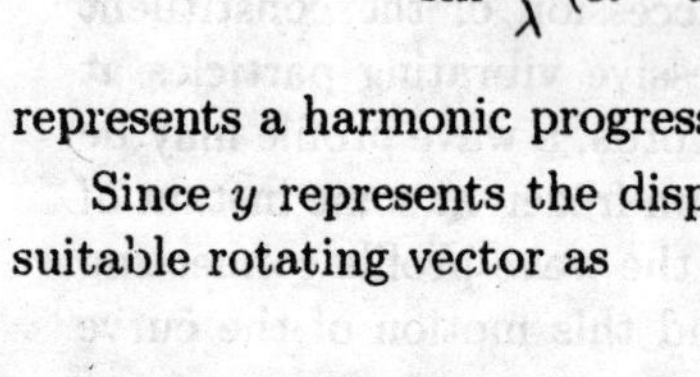

Fig 1.10

In case of harmonic disturbance, $\phi_s = \frac{2\pi x}{\lambda}$. The displacement rotating vector in eqn.(1.2.20) is shown in fig.(1.10). The rotating vector representing the displacement of another particle at a distance x' from the origin is written as

$$\vec{Y}' = ae^{j\left(\frac{\pi}{2} + \omega t - \phi'_s\right)}\vec{\nu}$$

The phase difference between these two vibrations is given by

$$\delta = \left(\omega t + \frac{\pi}{2} - \phi_s\right) - \left(\omega t + \frac{\pi}{2} - \phi'_s\right) = (\phi'_s - \phi_s).$$

In case of harmonic disturbance

$$\phi_s = \frac{2\pi x}{\lambda} \quad \text{and} \quad \phi'_s = \frac{2\pi x'}{\lambda}$$

$$\therefore \quad \delta = \text{phase diff.} = \phi'_s - \phi_s = \frac{2\pi}{\lambda}(x' - x) = \frac{2\pi}{\lambda} \times (\text{Path diff.})$$

Thus we get a very important relation

$$\text{Phase diff.} = \frac{2\pi}{\lambda} \times (\text{Path diff.}) \tag{1.2.21}$$

So, from eqn.(1.2.21) we see that if the path difference is an even multiple of $\frac{\lambda}{2}$, the corresponding phase difference is

$$\delta_{\text{even}} = \frac{2\pi}{\lambda} \times 2n \cdot \frac{\lambda}{2} = n(2\pi), \text{ where } n = 0, 1, 2, \text{ etc.} \tag{1.2.22}$$

When $n = 0$, the two rotating vectors coincide meaning phase difference is zero.

When $n = 1, \delta_{\text{even}} = 2\pi$, in this case the first particle which is nearer and leading in phase makes one complete oscillation and comes to the same state of motion as that of the second particle having a relative separation equal to λ.

But, when path diff. is an odd multiple of $\frac{\lambda}{2}$

$$\delta_{\text{odd}} = \frac{2\pi}{\lambda} \times (2n + 1)\frac{\lambda}{2} = (2n + 1)\pi, \ n = 0, 1, 2, \text{ etc.} \tag{1.2.23}$$

Thus we see in this case that the two particles having relative separation equal to $\frac{\lambda}{2}$ are in opposite state of vibration. These inferences will be of considerable importance in the consideration of the phenomenon of interference as shall see presently.

References

1. Advanced Textbook on Electricity and Magnetism, Vol.II, *P.K. Chakrabarti*, Modern Book Agency, Kolkata.
2. A Textbook on Waves and Acoustics, *P.K. Chakrabarti & S.B. Chowdhury*, New Central Book Agency (P) Ltd., Kolkata.
3. Optics, *C.J. Smith*, Ind. Edition, Radha Publishing House, Kolkata.
4. A Textbook on Light, *Barton A.W.*, Longmans Green & Co., London.

Chapter 2

Wave Theory and Laws of Optics

2.1 Introduction

In the preceding chapter we have gathered some knowledge about a wave and one theory about the nature of light is that it is a wave. Such a proposition came through Young, Fresnel and Huygens. We will see through subsequent chapters how this new concept about the nature of light could explain all the basic phenomena, one after another, with amazing success but even after all these successes it came to a halt when an awkward question was raised — *How could light (Sunlight) travel through vast void?* This was not possible to answer from the wave theory Young and others proposed but this could be answered through a different concept about light wave in the form of electromagnetic wave. The older wave theory is just a mechanical wave theory but light is not a mechanical wave like acoustic disturbance; it is a wave, true, but it is an *electromagnetic disturbance* — the nature of which was fully developed and explored by Maxwell (elaborately discussed in chapter 6).

2.2 Huygens' idea about Propagation of Wavefront

A wavefront is defined as the surface on which the particles vibrate in phase. If we consider a point source of light placed in an isotropic medium, then disturbances propagate in all directions equally, hence if we consider a sphere with the point source as centre, this spherical surface would represent a wavefront at a certain instant.

This inference can be drawn w.r.t. fig.(2.1) where we have shown two particles P_1 and P_2 on the surface of the sphere. The displacement of the particle P_1 due to wave from S to P_1 distant r from S is represented by

$$\vec{Y}_1 = ae^{j(\omega t + \frac{\pi}{2} - \phi_s)}\vec{\nu}$$

where $\phi_s = \frac{2\pi}{\lambda}r$. The displacement of the particle P_2 at the same instant is represented by a rotating vector

$$\vec{Y}_2 = ae^{j(\omega t + \frac{\pi}{2} - \phi'_s)}\vec{\nu}$$

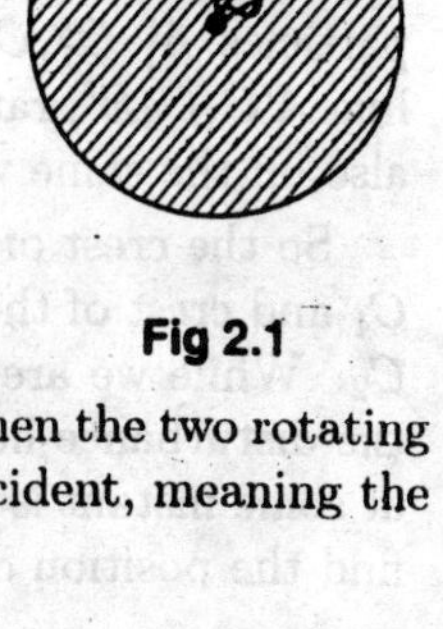

Fig 2.1

where $\phi'_s = \frac{2\pi}{\lambda}r$, because P_2 is also at the same distance w.r.t. the source. Then the two rotating vectors representing the displacements of the two particles are always coincident, meaning the

two particles are vibrating in phase. Thus all the particles which happen to lie on the surface of this sphere do vibrate in phase.

In case of a linear source of light, the particles lying on the surface of a cylinder with the linear source as axis vibrate in phase. Thus cylindrical surfaces represent wavefronts in this case. Whatever may be the nature of wavefronts, all the particles lying on any wavefront surface vibrate in unison and according to Huygens, they become the sources of secondary disturbance. So secondary wavelets generate with the vibrating particles as generators. So Huygens says that if a wavefront at any instant is given, subsequent positions of the wavefronts as the disturbance propagates can be constructed.

Let us consider fig.(2.2) where AB is the trace of a wavefront at the instant t. We want to locate the position of the wavefront at $(t + \delta t)$. Huygens' idea is that the different particles are taken as the centres and spheres are drawn with radius equal to $c\delta t$, where c is velocity of the disturbance. A common tangent plane $A'B'$ touching these spheres represents the subsequent position of the wavefront at $(t + \delta t)$. If we can construct the new wavefront, repeating the argument, we can construct the succeeding wavefronts.

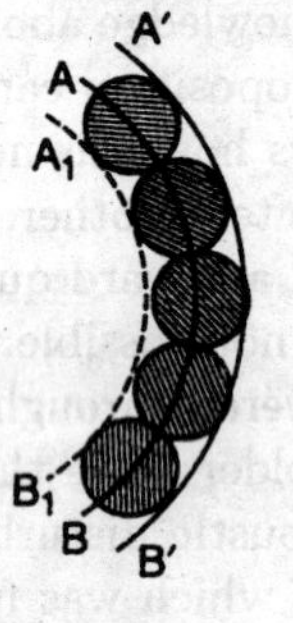

Fig 2.2

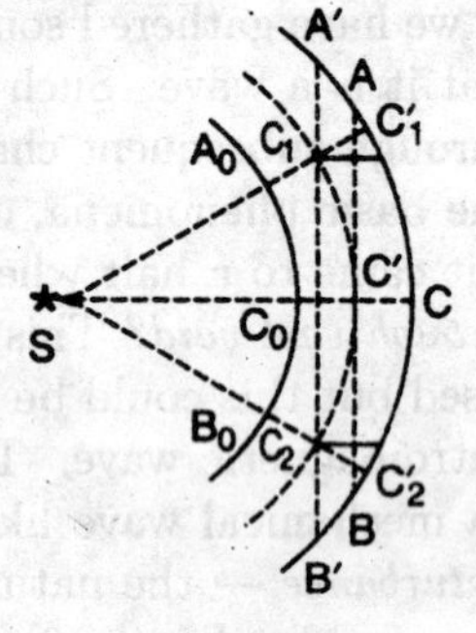

Fig 2.3

In this construction one basic difficulty we are facing is that we can construct another tangent plane common to all the spherical surfaces like A_1B_1 to the left. This would, in other words, imply that there is possibility of back propagation leading to back-waves. But we know that disturbance from the source has only the forward motion. How then to reconcile with this fact? Possibly a very simple explanation could be offered with reference to fig.(2.3). Suppose we have two wavefronts at $A_0C_0B_0$ and ACB such that the separation between them is λ, the wavelength of the disturbance. The two particles A, B on the advancing wavefront are in phase, but if we consider a vertical line AB, the particles A, B lying on one wavefront are not in line with particle C but they are in line with the particle C' behind C by a distance equal $\frac{\lambda}{2}$. So, when the secondary source C produces secondary wavecrest trying to expand back, C' particle produces a trough, thus the crest and trough coincide reducing the amplitude of the back-wave to zero value. Similarly, if the vertical line $A'C_1C_2B'$ is considered, then the particles A'_1, C'_1, C'_2 and B' are lying on one wavefront but C'_1 and C'_2 are not on the vertical line $A'C_1C_2B'$, rather C_1 and C_2 are on it but C_1, C'_1 are on the same wave normal, C_2, C'_2 are also on the same wave normal.

So the crest of the back-wave due to secondary source C'_1 coincides with the trough due to C_1 and crest of the back-wave due to C'_2 coincides with the trough of the wave from the source C_2. While we are considering the annihilation of the back-wave, it is to be remembered that the disturbance has spread on to the spherical envelope ACB. Once the trace of an wavefront at some instant is available, and remembering that back-wave propagation not possible, we can find the position of the advancing wavefront in course of time.

It is important to remember that there is difference in the nature of vibration of a particle in a medium when it is oscillating in any manner and when it is oscillating under the influence of the incident wave. In the first case waves propagate in both forward and backward direction but when oscillation is due to incident wave on it, then the disturbance in the forward direction will be said to be due to the vibrating particle.

The absence of back-wave will be mathematically explained later with the help of Kirchhoff-Fresnel's integral to be developed in the chapter on Diffraction. It will be shown there that the amplitude at a point P having direction vector inclined w.r.t. the forward direction is proportional to $(1 + \cos\theta)$ where θ is the angle of inclination. Hence, when $\theta = 0$, amplitude is maximum, hence intensity; when $\theta = \pi$, $(1 + \cos\theta)$ becomes zero, implying amplitude, hence intensity, to be equal to zero. This explains absence of back-wave.

Thus the principle of Huygens can be stated as — *All the particles of wavefront vibrate in phase and these vibrating particles become the sources of secondary disturbance and generate secondary wavelets which spread in the forward direction.*

It is also to be noted that when a spherical wavefront expands to a large distance from the source, the spherical wavefront is transformed into a plane wavefront. The line along which propagation of the disturbance, hence energy transference takes place, is called a ray and a ray is normal to the wavefront. Thus, when the wavefront is a plane wavefront as in case of light rays coming from the sun or the moon, we represent the rays as a parallel beam as shown in fig.(2.4).

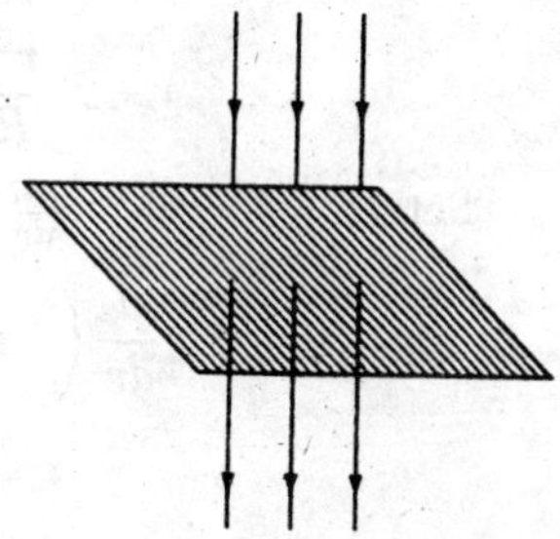

Fig 2.4

2.3 Wave Equation in Three Dimensions — Spherical Wave

The details of the wave eqns. in one-dimension as in the case of vibration of a string, two-dimensional wave equation as in the vibration of a plate or membrane, three-dimensional wave equation as in the case of a point vibrating source placed in an isotropic medium have been elaborately discussed elsewhere[1].

Here we will only consider three-dimensional wave equation represented as

$$\frac{\partial^2 u}{\partial t^2} = c^2 \left(\frac{\partial^2 u}{\partial x^2} + \frac{\partial^2 u}{\partial y^2} + \frac{\partial^2 u}{\partial z^2} \right). \tag{2.3.1}$$

This equation is transformed into one of single variable r, the radial distance. This transformation is done very simply noting that

$$\frac{\partial^2 u}{\partial x^2} + \frac{\partial^2 u}{\partial y^2} + \frac{\partial^2 u}{\partial z^2} = \nabla^2 u$$

Now, $$\nabla^2 u = \vec{\nu}\cdot(\vec{\nu}u) = \vec{\nu}\cdot\left(\frac{\partial u}{\partial r}\hat{r}\right) = \vec{\nu}\cdot\left(\frac{1}{r}\frac{\partial u}{\partial r}\vec{r}\right) = \frac{1}{r}\frac{\partial u}{\partial r}\vec{\nu}\cdot\vec{r} + \vec{r}\cdot\vec{\nu}\left(\frac{1}{r}\frac{\partial u}{\partial r}\right)$$

$$= \frac{3}{r}\frac{\partial u}{\partial r} + \frac{\partial^2 u}{\partial r^2} - \frac{1}{r}\frac{\partial u}{\partial r} = \frac{\partial^2 u}{\partial r^2} + \frac{2}{r}\frac{\partial u}{\partial r} = \frac{1}{r^2}\frac{\partial}{\partial r}\left(r^2\frac{\partial u}{\partial r}\right)$$

[1]See 'A Textbook of Mathematical Physics' by P.K. Chakrabarti and S.N. Kundu.

$$\therefore \quad \frac{\partial^2 u}{\partial t^2} = c^2 \cdot \frac{1}{r^2}\frac{\partial}{\partial r}\left(r^2 \frac{\partial u}{\partial r}\right) = c^2\left(\frac{\partial^2 u}{\partial r^2} + \frac{2}{r}\frac{\partial u}{\partial r}\right) \tag{2.3.2}$$

Set $u = R.T$; $\frac{\partial^2 u}{\partial t^2} = R\ddot{T}$, $\frac{\partial u}{\partial r} = T\dot{R}$; $\frac{\partial^2 u}{\partial r^2} = T\ddot{R}$

$$\therefore \quad R\ddot{T} = c^2\left(\ddot{R}T + \frac{2}{r}\dot{R}T\right)$$

Dividing by RT

$$\frac{\ddot{T}}{T} = c^2\left(\frac{\ddot{R}}{R} + \frac{2}{r}\frac{\dot{R}}{R}\right) = -\omega^2 \quad \Rightarrow \quad \ddot{T} = -\omega^2 T \tag{2.3.3}$$

$$\Rightarrow \quad \frac{\ddot{R}}{R} + \frac{2}{r}\frac{\dot{R}}{R} = -\frac{\omega^2}{c^2} = -k^2 \tag{2.3.4}$$

$$\therefore \quad \frac{r^2}{R}\ddot{R} + 2r\frac{\dot{R}}{R} + k^2 r^2 = 0 \quad \Rightarrow \quad \frac{d}{dr}(r^2\dot{R}) + k^2 r^2 R = 0 \tag{2.3.5}$$

Set $R = \frac{s}{r} \quad \Rightarrow \quad \frac{dR}{dr} = -\frac{s}{r^2} + \frac{1}{r}\frac{ds}{dr}$

$$\therefore \quad \frac{d}{dr}\left(-s + r\frac{ds}{dr}\right) + k^2 r^2 \cdot \frac{s}{r} = 0 \Rightarrow -\frac{ds}{dr} + \frac{ds}{dr} + r\frac{d^2 s}{dr^2} + rk^2 s = 0$$

$$\Rightarrow \frac{d^2 s}{dr^2} + k^2 s = 0 \tag{2.3.6}$$

Put $s = Ae^{\alpha r}$

$$\alpha^2 Ae^{\alpha r} + k^2 Ae^{\alpha r} = 0 \quad \Rightarrow \quad \alpha = \pm jk$$

$$\therefore \quad s = (A_1 e^{jkr} + A_2 e^{-jkr})$$

$$\therefore \quad R = \frac{A_1}{r}e^{jkr} + \frac{A_2}{r}e^{-jkr} \tag{2.3.7}$$

The time-dependent part is given by the solution of

$$\ddot{T} = -\omega^2 T \quad \Rightarrow \quad T = (C_1 e^{j\omega t} + C_2 e^{-j\omega t}) \tag{2.3.8}$$

$$\therefore \quad u = \left(\frac{A_1}{r}e^{jkr} + \frac{A_2}{r}e^{-jkr}\right)(C_1 e^{j\omega t} + C_2 e^{-j\omega t})$$

$$= \frac{A}{r}e^{j(\omega t + kr)} + \frac{B}{r}e^{-j(\omega t - kr)} + \frac{C}{r}e^{+j(\omega t - kr)} + \frac{D}{r}e^{-j(\omega t + kr)}.$$

Now, for a particular point as time passes on $u \neq \infty$, so $A = 0$, $C = 0$.

$$\therefore \quad u = \frac{B}{r}e^{-j(\omega t - kr)} + \frac{D}{r}e^{-j(\omega t + kr)} \tag{2.3.9}$$

So the eqn. of the spherical wave towards the right is given by

$$u = \frac{B}{r}\cos(\omega t - kr) = \frac{B}{r}\cos(kct - kr) = \frac{B}{r}\cos k(ct - r) \tag{2.3.10}$$

and towards the left,

$$u = \frac{D}{r}\cos(\omega t + kr) = \frac{D}{r}\cos k(ct + r). \tag{2.3.11}$$

Thus it shows that in case of spherical wave, the amplitude falls off as $\frac{1}{r}$, hence intensity as $\frac{1}{r^2}$.

Cylindrical Wave

In case of cylindrical disturbance as in case of linear source, since any section is a circle, $x^2 + y^2 = r^2$.

For a linear source of disturbance, the propagating disturbance is said to be cylindrical wave, and the wavefront at any instant is a circle defined by $r^2 = x^2 + y^2$.

$$\therefore \quad \frac{\partial^2 u}{\partial t^2} = c^2\left(\frac{\partial^2 u}{\partial x^2} + \frac{\partial^2 u}{\partial y^2}\right) = c^2\left(\frac{\partial^2 u}{\partial r^2} + \frac{1}{r}\frac{\partial u}{\partial r}\right)$$

Let $u = RT$.

$$\therefore \quad \frac{\partial^2 u}{\partial t^2} = R\ddot{T}, \quad \frac{\partial u}{\partial r} = \dot{R}T, \quad \frac{\partial^2 u}{\partial r^2} = \ddot{R}T.$$

So we get,

$$R\ddot{T} = c^2\left(\ddot{R}T + \frac{1}{r}\dot{R}T\right).$$

Dividing by RT, we get,

$$\frac{\ddot{T}}{T} = c^2\left(\frac{\ddot{R}}{R} + \frac{1}{r}\frac{\dot{R}}{R}\right) = -\omega^2 \text{ (say)} \quad \Rightarrow \quad \ddot{T} = -\omega^2 T \tag{2.3.12}$$

$$\Rightarrow \frac{\ddot{R}}{R} + \frac{1}{r}\cdot\frac{\dot{R}}{R} = -\frac{\omega}{c^2} = -k^2. \tag{2.3.13}$$

Multiplying (2.2.13) by r^2 we get,

$$r^2\frac{d^2R}{dr^2} + r\frac{dR}{dr} + k^2r^2R = 0. \tag{2.3.14}$$

Put $kr = x$, we get from eqn.(2.2.14)

$$\frac{d^2R}{dx^2} + \frac{1}{x}\cdot\frac{dR}{dx} + R = 0 \tag{2.3.15}$$

Eqn.(2.2.15) is the standard Bessel's equation of zero degree whose solution can be written as

$$R = a_0\left[1 - \frac{x^2}{2^2} + \frac{x^4}{2^2.4^2} - \frac{x^6}{2^2.4^2.6^2} + \cdots\right]. \tag{2.3.16}$$

If $a_0 = 1$, then

$$R = \sum_0^\infty (-1)^s\left(\frac{x}{2}\right)^{2s}\frac{1}{s!\Gamma(s+1)} = J_0(x), \quad \text{[where } J_0(0) = 1]$$

$$\therefore \quad u = A_1e^{ikct}J_0(kr) + A_2e^{-ikct}J_0(kr) \tag{2.3.17}$$

So the equation of the cylindrical disturbance to the right is given by

$$u = A_2e^{-ikct}J_0(kr)$$

and toward the left by

$$u = A_1e^{ikct}J - 0(kr). \tag{2.3.18}$$

2.4 Wave Theory and Phenomena of Reflection and Refraction

It has already been remarked that though Newton's corpuscular theory could explain the laws of reflection, it failed completely in explaining the refraction phenomenon. The wave theory supplemented by Huygens' idea about the propagation of the wavefront could explain both very satisfactorily.

Reflection from Wave Theory

Consider a plane wavefront moving towards the reflector fig.(2.5). At some instant the position of the wavefront is $A'D'E'$; as the wave advances, the wavefront shifts parallel to itself and the rays which are normal to the wavefront strike the reflector at different instants of time.

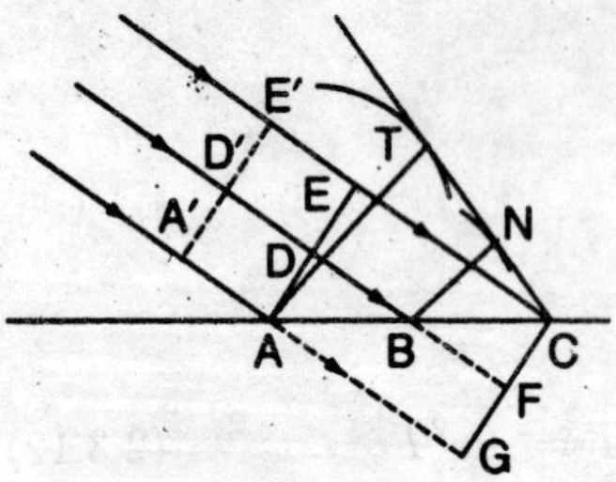

Fig 2.5

At the moment a ray strikes the reflector at A, the other two rays do not strike it; drop a normal ADE from the point A on the two other rays and a plane through it represents the wavefront at the moment of incidence. When the point A is struck by a ray, it becomes a source of secondary disturbance and disturbance spreads out with A as centre. Since we are here concerned with reflection phenomenon, we are to consider the disturbance in the same medium. The point A of the reflector is struck t secs earlier than the point C is struck such that $EC = AG = ct$, where c is the velocity of light. Taking A as the centre we describe a sphere of radius $R = EC = ct$, the section of the sphere by the plane of the paper is a circle, an arc of which is shown in the fig.(2.5). From C a tangent is drawn to the arc as CT such that AT is $\perp$ on to it. From B, the point where the ray corresponding to the point D strikes the reflector, a normal is dropped at N on CT. If we can prove that the disturbance which would have travelled to F in absence of the reflector has gone to N due to reflection, then we would conclude that CNT is the reflected wavefront.

Consider similar Δs CBN and ATC, we get

$$\frac{BN}{AT} = \frac{CB}{CA}.$$

From similar Δs CBF and CAG, we get

$$\frac{BF}{AG} = \frac{CB}{CA}; \text{ so, } \frac{BN}{AT} = \frac{BF}{AG}.$$

Since, $AT = AG$, we get, $BN = BF$.

It means that the disturbance which would have gone to F from B had there been no reflection has actually travelled to N due to reflection. Again, one reflected ray AT is normal to CT, another reflected ray BN is also normal to CT (by construction), hence the plane passing through CT represents the reflected wavefront.

ADE is the incident wavefront, hence $\angle EAC$ = angle of incidence = i and the angle ACT = angle of reflection = r.

Consider two Δs AEC and ATC; they are congruent, hence we get,

$$i = r \quad \text{(law of reflection)}.$$

Again since the incident wavefront, reflected wavefront and the normal at the point of incidence all lie in the plane of the paper, we conclude that the incident ray, the reflected ray and the normal at the point of incidence are coplanar.

Refraction from Wave Theory

As in the case of reflection, in this case also we consider a plane wavefront moving towards the surface of separation S between the two media of r.i. μ_1 and μ_2. The incident wavefront is represented by a surface passing through ABC. The disturbance from C takes a certain time t to reach the surface S at F given by, $t = (CF/v_1)$, where v_1 is the velocity of light in the incident medium. The point A of the surface is struck by the ray t secs earlier than the disturbance from C reaches F. Taking A as the centre, describe a sphere of radius $AN = v_2 t$, where v_2 is the velocity of light in the second medium. The sphere is cut by the plane of the diagram in a circle, whose element of arc length has been shown. Draw a tangent to this arc from F as FN; drop $\perp EN_1$ from E on the tangent FN.

If there were no surface of separation, i.e., if the first medium continued, then in the time t in which the disturbance from C comes to F, the disturbance from A would have gone to H, such that $AH = CF = v_1 t$; FGH would represent trace of the wavefront then.

Now, in similar Δs FEG and FAH, we get,

$$\frac{EG}{AH} = \frac{FE}{FA}.$$

Also, in similar Δs FEN_1 and FAN, we get,

$$\frac{EN_1}{AN} = \frac{FE}{FA}.$$

Therefore, we get,

$$\frac{EG}{AH} = \frac{EN_1}{AN} = \frac{EN_1}{v_2 t} \Rightarrow \frac{EG}{v_1 t} = \frac{EN_1}{v_2 t} \Rightarrow \frac{EG}{v_1} = \frac{EN_1}{v_2}. \tag{2.4.1}$$

This eqn.(2.4.1) shows that the disturbance that would have travelled to G from E if the first medium continued has actually gone from E to N_1 in presence of the second medium. Again, since EN_1 is dropped $\perp$ on to the tangent FN, we can say that a surface passing through FN is the refracted wavefront.

So the incident wavefront is ABC and the refracted wavefront is FN_1N; the angles of incidence i and refraction r are as shown in the fig.(2.6).

$$\therefore \quad \sin i = \frac{FC}{FA} \quad \text{and} \quad \sin r = \frac{AN}{AF}.$$

So,

$$\frac{\sin i}{\sin r} = \frac{FC}{AN} = \frac{v_1 t}{v_2 t} = \frac{v_1}{v_2}$$

Also,

$$\frac{c}{v_1} = \mu_1 \quad \text{and} \quad \frac{c}{v_2} = \mu_2$$

$$\therefore \quad \frac{\sin i}{\sin r} = \frac{c/\mu_1}{c/\mu_2} = \frac{\mu_2}{\mu_1} = \text{const.}$$

This is the second law of refraction, called Snell's law.

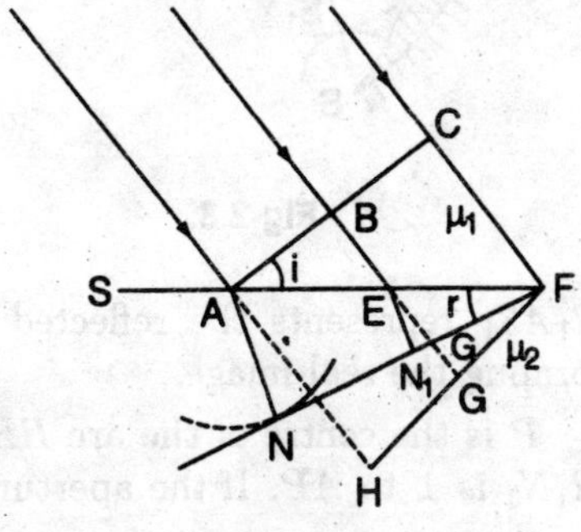

Fig 2.6

2.5 Reflection of Spherical Wave by a Plane Reflector

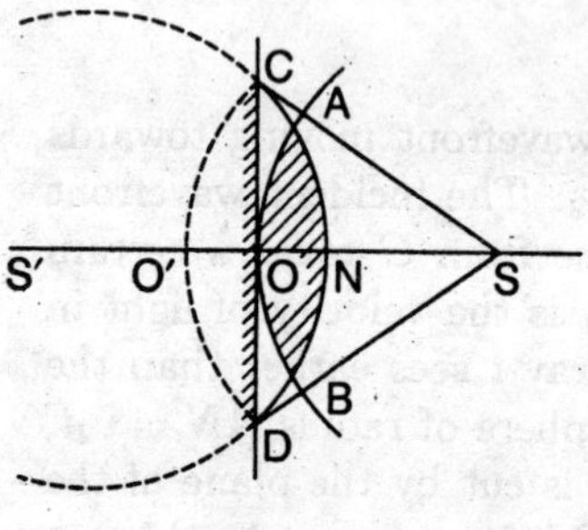

Fig 2.7

If we consider a point source S in an optically isotropic medium, the disturbance spreads out in the form of spherical wavefronts.

Consider the trace of the wavefront at some instant as AOB, the point O touching the mirror at O. In the time that the disturbance from A reaches C or from B reaches D, the disturbance from O would have travelled to O' in absence of the reflector, thus $CO'D$ would have been the wavefront at that instant. Due to presence of the reflector, the disturbance from O would come over to N such that $OO' = ON$. Draw a common tangent plane to these points, such that CND is the reflected spherical wavefront having centre at S' in fig.2.7.

Now, in the diagram, $OC^2 = ON\cdot(2S'N - ON) = 2S'N\cdot ON - ON^2 \simeq 2S'N\cdot ON.$
Also, $OC^2 = OO'(2SO' - OO') \simeq 2SO'\cdot OO'$

$$\therefore\ 2S'N\cdot ON = 2SO'\cdot OO' \quad \Rightarrow \quad S'N = SO' \text{ since } ON = OO'$$
$$\Rightarrow\ S'N - ON = SO' - OO' \quad \Rightarrow \quad S'O = SO.$$

2.6 Reflection of Spherical Wavefront from a Spherical Mirror

Let P be the point source of light spreading out waves in the form of spheres; the trace of the wavefront at some instant is represented as RBS. When disturbance from P strikes the mirror at R and S, similar disturbance lies at B only. In the time the disturbance from B reaches mirror at A, secondary spherical wavelets start travelling out into the same medium.

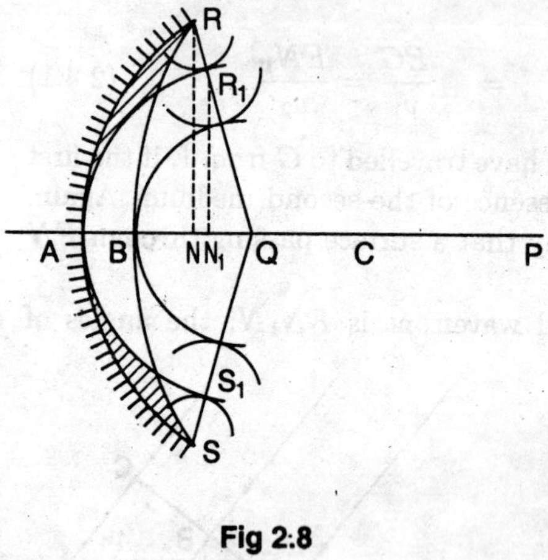

Fig 2.8

If time taken to travel from B to A be t, then, $BA = ct$, where c is the velocity of light in air. Take R and S as centres and describe spheres of radius $r = ct$, section of the sphere by the plane of the diagram will be circle and trace of the circle is shown in the diagram. When the secondary disturbance reaches R_1 reflected from R and S_1 reflected from S, the incident disturbance from B just reaches A of the reflector and just gets reflected. So the envelope R_1AS_1 represents the reflected wavefront which gradually converges to the point Q, thus forming the real image.

P is the centre of the arc RBS and C is the centre of the arc RAS. RN is $\perp$ to AP, also R_1N_1 is $\perp$ to AP. If the aperture of the mirror is small, i.e., for paraxial rays, $RR_1 \simeq NN_1$.

Now,
$$RN^2 = BN\cdot(2BP - BN) \simeq 2BN\cdot BP, \quad (BN^2 \text{ is very small})$$
$$R_1N_1^2 = AN_1\cdot(2AQ - AN_1) \simeq 2AN_1\cdot AQ, \quad (AN_1^2 \text{ is very small})$$

Also,

$$RN^2 = AN\cdot(2AC - AN) \simeq 2AN\cdot AC$$
$$AN_1 = AN + NN_1 = AN + RR_1 = AN + AB = AN + (AN - BN) = 2AN - BN$$
$$\Rightarrow AN_1 + BN = 2AN$$
$$\Rightarrow \frac{R_1N_1^2}{2AQ} + \frac{RN^2}{2BP} = 2 \times \frac{RN^2}{2AC}$$

Now, for small aperture, $R_1N_1 = RN, BP \simeq AP$.

$$\therefore \quad \frac{1}{v} + \frac{1}{u} = \frac{2}{r}, \quad \text{(mirror formula)}.$$

Phenomenon of Total Internal Reflection

Let AC be the incident plane wavefront, parallel rays in medium of r.i. $\mu_1 > \mu_2$ are incident at A on the surface of separation, the corresponding disturbance along the other ray at that instant reaches the point C. The disturbance from C strikes the surface at B after a time $t = (BC/v_1)$, where v_1 is the speed of light in medium 1. Taking A as centre and radius $R = v_2t = \frac{v_2}{v_1} \times BC$, describe a sphere, the section of the sphere by the plane of the diagram is a circle, whose trace is shown as an arc. From B tangent is drawn to the arc as BT, thus BT represents the wavefront t secs after the point A is struck by the incident disturbance.

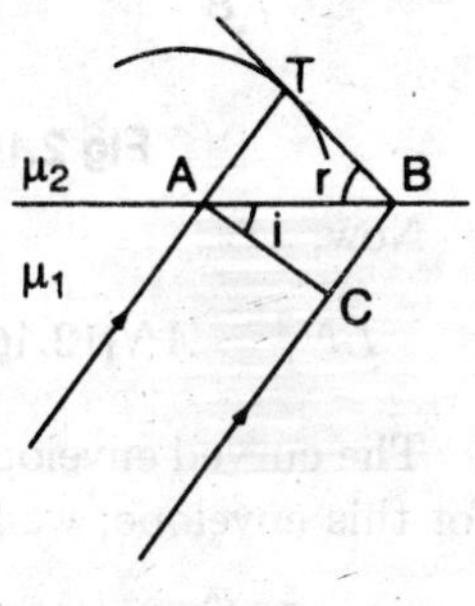

Fig 2.9

So, $\angle CAB = i, \angle ABT = r$.

$$\therefore \quad \frac{\sin i}{\sin r} = \frac{BC}{AB} \times \frac{AB}{AT} = \frac{v_1}{v_2}$$
$$\Rightarrow \quad BC = \frac{v_1}{v_2}\cdot AT = \frac{v_1}{v_2}AB\sin r, AT = \frac{v_2}{v_1}BC = \frac{v_2}{v_1}AB\sin i$$

Let the angle of incidence i be such that $(v_2/v_1)\sin i < 1$.

It then implies that $AT < AB$, hence $\sin r < 1$ or $r < \frac{\pi}{2}$, so there will be refracted ray in the second medium. Now, $\frac{c}{v_1} = \mu_1$ and $\frac{c}{v_2} = \mu_2$, and since $\mu_1 > \mu_2$, we get,

$$\frac{c}{v_1} > \frac{c}{v_2} \Rightarrow v_1 < v_2.$$

$$\therefore \quad BC < AT \quad \Rightarrow \quad AB\sin i < AB\sin r \quad \Rightarrow \quad \sin i < \sin r \quad \Rightarrow \quad i < r.$$

It means, the angle of refraction $r >$ angle of incidence i, that is, the ray bends away from the normal.

If $i = \theta$ be the angle of incidence such that $\frac{v_2}{v_1}\sin\theta = 1$, then $AT = AB \;\Rightarrow\; \sin r = 1$, it means $r = 90°$. So $i = \theta$ is the angle of incidence for which the angle of refraction is 90°. The angle of incidence for which the angle of refraction r is $\frac{\pi}{2}$ is called the *critical angle*.

Refraction Across Curved Refracting Surface

Spherical wavefront spreads out from the point source P and at some instant L_1AB_1 is the trace of the wavefront, the point A touching the pole of the curved refracting surface of radius

r separating two media of r.i. values μ_2 and $\mu_1(\mu_2 > \mu_1)$. The time taken by the disturbance to travel from L_1 to L or equivalently from B_1 to B is given by,

$$\frac{L_1L}{v_1} = \frac{BB_1}{v_1} = t$$

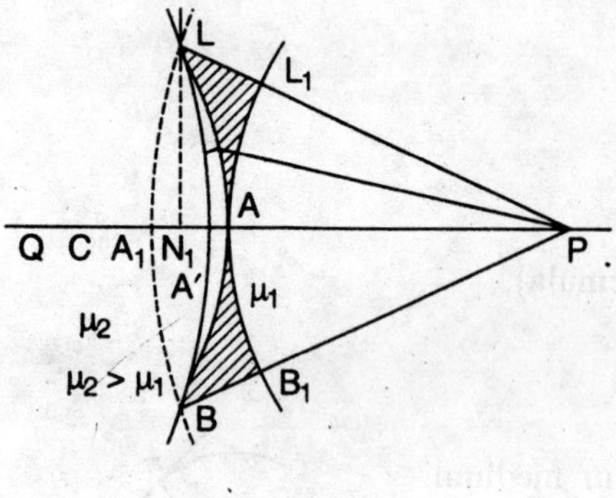

Fig 2.10

where v_1 is the velocity of the disturbance in the medium. Thus the point A of the refractor becomes the source of disturbance t secs earlier than the points L or B. So secondary disturbances travel out from A into the second medium. Take the point A as centre and draw a sphere of radius $AA' = v_2t$, where v_2 is the velocity in the second medium; the sphere is cut in a circle by the plane of the diagram, an arc is shown at A'. The envelope $LA'B$ represents refracted wavefront in the second medium. This wavefront converges to the point Q.

Now,

$$LN_1^2 = AN_1(2AC - AN_1) \simeq AN_1 \cdot 2AC, \quad \text{(considering the spherical surface } LAB\text{)}.$$

The curved envelope representing the refracted wavefront is also a part of a spherical surface. For this envelope, we have,

$$LN_1^2 = A'N_1(2A'Q - A'N_1) \simeq 2A'N_1 \cdot A'Q = 2AQA'N_1, \quad (\text{since } A'Q \simeq AQ)$$

Also, $LN_1^2 \simeq 2A_1P \cdot A_1N_1$

Now, $AA_1 = v_1t, AA' = v_2t$

$$\therefore \quad \frac{AA_1}{AA'} = \frac{v_1}{v_2} = \frac{\mu_2}{\mu_1} \Rightarrow \mu_1AA_1 = \mu_2AA'$$

$$\Rightarrow \mu_1(AN_1 + N_1A_1) = \mu_2(AN_1 - N_1A').$$

Now,

$$AN_1 = \frac{LN_1^2}{2AC} = \frac{LN_1^2}{-2r}; N_1A_1 = \frac{LN_1^2}{2A_1P} \simeq \frac{LN_1^2}{2AP} = \frac{LN_1^2}{2u}; N_1A' = \frac{LN_1^2}{2AQ} = \frac{LN_1^2}{-2v}$$

$$\therefore \mu_1\left(\frac{LN_1^2}{-2r} + \frac{LN_1^2}{2u}\right) = \mu_2\left(\frac{LN_1^2}{-2r} - \frac{LN_1^2}{-2v}\right) \quad \Rightarrow \quad \frac{\mu_2}{v} - \frac{\mu_1}{u} = \frac{\mu_2 - \mu_1}{r}$$

the usual formula for refraction at curved surface.

Spherical Wave Propagation through a Lens : Lens Formula

L_1AL_2 is the incident wavefront, diverging out from the point source P (object), $L'BL_2'$ is the emergent wavefront converging towards the point Q (image).

Now, $PA = PL_1$ and $QB = QL'$.

The optical path $L_1L' = \mu_1(L_1L + LL')$ where μ_1 is the r.i. of the surrounding medium. The optical path AB can be written as

$$\text{O.P.} \quad AB = \mu(AO + OB) \quad \therefore \quad \mu_1(L_1L + LL') = \mu(AO + OB).$$

If the lens be of small aperture, then,

$$L_1L = N_1O = N_1A + AO \quad \text{and} \quad LL' = ON_2 = OB + BN_2$$

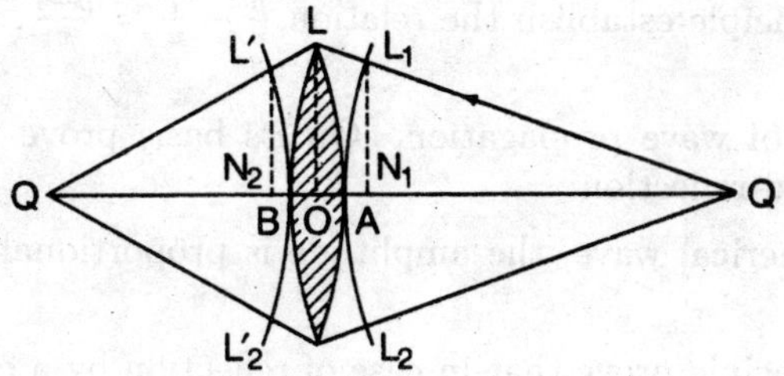

Fig 2.11

So, we get,

$$\mu_1\{(N_1A + AO) + (OB + BN_2)\} = \mu(AO + OB)$$

$$\Rightarrow (N_1A + BN_2) = \frac{\mu}{\mu_1}(AO + OB) - AO - OB$$

$$= \left(\frac{\mu}{\mu_1} - 1\right) AO + \left(\frac{\mu}{\mu_1} - 1\right) OB = \left(\frac{\mu}{\mu_1} - 1\right)(AO + OB) \qquad (2.6.1)$$

Now,

$$L_1N_1 = AN_1(2AP - AN_1) = AN_1 \cdot 2AP, AN_1 \text{ small;}$$
$$L'N_2^2 = BN_2(2BQ - BN_2) \simeq 2BN_2 \cdot BQ, BN_2 \text{ small;}$$
$$LO^2 = AO \cdot (-2r_1 - AO) \simeq AO \cdot (-2r_1), AO \text{ small.}$$

where r_1 is the radius of curvature of the right-hand face (−ve).

Also, $LO^2 = OB(2r_2 - OB) \simeq OB \cdot 2r_2$.

Now, $AP = u$ (object dist.) $\quad BQ = -v$ (image dist.).

So we get,

$$AN_1 = \frac{L_1N_1^2}{2u}; BN_2 = \frac{L'N_2^2}{-2v}; AO = \frac{LO^2}{-2r_1}; OB = \frac{LO^2}{2r_2}.$$

Putting these values in eqn.(2.6.1) we can get,

$$\frac{L_1N_1^2}{2u} + \frac{L'N_2^2}{-2v} = \left(\frac{\mu}{\mu_1} - 1\right)\left(\frac{LO^2}{2r_2} - \frac{LO^2}{2r_1}\right)$$

Since, $L_1N_1 \simeq L'N_2 \simeq LO$, we write,

$$\frac{1}{u} - \frac{1}{v} = \left(\frac{\mu}{\mu_1} - 1\right)\left(\frac{1}{r_2} - \frac{1}{r_1}\right) \quad \Rightarrow \quad \frac{1}{v} - \frac{1}{u} = \left(\frac{\mu}{\mu_1} - 1\right)\left(\frac{1}{r_1} - \frac{1}{r_2}\right)$$

the usual lens formula.

Exercises

1. Clearly explain Huygens' principle and deduce the law of refraction from it. What is the meaning of r.i.?
2. Using Huygens' principle establish the relation $\frac{\mu}{v} - \frac{1}{u} = \frac{\mu - 1}{r}$, the symbols having usual meaning.
3. State the principle of wave propagation. On its basis prove the law of refraction and explain total internal reflection.
4. Show in case of spherical wave, the amplitude is proportional inversely to the distance from the centre.
5. Using Huygens' principle prove that in case of reflection by a concave mirror $\frac{1}{v} + \frac{1}{u} = \frac{2}{r}$.
6. Establish the lens formula from Huygens' principle.

References

1. A Textbook of Mathematical Physics, *Chakrabarti, P.K.*, *Kundu, S.N.*, New Central Book Agency (P) Ltd., 8/1 Chintamoni Das Lane, Kolkata.
2. Treatise on Light, *Huygens, C.*, Dover Publication, New York (USA).
3. The Mathematical Theory of Huygens' Principle, *Baker, B.B.* and *Copson, E.J.*, Oxford University Press, London (England).
4. A Textbook on Light, *Majumder, K.G.*, Modern Book Agency, Kolkata.
5. Geometrical and Physical Optics, *Longhurst, R.S.*, Orient Longman.

Chapter 3

Interference of Light

3.1 Introduction

Newton with his corpuscular theory was still trying to stand, even after seeing that his theory could not quantitatively explain the refraction of light, the then existing phenomenon in optics. A new phenomenon was now observed in optics and even its existence could not be interpreted with the help of the corpuscular theory, not to speak about its qualitative explanation. This is the phenomenon called *interference*. However, this phenomenon is actually a characteristic of what we call a 'wave'.

What is the phenomenon? What is the constructional features of what is known as interference pattern? When and how to get the pattern?—These are the questions that will be our concern in this chapter.

3.2 What is Interference?

It has been observed that when two sources of light are simultaneously excited, then under certain conditions being fulfilled by the two sources, waves from these two sources superpose, producing on a screen alternate regions of darkness and brightness. This phenomenon is referred to as *interference* and the pattern consisting of alternate regions of darkness and brightness is called *interference pattern.*

In the description of the phenomenon we have mentioned two things which are of prime importance in this connection. Firstly, certain conditions to be fulfilled by the two sources and the second is superposition of waves from the two sources give rise to the phenomenon of interference. Let us now talk about these concepts elaborately.

3.3 Superposition of Waves and Principle of Superposition

The principle of superposition states that the displacement of a particle at a point under the influence of two waves is the algebraic sum of the displacements caused by the individual waves; or, in other words, it means that when superposition takes place, each wave retains its characteristics and produces its effect felt quite independent of the other.

This principle of superposition is valid, subject to the following conditions:

(1) The displacements y due to a wave must be small. This implies that the wave amplitude must not be large.

(2) Two waves must be monochromatic in nature.

The one-dimensional wave eqn. is represented as

$$\frac{\partial^2 y}{\partial t^2} = c^2 \frac{\partial^2 y}{\partial x^2} \tag{3.3.1}$$

Let y_1 and y_2 are two solutions of the wave eqn., then,

$$\frac{\partial^2 y_1}{\partial t^2} = c^2 \frac{\partial^2 y_1}{\partial x^2} \quad \text{and} \quad \frac{\partial^2 y_2}{\partial t^2} = c^2 \frac{\partial^2 y_2}{\partial x^2}.$$

If we set $y = y_1 + y_2$, then,

$$\frac{\partial^2}{\partial t^2}(y_1 + y_2) = \frac{\partial^2 y_1}{\partial t^2} + \frac{\partial^2 y_2}{\partial t^2} = c^2 \frac{\partial^2 y_1}{\partial x^2} + c^2 \frac{\partial^2 y_2}{\partial x^2} = c^2 \frac{\partial^2}{\partial x^2}(y_1 + y_2). \tag{3.3.2}$$

Eqn.(3.3.2) suggests that if y_1 and y_2 be two solutions of the wave eqn., then a linear combination of the two solutions also represents a solution of the wave eqn.

The solution of wave eqn.(3.3.1) can be written as

$$y = a_1(\omega t - kx) \quad \text{where} \quad k = \frac{2\pi}{\lambda}, \omega = 2\pi\nu = \frac{2\pi c}{\lambda}.$$

The acceleration of the particle at x is given by

$$\frac{\partial^2 y}{\partial t^2} = -a_1\omega^2 \cos(\omega t - kx) = -\omega^2 y.$$

Let us consider y_1 and y_2 as solutions representing displacement of a particle at x due to two waves. Then, if the displacement is not small, let us write

$$\frac{\partial^2 y_1}{\partial t^2} = -Ay_1 + By_1^2 + Cy_1^3 + Dy_1^4.$$

Similarly,

$$\begin{aligned} \frac{\partial^2 y_2}{\partial t^2} &= -Ay_2 + By_2^2 + Cy_2^3 + Dy_2^4 \\ \frac{\partial^2 y_1}{\partial t^2} + \frac{\partial^2 y_2}{\partial t^2} &= -A(y_1 + y_2) + B(y_1^2 + y_2^2) + C(y_1^3 + y_2^3) + D(y_1^4 + y_2^4). \end{aligned}$$

If the principle of superposition is to be valid, we must have

$$\frac{\partial^2 y_1}{\partial t^2} + \frac{\partial^2 y_2}{\partial t^2} = \frac{\partial^2}{\partial t^2}(y_1 + y_2) \tag{3.3.3}$$

$$-Ay_1 - Ay_2 = -A(y_1 + y_2) \tag{3.3.4}$$

$$B(y_1^2 + y_2^2) = B(y_1 + y_2)^2 \tag{3.3.5}$$

$$C(y_1^3 + y_2^3) = C(y_1 + y_2)^3, \text{ etc.} \tag{3.3.6}$$

Eqns. (3.3.5), (3.3.6) etc., are valid only if $B = 0$, $C = 0$, etc.

So we can conclude that the resultant displacement due to superposition of two waves is equal to the algebraic sum of the displacements due to them if y_1 and y_2 are small.

3.4 Application of the Principle of Superposition of Two Waves Travelling in Opposite Direction

Let the eqns. of the two waves be represented by

$$y_1 = a_1 \cos(\omega t - kx) \tag{3.4.1}$$
$$y_2 = a_2 \cos(\omega t + kx) \tag{3.4.2}$$

where y_1 and y_2 represent displacement of the particle at x due to an incident wave travelling towards the right and a wave travelling towards the left (reflected wave). The two displacements can be represented by two rotating vectors as

$$\vec{Y}_1 = a_1 e^{j(\omega t - kx)} \vec{\nu} \tag{3.4.3}$$
$$\vec{Y}_2 = a_2 e^{j(\omega t + kx)} \vec{\nu} \tag{3.4.4}$$

The resultant displacement is given by

$$\begin{aligned} \vec{Y} &= a_1 e^{j(\omega t - kx)} \vec{\nu} + a_2 e^{j(\omega t + kx)} \vec{\nu} = e^{j\omega t}(a_1 e^{-j\phi} + a_2 e^{j\phi}) \vec{\nu} \\ &= e^{j\omega t}(a_1 \cos\phi - j a_1 \sin\phi + a_2 \cos\phi + j a_2 \sin\phi) \vec{\nu} \\ &= e^{j\omega t}\left[(a_1 + a_2)\cos\phi - j\sin\phi(a_1 - a_2)\right] \vec{\nu} = (a - b_j) e^{j\omega t} \vec{\nu} \end{aligned} \tag{3.4.5}$$

where $a = (a_1 + a_2)\cos\phi$ and $b = (a_1 - a_2)\sin\phi$.

$$\therefore \quad \vec{Y} = \sqrt{a^2 + b^2} e^{-j\theta} \cdot e^{j\omega t} \vec{\nu} = \sqrt{a^2 + b^2} e^{j(\omega t - \theta)} \vec{\nu}$$

where

$$\tan\theta = \frac{(a_1 - a_2)\sin\phi}{(a_1 + a_2)\cos\phi} = \frac{a_1 - a_2}{a_1 + a_2} \tan\phi$$

$$\begin{aligned} \therefore \quad y &= \vec{Y} \cdot \vec{\nu} = \sqrt{a^2 + b^2} \cos(\omega t - \theta) \\ a^2 + b^2 &= (a_1 - a_2)^2 \cos^2\phi + \sin^2\phi (a_1 + a_2)^2 = a_1^2 + a_2^2 + 2a_1 a_2 \cos^2\phi - 2a_1 a_2 \sin^2\phi \\ &= a_1^2 + a_2^2 + 2a_1 a_2 \cos 2\phi \end{aligned}$$

If $a_1 = a_2 = A, \tan\theta = 0 \quad \Rightarrow \quad \theta = 0$

$$\sqrt{a^2 + b^2} = 2A^2(1 + \cos^2\phi) = \sqrt{4A^2 \cos^2\phi} = 2A\cos\phi$$
$$\therefore \quad y = 2A\cos\phi\cos(\omega t - \theta) = 2A\cos\phi\cos\omega t \tag{3.4.6}$$

Thus, the term independent of t and representing the amplitude is given by

$$A_0 = 2A\cos\phi = 2A\cos\frac{2\pi}{\lambda}x \tag{3.4.7}$$

$\therefore \quad y = A_0 \cos\omega t$ represents the equation of the resultant displacement due to superposition of the two waves. Since in this case the time and space occur in independent functional forms, by no adjustment of x and t we can repeat the displacement, so it does not represent a moving wave. Thus

$$y = 2A\cos\frac{2\pi x}{\lambda}\cos\omega t$$

represents a stationary wave and its amplitude is not constant at all points.

When $\cos(2\pi x/\lambda) = 1$, we get $A_0 = 2A$, maximum.

Thus, resultant wave pattern has its maximum displacement when

$$\frac{2\pi x}{\lambda} = n\pi, (n \in I) \quad \Rightarrow \quad x = n\frac{\lambda}{2}. \tag{3.4.8}$$

Thus, at $x = 0, \frac{\lambda}{2}, \lambda, 3\frac{\lambda}{2}$, the amplitude is maximum and these points are referred to as *antinodes.*

When $\cos\frac{2\pi x}{\lambda} = 0, y = 0$ for all times, thus it defines the points which are permanently at rest, called *nodes.*

$$\cos\frac{2\pi x}{\lambda} = 0 \quad \Rightarrow \quad \frac{2\pi x}{\lambda} = (2n+1)\frac{\pi}{2}, n = 0, 1, 2, \text{ etc.} \quad \Rightarrow \quad x = (2n+1)\frac{\lambda}{4}. \tag{3.4.9}$$

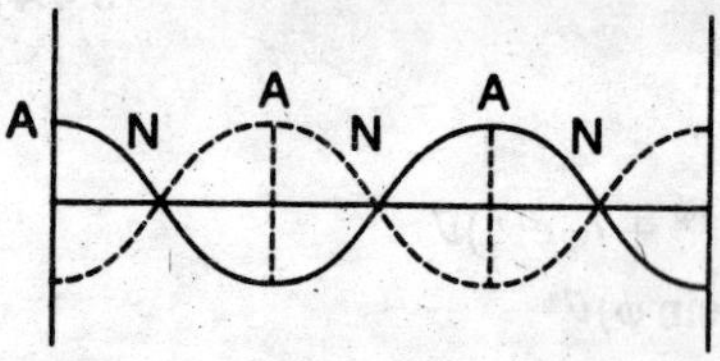

Fig 3.1

Thus, the nodes are situated at

$$x = \frac{\lambda}{4}, \frac{3\lambda}{4}, \frac{5\lambda}{4}, \text{ etc.}$$

The position of nodes (N) and antinodes (A) in case of stationary waves are as shown in fig. (3.1). It is to note that the nodal separation is $\frac{\lambda}{2}$; the antinodal separation is also $\frac{\lambda}{2}$. This wave pattern is confined to the region where superposition takes place. In case of a sound wave, the formation of a stationary wave takes place on the string of any stringed instrument or in the air column of an organ pipe. The formation of stationary wave in case of light was successfully demonstrated by Wiener.

3.5 Phenomenon of Interference

We are now in a position to discuss the phenomenon of Interference.

The first demonstration of the phenomenon of interference was due to Young using two slits, known as *Young's double-slit experiment,* which is really a modification of the experiment suggested first by Grimaldi as early as 1665.

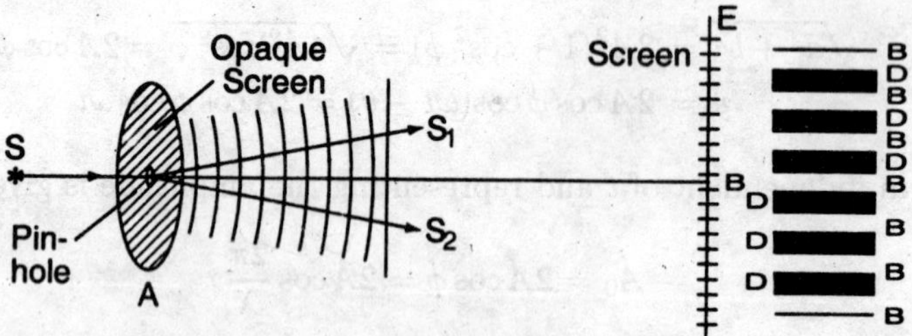

Fig 3.2

Two narrow slits S_1 and S_2 symmetrically situated w.r.t. the symmetric line SB are illuminated by light from the source S, passing through a pin-hole in an opaque screen A. The distance of the plane containing the slits is such that the spherical waves generated at the aperture O is nearly transformed into a plane wave, which at some instant occupies the plane of the slits.

From the plane of the slits an observation screen (which may be the field of view of a telescope) is placed at a considerably large distance apart. A typical experimental set-up has specified the dimension of the Young's experiment as given below—

Dimension of the aperture in the opaque screen in front of S = 0.25 *mm.*

Distance of slit plane from this pin-hole aperture = 1 *meter.*

Spatial separation between the slits S_1 *and* $S_2 = 0.7$ *mm.*

The linear-dimension of the slits S_1 *and* S_2 *both* = 0.1 *mm.*

The distance of the observation screen E w.r.t. the slit plane = 5 *meters.*

The wavelength used $\lambda = 0.5\ \mu m = 5000$Å.

Under these specifications we observe on the screen a pattern consisting of alternate regions of brightness (B) and darkness (D). The intensity distribution pattern on the screen is represented by intensity distribution curve, shown in fig. (3.3) to the extreme right.

3.5.1 Some Facts of Observation

- If the two slits S_1 and S_2 are widened out, gradually interference pattern disappears and the screen becomes illuminated under the influence of the source S_1 and S_2.
- If the two slits, keeping the dimension very small, are separated out, gradually we will find that the fringes are becoming closely spaced and loosing contrast.
- If a sheet of transparent material is placed over one of the slits, the fringe pattern gradually fades out with increasing thickness until at a certain thickness no interference fringe is visible.
- In case we get distinct fringes on the screen, we find that all the bright fringes are of equal thickness and intensity.

Any suitable theory should be able to explain the above facts of observation.

3.6 Theory of the Formation of Interference Pattern as in Young's Experiment

If we consider two sources S_1 and S_2 sending out waves in different directions, then at the point P_0 on the screen lying on the perpendicular bisector of the line joining the two sources, the two disturbances meet in phase or out of phase and this will be solely determined by the conditions of the two waves at the generation end, because the optical path connecting S_1 to

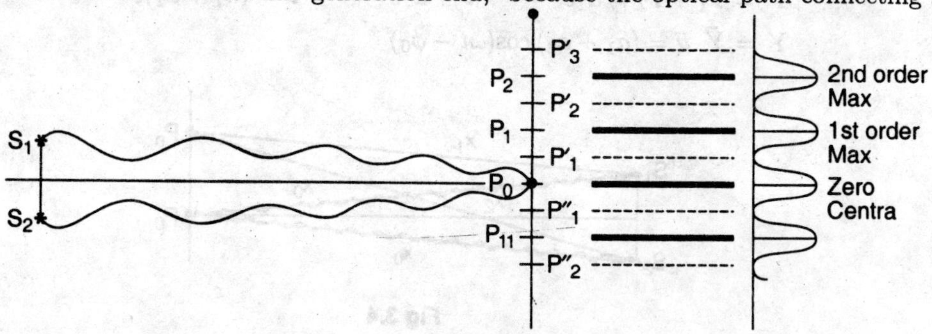

Fig 3.3

P_0 and S_2 to P_0 being equal does not introduce any phase difference. So, if the two sources are in phase, the waves coming from them superpose at P_0 in phase (crest-crest union or trough-trough union) producing constructive interference, hence brightness. On both sides of P_0 at the points P_1', P_1, P_2', etc., or at point P_1'', P_{11}, P_2'', etc., a path difference is introduced, hence a phase difference. So, when the waves start out from the native sources, they gradually suffer in phase difference due to difference in path followed. If the points on the screen are such that their path difference is an even multiple of $\frac{\lambda}{2}$, then the waves superpose in phase and due to such constructive interference, brightness is produced at these points. But, if the points are such that the path difference is an odd multiple of $\frac{\lambda}{2}$, the waves meet in opposition, destructive interference takes place producing darkness.

Thus the condition of the waves from the two sources at the time of superposition is determined by the condition both at the generation end and the meeting end. If at the generation end, there is a compromise between, the two sources to be always in phase, then the condition at the meeting end of the superposing waves is determined by the difference in optical path travelled by them. But, if there is so compromise in phase between the generator sources, then even points like P_0 on the screen whose optical path w.r.t. the two sources is the same or points like P_1, P_2, P_{11} whose optical path difference w.r.t. the two sources is an even multiple of $\frac{\lambda}{2}$, will not show any fixed character, but there will be rapid alternation between brightness and darkness, the frequency of alteration dependent on the frequency of the source. Since eye cannot respond to such fluctuation, every point appears to be bright and the fringe pattern disappears. Thus we understand that the two sources must maintain good relation so far as the phase is concerned. Such sources having the same phase always or having a predictable phase relation are said to be COHERENT sources. Thus coherency of the two sources is a crying need for the permanent interference pattern to be available. We will take up this question of coherency for further discussion in a very short while, but in the meantime let us put the above physical explanation on an analytical basis.

3.7 Analytical Method

Let the waves from the two sources S_1 and S_2 produce displacements of the particle at P_0 distant x_0 from the two sources given by,

$$\vec{Y}_1 = a_1 e^{j(\omega t - \phi_0)}\vec{\nu} \quad \text{and} \quad \vec{Y}_2 = a_2 e^{j(\omega t - \phi_0)}\vec{\nu} \text{ where } \phi_0 = \frac{2\pi}{\lambda}x_0$$

$$\therefore \quad \vec{Y} = \vec{Y}_1 + \vec{Y}_2 = a_1 e^{j(\omega t - \phi_0)}\vec{\nu} + a_2 e^{j(\omega t - \phi_s)}\vec{\nu} = (a_1 + a_2)e^{j(\omega t - \phi_s)}\vec{\nu}$$

$$\therefore \quad Y = \vec{Y}\cdot\vec{\nu} = (a_1 + a_2)\cos(\omega t - \phi_0) \tag{3.7.1}$$

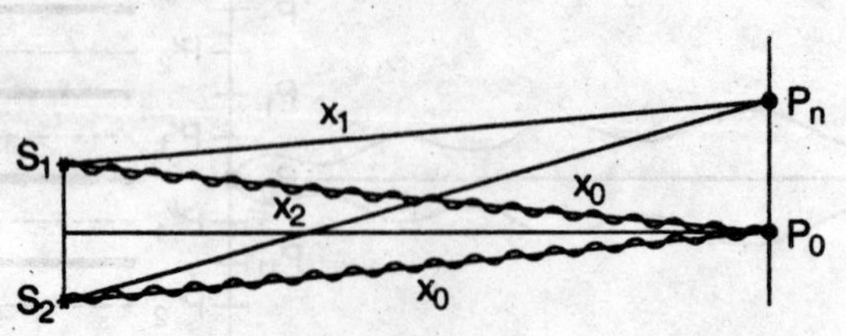

Fig 3.4

So the amplitude of vibration under the combined action of the waves from the two sources at the point P_0 is given by $A_0 = (a_1 + a_2)$, maximum.

But in case of the point like P_n, distant x_1 from S_1 and x_2 from S_2, the displacements of the particle at P_n due to them can be represented as

$$\vec{Y}_1 = a_1 e^{j(\omega t - \phi_s)}\vec{\nu} \quad \text{and} \quad \vec{Y}_2 = a_2 e^{j(\omega t - \phi'_s)}\vec{\nu} \quad \text{where } \phi_s = \frac{2\pi x_1}{\lambda},\ \phi'_s = \frac{2\pi x_2}{\lambda}$$

$$\therefore \quad \vec{Y} = a_1 e^{j(\omega t - \phi_s)}\vec{\nu} + a_2 e^{j(\omega t - \phi'_s)}\vec{\nu} = e^{j\omega t}\left(a_1 e^{-j\phi_s} + a_2 e^{-j\phi'_s}\right)\vec{\nu}$$

$$\Rightarrow \quad \vec{Y} = e^{j\omega t}\{(a_1\cos\phi_s + a_2\cos\phi'_s) - j(a_1\sin\phi_s + a_2\sin\phi'_s)\}\,\vec{\nu} = e^{j\omega t}(a - bj)\vec{\nu} \tag{3.7.2}$$

where $a = a_1\cos\phi_s + a_2\cos\phi'_s,\ b = a_1\sin\phi_s + a_2\sin\phi'_s$

$$\vec{Y} = \sqrt{a^2+b^2}\,e^{-j\theta}e^{j\omega t}\vec{\nu} = \sqrt{a^2+b^2}\,e^{j(\omega t-\theta)}\vec{\nu} = Ae^{j(\omega t-\theta)}\vec{\nu} \tag{3.7.3}$$

This represents a rotating vector whose magnitude A is a function of ϕ_s and ϕ'_s and at a certain instant it makes an angle $(\omega t - \theta)$ with $\vec{\nu}$; θ also depends on ϕ_s and ϕ'_s. Thus as a result of superposition of waves coming from the two sources, the particle at any point P_n may have maximum amplitude of vibration or minimum value depending on the position of the P_n on the screen. The resultant displacement eqn. is given by

$$y = \vec{Y}\cdot\vec{\nu} = A\cos(\omega t - \theta) \tag{3.7.4}$$

where

$$A^2 = a^2 + b^2 = a_1^2 + a_2^2 + 2a_1a_2\cos(\phi_s - \phi'_s) \tag{3.7.5}$$

Now, $\phi_s - \phi'_s$ represents the phase difference between the two component waves at the point of superposition.

Setting $(\phi_s - \phi'_s) = \delta$, we get

$$A^2 = a_1^2 + a_2^2 + 2a_1a_2\cos\delta \tag{3.7.6}$$

when $\delta = 2n\pi, A = \pm(a_1 + a_2)$.

The +ve sign before the bracket indicates crest-crest union of the component waves, a −ve sign indicates trough-trough union. But intensity being proportional to the square of amplitude, P_n is a bright point no matter whether crest-crest or trough-trough union takes place.

$$\delta = 2n\pi \quad \Rightarrow \quad \frac{2\pi}{\lambda}(x_2 - x_1) = 2n\pi \quad \Rightarrow \quad (x_2 - x_1) = 2n \times \frac{\lambda}{2} \tag{3.7.7}$$

Thus the points on the screen having path difference as even multiple of $\frac{\lambda}{2}$ become bright points.

Again, when $\delta = (2n+1)\pi$,

$$A^2 = (a_1 \sim a_2)^2 \quad \text{(minimum intensity)}$$

$$\delta = (2n+1)\pi \quad \Rightarrow \quad (x_2 - x_1) = (2n+1)\frac{\lambda}{2} \tag{3.7.8}$$

Thus, points on the screen having path difference as odd multiple of $\frac{\lambda}{2}$ become the points of minimum intensity.

If $a_1 = a_2 = a_0$, then for $\delta = 2n\pi$ intensity max. $I_m = 4a_0^2$, taking constant of proportionality equal to unity.

For $\delta = (2n+1)\pi, I = 0$, meaning darkness.

If $a_1 = a_2 = a_0$, the intensity distribution is given by

$$I = 2a_0^2 + 2a_0^2 \cos\delta = 2a_0^2(1+\cos\delta) = 4a_0^2\cos^2\frac{\delta}{2} = 4I_0\cos^2\frac{\delta}{2} \quad (3.7.9)$$

where $I_0 = a_0^2$.

A plot of intensity distribution function $y = \frac{I}{I_0}$ as given by eqn.(3.7.9) against δ is shown in fig. (3.5).

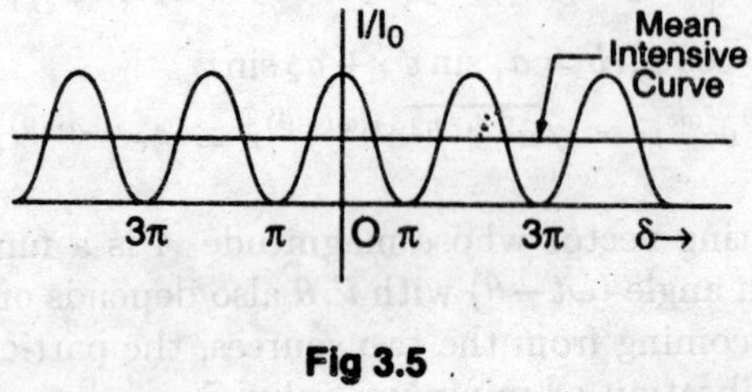

Fig 3.5

Actual intensity pattern is shown in fig. (3.6). This distribution of intensity is also sometimes referred to as $\cos^2$-pattern.

It is evident from the pattern that the thickness of the dark region is equal to that of bright region. The mean intensity is, however, a constant equal to $2I_0$.

3.8 COHERENCE is a Basic Need for Permanent Interference Pattern

We have seen while discussing Young's double-slit experiment that under certain conditions we do not get interference pattern. Thus we have stated that when the two slits are widened, the interference pattern disappears or that when a sheet of transparent material is placed over one of the slits, the pattern also disappears at a certain thickness of the sheet. Therefore, it is quite logical to conclude that certain conditions are required to be satisfied so that we can get permanent interference pattern. So we would now turn to the question : *What are the conditions of interference?*

In course of development of the intensity distribution function we have stated that when waves from the two sources superpose at certain point, the phase condition there is determined by two factors; one factor is the phase condition of the sources, the other factor is the phase relation developed in their journey towards the superposition spots. We introduce two terms, *spatial coherence* and *temporal coherence* to link up the formation of interference to these two factors ultimately deciding the fate of the waves when they superpose.

3.8.1 What is coherence?

Coherence is a property of light wave. Two waves are said to be coherent if they have a fixed and predictable phase relationship between them. In order that there is no abrupt phase change suffered by any of the waves, they must satisfy the condition of spatial coherence and temporal coherence.

3.8.2 What is Spatial Coherence?

Two independent sources are said to have zero correlation; even waves originating from the same source possess different degrees of spatial coherence depending on the extension of the source.

In Young's experiment, the two sources at S_1 and S_2 are derived from the same source S limited by the opening O in the opaque screen A. Keeping the two slits S_1 and S_2 in position if the opening O is widened, the fringes appear to lose sharpness or contrast or if the opening O is a pin-hole and the pin-hole slits S_1 and S_2 are widened or separated out, then also the fringes appear to lose in distinctiveness or contrast. In all these cases we say that spatial coherence is on the waste and if the two sources are independent, the waves from them are said to lose spatial coherence altogether and no interference is possible.

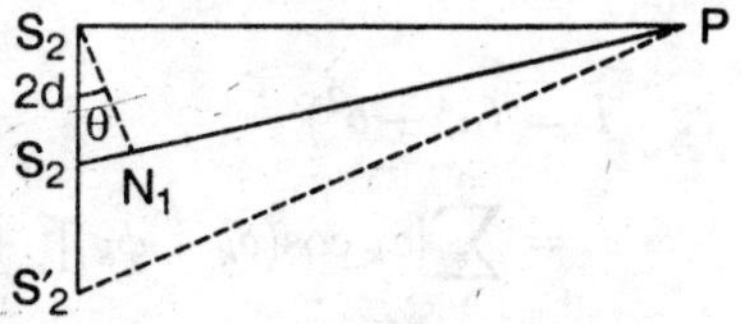

Fig 3.6

In order to understand the coherence property we must be familiar with the process of emission of light wave. When we consider an atomic emitter, a wave is emitted when the atom falls back from higher energy state to unexcited state. This takes a time of the order of 10^{-9} sec., called *emission time*. So two atomic emitters may be taken to be correlated in phase for 10^{-9} sec. only or less. This time increases as the nature of the source changes. So, if the two atomic emitters so positioned that the path difference of P w.r.t. them, i.e., $\Delta = 2d\sin\theta \le c \times 10^{-9}$, where c is the velocity of light, the two atomic sources are said to maintain *spatial coherence*, but as the two sources are drawn apart such that the paths S_1P and S_2P become S_1P and $S_2'P$, the spatial coherence becomes poorer. It is because, a wave from S_1 when it meets with a wave from S_2 at the point P, times of emission of these two superposing waves are different. The wave travelling along the path S_1P must have originated in the source S_1 at the time instant $(t - \frac{S_1P}{c})$; the wave along S_2P must have originated at $(t - \frac{S_2P}{c})$ in the source S_2.

$$\therefore \quad \Delta t = \left(t - \frac{S_1P}{c}\right) - \left(t - \frac{S_2P}{c}\right) = \frac{1}{c}(S_2P - S_1P). \tag{3.8.1}$$

Since c is a constant, Δt increases as $(S_2P - S_1P)$ increases only when the sources S_1 and S_2 are drawn apart.

If $\Delta t > 10^{-9}$ sec., the sources S_1 and S_2 lose spatial coherence; the waves from S_1 and S_2 meeting at P, though have constant path difference and hence phase difference, but there is no predictable phase relation at the time of birth.

So, for spatial coherence $\Delta t \le 10^{-9}$ sec.

$$\therefore \quad (S_2P - S_1P) \le c \times 10^{-9}.$$

If we take $c = 3 \times 10^{10}$ cm/s, then for spatial coherence, $(S_2P - S_1P) \le 30$ cm.

If $\Delta t \le 10^{-10}$ sec. $(S_2P - S_1P) \le 3$ cm. So spatial coherence depends on the linear separation between the two sources and also on the time of emission by the different sources.

3.8.3 If source is extended, no interference pattern is possible

If we consider an extended source, we have a number of atomic emitters which maintain predictable phase relation amongst themselves only for a very small time $t_c = 10^{-8}$ or 10^{-9} sec. or less, called *coherence time*. The atomic emitters may also have large linear separation. So the emitters though they belong to the same source suffer from absence of correlation in phase or suffer random phase change. So, at some point if waves from the extended source are superposed, we can write

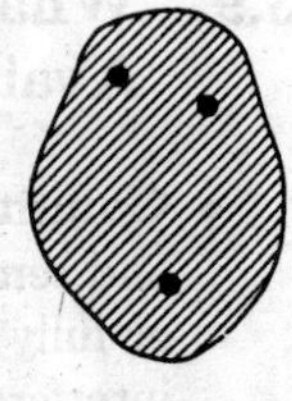

Fig 3.7

$$\vec{Y_1} = a_1 e^{j(\omega t + \delta_1 - \phi_1)}\vec{\nu}, \quad \vec{Y_2} = a_2 e^{j(\omega t + \delta_2 - \phi_2)}\vec{\nu}, \text{etc.}$$

So if we superpose,

$$\vec{Y} = \vec{Y}_1 + \vec{Y}_2 + \cdots = e^{j\omega t}\left[a_1 e^{j(\delta_1-\phi_1)} + a_2 e^{j(\delta_2-\phi_2)} + \cdots + e^{j(\delta_n-\phi_n)}\right]\vec{\nu}$$

$$= e^{j\omega t}[a_1\cos(\delta_1-\phi_1) + a_2\cos(\delta_2-\phi_2) + \cdots + j\{a_1\sin(\delta_1-\phi_1) + a_2\sin(\delta_2-\phi_2) + \cdots\}]\vec{\nu}$$

$$= (a+b_j)e^{j\omega t}\vec{\nu} = \sqrt{a^2+b^2}e^{j\theta}e^{j\omega t}\vec{\nu} = \sqrt{a^2+b^2}e^{j(\omega t+\theta)}\vec{\nu}.$$

$$\therefore \quad y = \vec{Y}\cdot\vec{\nu} = \sqrt{a^2+b^2}\cos(\omega t+\theta)$$

$$\therefore \quad I = (a^2+b^2)$$

$$= \sum_1^n[a_k\cos(\delta_k-\phi_k)]^2 + \sum_1^n[a_k\sin(\delta_k-\phi_k)]^2 + \sum_k^n\sum_{\substack{l \\ k\neq l}}^n a_k\cos(\delta_k-\phi_k)a_l\sin(\delta_l-\phi_l)$$

$$= \sum_1^n a_k^2 + \sum_k\sum_{\substack{l \\ k\neq l}} a_k a_l\cos(\delta_k-\phi_k)\cos(\delta_l-\phi_l) + a_k a_l\sin(\delta_k-\phi_k)\sin(\delta_l-\phi_l)$$

$$= \sum_1^n a_k^2 + \sum_k\sum_{\substack{l \\ k\neq l}} a_k a_l\cos(\delta_k-\phi_k-\delta_l+\phi_l).$$

Since δ_k and δ_ls are all randomly varying, the second term in all propability vanishes, hence we get,

$$I = \sum_l^n a_k^2 = a_1^2 + a_2^2 + \cdots + a_n^2.$$

If $a_1 = a_2 = a_3 = \cdots = a_n$, then, $I = na^2$. Hence no fringe pattern is available.

3.8.4 Temporal coherence

If the source is not extended and waves are derived from the same source, the path of one beam is so extended either by introduction of a transparent plate in one path or point of observation is so chosen that the path difference is exceeding the coherence length $l_c = ct_c$, then the interference fringes lose sharpness or contrast. We say that the beams lack in *temporal coherence.* At a certain value of the plate thickness or path difference, temporal coherence is completely lost and fringes disappear.

So when we get permanent fringe pattern with sufficient contrast we conclude that both spatial coherence and temporal coherence are maintained. So we can now summarise the conditions of getting permanent interference pattern.

3.9 What are the Conditions for Interference Pattern to be Available?

- **Condition 1 :** The interfering beams must be coherent w.r.t. both spatial coherence and temporal coherence. It is to be bourne in mind that beams cannot be fully coherent or fully non-coherent, so we talk about the degree of coherence. When we get permanent interference pattern with considerable contrast, we say that the beams have high degree of coherence.

- **Condition 2 :** The interfering waves must be having equal or comparable amplitudes so that the resultant amplitude $(a_1 + a_2)$ or $(a_1 - a_2)$ may be of any significance and hence the intensity variation can be recognised.
- **Condition 3 :** The waves must be having same frequency or in other words, monochromatic.
- **Condition 4 :** They must propagate along the same direction or if they have any inclination, that inclination must be small.
- **Condition 5 :** The linear distance of separation of the sources of waves and the distance of the screen are also to be small. Of course, this fact is a necessary demand of the spatial and temporal coherence.
- **Condition 6 :** In case of interference of polarised waves (the phenomenon of polarisation is discussed later), the plane of polarisation of the interfering beams must be same. In case the beams are polarised at right angles, no interference fringes are available.

3.10 Nature of the Fringes and Expression for the Fringe-width

If we refer to fig. (3.8), the coherent sources are taken along Y-axis, separation being d, the co-ordinates of S_1 are $(0, \frac{d}{2})$ and those of S_2 are $(0, -\frac{d}{2})$. Let us take a point P on the screen so that its co-ordinates are (x, y).

$$S_1P^2 = (x-0)^2 + \left(y - \frac{d}{2}\right)^2 = x^2 + \left(y - \frac{d}{2}\right)^2 \quad \Rightarrow \quad S_1P = \left[x^2 + \left(y - \frac{d}{2}\right)^2\right]^{\frac{1}{2}}$$

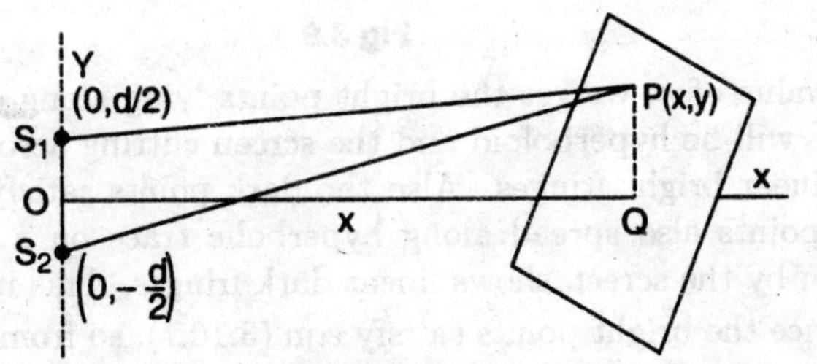

Fig 3.8

Similarly,

$$S_2P = \left[x^2 + \left(y + \frac{d}{2}\right)^2\right]^{\frac{1}{2}}$$

$$\therefore \quad S_2P - S_1P = n\lambda \quad \text{(for brightness at } P\text{)}$$

$$\Rightarrow \quad \left[x^2 + \left(y + \frac{d}{2}\right)^2\right]^{\frac{1}{2}} - \left[x^2 + \left(y - \frac{d}{2}\right)^2\right]^{\frac{1}{2}} = n\lambda$$

$$\Rightarrow \quad x^2 + \left(y + \frac{d}{2}\right)^2 = (n\lambda)^2 + x^2 + \left(y - \frac{d}{2}\right)^2 + 2n\lambda\left[x^2 + \left(y - \frac{d}{2}\right)^2\right]^{\frac{1}{2}}$$

$$\Rightarrow \quad 4y\cdot\frac{d}{2} = n^2\lambda^2 + 2n\lambda\left[x^2 + \left(y - \frac{d}{2}\right)^2\right]^{\frac{1}{2}}$$

$$\Rightarrow \quad (2yd - n^2\lambda^2)^2 = 4n^2\lambda^2\left[x^2 + \left(y - \frac{d}{2}\right)^2\right]$$

$$\Rightarrow \quad 4y^2d^2 - 4n^2\lambda^2 yd + n^4\lambda^4 = 4n^2\lambda^2\left(x^2 + y^2 - yd + \frac{d^2}{4}\right)$$

$$\Rightarrow \quad \frac{y^2}{\left(\frac{n^2\lambda^2}{4}\right)} - \frac{x^2}{\left(\frac{d^2 - n^2\lambda^2}{4}\right)} = 1 \qquad (3.10.1)$$

The eqn.(3.10.1) is the equation of a hyperbola, so the locus of the point P which happens to be a bright point is a hyperbola with O as centre. So the bright fringes are in the form of hyperbolic traces as shown in fig. (3.9).

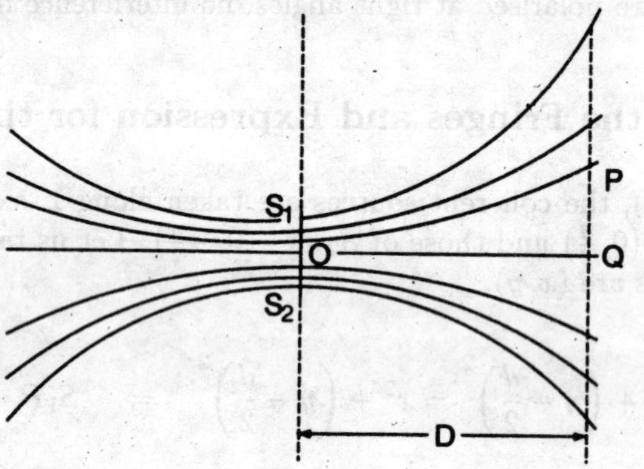

Fig 3.9

For a particular value of n, we get the bright points lying along a hyperbolic trace and in three-dimensions this will be hyperboloid and the screen cutting across the hyperboloid would give linear bright fringes. Also the dark points satisfy similar equation, hence dark points also spread along hyperbolic trace on a hyperboloid surface; the section by the screen shows linear dark fringes. This is shown in fig. (3.10).

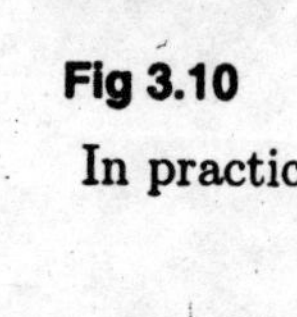

Fig 3.10

Since the bright points satisfy eqn.(3.10.1), so from this equation we can get the distance of the nth bright fringe w.r.t. the central line OQ.

So, let $x = D$ and $y = y_n$, locate the position of the nth bright fringe, then

$$\frac{4y_n^2}{n^2\lambda^2} - \frac{4D^2}{(d^2 - n^2\lambda^2)} = 1 \quad \Rightarrow \quad \frac{4y_n^2}{n^2\lambda^2} = \frac{(d^2 - n^2\lambda^2) + 4D^2}{(d^2 - n^2\lambda^2)}.$$

In practice, $D^2 \gg (d^2 - n^2\lambda^2)$ and also $d^2 \gg n^2\lambda^2$

$$\therefore \quad \frac{4y_n^2}{n^2\lambda^2} = \frac{4D^2}{d^2} \quad \Rightarrow \quad y_n^2 = \frac{D^2n^2\lambda^2}{d^2} \quad \therefore \quad y_n = \frac{Dn\lambda}{d}. \qquad (3.10.2)$$

This eqn.(3.10.2) locates the nth bright fringe position w.r.t. the central point. The position of the next bright fringe, i.e., $(n+1)$th bright fringe is given by

$$y_{n+1} = \frac{D(n+1)\lambda}{d}. \qquad (3.10.3)$$

From these two eqns. we get the thickness containing two consecutive bright fringes, hence the fringe width β is given as

$$\beta = y_{n+1} - y_n = \frac{D\lambda}{d}.$$

3.11 Shape of Interference Fringes Depending on the Disposition of the Slit-sources and Plane of the Screen

3.11.1 Show that the fringe shape obtained on a screen when the linear-dimension of the double-slits in Young's experiment is parallel to the plane of the screens

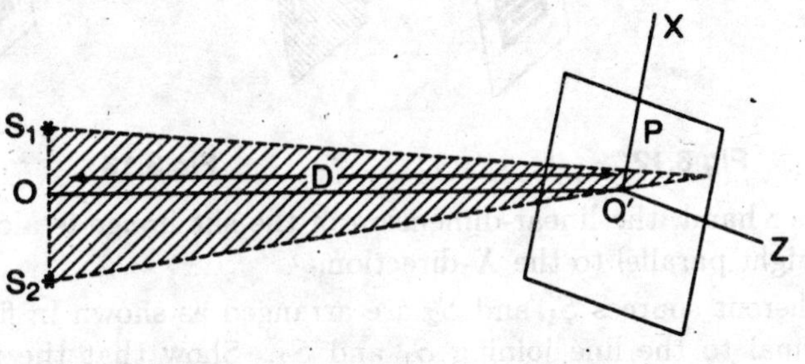

Fig 3.11

Let us consider two coherent sources lying in a plane parallel to the plane of the screen, along a line S_1S_2 in the X-direction; the receiving screen is placed at a distance D from O; the linear separation between S_1 and S_2 being d. We are to investigate the shape of the fringes as obtained on the screen. We have proved previously that with respect to the point O as the origin, the locus of the bright points is a hyperbola with the central line and the coherent source joining line as axes.

What will be the shape of the fringes (bright or dark) on the screen? We take O' as the origin. The bright point P then has the co-ordinate $(x, 0, z)$, the co-ordinates of the source points S_1 and S_2 are respectively $(\frac{d}{2}, -D, 0)$ and $(-\frac{d}{2}, -D, 0)$.

$$S_1P^2 = \left(x - \frac{d}{2}\right)^2 + D^2 + z^2 \quad \text{and} \quad S_2P^2 = \left(x + \frac{d}{2}\right)^2 + D^2 + z^2$$

$$\therefore \quad S_2P - S_1P = \sqrt{\left(x + \frac{d}{2}\right)^2 + D^2 + z^2} - \sqrt{\left(x - \frac{d}{2}\right)^2 + D^2 + z^2}.$$

Since P is the nth bright point, so $S_2P - S_1P = n\lambda$.

$$\therefore \quad n\lambda = \sqrt{\left(x + \frac{d}{2}\right)^2 + D^2 + z^2} - \sqrt{\left(x - \frac{d}{2}\right)^2 + D^2 + z^2}$$

$$\Rightarrow \quad n\lambda + \sqrt{\left(x - \frac{d}{2}\right)^2 + D^2 + z^2} = \sqrt{\left(x + \frac{d}{2}\right)^2 + D^2 + z^2}.$$

Squaring both sides

$$2n\lambda\sqrt{\left(x - \frac{d}{2}\right)^2 + D^2 + z^2} = (2xd - n^2\lambda^2).$$

Again, squaring and rearranging terms

$$4(n^2\lambda^2 - d^2)x^2 + 4n^2\lambda^2 z^2 = n^2\lambda^2(n^2\lambda^2 - d^2) - 4n^2\lambda^2 D^2$$

$$\text{or} \quad x^2 + \frac{n^2\lambda^2}{(n^2\lambda^2 - d^2)} z^2 = \frac{n^2\lambda^2}{4} - \frac{n^2\lambda^2}{(n^2\lambda^2 - d^2)} D^2$$

So, if $x \ll z$, i.e., the linear slit image as in Fresnel's biprism is along Z-axis, the fringe pattern will consist of straight fringes parallel to the X-axis, i.e., parallel to the linear-dimension of the slit image (cf. fig. 3.12).

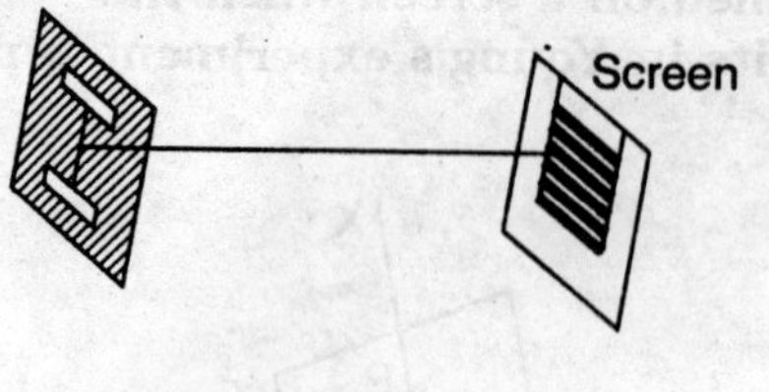

Fig 3.12

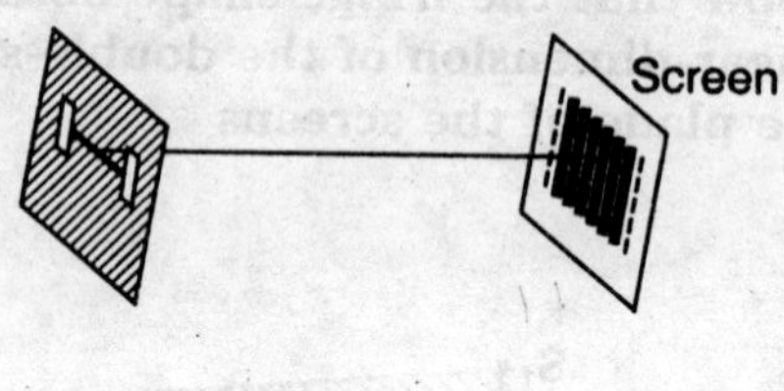

Fig 3.13

But if, on the other hand, the linear-dimension of the slit image is along the X-direction, the fringe will be straight parallel to the X-direction.

The two point coherent sources S_1 and S_2 are arranged as shown in fig. (3.13). Consider a plane which is normal to the line joining S_1 and S_2. Show that the interference pattern produced on this plane will consist of concentric circles and that the radius of the nth bright circle will be approximately $\sqrt{2}\sqrt{1 - \frac{n\lambda}{d}}D$, where we assume $D \gg d$ or x, the distances D, d and x are shown in fig. (3.14).

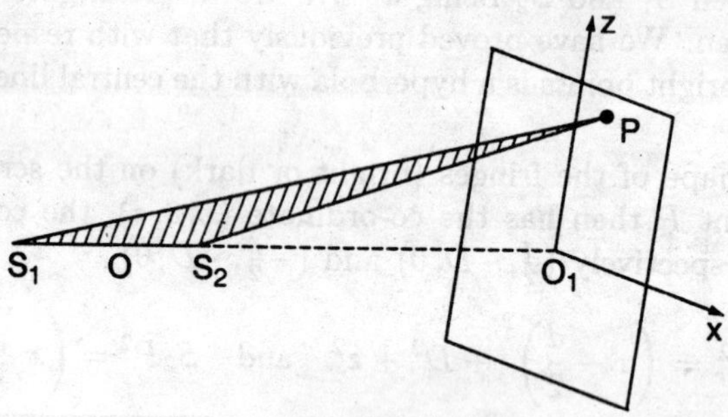

Fig 3.14

Consider origin at O_1, then any point P on the screen may be supposed to have co-ordinates $(x, 0, z)$ and the co-ordinates of the two sources S_1 and S_2 are

$$\left[0, \left(D + \frac{d}{2}\right), 0\right] \quad \text{and} \quad \left[0, \left(D - \frac{d}{2}\right), 0\right] \text{ respectively.}$$

Then, we can write

$$S_1P = \sqrt{x^2 + \left(D + \frac{d}{2}\right)^2 + z^2} \quad \text{and} \quad S_2P = \sqrt{x^2 + \left(D - \frac{d}{2}\right)^2 + z^2}$$

$$\therefore\ S_1P - S_2P = \sqrt{x^2 + \left(D + \frac{d}{2}\right)^2 + z^2} - \sqrt{x^2 + \left(D - \frac{d}{2}\right)^2 + z^2}.$$

For nth bright fringe at P, we have,

$$S_1P - S_2P = n\lambda$$

$$\therefore\ n\lambda = \sqrt{x^2 + \left(D + \frac{d}{2}\right)^2 + z^2} - \sqrt{x^2 + \left(D - \frac{d}{2}\right)^2 + z^2}.$$

Proceeding exactly as in 1, we get

$$x^2 + z^2 = \frac{D^2d^2}{n^2\lambda^2} + \frac{n^2\lambda^2}{4} - D^2 - \frac{d^2}{4} = D^2\left(\frac{d^2}{n^2\lambda^2} + \frac{n^2\lambda^2}{4D^2} - 1 - \frac{d^2}{4D^2}\right).$$

Since D is large, we get

$$x^2 + z^2 = D^2\left(\frac{d^2}{n^2\lambda^2} - 1\right) = \frac{D^2d^2}{n^2\lambda^2}\left(1 - \frac{n^2\lambda^2}{d^2}\right) = R_n^2$$

$$\therefore\quad R_n^2 = \frac{D^2d^2}{n^2\lambda^2}\left(1 + \frac{n\lambda}{d}\right)\left(1 - \frac{n\lambda}{d}\right)$$

$d \simeq n\lambda$, so the radius of the nth bright circle may be written as

$$R_n^2 = D^2 \times 2\left(1 - \frac{n\lambda}{d}\right) \quad\Rightarrow\quad R_n = D\sqrt{2}\left(1 - \frac{n\lambda}{d}\right)^{\frac{1}{2}}. \tag{3.11.1}$$

Thus we see that the bright fringes will be on different hyperboloids of revolution which are confocal with S_1 and S_2 as foci. Moreover we conclude from the above discussion that:

- If the screen is placed parallel to the line joining S_1 and S_2, we will get straight fringes parallel to the length of the slits at S_1 and S_2.
- If the screen be placed at right angles to the line joining S_1 and S_2, we will get a number of alternate bright and dark concentric circles round the point of intersection of the line joining S_1 and S_2 and the plane.
- If, however, the screen be placed at some angle with the line joining S_1 and S_2, the fringes will be a set of confocal ellipses.

All the above fringes are said to be *non-localised* because they will be available on the screen not at any specified position.

3.11.2 Two Important Questions

Q.1 *The dark region in the interference pattern means destruction of light energy. Is interference a violation of energy principle?*

Ans. It is true that dark region is a place deprived of light energy but that energy is not destroyed but simply it is redistributed or added to the bright region. When the fringes are not formed, the screen receives light energy from the two sources and is illuminated by light. The total energy tht the screen receives in absence of the fringe pattern is exactly equal to the sum total of energy associated with the bright fringes. Thus in interference there is redistribution of energy and not the destruction of energy.

If we consider two waves having amplitude of each equal to a, then at any point of the screen, the intensity is $(a^2 + a^2)$ i.e., $2a^2$. In the interference pattern also the average intensity of the energy distribution is $2a^2$. Thus the total energy is conserved.

Q.2 *Two candles or two torches can never produce interference pattern. Explain.*

Ans. A candle flame or a torch is an extended source; moreover, they are independent sources, so the light waves from them are completely lacking in coherence, both spatial and temporal. In such case of uncorrelated waves, interference pattern is impossible; what we get is illumination all over.

3.12 Classification of Interference Phenomenon

Depending on the manner how we get coherent sources, interference is divided into two classes (1) wavefront division class, (2) amplitude division class. To the first class belong a large number of experiments like *Fresnel Biprism*, Lloyd's single and double mirror, Billet's split lens, etc.; to the second also belong a series of experiments like Newton's rings, Michelson's interferometer, Jamin's interferometer, Fabry-Perot interferometer, etc.

3.12.1 Wavefront division class of interference

In this class of interference, a wavefront which represents the locus of isophasal particles is divided and the waves from these two parts are allowed to be taken along different paths and made to superpose giving rise to interference pattern. As the particles on a given wavefront vibrate in phase, they may be looked upon as coherent sources. But in order to obtain spatial coherence, only a limited portion of the wavefront is selected. If we refer to Young's experimental set-up, spherical disturbances set out from the small opening O, fall on the plane of the slits S_1 and S_2 becoming new sources of secondary disturbance and if these two slits are close-by, these secondary waves satisfy the condition of spatial coherence. If the screen is no placed that there are points on the screen whose path difference is less than the coherence length for the nature of the sources under consideration, then on the screen we would expect permanent interference pattern. In all such class of interference, only a very limited region of a wavefront is taken into consideration.

3.12.2 A few experimental set-up based on wavefront division

(A) Fresnel Biprism

Fig. 3.15 shows a Fresnel biprism experimental arrangement. The biprism consists of two acute-angled prisms with their bases put together. It is placed in front of a very narrow slit illuminated by monochromatic light. Rays from S after refraction in the upper half of the prism are rendered into a diverging beam spreading over the region PQ on the screen, all such rays as if come from S_1, the virtual image of S due to refraction in the upper half of the biprism;

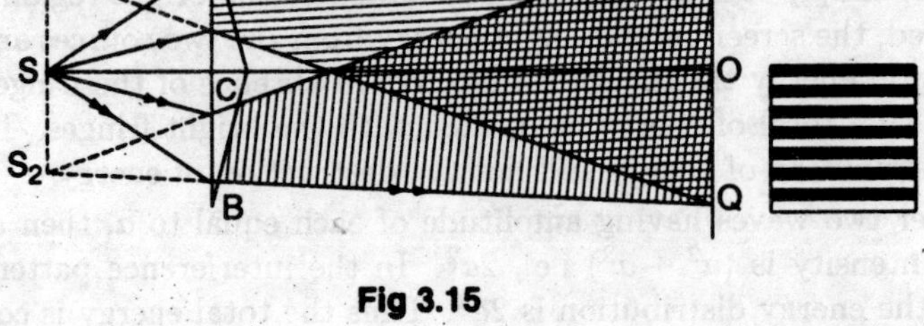

Fig 3.15

similarly, S_2 is the virtual image of the source S due to refraction at the lower half of the biprism, as if all the rays come from the virtual source S_2. Since both halves are symmetric, optical path introduced by the prism will be same for the two interferring beams. Again, since the virtual sources S_1 and S_2 are derived from the same source, any abrupt change in phase of S will equally affect the two sources S_1 and S_2. Thus the waves as if from the sources S_1 and S_2 continue to have constant phase relationship, so the waves interfere producing fringe system in the region PQ of the screen, the common region of overlap of the screen.

Theory of the experiment

The theory of the experiment we have already developed, assuming S_1 and S_2 as forming the two coherent sources having separation equal to d, the fringe width is given by

$$\beta = \frac{D\lambda}{d}.$$

So measuring the fringe width of the interference fringes, and also the value of D, we can determine λ from a knowledge of d.

Measurement of d

A convex lens of suitable focal length is introduced between the biprism and the screen and the position of the lens is adjusted so as to get sharp images of the sources S_1 and S_2 on the screen as S_1' and S_2' having separation equal to d_1; the lens is shifted to another position when again sharp images of the sources is obtained on the screen; the separation between the images formed now is d_2 (say).

In the first position magnification m_1 is given by

$$m_1 = \frac{\text{Image separation } S_2'S_1'}{\text{Object separation } S_1S_2} = \frac{d_1}{(d)}.$$

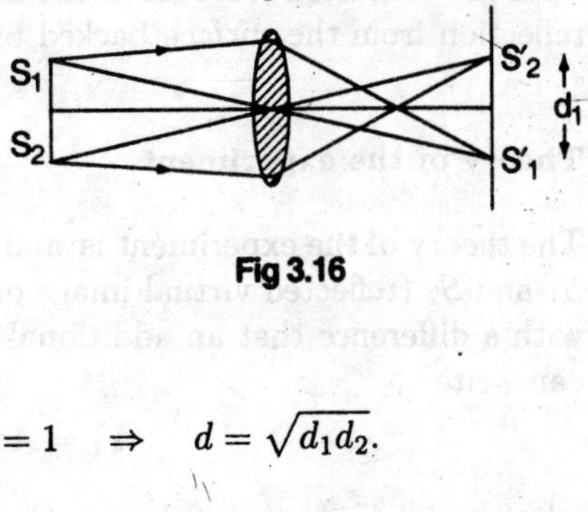

Fig 3.16

In the second position, magnification m_2 is given by

$$m_2 = \frac{\text{Image separation}}{\text{Object separation}} = \frac{d_2}{(d)} \quad \because \quad m_1 m_2 = 1 \quad \Rightarrow \quad d = \sqrt{d_1 d_2}.$$

Hence d can be evaluated[1].

(B) Lloyd's Mirror

This arrangement of obtaining interference pattern is due to waves coming from the two sources, one is the original point source (S_1), the other is the virtual source S_2 derived out of it due to reflection at the mirror MM_1. The waves directly coming from the source S_1 are confined within the region ADP and the waves reaching the screen through reflection are confined within the region MM_1QP, so superposition of these waves takes place in the position PQ of the screen, called the *region of overlap*. So fringe pattern will appear in this region of overlap.

[1]For experimental details see 'Advanced Textbook on Practical Physics', P.K. Chakrabarti and S.B. Chowdhury.

The fringe characteristics are (1) the point C lying on the perpendicular bisector of the line joining S_1 and S_2 called the *central line*, is equidistant from the sources S_1 and S_2. The fringes lie on one side of the central line OC.

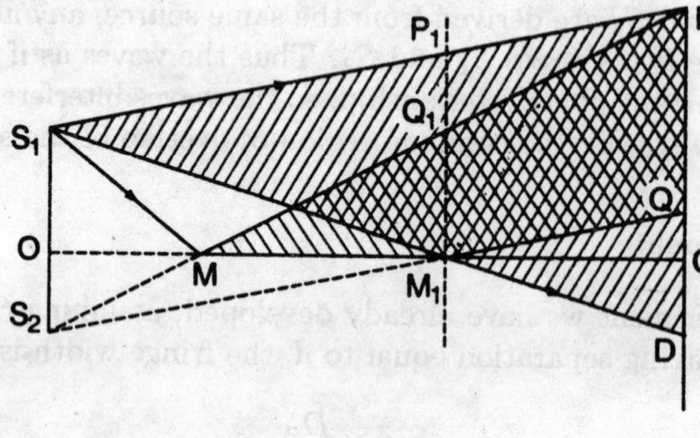

Fig 3.17

(2) In order to get the central fringe, the screen is to be shifted to the position $P_1Q_1M_1$ touching the edge of the mirror at M_1 or introducing a plate in the path of the direct beam so that the central point is shifted to the region PQ. It will be found that the point C which is equidistant w.r.t. the sources is a dark point in contradiction to our previous inference, according to which it should correspond to the central bright fringe. As we will see later, reflection from a surface is associated with a phase change of π which results in a path difference $\frac{\lambda}{2}$. This inference of the phase change due to reflection comes through a theorem due to Stokes and this experiment which shows that the central fringe is dark due to superposition of direct wave and reflected wave from the metallic or a dielectric mirror suggests that only in case of reflection from the surface backed by denser medium, phase change of π is introduced.

Theory of the experiment

The theory of the experiment is similar to what we have discussed earlier if we treat the sources S_1 and S_2 (reflected virtual image of the source S_1) as coherent sources producing interference with a difference that an additional phase difference is produced due to reflection. So here we can write

$$\vec{Y}_1 = Ae^{j(\omega t-\phi_s)}\vec{\nu}, \quad \vec{Y}_2 = Ae^{j(\omega t-\phi'_s)}\vec{\nu}$$

where $\phi_s = \frac{2\pi}{\lambda}x_1, \phi'_s = (\frac{2\pi}{\lambda}x_2 \pm \pi)$

$$\begin{aligned}
\therefore \quad \vec{Y} &= \vec{Y}_1 + \vec{Y}_2 = Ae^{j\omega t}\left(e^{-j\phi_s} + e^{-j\phi'_s}\right)\vec{\nu} \\
&= Ae^{j\omega t}(\cos\phi_s - j\sin\phi_s + \cos\phi'_s - j\sin\phi'_s)\vec{\nu} \\
&= Ae^{j\omega t}\left[(\cos\phi_s + \cos\phi'_s) - j(\sin\phi_s + \sin\phi'_s)\right]\vec{\nu} \\
&= Ae^{j\omega t}(a - bj)\vec{\nu} = A\sqrt{a^2+b^2}e^{j(\omega t-\theta)}\vec{\nu} \\
\therefore \quad y &= \vec{Y}\cdot\vec{\nu} = A\sqrt{a^2+b^2}\cos(\omega t - \theta) \\
\therefore \quad I &= A^2(a^2+b^2) \\
&= A^2(\cos^2\phi_s + \cos^2\phi'_s + \sin^2\phi_s + 2\cos\phi_s\cos\phi'_s + \sin^2\phi'_s + 2\sin\phi_s\sin\phi'_s) \\
&= 2A^2[1 + \cos(\phi'_s - \phi_s)] = 4A^2\cos^2\frac{\phi'_s - \phi_s}{2}
\end{aligned}$$

So the maximum intensity corresponds to the condition when

$$\cos\frac{\phi'_s - \phi_s}{2} = 1 \quad \Rightarrow \quad \phi'_s - \phi_s = 2n\pi.$$

Now, $\phi'_s - \phi_s = 2n\pi$

$$\Rightarrow \quad \frac{2\pi}{\lambda}x_2 - \frac{2\pi}{\lambda}x_1 \pm \pi = 2n\pi \quad \Rightarrow \quad \frac{2\pi}{\lambda}(x_2 - x_1) = (2n \pm 1)\pi$$

$$\Rightarrow \quad x_2 - x_1 = (2n \pm 1)\frac{\lambda}{2},$$

i.e., interference maximum occurs when the path difference is an odd multiple of $\frac{\lambda}{2}$.

Minimum of intensity occurs when $\cos\frac{\phi'_s - \phi_s}{2} = 0$

$$\Rightarrow \quad \frac{\phi'_s - \phi_s}{2} = (2n+1)\frac{\pi}{2} \quad \Rightarrow \quad \phi'_s - \phi_s = (2n+1)\pi$$

$$\Rightarrow \quad \frac{2\pi}{\lambda}(x_2 - x_1) = (2n+1)\pi \pm \pi = (2n+2)\pi, \quad \text{or,} \quad 2n\pi$$

$$\Rightarrow \quad (x_2 - x_1) = (2n+2)\frac{\lambda}{2}, \quad \text{or,} \quad 2n\frac{\lambda}{2}$$

It shows that when the path difference is an even multiple of $\frac{\lambda}{2}$, the intensity becomes $I = 0$. So when $x_2 = x_1$, we get the central fringe and this satisfies the condition of minimum.

(C) Fresnel's Double Mirror Arrangement

We have just described Lloyd's single mirror arrangement. In it interference phenomenon was of considerable interest to the then scientist, so was another arrangement due to Fresnel using two mirrors.

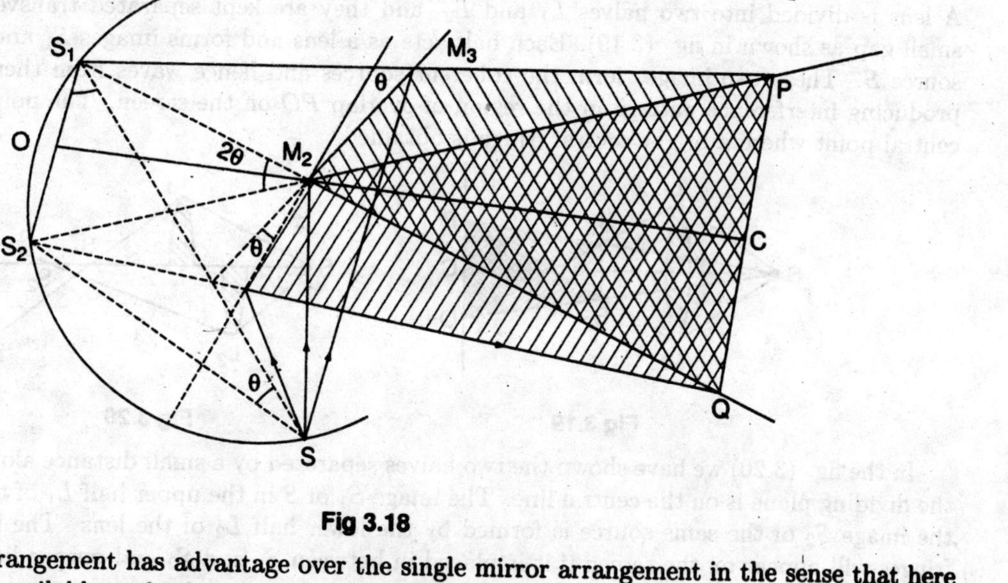

Fig 3.18

This arrangement has advantage over the single mirror arrangement in the sense that here fringes are available on both sides of the line of symmetry.

In this case, a very narrow slit S illuminated by monochromatic light sends out wavefronts, part of which is reflected from the mirror AB forming the virtual source S_1 due to reflection in the mirror AB; the other part of the wavefront is reflected in the mirror BC forming the virtual source S_2.

The waves reflected from the mirror M_1M_2 are spread over the region M_1M_2PQ; the waves reflected from the mirror M_2M_3 are spread over the region M_2M_3PQ; the region of overlap of these two sets of waves is represented by the cross thatched area and on the screen, fringes will appear in the region PQ. The central line is OC, the central point is C.

It is easy to see that

$$SM_2 = S_2M_2 = S_1M_2.$$

Hence S_1S_2 and S_1 all line on a circle of radius

$$SM_2 = S_2M_2 = S_1M_2 = R \quad \text{(say)}.$$

Also, $\angle S_1SS_2 = \theta = \frac{1}{2}\angle S_1M_2S_2$

$$\therefore \quad S_1S_2 = 2R\sin\theta \simeq 2R\theta.$$

Characteristics of the Fringe Pattern

Here central fringe is available and this fringe is bright. This is because the reflected waves from the two mirrors M_1M_2 and M_2M_3 suffer phase change by the same amount, whereas in case of single mirror the fringe pattern appears on one side of the central line and even if the central fringe is obtained on the screen by suitable adjustment of its position, it is found to be dark one.

Billet's Split Lens

A lens is divided into two halves L_1 and L_2, and they are kept separated transversely by a small gap as shown in fig. (3.19). Each half acts as a lens and forms images S_1 and S_2 of the source S. These two images form the coherent sources and hence waves from them interfere producing interference pattern in the region of overlap PQ on the screen. The point C is the central point where a bright central fringe is obtained.

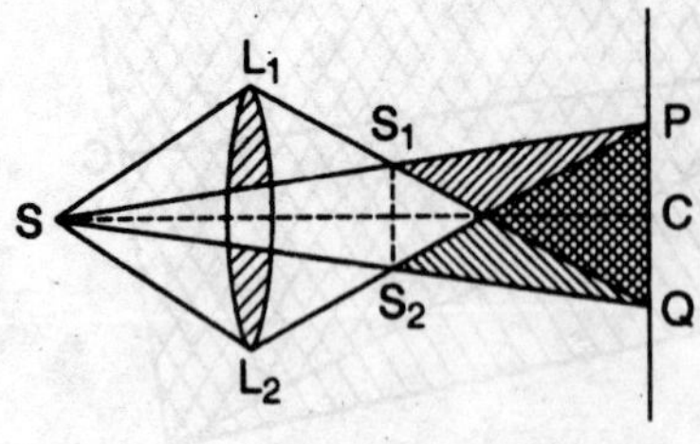

Fig 3.19

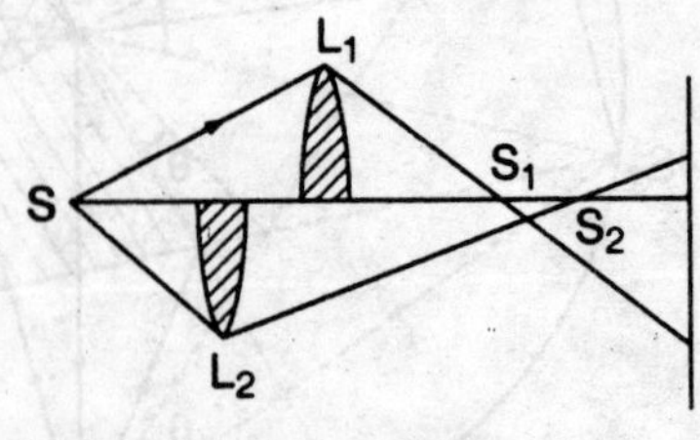

Fig 3.20

In the fig. (3.20) we have shown the two halves separated by a small distance along the axis, the dividing plane is on the central line. The image S_1 of S in the upper half L_1 of the lens and the image S_2 of the same source is formed by the lower half L_2 of the lens. The interference fringes will appear on the screen if it is placed in between S_1 and S_2 and perpendicular to the line joining S_1 and S_2.

3.13 Application of the Phenomenon of Interference

3.13.1 Finding the wavelength of the light used

The wavelength of light used in the biprism experiment can be determined as discussed earlier.

3.13.2 Finding the thickness of a thin film

The interferometric method may be applied to find out the thickness of a very thin film, which cannot be done by other conventional mechanical instruments.

In this method, a thin film is to be introduced in the path of one of the interfering beams, as a result, the fringe pattern shifts to that side of the central line OC where the film is placed. It is to be remembered that the thickness of the transparent film should not be such that temporal coherence no longer holds; if the thickness exceeds a certain value, the fringe system disappears due to absence of temporal coherence.

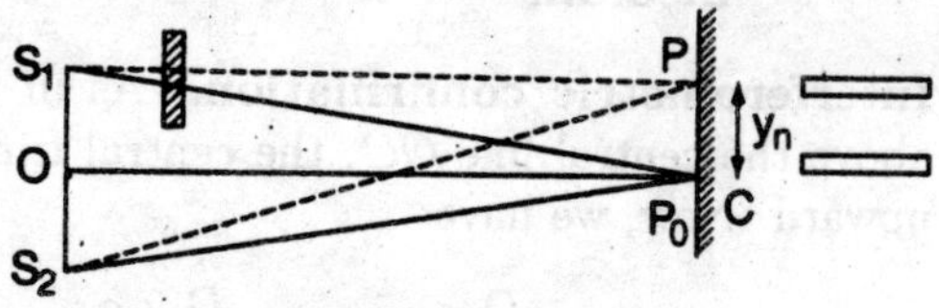

Fig 3.21

Now, as a result of the introduction of the thin film, the central fringe at P_0 characterised by zero optical path difference w.r.t. the source S_1 and S_2, is shifted to the position P, so now, the optical path to P from S_1 and S_2 must be equal.

$$\therefore\ (S_1P - l) + \mu l = S_2P \quad \Rightarrow \quad S_2P - S_1P = (\mu - 1)l.$$

In the original fringe pattern, P was the site of some order fringe, say nth bright fringe.

$$\therefore \quad (S_2P - S_1P) = n\lambda \quad \text{for } n\text{th bright fringe.}$$

If the central point P_0 is at a distance D w.r.t. O (cf. fig. 3.21), then

$$S_2P - S_1P = \left[D^2 + \left(y + \frac{d}{2}\right)^2\right]^{\frac{1}{2}} - \left[D^2 + \left(y - \frac{d}{2}\right)^2\right]^{\frac{1}{2}}.$$

Now,

$$\left[D^2 + \left(y + \frac{d}{2}\right)^2\right]^{\frac{1}{2}} = D\left[1 + \left(\frac{y + d/2}{D}\right)^2\right]^{\frac{1}{2}}$$

$$\simeq D\left[1 + \frac{1}{2}\frac{(y + d/2)^2}{D^2}\right], \text{ since } \frac{(y + d/2)}{D} \ll 1$$

Similarly,

$$\left[D^2 + \left(y - \frac{d}{2}\right)^2\right]^{\frac{1}{2}} = D\left[1 + \frac{1}{2}\frac{(y - d/2)^2}{D^2}\right]$$

$$\therefore \quad S_2P - S_1P = \frac{yd}{D} \tag{3.13.1}$$

Now, here we have taken P as the site of nth bright fringe at a distance y_n w.r.t. the central point C on the screen.

$$\therefore \quad \frac{y_n d}{D} = (S_2P - S_1P) = n\lambda \quad \Rightarrow \quad y_n = \frac{Dn\lambda}{d} \quad \therefore \quad (\mu - 1)l = n\lambda = \frac{y_n d}{D}.$$

This equation may be utilised to find out the thickness l of the film. If we can count the order number of fringe on which the central fringe falls due to displacement (i.e., if n value determined), or if the displacement of the central fringe, i.e., y_n, is found out, the value of l can be found out.

The above equation can also be thrown into a form containing the fringe width β, only if we remember that

$$\beta = \frac{D\lambda}{d} \quad \therefore \quad (\mu - 1)l = \frac{y_n\lambda}{\beta}.$$

3.13.3 The velocity of light in air is greater than the velocity of light in some medium

Interferometric confirmation. As in the application 2, if the transparent film is placed above the central line OC, the central fringe which was formed at C is found to be displaced upward. Now, we have

$$y_n = \frac{D}{d}(\mu - 1)l = \frac{D}{d}\left(\frac{c}{V} - 1\right)l \quad \because y_n \text{ is +ve, } \frac{c}{V} > 1 \quad \Rightarrow \quad c > V.$$

3.13.4 Measurement of the angle of the biprism

The deviation by each thin prism forming the biprism is given by, $\delta = (\mu - 1)A$, where δ is the deviation, A is the angle of the prism. So the linear separation of the two sources S_1 and S_2 may be written as

$$d = a \times 2\delta$$

where a is the separation between the source S and the biprism.

$$\therefore \quad d = a \times 2(\mu - 1)A \quad \Rightarrow \quad A = \frac{d}{2a(\mu - 1)}.$$

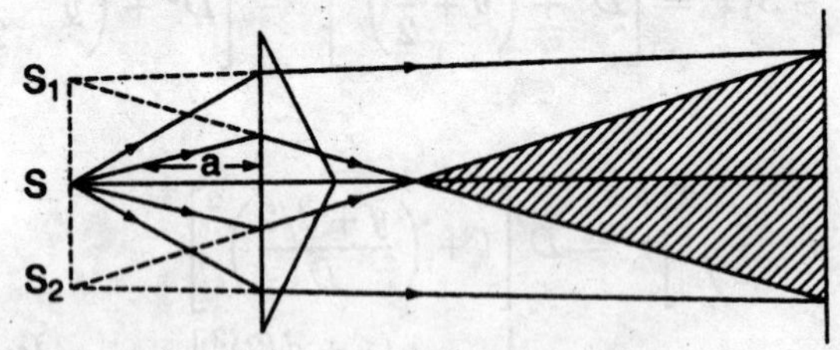

Fig 3.22

Hence, the value of A can be determined.

Theory Questions

1. What do you mean by interference? What is the basic requirement so as to obtain permanent interference pattern?

   Ans. Interference is the result of superposition of waves, due to which under certain conditions we get permanent pattern consisting of alternate regions of brightness and darkness; the bright region, called *bright fringe,* being due to constructive interference, the dark region, called the *dark fringe,* being due to destructive interference.

The phase condition when superposition of waves takes place depends on the phase relationship at the generation end as well as meeting end. If the sources of waves are in phase or have a predictable phase relationship, they are referred to as coherent sources. Coherency refers to spatial as well as temporal coherence. Spatial coherency demands that the linear separation between the sources must be small; temporal coherence demands that the path retardation introduced between the waves on their travel towards the meeting point must be less than the coherence length. So the most fundamental criterion to be fulfilled by the waves is the conditon of spatial coherence and temporal coherence.

2. **Why is it necessary to use narrow source for Fresnel's Biprism?**

   Ans. If we do not use a narrow source but an extended source, there will be no phase correlation amongst the different emitting sources, so condition of spatial coherence will not be satisfied.

3. **Explain why the fringes obtained in a Fresnel biprism experiment appear to be straight lines.**

   Ans. Actually the locus of the bright point or dark point is a hyperbola with the two sources S_1 and S_2 as foci, but on the screen placed at a certain distance, the line of intersection of the hyperboloidal surface with the screen becomes straight fringe.

4. **What is the nature of central fringe in Lloyd's mirror experiment? Explain the nature.**

   Ans. The fringe system in Lloyd's mirror experiment occurs on one side of the central line, so to get the central fringe either the screen is to be placed touching the edge of the mirror or placing a thin plate on the direct beam. Whatever may be the case, the central fringe is found to be dark.

   The reason is that at the central point which is equidistant w.r.t. the sources S_1 and S_2, the reflected beam suffers a phase change of π, thus at the central point the direct wave and reflected wave meet in opposition producing destructive interference, hence dark fringe.

5. **In Young's double-slit experiment find a relation between the linear-dimension and the spatial coherence of a source for a given separation between the slits.**

   Ans. Even if the slits S_1 and S_2 are pin-holes but separated by a small distance apart, the source S is an extended one such that AB is the linear separation between two atomic emitters. Then the waves arriving at S_1 from A and B will not maintain spatial coherence if $BS_1 - AS_1 > ct_c$, where t_c is the emission time.

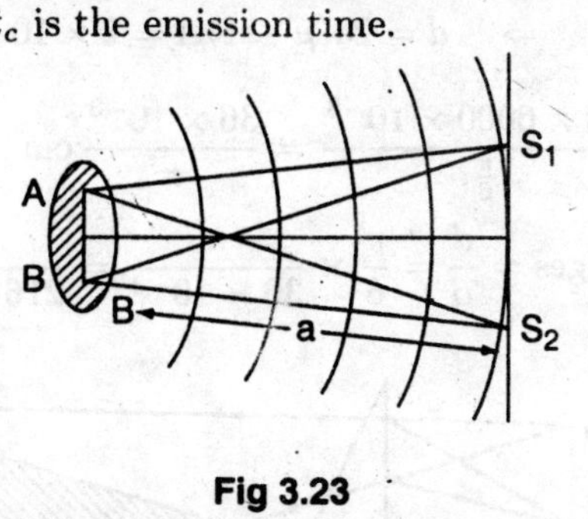

Fig 3.23

Now, take the separation between A and B as d, then

$$BS_1 - AS_1 = \frac{yd}{a}$$

$$\therefore\ (BS_1 - AS_1) \le ct_c \text{ for spatial coherence or, } \frac{yd}{a} \le ct_c \quad \Rightarrow \quad \left(d \le \frac{ct_c \times a}{y}\right.$$

Taking the values of the experimental arrangement the linear-dimension can be found.

6. What is the difference in the coherent portions of the virtual sources in Fresnel's biprism and Lloyd's mirror, and what effect would this produce in the appearance of the fringes?

Ans. In Fresnel's biprism the coherent sources are formed due to refraction in the two halves of the biprism but in Lloyd's mirror the direct waves and reflected waves form the coherent beams to superpose, giving rise to fringe pattern. Due to this difference in the process of getting the coherent waves, the fringe pattern in Lloyd's arrangement is usually brighter than that in Fresnel's biprism. Moreover, the fringe pattern in Fresnel's biprism extends both upwards and downwards w.r.t. the central line but in Lloyd's arragement the fringe pattern appears on one side of the central line. In Fresnel's biprism the central fringe is bright but in Lloyd's mirror the central fringe is dark due to an additional phase change of π due to reflection from the mirror.

7. It is said that two independent sources cannot produce interference pattern. Is the statement true whatever may be the nature of the sources?

Ans. Two independent sources of thermal origin cannot give interference pattern that can be observed because in such cases light emitted is composed of a large number of waves emitted randomly in space and time. Phase relation between two individual waves, even from one source, varies, not to speak about waves from two sources. But with the advent of laser light whose coherence length is much larger, it was possible to produce interference by two independent laser sources. Laser sources have very high degree of spatial coherence and temporal coherence which gives rise to interference fringes with two independent laser beams.

Numerical Problems

1. In Fresnel Biprism experiment,, the angle of the biprism is 3°, the r.i. of the material 1.5. Interference fringes are formed with a source of wavelength 6,000Å located 10 cm from the prism and the source-screen distance is 100 cm .Find the maximum number of fringes that can be observed.

Solution.

$$A = \frac{d}{2a(\mu-1)} \quad \Rightarrow \quad d = 2a(\mu-1)A = 2\times 10\times\frac{1}{2}\times\frac{3\pi}{180} = \frac{\pi}{6}$$

$$\beta = \frac{D\lambda}{d} = \frac{100\times 6000\times 10^{-8}}{\frac{\pi}{6}} = \frac{36\times 10^{-3}}{\pi}\text{ cm}$$

$$\therefore \text{ Number of fringes} = \frac{d}{\beta} = \frac{\pi}{6}\times\frac{\pi}{36\times 10^{-3}} = \frac{\pi^2}{216\times 10^{-3}} \simeq \frac{10}{216}\times 10^3 = 46$$

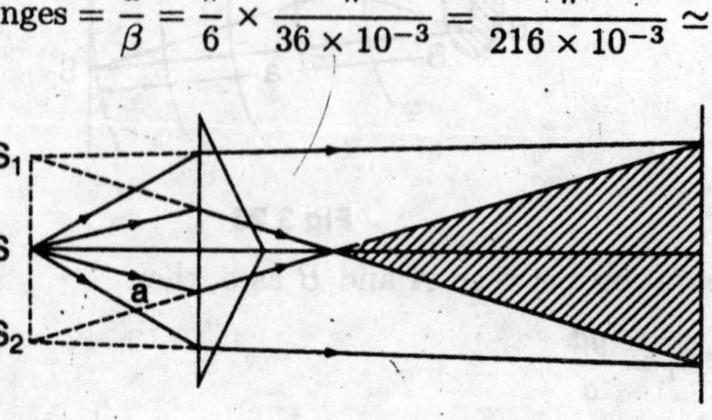

Fig 3.24

2. A source of wavelength 6000Å illuminates a Fresnel's biprism of angle 3° placed at a distance of 15 cm from it. Find the fringe width on a screen 85 cm from the biprism, the r.i. of the material being 1.5.

Solution.

$$A = \frac{d}{2a(\mu - 1)} \quad \Rightarrow \quad d = 2A(\mu - 1)a$$

$$\beta = \text{fringe width} = \frac{D\lambda}{d} = \frac{D\lambda}{2A(\mu-1)a} = \frac{85 \times 6000 \times 10^{-8}}{2 \times \frac{3\pi}{180} \times \frac{1}{2} \times 15}$$

$$= 6.49 \times 10^{-3}\,\text{cm} = 0.065\,\text{mm}$$

3. A Lloyd's mirror of length 5 cm is illuminated with light of wavelength 5,890Å from a narrow slit, the axis of which is parallel to the plane of the mirror. The height of the slit from the plane of the mirror is 0.1 cm while the foot of the ⊥ drawn from the slit on to the plane of the mirror is 5 cm away from the near edge of the mirror. Find the separation of the fringe at a distance of 1 m from the slit. Also determine the total width of the fringe system on the screen.

Solution. See fig. (3.25), here $OS_1 = 0.1$ cm so that $S_1S_2 = 0.2$ cm $D = 100$ cm.

$$\therefore \quad \beta = \frac{D\lambda}{d} = \frac{10^2 \times 5{,}890 \times 10^{-8}}{0.2} = .029\,\text{cm}$$

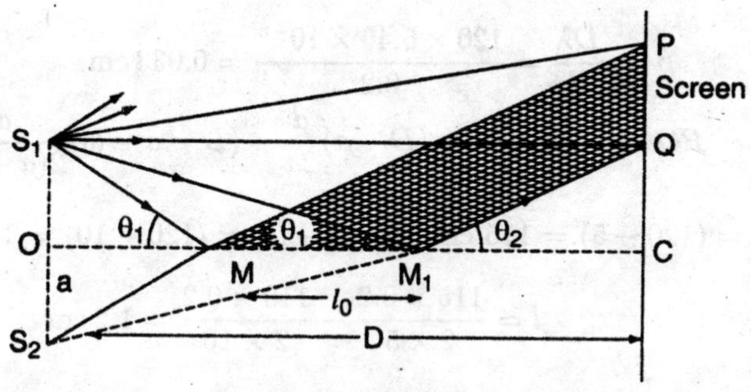

Fig 3.25

To find the total width of the fringe system refer to fig. (3.25) again.
The portion PQ is the region of overlap and hence portion containing the fringes.
Let $MO = a$, $MM_1 = l_0$ (mirror length), and θ_1 and θ_2 are the angles as indicated.
Then, $S_1S_2 = d$ (separation of the sources)

$$PQ = PC - QC = (D-a)\tan\theta_1 - (D - a - l_0)\tan\theta_2$$

$$= (D-a)\frac{d}{2a} - (D-a-l_0)\frac{a}{2(a+l_0)}.$$

We will use this relation to find the total width of the fringe system formed on the screen.
Here $D = 100$ cm, $a = 5$ cm, $l_0 = 5$ cm, $d = 0.2$ cm.

$$\therefore \quad PQ = \text{width} = (100-5) \times \frac{0.2}{2 \times 5} - (100 - 5 - 5) \times \frac{2}{10 \times 2}$$

$$= \frac{95 \times 0.2}{10} - \frac{90 \times 0.2}{10 \times 2} = 1\,\text{cm}$$

4. Fringes 0.010 cm apart are formed by a small biprism placed 28.7 cm in front of a narrow slit illuminated by monochromatic light. The screen on which the fringes are formed is 71.3 cm from the prism and the refracting angle of each half of the biprism is 1°. Find the wavelength of light used.

Solution.

$$A = \frac{d}{2a(\mu-1)}, \beta = \frac{D\lambda}{d} \quad \Rightarrow \quad \lambda = \frac{2a(\mu-1)A\beta}{D}$$

$$A = 1° = \frac{\pi}{180} \text{ rad}, \beta = .010 \text{ cm}, a = 28.7 \text{ cm},$$

$$D = 71.3 + 28.7 \text{ cm}, \mu = \frac{3}{2}$$

$$\therefore \quad \lambda = \frac{2 \times 28.7 \times \frac{1}{2}}{100.0} \times \frac{\pi}{180} \times .010 = 5.01 \times 10^{-5} \text{ cm}$$

5. A Lloyd's mirror of length 5.0 cm is illuminated with light of wavelength 5.46×10^{-5} cm from a narrow slit whose height w.r.t. the mirror-plane is 0.10 cm and which is at a distance of 5 cm measured in that from its nearer edge. Find the separation of the interference fringes at a distance of 120 cm from the slit and the total width of the pattern.

Solution. As in problem 3, we have here $d = 0.2$ cm.

$$\beta = \frac{D\lambda}{d} = \frac{120 \times 5.46 \times 10^{-5}}{0.2} = 0.033 \text{ cm}$$

$$PQ = \text{width} = l = (D-a)\frac{d}{2a} - (D-a-l_0)\frac{d}{2(a+l_0)}$$

Here, $(D-a) = (120-5) = 115$ cm $(D-a-l_0) = (120-10) = 110$ cm.

$$\therefore \quad l = \frac{115 \times 0.2}{2 \times 5} - \frac{110 \times 0.2}{2 \times 10} = 1.2 \text{ cm}$$

6. If the two elements of the double-slit were moved symmetrically apart with relative velocity V_R, calculate the rate at which fringes pass a point x cm from the centre of the fringe system formed on a screen y cm from the double-slit.

Solution. Let N be the number of fringes within the length x cm, then we have

$$\beta N = x \quad \Rightarrow \quad \frac{y\lambda}{d}N = x \quad \Rightarrow \quad N = \frac{xd}{\lambda y} \quad \therefore \quad \frac{\partial N}{\partial d} = \frac{x}{\lambda y}$$

Now,

$$\frac{\partial N}{\partial t} = \frac{\partial N}{\partial d} \cdot \frac{\partial d}{\partial t} = \frac{x}{\lambda y} V_R.$$

So, if V_R is +ve, $\frac{\partial N}{\partial t}$ is also +ve, hence, N increases, so the fringes move towards the centre of the pattern.

7. When a narrow monochromatic source of light was placed at distance of 50 cm from a biprism ($\mu = 1.5$), the distance between two consecutive bands formed on a screen placed at a distance of 100 cm from the biprism was found to be 0.012 cm Find the obtuse angle of the biprism. The wavelength of light used was 5,893Å.

Solution. Here $D = 150$ cm, $d = 2a(\mu - 1)A$

$$\beta = \frac{D\lambda}{d} \Rightarrow d = \frac{D\lambda}{\beta} = \frac{150 \times 5893 \times 10^{-8}}{0.012} = 0.737 \text{ cm}$$

$$\therefore A = \frac{d}{2a(\mu - 1)} = \frac{0.737}{2 \times 50 \times \frac{1}{2}} = 0.015 \text{ rad} = 0.015 \times \frac{180}{\pi} = 0.845 \text{ deg.}$$

$$\therefore \text{ Obtuse angle} = (180 - 2A) = (180 - 2 \times 0.845) \text{ deg} = 178.3°.$$

8. **Fringes are produced with monochromatic light of wavelength 5.45×10^{-5} cm A thin film of glass of r.i. 1.5 is placed normally in the path of one of the interfering rays the central bright fringe is found to move to a position occupied by the third bright fringe from the centre. Calculate the thickness of the film.**

Solution.

Here, $(\mu - 1)t = 3\lambda \quad \Rightarrow \quad t = \dfrac{3 \times 5.45 \times 10^{-5}}{\frac{1}{2}} = 32.7 \times 10^{-5}$ cm

9. **If light of $\lambda = 660 \times 10^{-9}$ m has wavetrains 20λ long, what is its (i) coherence length, (ii) coherence time?**

Solution. The coherence length is nothing but the length of wavetrains.

So here, the coherence length

$$\text{(i)} \quad l_c = 20 \times 660 \times 10^{-9} \text{ m} = 13.2 \ \mu\text{m}$$

$$\text{(ii)} \quad t_c = \frac{13.2 \times 10^{-6}}{3 \times 10^8} \text{ sec} = 4.4 \times 10^{-14} \text{ sec.}$$

10. **Light passes through two narrow slits 0.08 cm apart. If, on a screen 80 cm away, the distance between two second order maxima is 0.2 cm, what is the wavelength of light?**

Solution. For the solution of the problem refer to fig. (3.26)

$$S_2P - S_1P = n\lambda \text{ for } n\text{th bright fringe at } P.$$

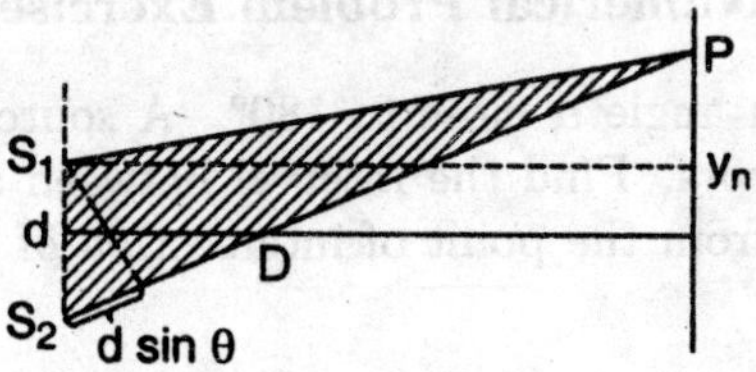

Fig 3.26

Again, $S_2P - S_1P \simeq d \sin\theta$.

$$\therefore \quad d\sin\theta = n\lambda, \text{ is the double-slit eqn.}$$

Now,

$$\sin\theta \simeq \theta \simeq \tan\theta. \quad \therefore \quad \left(y_n - \frac{d}{2}\right) = D\tan\theta.$$

Hence,

$$\frac{d}{D}\left(y_n - \frac{d}{2}\right) = n\lambda \quad \Rightarrow \quad Dn\lambda = y_n d - \frac{d^2}{4} \simeq y_n d.$$

Hence, $n = 2, y_n = \frac{0.2}{2} = 0.1$ cm.

$$\therefore \quad \lambda = \frac{y_n d}{2D} = \frac{0.1 \times .08}{2 \times 80} = 5 \times 10^{-5} \text{ cm}$$

11. Light of 600×10^{-9} m wavelength passes thro' a double-slit and forms interference fringes on a screen 1.2 m away. If the slits 0.02 cm apart, find the distance between the zeroth and (i) third order maximum, (ii) third order minimum.

Solution. As in previous example

$$n\lambda = d \sin\theta \text{ for } n\text{th bright fringe}$$

$$\left(n + \frac{1}{2}\right)\lambda = d \sin\theta \text{ for } n\text{th dark fringe}$$

Now, $\sin\theta \simeq \theta \simeq \tan\theta \simeq \frac{y}{D}$

$$\therefore \quad \text{(i)} \quad 3\lambda = \frac{yd}{D} \quad \Rightarrow \quad y = \frac{3\lambda \times D}{d} = \frac{3 \times 600 \times 10^{-7} \times 120}{0.02} = 1.08 \text{ cm}$$

$$\text{(ii)} \quad \frac{5}{2}\lambda = \frac{yd}{D} \quad \Rightarrow \quad y = \frac{5}{2}\frac{\lambda D}{d} = 0.90 \text{ cm}$$

12. A double-slit is illuminated by light containing two wavelengths, 450×10^{-9} m and 600×10^{-9} m What is the lowest order at which a maximum of one wavelength will fall exactly on a minimum of the other?

Solution. Let n be the lowest order, then

$$n\lambda_1 = \left(n + \frac{1}{2}\right)\lambda_2 \quad \Rightarrow \quad n = 2.$$

Numerical Problem Exercise

1. Two flat mirrors form an angle α close to 180°. A source of light S is placed at equal distances b from the mirrors. Find the interval between adjacent interference bands on a screen at a distance a from the point of intersection of the mirrors, the wavelength of light is λ. $\left[\beta = \frac{\lambda(a+b)}{2b\alpha}\right]$
2. With Fresnel's double mirror what must be the angle between the mirrors in degrees in order to produce Na fringes 0.5 mm apart if the slit is 30 cm from the intersection of the mirrors and the screen 120 cm from the slit. [179.83°]
3. On placing a thin sheet of mica of $\mu = 1.58$ on one of the interfering beams in a double-slit experiment, it was found that the central fringe on the screen is occupied by what used to be the seventh bright fringe. If $\lambda = 5{,}500$Å, find the thickness of the mica sheet. [6.6×10^{-5} mm]
4. Interference fringes were produced by Young's slit method, the wavelength of light being 6×10^{-7} m. When a film of material 3.6×10^{-3} cm thick was placed over one of the slits, the fringe pattern was displaced by a distance equal to 30 times that between two adjacent fringes. Calculate μ of the film. [$\mu = \frac{3}{2}$]

5. In Young's slit experiment, the separation of four bright fringes is 2.5 mm when wavelength is 6.2×10^{-7} m. The distance from the slits to the screen is 80 cm. Calculate the separation of two slits. [$d = 0.8$ mm.]
6. In a Young's experiment, the centres of the double-slits are 0.25 mm apart and the wavelength of light used is 6.0×10^{-4} mm. Calculate the angle θ subtended at the slits by the adjacent maxima of the fringe pattern. [2.4×10^{-2} rad]
7. If in example 3 it is found that the central fringe shifts by 0.2 cm, find the thickness of the sheet and the distance between the coherent sources $S_1S_2 = 0.1$ cm and the separation of the screen is given as 50 cm. [6.7×10^{-4} cm]
8. In worked out example 3, take $OS_1 = 0.1$ cm, $MO = 5$ cm, $MM_1 = 5$ cm and $M_1C = 190$ cm respectively. Determine the position of the region where the fringes will be visible and calculate the number of fringes. [2 cm, 40]
9. Assume a plane wave incident normally on a plane containing two holes separated by a distance 0.08 cm. If we place a convex lens of focal length 10 cm, find the fringe width as observed on the focal plane of the lens if the wavelength used is 6.2×10^{-7} m. [7.75×10^{-3} cm]

Question Exercise

1. How would you determine the wavelength of a monochromatic light with the help of Lloyd's mirror? What is the difference in the coherent portions of the virtual sources in Fresnel's biprism and Lloyd's mirror, and what effect would this produce in the appearance of the fringes?
2. Explain the method of finding the wavelength of light by Fresnel biprism.
3. State the conditions to be fulfilled for the production of interference fringes. Why is it necessary to use narrow source in Fresnel's biprism? Deduce the theory of Fresnel's biprism and work out mathematically the shape of the interference fringes in space. Explain why they appear to be straight lines in usual optical experiments.
4. Why two independent sources of thermal origin cannot produce interference? Can such interference ever occur if you replace the nature of the sources?
   Investigate the shape of interference fringes on a plane at right angles to the line joining the two coherent sources.
5. Explain the concepts of temporal and spatial coherence of a light source. With the help of Young's double-slit experiment find a relation between the linear-dimension and spatial coherence of a source for a given separation between the slits.
6. Show mathematically that the intensity on a screen due to an extended source is given by $I = \sum_{1}^{n} a_k^2$, where a_k is the amplitude due to the kth atomic emitter.
7. In an Young's double-slit experiment, the interference fringes are obtained on a screen. Discuss the effect on the fringes when
   (a) the double-slits are moved apart along the line joining them,
   (b) the slit-dimension is increased,
   (c) the screen is moved away from the double slit.

8. Describe the formation of fringes in Billet split lens arrangement. Investigate the nature of fringes on a screen when the two split halves have relative separation.
9. Investigate the formation of interference fringes in Fresnel's double mirror arrangement. Is it an improvement of Lloyd's single mirror arrangement?
10. On a cardboard a number of slits of the same width but an random distances are cut. This is placed in front of a monochromatic light source. Investigate the pattern.

Reference

1. Advanced Textbook on Electricity and Magnetism, Vol. II, *Chakrabarti P.K.*, Modern Book Agency, Kolkata.
2. Optics, *Ghatak Ajoy*, Tata McGraw-Hill Publishing Co. Ltd., New Delhi.
3. A Textbook on Light, *Majumder K.G.*, Modern Book Agency, Kolkata.
4. Principles of Optics, *Born M. and Wolf E.*, Pergamon Press, Oxford (England).
5. Geometrical and Physical Optics, *Longhurst R.S.*, Orient Longman.

Chapter 4

Interference Based on Amplitude Division

4.1 Introduction

In the last chapter we discussed the interference phenomenon based on division of wavefront, where we got coherent sources by the division of wavefront; now in this chapter we will introduce another class of interference phenomenon, where we will produce coherent waves by division of amplitude. In the production of the coherent waves, in these two classes of interference phenomena difference lies in the fact that in the former class, a narrow source is required but in the latter case, an extended source is taken, a single ray produces a series of coherent waves as a result of multiple reflections and refractions. To this class belongs a number of very important arrangements like Newton's rings, interference in thin films, and a number of interferometers like Michelson's interferometer, Jamin's interferometer and Fabry-Perot interferometer, etc. Since in all these interference phenomena an extended source can be used, the fringes produced here are brighter than those in case of wavefront division class of interference.

4.2 A Theorem of Importance in the Consideration of the Fringes—Stokes' Theorem

This theorem is of considerable importance in deriving the condition of constructive and destructive interference.

Let a wave of wave amplitude a is incident on the dividing surface, a part is reflected along OB and a part is transmitted into the second medium along OC. We introduce a few terms in this connection.

(a) Reflection Co-efficient r

When light travelling through the first medium meets the surface of separation backed by a denser medium (i.e., medium 2 is denser), the ratio of the wave amplitude reflected to the incident wave amplitude is known as *reflection co-efficient* r. Thus

$$\frac{\text{The reflected wave amplitude along } OB}{\text{Incident wave amplitude } (a)} = r$$

$\Rightarrow$ The reflected wave amplitude along $OB = ar$

(b) Transmission Co-efficient, t, for Transmission from Medium 1 (Rarer) to Medium 2 (Denser)

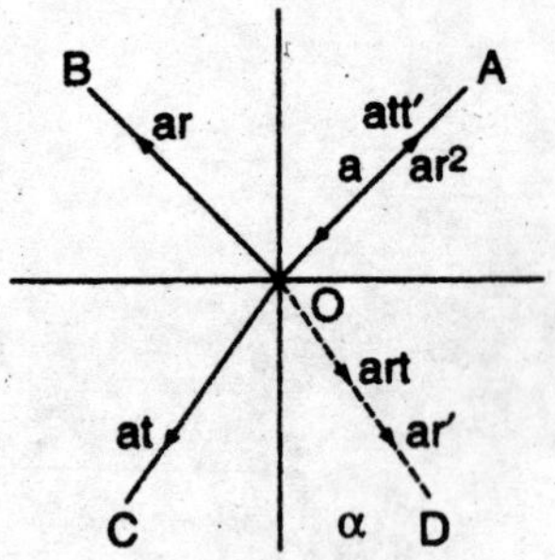

Fig 4.1

$$t = \frac{\text{Wave amplitude transmitted to the second medium along } OC}{\text{Incident wave amplitude } (a)}$$

$\Rightarrow$ transmitted wave amplitude along $OC = at$

The corresponding quantities when light travels from denser to rarer medium are denoted by r' and t'. Since, light path is reversible, if the reflected wave along OB is reversed, it splits into two parts of wave amplitude ar^2 along OA, and art along OD; similarly, if the transmitted wave along OC be reversed, it splits into atr' along OD and att' along OA.

$$\therefore \quad att' + ar^2 = a \Rightarrow r^2 + tt' = 1 \Rightarrow tt' = (1 - r^2) \qquad (4.2.1)$$

$$\text{and} \quad art + atr' = 0 \Rightarrow r = -r' \qquad (4.2.2)$$

Eqn.(4.2.2) suggests that the reflection co-efficient in both cases of reflection is same so far as the magnitude is concerned but the negative sign suggests that in one of the two cases of reflection, there is a phase change of π. But in which case? Is it in the case when light travelling through a rarer medium gets reflected from the surface backed by a denser medium, or in the case when light travelling through a denser medium gets reflected from the surface backed by a rarer medium?

The definite conclusion comes with reference to Lloyd's mirror arrangement where we have seen that the central fringe is a dark fringe instead of a bright fringe. This is because here direct waves and reflected waves superpose giving the interference pattern, and reflection takes place from a metallic or a dielectric reflecting surface. Hence we conclude—

In case of reflection from a surface backed up a denser medium, there is a phase change of π.

This inference can also be drawn mathematically from Maxwell's field equation from which we get (from Chapter 6)

$$a_r = \text{reflected wave amplitude} = \left(\frac{n_2 - n_1}{n_2 + n_1}\right) \quad \text{and} \quad a_t = \frac{n_2}{n_1}\left(\frac{2n_1}{n_1 + n_2}\right).$$

In this case, light is travelling through a medium of r.i. n_1, into the medium of r.i. n_2. If the incident medium be n_2, then,

$$a'_r = \frac{n_1 - n_2}{(n_1 + n_2)}.$$

So, if r and r' be the reflection co-efficients in the two cases, we get,

$$a_r = ar = \frac{n_2 - n_1}{n_2 + n_1} = +\text{ve for } n_2 > n_1$$

$$a'_r = ar' = \frac{n_1 - n_2}{n_1 + n_2} = -\text{ve but equal to } ar$$

$$\therefore \quad ar = -ar' \quad \Rightarrow \quad r = -r'$$

4.3 Interference in an Wedge-shaped Film—Multiple Beam Interference

In case of two-beam interference we saw that an extended source does not produce interference fringes because for an extended source spatial coherence is not maintained. This can be explained if we refer to a biprism experiment and consider two slit-images of the single slit source formed due to refraction in the two halves of the biprism. Let the two images AB and CD be very narrow initially, so that distinct fringes are obtained on the screen. Now, as the dimension of the linear slit is increased, the slit-images AB and CD also increase in width and at a certain value, the fringe pattern disappears. We can recognise conjugate pairs of points of the slits; A and C are one extreme pair of conjugate points, also B, D form another extreme pair of conjugate points. Each pair of points forms an interference pattern whose intensity follows the $\cos^2$-distribution. The interference fringe due to conjugate pair A and C is represented by the continuous curve and the interference pattern due to the conjugate pair B, D is represented by the dashed curve. P is the central bright band (fig.4.3) of the interference pattern due to the conjugate pair (A, C), while Q is the central bright band of the interference pattern due to the other extreme conjugate pair (B, D). So there is a separation between these central

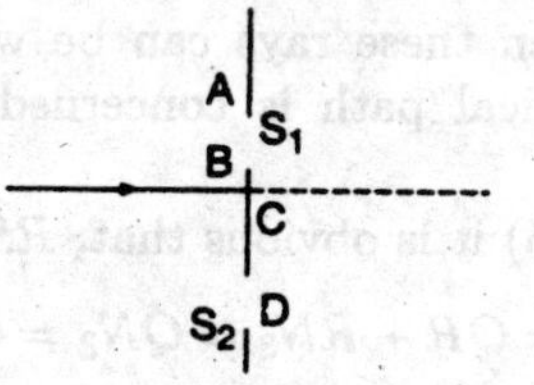

Fig 4.2

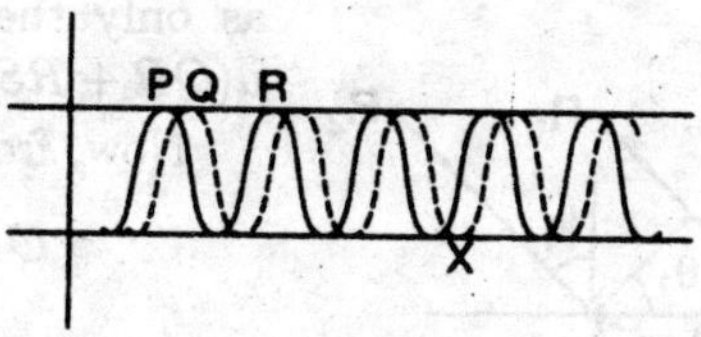

Fig 4.3

bright band positions and is referred to as the breadth of a band. This breadth or separation depends on the dimension of the slit-images. If it so happens that Q falls on R, the first order bright band of the first extreme conjugate pair (A, C), the interference pattern disappears and we will get just illumination. So we say that in case of two-beam interference, the slit source, giving rise to two coherent image-slit sources must be very narrow. But in case of multiple-beam interference, this restriction is not needed, because here a single ray from a source splits up into a number of reflected rays (also a system of transmitted rays available) which form a coherent beam and they will superpose giving interference fringes. Such splitting into reflected components and transmitted components occur when rays from an extended source fall on a thin film or glassplate, either parallel or wedge-shaped.

In such interference phenomenon individual incident ray need not be coherent w.r.t. each other; what happens is the formation of a number of coherent reflected rays (or transmitted rays) derived from a single incident ray. So, in order that a bright fringe is obtained in a particular direction, a large number of point sources is needed, hence an extended source is required for producing sufficiently bright fringe pattern.

Path Retardation in case of Parallel Plate

Let us consider a point source P from which a ray is incident at Q at some angle ϕ_1; reflects along QR_1, refracts along QR, a part of it transmits into the lower medium along RT_1, a part reflects along RS which finally goes back into the incident medium along SR_2, parallel to the ray QR_1. In this way a series of parallel rays in the incident medium like QR_1, SR_2, S_2R_3, etc., are available; also a series of transmitted parallel rays like RT_1, S_1T_2, etc., are available fig.(4.4). Either the reflected parallel beam or transmitted parallel beam is focussed by a convex lens producing bright or dark fringes depending on the path difference between rays.

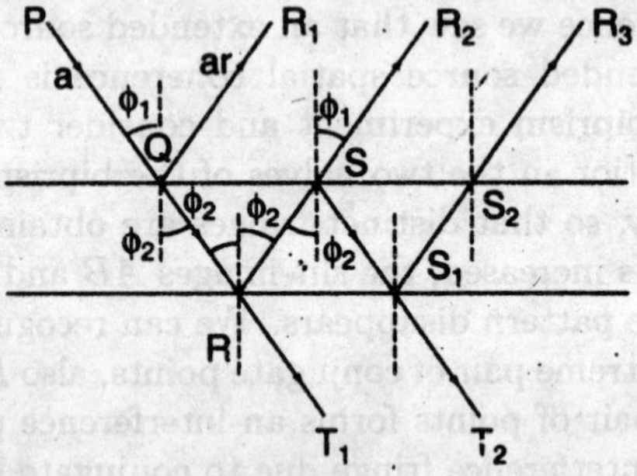

Fig 4.4

Calculation of Path Difference between two Successive Reflected Rays

The incident ray PQ reflects along QR_1 from the top surface, refracts along QR, again a part reflects along RS and comes back along SR_2 in the incident medium. If we consider these two rays only, then the path difference between these rays can be written as (so far as only the geometrical path is concerned) fig.(4.5), $\Delta = \mu(QR + RS) - QN_1$.

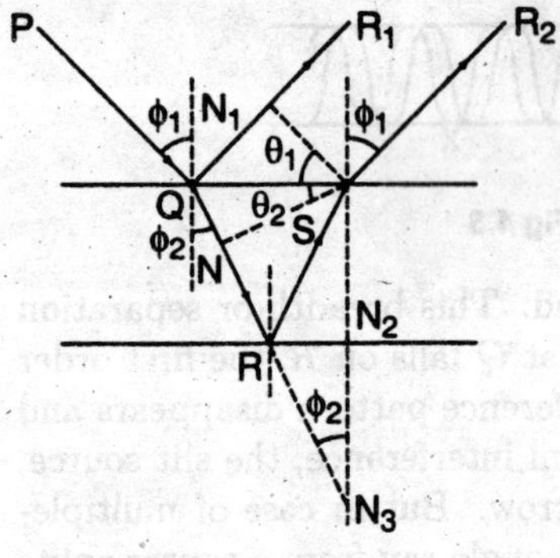

Fig 4.5

Now, from fig.(4.5) it is obvious that, $RS = RN_3$

$$\therefore \quad QR + RS = QR + RN_3 = QN_3 = QN + NN_3.$$

Again, $\sin\phi_1 = \mu \sin\phi_2 \quad \because \angle N_1SQ = \theta_1 = \phi_1$

$$\Rightarrow \quad \frac{QN_1}{QS} = \mu \times \frac{QN}{QS} \quad \Rightarrow \quad QN_1 = \mu QN$$

So the path difference becomes

$$\begin{aligned} \Delta &= \mu(QN + NN_3) - QN_1 \\ &= \mu QN + \mu NN_3 - QN_1 = \mu NN_3, \\ &\because \quad QN_1 = \mu QN. \end{aligned}$$

Now,

$$\frac{NN_3}{SN_3} = \cos\phi_2 \quad \Rightarrow \quad NN_3 = SN_3 \cos\phi_2 = 2e\cos\phi_2$$

where e is the thickness of the plate.

$$\therefore \quad \Delta = 2\mu e \cos\phi_2 \tag{4.3.1}$$

The two rays are obtained through reflection process, one reflection takes place from the top surface, the other takes place at the lower surface. According to Stokes' law, when reflection takes place from the surface backed by a denser medium (here reflection at the point Q), a phase change of π occurs which is equivalent to a path difference $\frac{\lambda}{2}$.

So the total path difference between the two rays now becomes

$$\begin{aligned} \Delta_T &= \Delta \text{ (geometrical path diff.)} \pm \text{path diff. due to reflection} \\ &= \Delta \pm \frac{\lambda}{2} = 2\mu e \cos\phi_2 \pm \frac{\lambda}{2}. \end{aligned}$$

If $\Delta_T = 2n\cdot\frac{\lambda}{2}$, we get brightness; $\Delta_T = (2n \pm 1)\frac{\lambda}{2}$, we get darkness.

So, for bright fringe,

$$2\mu e\cos\phi_2 \pm \frac{\lambda}{2} = 2n\cdot\frac{\lambda}{2} \quad \Rightarrow \quad 2\mu e\cos\phi_2 = (2n \mp 1)\frac{\lambda}{2} \quad \text{BRIGHTNESS.} \tag{4.3.2}$$

Dark fringe condition is

$$2\mu e\cos\phi_2 = 2n\cdot\frac{\lambda}{2} \quad \text{or,} \quad (2n+2)\frac{\lambda}{2} \quad \text{DARKNESS.} \tag{4.3.3}$$

Thus we conclude that those rays, such that the two adjacent members have geometrical path difference as an odd multiple of $\frac{\lambda}{2}$, will interfere constructively, producing bright fringe and those having geometrical path difference as an even multiple of $\frac{\lambda}{2}$, will interfere destructively, producing dark fringe.

It is also clear that since μ, e are constants here, the geometrical path difference will vary only when ϕ_2 varies. This implies, the angle of incidence varies. So rays from the source P incident on the plate at different angles may have geometrical path difference either given by eqn.(4.2.2) or by eqn.(4.3.1) and will interfere constructively or destructively producing respectively bright fringe or dark fringe. Though it is true that a large number of reflected rays is available from a single ray, the intensity rapidly falls off with multiple reflection. Hence only a few reflections need be considered.

Intensity at a Point due to Superposition of Multiple Reflected Rays

In fig.(4.6) we have represented a system of parallel reflected rays back into the incident medium through division of amplitude. If the reflection coefficient from a surface backed by a denser medium be r and that from a surface backed by a rarer medium as r' (where $r = r'$ according to Stokes' law) and transmission co-efficient from rarer to denser as t and t' for the reverse case, then for a wave amplitude a incident on the top surface, a system of parallel rays is available in the incident medium with wave amplitudes ar, $atr't'$, $att'r'^3$, etc. Also a series of parallel rays transmitted to the lower medium are available with amplitudes att', $att'r'^2$, $att't'^5$, etc.

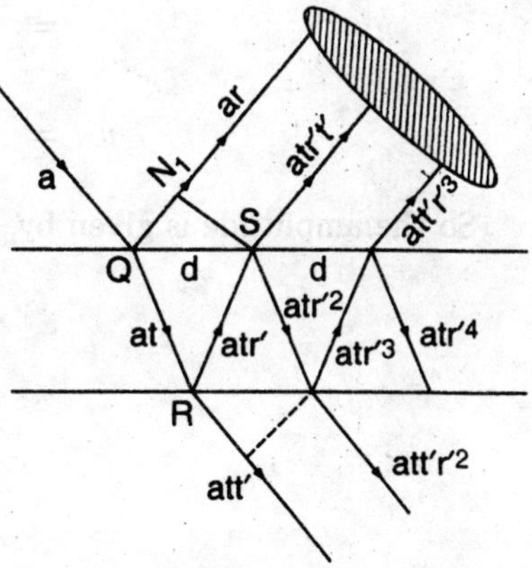

Fig 4.6

If we consider the superposition of the wave amplitudes in the incident medium, we have to superpose a number of simple harmonic waves having constant phase difference w.r.t. each other. Thus the first wave has wave amplitude ar, it can be represented as,

$$\vec{Y}_1 = are^{j(\omega t - \phi_s)}\vec{\upsilon},$$

where $\phi_s = \frac{2\pi}{\lambda}x$, where x is the distance of the point of convergence w.r.t. the point N_1, the foot of the $\perp$ from S to the ray from Q.

The second wave has travelled from Q to R, then from R to S, it has wave amplitude $att'r'$ and travels a different path-length to the point of convergence after its origin at Q. The point of convergence w.r.t. N_1 and S is same and a distance equal to x, hence displacement of the particle at the point of convergence due to this wave can be represented as,

$$\vec{Y}_2 = att'r'e^{j(\omega t-\phi'_s)}\vec{\nu}$$

where

$$\phi'_s = \frac{2\pi}{\lambda}(x+\Delta) = \frac{2\pi}{\lambda}x + \frac{2\pi}{\lambda}\times 2\mu e\cos\phi_2 = (\phi_s+\delta), \delta = \frac{2\pi}{\lambda}\times 2\mu e\cos\phi_2.$$

Similarly, $\vec{Y}_3 = att'r'^3 e^{j(\omega t+\overline{\phi_s+2\delta})}\vec{\nu}$, etc.

$$\begin{aligned}
\therefore \quad \vec{Y} &= \vec{Y}_1+\vec{Y}_2+\vec{Y}_3+\cdots \\
&= are^{j(\omega t-\phi_s)}\vec{\nu} + att'r'e^{j(\omega t-\overline{\phi_s+\delta})}\vec{\nu} + att'r'^3e^{j(\omega t-\overline{\phi_s+2\delta})}\vec{\nu}+\cdots \\
&= are^{j(\omega t-\phi_s)}\vec{\nu} + att'r'e^{j(\omega t-\phi_s+\delta)}\vec{\nu} + att'r'^3e^{j(\omega t-\phi_s-2\delta)}\vec{\nu}+\cdots \\
&= ae^{j(\omega t-\phi_s)}\vec{\nu}\left[r+tt'r'e^{-j\delta}+tt'r'^3e^{-2j\delta}+\cdots\right] \\
&= ae^{-j(\omega t-\phi_s)}\vec{\nu}\left[r+r'tt'\left\{e^{-j\delta}+(r')^2e^{-2j\delta}+\cdots(n-1)\text{ terms}\right\}\right] \qquad (4.3.4)\\
&= ae^{-j(\omega t-\phi_s)}\vec{\nu}\left[r+r'\left\{tt'e^{-j\delta}+tt'(r')^2e^{-j2\delta}+\cdots(n-1)\text{ terms}\right\}\right] \\
&= rae^{-j(\omega t-\phi_s)}\vec{\nu}\left[1-tt'e^{-j\delta}\left\{1+r^2e^{-j\delta}+r^4e^{-2j\delta}+\cdots(n-1)\text{ terms}\right\}\right] \\
&= rae^{j(\omega t-\phi_s)}\vec{\nu}\left[1-tt'e^{j\delta}\times\frac{1-(r^2e^{-j\delta})^{n-1}}{1-r^2e^{-j\delta}}\right]
\end{aligned}$$

As n tends to a large value $(r^2)^{n-1}\to 0$

$$\begin{aligned}
\therefore \quad \vec{Y} &= are^{j(\omega t-\phi_s)}\vec{\nu}\left[1-tt'e^{-j\delta}\times\frac{1}{1-r^2e^{-j\delta}}\right] \\
&= are^{j(\omega t-\phi_s)}\vec{\nu}\left[1-(1-r^2)\frac{e^{-j\delta}}{(1-r^2e^{-j\delta})}\right] \qquad (4.3.5)\\
&= are^{j(\omega t-\phi_s)}\vec{\nu}\left(\frac{1-e^{-j\delta}}{1-r^2e^{-j\delta}}\right) \\
y &= \vec{Y}\cdot\vec{\nu} = ar\left(\frac{1-e^{-j\delta}}{1-r^2e^{-j\delta}}\right)\cos(\omega t-\phi_s).
\end{aligned}$$

So the amplitude is given by,

$$A_r = ar\left(\frac{1-e^{-j\delta}}{1-r^2e^{-j\delta}}\right)$$

$$\therefore \quad A_rA_r^* = a^2r^2\frac{(1-e^{+j\delta})(1-e^{-j\delta})}{(1-r^2e^{j\delta})(1-r^2e^{-j\delta})}$$

$$\Rightarrow \quad A^2 = a^2r^2\frac{[1-(e^{j\delta}+e^{-j\delta})+1]}{1-r^2(e^{j\delta}+e^{-j\delta})+r^4}$$

Now, $e^{j\delta}+e^{-j\delta} = \cos\delta + j\sin\delta + \cos\delta - j\sin\delta = 2\cos\delta$

$$\begin{aligned}
\therefore \quad A^2 &= a^2r^2\frac{2(1-\cos\delta)}{1-2r^2\cos\delta+r^4} = 4a^2r^2\frac{\sin^2\delta/2}{(1-r^2)^2+2r^2(1-\cos\delta)} \\
&= 4a^2r^2\frac{\sin^2\delta/2}{(1-r^2)^2+4r^2\sin^2\delta/2} = \frac{4a^2r^2\sin^2\delta/2}{(1-r^2)^2+4r^2\sin^2\delta/2} \qquad (4.3.6)
\end{aligned}$$

The normalised reflected intensity I_r is written as,

$$(I_r)_N = \frac{I_r}{I_0} = \frac{A^2}{a^2} = \frac{4r^2 \sin^2 \delta/2}{(1-r^2)^2 + 4r^2 \sin^2 \delta/2}.$$

Introducing a new parameter $g^2 = \frac{4r^2}{(1-r^2)^2}$, we get,

$$I_r = \frac{\sin^2 \delta/2}{\frac{1}{g^2} + \sin^2 \delta/2} = \frac{g^2 \sin^2 \delta/2}{1 + g^2 \sin^2 \delta/2}. \tag{4.3.7}$$

In case of transmitted light, the transmitted waves produce displacements at the point of convergence of the transmitted rays which can likewise be represented by,

$$\begin{aligned}
\vec{Y}' &= att' e^{j(\omega t - \phi_s)} \vec{\nu} \\
\vec{Y}'' &= att' r'^2 e^{j(\omega t - \phi'_s)} \vec{\nu} = att' r'^2 e^{j(\omega t - \phi_s + \delta)} \vec{\nu}, \text{ etc.} \\
\therefore \quad \vec{Y}_t &= att' e^{j(\omega t - \phi_s)} + att' r'^2 e^{j(\omega t - \phi_s + \delta)} \vec{\nu} + att' r^4 e^{j(\omega t - \phi_s + 2\delta)} \vec{\nu} + \cdots \\
&= att' e^{j(\omega t - \phi_s)} \vec{\nu} (1 + r^2 e^{j\delta} + r^4 e^{2j\delta} + \cdots) = att' e^{j(\omega t - \phi_s)} \vec{\nu} \times \frac{1 - (r^2 e^{j\delta})^n}{1 - r^2 e^{j\delta}}
\end{aligned}$$

When $n \to$ large value, $(r^2)^n \to 0$, we have,

$$\vec{Y}_t = \frac{att'}{1 - r^2 e^{j\delta}} e^{j(\omega t - \phi_s)} \vec{\nu} \quad \therefore \quad y_t = \vec{Y}_t \cdot \vec{\nu} = \frac{att'}{1 - r^2 e^{j\delta}} \cos(\omega t - \phi_s).$$

So the resultant complex transmitted amplitude is given by,

$$A_t = \frac{att'}{1 - r^2 e^{j\delta}}$$

$$\begin{aligned}
\therefore \quad |A_t|^2 &= A_t A_t^* = \frac{(att')^2}{(1 - r^2 e^{j\delta})(1 - r^2 e^{-j\delta})} = \frac{(att')^2}{1 - r^2(e^{j\delta} + e^{-j\delta}) + r^4} \\
&= \frac{a^2(1-r^2)^2}{1 - 2r^2 \cos\delta + r^4} = \frac{a^2(1-r^2)^2}{(1-r^2)^2 + 4r^2 \sin^2 \delta/2} \\
\therefore \quad (I_t)_N &= \frac{|A_t|^2}{a^2} = \frac{(1-r^2)^2}{(1-r^2)^2 + 4r^2 \sin^2 \delta/2} \\
\text{or,} \quad (I_t)_N &= \frac{I_t}{I_0} = \frac{1}{1 + g^2 \sin^2 \delta/2}
\end{aligned} \tag{4.3.8}$$

Eqns.(4.3.5) and (4.3.6) show that $I_r + I_t = 1$ which suggests that when I_r is maximum, I_t is minimum and vice versa.

Intensity Distribution Functions (IDF)

The variation of intensity with δ is known as *intensity distribution* and the function representing such variation is known as IDF.

Reflected IDF

The intensity variation in case of reflected beam is given by eqn.(4.3.5) and can be written as,

$$I_r = \frac{4a^2r^2\sin^2\delta/2}{(1-r^2)^2+4r^2\sin^2\delta/2} = \frac{4r^2\sin^2\delta/2}{(1-r^2)^2+4r^2\sin^2\delta/2}I_0$$

where I_0 represents the intensity of the incident ray. Writing I_r/I_0 as $(I_r)_N$, called *normalised intensity* and $g^2 = \frac{4r^2}{(1-r^2)^2}$, called *finesse factor*, we can write the normalised distribution function as,

$$(I_r)_N = \frac{I_r}{I_0} = \frac{g^2\sin^2\delta/2}{1+g^2\sin^2\delta/2}.$$

Now, $\frac{\delta}{2} = \frac{1}{2}\cdot\frac{2\pi}{\lambda}\times 2\mu e\cos\phi_2$ and in case of normal incidence $\phi_2 = 0$, hence,

$$\frac{\delta}{2} = \frac{2\pi\mu e}{\lambda} = \text{const.}$$

In case of parallel incidence also, the angle ϕ_2 is fixed. Again,

$$(I_r)_N = \frac{g^2}{g^2+\frac{1}{\sin^2\delta/2}}$$

shows that for a particular value of g^2,

$$(I_r)_N \to \text{max as } \sin\frac{\delta}{2}\to\pm1 \quad\Rightarrow\quad \frac{\delta}{2} = (2n+1)\frac{\pi}{2}.$$

So, for constructive interference of reflected rays forming brightness, we get,

$$\frac{\delta}{2} = (2n+1)\frac{\pi}{2} \quad\Rightarrow\quad \frac{2\pi\mu\cos\phi_2}{\lambda} = (2n+1)\frac{\pi}{2} \quad\Rightarrow\quad 2\mu e\cos\phi_2 = (2n+1)\frac{\lambda}{2}. \qquad (4.3.9)$$

This inference is already drawn in eqn.(4.2.2).
So condition of BRIGHTNESS, we have,

$$\delta = \pi, 3\pi, 5\pi, \text{ etc.}$$

The minimum brightness corresponds to

$$\delta = 2n\pi \quad\Rightarrow\quad \delta = 0, 2\pi, 4\pi, 6\pi, \text{ etc.}$$

$\delta = 0$ is possible only if $e = 0$, i.e., the plate or film is of vanishingly small thickness where $(I_r)_N = 0$. So a very thin oil or water layer appears to be dark.

Transmitted IDF

The intensity distribution in case of transmitted light is given by eqn.(4.3.6) as

$$(I_t)_N = \frac{I_t}{I_0} = \frac{1}{1+g^2\sin^2\delta/2}.$$

It shows that for a particular value of g^2, $(I_t)_N$ will be maximum if

$$\sin\frac{\delta}{2} = 0 \quad \Rightarrow \quad \frac{\delta}{2} = s\pi \quad \Rightarrow \quad \delta = 2s\pi, s = 0, 1, 2, \text{ etc.}$$

$$\therefore \quad \delta = 0, 2\pi, 4\pi, 6\pi, \text{ etc.} \quad \text{BRIGHTNESS}$$

The maximum value of $(I_t)_N = 1$.

The minimum of transmitted intensity corresponds to $\sin\frac{\delta}{2} = \pm 1$ and its value is given as

$$[(I_t)_N]_{\min} = \frac{1}{1+g^2} \to 0 \text{ as } g \to \text{ large value.}$$

$$\sin\frac{\delta}{2} = \pm 1 \quad \Rightarrow \quad \delta = (2n+1)\pi \quad \therefore \quad \delta = \pi, 3\pi, 5\pi, \text{ etc.} \quad \text{DARKNESS}$$

Before we plot the IDFs, we must know the behavioural pattern of the distribution functions.

Reflected IDF Behaviour

$$(I_r)_N = \frac{g^2 \sin^2 \delta/2}{1 + g^2 \sin^2 \delta/2}$$

$$[(I_r)_N]_{\max} = \frac{g^2}{1+g^2} \to 1 \text{ as } g^2 \to \text{very large value.}$$

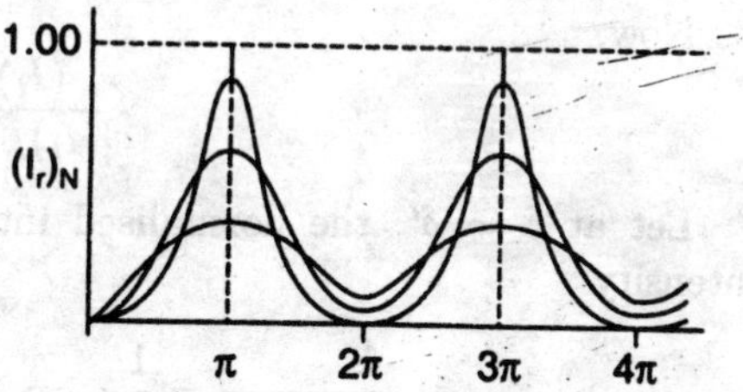

Fig 4.7

So, in the reflected light, the normalised maximum is less than unity and $[(I_r)_N]_{\max}$ increases as g^2 value increases (fig.4.7).

The sharpness of the fringes in the reflected light is determined by the peakiness of the curve; whether the curve is steep or flat depends on the value of g^2. If the curve is flat, its spread is greater; if it is steep, its spread is small, hence narrower. Narrower the curve is, sharper is the fringe. Narrowness or broadness is expressed through a factor, called *half-intensity width*, i.e., phase-angle separation $\Delta\delta$ where the intensity is half of the normalised maximum.

Let us consider once again the reflected IDF given as

$$(I_r)_N = \frac{g \sin^2 \delta/2}{1 + g^2 \sin^2 \delta/2}.$$

Let again at $\delta = \delta_1$, $(I_r)_N$ is half of the normalised maximum $[(I_r)_N]_{\max}$.

$$\therefore \quad \frac{(I_r)_N}{[(I_r)_N]_{\max}} = \frac{1}{2} \quad \Rightarrow \quad \frac{g^2 \sin^2(\delta_1/2)}{1 + g^2 \sin^2 \frac{\delta_1}{2}} \times \frac{(1+g^2)}{g^2} = \frac{1}{2}$$

$$\Rightarrow \quad \frac{(1+g^2)\sin^2 \delta_1/2}{1 + g^2 \sin^2 \delta_1/2} = \frac{1}{2} \quad \Rightarrow \quad (2+g^2)\sin^2\frac{\delta_1}{2} = 1$$

$$\Rightarrow \sin^2\frac{\delta_1}{2} = \frac{1}{2+g^2} \tag{4.3.10}$$

If the fringe be narrow, $\frac{\delta_1}{2}$ must be near zero or near multiple of π $\Rightarrow$ left-hand side becomes small, and for this smallness and hence narrowness of the fringe, g^2 requires to increase. Narrowness of a fringe also means sharpness of a fringe, hence sharpness increases as g^2 increases; hence, g^2 is referred to as finesse co-efficient. Of course, the finesse co-efficient is related to reflection co-efficient of the plate.

Transmitted IDF Behaviour

$$(I_t)_N = \frac{1}{1+g^2\sin^2\delta/2},$$

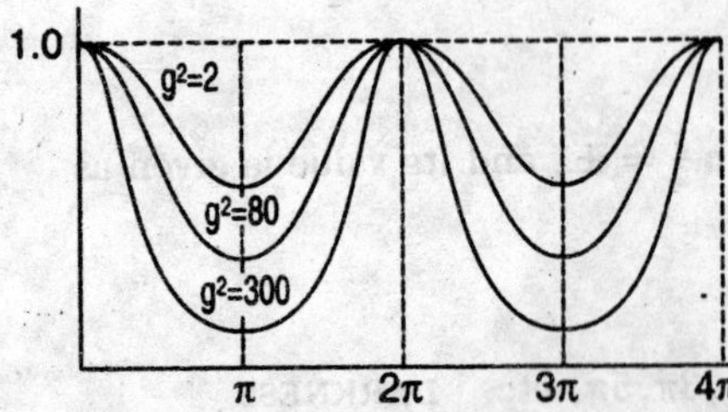

Fig 4.8

the maximum occurs at $\delta = 0, 2\pi, 4\pi$, etc. and its value is equal to unity.

$\therefore \quad [(I_t)_N]_{max} = 1$. The minimum occurs at $\delta = \pi, 3\pi, 5\pi$, etc. and its value is

$$[(I_t)_N]_{min} = \frac{1}{1+g^2},$$

this minimum becomes smaller greater the value of g^2 (fig.4.8).

The spread of the intensity curve is determined by the half-intensity width, smaller its value, greater is the sharpness of the fringe.

Now,

$$\frac{(I_t)_N}{[(I_t)_N]_{max}} = \frac{1}{1+g^2\sin^2\delta/2}.$$

Let at $\delta = \delta'$, the normalised intensity is reduced to half of the maximum normalised intensity.

$$\therefore \quad \frac{1}{1+g^2\sin^2\delta'/2} = \frac{1}{2} \quad \Rightarrow \quad 1+g^2\sin^2\frac{\delta'}{2} = 2 \quad \Rightarrow \quad \sin^2\frac{\delta'}{2} = \frac{1}{g^2}. \qquad (4.3.11)$$

In order that the fringes are narrow, $\frac{\delta'}{2}$ must be near zero or near multiple of π. For g^2 large, $\frac{\delta'}{2}$ is near zero or near multiple of π; hence, fringes become sharply peaked.

Difference in the Two IDFs

If we compare the two NIDFs, a few things become at once clear—

(1) The $[(I_r)_N]_{max}$ is always less than $[(I_t)_N]_{max}$. It is no wonder because if we consider amplitudes of the first few reflected and transmitted rays, we get

	First ray	Second ray	Third ray
Reflected amplitude	$ar = (A_1)$	$att'r' = (A_2)$	$att'r'^3 = (A_3)$
Transmitted amplitude	$att'(= A_1')$	$att'r'(= A_2')$	$att'r^2(= A_3')$

Now, $|r| = |r'|, tt' = (1-r^2)$

$$\therefore \quad I_1 = a^2r^2 = \left(\frac{n_2-n_1}{n_2+n_1}\right)^2, I_1' = a^2(1-r^2)^2 = \frac{n_2^2}{n_1^2}\left(\frac{2n_1}{n_1+n_2}\right)^2$$

$$\therefore \quad \frac{I_1}{I_1'} = \frac{(n_2-n_1)^2}{(n_2+n_1)^2}\cdot\frac{n_1^2}{n_2^2}\frac{n_1^2(n_1+n_2)^2}{4n_1^2} = \left(\frac{n_2-n_1}{2n_2}\right)^2$$

Now, if $n_1 = 1, n_2 = 3/2$, then,

$$\left(\frac{n_2-n_1}{2n_2}\right) = \left(\frac{3/2-1}{2\times 3/2}\right) = \frac{1}{2}\times\frac{1}{3} = \frac{1}{6} \quad \therefore \quad \frac{I_1}{I_1'} < 1 \quad \Rightarrow \quad I_1' > I_1$$

(2) The maximum intensity of the fringes in the transmitted light is greater than the maximum intensity of the fringes in reflected light and the minimum of the intensity of the fringes in transmitted and reflected light is not same, the minimum value is zero in reflected light but it is not zero in transmitted light. But, however, the total variation of intensity from maximum to minimum is same in both the cases.

(3) The fringe pattern in reflected light is complementary to the fringe pattern in transmitted light. So the rays incident normal to the plate give rise to central circular fringe region which is bright in case of reflected intensity pattern but dark in transmitted intensity pattern.

Nature of the Fringes

Consider fig.(4.9) where the screen is placed at the focal plane of the lens. Let two successive reflected rays be brought to focus $P(-x, o, z)$; the points S_1 and S_2 are the origin of the reflected rays and have co-ordinates $\{0, (D + d/2), 0\}$ and $\{0, (D - d/2), 0\}$ respectively,

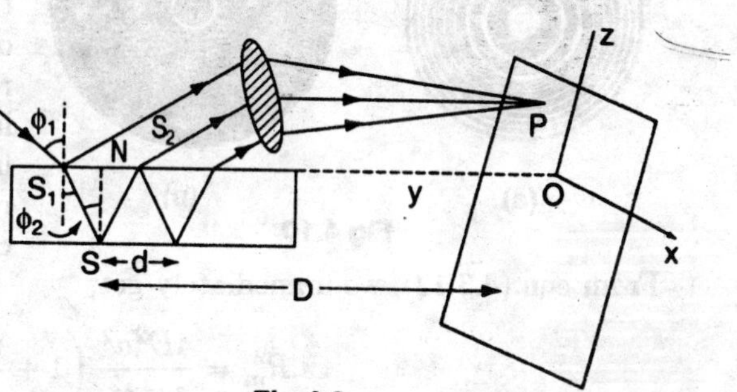

Fig 4.9

$$\therefore \quad S_1P = \sqrt{x^2 + (D + d/2)^2 + z^2},$$
$$S_2P = \sqrt{x^2 + (D - d/2)^2 + z^2}.$$

For brightness at P, we must have,

$$\Rightarrow \quad S_1P - S_2P = (2n+1)\lambda/2 = m \times \frac{\lambda}{2}, \quad \text{where } m \text{ is an odd integer.}$$

$$\therefore \quad \frac{m\lambda}{2} = \sqrt{x^2 + \left(D + \frac{d}{2}\right)^2 + z^2} - \sqrt{x^2 + \left(D - \frac{d}{2}\right)^2 + z^2}$$

$$\Rightarrow \quad \frac{m\lambda}{2} + \sqrt{x^2 + \left(D - \frac{d}{2}\right)^2 + z^2} = \sqrt{x^2 + \left(D + \frac{d}{2}\right)^2 + z^2}$$

Squaring both sides, we get,

$$\frac{m^2\lambda^2}{4} + m\lambda\sqrt{x^2 + \left(D - \frac{d}{2}\right)^2 + z^2} + x^2 + \left(D - \frac{d}{2}\right)^2 + z^2 = x^2 + \left(D + \frac{d}{2}\right)^2 + z^2$$

$$\Rightarrow \left(D + \frac{d}{2}\right)^2 - \left(D - \frac{d}{2}\right)^2 = \frac{m^2\lambda^2}{4} + m\lambda\sqrt{x^2 + \left(D - \frac{d}{2}\right)^2 + z^2}$$

$$\Rightarrow 2Dd - \frac{m^2\lambda^2}{4} = m\lambda\sqrt{x^2 + \left(D - \frac{d}{2}\right)^2 + z^2}$$

$$\Rightarrow 4D^2d^2 + \frac{m^4\lambda^4}{16} - Ddm^2\lambda^2 = m^2\lambda^2x^2 + m^2\lambda^2\left(D - \frac{d}{2}\right)^2 + m^2\lambda^2z^2$$

$$\Rightarrow \frac{4D^2d^2}{m^2\lambda^2} + \frac{m^2\lambda^2}{16} - Dd = x^2 + z^2 + \left(D - \frac{d}{2}\right)^2$$

$$\Rightarrow x^2 + z^2 = \frac{4D^2d^2}{m^2\lambda^2} + \frac{m^2\lambda^2}{16} - D^2 + Dd - \frac{d^2}{4} - Dd = D^2\left(\frac{4d^2}{m^2\lambda^2} + \frac{m^2\lambda^2}{16D^2} - 1 - \frac{d^2}{4D^2}\right)$$

Since, $\frac{m^2\lambda^2}{16D^2}, \frac{d^2}{4D^2}$ are all very small quantities,

$$x^2 + z^2 = D^2\left(\frac{4d^2}{m^2\lambda^2} - 1\right) = \frac{4D^2d^2}{m^2\lambda^2}\left(1 - \frac{m^2\lambda^2}{4d^2}\right) = R_m^2. \tag{4.3.12}$$

In the same way we can show that the bright fringes in transmitted light also form concentric circles. The two systems of fringes, one concentric circular fringes fig.4.10(a) for reflected light and another concentric circular fringes are as shown in fig.4.10(b) for transmitted light. It is evident from the pattern that they are complementary to each other, thus the pattern in the reflected light is a system of concentric circular fringes with the central region bright circle, surrounded by dark annular region, bright annular region, etc. whereas in the transmitted light we have dark central region surrounded by alternate bright and dark circular fringes.

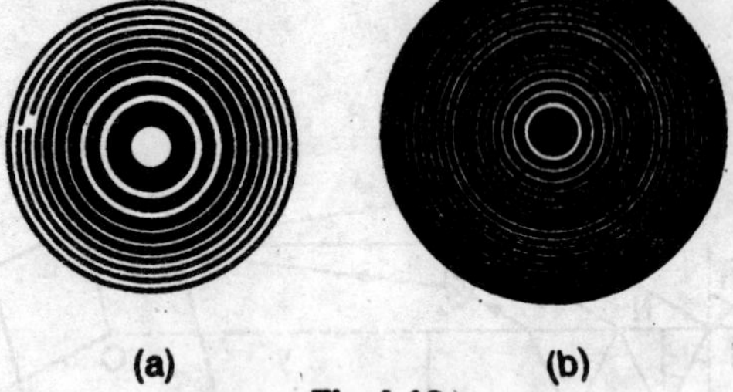

Fig 4.10

From eqn.(4.3.12), we immediately get,

$$R_m^2 = \frac{4D^2d^2}{m^2\lambda^2}\left(1 + \frac{m\lambda}{2d}\right)\left(1 - \frac{m\lambda}{2d}\right)$$

If $m\lambda \simeq d$, $R_m^2 = 4D^2(3/2)\left(1 - \frac{m\lambda}{2d}\right) = 6D^2\left(1 - \frac{m\lambda}{2d}\right)$.

Moreover, $R_m^2 = \frac{4D^2d^2}{\lambda^2}\left(\frac{1}{m^2} - \frac{\lambda^2}{4d^2}\right)$ suggests that as m increases, R_m decreases; it means that the circular fringes are more crowded towards the edge of the pattern.

Interference Fringes in Wedge-shaped Film

We so long considered fringes formed in the reflected and transmitted light when the plate or film is plane-parallel. But we will now consider the phenomenon when the film is wedge-shaped, i.e., the surfaces enclosing the film are inclined at an angle α.

PS_1 is the incident ray, reflected part goes along S_1O, the transmitted part goes along S_1S_3 at an angle θ', gets reflected along S_3S_2, partly gets back in the incident medium along S_2O, such reflected rays unite producing constructive or destructive interference. S_2N_1, S_2N_2, S_2N_3 are all perpendiculars dropped from S_2 on to ray S_1S_3, to ray S_1O and to the surface CD. The angles i, r, θ' and α are as shown in the fig.(4.11).

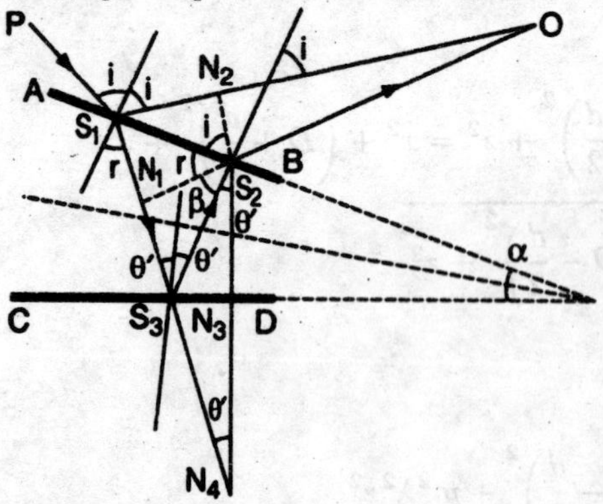

Fig 4.11

From the fig.(4.11), we get,

$$\beta + r = (90 - \theta') + \alpha$$

Again, $\beta + 2\theta' = 90°$

$$\therefore \quad 90° - 2\theta' + r = (90 - \theta') + \alpha \quad \Rightarrow \quad \theta' = (r - \alpha) \tag{4.3.13}$$

The path difference between the reflected rays

$$\begin{aligned}\Delta &= \mu(S_1S_3 + S_3S_2) - S_1N_2 = \mu(S_1S_3 + S_3N_4) - S_1N_2\\ &= \mu S_1N_4 - S_1N_2 = \mu(S_1N_1 + N_1N_4) - S_1N_2 = \mu S_1N_1 + \mu N_1N_4 - S_1N_2.\end{aligned}$$

Now, $\sin i = \mu \sin r$

$$\Rightarrow \quad \frac{S_1N_2}{S_1S_2} = \mu\frac{S_1N_1}{S_1S_2} \quad \Rightarrow \quad S_1N_2 = \mu S_1N_1$$

$$\therefore \quad \Delta = \mu N_1N_4 = \mu S_2N_4 \cos\theta' = \mu \times 2e\cos(r-\alpha)$$

where e represents thickness of the film at S_2.

Considering the path-difference introduced due to reflection, we can write,

$$\Delta_T = \Delta \pm \frac{\lambda}{2} = 2\mu e\cos(r-\alpha) \pm \frac{\lambda}{2}$$

So **brightness**, if

$$\Delta_T = 2\mu e\cos(r-\alpha) \pm \frac{\lambda}{2} = 2s\cdot\frac{\lambda}{2},$$

darkness, if

$$\Delta_T = 2\mu e\cos(r-\alpha) \pm \frac{\lambda}{2} = (2s+1)\frac{\lambda}{2}.$$

So, if $2\mu e\cos(r-\alpha) = (2s \pm 1)\frac{\lambda}{2}$, i.e., odd multiple of $\frac{\lambda}{2}$, we get brightness; if $2\mu e\cos(r-\alpha)$ is an even multiple of $\lambda/2$, we get darkness.

Interference in Transmitted Light through an Wedge-shaped Film

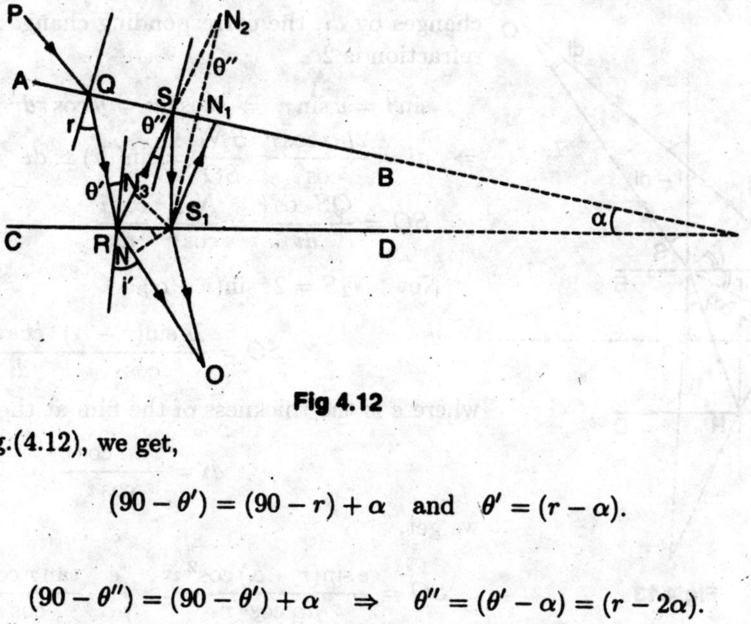

Fig 4.12

From the fig.(4.12), we get,

$$(90-\theta') = (90-r) + \alpha \quad \text{and} \quad \theta' = (r-\alpha).$$

Again,

$$(90-\theta'') = (90-\theta') + \alpha \quad \Rightarrow \quad \theta'' = (\theta'-\alpha) = (r-2\alpha).$$

So the path difference between the two transmitted rays RO and S_1O is given by

$$\Delta_t = \mu(RS + SS_1) - RN = \mu(RS + SN_2) - RN$$
$$= \mu RN_2 - RN = \mu(RN_3 + N_3N_2) - RN.$$

Now, $\sin i' = \mu \sin\theta'$

$$\Rightarrow \quad \frac{RN}{RS_1} = \mu\frac{RN_3}{RS_1} \quad \Rightarrow \quad RN = \mu RN_3 \quad \therefore \quad \Delta_t = \mu N_3N_2.$$

Again,

$$\frac{N_3N_2}{2e} = \cos\theta'' \Rightarrow N_3N_2 = 2e\cos\theta'' \quad \therefore \quad \Delta_t = 2\mu e\cos\theta'' = 2\mu e\cos(r - 2\alpha).$$

Here both the reflections take place from a surface backed by a rarer medium, hence no phase change of π is associated with the reflection process. Hence the total path difference is also $2\mu e\cos(r - 2\alpha)$.

$$\therefore \quad (\Delta_t)_T = 2\mu e\cos(r - 2\alpha) = 2n \times \frac{\lambda}{2} \text{ for Brightness}$$
$$= (2n + 1)\frac{\lambda}{2} \text{ for Darkness.}$$

If α is small, the path difference becomes $(\Delta_t)_T = 2\mu e\cos r$, same as in case of parallel plate.

Location of Fringes Formed in an Wedge-shaped Film : Its Fringe-width

The fig.(4.13) shows the formation of the fringes in reflected light by the wedge-shaped film, the wedge-angle being α. The different angles of our interest are as shown in the figure. The angle of incidence at Q is i; the corresponding refraction angle is r; at S, the emerging angle is $(i - di)$, say, the corresponding angle in the film is $(r - 2\alpha)$. So, when the angle of incidence changes by di, the corresponding change in the angle of refraction is 2α.

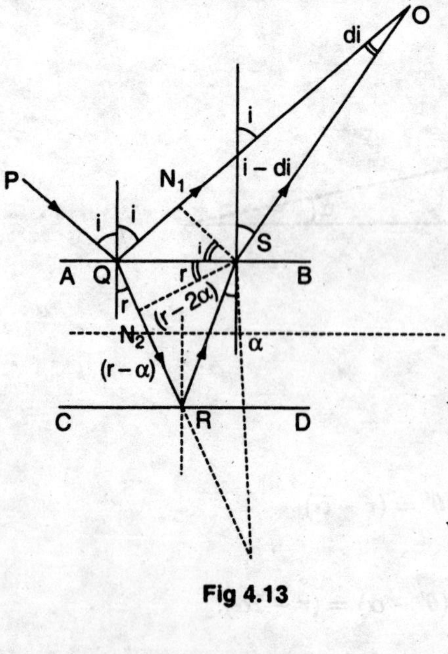

Fig 4.13

$$\therefore \quad \sin i = \mu\sin r \;\Rightarrow\; \cos i\,di = \mu\cos r\,dr = \mu\cos r\cdot 2\alpha$$

$$\Rightarrow \; di = \frac{2\mu\alpha\cos r}{\cos i}, \; \frac{SN_1}{SO} = \sin(di) \simeq di \;\Rightarrow\; SO = \frac{SN_1}{di}$$

$$\therefore \quad SO = \frac{QS\cos i}{di} = \frac{N_2S}{\cos r}\cdot\frac{\cos i}{di}.$$

Now, $N_2S = 2e\sin(r - \alpha)$.

$$\therefore \; SO = \frac{2e\sin(r-\alpha)}{\cos r}\cdot\frac{\cos i}{di}$$

where e is the thickness of the film at the point S.

$$\because \; di = \frac{2\mu\alpha\cos r}{\cos i},$$

we get,

$$SO = \frac{e\sin(r-\alpha)\cos^2 i}{\mu\alpha\cos^2 r} \simeq \frac{e}{\mu\alpha}\cdot\frac{\tan r\cos^2 i}{\cos r}. \quad (4.3.14)$$

Let the distance of S w.r.t. The apex of film be x_n, then,

$$e = x_n\alpha \quad \therefore \quad SO = \frac{x_n}{\mu} \cdot \frac{\tan r \cos^2 i}{\cos r} \tag{4.3.15}$$

From eqn.(4.3.14 or 4.3.15) we can draw the following important conclusions—

Conclusion 1

The location of nth dark fringe is given by

$$2\mu e \cos r = n\lambda.$$

So, $e = 0 \quad \Rightarrow \quad n = 0$, thus the central fringe is a dark fringe and is situated at the apex of the film, i.e., $e = 0$. From eqn.(4.3.14) we get,

$$SO = 0, \text{ because } x_n = 0.$$

Conclusion 2

As the point S shifts along the film surface, x_n increases, and so SO also increases. So the fringes are located further and further away from the film surface as we proceed from the apex towards the thicker part of the film, but always localised, meaning at a certain distance w.r.t. the surface of the film.

Conclusion 3

If the bounding surfaces are plane parallel, then $\alpha = 0$, it implies $SO = \infty$, evident from eqn.(4.3.14); so in order to locate or see these fringes, the eye or the receiving telescope must be adjusted for normal vision.

Thus, the fringes formed in thin film are localised fringes whereas the fringes obtained in Fresnel's biprism, Lloyd's mirror and practically in all cases of wavefront division class of interference, fringes are non-localised.

Fringe-width β

The distance x_n of the nth dark fringe is given by,

$$2\mu e \cos r = n\lambda, \text{ where } e = x_n\alpha \quad \Rightarrow \quad 2\mu x_n \alpha \cos r = n\lambda. \tag{4.3.16}$$

So, if we consider a parallel beam of rays falling normally, then $r = 0$, and we get,

$$2\mu x_n \alpha = n\lambda \quad \Rightarrow \quad x_n = \frac{n\lambda}{2\mu\alpha} \quad \therefore \quad x_{n+1} = \frac{(n+1)\lambda}{2\mu\alpha}$$

$$\therefore \quad \beta = x_{n+1} - x_n = \frac{\lambda}{2\mu\alpha} = \text{constant.} \tag{4.3.17}$$

For bright fringe, we would have,

$$2\mu x'_n \alpha = (2n+1)\frac{\lambda}{2}$$

$$\Rightarrow \quad x'_n = \frac{(2n+1)\lambda}{4\mu\alpha}, \quad x'_{n+1} = \frac{(2n+3)\lambda}{4\mu\alpha}$$

$$\therefore \quad \beta = x'_{n+1} - x'_n = \frac{\lambda}{2\mu\alpha}.$$

Thus, the dark fringe-width or bright fringe-width is the same, which is the characteristic feature of interference fringes.

4.4 Fringes of Equal Width

The maxima and minima of interference in reflected light from a thin wedge shaped film are given respectively by,

$$2\mu e \cos r = (2n+1)\tfrac{\lambda}{2} \quad \text{— Maxima}$$

$$2\mu e \cos r = n\lambda \quad \text{— Minima}$$

Thus, if μ, r, λ are held constant as in the case of parallel beam of monochromatic light, maxima or minima represent the locus of the points where the thickness of the film is constant. Such fringes are fringes of *equal width or thickness.* Practical realisation of this kind of fringe is found in the Newton's rings experimental arrangement where we get circular fringes (detailed discussion later on).

4.5 Fringes of Equal Inclination

This is another type of fringe to be available in case of a thin air-film. If e is constant, μ, λ are also held constant, then different order numbers n satisfying maxima or minima will be determined by the values of r and hence i, the angle of incidence.

Thus an air-film enclosed between two perfectly plane and parallel glass-plates is illuminated by monochromatic extended source. Thus, if we consider a point source S from which divergent rays fall on the air-film, then the rays falling on the circumference of a circle with centre at A and radius equal to AN_1 or AN_2, will have the same angle of incidence. Hence, these points will be either bright or dark according as the condition of brightness or darkness is satisfied.

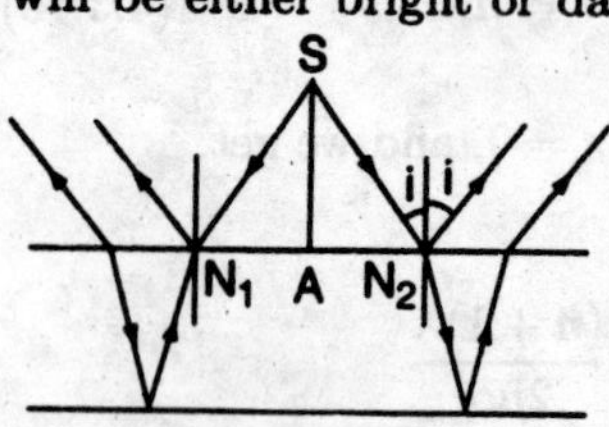

Fig 4.14

Thus, here also we get alternate bright and dark rings with the point like A as centre. Since i is remaining same for a particular order number n, these fringes are referred to as *fringes of equal inclination.* They are also referred to as Haidinger's fringes. These fringes may be observed in reflected light or transmitted light. If the bounding surfaces of the plate are perfectly parallel, perfect circular fringes will be obtained. Distortion in the circular shape will come due to non-parallelism of the two surfaces.

4.6 Newton's Rings : Fringes of Equal Width

The formation of Newton's rings is an example of interference in an wedge-shaped film of variable thickness and the fringes obtained are circular fringes which are fringes due to reflection from the points where the thickness of the film is constant.

Experimental Arrangement

Na-vapour lamp placed at the focus of the lens L_1 sends parallel beam of rays to fall on an inclined glass-plate G_2, inclined at an angle of 45°.

The rays after reflection are directed vertically downwards on to a plano-convex lens AB of large radius of curvature and placed on another glass-plate G_1. The microscope M placed above the glass-plate G_2 is focussed on to the circular rings formed. The arrangement is shown in fig.(4.16), (a) showing the complete set-up (b) shows an wedge-shaped air-film of variable thickness formed between the curved surface of the lens and the glass-plate G_1. The observed rings are shown in fig.(4.16) which is an actual photograph of the fringe-pattern.

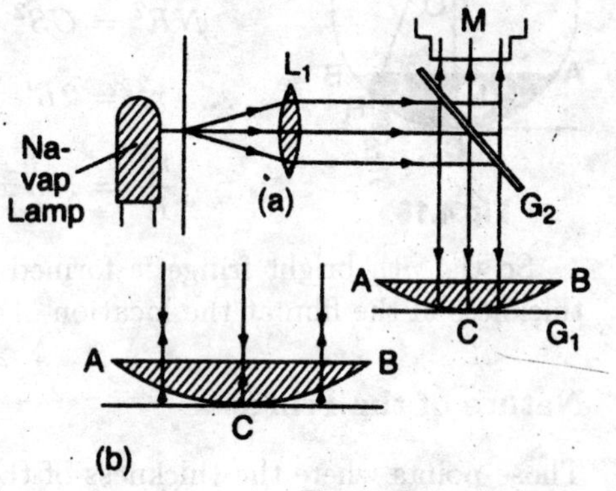

Fig 4.15

Theory

The formation of Newton's rings belongs to fringes of equal width formed in reflected light from a wedge-shaped film of variable thickness. If the lens be of large radius of curvature, the curved surface of the lens approximates to a plane surface, thus a very narrow air-film is enclosed between the lens surface and the glass-plate G_1, having the apex at C.

Fig 4.16

Consider a ray PQ directed vertically downwards, proceeds along QR and strikes the upper surface of wedge-shaped film at R almost normally; a part reflects from R, the other part goes through the air-film and strikes the glass-plate G_1 at S and gets reflected. The path difference is $2e$. The first reflection takes place from the surface backed by a rarer medium whereas the second reflection takes place from a surface backed by a denser medium so in this case the second reflection introduces a phase change of π or path diff. $\frac{\lambda}{2}$.

Total path difference between the reflected rays $= 2e \pm \frac{\lambda}{2}$.

So,

$$\left.\begin{aligned} \left(2e \pm \frac{\lambda}{2}\right) &= 2n \cdot \frac{\lambda}{2} \quad \Rightarrow \quad \text{Brightness} \\ \left(2e \pm \frac{\lambda}{2}\right) &= (2n+1)\frac{\lambda}{2} \quad \Rightarrow \quad \text{Darkness} \end{aligned}\right\} \tag{4.6.1}$$

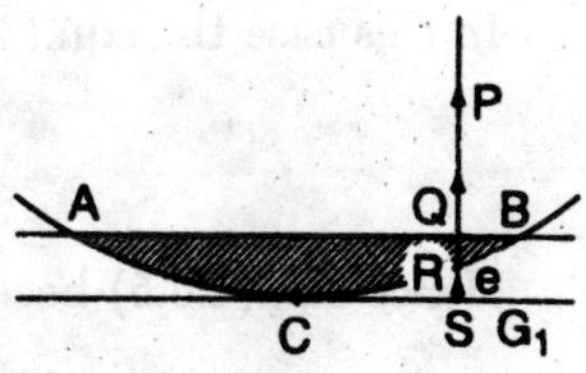

Fig 4.17

From the first condition, it is evident that

$$2e = (2n \pm 1)\frac{\lambda}{2}, \quad \text{i.e., odd multiple of } \tfrac{\lambda}{2} \text{ gives brightness} \tag{4.6.2}$$

$$2e = 2n\cdot\frac{\lambda}{2}, \quad \text{i.e., even multiple of } \tfrac{\lambda}{2} \text{ gives darkness} \tag{4.6.3}$$

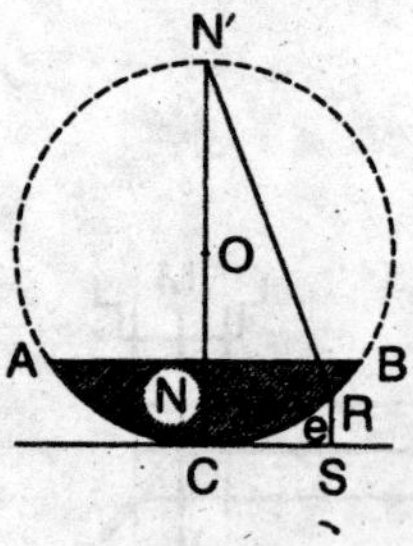

Fig 4.18

Now, thickness e can be expressed in terms of the radius of curvature of the surface.

Considering fig.(4.18) where $RS = e, CS = r_n$, then from the figure,

$$NR^2 = CS^2 = N'N\cdot NC = (N'C - NC)NC = (2R' - e)e \simeq 2R'e$$

$$\therefore \quad r_n^2 = 2R'e \quad \Rightarrow \quad 2e = \frac{r_n^2}{R'}$$

$$\therefore \quad \frac{r_n^2}{R'} = (2n \pm 1)\frac{\lambda}{2} \quad \Rightarrow \quad r_n^2 = R'(2n \pm 1)\frac{\lambda}{2} \tag{4.6.4}$$

So the nth bright fringe is formed at a distance r_n from the point of contact C and the thickness of the film at the location of the bright fringe is e.

Nature of the Fringe

Those points where the thickness of the film is e and e is such that it satisfies eqn.(4.6.1), we would get brightness. Thus bright points will lie along the circumference of a circle with C as the centre, hence fringes available here are circular in nature. The radius of the nth bright circular fringe is given by eqn.(4.6.4).

Now, if we denote D_n as the diameter of the nth bright ring, then,

$$D_n^2 = 4r_n^2 = 4R'(2n \pm 1)\frac{\lambda}{2}.$$

Consider again the $(n+s)$th bright ring diameter, denoted as D_{n+s} given by,

$$D_{n+s}^2 = 4r_{n+s}^2 = 4R'\left(\overline{2n+s} \pm 1\right)\frac{\lambda}{2}$$

$$\therefore \quad D_{n+s}^2 - D_n^2 = 4R'\cdot 2S\cdot\frac{\lambda}{2} \quad \Rightarrow \quad \lambda = \frac{D_{n+s}^2 - D_n^2}{4R'S} \tag{4.6.5}$$

Here, λ = wavelength of light used, R' = radius of curvature of the lens.

When a liquid is taken to form the wedge-shaped film

When a liquid is taken to form the film, we get the circular fringes once again, but the fringe-width decreases compared to the fringe-width with air-film.

In this case the eqn.(4.6.1) is given by,

$$2\mu e = (2n \pm 1)\frac{\lambda}{2} \quad \text{for bright fringe.}$$

Hence eqn.(4.6.5) becomes,

$$\lambda = \frac{D_{n+s}^2 - D_n^2}{4\mu R'_s} \tag{4.6.6}$$

Fringe-width β

$$D_n^2 = 4R'(2n \pm 1)\frac{\lambda}{2}, \quad D_{n+1}^2 = 4R'\left(\overline{2n+1} \pm 1\right)\frac{\lambda}{2}$$

$$\therefore \quad \beta = (D_{n+1} - D_n) = \frac{(D_{n+1}^2 - D_n^2)}{(D_{n+1} + D_n)} = \frac{4R'\lambda}{(D_{n+1} + D_n)}$$

Now, $D_n + D_{n+1} \simeq 2D_n$

$$\therefore \quad \beta = \frac{4R'\lambda}{2D_n} = \frac{2R'\lambda}{D_n}. \tag{4.6.7}$$

Now, D_n increases with n, hence β decreases as the order number increases, it means there is more crowding of the circular rings as we move outwards from centre of the pattern.

When a liquid is taken to form the film, the fringe-width is given by,

$$\beta' = \frac{2R'\lambda}{\mu D_n}. \tag{4.6.8}$$

Thus, $\beta' < \beta$ since $\mu > 1$. Thus the fringe-width decreases.

Nature of Central Fringe

In the arrangement shown in fig.(4.15), the central fringe formed at the point of contact is a dark fringe. This fringe is characterised by $n = 0$ and here $e = 0$. Thus, thickness of the film is zero, still there is a molecular gap where dut to reflection, a phase change of π occurs, hence the reflected rays always meet in opposition producing dark fringe. But, of course, the central fringe may be bright depending on the disposition of the lens.

Newton's Rings under Different Configurations of the Lens

Case I

When the point C of the lens is at a height d from the glass plate G_1, the thickness of the film at R is given by, $(e + d) = e'$. So the condition of brightness is given as,

$$2e' = (2n \pm 1)\frac{\lambda}{2} \text{ in case of air-film}$$

$$\text{and } \quad 2\mu e' = (2n \pm 1)\frac{\lambda}{2} \text{ in case of liquid-film.}$$

$$\therefore \quad 2\mu e' = (2n \pm 1)\frac{\lambda}{2} \Rightarrow 2\mu(e + d) = (2n \pm 1)\frac{\lambda}{2}$$

$$\Rightarrow \quad 2\mu e + 2\mu d = (2n \pm 1)\frac{\lambda}{2} \Rightarrow \left(\mu\frac{r_n^2}{R'} + 2\mu d\right) = (2n \pm 1)\frac{\lambda}{2}$$

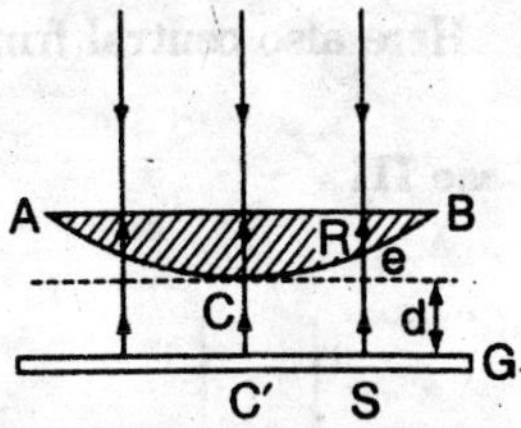

Fig 4.19

The radius of the $(n + s)$th ring is given by,

$$\left(\mu\frac{r_{n+s}^2}{R'} + 2\mu d\right) = \left(\overline{2n+s} \pm 1\right)\frac{\lambda}{2}$$

$\therefore$ Taking the difference we get,

Fig 4.20

$$\frac{\mu(r_{n+s}^2 - r_n^2)}{R'} = s\lambda \quad \therefore \quad \lambda = \frac{\mu(r_{n+s}^2 - r_n^2)}{R's} = \frac{\mu(D_{n+s}^2 - D_n^2)}{4R'S} \tag{4.6.9}$$

This equation is similar to eqn.(4.6.6) derived when the lens and glass-plate are in contact. But in this case the central fringe may be bright depending on the value of d. Thus, if

$$2\mu d = \frac{\lambda}{2} \quad \Rightarrow \quad \mu d = \frac{\lambda}{4}, \quad \text{the central fringe becomes bright.}$$

Case II

When the lens is placed on a concave glass-plate.

In this case, $e = RS - R_1S$. Now, $RS = QC$. Also we have,

$$(2R' - QC)QC = CS^2 \quad \Rightarrow \quad 2R'QC = CS^2 \quad \Rightarrow \quad QC = \frac{CS^2}{2R'}.$$

Also, $R_1S = \frac{CS^2}{2R''}$, where R' is the radius of curvature of the lens and R'' is the radius of curvature of the concave glass-plate.

$$\therefore \quad e = \frac{CS^2}{2}\left(\frac{1}{R'} - \frac{1}{R''}\right) = \frac{r_n^2}{2}\left(\frac{1}{R'} - \frac{1}{R''}\right).$$

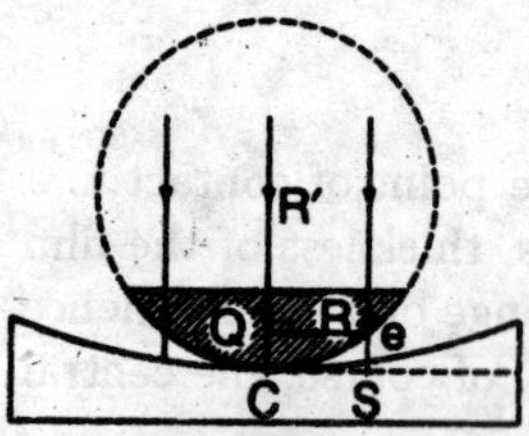

Fig 4.21

So, for brightness, we would have,

$$2\mu e = (2n \pm 1)\frac{\lambda}{2} \quad \Rightarrow \quad \mu r_n^2\left(\frac{1}{R'} - \frac{1}{R''}\right) = (2n \pm 1)\frac{\lambda}{2}.$$

Similarly, the radius of the $(n+s)$th ring is given by,

$$\mu r_{n+s}^2\left(\frac{1}{R'} - \frac{1}{R''}\right) = (2n + s \pm 1)\frac{\lambda}{2}.$$

$\therefore$ Taking difference we get,

$$\mu\left(\frac{1}{R'} - \frac{1}{R''}\right)(r_{n+s}^2 - r_n^2) = s\lambda$$

$$\lambda = \mu\frac{(r_{n+s}^2 - r_n^2)}{s}\left(\frac{1}{R'} - \frac{1}{R''}\right) = \mu\frac{(D_{n+s}^2 - D_n^2)}{4s}\left(\frac{1}{R'} - \frac{1}{R''}\right) \tag{4.6.10}$$

Here also central fringe is a dark fringe.

Case III

When the lens is placed on a convex glass-plate.

Here, $e = RS + SR_1$

Now, $RS = QC = CS^2/2R'$

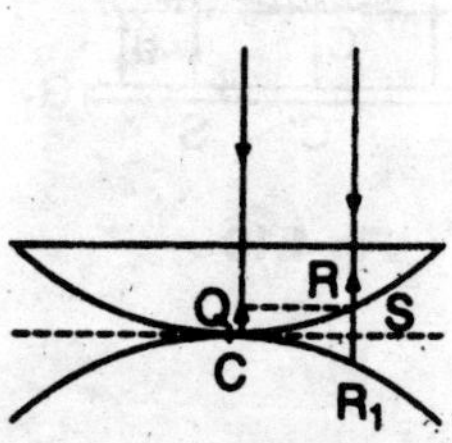

Fig 4.22

$$SR_1 = \frac{CS^2}{2R''} \quad \therefore \quad e = \frac{CS^2}{2}\left(\frac{1}{R'} + \frac{1}{R''}\right).$$

The condition for bright fringe at R is

$$2\mu e = (2n \pm 1)\frac{\lambda}{2}$$

$$\mu \cdot CS^2\left(\frac{1}{R'}+\frac{1}{R''}\right)=(2n\pm 1)\frac{\lambda}{2} \quad\Rightarrow\quad \mu r_n^2\left(\frac{1}{R'}+\frac{1}{R''}\right)=(2n\pm 1)\frac{\lambda}{2}$$

The radius of the $(n+s)$th bright fringe is given by,

$$\mu r_{n+s}^2\left(\frac{1}{R'}+\frac{1}{R''}\right)=(2n+s\pm 1)\frac{\lambda}{2}$$

$$\therefore \quad \mu\left(\frac{1}{R'}+\frac{1}{R''}\right)(r_{n+s}^2-r_n^2)=s\lambda \quad\Rightarrow\quad \lambda=\mu\frac{(r_{n+s}^2-r_n^2)}{s}\cdot\left(\frac{1}{R'}+\frac{1}{R''}\right) \quad (4.6.11)$$

Application of Newton's Rings

(1) To determine the wavelength

Finding the diameter of the nth bright fringe and $(n+s)$th bright fringe obtained according to the arrangement shown in fig.(4.15) and using eqn.(4.6.5) we can find the wavelength λ given as,

$$\lambda=\frac{D_{n+s}^2-D_n^2}{4R's}.$$

The radius of curvature of the plano-convex lens is available from a knowledge of the focal length of the lens as,

$$\frac{1}{f}=(\mu_L-1)\left(\frac{1}{R'}\right),$$

where μ_L is the r.i. of the lens material.

(2) To determine the r.i. of a liquid

The space between the lens and the glass-plate G_1 is filled with the given liquid and a liquid film is formed. Newton's rings are obtained in the usual arrangement as shown in fig.(4.15). Then from eqn.(4.6.6) we can write,

$$\lambda=\frac{D_{n+s}^2-D_n^2}{4\mu R'S},$$

thus measuring the diameter of $(n+s)$th and nth ring, and knowing λ and R', we can find out the value of μ of the liquid.

4.7 Wiener and Ives Experiment

Stationary Light Waves : Colour Photography by Lippmann

Stationary waves are formed due to superposition of direct waves and reflected waves as in case of stringed instruments in sound. It is a characteristic feature of a wave. Light is also a wave, so stationary wave pattern must also be available in light. But experiment to show stationary wave in case of light is difficult because of extremely short wavelength.

Ives allowed a parallel beam of light on the glass side of a photographic plate while the emulsion side was placed on a mirror. In fig.4.22(b) is shown a magnified version of the arrangement and in fig.4.22(c) the formation of the stationary pattern in the emulsion is shown.

The direct waves of monochromatic light after reflection at the Hg surface of the mirror superpose giving rise to the stationary wave formation with nodal and antinodal planes as shown. As the film (i.e., the emulsion) is not affected at the nodal planes, on development it will exhibit a number of transparent parallel layers of gelatine separated by thin dark parallel layers of silver (antinodal layers). The distance of separation between two consecutive antinodal or nodal layers is equal to $\frac{\lambda}{2}$. As this distance is very small Wiener suggested a modification to take the photographic film at an angle and thus increasing considerably on the film the distance of separation of the nodal planes.

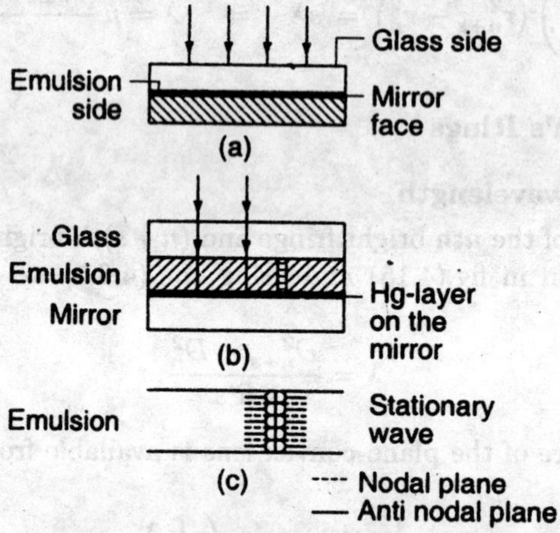

Fig 4.23

If now white light be made incident on the film, the light of wavelength λ will be reflected from the silver layers and will differ in path by multiples of $\lambda \left(= \frac{\lambda}{2} + \frac{\lambda}{2}\right)$. Thus we get reflected light of wavelength λ only while other wavelengths are absorbed. Based on this fact Lippmann introduced colour photography. In the camera he placed a photographic plate with the glass side towards the objective while the film side was backed by a layer of Hg. When the plate was exposed and then developed thin parallel layers of Ag (the antinodal layers) are formed, each of the incident wavelength forming its own system of characteristic antinodal layers. Thus original colour will be reproduced.

Numerical Problems

1. In an experiment with Newton's rings, the diameters of 6th and 20th rings formed by light of wavelength 5,000Åare 1.76 and 3.22 mm. Calculate the radius of curvature of the convex surface.

Solution.

$$\lambda = \frac{D_{n+s}^2 - D_n^2}{4R'S}$$

$$\lambda = 5000 \times 10^{-8}\,\text{cm}, D_n = D_6 = 1.76 \times 10^{-1}\,\text{cm}, D_{n+s} = D_{20} = 3.22 \times 10^{-1}$$

$$\therefore\ 5000 \times 10^{-8} = \frac{(3.22^2 - 1.76^2) \times 10^{-2}}{4R' \times 14} \Rightarrow R' = \frac{(3.22^2 - 1.76^2) \times 10^{-2}}{5000 \times 10^{-8} \times 4 \times 14} = 32.7\,\text{cm}.$$

2. Newton's rings by reflected light of wavelength 5,000Åare formed between two biconvex lenses of equal radii of curvature. If the distance between the 5th and 15th dark rings is 0.085 cm, calculate the radius of either lens.

Solution. According to Case III, we get,

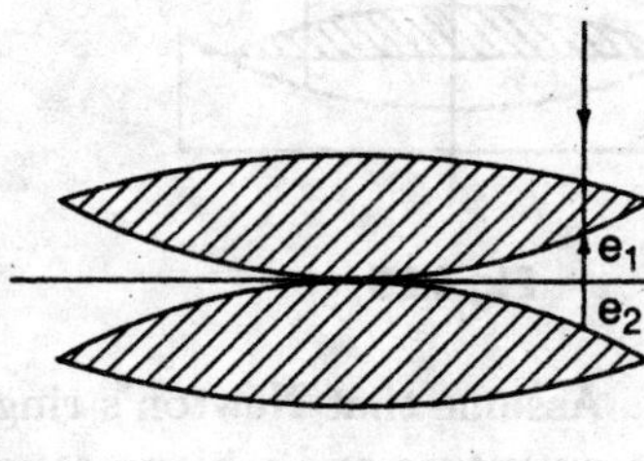

Fig 4.24

$$r_n^2\left(\frac{1}{R'}+\frac{1}{R''}\right)=2n\cdot\frac{\lambda}{2}=n\lambda,\ \text{for dark fringe}$$

$$\Rightarrow\ \frac{2}{R}\cdot r_n^2=n\lambda\ \Rightarrow\ r_n^2=\frac{n\lambda}{(2/R)}$$

$$\therefore\ r_5^2=\frac{5\lambda}{(2/R)}\ \Rightarrow\ r_5=\sqrt{\frac{5\lambda}{(2/R)}}$$

Similarly,

$$r_{15}=\sqrt{\frac{15\lambda}{(2/R)}}$$

$$\therefore\ r_{15}-r_5=\sqrt{\frac{15\lambda}{(2/R)}}-\sqrt{\frac{5\lambda}{(2/R)}}=.085$$

$$\Rightarrow\ \frac{1}{\sqrt{(2/R)}}=\frac{.085}{\sqrt{15\lambda}-\sqrt{5\lambda}}=\frac{0.085}{\sqrt{5\lambda}(\sqrt{3}-1)}$$

$$\Rightarrow\ \sqrt{R/2}=\frac{.085}{\sqrt{5\lambda}(\sqrt{3}-1)}\ \Rightarrow\ R=107.9\,\text{cm}.$$

3. Newton's rings by reflection are formed between two convex lenses having equal radii of curvature being 100 cm each. Calculate the distance between the 5th and 15th dark rings using monochromatic light of wavelength 5,400Å. [0.0839 cm.]

4. Newton's rings are formed by reflection in the air-film between a plane surface and a spherical surface of radius 50 cm. If the diameter of the third bright ring is 0.181 cm and that of the twenty-third bright ring is 0.501 cm, what is the wavelength of the light used?

Solution. $2e\cos r=(2n\pm1)\frac{\lambda}{2}$. Again, $2e=r_n^2/R'$

$$\therefore\ \frac{r_n^2}{R'}\cos r=(2n\pm1)\frac{\lambda}{2}\ \Rightarrow\ \frac{4r_n^2}{R'}\cos r=4(2n\pm1)\frac{\lambda}{2}\ \Rightarrow\ \frac{D_n^2}{R'}\cos r=4(2n\pm1)\frac{\lambda}{2}$$

$$\therefore\ \frac{D_{n+s}^2}{R'}\cos r=4\left(2n+s\pm1\right)\frac{\lambda}{2}\ \Rightarrow\ \frac{D_{n+s}^2-D_n^2}{R'}\cos r=4\times2s\cdot\frac{\lambda}{2}=4s\lambda$$

Here, $r=0,\cos r=1,n=3,n+s=23.$

Hence, $\lambda=\frac{D_{n+s}^2-D_n^2}{4sR'}=\frac{0.501^2-0.181^2}{4\times20\times50}=5456\text{Å}.$

5. A soap film of r.i. $\frac{4}{3}$ and of thickness 1.5×10^{-4} cm is illuminated by white light incident at an angle of 45°. The light reflected by it is examined spectroscopically in which it is found that a dark band corresponds to a wavelength of 5×10^{-5} cm. Calculate the order of the dark band.

Solution. $2\mu e\cos r = 2n\cdot\frac{\lambda}{2}$.

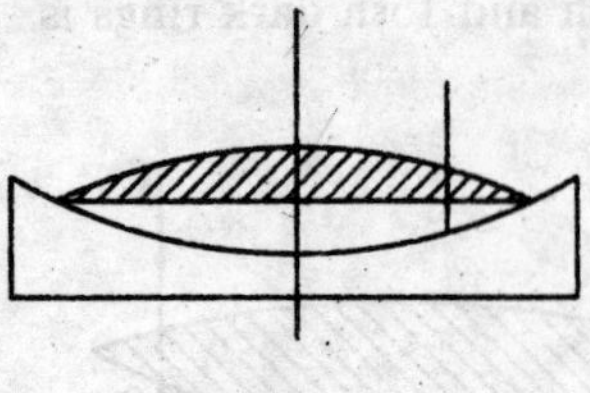

Fig 4.25

Now, $\sin i = \mu \sin r$

$$\Rightarrow \sin r = \frac{\sin i}{\mu} = \frac{1/\sqrt{2}}{4/3} = \frac{3}{4\sqrt{2}} = 0.5303 \Rightarrow r = 32°1'$$

$$\therefore\ 2\times\frac{4}{3}\times 1.5\times 10^{-4}\times .8479 = n\times 5\times 10^{-5}$$

$$\Rightarrow n = 6.8, \text{ hence 7th order.}$$

6. Assume that Newton's rings are formed between a plano-convex lens of 50 cm radius of curvature and a plano-concave lens, placed so as to enclose an air-film. If the 20th dark ring seen in reflected light of 5,500Åwavelength, has a diameter 1.2 cm, what is the radius of curvature of the concave surface?

Solution. When the lower surface of the lens is the plane surface, the radius of the nth ring is given by

$$r_n^2 = R'\times 2n\cdot\frac{\lambda}{2} \text{ for dark ring} \Rightarrow D_n^2 = 4R''\times n\lambda$$

$$R'' = \frac{D_n^2}{4n\lambda} = \frac{4.22^2}{4\times 20\times 5500\times 10^{-8}} = 338.3\,\text{cm}\,.$$

7. In Newton's ring experiment, the radius of the nth dark ring is 4 mm and the radius of the $(n+5)$th dark ring is 6 mm. If the radius of curvature of the lower surface of the lens is 10 meters, find the wavelength of light and the ring number n.

Solution.

$$\frac{D_{n+s}^2 - D_n^2}{R'} = 4\times 2s\cdot\frac{\lambda}{2} = 4s\lambda$$

$$\Rightarrow \frac{D_{n+s}^2 - D_n^2}{R'} = 4s\lambda = 4\times 5\times\lambda$$

$$\Rightarrow \lambda = \frac{(2\times 6\times 10^{-3})^2 - (2\times 4\times 10^{-3})^2}{20\times 10} = 4\times 10^{-5}\text{ cm}$$

Again, $r_n^2 = R/n\lambda \Rightarrow (4\times 10^{-3})^2 = 10\times n\times 4\times 10^{-7} \Rightarrow n = 4.$

8. Two rectangular glass-plates are laid one upon another with a thin wire between them at one edge, so as to enclose a thin wedge-shaped air-film. The plates are illuminated normally by sodium light ($\lambda = 5893$Å). Bright and dark fringes are formed, there being 10 of each per cm length of the wedge, measured normal to the edge in contact. Find the angle of the wedge.

Solution. The central dark fringe forms at the edge of the wedge. Let x_n be the distance of the nth dark fringe, x_n being measured from the apex of the film. Then, if e represents the thickness of the air-film, $2e = n\lambda$. Also, $e = x_n\alpha \Rightarrow 2x_n\alpha = n\lambda$. $\therefore\ 2x_{n+1}\alpha = (n+1)\lambda\cdot(n+1)\lambda$. So the fringe-width $\beta = x_{n+1} - x_n = \frac{\lambda}{2\alpha}$.
Here, $\beta = \frac{1}{10}\,\text{cm}, \lambda = 5893\times 10^{-8}\,\text{cm}$.

$$\therefore\ \frac{1}{10} = \frac{5893\times 10^{-8}}{2\times\alpha} \Rightarrow \alpha = \frac{5893\times 10^{-8}}{2\times 10} = 2.947\times 10^{-4}\,\text{rad} = 1'47.4''.$$

9. Similar problem:

Two similar rectangular plates are placed in contact along one edge and separated by a strip of paper along the opposite edge, thus forming air wedge of very small angle between them. When the wedge is illuminated normally by light from a sodium lamp, it appears to be crossed by bright bands with a spacing of 1.20 mm. Calculate the angle of the wedge. [0.84′]

10. Two wires, with diameters 0.010 cm and 0.020 cm respectively, are laid parallel to each other and 2.00 cm apart on a piece of plane glass; another piece of plane glass is then laid on top. If the light used is of wavelength 5.46×10^{-5} cm, calculate the distance between consecutive bright fringes.

Solution. $\beta = \frac{\lambda}{2\alpha}$

$$\tan\frac{\alpha}{2} = \frac{.010/2}{2} \Rightarrow \frac{\alpha}{2} = \frac{.010}{4} \Rightarrow 2\alpha = .010$$

$$\therefore \beta = \frac{5.46 \times 10^{-5}}{.010} = 0.0054\,\text{cm}.$$

11. The convex surface (radius of curvature 400 cm) of a plano-convex lens rests on a concave surface of 500 cm radius and Newton's rings are viewed by reflected light of wave length 5.9×10^{-5} cm. Calculate the diameter of the 5th bright ring seen.

Solution. In this case,

$$r_n^2\left(\frac{1}{R'} - \frac{1}{R''}\right) = (2n \pm 1)\frac{\lambda}{2}.$$

Here, $n = 5, R' = 400\,\text{cm}, R'' = 500\,\text{cm}$

$$\therefore r_5^2\left(\frac{1}{400} - \frac{1}{500}\right) = 11 \times \frac{\lambda}{2}, \text{ or, } 9 \times \frac{\lambda}{2}$$

$$\Rightarrow r_5^2 \times \frac{1}{2000} = 11 \times \frac{\lambda}{2}, \text{ or, } 9 \times \frac{\lambda}{2}$$

$$\therefore D_5 = 2r_5 = 2\sqrt{\frac{11 \times \lambda}{2} \times 2000}$$

$$\text{or, } 2\sqrt{\frac{9\lambda}{2} \times 2000} = 1.61\,\text{cm or } 1.46\,\text{cm}$$

The minimum diameter is 1.46 cm.

Short Theory Questions with Answers

1. What is interference by division of amplitude?

Ans. In the class of interference termed amplitude division class, coherent disturbances are produced due to multiple reflection from or transmission through a thin film or glass plate with phase difference continuously varying. These reflected disturbances or transmitted disturbances interfere constructively or destructively producing brightness or darkness.

2. To which class of interference does the biprism experiment belong—the Newton's ring? In biprism experiment a narrow source is used but in Newton's rings experiment a broad source is used—why?

   Ans. There are two classes of interference, one based on division of wavefront and the other on division of amplitude, depending on how we obtain coherent sources. Biprism experiment belongs to interference phenomena based on division of wavefront and Newton's ring experiment on division of amplitude.

   In biprism experiment a very narrow source is necessary because it produces two virtual images S_1 and S_2 of the source S. If the source is a narrow slit, the virtual image sources will also be slit-like having corresponding points like (A, C) and (B, D). Each pair of conjugate points will produce interference pattern of its own. The relative displacement of the corresponding maxima of intensity pattern due to interference of the two sources like (A, C) and (B, D) depend on the extension of the slit source S. For a certain width, the central maximum of one pattern may coincide with the first order maximum of the other. In this case interference fringes vanish. We can also say that if the source S is extended, temporal coherence condition is not satisfied, hence no interference. But in Newton's ring experiment, a ray gives rise to multiple rays by reflection or transmission. It is the interference of these rays that give rise to Newton's ring. In this case then phase relationship of the incident rays is of no concern. So the incident rays may not be derived from an infinitely narrow source as is necessary in a biprism experiment. So a broad source is used in Newton's experiment for the sake of sufficient light intensity.

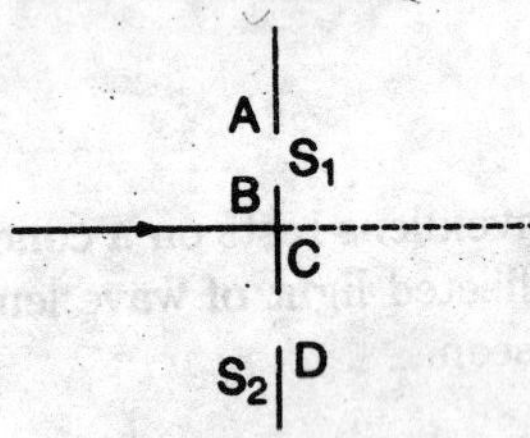

Fig 4.26

3. How does colour appear in thin films?

   Ans. If a parallel beam of white light be incident on an wedge-shaped film, μ, λ and r will be different for different colours. Where the thickness e of the film is practically zero, all colours suffer a phase change of π or path diff. equal to $\frac{\lambda}{2}$, so that thin edge becomes perfectly dark. But as e value increases and since $\lambda_v < \lambda_r$, the first order bright fringe for violet will be formed at a smaller thickness than the corresponding bright fringe for red, producing differently coloured fringes at different thicknesses. Beyond this thickness of the film, we arrive at some value of thickness where the condition for maximum is satisfied by two or more colours, thus getting a colour band due to overlapping of coloured bright fringes. If overlapping takes place to such an extent and this takes place at sufficient value of thickness that we get uniform illumination.

4. Why a very thin film appears to be black when illuminated by white light while a thin film provides beautiful colouration?

   Ans. When a very thin film is used e is practically zero at any place and so all the colours suffer a phase change of π on reflection from the film, thus the film appears to be dark. For the second part, refer to the answer given in Question 3.

Exercise

1. Explain the colours in thin films. What is Stokes' formula? Has it anything to do with the interference of the reflected rays from a thin film?
2. What do you mean by fringes of equal width? Give a practical arrangement to realise such fringes.

3. What are fringes of equal inclination (also called *Haidinger's fringes*)? What is the application of such fringes?
4. Give the theory of Newton's rings and show how, from their study, the wavelength of monochromatic light can be determined.
5. What do you mean by Finesse factor? Derive the intensity distribution functions for the reflected and transmitted light when a ray is incident on a parallel film.
6. A thin film of air is contained between two parallel-faced glass plates, the distance and angle between which can be varied. How would you adjust the plates in order to obtain—(a) St. fringes, (b) Circular fringes?
   Describe with the help of a diagram the method of illumination and of viewing the fringes.
7. Show that the interference pattern in the reflected and transmitted lights are complementary. **[CU 1998]**

Numerical Problems

1. In Newton's ring expt., the diameter of the nth and $(n+8)$th bright rings are 4.2 mm and 7 mm respectively. Radius of curvature of the lower surface of the lens is 2 meters. Determine the wavelength of light and the ring number. **[λ = 49,00Å, $n = 4$]**
2. Light of wavelength 6×10^{-5} cm falls normally on two glass plates which enclose a thin wedge-shaped film of air. The plates touch each other at one end and separated by a wire of diameter 0.05 mm placed at a point whose distance from the touched end is 15 cm. Find the width between any two bright fringes.
   [Hints. $2(\alpha x_{n+1} - \alpha x_n) = \lambda \Rightarrow 2\alpha\beta = \lambda$; here, $\alpha = 0.005/15$, so β can be calculated.**]**
3. A drop of oil ($\mu = 1.58$) is placed between a plano-convex lens and a plane glass plate on which the former rests. The lens and the plate are made of the same r.i. value equal to 3/2. Calculate the diameter of the 5th bright ring if the radius of curvature of the lens surface is 200 cm and light of wavelength 5900Åis reflected normally from the system. What effect will be observed when the plate is replaced by another of r.i. 1.65?
   [0.4054 cm rings with bright centre]
4. An oil film ($\mu = 1.2$) on water ($\mu = 1.33$) is viewed from directly above with light of wavelength 6000Åin air. The film appears circular and has a centre-thickness 1 μm and zero thickness at the edge. Explain whether the edge will appear bright or dark. How many dark rings appear in the fringe? **[C.U. 1998]**

References

1. **Born, M. and Wolf, E.**—*Principles of Optics,* Pergamon Press.
2. **Ditchburn**—*Light,* Blackie & Son Ltd., London.
3. **Jenkins and White**—*Fundamentals of Optics,* McGraw-Hill Book Co.
4. **Mazumder, K.G.**—*A Textbook of light,* Modern Book Agency, Calcutta.

Chapter 5

Diffraction

5.1 Introduction

We have already remarked that the geometrical optics based on the rectilinear propagation of light is only an approximate truth because when light passes across a sharp obstacle, viz. edge of a blade, a narrow slit, etc., light ray bends so that the shadow of the obstacle loses in sharpness of the boundary of the shadow. If light travelled strictly in straight path, we would get a shadow with well-defined boundary, but it has been found that when the irregularities of the surface across which light wave travels are comparable to the wavelength of the wave, appreciable bending of light takes place. As a result, we observe that to some depth within, or some distance beyond the geometrical boundary of the shadow, there are alternate regions of darkness and brightness. The presence of the bright region clearly shows that light has reached there, showing thereby the bending of light beam. This bending of light over a sharp obstacle is referred to as *diffraction phenomenon.*

The occurrence of alternate regions of darkness and brightness to some depth within the geometrical boundary of the shadow, is definitely the superposition of waves as in case of *interference*, but there is basic difference between these two phenomena with regard to both genesis as well as structural features. In interference pattern, the alternate regions of darkness and brightness are of equal thickness, and all the bright regions, called the *bright fringes,* are of equal intensity. But the diffraction fringes are not of equal thickness nor is the intensity of all these bright regions same. This is because the interference fringes (dark or bright) are due to superposition of two coherent primary waves, whereas the diffraction fringes are due to superposition of coherent secondary wavelets. Since ultimate diffraction pattern is obtained through interference, a big slit which consists of secondary sources widely separated, cannot produce interference, as a result we do not get diffraction maxima or minima. But if the slit is narrow, we can realise a large number of sources in phase, satisfying the condition of spatial coherence and hence diffraction maxima and minima are available. Thus when light passes through a wide aperture, we get illumination on a screen, but if light passes through a narrow slit, we get an illuminated region followed by alternate regions of darkness and brightness.

The fact of observation is represented as in fig. 5.1(a), (b) and (c).

If the aperture is of width $a \gg \lambda$, then on a screen we will get uniformly illuminated region, no diffraction fringes appear fig. 5.1(b).

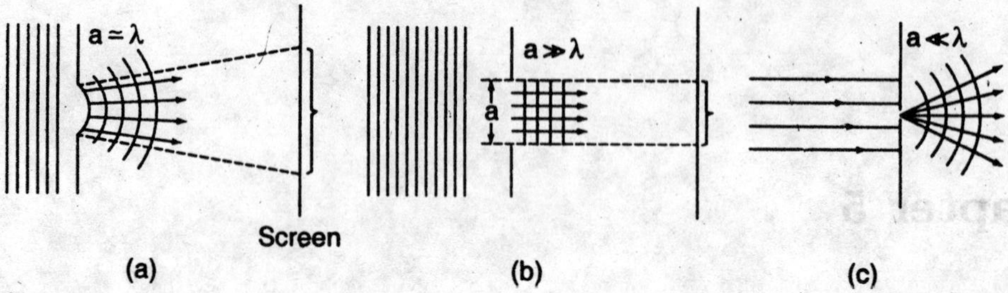

Fig 5.1

In case fig. 5.1(a) the width of the aperature is of the order of the wavelength, i.e., $a \simeq \lambda$, hence diffraction fringes appear. But when $a \ll \lambda$, as in fig. 5.1(c), the screen is non-uniformly illuminate and fringes do not appear.

5.2 Huygens' Principle—Fresnel's Extension

The propagation of wavefront was explained very simply by Huygens' principle which we have already come across. Though very simply this principle could be led to explain the propagation of the wave, it could not explain many of the facts of observation like why there is no back wave propagation, why the diffraction fringes would be produced, etc. However, it should be remembered that at the time of Huygens, the idea was that light is a disturbance in a medium called *ether* which may be supposed to consist of closely packed spheres. So when a disturbance travels, the momentum is transferred in the forward direction.

Huygens' principle really applies to isolated waves or pulses. Fresnel extended this principle to give wider coverage, to the extended train of simple harmonic waves.

Fresnel's Theory of Spherical Wave Propagation

For the basic assumptions in the development of his theory, Fresnel took the following:

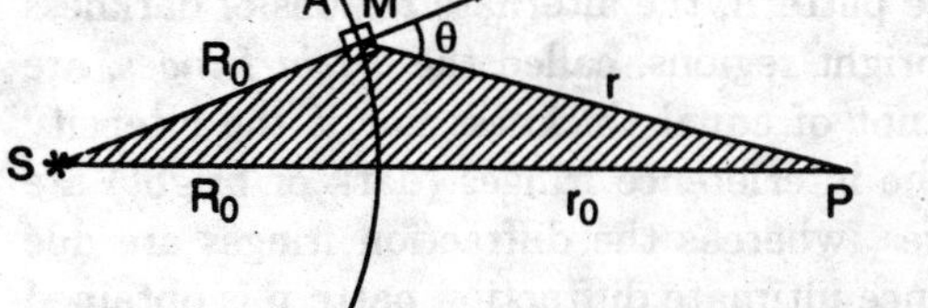

Fig 5.2

1. Every element of an unobstructed wavefront through an aperture acts as a secondary source of disturbance.

2. The secondary disturbances from these elements superpose at any point P to give the resultant disturbance to be felt there.

3. The amplitude of the disturbance at P due to the sources in any elementary area of the primary wavefront is

(a) proportional to the amplitude of the disturbance at these elements,

(b) proportional to the area of the element,

(c) inversely proportional to the distance of the element w.r.t. P,

(d) dependent on the angle θ between the wave normal at the element under consideration, and the line joining the element and the field point P. With these assumptions in the background, let us take a source S sending out disturbances as spherical wavefront, a part of this unobstructed wavefront being shown as AB. If the wave amplitude be a, then the displacement at the element of area $d\sigma$ at M is given as

$$y_0 = \frac{a}{R_0} e^{ikR_0}.$$

This is the primary wave acting as the forcing system on the element of the aperture, which contains the portion of the wavefront. Each such element behaves as the forced system and so every element vibrates as $y_0 e^{i\alpha}$, α representing the phase difference introduced. So the displacement due to the disturbance reaching P is now given by

$$d\psi = \frac{y_0}{r} e^{i\alpha} \cdot e^{ikr} f(\theta) d\sigma,$$

where $f(\theta)$ takes into account the obliquity.

$$\therefore \quad d\psi = \frac{af(\theta)}{R_0} e^{i\alpha} \cdot e^{ikR_0} \frac{e^{ikr}}{r} d\sigma = \frac{a}{R_0} f(\theta) e^{i\alpha} \frac{e^{ik(R_0+r)}}{r} d\sigma.$$

Total amplitude at P is given by

$$\psi_p = \int \frac{a}{R_0} f(\theta) e^{i\alpha} \frac{e^{ik(R_0+r)}}{r} d\sigma. \tag{5.2.1}$$

The integration is over the whole aperture.

This integral is to be evaluated so as to get the effect of the spherical wavefront at P.

This integral is known as *Fresnel integral*. This can be utilized to get the intensity at some point ahead of a spherical wavefront.

5.3 Fresnel's Half-period Zones

To evaluate the integral, Fresnel considers that the spherical wavefront be divided into spherical zones or strips. The line SA_0P is the central line and the point A_0 on the wavefront is called the *pole of the wavefront* w.r.t. P. If we draw circles taking P as centre and

$$r_0, r_0 + \frac{\lambda}{2}, \left(r_0 + 2 \cdot \frac{\lambda}{2}\right) \cdots \left(r_0 + \overline{n-1}\frac{\lambda}{2}\right)$$

as centre, then we cut out from the spherical wavefront surface circular zones, first between A_0 and A_1, then between A_1 and A_2; thus the nth zone area becomes

$$d\sigma = 2\pi R_0 \sin\alpha . R_0 \, d\alpha = 2\pi R_0^2 \sin\alpha \, d\alpha.$$

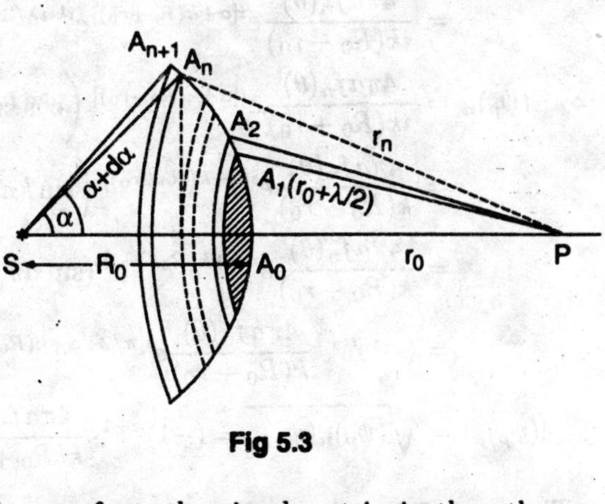

Fig 5.3

So, the amplitude at P due to the disturbances from the circular strip in the nth zone between r and $r + dr$ is given as

$$d\psi_p = \frac{a}{R_0} e^{i\alpha} f_n(\theta) \frac{e^{ik(r+R_0)}}{r} \cdot 2\pi R_0^2 \sin\alpha \, d\alpha.$$

Now, $r^2 = (R_0 + r_0)^2 + R_0^2 - 2R_0(R_0 + r_0)\cos\alpha$

$$\Rightarrow \quad r\,dr = R_0(R_0 + r_0)\sin\alpha\,d\alpha \quad \Rightarrow \quad \sin\alpha\,d\alpha = \frac{r\,dr}{R_0(R_0 + r_0)}$$

$$\therefore \quad d\psi_p = \frac{a}{R_0}e^{i\alpha}f_n(\theta)\frac{e^{ik(r+R_0)}}{r}\cdot 2\pi R_0^2\cdot\frac{r\,dr}{R_0(R_0 + r_0)} = \frac{2\pi a e^{i\alpha}f_n(\theta)}{(R_0 + r_0)}e^{ik(r+R_0)}dr.$$

So the amplitude due to the nth zone at P

$$\begin{aligned}(\psi_p)_n &= \frac{2\pi a e^{i\alpha}}{(R_0 + r_0)}f_n(\theta)\int_{r_{n-1}}^{r_n} e^{ik(r+R_0)}dr\\ &= \frac{2\pi a e^{i\alpha}f_n(\theta)}{(R_0 + r_0)}\cdot e^{ikR_0}\int_{r_{n-1}}^{r_n} e^{ikr}dr\\ &= \frac{2\pi a f_n(\theta)}{ik(R_0 + r_0)}e^{i(\alpha + kR_0)}\left[e^{ikr_n} - e^{ikr_{n-1}}\right].\end{aligned}$$

Now, $r_n = r_0 + n\frac{\lambda}{2}; r_{n-1} = r_0 + (n-1)\frac{\lambda}{2}$

$$\begin{aligned}\therefore \quad (\psi_p)_n &= \frac{2\pi a f_n(\theta)}{ik(R_0 + r_0)}e^{i(\alpha + kR_0)}\left[e^{ik(r_0 + n\lambda/2)} - e^{ik\left(r_0 + \overline{n-1}\lambda/2\right)}\right]\\ &= \frac{2\pi a f_n(\theta)}{ik(R_0 + r_0)}e^{i(\alpha + kR_0)}\cdot e^{ik(r_0 + n\lambda/2)}\left[1 - e^{-ik\lambda/2}\right]\\ &= \frac{2\pi a f_n(\theta)}{ik(R_0 + r_0)}e^{i(\alpha + kR_0)}\cdot e^{ik(r_0 + n\lambda/2)}\left[1 - \cos k\frac{\lambda}{2} + i\sin k\frac{\lambda}{2}\right].\end{aligned}$$

Now, $k = \frac{2\pi}{\lambda} \quad \Rightarrow \quad \frac{k\lambda}{2} = \pi \quad \therefore \quad \cos k\frac{\lambda}{2} = -1, \sin k\frac{\lambda}{2} = 0$

$$\begin{aligned}(\psi_p)_n &= \frac{4\pi a f_n(\theta)}{ik(R_0 + r_0)}e^{i(\alpha + kR_0)}\cdot e^{ik(r_0 + n\lambda/2)}\\ &= \frac{4\pi a f_n(\theta)}{ik(R_0 + r_0)}e^{i[\alpha + k(R_0 + r_0)]}e^{ikn\lambda/2}\\ \Rightarrow \quad (\psi_p)_n &= \frac{4\pi a f_n(\theta)}{ik(R_0 + r_0)}e^{i[\alpha + k(R_0 + r_0)]}\left(\cos kn\frac{\lambda}{2} + i\sin kn\frac{\lambda}{2}\right)\\ &= \frac{4\pi a f_n(\theta)}{k(R_0 + r_0)}e^{i[\alpha + k(R_0 + r_0)]}\left(\sin kn\frac{\lambda}{2} - i\cos kn\frac{\lambda}{2}\right)\\ &= \frac{4\pi a f_n(\theta)}{k(R_0 + r_0)}e^{i[\alpha + k(R_0 + r_0)]}\left(\sin n\pi + i(-1)^{n+1}\right)\\ &= (-1)^{n+1}\frac{4\pi a f_n(\theta)}{k(R_0 + r_0)}e^{i[\pi/2 + \alpha + k(R_0 + r_0)]}, \left\{\sin n\pi = 0, i = e^{i\pi/2}, \cos n\pi = (-1)^n\right\}\end{aligned}$$

$$\therefore \quad |(\psi_p)_n| = \sqrt{(\psi_p)_n(\psi_p)_n^*} = (-1)^{n+1}\cdot\frac{4\pi a f_n(\theta)}{k(R_0 + r_0)} = (-1)^{n+1}\frac{2f_n(\theta)a\lambda}{(R_0 + r_0)} \qquad (5.3.1)$$

The eqn. (5.3.1) shows that amplitude of the disturbance at the point P due to different zones are alternately positive and negative. Thus

$$(\psi_p)_1 = \frac{2f_1(\theta)a\lambda}{(R_0 + r_0)} = \phi_1; (\psi_p)_2 = -\frac{2f_2(\theta)a\lambda}{(R_0 + r_0)} = \phi_2; (\psi_p)_3 = \frac{2f_3(\theta)a\lambda}{(R_0 + r_0)} = \phi_3, \quad \text{etc.}$$

So the resultant amplitude at P due to a number of zones into which the spherical wavefront is divided is given by

$$\psi_p = (\psi_p)_1 + (\psi_p)_2 + (\psi_p)_3 + \cdots = \phi_1 - \phi_2 + \phi_3 - \phi_4 \cdots n \text{ terms.} \tag{5.3.2}$$

Now, $f(\theta)$ is a slowly varying function, whose value decreases with increase in the value of n. So the values of ϕ_1, ϕ_2, ϕ_3, etc., are of decreasing sequence. So the series sum in eqn. (5.3.2) can be computed depending on whether n is even or odd.

When n is Odd

$$\psi_p = \frac{\phi_1}{2} + \left(\frac{\phi_1 + \phi_3}{2} - \phi_2\right) + \cdots + \left(\frac{\phi_{n-2} + \phi_n}{2} - \phi_{n-1}\right) + \frac{\phi_n}{2}.$$

Depending on the variation of $f(\theta)$, we can have

$$\frac{\phi_{j-1} + \phi_{j+1}}{2} = \phi_j \quad \therefore \psi_p = \frac{\phi_1}{2} + \frac{\phi_n}{2}. \tag{5.3.3}$$

If n is Even

$$\psi_p = \frac{\phi_1}{2} + \left(\frac{\phi_1 + \phi_3}{2} - \phi_2\right) + \left(\frac{\phi_3 + \phi_5}{2} - \phi_4\right) + \frac{\phi_5}{2} + \cdots$$
$$+ \left(\frac{\phi_{n-3} + \phi_{n-1}}{2} - \phi_{n-2}\right) + \frac{\phi_{n-1}}{2} - \phi_n.$$

If $\frac{\phi_{j-1}+\phi_{j-3}}{2} = \phi_{j-2}$, we get

$$\psi_p = \frac{\phi_1}{2} + \frac{\phi_{n-1}}{2} - \phi_n. \tag{5.3.4}$$

When n is sufficiently large, both ϕ_n and ϕ_{n-1} become very small compared to ϕ_1, hence, $\Psi_p = \frac{\phi_1}{2}$, i.e., the resultant displacement is equal to half of the displacement due to the first-half-period zone.

When the Wavefront is a Plane Wavefront

From a point source, spherical waves spread out and a spherical wavefront transforms into a plane wavefront, when it has spread to a large distance from the source. We will now investigate the effect of a plane wavefront at a point ahead of it.

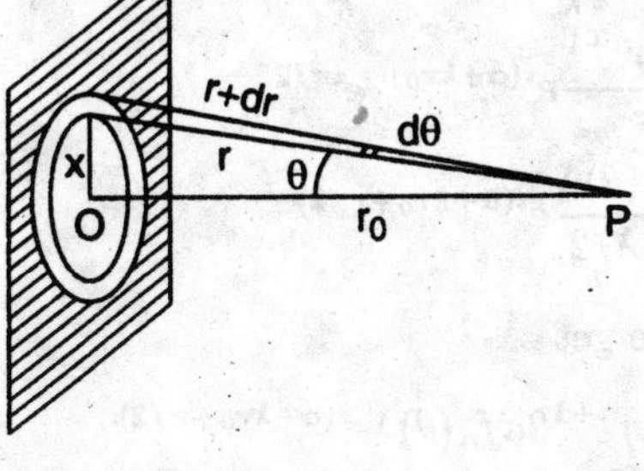

Fig 5.4

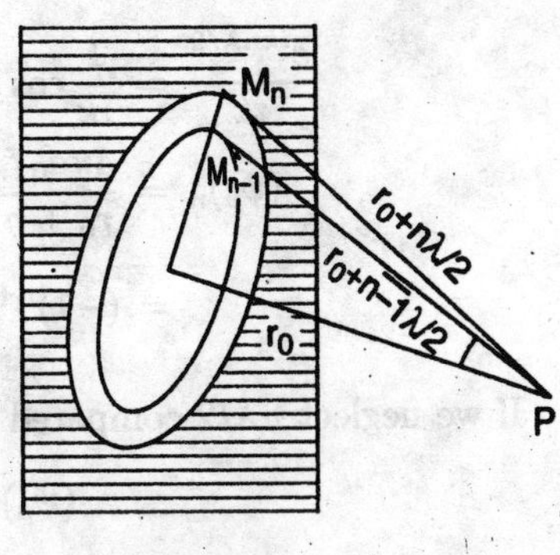

Fig 5.5

The incident vibration may be taken to have amplitude a, then each element of the wavefront vibrates as $ae^{i\alpha}$ and the displacement at P due to the zone between radii r and $r+dr$ is given as

$$d\psi = ae^{i\alpha}.\frac{e^{ikr}}{r}f(\theta)d\sigma$$

where $d\sigma = 2\pi r^2 \sin\theta \, d\theta$

$$\therefore \quad d\psi = ae^{i\alpha}.\frac{e^{ikr}f(\theta)}{r} \times 2\pi r^2 \sin\theta \, d\theta.$$

Now, $\frac{r_0}{r} = \cos\theta \quad \Rightarrow \quad \frac{r_0}{r^2}dr = \sin\theta \, d\theta$

$$\therefore \quad d\psi = 2\pi a r_0 e^{i\alpha} f(\theta)\frac{e^{ikr}}{r}dr.$$

Consider the nth Fresnel zone. The displacement at P due to the nth Fresnel zone is given by

$$(\psi_p)_n = 2\pi a f_n(\theta)e^{i\alpha}r_0 \int_{r_{n-1}}^{r_n} \frac{e^{ikr}}{r}dr.$$

Now, $r_n = r_0 + n\frac{\lambda}{2}; r_{n-1} = r_0 + \overline{n-1}\frac{\lambda}{2}$

$$\therefore \quad (\psi_p)_n = 2\pi a f_n(\theta)e^{i\alpha}r_0 \left[\frac{1}{ik}\frac{e^{ikr}}{r}\bigg|_{r_{n-1}}^{r_n} + \int_{r_{n-1}}^{r_n}\frac{1}{r^2}.\frac{e^{ikr}}{ik}dr\right] \tag{5.3.5}$$

The second term and resulting terms available through integration by parts, will contain higher powers of r (i.e., r^3, r^4, etc.) in the denominator; hence, the first term can be retained in preference to other terms

$$\begin{aligned}\therefore \quad (\psi_p)_n &= \frac{2\pi a r_0 f_n(\theta)}{ik}e^{i\alpha}\left[\frac{e^{ik(r_0+n\lambda/2)}}{r_0+n\lambda/2} - \frac{e^{ik(r_0+\overline{n-1}\lambda/2)}}{r_0+n\lambda/2-\lambda/2}\right] \\ &= \frac{2\pi r_0 a f_n(\theta)}{ik}\frac{e^{i(\alpha+kr_0)}}{r_0+n\lambda/2}.e^{ikn\lambda/2}\left(1-e^{-ik\lambda/2}\right) \\ &= \frac{4\pi a r_0 f_n(\theta)}{r_0+n\lambda/2}e^{i(\alpha+kr_0)}\frac{e^{ikn\lambda/2}}{ik} \quad \text{assuming } (n-1)\tfrac{\lambda}{2} \simeq n\tfrac{\lambda}{2}.\end{aligned}$$

Now,

$$\frac{e^{ikn\lambda/2}}{ik} = \frac{1}{ik}\cos kn\frac{\lambda}{2} + \frac{\sin kn\lambda/2}{k} = i\frac{(-1)}{k}\cos n\pi + 0 = \frac{(-1)^{n+1}}{k}i$$

$$\begin{aligned}\therefore \quad (\psi_p)_n &= \frac{4\pi a_0 f_n(\theta)}{r_0+n\lambda/2}\frac{(-1)^{n+1}}{k}e^{i(\alpha+kr_0)}.e^{i\pi/2} \\ &= (-1)^{n+1}\frac{2ar_0 f_n(\theta)\lambda}{r_0+n\lambda/2}e^{i(\alpha+kr_0+\pi/2)}\end{aligned}$$

If we neglect $n\lambda/2$ compared to r_0, we get

$$(\psi_p)_n = (-1)^{n+1}2af_n(\theta)\lambda e^{i(\alpha+kr_0+\pi/2)}$$

$$\therefore \quad |(\psi_p)_n| = \sqrt{(\psi_p)_n(\psi_p)_n^*} = (-1)^{n+1}2af_n(\theta)\lambda \tag{5.3.6}$$

5.4 Classes of Diffraction

There are two classes of diffraction : (1) *Fraunhofer class,* (2) *Fresnel class.*

In Fraunhofer class of diffraction the source and observation plane are both at infinite distance of separation with respect to the diffracting aperture, and as a result, both the incident wavefront and the diffracted wavefront are planar. But in Fresnel case the source and the observation plane are at a finite distance w.r.t. the diffracting aperture. There may also be a mixture of these two classes, i.e., the source may be at infinite distance w.r.t. the diffracting aperture (Fraunhofer) but the observation plane is at a finite distance (Fresnel). It is also to be noted that in Fraunhofer diffraction, the obliquity factor remains constant over the part of the wavefront utilised for the diffraction, but this is not the case with Fresnel diffraction.

5.5 Kirchhoff's Fresnel Integral

Fresnel integral that we have just seen, though explains the diffraction intensity in some simple cases very well, leaves scope for criticism so far as the phase condition of the disturbance, obliquity factor, etc., are concerned. Kirchhoff, purely on the basis of the wave equation, derived an equation involving an integral, called *Kirchhoff-Fresnel Integral* which evaluates the displacement at any point due to a wavefront in correct phase and magnitude. This equation is also called *Kirchhoff-Fresnel Diffraction Integral.*

This integral can be evaluated very conveniently to get the correct solution of diffraction problems.

The formulation of the diffraction integral is nothing but the solution of the homogeneous wave equation written as

$$\nabla^2\phi - \frac{l}{c^2}\frac{\partial^2\phi}{\partial t^2} = 0 \tag{5.5.1}$$

where $\nabla^2 \equiv$ Laplacian operator $= \frac{\partial^2}{\partial x^2} + \frac{\partial^2}{\partial y^2} + \frac{\partial^2}{\partial z^2}$ in Cartesian co-ordinates, $\phi(x, y, z, t)$ is a space and time-dependent function representing some disturbance.

If we try to solve eqn. (5.5.1) by the method of separation of variables, we put

$$\phi = \psi(x, y, z)\xi(t)$$

$$\frac{\partial^2\phi}{\partial x^2} + \frac{\partial^2\phi}{\partial y^2} + \frac{\partial^2\phi}{\partial z^2} = \left(\frac{\partial^2\psi}{\partial x^2} + \frac{\partial^2\psi}{\partial y^2} + \frac{\partial^2\psi}{\partial z^2}\right)\xi(t); \quad \frac{\partial^2\phi}{\partial t^2} = \psi\frac{d^2\xi}{dt^2}$$

$$\therefore \quad \nabla^2\psi\xi(t) = \frac{\psi}{c^2}\frac{d^2\xi}{dt^2} \quad \Rightarrow \quad \frac{c^2}{\psi}\nabla^2\psi = \frac{1}{\xi}\frac{d^2\xi}{dt^2} \tag{5.5.2}$$

The left-hand side of eqn. (5.5.2) is purely space-dependent part, while the right-hand side is a time-dependent function, hence we can write

$$\frac{c^2}{\psi}\left(\frac{\partial^2\psi}{\partial x^2} + \frac{\partial^2\psi}{\partial y^2} + \frac{\partial^2\psi}{\partial z^2}\right) = \frac{1}{\xi}\frac{d^2\xi}{dt^2} = -\omega^2 \tag{5.5.3}$$

where ω^2 is a constant independent of space and time.

$$\therefore \quad \frac{1}{\psi}\left(\frac{\partial^2\psi}{\partial x^2} + \frac{\partial^2\psi}{\partial y^2} + \frac{\partial^2\psi}{\partial z^2}\right) = -\frac{\omega^2}{c^2} \tag{5.5.4}$$

$$\frac{1}{\xi}\frac{d^2\xi}{dt^2} = -\omega^2 \tag{5.5.5}$$

Now, if we set $\frac{\omega^2}{c^2} = k^2$ and $\omega = 2\pi\nu$, then c being equal to $\nu\lambda$, we get

$$k^2 = \frac{\omega^2}{c^2} = \frac{4\pi^2\nu^2}{\nu^2\lambda^2} = \frac{4\pi^2}{\lambda^2} \quad \Rightarrow \quad k = \frac{2\pi}{\lambda}, \text{ called the } \textit{propagation constant.}$$

The space part satisfies the eqn. (5.4) given as

$$\nabla^2\psi = -\frac{\omega^2}{c^2}\psi = -k^2\psi. \tag{5.5.6}$$

Consider an incoming wave satisfying eqn. (5.5.6) and we are to find its effect at some point P. Let P be within a volume V (fig. 5.6) surrounded by the surface S, and the function ψ possesses continuous first and second order derivatives within and on the surface. Also there is another function ψ' possessing properties similar to those of ψ, then by applying Green's theorem we get

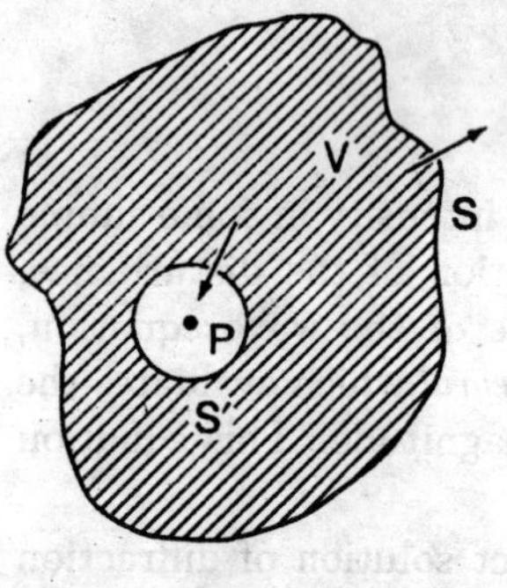

Fig 5.6

$$\iiint (\psi'\nabla^2\psi - \psi\nabla^2\psi')\,dv = \iint \left(\psi'\frac{\partial\psi}{\partial n} - \psi\frac{\partial\psi'}{\partial n}\right) ds. \tag{5.5.7}$$

If ψ' also satisfies the time-independent wave equation, then

$$\Delta^2\psi' = -k^2\psi'.$$

If we take ψ' to be the spherically symmetric solution of the wave equation, then

$$\psi' = \frac{e^{-ikr}}{r}$$

where r is the radial distance measured from P. There is singularity of ψ' at $P(r = 0)$, so P is to be excluded from the domain of integration. Consider a small spherical surface S' of radius r_0 about the point P. The volume V is now bounded externally by S and internally by S' and the unit vectors outward from S and S' are $\bar{n}$ and $\bar{n}'(\bar{n} = -\bar{n}')$, see fig. 5.6. Then eqn. (5.5.7) can be written as

$$\iiint_V (\psi'\nabla^2\psi - \psi\nabla^2\psi')\,dv = \iint_S \left(\psi'\frac{\partial\psi}{\partial n} - \psi\frac{\partial\psi'}{\partial n}\right) ds + \iint_{S'} \left(\psi'\frac{\partial\psi}{\partial n'} - \psi\frac{\partial\psi'}{\partial n'}\right) ds. \tag{5.5.8}$$

Since, $\nabla^2\psi = -k^2\psi$ and $\nabla^2\psi' = -k^2\psi'$, the integrand in the left-hand side of eqn. (5.5.8) reduces to zero at every point within the volume v, hence

$$\iint \left(\psi'\frac{\partial\psi}{\partial n} - \psi\frac{\partial\psi'}{\partial n}\right) ds = -\iint_{S'} \left(\psi'\frac{\partial\psi}{\partial n'} - \psi\frac{\partial\psi'}{\partial n'}\right) ds.$$

Again, $\frac{\partial}{\partial n} = -\frac{\partial}{\partial n'}$ because of the direction of unit vectors.

$$\therefore \quad \iint_S \left(\psi'\frac{\partial\psi}{\partial n} - \psi\frac{\partial\psi'}{\partial n}\right) ds = -\iint_{S'} \left(-\psi'\frac{\partial\psi}{\partial n} + \psi\frac{\partial\psi'}{\partial n}\right) ds$$

$$\Rightarrow \iint_S \left[\frac{e^{-ikr}}{r}\frac{\partial\psi}{\partial n} - \psi\frac{\partial}{\partial n}\left(\frac{e^{-ikr}}{r}\right)\right] ds$$

$$= \iint_{S'} \left[\frac{e^{-ikr}}{r}\frac{\partial\psi}{\partial n} - \psi\frac{\partial}{\partial n}\left(\frac{e^{-ikr}}{r}\right)\right] ds$$

$$= \iint_{S'} \left[\frac{e^{-ikr}}{r}\left(-\frac{\partial\psi}{\partial r}\right) + \psi\frac{\partial}{\partial r}\left(\frac{e^{-ikr}}{r}\right)\right] ds, \quad \because \frac{\partial}{\partial n} = -\frac{\partial}{\partial r}$$

(So far as the surface S' is concerned.)

$$= \iint_{S'} \left[\frac{e^{-ikr}}{r}\left(-\frac{\partial\psi}{\partial r}\right) + \psi\frac{\partial}{\partial r}\left(\frac{e^{-ikr}}{r}\right)\right] ds.$$

$$= \iint_{S'} \left[\frac{e^{-ikr}}{r}\left(-\frac{\partial\psi}{\partial r}\right) + \psi\left(-ik\frac{e^{-ikr}}{r} - \frac{e^{-ikr}}{r^2}\right)\right] ds$$

$$= \int_{\omega} \left[re^{-ikr}\left(-\frac{\partial\psi}{\partial r}\right) - \psi ikre^{-ikr} + \psi e^{-ikr}\right] d\omega$$

where $d\omega = r^2 ds$ is the solid angle subtended by ds on S' at P. The last term in the integrand is different from zero as $r \to 0$. If the surface S' collapses round P, the value of ψ on, and inside S' is a constant and equal to ψ_p. Then, the right-hand side can be written as

$$\lim_{r\to 0}\int \psi e^{-ikr} d\omega = +\psi_p \int_{r\to 0} e^{-ikr} d\omega = +\psi_p \int d\omega = +\psi_p \times 4\pi$$

because total solid angle about a point is 4π.

$$\therefore \quad 4\pi\psi_p = +\iint_S \left[\frac{e^{-ikr}}{r}\frac{\partial\psi}{\partial n} - \psi\frac{\partial}{\partial n}\left(\frac{e^{-ikr}}{r}\right)\right] ds$$

$$\Rightarrow \quad \psi_p = \frac{1}{4\pi}\iint_S \left[\frac{e^{-ikr}}{r}\frac{\partial\psi}{\partial n} - \psi\frac{\partial}{\partial n}\left(\frac{e^{-ikr}}{r}\right)\right] ds$$

$$= \frac{1}{4\pi}\iint_S \left[\frac{e^{-ikr}}{r}\frac{\partial\psi}{\partial n} - \psi e^{-ikr}\frac{\partial}{\partial n}\left(\frac{1}{r}\right) + ik\psi\frac{e^{-ikr}}{r}\frac{\partial r}{\partial n}\right] ds$$

$$\therefore \quad \phi_p = \psi_p e^{ikct}$$

$$= \frac{1}{4\pi}\iint_S \left[\frac{e^{ik(ct-r)}}{r}\frac{\partial\psi}{\partial n} - \psi e^{ik(ct-r)}\frac{\partial}{\partial n}\left(\frac{1}{r}\right) + \frac{ik\psi}{r}e^{+ik(ct-r)}\frac{\partial r}{\partial n}\right] ds$$

$$= \frac{1}{4\pi}\iint_S \left[e^{ikc(t-r/c)}\frac{\partial\psi}{\partial n} - \psi e^{ikc(t-r/c)}\frac{\partial}{\partial n}\left(\frac{1}{r}\right) + \frac{ik\psi}{r}e^{+ikc(t-r/c)}\frac{\partial r}{\partial n}\right] ds \quad (5.5.9)$$

Now, $\psi e^{ikc(t-r/c)} = \psi e^{ikct'} = \phi\big|_{t=t'}$, so it represents the value of ϕ at the element ds at time $t' = t - \frac{r}{c}$, i.e., $\frac{r}{c}$ seconds earlier than the instant it arrives at P. So it may be denoted as $\{\phi\}_{t-r/c}$.

In the same way we can write

$$\frac{e^{ikc(t-r/c)}}{r}\frac{\partial\psi}{\partial n} = \frac{1}{r}\left\{\frac{\partial\phi}{\partial n}\right\}_{t-r/c}$$

$$\frac{ik\psi}{r}e^{ikc(t-r/c)} = \frac{1}{cr}\frac{\partial}{\partial t}\left\{\psi e^{ikc(t-r/c)}\right\} = \frac{1}{cr}\frac{\partial}{\partial n}\{\phi\}_{t-r/c} = \frac{1}{cr}\left\{\frac{\partial\phi}{\partial t}\right\}_{t-r/c}.$$

With these notations, eqn. (5.5.9) takes the form

$$\psi_p = \frac{1}{4\pi}\iint_S \left[\frac{1}{r}\left\{\frac{\partial\phi}{\partial n}\right\}_{t-r/c} - \frac{\partial}{\partial n}\left(\frac{1}{r}\right)\{\phi\}_{t-r/c} + \frac{1}{cr}\left\{\frac{\partial\phi}{\partial t}\right\}_{t-r/c}\frac{\partial r}{\partial n}\right] ds \qquad (5.5.10)$$

All the quantities within the brace { } refer to the retarded value. Eqn. (5.5.10) is known as *Kirchhoff-Fresnel diffraction integral.* This integral when evaluated reproduces disturbance at some point in proper phase and amplitude. The obliquity factor which was introduced on an ad hoc basis in the form of $f_n(\theta)$ in Fresnel's consideration has been theoretically established.

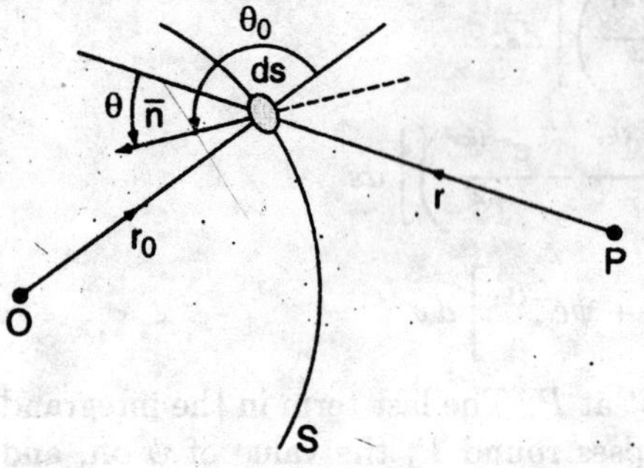

Fig 5.7

Let us consider any surface S enclosing the source O; $\bar{n}$ is the unit outward drawn normal vector from the side of the field point P. Then

$$\frac{\partial r_0}{\partial n} = \cos\theta_0, \frac{\partial r}{\partial n} = \cos\theta,$$

where θ_0 and θ are the angles as indicated in fig. 5.7. Let us take a spherical wave given by

$$\phi = \frac{A}{r_0}e^{ik(ct-r_0)} = \frac{A}{r_0}e^{ikct}.\, e^{-ikr_0} = \psi e^{ikct}$$

$$\therefore \quad \psi = \frac{A}{r_0}e^{-ikr_0} \quad \text{(space part of the wave).}$$

$$\frac{\partial\psi}{\partial n} = \frac{\partial\psi}{\partial r_0}\cdot\frac{\partial r_0}{\partial n} = \cos\theta_0\frac{\partial\psi}{\partial r_0} = \cos\theta_0\left[-\frac{A}{r_0^2}e^{-ikr_0} - ik\frac{A}{r_0}e^{-ikr_0}\right], \text{ also } \frac{\partial r}{\partial n} = \cos\theta.$$

With these values we at once get

$$\frac{1}{r}e^{ik(ct-r)}\frac{\partial\psi}{\partial n} = -\frac{A\cos\theta_0}{rr_0}e^{+ik(ct-\overline{r+r_0})}\left(\frac{1}{r_0}+ik\right)$$

$$\psi e^{ik(ct-r)}\frac{\partial}{\partial n}\left(\frac{1}{r}\right) = \frac{A}{r_0}e^{ik(ct-\overline{r+r_0})}\left(-\frac{1}{r^2}\right)\cos\theta.$$

So eqn. (5.5.9) can be written as

$$\phi_p = \frac{1}{4\pi}\iint_S \left[-\frac{A\cos\theta_0}{rr_0}e^{+ik(ct-\overline{r+r_0})}\left(\frac{1}{r_0}+ik\right)\right.$$

$$\left. + \frac{A}{r_0 r^2}e^{ik(ct-\overline{r+r_0})}\cos\theta + ik\frac{A}{rr_0}e^{ik(ct-\overline{r+r_0})}\cos\theta\right] ds$$

$$= \frac{1}{4\pi}\iint_S \frac{A}{rr_0}e^{ik(ct-\overline{r+r_0})}\left\{-\left(ik+\frac{1}{r_0}\right)\cos\theta_0 + \cos\theta\left(ik+\frac{1}{r}\right)\right\} ds$$

Now, $k = \frac{2\pi}{\lambda}$, so if r and r_0 both $\gg \lambda$, $\frac{1}{r_0}$ and $\frac{1}{r}$ both $\ll \frac{1}{\lambda}$, hence

$$\left\{\left(ik + \frac{1}{r}\right)\cos\theta - \left(ik + \frac{1}{r_0}\right)\cos\theta_0\right\} \to ik(\cos\theta - \cos\theta_0)$$

$$\phi_p = \frac{1}{4\pi}\iint_S \frac{Aik}{rr_0} e^{ik(ct - \overline{r + r_0})}(\cos\theta - \cos\theta_0)ds. \tag{5.5.11}$$

Eqn. (5.5.11) is the standard diffraction equation, called *Kirchhoff-Fresnel diffraction equation.* All diffraction problems can be solved using this Kirchhoff-Fresnel integral.

Application of Kirchhoff-Fresnel Integral

Example 1

Consider a surface S containing an aperture. In fig. 5.8, $\cos\theta_0 = \cos(180 - \theta') = -\cos\theta'$.

If $\theta = \theta'$, then

$$\psi_p = \frac{1}{4\pi}\iint_S \frac{Aik}{rr_0} e^{ik(-\overline{r + r_0})}(\cos\theta + \cos\theta')ds$$

$$= \frac{Aik}{4\pi}\iint_S \frac{1}{r^2} e^{-ik2r} \cdot 2\cos\theta ds$$

$$= \frac{Ai}{\lambda}\iint_S \frac{1}{r^2} e^{-ik2r}\cos\theta ds \text{ if } r = r_0.$$

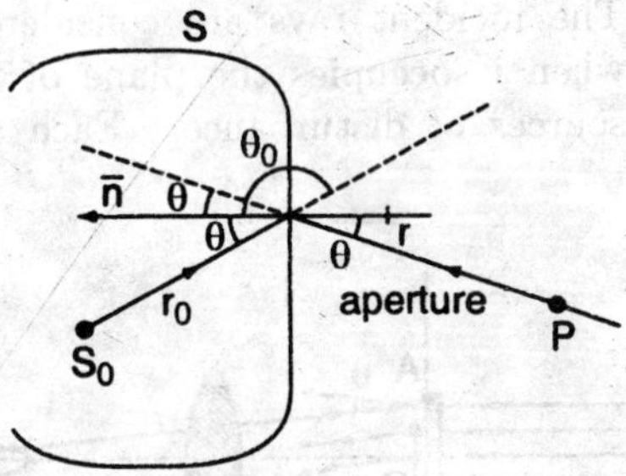

Fig 5.8

More applications are given with the respective diffraction patterns.

Fraunhofer Diffraction

Single-Slit Diffraction Pattern
Double-Slit Diffraction Pattern
Diffraction Pattern of N-Slits—Diffraction Grating

5.6 Introduction

In the previous section we have considered the Kirchhoff-Fresnel diffraction integral which reproduces the correct phase and obliquity factor of the disturbance at some point due to an advancing wavefront. We have also seen the effect of a wavefront when it falls on an aperture on a screen. We have said that there are two classes of diffraction phenomena—one is called *Fraunhofer,* the other is called *Fresnel diffraction.* In Fraunhofer class, the source and the observation plane are both at infinite distance from the diffracting aperture, but in Fresnel class, one or both are at finite distance.

It is also to remember that the diffraction pattern available on the screen is due to interference of waves. What is then the difference between interference phenomenon and diffraction phenomenon? We should bear in our mind that though the diffraction intensity is obtained through constructive and destructive interference, diffraction pattern is due to interference of secondary waves, whereas the interference phenomenon is due to superposition of primary waves coming from two coherent sources. Structurally, the interference maxima are all of equal intensity and width. Whereas the diffraction central maximum is most intense and wide, the secondary maxima are less intense and of less width than the central maximum.

The Kirchhoff-Fresnel integral may be utilised to study the intensity pattern in case of both classes of diffraction phenomena. We will first consider the diffraction phenomena belonging to the Fraunhofer class, then we will take up the problem of Fresnel class.

5.7 Diffraction Due to a Single Slit

A narrow slit is illuminated by monochromatic light of wavelength λ; O is the centre of the slit, and points symmetrically situated w.r.t. the centre, will be called *corresponding points.* The incident rays are considered to be parallel, and hence the wavefront is planar, and when it occupies the plane of the slit, all the particles in the aperture become secondary sources of disturbance. Each such vibrating particle sends out waves, some disturbances continue to move in the original direction, some are deviated or diffracted at some angle. The secondary disturbance undeviated will be brought to a point of focus by the lens, and this forms the central maximum; those diffracted through a certain angle θ (say) will form another parallel beam which will be brought to another point of focus on the screen. Now, of the parallel rays diffracted at an angle θ, the point of focus will be such that some of these disturbances will reinforce, some will destroy. Let us consider a very small length dy of the slit whose linear dimension is large compared to the width.

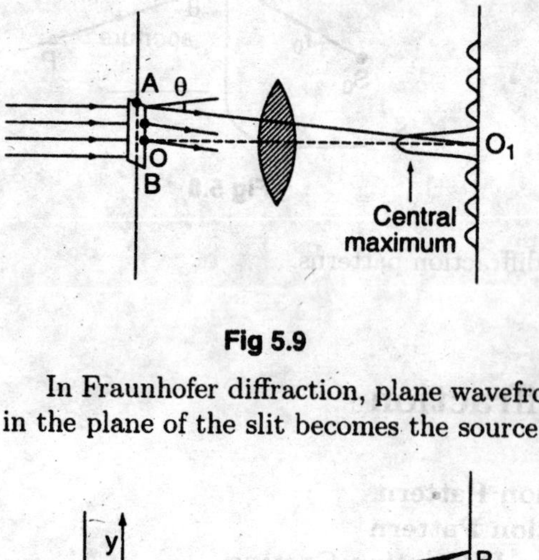

Fig 5.9

In Fraunhofer diffraction, plane wavefront is incident on the slit and each vibrating particle in the plane of the slit becomes the source of secondary disturbances.

As the wavefront is planar, the energy passing through unit area of the wavefront per sec is constant, and so the intensity C^2 associated with the incident wavefront is remaining constant over the entire slit aperture. So the vibration of the secondary sources at the origin (and the mid-point of the slit) and also at all points in the slit can be written as

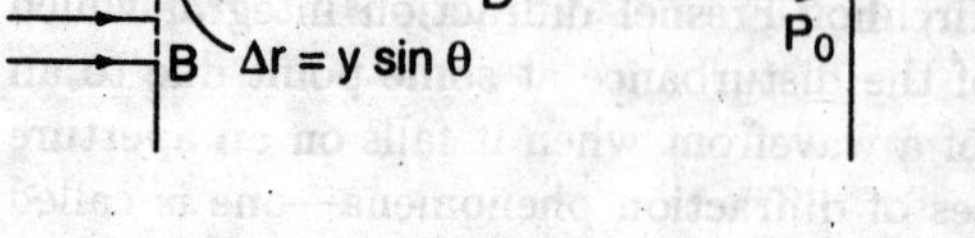

Fig 5.10

$$\phi = \left(\frac{C}{a}\right) e^{i\omega t}$$

where $\frac{C}{a}$ is the amplitude of vibration of the secondary source particle at the origin.

If we consider a linear element dy at a distance y, the displacement at P, distant r_1 from the source, is given by

$$d\phi_p = \frac{(C/a)}{r_1} e^{i(\omega t - kr_1)} dy. \qquad (5.7.1)$$

Now, $r = r_1 + y\sin\theta$. Since y and $\sin\theta$ are both small, the factor r_1 in the denominator can be replaced by r but not in the exponential factor.

$$\therefore \; d\phi_p = \frac{(C/a)}{r} e^{i(\omega t - kr)} \cdot e^{+iky\sin\theta} dy \tag{5.7.2}$$

$$\therefore \; \text{Total displacement } \phi_p = \frac{(C/a)}{r} e^{i(\omega t - kr)} \int_{-a/2}^{a/2} +e^{iky\sin\theta} dy.$$

Put $k\sin\theta = \frac{2\pi}{\lambda}\sin\theta = \beta$.

$$\therefore \; \int_{-a/2}^{a/2} +e^{+i\beta y} dy = +\frac{1}{i\beta} e^{+i\beta y}\Big|_{-a/2}^{a/2} = \frac{1}{i\beta}\left(e^{+i\beta a/2} - e^{-i\beta a/2}\right).$$

Now

$$e^{i\beta a/2} = \cos\left(\frac{\beta a}{2}\right) + i\sin\left(\beta\frac{a}{2}\right); e^{-i\beta a/2} = \cos\left(\frac{\beta a}{2}\right) - i\sin\left(\beta\frac{a}{2}\right)$$

$$\therefore \; e^{i\beta a/2} - e^{-i\beta a/2} = 2i\sin\left(\beta\frac{a}{2}\right) \quad \Rightarrow \quad \frac{e^{i\beta a/2} - e^{-i\beta a/2}}{2i} = \sin\left(\beta\frac{a}{2}\right)$$

$$\therefore \; \int_{-a/2}^{a/2} e^{-i\beta y} dy = \frac{2\sin\left(\beta\frac{a}{2}\right)}{\beta} = a\frac{\sin\left(\beta\frac{a}{2}\right)}{\left(\beta\frac{a}{2}\right)}$$

$$\therefore \; \phi_p = \frac{(C/a)}{r} e^{i(\omega t - kr)} a \cdot \frac{\sin\alpha}{\alpha} = \frac{C}{r} e^{i(\omega t - kr)} \frac{\sin\alpha}{\alpha} \tag{5.7.3}$$

Its complex conjugate ϕ_p^* is given by

$$\phi_p^* = \frac{C}{r} e^{-i(\omega t - kr)} \frac{\sin\alpha}{\alpha}$$

where $\alpha = \beta\frac{a}{2} = k\sin\theta \times \frac{a}{2} = \pi a\frac{\sin\theta}{\lambda}$. The intensity at any point P due to wavelets diffracted at an angle θ is, therefore, given by

$$I_p = \phi_p \cdot \phi_p^* = \frac{C^2}{r^2}\frac{\sin^2\alpha}{\alpha^2}.$$

Now, when $\alpha \to 0$, $\lim_{\alpha\to 0} \frac{\sin\alpha}{\alpha} = 1$.

$$\therefore \; \lim_{\alpha\to 0} I_p = I_0 = \frac{C^2}{r^2} \simeq \frac{C^2}{D^2}.$$

The distance of the receiving screen is so large compared to the dimension of the slit that in the region of the screen receiving diffraction pattern we can consider $r \simeq D$ at all points of observation.

$$\therefore \; I_p = I_0\frac{\sin^2\alpha}{\alpha^2} = I_0\frac{\sin^2\left(\frac{\pi a\sin\theta}{\lambda}\right)}{\left(\frac{\pi a\sin\theta}{\lambda}\right)} \tag{5.7.4}$$

Eqn. (5.7.4) shows that the intensity depends on the angle of diffraction θ. Eqn. (5.7.4) may be termed as single-slit diffraction intensity function. It has maximum value equal to I_0

corresponding to $\alpha = 0 \Rightarrow \theta = 0$. Eqn. (5.7.4) can be put in another form w.r.t. fig. 5.10 $\sin\theta \simeq \theta \simeq \tan\theta = \frac{Y}{D}$.

$$\therefore \quad \frac{\pi a \sin\theta}{\lambda} = \frac{2\pi}{\lambda}\cdot\left(\frac{a}{2}\right)\left(\frac{Y}{D}\right)$$

$$\therefore \quad I_p = I_0\left[\frac{\sin\{2\pi/\lambda\cdot(a/2)(Y/D)\}}{2\pi/\lambda\cdot(a/2)(Y/D)}\right]^2 \tag{5.7.5}$$

5.8 Derivation of Single-Slit Intensity Function from Kirchhoff-Fresnel Integral

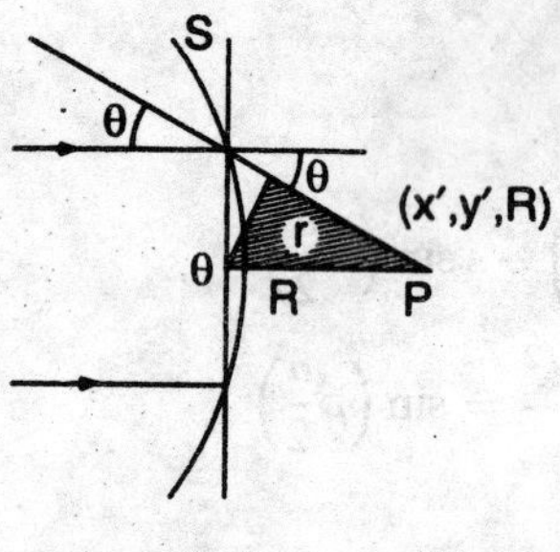

Fig 5.11

The single-slit intensity function can be deduced by using Kirchhoff-Fresnel integral. Let us consider a surface S containing the aperture now

$$\begin{aligned}\phi_p &= \frac{1}{4\pi}\iint \frac{Aik}{rr_0}e^{ik(ct-\overline{r+r_0})}(\cos\theta - \cos\theta_0)ds \\ &= \frac{Ak}{4\pi}\iint_{\text{aperture}} \frac{1}{rr_0}e^{ik(ct-r_0)}\cdot e^{-ikr}(\cos\theta-\cos\theta_0)ds\cdot e^{i\pi/2} \\ &= \frac{Ak}{4\pi}\iint_{\text{aperture}} \frac{1}{rr_0}e^{ik(ct-r_0+\pi/2)}e^{-ikr}(\cos\theta-\cos\theta_0)ds\end{aligned}$$

$\frac{1}{rr_0}e^{ik(ct-r_0)}$ may be taken to be a constant so far as the integration is concerned, as also the inclination factor.

$$\therefore \quad \phi_p = \text{Constant}\ \frac{Ak}{4\pi}\iint e^{-ikr}ds. \tag{5.8.1}$$

Now, $ds = dx\ dy$.

Let us consider $ABCD$ as representing aperture having $AB = b$ and $AD = a$. Consider a surface S which contains the aperture and at some instant, the plane wavefront falls on the aperture, and then $\iint$ would mean integration over the aperture $ABCD$.

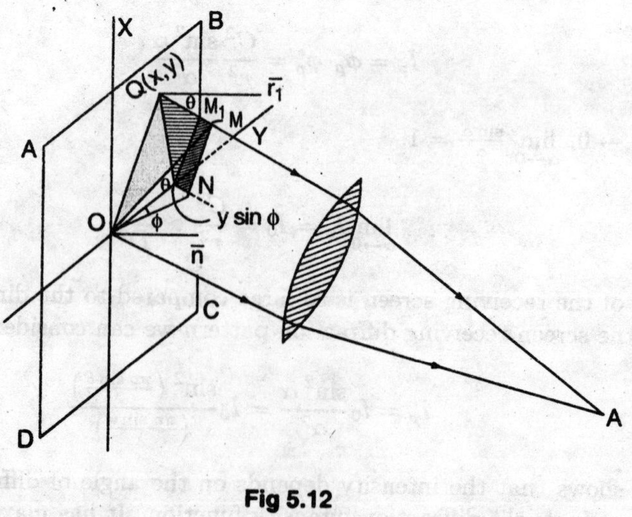

Fig 5.12

Now w.r.t. the centre O of the aperture, the field point P is at a distance R; Q is a point on the plane $ABCD$ and its co-ordinates are (x, y) such that $QP = r$. From O drop a $\perp$ on the line QP, the foot of the perpendicular being M, so

$$r = MP + QM \simeq OP + QM = R + QM.$$

From fig. 5.12 which is self-explanatory, we get $MM_1 = NQ' = y\sin\phi$, where ϕ is $\angle Q'ON$. Also, $QM_1 = x\sin\theta$ where θ is the angle of diffraction.

$$\therefore\ QM = MM_1 + M_1Q = y\sin\phi + x\sin\theta$$
$$\therefore\ r = R + x\sin\theta + y\sin\phi$$

ds is an elementary area at $Q(x, y)$. So eqn. (5.8.1) can be written as

$$\begin{aligned}\phi_p &= \text{Constant} \times \frac{Ak}{4\pi}\int e^{-ik(R+x\sin\theta+y\sin\phi)}dxdy \\ &= \text{Constant} \times \frac{A}{2\lambda}e^{-ikR}\int_{-a/2}^{a/2} e^{-ikx\sin\theta}dx\int_{-b/2}^{b/2} e^{-iky\sin\phi}dy \\ &= C_1 \times \frac{A}{2\lambda}\int_{-a/2}^{a/2} e^{-ikx\sin\theta}dx\int_{-b/2}^{b/2} e^{-iky\sin\phi}dy \qquad (5.8.2)\end{aligned}$$

The width of the slit is so small compared to the linear dimension (it is a linear slit) that the second integral in eqn. (5.8.2) practically remains constant over the domain of integration and so in this case, we get

$$\begin{aligned}\phi_p &= C_2 \times \frac{A}{2\lambda}\left[-\frac{1}{ik\sin\theta}e^{-ikx\sin\theta}\Big|_{-a/2}^{a/2}\right] \text{ where } C_2 = C_1\int_{-b/2}^{b/2} e^{-iky\sin\phi}dy \\ &= C_2 \times \frac{A}{2\lambda}\left[-\frac{1}{ik\sin\theta}\left(e^{-ik\sin\theta\times a/2} - e^{ik\sin\theta\times a/2}\right)\right] \\ &= C_2\frac{A}{2\lambda}\left[\frac{e^{ik\sin\theta\times a/2} - e^{-ik\sin\theta\times a/2}}{ik\sin\theta}\right]\end{aligned}$$

$$k\sin\theta \times a/2 = \frac{2\pi}{\lambda}\sin\theta \times a/2 = \frac{\pi a\sin\theta}{\lambda} = \alpha \Rightarrow k\sin\theta = \frac{2\pi}{\lambda}\sin\theta = \frac{2\alpha}{a}$$

$$\begin{aligned}\therefore\quad \phi_p &= C_2 \times \frac{A}{2\lambda}\left[\frac{(e^{i\alpha} - e^{-i\alpha})}{i2\alpha/a}\right] \\ &= C_2 \times \frac{A}{2\lambda}\cdot a\left[\frac{e^{i\alpha} - e^{-i\alpha}}{i2\alpha}\right] = C_2\frac{A}{2\lambda}a\cdot\frac{\sin\alpha}{\alpha}. \qquad (5.8.3)\end{aligned}$$

Now

$$C_1 = \frac{e^{ik(ct-r_0+\frac{\pi}{2})}}{rr_0}\cdot e^{-ikR}(\cos\phi - \cos\phi_0)$$

$$C_2 = C_1 \times \int_{-b/2}^{b/2} e^{-iky\sin\phi}dy = C_1 \times \frac{\sin\beta}{\beta}$$

where $\beta = \frac{\pi b \sin\phi}{\lambda} = C_1 \times 1$ because $\beta \to 0$ as $b \to 0$

$$|\phi_p|^2 = \phi_p \cdot \phi_p^* = \frac{A^2}{4\lambda^2} a^2 \frac{\sin^2\alpha}{\alpha^2} \cdot \frac{1}{r^2 r_0^2} (\cos\phi - \cos\phi_0)^2$$

$$\phi_p = \frac{Aa}{2\lambda} \frac{\sin\alpha}{\alpha} \cdot \frac{1}{rr_0} (\cos\phi - \cos\phi_0)$$

$$\therefore\quad I = \phi_p^2 = \frac{A^2 a^2}{4\lambda^2 r^2 r_0^2} (\cos\phi - \cos\phi_0)^2 \frac{\sin^2\alpha}{\alpha^2} \tag{5.8.4}$$

$I \to I_{max} = I_0$ when $\alpha \to 0 \Rightarrow \theta \to 0$.

$$\therefore\quad I_0 = \frac{A^2 a^2}{4\lambda^2 r^2 r_0^2} (\cos\phi - \cos\phi_0)^2$$

$$\therefore\quad I = I_0 \frac{\sin^2\alpha}{\alpha^2} = I_0 \frac{\sin^2\left(\frac{\pi a \sin\theta}{\lambda}\right)}{\left(\frac{\pi a \sin\theta}{\lambda}\right)^2} \tag{5.8.5}$$

Thus we get the same intensity function.

5.9 Phenomenon of Diffraction can be looked upon as Fourier Transform of a Suitable Slit Function

We have derived the diffraction eqn. (5.7.2) due to diffraction in a single-slit. The equation gives us

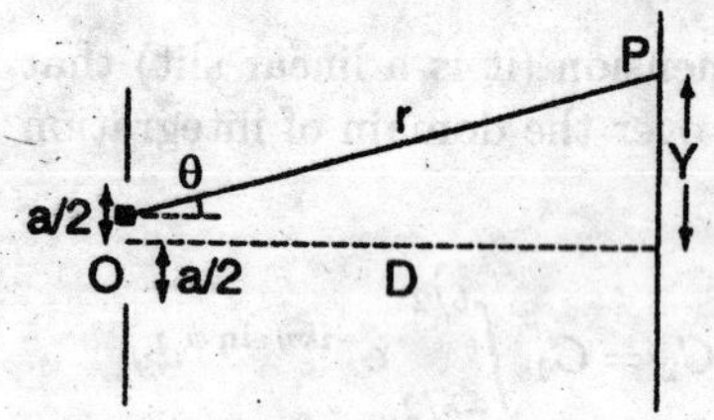

Fig 5.13(a)

$$d\phi_p = \frac{(C/a)}{r} e^{i(\omega t - kr)} e^{iky\sin\theta} dy.$$

For small angle diffraction, we can write (cf. fig. 5.13(a)), $\tan\theta \simeq \theta = \sin\theta = \frac{Y}{D} = \frac{Y}{r}$.

$$\therefore\quad d\phi_p = \frac{C}{aD} e^{i(\omega t - kr)} e^{ikyY/D} dy.$$

So the total displacement at P can be written as

$$\phi_p(Y) = \bar{C} \int_{-a/2}^{a/2} e^{ikyY/D} dy, \bar{C} = \frac{C}{aD} e^{i(\omega t - kr)}.$$

If we define a function called *slit function* $s(y)$ such that

$$s(y) = \begin{cases} 1, & \text{for } -\frac{a}{2} \le y \le \frac{a}{2} \\ 0, & \text{elsewhere} \end{cases}$$

Then

$$\phi_p(Y) = \bar{C} \int_{-\infty}^{\infty} e^{ikyY/D} s(y) dy. \tag{5.9.1}$$

In this formulation, therefore, $\phi_p(Y)$ is the Fourier Transform of the slit function $s(y)$ and is called the *displacement function*. Then the amplitude function $\psi_p(Y)$ can be obtained as

$$\psi_p(Y) = \int_{-\infty}^{\infty} e^{ikyY/D} s(y) dy = \tilde{F}[s(y)]. \tag{5.9.2}$$

So, if $s(y)$ is defined, we can calculate its Fourier Transform to get the amplitude function $\psi_p(Y)$. Reciprocally, if $\psi_p(Y)$ is given, its corresponding slit function can be obtained by inverse transformation given by

$$s(y) = \int_{-\infty}^{\infty} \psi_p(Y) e^{-i2\pi yY/D} dy.$$

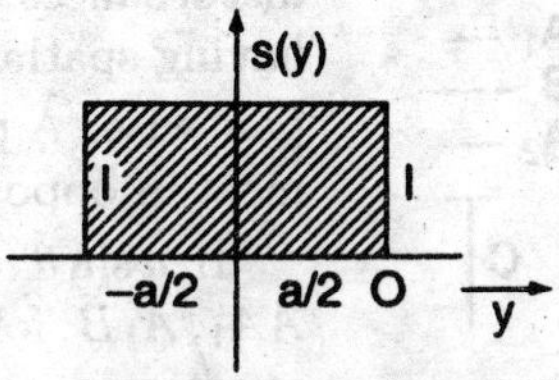

Fig 5.13(b)

In case of a single-slit, the slit function $s(y)$ is a unit step function defined as above and represented as in fig. 5.13(b). So the amplitude function is given as the F.T. of $s(y)$ and written as

$$\psi_p(Y) = \tilde{F}[s(y)] = \int_{-\infty}^{\infty} s(y) e^{ikyY/D} dy$$

$$= \int_{-\infty}^{-a/2} s(y) e^{ikyY/D} dy + \int_{-a/2}^{a/2} s(y) e^{ikyY/D} dy + \int_{a/2}^{\infty} s(y) e^{ikyY/D} dy$$

$$= 0 + \frac{e^{ikyY/D}\Big|_{-a/2}^{a/2}}{ikY/D} + 0 \text{ the first and the last integrals are zero}$$

because $s(y)$ is a step function

$$= \frac{e^{ikaY/2D} - e^{-ikaY/2D}}{ik\frac{Y}{D}} = \frac{2\sin\frac{ka}{2}\frac{Y}{D}}{k\frac{Y}{D}} = \frac{2\sin\left(\frac{\pi a \sin\theta}{\lambda}\right)}{\left(\frac{2\pi\sin\theta}{\lambda}\right)} = a\frac{\sin\alpha}{\alpha}$$

$$\therefore \quad I_p = I_0 \frac{\sin\alpha}{\alpha}. \tag{5.9.3}$$

5.10 Discussion on the Intensity Function

$$I_p = I_0 \frac{\sin\alpha}{\alpha^2}$$

So the maximum or minimum is given by

$$\frac{dI_p}{d\alpha} = 0 = I_0\left(\frac{1}{\alpha^2} 2\sin\alpha\cos\alpha - \frac{2\sin^2\alpha}{\alpha^3}\right)$$

$$\Rightarrow \sin\alpha\left(\cos\alpha - \frac{\sin\alpha}{\alpha}\right) = 0 \quad \Rightarrow \quad \sin\alpha = 0 \quad \text{or,} \quad \alpha = \tan\alpha$$

$$\frac{d^2I_p}{d\alpha^2} = I_0\left(-\frac{4\sin\alpha\cos\alpha}{\alpha^3} + \frac{2}{\alpha^2}\cos^2\alpha - \frac{2}{\alpha^2}\sin^2\alpha + \frac{6\sin^2\alpha}{\alpha^4} - \frac{4\sin\alpha\cos\alpha}{\alpha^3}\right)$$

If $\sin\alpha = 0$, $\frac{d^2I_p}{d\alpha^2}$ = +ve, hence I_p will be minimum. If $\alpha = \tan\alpha$, $\frac{d^2I_p}{d\alpha^2}$ is −ve, hence $\alpha = \tan\alpha$ is the condition for maximum.

Intensity Minima

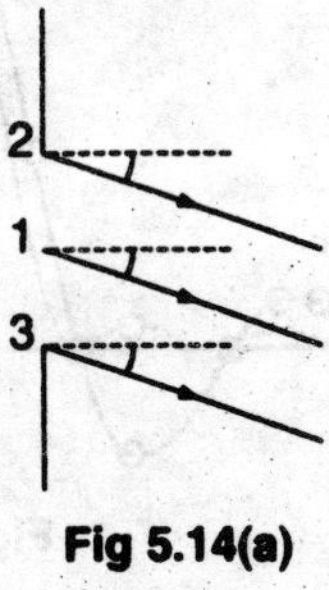

Fig 5.14(a)

$\sin\alpha = 0$ defines minima of intensity

$\Rightarrow \alpha = s\pi, s \in I_n$ (positive or negative integer, zero excluded)

$\Rightarrow \dfrac{\pi a\sin\theta}{\lambda} = s\pi$

$$\Rightarrow a\sin\theta = s\lambda \tag{5.10.1}$$

Thus eqn. (5.10.1) gives the angular positions of the maxima.

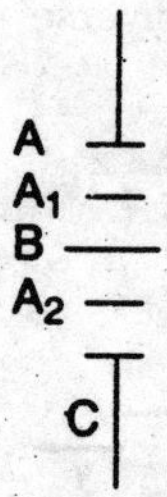

Fig 5.14(b)

To understand the physical content of eqn. (5.10.1), we refer to fig. 5.14.

When $a \sin\theta = \lambda$, we will consider the slit to be divided into two halves. The disturbances from two points, one in the upper half, the other in the lower half, having spatial separation $\frac{a}{2}$ and thus having the path difference $\frac{a}{2}\sin\theta = \frac{\lambda}{2}$ cancel pair-wise. A path difference $\frac{\lambda}{2}$ is equal to phase difference π, thus such disturbances meet in opposite phase.

If $a \sin\theta = 2\lambda$, the slit is to be considered divided into four equal parts—AA_1, A_1B forming the upper two quarters, BA_2, A_2C forming the lower two quarters.

Consider pair disturbances from two points having spatial separation $\frac{a}{4}$, thus the path difference is given as $\frac{a}{4}\sin\theta = \frac{\lambda}{2}$, leading to pairwise cancellation. Thus all the disturbances originating from the slit points cancel out, producing zero intensity on the screen. So the eqn. (5.10.1) is the equation of the minima in the intensity pattern.

Thus,

$$a\sin\theta = s\lambda \longrightarrow \text{Minima},$$
$$s \in I_n \quad \text{(all integers, other than zero)}.$$

Maxima

The maxima are defined by the equation $\alpha = \tan\alpha$.

This equation is a transcendental equation, the solution can be obtained through graphical method. To solve the equation of maxima, we get

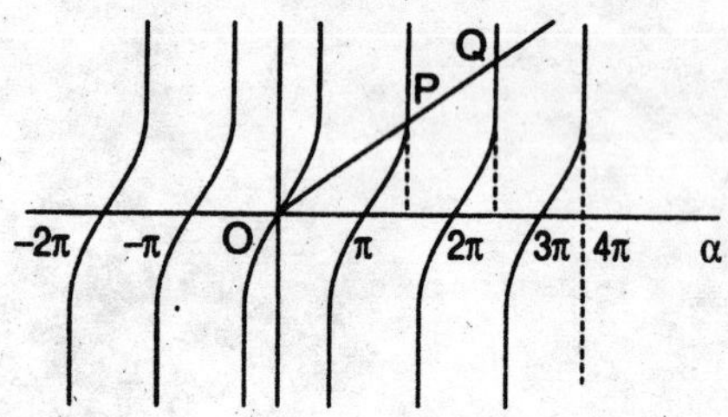

Fig 5.14(c)

$$y = \tan\alpha \tag{5.10.2}$$
$$y = \alpha \tag{5.10.3}$$

The eqn. (5.10.3) is a straight line making an angle of 45° with x-axis along which α is plotted, this is represented in fig. 5.14(c). Eqn. (5.10.2) is simply the $\tan\alpha$ graph. The straight line intersects the $\tan\alpha$ vs. α graph at the origin and at P, Q, R, etc., where the α values are $0, 1.430\pi$, 2.459π and 3.471π, etc.

The intensity function I_P is given by

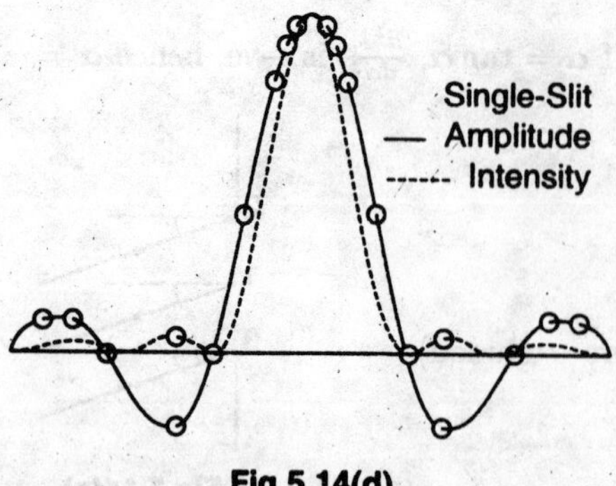

Fig 5.14(d)

$$I_P = I_0 \frac{\sin^2\alpha}{\alpha^2}$$
$$\text{when,}\quad \alpha = 0, I_P = I_0$$
$$\text{when,}\quad \alpha = 1.430\pi, I_P = 0.0496 I_0$$
$$\text{when,}\quad \alpha = 2.459\pi, I_P = 0.0168 I_0$$

Thus for $\alpha = 0$ defining the central maximum, the second maximum occurring at $\alpha = 1.430\pi$, the intensity drops to 4.96% of the central maximum intensity and the third maximum is only 1.68% of the central maximum. The intensity distribution and amplitude distribution is shown in fig. 5.14(d).

The intensity distribution as well as location of maxima is shown in fig. 5.15. The actual photograph of a single-slit pattern is shown in fig. 5.15(a).

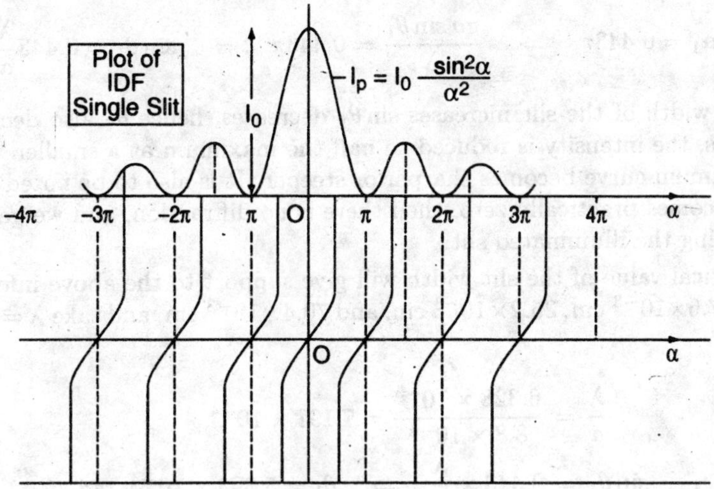

Fig 5.15

Fig 5.15(a)

Half-intensity Band-width

I_0 is the central maximum intensity of the single-slit diffraction pattern; the half-intensity maximum occurs at two values of α given by the equation

$$\frac{1}{2} = \frac{I_P}{I_0} = \frac{\sin^2\alpha}{\alpha^2} \quad \Rightarrow \quad \sin\alpha = \pm\frac{1}{\sqrt{2}}\alpha. \quad (5.10.4)$$

This is another transcendental equation and also can be solved graphically plotting two groups

(i) $y = \pm\frac{1}{\sqrt{2}}\alpha$, (ii) $y = \sin\alpha$.

Finding the points of intersection of two graphs we get

$$\alpha_1 = \pm 0.443\pi.$$

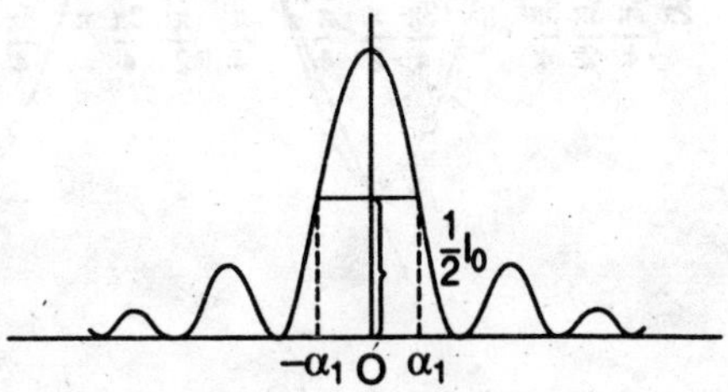

Fig 5.16

So the curve is symmetric about $\alpha = 0$ line and the intensity is reduced to half of the maximum at an angle α_1 given by

$$\alpha_1 = 0.443\pi \quad \Rightarrow \quad \frac{\pi a \sin\theta_1}{\lambda} = 0.443\pi \quad \Rightarrow \quad \sin\theta_1 = 0.443\frac{\lambda}{a}.$$

So, if a, the width of the slit increases $\sin\theta_1$ decreases, hence θ_1 also decreases. It means when a increases, the intensity is reduced to half the maximum at a smaller value of α, hence the central maximum curve becomes sharper or steeper. It is also to be noted that at a certain value of a, θ_1 becomes practically zero, then there is no diffraction, and we get only a uniform region representing the illuminated slit.

A few numerical value of the slit width will give support to the above inference. Take $a = 8.8\times10^{-3}$ cm, 17.6×10^{-3} cm, 35.2×10^{-3} cm, and 70.4×10^{-3} cm, and take $\lambda = 6.328\times10^{-5}$ cm.

When

$$\frac{\lambda}{a} = \frac{6.328 \times 10^{-5}}{8.8 \times 10^{-4}} = 7.191 \times 10^{-4}$$

$$\sin\theta_1 = 0.443 \times \frac{\lambda}{a} \quad \Rightarrow \quad \theta_1 = 5.393 \times 10^{-4} \text{ rad}.$$

When

$$\frac{\lambda}{a} = \frac{6.328 \times 10^{-5}}{2 \times 8.8 \times 10^{-3}} = 3.595 \times 10^{-3}$$

$$\sin\theta_1' = 0.443 \times \frac{\lambda}{a} \quad \Rightarrow \quad \theta_1' = 2.697 \times 10^{-3} \text{ rad}.$$

When

$$\frac{\lambda}{a} = \frac{6.328 \times 10^{-5}}{3 \times 8.8 \times 10^{-3}} = 1.798 \times 10^{-3}$$

$$\sin\theta_1'' = 0.443 \times \frac{\lambda}{a} \quad \Rightarrow \quad \theta_1'' = 1.348 \times 10^{-3} \text{ rad}.$$

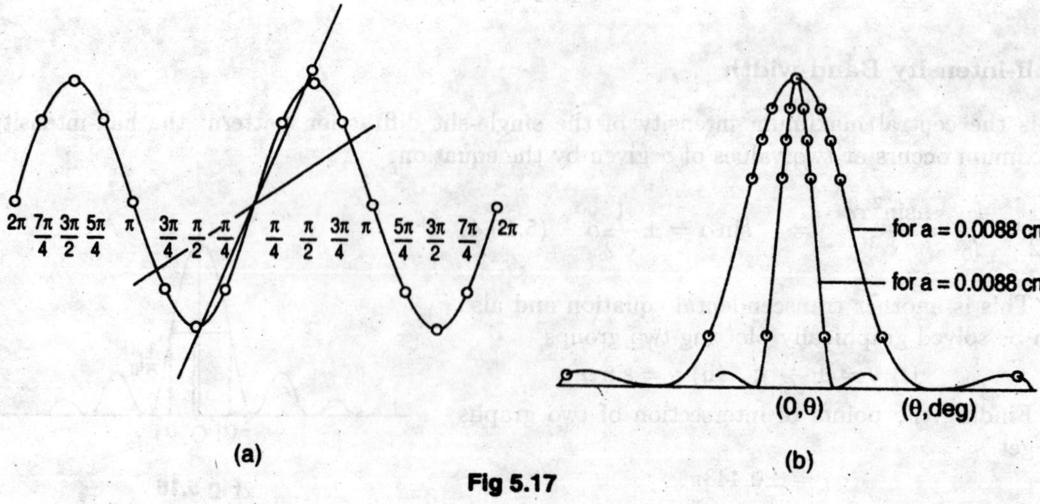

Fig 5.17

When

$$\frac{\lambda}{a} = \frac{6.328 \times 10^{-5}}{4 \times 8.8 \times 10^{-3}} = 9.063 \times 10^{-3}$$

$$\sin\theta_1''' = 0.443 \times \frac{\lambda}{a} \quad \Rightarrow \quad \theta_1''' = 6.797 \times 10^{-3} \text{ rad}$$

Thus as the slit width increases, the half-intensity occurs at smaller values of α, and also for θ. This is confirmed by fig. 5.17(a) and 5.17(b).

The first minimum for these cases occur at

$$\sin\theta_m = \frac{\lambda}{a} = 7.191 \times 10^{-3} \quad \Rightarrow \quad \theta_m = 7.191 \times 10^{-3} \text{ rad}$$

$$\sin\theta_m' = 3.595 \times 10^{-3} \quad \Rightarrow \quad \theta_m' = 3.595 \times 10^{-3} \text{ rad}$$

So the first minimum occurs at a smaller angle and it will vanish altogether. When this occurs, no diffraction pattern is observed, only an illuminated slit image appears.

A few Points to Remember

IDF (Intensity Distribution Function)

$$I_P = I_0 \frac{\sin^2\alpha}{\alpha^2}$$

$$\text{Minima}: \quad \alpha = \pm\pi, \ \pm 2\pi, \ \pm 3\pi, \text{ etc.}$$

$$\text{Maxima}: \quad \alpha = 0, \ 1.43\pi, \ 2.46\pi, \text{ etc.}$$

Important Note

As the ratio $\frac{\lambda}{a}$ decreases with the increase of the slit width, the central maximum narrows down, and at a certain value of this ratio, the diffraction pattern disappears.

Example. A convex lens of focal length 20 cm is placed behind a slit of width 0.6 mm. If a plane wave of wavelength 6000Åfalls normally on the slit, calculate the separation between the second minima on either side of the central maxima.

The minima are given by the equation

$$\alpha = \pm s\pi \quad \Rightarrow \quad \frac{\pi a \sin\theta}{\lambda} = \pm s\pi \quad \Rightarrow \quad \sin\theta = \pm\frac{s\lambda}{a}.$$

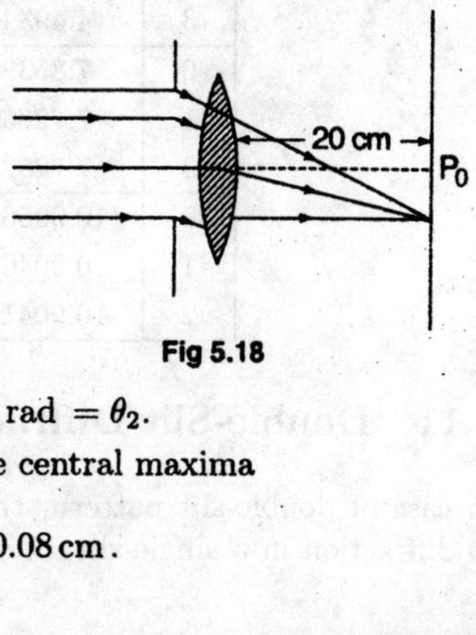

Fig 5.18

Now

$$\frac{\lambda}{a} = \frac{6000 \times 10^{-8}}{6 \times 10^{-2}} = 1 \times 10^{-3}$$

$\therefore$ The second minima will occur at $\sin\theta_2 = \frac{2\lambda}{a} = 2 \times 10^{-3} \text{ rad} = \theta_2$.

So the separation of the second minima on either side of the central maxima

$$= 2 \times 2 \times 10^{-3} \times 20 = 80 \times 10^{-3} \text{ cm} = 0.08 \text{ cm}.$$

Solution of the Equation defining the Secondary Maxima in Case of Single-Slit Diffraction

The equation for the secondary maxima is given by

$$\alpha = \tan\alpha \quad \Rightarrow \quad \alpha - \frac{\sin\alpha}{\cos\alpha} = 0 \quad \Rightarrow \alpha\cos\alpha - \sin\alpha = 0.$$

The values of α satisfying the equation $\alpha = \tan\alpha$ can be obtained graphically by the intersection of two graphs $y = \alpha$ and $y = \tan\alpha$. The approximate value of α obtained from this graphical solution may be corrected with the help of Newton-Raphson's method for the transcendental equation.

The equation $\alpha = \tan\alpha$ may be put in an equivalent form as

$$f(\alpha) = \alpha\cos\alpha - \sin\alpha = 0.$$

The approximate solutions as evident from the graphical method and may be taken as $\alpha_0 = 1.5\pi$, 2.5π, 3.5π, etc. By the iteration method of Newton-Raphson, the correct root of the equation may be successively approached.

According to this method, the $(n+1)$th approximation is given as

$$\alpha_{n+1} = \alpha_n - \frac{f(\alpha_n)}{f'(\alpha_n)}$$

where

$$f'(\alpha_n) = \frac{d}{d\alpha} f(\alpha\Big|_{\alpha=\alpha_n} = -\alpha_n \sin\alpha_n$$

$$\therefore \quad \alpha_{n+1} = \alpha_n + \frac{(\alpha_n\cos\alpha_n - \sin\alpha_n)}{\alpha_n\sin\alpha_n}.$$

With $\alpha_0 = 1.5\pi, 2.5\pi, 3.5\pi$, the successive iterates are given in the table 1.

Table 1

n	α_n	$f(\alpha_n)$	α_{n+1}	Convergent
0	4.71239	—	4.50018	
1	4.50018	—	4.49342	
2	4.49342	—	4.49341	4.49341
3	4.49341	—	4.49341	$= 1.4303\pi$
0	7.85398	—	7.72666	
1	7.72666	—	7.72525	7.72525
3	7.72525	—	7.72525	$= 2.4590\pi$
0	10.9956	—	10.9046	
1	10.9046		10.9041	10.9041
2	10.9041		10.9041	$= 3.4708\pi$

5.11 Double-Slit Diffraction

In case of double-slit pattern, the intensity distribution will be due to two factors, one due to diffraction in a single-slit, and the other is the interference between the disturbances from the two slits. So the resultant pattern will be the superposition of two intensity patterns—diffraction intensity and interference intensity.

Consider fig. 5.19 where we have shown two slits of width a each and separated by a distance b of opaque space. The two slits are AB and CD, and OP_0 is the central line such that $AO = OD$. Any two points equidistant from A and C in the slits are called *corresponding points* of the slits, and the distance $d = (a+b)$ of these corresponding points will be called *double-slit constant.*

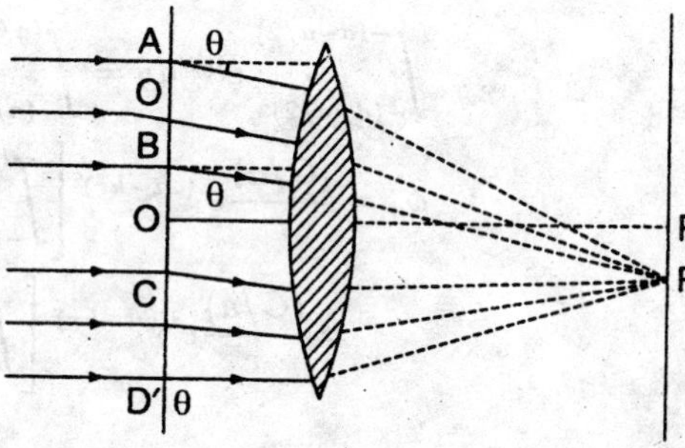

Fig 5.19

Let a plane wave be incident on the slit plane and the disturbances or rays diffracted by an angle θ are considered. So at the point P_1 on the screen each slit will produce its diffraction intensity through interference of the diffracted disturbances, from a slit, also intensity at P_1 will be produced due to interference between the disturbances from the two slits. Corresponding points of the two slits separated by the distance equal to $(a+b)$ become coherent sources and hence capable of producing interference.

In fig. 5.20 we have shown two slits of equal widths a and separated by an opaque space b. Set up the co-ordinate with origin at O; consider a line element dy at a distance y w.r.t. the centre. The plane wave-front occupies the plane of the slit and the energy C^2 associated with the wavefront is uniformly distributed over both the slits and so the vibration of the secondary sources in the slit-plane can be represented as $\phi = (C/a)e^{i\omega t}$, where (C/a) is the amplitude of vibration of the secondary sources. The displacement at P due to the secondary source of length dy at y w.r.t. the centre O is given by

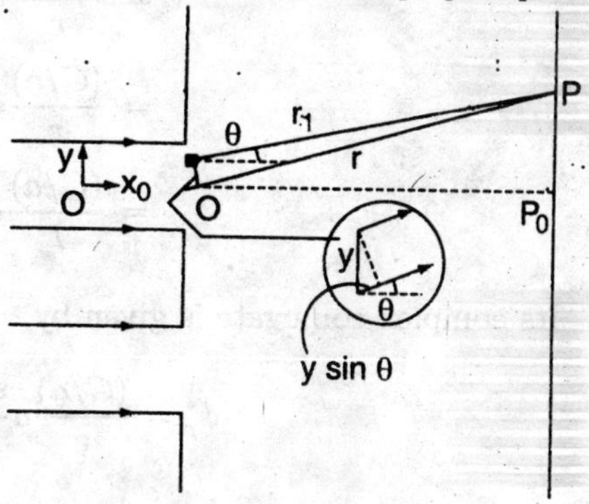

Fig 5.20

$$d\phi = \frac{(C/a)}{r_1} e^{i(\omega t - kr_1)} dy.$$

Now, $r = r_1 + y\sin\theta \Rightarrow r_1 = r - y\sin\theta$, since the length of the slit a is small compared to r, r_1 in the denominator can be replaced by r but not in the exponential term.

$$\therefore \quad d\phi = \frac{(C/a)}{r} e^{i\{\omega t - k(r - y\sin\theta)\}} dy = \frac{(C/a)}{r} e^{i(\omega t - kr)} . \, e^{iky\sin\theta} dy. \tag{5.11.1}$$

To find the displacement at P due to all the secondary sources, we are to integrate eqn. (5.11.1) for all possible values of y so as to include all the secondary sources included in the two slits. Over the first slit y varies from $y = -a/2$ to $a/2$ and for the second slit y varies from $y = -\left(d + \frac{a}{2}\right)$ to $-\left(d - \frac{a}{2}\right)$. So the total displacement at P is given by

$$\phi_p = \frac{(C/a)}{r} e^{i(\omega t - kr)} \left[\int_{-(a/2)}^{(a/2)} e^{iky\sin\theta} dy + \int_{-(d+a/2)}^{-(d-a/2)} e^{iky\sin\theta} dy \right]$$

Putting $k\sin\theta = \beta$, we get

$$\phi_p = \frac{(C/a)}{r} e^{i(\omega t - kr)} \left[\int_{-(a/2)}^{(a/2)} e^{i\beta y} dy + \int_{-(d+a/2)}^{-(d-a/2)} e^{i\beta y} dy \right]$$

The second integral can be suitably transformed by putting $\eta = y + d$. When $y = -\left(d + \frac{a}{2}\right)$, $\eta = -\frac{a}{2}$; when $y = -\left(d - \frac{a}{2}\right), \eta = \frac{a}{2}$

$$\therefore \int_{-(d+a/2)}^{-(d-a/2)} e^{i\beta y} dy = \int_{-(a/2)}^{(a/2)} e^{i\beta(\eta-d)} d\eta$$

$$\therefore \phi_p = \frac{(C/a)}{r} e^{i(\omega t - kr)} \left[\int_{-(a/2)}^{(a/2)} e^{i\beta y} dy + \int_{-(a/2)}^{(a/2)} e^{i\beta(\eta-d)} d\eta \right]$$

$$= \frac{(C/a)}{r} e^{i(\omega t - kr)} \left[\int_{-(a/2)}^{(a/2)} e^{i\beta y} dy + e^{-i\beta d} \int_{-(a/2)}^{(a/2)} e^{i\beta\eta} d\eta \right]$$

$$\int_{-(a/2)}^{(a/2)} e^{i\beta\eta} d\eta = \int_{-(a/2)}^{(a/2)} e^{i\beta y} dy$$

$$\therefore \phi_p = \frac{(C/a)}{r} e^{i(\omega t - kr)} \left(1 + e^{-i\beta d}\right) \int_{-(a/2)}^{(a/2)} e^{i\beta y} dy$$

$$= \frac{(C/a)}{r} e^{i(\omega t - kr)} \left(1 + e^{-i\beta d}\right) \left(\frac{e^{i\beta a/2} - e^{-i\beta a/2}}{i\beta} \right)$$

$$= \frac{(C/a)}{r} e^{i(\omega t - kr)} \frac{2 \sin\left(\beta \frac{a}{2}\right)}{\beta} \left(1 + e^{-i\beta d}\right)$$

$$= \frac{(C/a)}{r} \cdot a \frac{\sin\left(\beta \frac{a}{2}\right)}{\left(\beta \frac{a}{2}\right)} e^{i(\omega t - kr)} \left(1 + e^{-i\beta d}\right)$$

Its complex conjugate is given by

$$\phi_p^* = \frac{(C/a)}{r} a \frac{\sin\left(\beta \frac{a}{2}\right)}{\left(\beta \frac{a}{2}\right)} \cdot e^{-i(\omega t - kr)} \left(1 + e^{i\beta d}\right). \tag{5.11.2}$$

Intensity I is given by

$$I = \phi_p \phi_p^* = \frac{C^2}{r^2} \frac{\sin^2\left(\beta \frac{a}{2}\right)}{\left(\beta \frac{a}{2}\right)^2} \left(1 + e^{i\beta d}\right)\left(1 + e^{-i\beta d}\right)$$

$$= \frac{C^2}{r^2} \frac{\sin^2\left(\beta \frac{a}{2}\right)}{\left(\beta \frac{a}{2}\right)^2} \left[2 + \left(e^{i\beta d} + e^{-i\beta d}\right)\right]$$

$$= \frac{C^2}{r^2} \frac{\sin^2\left(\beta \frac{a}{2}\right)}{\left(\beta \frac{a}{2}\right)^2} \left[2 + 2\cos \beta d\right]$$

$$= \frac{C^2}{r^2} \frac{\sin^2\left(\beta \frac{a}{2}\right)}{\left(\beta \frac{a}{2}\right)^2} 4\cos^2\left(\beta \frac{d}{2}\right) \tag{5.11.3}$$

Taking

$$\frac{\beta a}{2} = \frac{\pi a \sin\theta}{\lambda} = \alpha$$

$$\frac{\beta d}{2} = \frac{\pi d \sin\theta}{\lambda} = \gamma \quad \text{eqn. (5.11.3) becomes}$$

$$I = \frac{C^2}{r^2} \cdot 4 \frac{\sin^2 \alpha}{\alpha^2} \cos^2 \gamma$$
$$= 4I_0 \frac{\sin^2 \alpha}{\alpha^2} \cos^2 \gamma \quad \Rightarrow \quad \frac{I}{I_0} = 4\frac{\sin^2 \alpha}{\alpha^2} \cos^2 \gamma, \tag{5.11.4}$$

where $I_0 = \frac{C^2}{r^2}$ is the maximum intensity of the single-slit diffraction pattern.

5.12 Discussion on Intensity Distribution Function (IDF)

The intensity distribution function for a double-slit is given by eqn. (5.11.4) and it is a product of two functions. One represents the diffraction pattern due to a single-slit, where interference amongst the secondary sources contained in a slit takes place; this effect is contained in the first function, namely $\frac{\sin^2 \alpha}{\alpha^2}$. The other function $\cos^2 \gamma$ takes into account the interference effect between the secondary sources in the two slits. If the slit widths a are very small so that $\frac{\sin^2 \alpha}{\alpha^2}$ remains practically constant, the intensity distribution is predominantly determined by the $\cos^2 \gamma$ term and in this case we get the interference pattern due to two point sources (essence of Young's double-slit experiment).

Now, $d = a + b$, if $a \simeq 0$, $d = b$, the separation between the point holes. So in this case

$$I = I_0 \lim_{a \to 0} \frac{\sin^2 \alpha}{\alpha} \cos^2 \gamma = I_0 \cos^2 \gamma = I_0 \cos^2 \left(\frac{\pi b \sin \theta}{\lambda} \right).$$

So I is max when

$$\cos \frac{\pi b \sin \theta}{\lambda} = \pm 1 = \cos n\pi \quad \Rightarrow \quad \frac{\pi b \sin \theta}{\lambda} = n\pi$$
$$b \sin \theta = n\lambda \quad \Rightarrow \quad b\frac{Y_n}{D} = n\lambda$$
$$\therefore \quad Y_n = \frac{Dn\lambda}{b}, \quad Y_{n+1} = \frac{D(n+1)\lambda}{b}$$

$\therefore$ Fringe width $\omega_d = Y_{n+1} - Y_n = \frac{D\lambda}{b}$, hence all the bright fringes are of equal width, a characteristic feature of the interference pattern.

Plot of Eqn. (5.11.4) with a given value of a and $(a + b) = d$

The plot is shown in fig. 5.21.

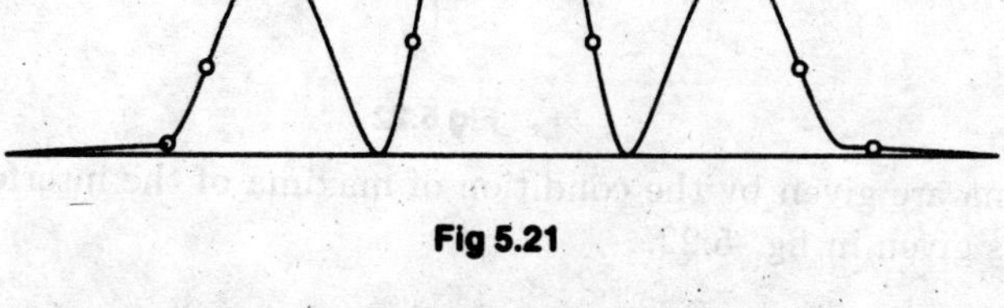

Fig 5.21

Minima

$I = 0$ when either of the two factors is equal by zero.

If

$$\cos\gamma = 0, \gamma = \frac{\pi d \sin\theta}{\lambda} = (2n+1)\frac{\pi}{2}, n \in \text{ integers} \quad \Rightarrow \quad d\sin\theta = (2n+1)\frac{\lambda}{2} \qquad (5.12.1)$$

The minima determined by eqn. (5.12.1) are the interference minima.

The second factor giving minima of the double-slit diffraction pattern is given by

$$\frac{\sin\alpha}{\alpha} = 0 \quad \Rightarrow \quad \sin\alpha = 0 \quad \Rightarrow \quad \alpha = \pm s\pi \ (s \text{ is any integer} \neq 0)$$

$$\Rightarrow \quad \frac{\pi a \sin\theta}{\lambda} = s\pi \quad \Rightarrow \quad a\sin\theta = s\lambda \qquad (5.12.2)$$

These minima are the diffraction minima.

Maxima

The maxima of the double-slit pattern are determined by the product of two factors as given in eqn. (5.11.4). If the slit width a is very small and the intensity near the centre of the pattern is considered, the first factor, *viz.*, $\frac{\sin^2\alpha}{\alpha^2}$ remains practically constant and so the maxima is controlled by the second factor, i.e., $\cos^2\gamma$. So the maxima are given by

$$\cos\gamma = 1 \quad \Rightarrow \quad \gamma = n\pi, n \in I_n, \quad \text{(including zero)}$$

$$\Rightarrow \quad \frac{\pi d \sin\theta}{\lambda} = n\pi \quad \Rightarrow \quad d\sin\theta = n\lambda \qquad (5.12.3)$$

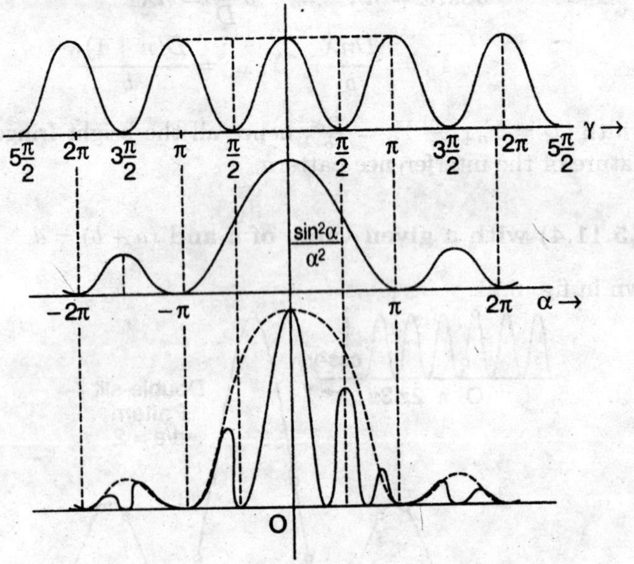

Fig 5.22

So the maxima are given by the condition of maxima of the interference pattern. A plot of $\cos^2\gamma$ versus γ is given in fig. 5.22.

Missing Orders

For small value of the slit width, and for observation restricted to the neighbourhood of the central region of the pattern, the maxima are predominantly controlled by the interference factor. So it might so happen that the maxima of the interference pattern coincide with the minima of the diffraction pattern for a certain value of θ. Then those interference maxima will be missing (cf. fig. 5.22).

The maxima are given by

$$d \sin\theta = n\lambda.$$

The diffraction minima are given by

$$a \sin\theta = s\lambda \quad \therefore \quad \frac{s}{n} = \frac{d}{a} \quad \Rightarrow \quad s = n\frac{d}{a} \tag{5.12.4}$$

For $d/a = 2, n = 2s \;\Rightarrow\; n = 2, 4, 6$, etc. corresponding $s = 1, 2, 3$ etc. are missing. Thus for $d = 2a$, the second, fourth, sixth order maxima will be missing.

For $d/a = 3, n = 3s \;\Rightarrow\; n = 3, 6, 9$, etc. will be missing (see fig. 5.22).

Thus, the maxima in the double-slit diffraction pattern will appear as uneven humps enveloped within the single-slit diffraction pattern. The actual photograph of a double-slit pattern is shown in fig. 5.23.

Fig 5.23

5.13 Double-slit Diffraction through Fourier Transform

As in single-slit diffraction, double-slit diffraction amplitude function $\psi_P(Y)$ can also be obtained through Fourier transformation of suitable slit function. In this case the slit function $s(y)$ can be defined as

$$s(y) = 1 \text{ for } -\frac{a}{2} \le y \le \frac{a}{2} \text{ and for } -\left(d+\frac{a}{2}\right) \le y \le -\left(d-\frac{a}{2}\right)$$
$$= 0 \text{ elsewhere}$$

$$\therefore \quad \psi_P(Y) = \int_{-\infty}^{\infty} s(y) e^{iky(Y/D)} dy = \int_{-(a/2)}^{(a/2)} e^{iky(Y/D)} dy + \int_{-(d+a/2)}^{-(d-a/2)} e^{iky(Y/D)} dy$$

Put $\eta = y + d \Rightarrow \eta = -(a/2)$ when $y = -(d + a/2)$ and $\eta = \frac{a}{2}$ when $y = -(d - a/2)$

$$\therefore \quad \psi_p(Y) = \int_{-(a/2)}^{(a/2)} e^{iky(Y/D)} dy + \int_{-(a/2)}^{(a/2)} e^{ik(\eta-d)(Y/D)} d\eta$$

$$= \int_{-(a/2)}^{(a/2)} e^{iky(Y/D)} dy + e^{-ikd(Y/D)} \int_{-(a/2)}^{(a/2)} e^{ik\eta} d\eta$$

$$= \left[1 + e^{-ikd(Y/D)}\right] \int_{-(a/2)}^{(a/2)} e^{iky(Y/D)} dy$$

Hence magnitude of $\psi_p(Y)$ can be found out.

5.14 Diffraction Pattern in a Multi-Slit : Theory of Amplitude Diffraction Grating

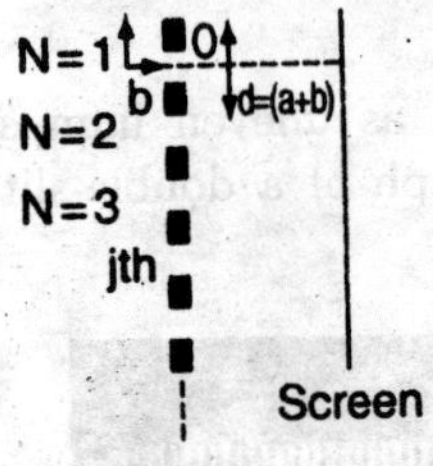

Fig 5.24

A grating consists of a large number of equi-spaced slits, each of width a and separated by a distance d, such that the grating constant is $(a + b) = d$; this array of the slits is shown in fig. 5.24.

When such an arrangement of slits is illuminated by parallel rays or when plane wavefront occupies the plane of the slits, the secondary disturbances from all the N-openings interfere with each other to produce the resultant pattern on the observation screen.

Construction of Grating : Replica Grating

A replica grating is usually used in the laboratory to find the unknown wavelength. On a glass plate a large number of equidistant parallel lines are drawn by a diamond point of a precision instrument and this is called the *master grating.* This is too costly. A replica of this master grating is usually made by pouring *cellulose acetate* properly diluted on to the surface of the master grating, and when taken under water, a thin tough film containing the impression of the rulings on the master grating separates out. This film mounted on a plane glass plate constitutes what is known as *replica grating.* In this case, each ruling serves as the opaque space of width b and the transparent portion between two consecuting rulings serves as the opening of width a (slit width).

Theory of the Grating

As shown in the fig. 5.24, there are N-slits with repeat distance $d = (a + b)$, also called *grating constant.* If the centre of the first slit is taken as the origin, then any slit element at a distance y, of length dy, will produce displacement at some point P on the observation screen given by

$$d\phi_p = \frac{(C/a)}{r} e^{i(\omega t - kr)} \cdot e^{iky \sin\theta} dy.$$

To find the total displacement at P we have to integrate this expression for all the slit elements in the different slits.

For the first slit, the integration limit is from $-(a/2)$ to $(a/2)$.
For the second slit, from $-(d+a/2)$ to $-(d-a/2)$.
For the third slit, from $-(2d+a/2)$ to $-(2d-a/2)$.
For the jth slit, from $-[(j-1)d+(a/2)]$ to $-[(j-1)d-(a/2)]$.
$\therefore$ The resultant displacement is

$$\phi_p = \frac{(C/a)}{r}e^{i(\omega t-kr)}\sum_j \int_{-[(j-1)d+a/2]}^{-[(j-1)d-a/2]} .e^{iky_j \sin\theta}dy_j$$

$$= \frac{(C/a)}{r}e^{i(\omega t-kr)}\sum_j \int_{-[(j-1)d+a/2]}^{-[(j-1)d-a/2]} e^{i\beta y_j}dy_j. \tag{5.14.1}$$

If we put $\eta_j = y_j + (j-1)d$, then

$$\eta_j = \begin{cases} +a/2, & \text{when } y_j = -[(j-1)d-a/2] \\ -a/2, & \text{when } y_j = -[(j-1)d+a/2] \end{cases}$$

So we can write

$$\phi_p = \frac{(C/a)}{r}e^{i(\omega t-kr)}\sum_j \int_{-(a/2)}^{(a/2)} e^{i\beta\{\eta_j-(j-1)d\}}d\eta_j$$

$$= \frac{(C/a)}{r}e^{i(\omega t-kr)}\sum_j e^{-i\beta(j-1)d}\int_{-(a/2)}^{(a/2)} e^{i\beta\eta_j}d\eta_j$$

$$= \frac{(C/a)}{r}e^{i(\omega t-kr)}\sum_j e^{-i\beta(j-1)d}\times\frac{(e^{i\beta a/2}-e^{-i\beta a/2})}{i\beta}$$

$$= \frac{(C/a)}{r}e^{i(\omega t-kr)}\frac{2\sin\beta\frac{a}{2}}{\beta}\sum_j e^{-i\beta(j-1)d}$$

$$= \frac{(C/a)}{r}e^{i(\omega t-kr)}a\frac{\sin\beta\frac{a}{2}}{(\beta a/2)}\sum_j e^{-i\beta(j-1)d}$$

$$= \frac{C}{r}e^{i(\omega t-r)}\frac{\sin\alpha}{\alpha}(1+e^{-i\beta d}+e^{-2i\beta d}+\cdots N \text{ terms})$$

$$= \frac{C}{r}e^{i(\omega t-r)}\frac{\sin\alpha}{\alpha}\frac{(1-e^{-Ni\beta d})}{(1-e^{-i\beta d})} \tag{5.14.2}$$

$$\therefore\quad I = \phi_p\phi_p^* = \frac{C^2}{r^2}\frac{\sin^2\alpha}{\alpha^2}\frac{(1-e^{+Ni\beta d})}{(1-e^{+i\beta d})}\frac{(1-e^{-Ni\beta d})}{(1-e^{-i\beta d})} \tag{5.14.3}$$

Now

$$(1-e^{-i\beta d})(1-e^{i\beta d}) = 2[1-(e^{i\beta d}+e^{-i\beta d})] = 2[1-2\cos\beta d] = 4\sin^2\beta\frac{d}{2}$$

$$(1-e^{-Ni\beta d})(1-e^{+Ni\beta d}) = 4\sin^2\left(N\beta\frac{d}{2}\right)$$

$$\therefore\ I = \frac{C^2}{r^2}\frac{\sin^2\alpha}{\alpha^2}\cdot\frac{\sin^2(N\beta\frac{d}{2})}{\sin^2(\beta\frac{d}{2})} = \frac{C^2}{r^2}\frac{\sin^2\alpha}{\alpha^2}\cdot\frac{\sin^2 N\gamma}{\sin^2\gamma}$$

where $\alpha = \beta\frac{a}{2} = \frac{\pi a\sin\theta}{\lambda}$; $\gamma = \beta\frac{d}{2} = \frac{\pi d\sin\theta}{\lambda}$.

When $\alpha \to 0, \gamma \to 0$ and $I = I_0 = \frac{C^2}{r^2}$,

$$\therefore \quad I = I_0 \frac{\sin^2 \alpha}{\alpha^2} \cdot \frac{\sin^2 N\gamma}{\sin^2 \gamma}. \tag{5.14.4}$$

Eqn. (5.14.4) is the IDF in multi-slit diffraction pattern. From this equation we at once get the intensity distribution in case of double-slit by putting $N = 2$ in eqn. (5.14.4). Thus

$$I = I_0 \frac{\sin^2 \alpha}{\alpha^2} \cdot \frac{\sin^2 \gamma}{\sin^2 \gamma} = I_0 \frac{\sin^2 \alpha}{\alpha^2} \cdot 4 \cos^2 \gamma.$$

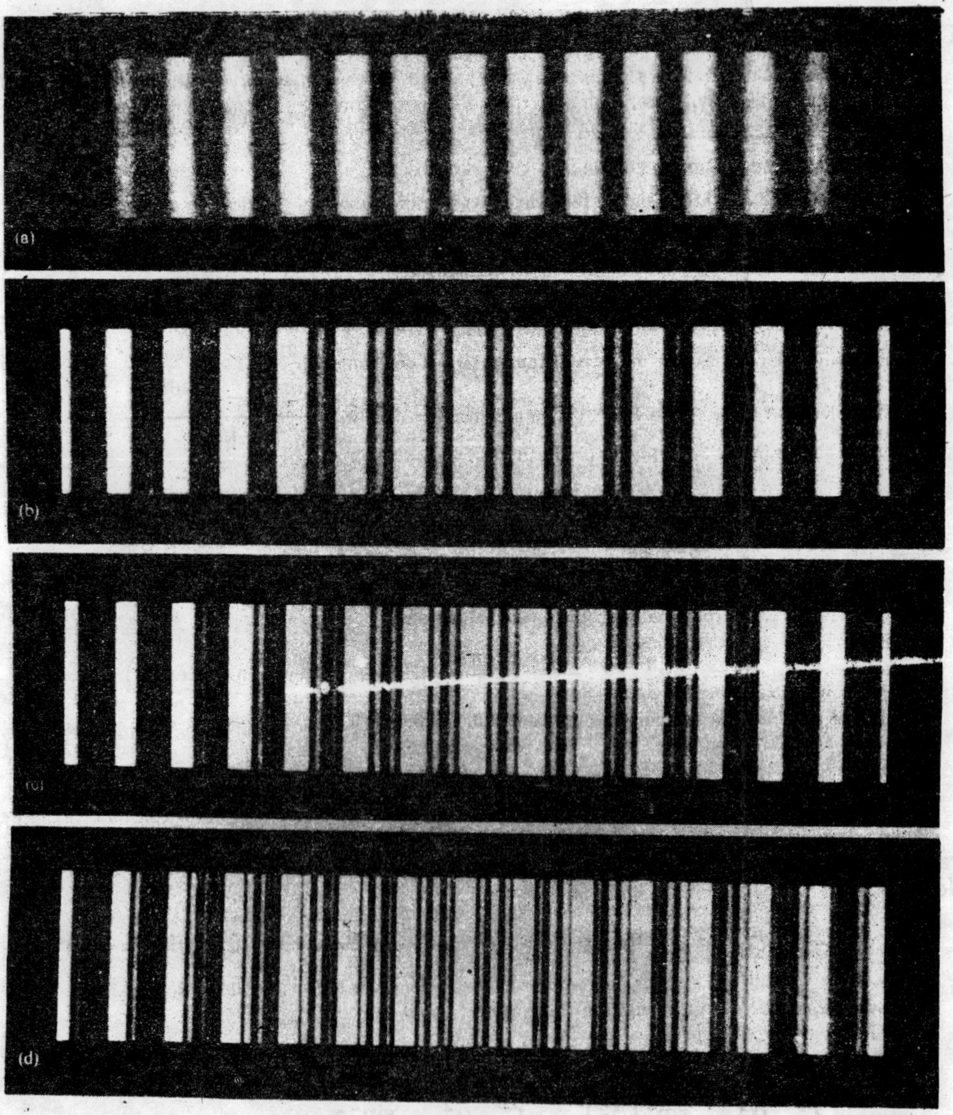

Fig 5.25

Discussion of the IDF

We have seen previously in case of double-slit pattern that as the slit width decreases, the diffraction intensity due to single-slit effect remains practically constant and the intensity pattern is predominantly controlled by the interference factor $\cos^2\gamma$. Also, if we compare the single-slit diffraction pattern with the double-slit pattern, one fact becomes evident that the central maximum of double-slit pattern is sharper than that in single-slit pattern. So as N increases, the central maximum narrows down, becoming steeper and sharper. This is confirmed by the photographs of diffraction patterns due to two, three, four and five slits (cf. fig. 5.25(a)).

Principal Maxima in Grating Pattern

The IDF for a grating shows that the intensity of the pattern depends on two factors, one representing the single-slit diffraction effect, that is, $I_0\frac{\sin^2\alpha}{\alpha^2}$; the other representing the interference of the secondary disturbances for N-slits, that is, $\frac{\sin^2 N\gamma}{\sin^2\gamma}$. Now, if the slit-width a is very small and observation is confined to the neighbourhood of the centre of the pattern, the first factor remains practically constant. So the maxima of the pattern is determined by the factor $\frac{\sin^2 N\gamma}{\sin^2\gamma}$.

Now, let $I_2 = \frac{\sin^2 N\gamma}{\sin^2\gamma}$, so I_2 is maximum when $\gamma = n\pi$, ($n \in$ integers, including 0). $\frac{\sin N\gamma}{\sin\gamma} = \frac{0}{0}$, hence indeterminate. So applying L' Hospital's rule, as $\gamma \to n\pi$, we get

$$\lim_{\gamma\to 0}\frac{\sin N\gamma}{\sin\gamma} = \lim_{\gamma\to 0}\frac{\frac{d}{d\gamma}(\sin N\gamma)}{\frac{d}{d\gamma}(\sin\gamma)} = N\cdot\lim_{\gamma\to 0}\frac{\cos N\gamma}{\cos\gamma} = N \quad \therefore \quad I_2 = N^2.$$

These maxima are called *principal maxima.*

$$\therefore \quad \gamma = n\pi \quad \Rightarrow \quad \frac{\pi d\sin\theta}{\lambda} = n\pi \quad \Rightarrow \quad d\sin\theta = n\lambda,$$

principal maxima corresponding to values $n = 0, 1, 2, 3$, etc. Thus

$$d\sin\theta = n\lambda. \tag{5.14.5}$$

If m be the number of rulings per unit length, then since $(a+b)$, called the *grating constant* contains one ruling, we can write

$$m = \frac{1}{(a+b)} = \frac{1}{d}.$$

So eqn. (5.14.5) becomes

$$\sin\theta = mn\lambda. \tag{5.14.6}$$

Eqn. (5.14.6) is called the *grating equation* where n is the order of the principal maxima and m is the number of rulings per unit length of the grating.

The intensity of the principal maxima is given by

$$I_{pm} = I_1 \times N^2 \tag{5.14.7}$$

Eqn. (5.14.7) shows that the secondary principal maxima increases with the value of N, the number of rulings. It means that with increase of N, the principal maxima sharpen but for a particular value of N, since the pattern also is controlled by the diffraction factor and since for greater value of θ, this factor decreases, the intensity of higher order principal maxima is on the wane.

Secondary Maxima and the Number of Secondary Maxima between two Principal Maxima

The factor I_2 depends on γ, hence for maxima $\frac{dI_2}{d\gamma} = 0$.

Now

$$I_2 = \frac{\sin^2 N\gamma}{\sin^2\gamma}$$

$$\therefore \quad \frac{dI_2}{d\gamma} = 2N\frac{\sin N\gamma\cos N\gamma}{\sin^2\gamma} - \frac{2\sin^2 N\gamma\cos\gamma}{\sin^3\gamma}$$

$$= 2N\frac{\sin^2 N\gamma}{\sin^2\gamma}\cdot\cot N\gamma - \frac{2\sin^2 N\gamma}{\sin^2\gamma}\cot\gamma = 2I_2(N\cot N\gamma - \cot\gamma) \qquad (5.14.8)$$

So for maxima

$$N\cot N\gamma = \cot\gamma \quad \Rightarrow \quad N\tan\gamma = \tan N\gamma. \qquad (5.14.9)$$

Calculating $\frac{d^2 I_2}{d\gamma^2}$ we get

$$\frac{d^2 I_2}{d\gamma^2} = \frac{2dI_2}{d\gamma}(N\cot N\gamma - \cot\gamma) + 2I_2(1 - N^2) + \cot^2\gamma - N^2\cot^2 N\gamma. \qquad (5.14.10)$$

Using eqn. (5.14.9) in eqn. (5.14.10), we get $\frac{d^2 I_2}{d\gamma^2} = 2I_2(1 - N^2) =$ –ve quantity because $n > 1$, hence eqn. (5.14.9) defines maxima.

Now, the solution of (5.14.9) other than $\gamma = n\pi$ defining the principal maxima represents secondary maxima.

Eqn. (5.14.9) can be written as

$$\frac{N\sin\gamma}{\cos\gamma} = \frac{\sin N\gamma}{\cos N\gamma} \Rightarrow \frac{N\cos N\gamma}{\cos\gamma} = \frac{\sin N\gamma}{\sin\gamma}$$

$$\Rightarrow \frac{N^2\cos^2 N\gamma}{\cos^2\gamma} = \frac{N^2\sin^2 N\gamma}{N^2\sin^2\gamma}$$

$$\Rightarrow \frac{N^2(1 - \sin^2 N\gamma)}{(1 - \sin^2\gamma)} = \frac{N^2\sin^2 N\gamma}{N^2\sin^2\gamma} = \frac{N^2}{1 + (N^2 - 1)\sin^2\gamma}$$

$$\therefore \quad \frac{\sin^2 N\gamma}{\sin^2\gamma} = \frac{N^2}{1 + (N^2 - 1)\sin^2\gamma}$$

So the intensity of the secondary maxima is given by

$$I_{sm} = I_1 \times \frac{\sin^2 N\gamma}{\sin^2\gamma}$$

$$= I_1 \times \frac{N^2}{1 + (N^2 - 1)\sin^2\gamma} = \frac{I_{pm}}{1 + (N^2 - 1)\sin^2\gamma} \qquad (5.14.11)$$

where

$$I_{pm} = I_1 \times N^2$$

$$\therefore \quad \frac{I_{pm}}{I_{sm}} = 1 + (N^2 - 1)\sin^2\gamma \qquad (5.14.12)$$

From eqn. (5.14.12) we conclude that as N increases, the intensity of the secondary maximum decreases. To find the number of secondary maxima, we have to see the position of the secondary minima.

Secondary Minima

$I_2 = 0$ when $\sin N\gamma = 0$.

Now

$$\frac{dI_2}{d\gamma} = \frac{2N \sin N\gamma \cos N\gamma}{\sin^2 \gamma} - \frac{2 \sin^2 N\gamma \cos \gamma}{\sin^3 \gamma}.$$

So when $\sin N\gamma = 0$, $\frac{dI_2}{d\gamma} = 0$ defining minima, called *secondary minima*. So the secondary minima are defined by

$$\sin N\gamma = 0 \quad \Rightarrow \quad N\gamma = s\pi \quad \Rightarrow \quad \frac{\pi d \sin\theta}{\lambda} = \frac{s}{N}\pi \quad \Rightarrow \quad d\sin\theta = \frac{s}{N}\lambda \qquad (5.14.13)$$

where s may be any integer other than 0, N, $2N$, etc., corresponding to which we get central, first order, second order principal maxima, etc.

Thus between $s = 0$ and $s = N$, we can think of integer values $1, 2, 3, \cdots, (N-1)$, thus showing $(N-1)$ secondary minima. This inference is confirmed in fig. 5.26 where we have taken $N = 6$.

Number of Secondary Maxima between two Consecutive Principal Maxima

Between two secondary minima, there is one (at least one) secondary maxima. Thus between two principal maxima, since there are $(N-1)$ secondary minima, it implies $(N-2)$ secondary maxima. Thus, in fig. 5.26 again we see four secondary minima with $N = 6$.

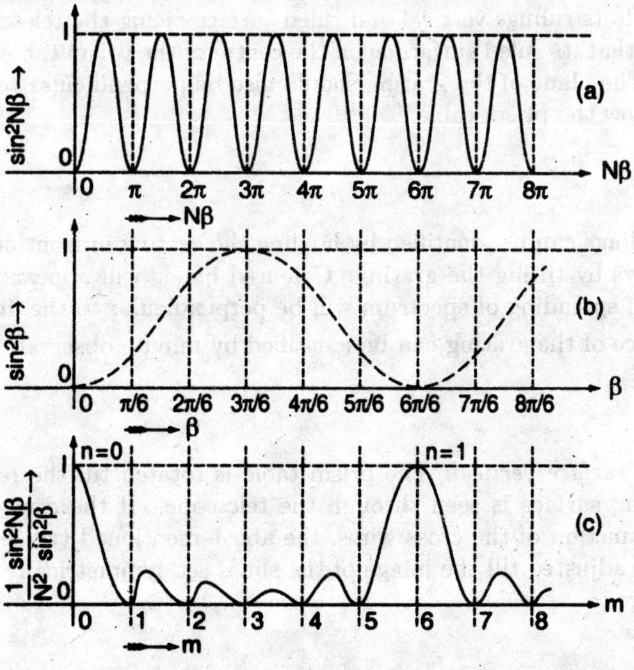

Fig 5.26

Whatever may be the number of secondary maxima, since the intensity is also controlled by the first factor I_1, it falls off for larger value of θ up to the mid-region between two consecutive principal maxima. Also they are not symmetrically situated w.r.t. the positions of the secondary minima. This asymmetry is more pronounced in the neighbourhood of a principal maximum and the secondary peaks shift towards the position of the next principal maximum (see fig. 5.26).

Experimental Procedure

Eqn. (5.14.6) may be employed to find the wavelength of radiation.

Step 1

The spectrometer is adjusted for parallel rays.

Step 2

Setting the telescope perpendicular to the collimator axis: The telescope is set to receive the direct light from the collimator and is adjusted till the vertical cross wire lies along the central line of the image of the slit. The reading of one of the verniers is noted for this position of the telescope. The telescope is then rotated by exactly 90° towards left or right by noting the reading of the same vernier and is clamped at this position.

Step 3

Mounting of grating on the prism table: The grating is placed on its holder and is mounted on the prism table with its rulings vertical and ruled surface facing the telescope. The position of the grating is such that its ruled surface is on the centre of the prism table and extends equally on its both sides. The plane of the grating should also be perpendicular to the line joining two levelling screws below the prism table.

Note

The direction of rulings can be identified by holding the grating in front of a lamp and looking for the reflected rays by tilting the grating. Coloured bands will appear if it is held properly and the direction of spreading of spectrum will be perpendicular to the line of the rulings.

The ruled surface of the grating can be identified by minute observation.

Step 4

Setting the grating surface vertical: The prism table is rotated till the reflected image of the slit from the grating surface is seen through the telescope. If the centre of the image does not coincide with junction of the cross wires, the above-mentioned two levelling screws below the prism table are adjusted till the image of the slit is set symmetrically on both sides of the horizontal cross wire.

Note

In the above position another reflected image will be seen on slight rotation of the prism table. Of the two reflected images, the brighter one is due to reflection from the front surface and the other due to reflection from the back surface of the grating. The ruled surface can be identified by holding an electric lamp behind the sodium flame. The image formed due to reflection from the ruled surface will appear coloured. Thus whether the front or the back surface is ruled can be verified.

Step 5

Setting the grating surface normal to the incident rays: The reading of the position of the prism table is noted for the above adjustment where the reflected image of the slit from the front surface of the grating coincides with the vertical cross wire of the telescope set at right angles to the axis of the collimator. Grating surface in this position obviously makes an angle 45° with the incident rays. The prism table is now rotated through an angle 45° in the appropriate direction by noting the scale reading so that rays from the collimator may be incident normally on the unruled surface of the grating. Now the grating surface is set normal to the rays with its ruled surface away from the collimator. The prism table is fixed in this position.

Step 6

Setting the grating with the rulings vertical: The telescope is rotated to see the diffraction fringes on both sides of the central band. If the rulings of the grating are not vertical, the diffracted images on one side will appear displaced upwards and on the other side displaced downwards. The third screw under the prism table is rotated till the centres of the diffracted images on both sides coincide with the junction of the cross wires. This screw rotates the grating in its own plane and this adjustment sets the rulings vertical.

After all these adjustment for the grating, the collimator slit is made very narrow so that sharp principal maxima are seen on both sides of the central maximum. The position of the source is also adjusted to get fringes on both sides equally bright.

Step 7

Measurement of angles of diffraction (for determination of m): The telescope is rotated and is set with its vertical cross wire on the extreme distinct principal maximum on the left side. The readings of the two verniers are noted. The order number of the principal maximum is also noted. The telescope is turned slowly towards right and its vertical cross wire is set on the next lower order bright fringe, and the corresponding readings of the two verniers and order number of the fringe are noted. This process of recording the positions of the telescope by the two vernier readings and corresponding order number of the principal maxima are continued till the extreme right bright band of the same order as the extreme left bright band is reached. If the readings of the two verniers for any particular order differ by 180°, the recordings are proper. The above recordings are repeated from right end to the left end.

The difference between the two readings of the same vernier, for a particular order principal maximum on opposite sides, gives twice the angle of diffraction for that order number. This is calculated separately for the two verniers for both the sets. The principal maximum of each order and the mean value θ is noted each case.

Measuring θ and knowing n and λ, m is calculated from the grating relation and the mean value of m is used for the next part of the experiment.

Note

A graph may be plotted with $\sin\theta$ along y-axis and order number along x-axis. On plotting the points, a mean straight is drawn passing through the origin. Dividing the slope of the straight line by λ, the mean value of m may be obtained.

Step 8

Measurement of angles of diffraction (for determination of λ*)*: The sodium light is now replaced by the source with the known wavelength and the above procedure as in Step 7 is repeated. Measuring θ and knowing m and n, λ is calculated.

Note

Mean λ may be obtained graphically as above by plotting $\sin\theta$ along y-axis and n along x-axis.

Grating Spectra

The maxima corresponding to wavelength λ is given by the grating equation

$$\sin\theta = mn\lambda$$

where m is the number of rulings per unit length and n is the order number. So the principal maxima for wavelength λ is given by

$$\sin\theta_1 = m\lambda \text{ 1st order;} \quad \sin\theta_2 = 2m\lambda \text{ 2nd order; etc.}$$

If the incident light consists of a number of wavelengths λ, λ_1, λ_2, etc., then the diffraction angles for the 1st order principal maxima for the different wavelengths are given by

$$\sin\theta_1 = m\lambda \quad \sin\theta_1' = m\lambda_1 \quad \sin\theta_1'' = m\lambda_2, \text{ etc.}$$

Thus we get a spectrum for wavelengths in the first order, called the *grating spectrum*.

The Grating Spectrum is Normal

The spectrum is said to be normal if the rate of change of diffraction angle w.r.t. wavelength is a constant, or if the angular separation $\delta\theta$ between the corresponsing principal maxima of two wavelengths differing by $\delta\lambda$ is proportional to $\delta\lambda$. $\lim\limits_{\delta\lambda\to 0} \frac{\delta\theta}{\delta\lambda} = \frac{d\theta}{d\lambda}$ is called the *dispersion between the two wavelengths*, so if $\frac{d\theta}{d\lambda} = A$(constant), the spectrum is said to be normal.

In case of grating nth order principal maximum for wavelength λ is given by the grating equation

$$\sin\theta = mn\lambda \quad \Rightarrow \quad \cos\theta\, d\theta = mnd\lambda \quad \Rightarrow \quad \frac{d\theta}{d\lambda} = \frac{mn}{\cos\theta}. \tag{5.14.14}$$

For small value of θ, the right-hand side is a constant and hence the grating spectrum is a *normal* one. In this case the angular dispersion is proportional to m, the number of rulings per unit length and also to the order number n.

Width of the Principal Maxima

The equation giving the position of the principal maxima is given by

$$d \sin \theta = n\lambda.$$

The first minima after the nth maxima is given by

$$d \sin(\theta + \Delta\theta) = n\lambda \pm \frac{\lambda}{N}$$

$$\Rightarrow \quad d\{\sin\theta \cos\Delta\theta + \sin\Delta\theta \cos\theta\} = n\lambda \pm \frac{\lambda}{N}$$

$$\Rightarrow \quad d \times (\sin\theta + \Delta\theta\cos\theta) = n\lambda \pm \frac{\lambda}{N}$$

$$\Rightarrow \quad \Delta\theta\cos\theta = \pm\frac{\lambda}{Nd}$$

$$\Rightarrow \quad \Delta\theta = \pm\frac{\lambda}{Nd\cos\theta}. \qquad (5.14.15)$$

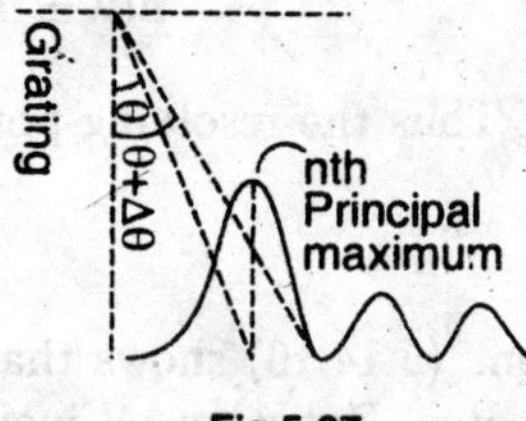

Fig 5.27

So the angular half-width of the nth order principal maxima is given by

$$\omega_h = \Delta\theta = \frac{\lambda}{Nd\cos\theta} = \frac{1}{Nn\cot\theta}.$$

It implies that as N increases, the half-width decreases, hence the principal maxima get sharpened as N increases.

Resolving Power and Resolution by Grating

The resolving power and its application in different optical devices has been elaborately discussed in a separate chapter. Here we will consider the problem of resolution of a grating.

According to Rayleigh's criterion, two spectral lines of wavelengths λ and $\lambda \pm d\lambda$ will be just resolved, that is identified as separate, if the nth principal maximum of wavelength $(\lambda \pm d\lambda)$ just falls on the first minimum after the nth principal maximum of the other component wavelength λ. Such case of resolution of two wavelengths very close to each other is referred to as *chromatic resolution* and is denoted as $\frac{\lambda}{d\lambda}$; smaller the value of $d\lambda$, greater is the chromatic resolution (more details is given elsewhere). Here we just mention the physical significance of the criterion. Whatever may be the optical instruments, the ultimate receptor is our eye which cannot differentiate between two regions if the contrast between them is less than 20% approximately. So if two spectral lines through diffraction produce diffraction patterns such that the overlap of the two patterns occurs to such an extent that in between the positions of the peak of intensity of the two spectral lines, the resultant intensity drops at least by 20%, then the two peaks could be distinguished, hence the spectral lines giving the peaks could be distinguished or resolved.

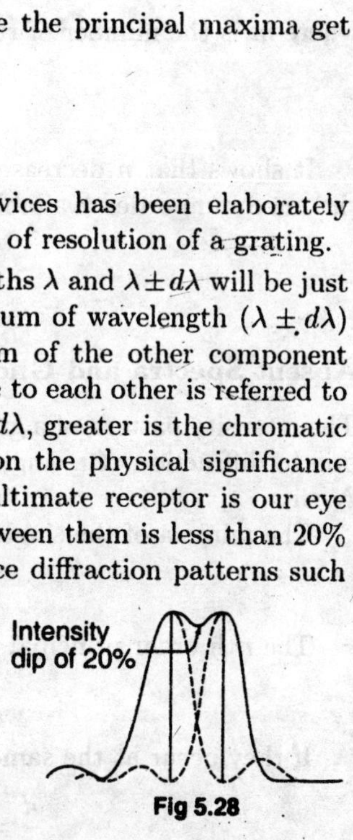

Fig 5.28

Consider the diffraction pattern formed by a grating. There will be formation of principal maxima due to $\lambda + d\lambda$ and λ as shown in fig. 5.28. The wavelength difference is such that the intensity dip at the mid-way between the peaks due to wavelength λ and $\lambda + d\lambda$ is about 20%, hence the two wavelengths can be identified as separate.

According to the theory of grating, the nth principal maximum of wavelength $(\lambda \pm d\lambda)$ occurs at a diffraction angle θ given by

$$d \sin \theta' = n(\lambda \pm d\lambda).$$

The first minimum on either side of the nth principal maximum of wavelength λ is given by

$$d \sin \theta = n\lambda \pm \frac{\lambda}{N}.$$

If they are to coincide in position, we must have,

$$n(\lambda \pm d\lambda) = n\lambda \pm \frac{\lambda}{N} \quad \Rightarrow \quad \frac{\lambda}{d\lambda} = \text{resolving power} = nN.$$

Thus the resolving power of a grating is given by,

$$RP = \frac{\lambda}{d\lambda} = nN. \tag{5.14.16}$$

Eqn. (5.14.16) shows that for greater resolution by a grating, the number of rulings must be greater. But when N increases, the order of the spectrum decreases.

Maximum Order available from a given grating

We write the grating equation as

$$\sin \theta = mn\lambda.$$

where m is the number of rulings per unit length, n is the order number.

$$\therefore \quad n = \frac{\sin \theta}{m\lambda}.$$

It shows that n decreases as m increases. If we take $m = \frac{7620}{2.54}$, $\lambda = 5000 \times 10^{-8}$ cm, then the maximum order available is given by

$$(n)\max = \frac{1}{\left(\frac{7620}{2.54}\right) \times 5000 \times 10^{-8}} \simeq 7.$$

Absent Spectra and Ghost Line

The intensity pattern of a grating consisting of N slits is the resultant of two contributions, one due to diffraction of a single-slit and the other due to superposition of the disturbances from N-slits.

The minima of the single-slit pattern are defined by the diffraction angle θ given as

$$a \sin \theta = p\lambda. \tag{5.14.17}$$

The nth order principal maxima is given by

$$d \sin \theta = n\lambda.$$

If they occur at the same position, nth order spectrum would be absent, thus,

$$\frac{d}{a} = \frac{n}{p} \quad \Rightarrow \quad \frac{a+b}{a} = \frac{n}{p} \quad \Rightarrow \quad \frac{b}{a} = \frac{n}{p} - 1. \tag{5.14.18}$$

If $\frac{b}{a} = 1, n = 2p$, hence $n = 2, 4, 6$ order spectrum would be absent. If $\frac{b}{a} = 2, n = 3p$, hence, $n = 3, 6, 9$ order spectrum would be absent.

Ghost Line

The ghost line in the grating spectrum is due to an error in the construction of the grating. In a good grating, there is periodicity of clear spaces; if there is irregularity in the periodicity of the clear spaces, additional lines other than the actual line will appear. These additional lines are referred to as *ghost line*. Thus if d be the grating constant in one part of the grating and d' be the constant in some other part, then nth order principal maximum for wavelength λ is formed at

$$d\sin\theta = n\lambda; \quad d'\sin\theta' = n\lambda \quad \Rightarrow \quad d\sin\theta = d'\sin\theta'. \tag{5.14.19}$$

Thus we get ghost line due to irregularities in the periodicity of the clear spaces.

Other errors due to irregularities and their effects on the Grating Spectra

1. **Random error in which there is a background illumination.**
2. **Progressive errors due to which the spectral lines are sharper in planes other than the focal plane of the optical system.**

Overlapping of Spectral Lines in the Grating Spectrum

The principal maxima are given by,

$$d\sin\theta = n\lambda \quad \Rightarrow \quad (a+b)\sin\theta = n\lambda \quad \Rightarrow \quad \sin\theta = \frac{n\lambda}{(a+b)} \tag{5.14.20}$$

So θ will be same when $n\lambda$ is constant. Thus in the first order, a wavelength λ_1 may fall on the second order for λ_2 satisfying relation $\lambda_1 = 2\lambda_2$. Thus the first order maximum of light of wavelength λ_1, in the extreme red in Helium of wavelength 7065Å, may come very close to coincide with extreme violet of wavelength 4026 in the second order. This overlapping causes confusion about the order of the spectrum.

If the Incidence is not Normal but Oblique

In this case

$$\gamma = \frac{\pi}{\lambda}d(\sin i \pm \sin\theta).$$

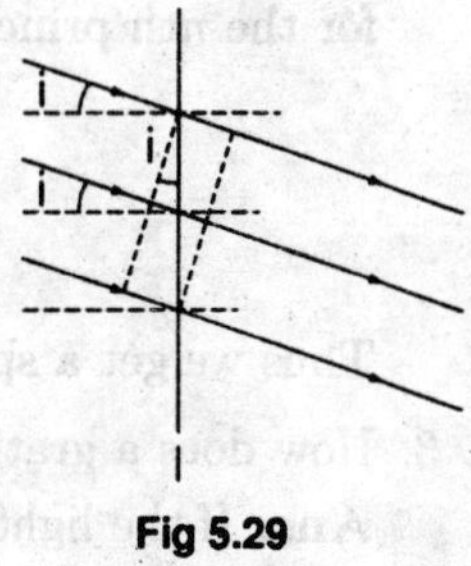

Fig 5.29

$\therefore$ The principal maxima are defined by

$$\frac{\pi d}{\lambda}(\sin i \pm \sin\theta) = n\pi \quad \Rightarrow \quad d(\sin i \pm \sin\theta) = n\lambda. \tag{5.14.21}$$

The Condition for Minimum Deviation in case of Grating Diffraction

For oblique incidence at an angle i, the deviation suffered for an angle of diffraction θ is given by,

$$D = i \pm \theta$$

$$\frac{dD}{di} = 1 \pm \frac{d\theta}{di}, \text{ for minimum, } \frac{d\theta}{di} = \pm 1.$$

For this angle of diffraction, let the nth order principal maximum be obtained. The condition for maximum is given by,

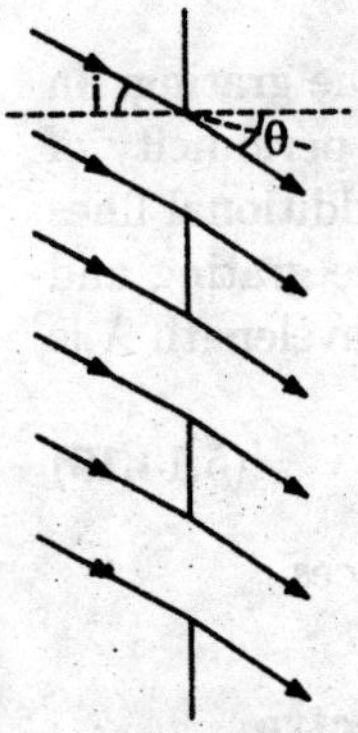

$$d(\sin i \pm \sin\theta) = n\lambda \quad \text{(cf. eqn.5.14.21)}$$

If we differentiate this equation w.r.t. i, we get

$$d\left(\cos i - \cos\theta\frac{d\theta}{di}\right) = 0 \Rightarrow d(\cos i - \cos\theta) = 0$$

$$\Rightarrow \cos i = \cos\theta$$

$$\Rightarrow i = \theta.$$

Fig 5.30

Thus, for minimum deviation, the angle of incidence is equal to the angle of diffraction.

Short Theory Questions with Answers

1. If the distance between the rulings, i.e., the clear space, is very large, what will be the effect observed?

   Ans. According to the grating equation we have, $d\sin\theta = n\lambda$ for nth order principal maximum. If we take $\sin\theta = 1$, then $mn\lambda = 1$.

   Now, $m = \frac{1}{d}$, so as d increases with the increase in the value of b, m decreases and n becomes large. So a large number of bands separated by small angles will be obtained.

2. How do you explain the formation of grating spectrum?

   Ans. The grating equation gives us $\sin\theta = mn\lambda$.

   So the nth order principal maxima for wavelength λ will be formed corresponding to the angle θ. If the light consists of other wavelengths λ_1, λ_2, etc., the corresponding θ values for the nth principal maxima would be given by

$$\sin\theta_1 = mn\lambda_1$$

$$\sin\theta_2 = mn\lambda_2, \text{ etc.}$$

   Thus we get a spectrum of nth order.

3. How does a grating spectrum differ from a prism spectrum?

   Ans. If the light used is of composite nature, then in both grating and prism we will get a spectrum of colours corresponding to different wavelengths present. Separation of two colours is said to be the dispersion between them.

   In grating we have shown that $\frac{d\theta}{d\lambda} = \frac{mn}{\cos\theta}$, so for small value of θ, the right-hand side is a constant, hence the angular separation between two colours is proportional to the wavelength difference.

   In prismatic spectrum, we have shown previously that the angular dispersion denoted as $\frac{d\delta_m}{d\lambda}$ where δ_m is the minimum deviation is given by

$$\frac{d\delta_m}{d\lambda} = \frac{d\delta_m}{d\mu}\cdot\frac{d\mu}{d\lambda}.$$

Taking $\mu = \frac{\sin(A-\delta_m)/2}{\sin A/2}$ and $\mu = A + \frac{B}{\lambda^2}$ and calculating $\frac{d\delta_m}{d\mu}$ and $\frac{d\mu}{d\lambda}$, we get dispersion in case of a prism as

$$\text{Dispersion} = \frac{d\delta_m}{d\lambda} = -\frac{4B \sin A/2}{\lambda^3\sqrt{1-\mu^2 \sin^2 A/2}}.$$

So there will be some points of similarity and dissimilarity in case of the two spectra.

Prismatic spectrum	Grating spectrum
(1) Only one spectrum obtained and hence intensity of the lines high.	(1) A number of spectrum obtained and as the incident intensity is distributed over a number of spectrum formed, its intensity is lower.
(2) Angular dispersion is not normal, it is inversely proportional to the cube of the wavelength, hence the spectrum is not normal.	(2) Grating spectrum is normal where $d\theta$ is proportional to $d\lambda$.
(3) Dispersion in prism spectrum depends on the material of the prism.	(3) Dispersion depends on the order number and number of rulings but not on the material.
(4) Resolving power is less than that of grating.	(4) Grating has greater resolving power than that of prism.
(5) Purity of the spectrum is less than that of the grating.	(5) Grating spectrum is more pure than that of the prismatic spectrum.

4. Explain the changes you will observe in the grating spectra as you pass from one order to the other, if the radiation is not strictly monochromatic?

Ans. With increase in the number of order in the grating spectrum both the angular dispersion and resolving power increases; it means two lines in the radiation will have greater separation as the order number of spectrum considered is greater.

Numerical Problems

1. A screen is placed 2 meters away from a narrow slit. Find the slit width if the first minimum lies 5 mm on either side of the central maximum when plane waves (of λ = 5000Å) are incident on the slit.

Solution. The first minimum is given by

$$\alpha = \pm\pi \quad \Rightarrow \quad \frac{\pi a \sin\theta}{\lambda} = \pm\pi$$

$$\Rightarrow \ a\sin\theta = \pm\lambda = \pm 5000 \times 10^{-8} \quad \Rightarrow \quad a \times \frac{\pm y}{D} = \pm 5000 \times 10^{-8}$$

$$\Rightarrow \ a = \frac{D \times 5000 \times 10^{-8}}{5 \times 10^{-1}} = \frac{2 \times 10^2 \times 5000 \times 10^{-8}}{0.5} = 2 \times 10^{-2} = 0.02\,\text{cm}.$$

2. A slit 0.025 cm wide is placed in front of a convex lens and illuminated by a parallel beam of light of wavelength 5896×10^{-8} cm. In the diffraction pattern formed at the focal plane

of the lens, the distance between the third minimum on the left and the third minimum on the right is 3 mm. What is the focal length of the lens?

Solution. The third minimum is given by

$$\alpha = \pm 3\pi \quad \Rightarrow \quad a\sin\theta = \pm 3\lambda \quad \Rightarrow \quad a\frac{(\pm y)}{f} = \pm 3\lambda$$

$$\Rightarrow \quad f = \frac{ay}{3\lambda} = \frac{0.025 \times \frac{1}{2} \times .3}{3 \times 5896 \times 10^{-8}} = 21.2\,\text{cm}.$$

3. Light of wavelengths 5000Åand 5200Åfalls on a grating having 5000 lines/cm. If a lens of focal length 2 meters is used to form spectra on a screen, find the distance between the lines in (a) the first order, (b) the third order.

Solution. The nth principal maximum of wavelength λ is given by,

$$\sin\theta = mn\lambda.$$

For wavelength $\lambda_1, \sin\theta_1 = m\lambda_1$. Here, $m = 5000, \lambda = 5200 \times 10^{-8}\,\text{cm}, \lambda_1 = 5000 \times 10^{-8}$

$$\sin\theta = 5000 \times 5200 \times 10^{-8} = 0.26, \quad \theta = 0.263\,\text{rad}$$

$$\sin\theta_1 = 5000 \times 5000 \times 10^{-8}, \quad \theta_1 = 0.253\,\text{rad}$$

$$\therefore \quad \Delta y = 2 \times 10^2 \times (0.263 - 0.253) = 2 \times 10^2 \times 0.100 = 2\,\text{cm}.$$

For 3rd order,

$$\sin\theta' = 3 \times 5200 \times 10^{-8} \times 5000 = 0.78, \quad \theta' = 0.895(\text{rad})$$

$$\sin\theta'' = 3 \times 5000 \times 5000 \times 10^{-8} = 0.75, \quad \theta'' = 0.848$$

$$\therefore \quad \Delta y' = 2 \times 10^2 \times (0.895 - 0.848) = 9.4\,\text{cm}.$$

4. A plane transmission grating having 3000 lines/cm gives an angle of diffraction of a line by 30° in the third order. Find the wavelength.

Solution.

$$\sin\theta = mn\lambda \quad \Rightarrow \quad \lambda = \frac{\sin\theta}{mn} = \frac{\sin\frac{\pi}{6}}{3000 \times 3} = 5556\text{Å}.$$

5. The double-slit arrangement is illuminated with light of Hg-vapour lamp so filtered that only strong green line ($\lambda = 5460$Å) is effective, the slits are 0.10 mm wide and the screen on which the interference pattern appears is 20 cm away. What is the angular position (i) *of the first minimum* (ii) *of the tenth maximum?*

Solution. The condition for the diffraction minima in double-slit pattern is given by

$$a\sin\theta = s\lambda.$$

Here, $s = 1, \lambda = 5460\text{Å} = 5460 \times 10^{-8}\,\text{cm}, a = 0.01\,\text{cm}.$

$$\therefore \quad \sin\theta = \frac{\lambda}{a} = \frac{5460 \times 10^{-8}}{.01}.$$

6. A parallel beam of light of wavelength 5.2×10^{-7} m is normally incident on a diffraction grating, the first order spectrum is formed in a direction making an angle θ_1 with the normal such that $\sin\theta_1 = 0.25$. Calculate (i) the number of lines/m on the grating, (ii) the angle θ_2 between the normal and the second order spectrum of the same light, (iii) highest order.

Solution.

$$\sin\theta_1 = mn\lambda = m\lambda$$

$$\Rightarrow \quad m = \frac{\sin\theta_1}{5.2 \times 10^{-7}} = \frac{0.25}{5.2 \times 10^{-7}} = 4.8 \times 10^{-5}/m = 5 \times 10^5/m$$

$$\sin\theta_2 = m \times 2\lambda = 5.0 \times 10^5 \times 2 \times 5.2 \times 10^{-7} \quad \Rightarrow \quad \theta_2 = 31°$$

Take $\sin\theta = 1$, then the highest order is

$$n = \frac{1}{5 \times 10^5 \times 5.2 \times 10^{-7}} \simeq 4.$$

Three-Dimensional Grating: X-ray Crystal Diffraction

The grating that we have just considered is one-dimensional one, consisting of periodic arrangement of clear spaces, called *slits*. We have also seen that when the dimension of the clear space is comparable to the wavelength λ, diffraction effect is appreciable. So in an optical grating, the transparent portion of the film between two rulings is comparable to the visible wavelength but X-ray, though an electromagnetic wave, has wavelength of the order of 1Åonly, so diffraction of X-ray by ordinary grating is not possible. The grating equation that we have derived is given by

$$\sin\theta = mn\lambda. \tag{5.14.22}$$

If we take $\sin\theta = 1, n = 1$, then

$$1 = m\lambda = m \times 10^{-8}\text{ cm} \quad \Rightarrow \quad m = 10^8 \text{ lines/cm.}$$

So for diffraction effect to be observed, the optical grating should consist of 10^8 lines/cm. So it is not possible to draw by any precision instrument, such a large number of lines/cm, or number of rulings comparable to this number.

Actually, however, the fact that a crystal consists of a regular arrangement of atoms or ions in the crystal lattice and the interatomic or interionic separation is of the order of 1Å, aroused in the mind of Von Laue the idea of demonstrating the diffraction phenomenon with X-ray.

A crystal is built up with repetition of some unit, called *unit cell,* throughout the bulk of the body, and depending on the class to which a crystal belongs, atoms or ions are arranged in a three-dimensional array to form the crystal. Since the arrangement is a repetition of a fundamental grouping, it is not necessary to consider the entire array, rather it is better and more convenient to consider a region defined by three vectors $\vec{a}$, $\vec{b}$ and $\vec{c}$ which, on translation by integral multiple, reproduces similar region. These vectors are called *basis vectors.* The region of volume $v = |\bar{a} \cdot \bar{b} \times \bar{c}|$ is called a *unit cell.*

The faces of a crystal are conveniently represented by Miller index in which the ratio of the reciprocals of the reduced intercepts is taken and the fraction cleared off. A plane represented by (121) in Miller representation has reduced intercepts 2, 1, 2 on the three axes. In fig. 5.31 we have shown a face ABC which cuts the three axes at

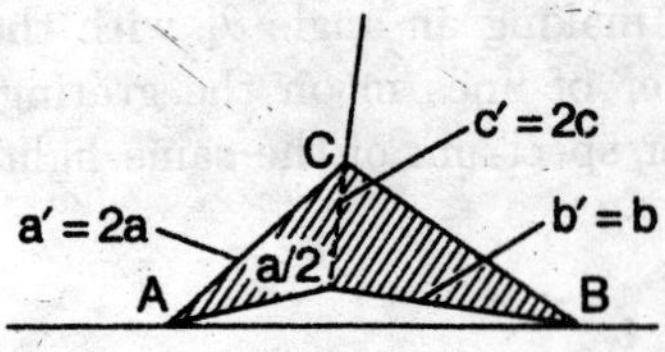

Fig 5.31

$$a' = 2a, b' = 1b, c' = 2c \quad \Rightarrow \quad \frac{a'}{a} = 2, \frac{b'}{b} = 1, \frac{c'}{c} = 2, \frac{a'}{a}, \text{ etc.}$$

are called *reduced intercepts*. Taking reciprocals of the reduced intercepts we get, $\frac{1}{2}, 1, \frac{1}{2}$; hence Miller's index for such a face is (121) after clearing off the fraction.

A cubic crystal may be of three types:

- Simple cube (SC) where the atoms occupy the corners of the repetitive units, called *cells* in the form of cubes (shown in fig. 5.32(a)).
- Body-centred cube (BCC) where there is an additional atom at body centre (fig. 5.32(b)).
- Face-centred cube (FCC) where at the centre of each face there is an additional atom (fig. 5.32(c)).

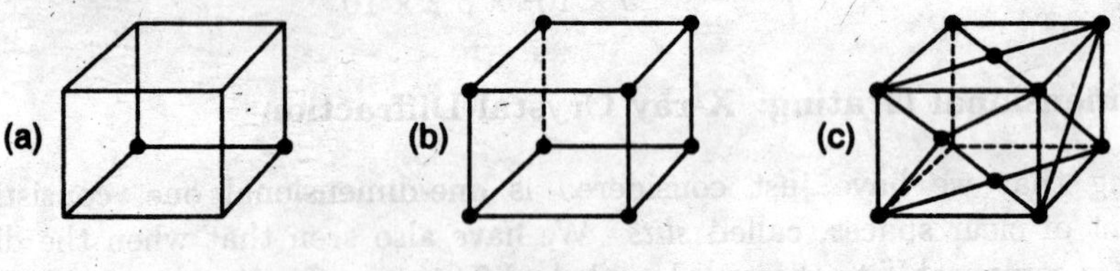

Fig 5.32

The nearest distance between two atoms in simple cube is a, the side of the cube; in BCC it is $\frac{\sqrt{3}a}{2}$, in FCC it is $\frac{\sqrt{2}a}{2}$. NaCl belongs to simple cube formation; Li, Na, Fe belong to BCC formation; Ag belongs to FCC formation.

Thus in BCC ion, the density is 7.9 gm/cc, the atomic weight is 56, and there are two atoms in a unit cell, and the nearest neighbour distance is $\frac{\sqrt{3}a}{2}$.

So the mass associated with the unit cell is $m = \frac{M}{N} \times 2$, where N is the Avogadro number, ρ being the density, and volume of the unit cell being a^3, we can easily write $\rho a^3 = m$.

$$\therefore \quad \rho a^3 = \frac{M}{N} \times 2 \quad \Rightarrow \quad \frac{\sqrt{3}a}{2} = \text{(nearest neighbour distance)}$$

$$= \frac{\sqrt{3}}{2}\left(\frac{2M}{NP}\right)^{1/3} = \frac{\sqrt{3}}{2}\left(\frac{2 \times 56}{6.02 \times 10^{23} \times 7.9}\right)^{1/3} = 2.52 \times 10^{-8} \text{ cm}.$$

So we see that the atoms in the crystal lattice may serve as the diffracting centres. This suggestion of Von Laue was put to experimental test by Knipping and Friedrich and his idea was confirmed. The crystal grating, however, differs from an optical grating in one respect that in the former, the atomic centres of diffraction are not all in one plane but are distributed in space, whereas in optical grating they are in one plane. A crystal grating may be looked upon as a three-dimensional space grating rather than one-dimensional plane grating.

Experimental Arrangement of Laue, Friedrich and Knipping

The experimental arrangement is shown in fig. 5.33. A narrow beam of X-rays collimated by two slits S_1 and S_2 is allowed to pass through a thin crystal of Zn and the transmitted beam is received on a photographic plate placed behind the crystal. After many hours of exposure, the photograpic plate is developed and it shows a pattern as shown in the figure separately. It contains a central black spot corresponding to the incidence of the direct beam surrounded by a series of fainter black spot arranged regularly, corresponding to the diffracted rays from various crystal planes. The arrangement of secondary Laue's spot around the central spot speaks of the arrangement of the crystal lattice.

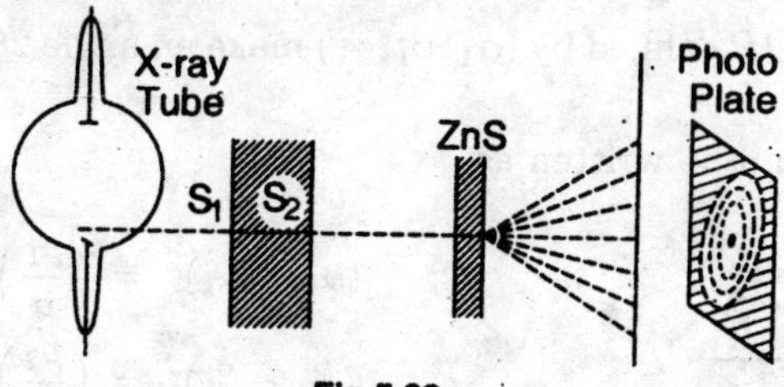

Fig 5.33

Theory of the Grating

Let us consider a linear arrangement of diffracting centres as shown.

The path difference is given by

$$\Delta = CB - AD = \bar{r}\cdot\bar{n}_0 - \bar{r}\cdot\bar{n}_1$$
$$= AB(\cos\theta_0 - \cos\theta_1) = AB(\alpha_0 - \alpha_1) \quad (5.14.23)$$

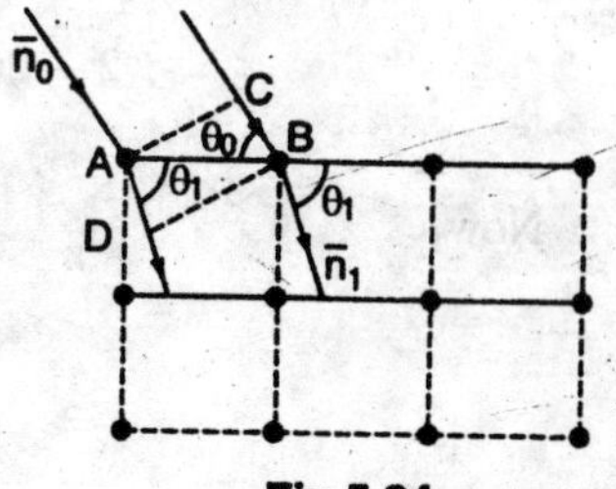

Fig 5.34

where α_0 and α_1 are substituted for $\cos\theta_0$ and $\cos\theta_1$. Due to diffraction at A and B, the path difference, in general, for arbitrary orientation of $\bar{r}$ is given by

$$\Delta = \bar{r}\cdot(\bar{n}_0 - \bar{n}_1). \quad (5.14.24)$$

If we write $\bar{r} = a\hat{a} + b\hat{b} + c\hat{c}$, where $\hat{a}, \hat{b}$ and $\hat{c}$ are unit vectors along the crystallographic axes, then

$$\Delta = \bar{r}\cdot(\bar{n}_0 - \bar{n}_1) = (a\hat{a} + b\hat{b} + c\hat{c})\cdot(\bar{n}_0 - \bar{n}_1)$$
$$= a(\hat{a}\cdot\bar{n}_0 - \hat{a}\cdot\bar{n}_1) + b(\hat{b}\cdot\bar{n}_0 - \hat{b}\cdot\bar{n}_1) + c(\hat{c}\cdot\bar{n}_0 - \hat{c}\cdot\bar{n}_1)$$
$$= a(\cos\theta_0 - \cos\theta_1) + b(\cos\phi_0 - \cos\phi_1) + c(\cos\psi_0 - \cos\psi_1) \quad (5.14.25)$$

The unit vectors $\bar{n}_0$ and $\bar{n}_1$ makes angles $(\theta_0, \phi_0, \psi_0)$ and $(\theta_1, \phi_1, \psi_1)$ respectively with $\hat{a}, \hat{b}, \hat{c}$. If the direction cosines of $\bar{n}_0$ and $\bar{n}_1$ be written as $\cos\theta_0 = \alpha_0, \cos\phi_0 = \beta_0, \cos\psi_0 = \gamma_0$ and $\cos\theta_1 = \alpha_1, \cos\phi_1 = \beta_1$ and $\cos\psi_1 = \gamma_1$, then the conditions for diffraction maxima, in accordance with the optical grating consideration, become

$$a(\alpha_0 - \alpha_1) = p_1\lambda \quad (5.14.26)$$
$$b(\beta_0 - \beta_1) = p_2\lambda \quad (5.14.27)$$
$$c(\gamma_0 - \gamma_1) = p_3\lambda. \quad (5.14.28)$$

where p_1, p_2, p_3 are integers

$$\therefore \quad (\alpha_0 - \alpha_1) = \frac{p_1}{a}\lambda \quad (5.14.29)$$
$$(\beta_0 - \beta_1) = \frac{p_2}{b}\lambda \quad (5.14.30)$$
$$(\gamma_0 - \gamma_1) = \frac{p_3}{c}\lambda \quad (5.14.31)$$

This triplet relations form what is known as *Laue's equation.*

These equations show that for a given direction of incidence defined by the direction cosines $(\alpha_0, \beta_0, \gamma_0)$, the diffraction maxima occur in a direction defined by $(\alpha_1, \beta_1, \gamma_1)$ for a set of integers (p_1, p_2, p_3) for a particular wavelength.

To find the angle of diffraction and derive the grating equation, we make the following construction.

Let the diffracted ray OP defined by $(\alpha_1, \beta_1, \gamma_1)$ make an angle 2θ with the incident direction $(\alpha_0, \beta_0, \gamma_0)$.

The triplet relations can be written as

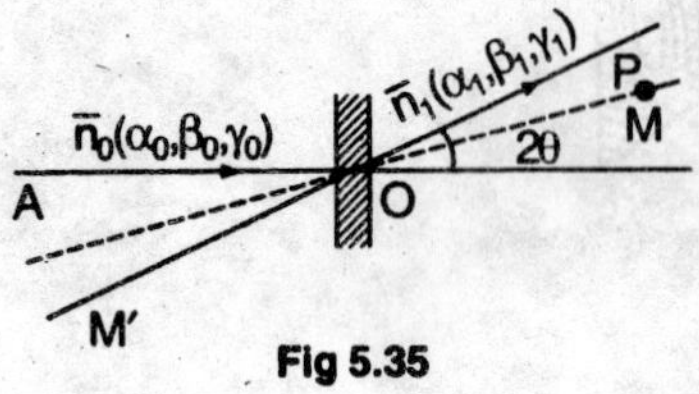

Fig 5.35

$$(\alpha_0 - \alpha_1)^2 = \left(\frac{p_1}{a}\right)^2 \lambda^2$$
$$(\beta_0 - \beta_1)^2 = \left(\frac{p_2}{b}\right)^2 \lambda^2$$
$$(\gamma_0 - \gamma_1)^2 = \left(\frac{p_3}{c}\right)^2 \lambda^2$$

Adding we get, (remembering $\alpha_0^2 + \beta_0^2 + \gamma_0^2 = \alpha_1^2 + \beta_1^2 + \gamma_1^2 = 1$)

$$2 - 2(\alpha_1\alpha_0 + \beta_1\beta_0 + \gamma_1\gamma_0) = \lambda^2 \left(\frac{p_1^2}{a^2} + \frac{p_2^2}{b^2} + \frac{p_3^2}{c^2}\right)$$

Now,

$$\bar{n}_0 = \alpha_0\hat{a} + \beta_0\hat{b} + \gamma_0\hat{c};\ \bar{n}_1 = \alpha_1\hat{a} + \beta_1\hat{b} + \gamma_1\hat{c}$$
$$\Rightarrow \bar{n}_0 \cdot \bar{n}_1 = \alpha_0\alpha_1 + \beta_0\beta_1 + \gamma_0\gamma_1$$
$$\therefore\ 2 - 2\bar{n}_0 \cdot \bar{n}_1 = \lambda^2 \left(\frac{p_1^2}{a^2} + \frac{p_2^2}{b^2} + \frac{p_3^2}{c^2}\right)$$
$$\Rightarrow 2(1 - \cos 2\theta) = \lambda^2 \left(\frac{p_1^2}{a^2} + \frac{p_2^2}{b^2} + \frac{p_3^2}{c^2}\right)$$
$$\Rightarrow 4\sin^2\theta = \lambda^2 \left(\frac{p_1^2}{a^2} + \frac{p_2^2}{b^2} + \frac{p_3^2}{c^2}\right)$$
$$\Rightarrow \sin\theta = \frac{\lambda}{2}\left(\frac{p_1^2}{a^2} + \frac{p_2^2}{b^2} + \frac{p_3^2}{c^2}\right)^{1/2} \qquad (5.14.32)$$

Consider a bisector plane MM' bisecting the angle 2θ; since this plane makes equal angles with the incident ray and the diffracted ray, the diffraction may be looked upon as though the reflection is taking place from this plane, called *net plane.* Other net planes parallel to this just indicated, will also reflect the incident ray as it penetrates the crystal. So diffraction may be looked upon as space reflection from the set of parallel net planes (Bragg's concept).

Let the position vector of the point $M(x, y, z)$ on the bisector of the angle of diffraction be denoted as $\bar{s}$, so we can write

$$\bar{s} \cdot \bar{n}_0 = \bar{s} \cdot \bar{n}_1$$
$$\Rightarrow (x\hat{a} + y\hat{b} + z\hat{c}) \cdot (\alpha_0\hat{a} + \beta_0\hat{b} + \gamma_0\hat{c}) = (x\hat{a} + y\hat{b} + z\hat{c}) \cdot (\alpha_1\hat{a} + \beta_1\hat{b} + \gamma_1\hat{c})$$
$$\Rightarrow (\alpha_0 - \alpha_1)x + (\beta_0 - \beta_1)y + (\gamma_0 - \gamma_1)z = 0$$

From Laue's equation put the values of $(\alpha_0 - \alpha_1)$, $(\beta_0 - \beta_1)$ and $(\gamma_0 - \gamma_1)$, and we obtain

$$\frac{p_1\lambda}{a}x + \frac{p_2\lambda}{b}y + \frac{p_3\lambda}{c}z = 0 \quad \Rightarrow \quad \frac{x}{\left(\frac{a}{p_1}\right)} + \frac{y}{\left(\frac{b}{p_2}\right)} + \frac{z}{\left(\frac{c}{p_3}\right)}. \tag{5.14.33}$$

So the plane MM' makes intercepts on the three axes given as

$$a' = \frac{a}{p_1}, b' = \frac{b}{p_2}, c' = \frac{c}{p_3}.$$

The reduced intercepts are

$$\frac{a'}{a} = \frac{1}{p_1}, \frac{b'}{b} = \frac{1}{p_2}, \frac{c'}{c} = \frac{1}{p_3}.$$

The reciprocals of the reduced intercepts are p_1, p_2, p_3 and if $p_1 = nh$, $p_2 = nk$, $p_3 = nl$, then Miller index for such a plane is (h, k, l).

The distance of separation between a set of net planes is given by

$$d = \frac{1}{\sqrt{\left(\frac{h^2}{a^2}\right) + \left(\frac{k^2}{b^2}\right) + \left(\frac{l^2}{c^2}\right)}} = \frac{n}{\sqrt{\left(\frac{p_1^2}{a^2}\right) + \left(\frac{p_2^2}{b^2}\right) + \left(\frac{p_3^2}{c^2}\right)}}$$

$$= \frac{n\lambda}{2\sin\theta}, \text{ using relation (5.1.32)}$$

$$\Rightarrow 2d\sin\theta = n\lambda. \tag{5.14.34}$$

Eqn. (5.14.32) is the crystal grating equation, also called *Bragg's law,* where 2θ may be taken to define the diffraction angle, θ may be considered the angle of incidence w.r.t. the net planes, and d is the separation of the atomic planes.

5.15 Fraunhofer Diffraction in a Rectangular Aperture

In this case the width of the slit is not negligible compared to the linear dimension of the slit. The incident plane wavefront at some instant occupies the plane of the aperture and the intensity is spread uniformly all over the aperture; hence the vibration of the source particles in the aperture is given by

$$\phi_s = \frac{C}{ab} \cdot e^{i\omega t}.$$

Consider an elementary area $ds = dxdy$ at the point $Q(x, y)$, see fig. 5.12, then the displacement at P, the field point, due to sources in ds is written as

$$d\phi_p = \frac{C}{abR} \cdot e^{i(\omega t - kr)} dxdy$$

Now, $r = R + x\sin\theta + y\sin\phi$ (cf. fig. 5.12)

$$\therefore \quad d\phi_p = \frac{C}{abR} \cdot e^{i(\omega t - kR - kx\sin\theta - ky\sin\phi)} dxdy$$

$$= \frac{C}{ab} \frac{e^{i(\omega t - kR)}}{R} e^{-ikx\sin\phi} dx e^{-iky\sin\theta} dy \tag{5.15.1}$$

So the resultant displacement at P due to all the sources contained in the rectangular aperture is given by,

$$\phi_p = \frac{C}{abR} e^{i(\omega t - kR)} \int_{-a/2}^{a/2} e^{-ikx\sin\theta} dx \int_{-b/2}^{b/2} e^{-iky\sin\theta} dy.$$

Putting $k\sin\theta = \alpha_0, k\sin\phi = \beta_0$, we get

$$\int_{-b/2}^{b/2} e^{-iky\sin\theta} dx = \int_{-a/2}^{a/2} e^{-ix\alpha_0} dx = -\frac{e^{-ia\alpha_0/2} - e^{-ia\alpha_0/2}}{i\alpha_0}$$

$$= \frac{2\sin\alpha_0 a/2}{\alpha_0} = a\cdot\frac{\sin\alpha_0 a/2}{(\alpha_0 a/2)} = a\cdot\frac{\sin\alpha}{\alpha}$$

Similarly,

$$\int_{-b/2}^{b/2} e^{-iky\sin\phi} dy = \int_{-b/2}^{b/2} e^{-iy\beta_0} dy = b\frac{\sin\beta}{\beta}$$

where, $\alpha = \alpha_0 a/2 = (\pi a\sin\theta)/\lambda, \beta = \beta_0 b/2 = (\pi b\sin\phi)/\lambda$.

$$\therefore \quad \phi_p = \frac{(C/ab)}{R} ab\frac{\sin\alpha}{\alpha}\cdot\frac{\sin\beta}{\beta}\cdot e^{i(\omega t - kR)}.$$

Its complex conjugate is ϕ_p^* given by

$$\phi_p^* = \frac{(C/ab)}{R} ab\frac{\sin\alpha}{\alpha}\cdot\frac{\sin\beta}{\beta}\cdot e^{-(\omega t - kR)}$$

$$\therefore \quad I = \phi_p\phi_p^* = \frac{C^2}{R^2}\cdot\frac{\sin^2\alpha}{\alpha^2}\cdot\frac{\sin^2\beta}{\beta^2}. \tag{5.15.2}$$

If the intensity is such that $I = I_0 = C^2/R^2$, then

$$I = I_0\frac{\sin^2\alpha}{\alpha^2}\cdot\frac{\sin^2\beta}{\beta^2}. \tag{5.15.3}$$

Discussion on the Intensity Distribution Function (IDF) for a Rectangular Aperture

Minima of the pattern

The intensity of the pattern is controlled by two factors, one factor controlling the intensity along x-axis, the other controlling the intensity along y-axis. The first factor is $\frac{\sin^2\alpha}{\alpha^2}$ and the second factor is $\frac{\sin^2\beta}{\beta^2}$.

So for minimum, either of the two factors becomes zero, i.e.,

$$\sin\alpha = 0 \Rightarrow \alpha = n\pi,\ n \in \text{all integers, excluding zero}$$
$$\text{or, } \sin\beta = 0 \Rightarrow \beta = s\pi,\ s \in \text{all integers, excluding zero.}$$

Now,

$$\alpha = n\pi \quad \Rightarrow \quad \frac{\pi a\sin\theta}{\lambda} = n\pi \quad \Rightarrow \quad a\sin\theta = n\lambda \tag{5.15.4}$$

$$\beta = s\pi \quad \Rightarrow \quad \frac{\pi b\sin\phi}{\lambda} = s\pi \quad \Rightarrow \quad b\sin\phi = s\lambda \tag{5.15.5}$$

Central Maximum

The central maxima are given by the conditions $\alpha \to 0, \beta \to 0$. Under these conditions, $\frac{\sin\alpha}{\alpha} = 1$ and $\frac{\sin\beta}{\beta} = 1$; hence we get, $I = I_0 = C^2/R^2$. The diffraction pattern obtained is shown in fig. 5.36.

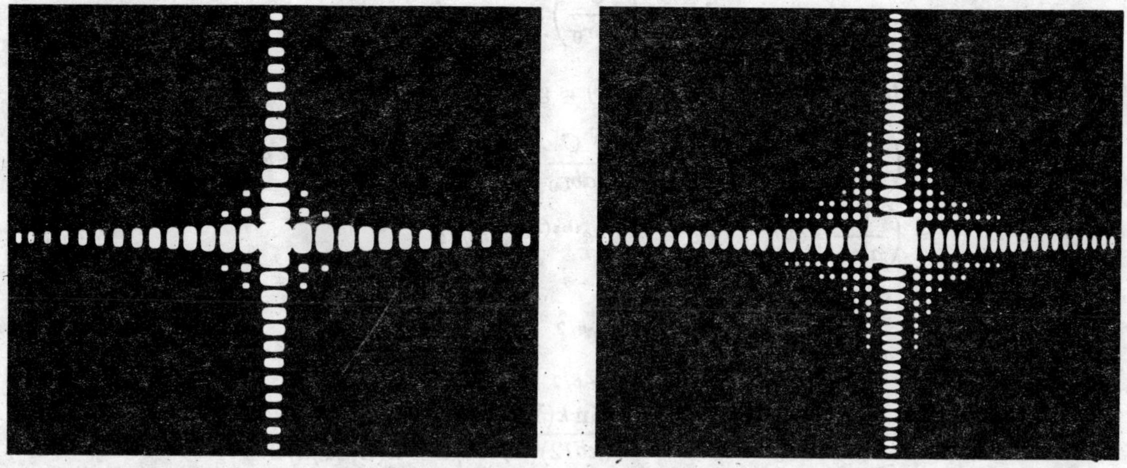

Fig 5.36

Another method to find the IDF for a Rectangular Aperture

In fig. 5.37 we have shown a rectangular aperture $AOCD$ of length b along the y-axis and width a along the z-axis. P is a point on the screen having co-ordinates (X, Y, Z). Consider an elementary area $dydz$ at $M(o, y, z)$, then

$$\begin{aligned} O_1P^2 = r_0^2 &= X^2 + Y^2 + Z^2 \simeq X^2 \\ MP^2 = r^2 &= X^2 + (Y - y)^2 + (Z - z)^2 \\ &= X^2 + Y^2 + Z^2 - 2Yy - 2Zz \\ &= r_0^2 - 2Yy - 2Zz \\ \Rightarrow \quad r &= r_0 - \frac{Yy + Zz}{r_0} \\ &\simeq r_0 - y(Y/X) - z(Z/X) \end{aligned}$$

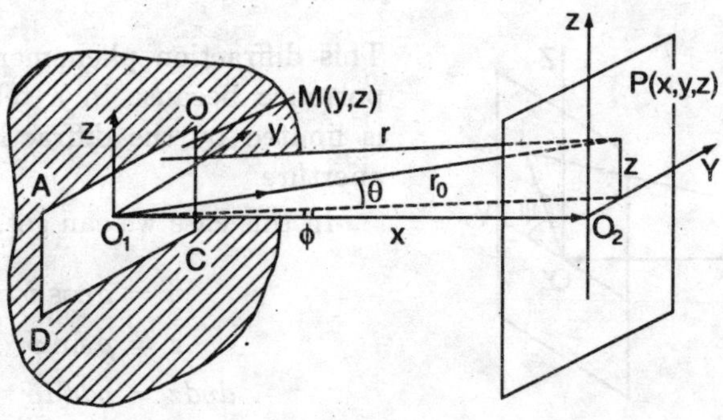

Fig 5.37

When the plane wave occupies the plane of the aperture, every particle contained in the aperture becomes the source of the secondary disturbances.

The energy C^2 is spread uniformly throughout the clear space of the aperture, hence the vibration of the source particles in the area $dydz$ at M can be written as

$$d\phi_s = \left(\frac{C}{abr_0}\right) dydz\cdot e^{i\omega t} \tag{5.15.6}$$

The vibration of the particle at $P(X, Y, Z)$ is given as

$$\begin{aligned} d\phi_p &= \left(\frac{C}{abr_0}\right) e^{i(\omega t - kr)} dydz = \left(\frac{C}{abr_0}\right) e^{i[\omega t - k\{r_0 - y(Y/X) - z(Z/X)\}]} dydz \\ &= \left(\frac{C}{abr_0}\right) e^{i(\omega t - kr_0)} e^{iky(Y/X)} e^{ikz(Z/X)} dydz \end{aligned} \tag{5.15.7}$$

$$\begin{aligned} \therefore \quad \phi_p &= \left(\frac{C}{abr_0}\right) e^{i(\omega t - kr_0)} \int_{-b/2}^{b/2} e^{iky(Y/X)} dy \int_{-a/2}^{a/2} e^{ikz(Z/X)} dz \\ &= \frac{C}{abr_0} e^{i(\omega t - kr_0)} (ab) \frac{\sin k(Y/X)b/2}{(kb/2)(Y/X)/2} \cdot \frac{\sin k(Z/X)a/2}{k(Z/X)a/2}. \end{aligned}$$

$$\therefore \quad I_p = \phi_p \phi_p^* = \left(\frac{C}{r_0}\right)^2 \frac{\sin^2 k(b/2)(Y/X)}{[k(b/2)(Y/X)]^2} \cdot \frac{\sin^2 k(a/2)(Z/X)}{[k(a/2)(Z/X)]^2} \tag{5.15.8}$$

Now,

$$\frac{Y}{X} \tan\phi \sim \phi \sim \sin\phi, \quad \frac{Z}{X} = \frac{Z}{r_0} = \sin\theta \sim \theta.$$

$$\therefore \quad I_p = C^2 \frac{\sin^2\alpha}{\alpha^2} \cdot \frac{\sin^2\beta}{\beta^2} \tag{5.15.9}$$

where

$$\alpha = \frac{kb}{2}\sin\phi = \frac{\pi b \sin\phi}{\lambda}, \beta = \frac{ka}{2}\sin\theta = \frac{\pi a \sin\theta}{\lambda}.$$

Thus we get IDF for a rectangular aperture.

Fraunhofer Diffraction in a Circular Aperture

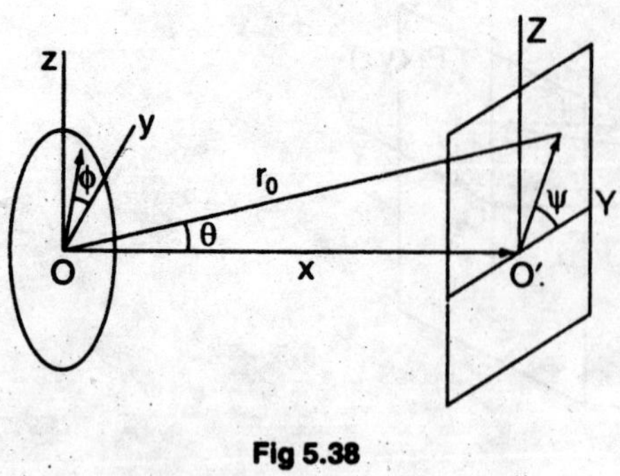

Fig 5.38

This diffraction phenomenon is of practical importance because in a telescope, the resolution is limited by the diffraction effect in a circular aperture.

In this case we can get,

$$\begin{aligned} y &= \rho\cos\phi \quad Y = \rho'\cos\psi \\ z &= \rho\sin\phi \quad Z = \rho'\sin\psi \\ dydz &= \rho d\phi d\rho \end{aligned} \tag{5.15.10}$$

$$
\begin{aligned}
d\phi_p &= \frac{C}{\pi a^2 r_0} e^{i(\omega t - kr_0)} e^{ik\rho\cos\phi\rho'\cos\psi} \cdot e^{ik\rho\sin\phi\rho'\sin\psi}\rho d\phi d\rho \\
&= \frac{C}{\pi a^2 r_0} e^{i(\omega t - kr_0)} e^{ik\rho\rho'\cos(\phi-\psi)}\rho d\phi d\rho \\
\therefore \quad \phi_p &= \frac{C}{\pi a^2 r_0} e^{i(\omega t - kr_0)} \int_0^a \int_0^{2\pi} e^{ik\rho\rho'\cos(\phi-\psi)}\rho\, d\rho d\phi \\
&= \frac{C}{\pi a^2 r_0} e^{i(\omega t - kr_0)} \int_0^a \left[\int_0^{2\pi} e^{ik\rho\rho'\cos(\phi-\psi)} d\phi\right]\rho\, d\rho
\end{aligned}
\tag{5.15.11}
$$

Now, $J_0(x)$, called *Bessel function* of zero order is defined as

$$
J_0(x) = \frac{1}{2\pi}\int_0^{2\pi} e^{ix\cos(\phi-\psi)} d\phi. \tag{5.15.12}
$$

The value of $J_0(x)$ can be obtained as a power series in x from the general Bessel function of order n denoted as $J_n(x)$ and given by,

$$
J_n(x) = \sum_{r=0}^{\infty}(-1)^r\left(\frac{x}{2}\right)^{2r+n}\frac{1}{\Gamma(r+1)\Gamma(n+r+1)} \tag{5.15.13}
$$

$$
\therefore \quad J_0(x) = \sum_0^{\infty}\left(\frac{x}{2}\right)^{2r}(-1)^r\frac{1}{\Gamma(r+1)\Gamma(r+1)} = 1 - \left(\frac{x}{2}\right)^2 + \left(\frac{x}{2}\right)^4 \cdot \frac{1}{[\Gamma(3)]^2} + \cdots \tag{5.15.14}
$$

$$
\therefore \quad \int_0^{2\pi} e^{ik\rho\rho'\cos(\phi-\psi)} d\phi = J_0(x),
$$

where $x = (k\rho\rho'/X) = 2\pi\left(\frac{\rho}{\lambda}\right)\left(\frac{\rho'}{X}\right)$.

So,

$$
\begin{aligned}
\phi_p &= \frac{C}{\pi a^2 r_0} e^{i(\omega t - kr_0)} \int_0^a \left[\int_0^{2\pi_0} e^{ix\cos(\phi-\psi)} d\phi\right]\rho\, d\rho \\
&= \left(\frac{C}{\pi a^2 r_0}\right) e^{i(\omega t - kr_0)} \cdot 2\pi \int_0^{2\pi a\rho'/\lambda X} J_0(x) x\, dx \cdot \left(\frac{\lambda X}{2\pi\rho'}\right)^2 \\
&= \frac{2\pi C}{\pi a^2 r_0} e^{i(\omega t - kr_0)} \cdot \left(\frac{\lambda X}{2\pi\rho'}\right)^2 \int_0^{(2\pi a\rho'/\lambda X)} J_0(x) x\, dx \\
&= (2C/r_0) e^{i(\omega t - kr_0)} (\lambda X / 2\pi\rho' a)^2 \int_0^{(2\pi a\rho'/\lambda X)} J_0(x) x\, dx
\end{aligned}
$$

Again, from the property of the Bessel function[1]

$$
\int_0^{(2\pi a\rho'/\lambda x)} x J_0(x) dx = \int_0^{x'} x J_0(x) dx = J_1(x') x' \tag{5.15.15}
$$

[1] For further details, see *A Text-book of Mathematical Physics* by Chakrabarti and Kundu.

where $J_1(x')$ is the first order Bessel function.

$$\therefore \quad \phi_p = \left(\frac{2C}{r_0}\right) e^{i(\omega t - kr_0)} \left(\frac{\lambda X}{2\pi\rho' a}\right)^2 \cdot \left(\frac{2\pi a\rho'}{\lambda X}\right) J_1\left(\frac{2\pi a\rho'}{\lambda X}\right)$$

$$\Rightarrow \quad \phi_p = \left(\frac{2C}{r_0}\right) e^{i(\omega t - kr_0)} \cdot \left(\frac{\lambda X}{2\pi\rho' a}\right) J_1\left(\frac{2\pi a\rho'}{\lambda X}\right)$$

$$= \left(\frac{2C}{r_0}\right) e^{i(\omega t - kr_0)} \frac{J_1(\nu)}{\nu} \tag{5.15.16}$$

where $\nu = \left(\frac{2\pi a\rho'}{\lambda X}\right) = \frac{2\pi a}{\lambda}\sin\theta$.

$$\therefore \quad I_p = \phi_p\phi_p^* = \left(\frac{2C}{r_0}\right)^2 \left[\frac{J_1(\nu)}{\nu}\right]^2 = \left(\frac{C}{r_0}\right)^2 \left[\frac{2J_1(\nu)}{\nu}\right]^2. \tag{5.15.17}$$

Now,

$$J_1(\nu) = \sum_0^\infty (-1)^r \left(\frac{\nu}{2}\right)^{2r+1} \frac{1}{\Gamma(r+1)\Gamma(r+2)}$$

$$= \left(\frac{\nu}{2}\right) - \left(\frac{\nu}{2}\right)^3 \frac{1}{\Gamma(2)\Gamma(3)} + \left(\frac{\nu}{2}\right)^5 \frac{1}{\Gamma(3)\Gamma(4)} - \cdots$$

$$= \left(\frac{\nu}{2}\right) - \frac{1}{2}\left(\frac{\nu}{2}\right)^3 + \left(\frac{\nu}{2}\right)^5 \frac{1}{2}\cdot\frac{1}{3}\cdot\frac{1}{2} - \cdots$$

$$= \frac{\nu}{2}\left[1 - \frac{1}{2}\left(\frac{\nu}{2}\right)^2 + \frac{1}{12}\left(\frac{\nu}{2}\right)^4 - \cdots\right] \tag{5.15.18}$$

$$I_p = \left(\frac{C}{r_0}\right)^2 \left[\frac{J_1(\nu)}{(\nu/2)}\right]^2 \tag{5.15.19}$$

$$= \left(\frac{C}{r_0}\right)^2 \left[\frac{(\nu/2)}{(\nu/2)}\left\{1 - \frac{1}{2}\left(\frac{\nu}{2}\right)^2 + \frac{1}{12} + \left(\frac{\nu}{2}\right)^4 - \cdots\right\}\right]^2$$

As $\nu \to 0$, $\frac{J_1(\nu)}{\nu} = \frac{1}{2}$ by applying L'Hospital's theorem.

$$\therefore \quad I_p = I_0 = \left(\frac{C}{r_0}\right)^2.$$

So the intensity distribution is given by

$$I_p = I_0 \left[\frac{J_1(\nu)}{\nu/2}\right]^2. \tag{5.15.20}$$

Principal Maxima

The principal maxima corresponds to $\nu = 0$, when $I = I_0$ (maximum). Now, $\nu = \frac{2\pi a \sin\theta}{\lambda}$, so when $\nu = 0, \theta = 0$. So the undiffracted rays go to form the central bright disc, called *Airy's disc*.

Secondary Maxima

The secondary maxima would be given by the condition

$$\frac{d}{d\nu}\left[\frac{J_1(\nu)}{\nu}\right] = 0 \tag{5.15.21}$$

where $\nu \neq 0$.

Now, the property of Bessel function is that

$$\frac{d}{dx}[x^{-n}J_n(x)] = -x^{-n}J_{n+1}(x). \tag{5.15.22}$$

So taking $n = 1$, we get from above

$$\frac{d}{dx}[J_1(x)/x] = -\frac{1}{x}J_2(x) \tag{5.15.23}$$

$\therefore$ Eqn. (5.15.23) for secondary maxima thus reduces to

$$-\frac{1}{\nu}J_2(\nu) = 0 \quad \Rightarrow \quad J_2(\nu) = 0. \tag{5.15.24}$$

So the values of ν satisfying eqn. (5.15.24) defines the position of the secondary maxima.

Minima

The intensity will be zero at those points defined by the equation given as

$$J_1(\nu) = 0 \tag{5.15.25}$$

Now, $\nu = 0$ defines the principal maximum where the normalised intensity becomes equal to 1. The value of ν is defined as $\nu = \frac{2\pi a \sin\theta}{\lambda}$.

So, $\nu = 0 \Rightarrow \sin\theta = 0 \Rightarrow \theta = 0$ giving the central maximum. The other maxima are given by the solution of $J_2(\nu) = 0$.

From the table of the Bessel functions, the solutions of the above equation can be written as

$$\frac{\nu}{2} = 0.819\pi,\ \ 1.333\pi,\ \ 1.847\pi,\ \ 2.361\pi,\ \ \text{etc.} \tag{5.15.26}$$

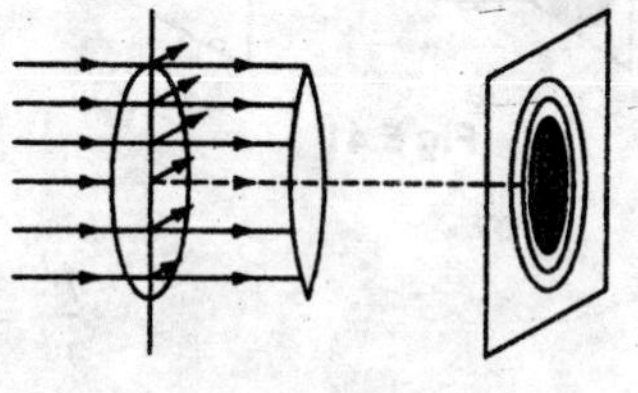

Fig 5.39

The solutions of eqn. (5.15.25) giving the secondary minima are given as

$$\frac{\nu}{2} = 0.610\pi,\ \ 1.116\pi,\ \ 1.619\pi,\ \ 2.1290\pi,\ \ \text{etc.} \tag{5.15.27}$$

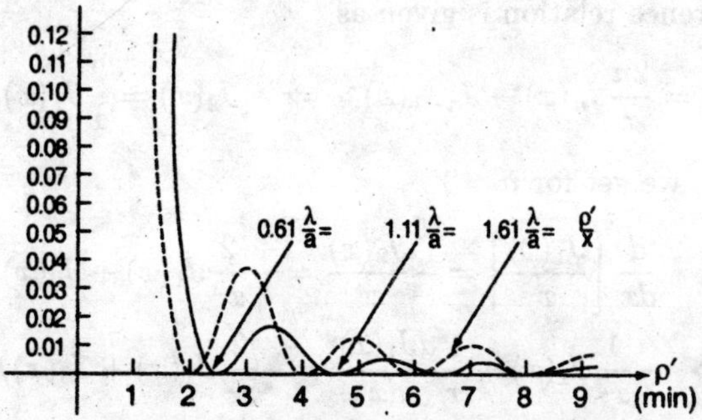

Fig 5.40

The plot of IDF is shown in fig. 5.40. In this figure normalised intensity (I_p/I_0) is plotted against ρ' (in mm) with specified parameter values of radius $a = 0.5$ mm, $\lambda = 500$ nm and $X = 4$ nm. The broken lines represent IDF for a slit of length equal to $2a = 1$ mm. The first three zeros of intensity are shown to be at

$$\frac{\nu}{2} = \frac{\pi a \sin\theta}{\lambda} = .610\pi \quad \text{1st Minimum}$$

$$\Rightarrow \quad \sin\theta = .610\frac{\lambda}{a} \quad \Rightarrow \quad \left(\frac{\rho'}{X}\right)_1 = 0.610\frac{\lambda}{a}.$$

The second minimum occurs at

$$\left(\frac{\rho'}{X}\right)_2 = 1.116\lambda/a \quad \text{2nd Minimum}$$

$$\left(\frac{\rho'}{X}\right)_3 = 1.619\lambda/a \quad \text{3rd Minimum, etc.}$$

So the radius of the 1st dark ring is $(\rho')_1 = .610\frac{\lambda}{a}X = .610\lambda af$, where $f = X$ is the focal length of the lens. Similarly, we can find the other radii.

The fraction of the total incident energy contained within a circle of (angular) radius θ in the image plane, centered on the geometrical image can be calculated from the intensity expression. From these calculations we find that about 90% of the energy is contained within the circle bounded by the second dark ring.

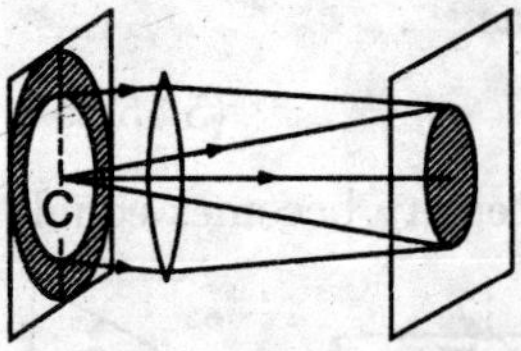

Fig 5.41

If $F(R)$ represents the fractional energy contained in a circle of radius R, then

$$F(R) = \frac{\int_0^R I(\rho') \times 2\pi\rho' d\rho'}{\int_0^\infty I(\rho') \times 2\pi\rho' d\rho'} \tag{5.15.28}$$

Now,

$$\frac{\rho'}{r'} = \sin\theta, \quad \nu = ka\sin\theta = \frac{ka\rho'}{r'} \quad \Rightarrow \rho' = \frac{r'\nu}{ka}.$$

$$\therefore \quad F(\nu) = \frac{\int_0^\nu [2J_1(\nu)/\nu]^2 \nu d\nu}{\int_0^\infty [2J_1(\nu)/\nu]^2 \nu d\nu}. \tag{5.15.29}$$

An important recurrence relation is given as

$$J_{n+1}(x) = \frac{2n}{x}J_n(x) - J_{n-1}(x) \quad \Rightarrow \quad J_2(x) = \frac{2}{x}J_1(x) - J_0(x).$$

From eqn. (5.15.22), we get for $n = 1$

$$\frac{d}{dx}\left[\frac{J_1(x)}{x}\right] = -\frac{J_2(x)}{x} = -\frac{2}{x^2}J_1(x) + J_0(x)/x \tag{5.15.30}$$

$$\Rightarrow \quad -\frac{1}{x^2}J_1(x) + \frac{1}{x}\frac{dJ_1(x)}{dx} = -\frac{2}{x^2}J_1(x) + J_0(x)/x$$

$$\Rightarrow \quad \frac{J_1(x)}{x^2} = J_0(x)/x - \frac{1}{x}\frac{dJ_1(x)}{dx}$$

$$\Rightarrow \quad \frac{J_1^2(x)}{x} = J_1(x)\left[J_0(x) - \frac{dJ_1(x)}{dx}\right] = \left[-J_0(x)\frac{dJ_0(x)}{dx} - \frac{dJ_1}{dx}J_1(x)\right]$$

$$= -\frac{1}{2}\frac{d}{dx}[J_1^2(x) + J_0^2(x)] \qquad (5.15.31)$$

Using eqn. (5.15.30) in eqn. (5.15.31), we get

$$F(\nu) = \frac{\int_0^{\nu} \frac{d}{d\nu}[J_0^2(\nu) + J_1^2(\nu)]d\nu}{\int_0^{\infty} \frac{d}{d\nu}[J_0^2(\nu) + J_1^2(\nu)]d\nu} = \frac{[J_0^2(\nu) + J_1^2(\nu)]|_0^{\nu}}{[J_0^2(\nu) + J_1^2(\nu)]|_0^{\infty}}$$

$$= \frac{J_0^2(\nu) + J_1^2(\nu) - J_0^2(0) - J_1^2(0)}{J_0^2(\infty) + J_1^2(\infty) - J_0^2(0) - J_1^2(0)} = 1 - J_0^2(\nu) - J_1^2(\nu)$$

because $J_0(0) = 1$, $J_0(\infty) = 0$ and $J_1(0) = J_1(\infty) = 0$.

$$\therefore \; F(\nu) = 1 - J_0^2(\nu) - J_1^2(\nu). \qquad (5.15.32)$$

A plot of $F(\nu)$ against ν according to the relation (5.15.32) is shown in fig. 5.42. From the figure (5.42), it is evident that about 90% of the energy is contained within the circle bounded by the second dark ring.

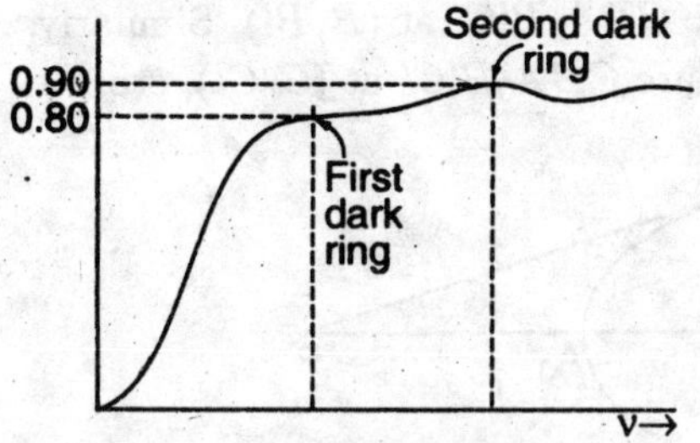

Fig 5.42

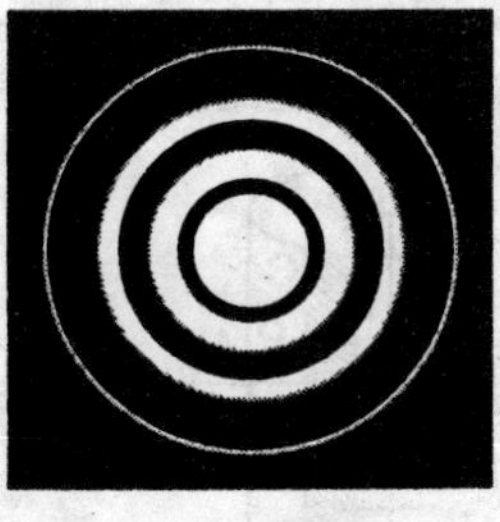

Fig 5.43

The diffraction pattern due to a circular aperture is shown in fig. 5.43 which clearly indicates the formation of Airy's disc, surrounded by alternate bring and dark rings.

The first minimum is defined by

$$\sin\theta = 0.610\frac{\lambda}{a} \quad \Rightarrow \quad \theta = 0.610\frac{\lambda}{a} = 1.22\frac{\lambda}{D},$$

where D is the diameter of the circular aperture. The rays diffracted at angle θ make a cone (see fig. 5.41) with the semi-vertical angle θ. So the diffraction fringes minima or maxima will be circular. The central circular bright disc formed by the rays characterised by $\theta = 0$ (i.e., undiffracted rays) is the brightest, it is called *Airy's disc* it is surrounded by alternate dark and bright rings of rapidly decreasing intensity. If the circular aperture is not sufficiently illuminated or if the exposure to the photographic plate is not of long duration, only the Airy's disc is available. By long exposure or for a strong source, two or three secondary maxima are also available.

In fig. 5.40, broken lines indicate the intensity of a slit of linear dimension $l = 2a = D$ (diameter of the circular aperture). From the graphical representation it is evident that minima occur at smaller value of θ in case of slit diffraction.

Eqn. (5.15.27) defining the positions of the minima is very important because it will be used in the consideration of resolving power of a telescope (discussed in chapter entitled Resolving Power).

Fresnel Diffraction

5.16 Introduction

At the beginning of our discussion on the phenomenon of diffraction we introduced the difference between the Fresnel and Fraunhofer type of diffraction pattern. The different diffraction patterns (single-slit, double-slit, grating, etc.) belonging to Fraunhofer class have been discussed elaborately in the preceding sections. At present we will consider the Fresnel class of diffraction.

The essential difference between these two classes of diffraction lies in the positions of the light source, screen with respect to the diffracting centre. In Fraunhofer class we considered the incident wavefront and the diffracted wavefront both the plane wavefronts, meaning that both the source and the receiving screen are virtually at infinity. But in Fresnel's diffraction either the source or the screen or both are at finite distance from the diffracting element. In this case the wavefronts are either spherical or cylindrical.

To explain the Fresnel diffraction at straight edge, small slit, a disc, etc., the simplest approach is to consider the Fresnel half-period zones as outlined at the beginning of this chapter. To recapitulate the consideration, we take the cylindrical wavefront emerging out of a narrow linear slit, $PA_0 = b$ is the shortest distance of P w.r.t. the wavefront, A_0 being referred to as the pole w.r.t. P.

Taking $PB = b + \frac{\lambda}{2}$ as radius draw arcs which cut the arc $CBA_0B'C'$ at (B, B'). Similarly, take $PC = b + 2\frac{\lambda}{2}$ and construct similar arcs which cut the arc $CBA_0B'C'$ at (C, C'), etc.

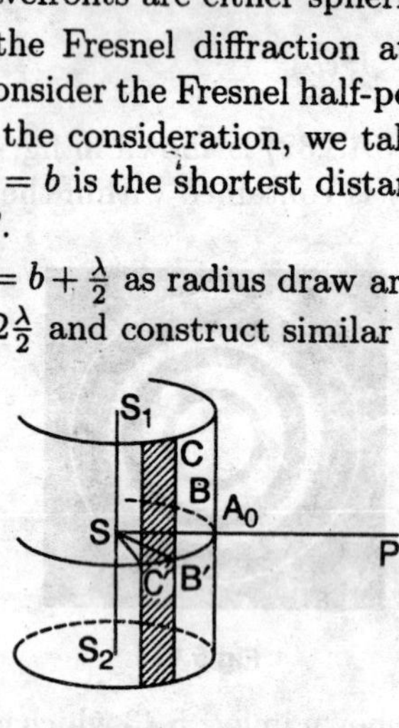

Fig 5.44

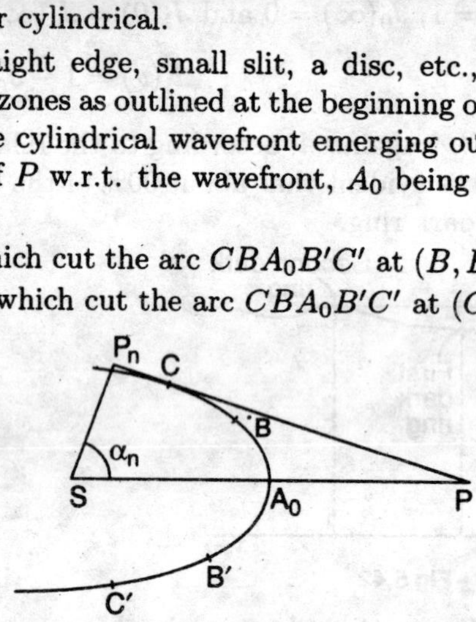

Fig 5.45

The strips of the wavefront of length $S_1S_2 = L$ and width A_0B or A_0B' are called the ***first half-period strips*** or ***elements of the wavefront.*** The strips of width BC or B_1C_1 are the second half-period element and so on.

Let α_n be the angle subtended at S between the end P_n of the nth half-period element and the line SA_0P. The length $PP_n = b + n\frac{\lambda}{2}$.

$$\therefore \quad (PP_n)^2 = SP_n^2 + SP^2 - 2SP_n \cdot SP\cos\alpha_n = a^2(a+b)^2 - 2a(a+b)\cos\alpha_n$$

$$\Rightarrow \quad \left(b + n\cdot\frac{\lambda}{2}\right)^2 = a^2 + a^2 + 2ab + b^2 - 2a(a+b)\cos\alpha_n$$

$$\Rightarrow \quad b^2 + bn\lambda + n^2\frac{\lambda^2}{4} = 2a^2 + 2ab + b^2 - 2a(a+b)\cos\alpha_n$$

$$\Rightarrow \quad bn\lambda = 2a^2 + 2ab - 2a(a+b)\left(1 - \frac{\alpha_n^2}{2} + \cdots\right) = a(a+b)\alpha_n^2, \text{ neglecting } n^2\frac{\lambda^2}{4}$$

$$\Rightarrow \quad \alpha_n = \sqrt{\frac{bn\lambda}{(a(a+b)}} = k\sqrt{n}, \text{ where } k \text{ is a constant} = \sqrt{\frac{b\lambda}{a(a+b)}}.$$

The area of the nth strip

$$= La(\alpha_n - \alpha_{n-1}) = Lak(\sqrt{n} - \sqrt{n-1}) = Lak\left[\sqrt{n} - \sqrt{n}\left(1 - \frac{1}{n}\right)^{\frac{1}{2}}\right] = \frac{Lak}{2\sqrt{n}}.$$

Thus the area of the strip slightly decreases with the increase in the order number.

The phase difference between the waves reaching the point P_0 from A_0 and B is $\delta = \frac{2\pi}{\lambda} \times$ path diff. $= \frac{2\pi}{\lambda} \times \frac{\lambda}{2} = \pi$. So if the phase condition of the wavelet from A_0 reaching P is taken as reference, that of the wavelet from B is π; hence any wavelet reaching P from any point in the first half-period strip has on the average a phase $\frac{\pi}{2}$; similarly, the wavelet from B reaching P has phase π, the wavelet from C reaching phase 2π w.r.t. the wavelet reaching P from A_0; so the average phase for any wavelet from the second half-period strip is $\frac{3\pi}{2}$, thus wavelets from the first strip and second strip reaching P differ in phase by π (opposite in phase). Thus, if the displacement due to the wavelets from the first half-period element (considering both halves of the half-period strip with respect to A_0) be D_1, that due to those from the second element D_2 and so on, the resultant displacement

$$D = D_1 - D_2 + D_3 - \cdots.$$

Fresnel Half-period Zones in case of Spherical Wavefronts

We have already discussed the construction of Fresnel half-period zones in case the wavefront is spherical. In this case also we have the resultant displacement given as

$$D = d_1 - d_2 + d_3 - \cdots.$$

So when n is large, in both the cases the resultant displacement D is given by

$$D = \frac{D_1}{2} \quad \text{or} \quad \frac{d_1}{2}.$$

So Frésnel diffraction phenomena where either cylindrical or spherical wavefronts are involved, can be given simple explanation in terms of Fresnel half-period zones of half-period strips. More rigorous analysis, of course, comes through the formulation of Fresnel integrals and intensity distribution obtained through the plot of Cornu spiral. We will first use the concept of half-period elements to explain some of the basic problems of Fresnel pattern and then utilise the rigorous analytical approach involving Fresnel's integral.

In case of Plane Wavefronts

In case of plane wavefront also it is shown that the wavelets coming from the first half-period zone and second half-period zone differ in phase on the average by an amount equal to π. Thus the resultant displacement at a point P ahead of the wavefront is given by

$$D = d_1 - d_2 + d_3 - \cdots.$$

The resultant displacement as shown at the beginning of our discussion is $D = \frac{D_1}{2}$, that is, half the displacement due to the first half-period element.

From the consideration of the half-period elements we can explain some of the basic facts of Fresnel diffraction.

Explanation of Rectilinear Propagation

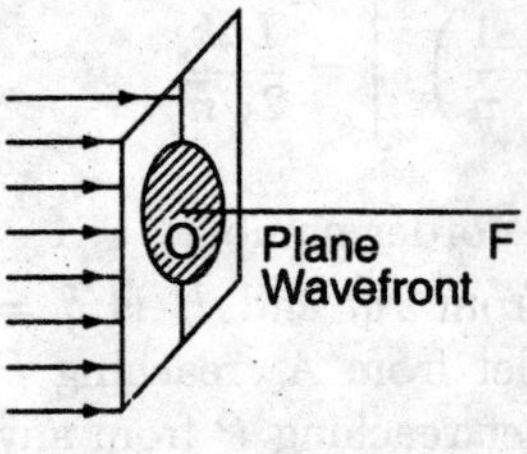

Fig 5.46

Suppose in the path of a parallel beam of rays an opaque disc is placed. To find the intensity at P we consider the half-period elements in the plane wavefront. If the size of the disc is such that it covers the first half-period zone, then from the second half-period zones onwards all other zones are unobstructed hence, the resultant displacement is given by

$$D = \left(\frac{d_2}{2}\right).$$

The area of the half-period zones depends on the value of b, the distance of P w.r.t. the pole O, thus for a nearer point, the area of the half-period zones becomes so small that the opaque disc covers a large number of half-period zones. So when P is very close, intensity at P will be very small. If P is not far off, even a very small obstacle is sufficient to cover up a large number of half-period zones resulting in small intensity at P. This explains rectilinear propagation.

A Circular aperture in the Path of Parallel Rays

When the plane wavefront occupies the plane of the aperture, depending on the size of the aperture and the distance of the point w.r.t. O, it might so happen that the aperture transmits only the first half-period zone, so the displacement is $D = d_1$. If the first two half-period zones are transmitted, the displacement will be given by $D' = (d_1 - d_2)$ which is very nearly equal to zero. If the aperture transmits first three zones, the displacement is $D'' = (d_1 - d_2 + d_3) \simeq d_1$, so the intensity again maximum. So at the point P, the intensity will be either maximum or minimum depending on whether the aperture transmits odd or even number of half-period zones.

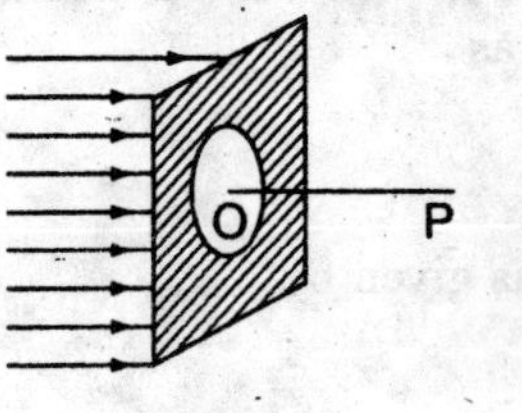

Fig 5.47

Zone Plate—its construction and uses

Before we go over to the rigorous approach to explain fresnel diffraction pattern in terms of Fresnel integral, we finish off the discussion of the zone plate based on the consideration of Fresnel's half-period zones.

Zone plate may be considered as a practical outcome of the Fresnel's concept about half-period zones. We have seen all through our previous simple consideration of Fresnel's diffraction problems that the wavefronts whether plane, spherical or cylindrical can be divided into Fresnel's half-period zones or strips such that phase condition of wavelets reaching certain point P from all odd or all even half-period zones is same. Thus we have seen the resultant displacement at P due to a wavefront divided into Fresnel zones is given by

$$\begin{array}{ccccccc} D = & d_1 & - & d_2 & + & d_3 & - d_4 + \cdots \\ & \downarrow & & \downarrow & & \downarrow & \\ & \text{First half-zone} & & \text{Second half-zone} & & \text{Third half-zone,} & \text{etc.} \end{array}$$

If all the wavefront is unobstructed, the series displacement sum tends to $\left(\frac{d_1}{2}\right)$ approximately. But if only the first half-period zone is unobstructed while all other zones are locked, then the resultant displacement is

$$D = d_1 \quad \Rightarrow \quad \text{Intensity} \propto d_1^2.$$

But for the whole wavefront, the intensity is proportional to $\left(\frac{d_1^2}{4}\right)$ which is four times less compared to (d_1^2). So it is evident from this consideration that if the odd or even zones are blocked, the resultant intensity at some point will be much higher than the intensity available when the whole wavefront is allowed to be transmitted. This consideration is utilized in the construction of the zone plate.

Theory of Zone Plate

Consider a plane $\perp$ to the line joining O and S through the point P on the line OS. Let $SP = u$, $PO = v$. Take points M_1, M_2, M_3, M_4, etc. on the plane $ABCD$ such that

$$SM_1 + M_1O = SP + PO + \frac{\lambda}{2} = u + v + \frac{\lambda}{2}$$

$$SM_2 + M_2O = u + v + 2\frac{\lambda}{2}$$

....

$$SM_n + M_nO = u + v + n\frac{\lambda}{2}.$$

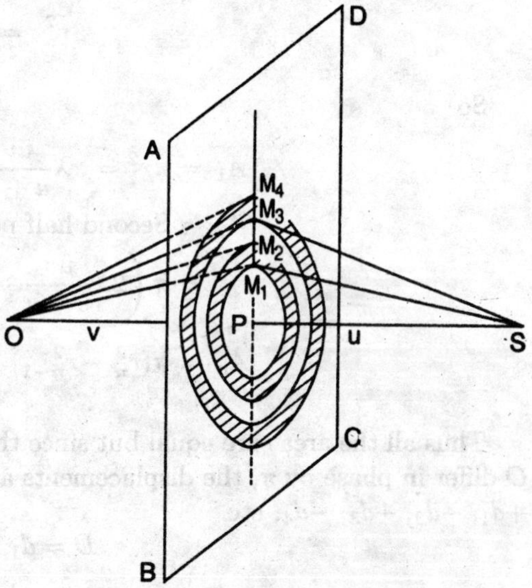

Fig 5.48

The radius PM_1 is the radius of the first half-period zone and let $PM_1 = r_1$, also let $PM_2 = r_2$.

$\therefore$ The first circular zone is of area $A_1 = \pi r_1^2$.

From fig. 5.48

$$SM_1^2 = r_1^2 + u^2 = u^2\left(1 + \frac{r_1^2}{u^2}\right), OM_1^2 = r_1^2 + v^2 = v^2\left(1 + \frac{r_1^2}{v^2}\right)$$

$$\Rightarrow SM_1 = u\left(1 + \frac{r_1^2}{u^2}\right)^{\frac{1}{2}} = u\left(1 + \frac{1}{2}\frac{r_1^2}{u^2}\right), OM_1 = v\left(1 + \frac{r_1^2}{v^2}\right)^{\frac{1}{2}} = v\left(1 + \frac{1}{2}\frac{r_1^2}{v^2}\right)$$

$$\Rightarrow SM_1 + OM_1 = (u + v) + \frac{1}{2}r_1^2\left(\frac{1}{u} + \frac{1}{v}\right)$$

$$\Rightarrow SM_1 + OM_1 - (u + v) = \frac{1}{2}r_1^2\left(\frac{1}{v} + \frac{1}{u}\right) \quad \Rightarrow \quad \frac{\lambda}{2} = \frac{1}{2}r_1^2\left(\frac{1}{v} + \frac{1}{u}\right)$$

Similarly,

$$SM_n + OM_n - (u + v) = \frac{1}{2}r_n^2\left(\frac{1}{v} + \frac{1}{u}\right) \Rightarrow n\frac{\lambda}{2} = \frac{1}{2}r_n^2\left(\frac{1}{v} + \frac{1}{u}\right)$$

$$\Rightarrow r_n^2\left(\frac{1}{v} + \frac{1}{u}\right) = n\lambda \qquad (5.16.1)$$

Putting $n = 1, 2, 3$, etc. we get r_1, r_2, r_3, etc.

$$\begin{aligned} r_1^2 &= \lambda \frac{uv}{u+v}, \\ r_2^2 &= 2\lambda \frac{uv}{u+v}, \\ r_3^2 &= 3\lambda \frac{uv}{u+v} \\ &\dots\dots\dots \\ r_{n-1}^2 &= (n-1)\lambda \frac{uv}{u+v}, \\ r_n^2 &= n\lambda \cdot \frac{uv}{u+v} \end{aligned}$$

So

$$\begin{aligned} A_1 &= \pi r_1^2 = \pi\lambda \frac{uv}{u+v} \\ A_2 &= \text{Second half-period zone area} = \pi(r_2^2 - r_1^2) \\ &= \pi\left(2\lambda \frac{uv}{u+v} - \lambda \frac{uv}{u+v}\right) = \pi \frac{\lambda uv}{(u+v)} = A_1 \\ A_n &= \pi(r_n^2 - r_{n-1}^2) = \pi \frac{\lambda uv}{(u+v)} = A_1 \end{aligned}$$

Thus all the areas are equal but since the wavelets in the alternate zones reaching the point O differ in phase by π, the displacements at O due to the wavelets from the zones are given as $+d_1$, $-d_2$, $+d_3$, $-d_4$, etc.

$$\therefore \quad D = d_1 - d_2 + d_3 - d_4 + \cdots.$$

But though the areas of the zones are same, the values of d_1, d_2, d_3, etc., will form a decreasing sequence because of the obliquity of the different zones.

If the alternate zones are blocked, may be either even numbered or odd numbered zones are blocked, then the resultant displacement will be many times large than the displacement due to the wavelets from all zones. Thus the point O will be a point of maximum illumination and so light from S is said to be focussed at O.

From eqn. (5.16.1) above, we get

$$\frac{1}{v} + \frac{1}{u} = \frac{1}{(r_n^2/n\lambda)} \tag{5.16.2}$$

This result is similar to the action of a converging lens whose focal length is given by

$$\frac{1}{f} = \frac{n\lambda}{r_n^2}.$$

Similarity and Dissimilarity between a Convex Lens and a Zone Plate

So far as the focussing action of the zone plate is concerned, it satisfies the same equation as that for a convex lens (c.f. eqn. 5.16.2). But there are points of differences also. A convex lens has a fixed focl length for a particular wavelength but a zone plate has a number of focii. Consider fig. 5.49, here the circles are drawn with radii $\sqrt{1k}$, $\sqrt{2k}$, $\sqrt{3k}$, etc., where $k = \frac{\lambda uv}{u+v}$, where u is the source point and v is the point of focus or bright point. If the source point S is at a distance equal to $\frac{k^2}{\lambda}$, then

$$r_1^2 = \frac{\lambda uv}{u+v}, \quad r_2^2 = \frac{2\lambda uv}{u+v}, \text{ etc.}$$

such that blackened rings correspond to second, fourth, sixth, etc., half-period zones and the first, third, fifth, etc., half-period zones are unobstructed. So at O, the displacement is given by,

$$D = d_1 + d_3 + d_5 + \cdots$$

hence O is a bright point.

Fig 5.49

Thus at the first focus for the zone plate shown in fig. 5.49 where the displacement is given by

$$D = d_1 + d_3 + d_5 + \cdots,$$

odd zone contributions sum, hence a maximum. Again, if $v = v_1$ is such that the first half-period zone which is the first unobstructed zone for v contains first three half-period zones and similarly for the other unobstructed half-period zones, then the contributions by two will cancel leaving the contribution from the third; this is true for any of the unobstructed half-period zone. Thus we get another bright point.

Now, the area of a half-period zone for a particular value of u and v is given by

$$A = \frac{\pi uv\lambda}{(u+v)} = \frac{\pi u\lambda}{\left(1+\frac{v}{u}\right)}. \tag{5.16.3}$$

So for a particular source, if v changes, the area of the half-period zone also varies; when v decreases, A increases which means that the previous first half-period zones now may include three half-period zones each having an area

$$A' = \frac{1}{3}\frac{\pi uv\lambda}{(u+v)} = \frac{\pi v'\lambda u}{(u+v')} \quad \Rightarrow \quad v' = \frac{uv}{(2u+2v)},$$

so the second focus is at a distance v'. Also, $r_n^2 = n\lambda\frac{uv}{u+v}$

$$r_n'^2 = n\lambda\frac{uv'}{u+v'} n\lambda \times \frac{1}{3}\frac{uv}{u+v} \quad \therefore \quad r_n'^2 = \frac{1}{3}r_n^2.$$

So the first focal length is given by,

$$\frac{1}{f} = \frac{n\lambda}{r_n^2}, \frac{1}{f'} = \frac{n\lambda}{r_n'^2} = \frac{3n\lambda}{r_n^2} \quad \Rightarrow \quad f' = \frac{r_n^2}{3n\lambda}$$

giving the second focal length. Thus the zone plate may be looked upon as a multiple focus lens. So the zone plate formula can be written as,

$$\frac{1}{V} + \frac{1}{U} = (2n+1)\frac{\lambda}{r_n^2} = \frac{1}{f_n}. \tag{5.16.4}$$

Again, as λ increases f decreases, so in a zone plate red-focus will be nearer to the zone plate than the violet-ray focus. But in case of a lens since $\frac{1}{f} = (\mu - 1)k$ and $\mu_v > \mu_R$, $f_v < f_r$.

Zone-Plate Construction

To construct a zone plate on a paper draw a number of circle whose radii are proportional to the square root of the natural numbers. Alternate circular zones are blackened. A reduced photograph of this drawing pasted on a thin glass plate constitutes what we call a zone plate. If the first half-period zone is transparent and the even zones are blackened the zone plate is said to be a *positive zone plate*; if the centre is blackened and all other odd zones are blackened, the zone plate is said to be a negative zone plate (fig. 5.50(b) and (a)).

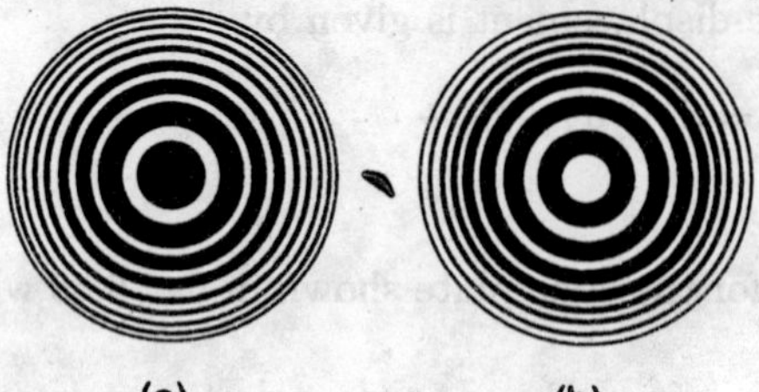

(a) (b)

Fig 5.50

Phase Reversal Zone-Plate

In case of the zone plate just considered, alternate zones are blackened. So when such a zone plate is exposed to parallel beam of rays, half the incident intensity is transmitted and utilised in the formation of the bright focus. In a zone plate, wavelets from two successive zones differ in phase by π, so if we can reverse the phase of the wavelets from every alternate zones, there is no question of blocking alternate zones; all the zones may be utilised for transmission and hence intensity of bright focus points will be greatly increased. This idea was first put to practical realisation of what is known as *phase reversal* zone plate by R.W. Wood. He took a photograph of a usual zone plate on a photographic plate containing a layer of gelatine mixed with bichromate of potash. The portion of gelatine receiving light has chemical reaction and here solubility decreases. Such a photographic plate when developed under warm water will show differential thickness of gelatine; the circular zones receiving light will have greater thickness of gelatine than the circular zones which did not receive light (see fig. 5.51).

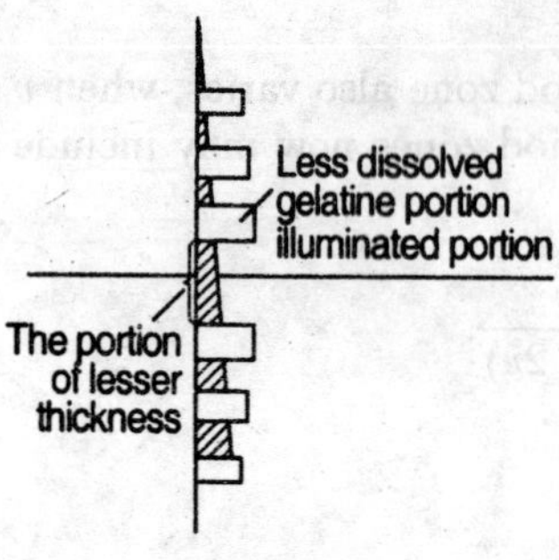

Fig 5.51

By controlling the thickness of the gelatine layer, light passing through two successive half-period zones may be rendered almost in phase; hence with such a zone plate it is possible to obtain sharper points of focus.

Babinet's Principle

This principle is a very important and basic principle applicable to the diffraction pattern in many cases.

In order to understand the principle we must first understand the meaning of *complementary diffraction screen*.

Two diffraction screens are shown in fig. 5.52; fig. (a), is a metal plate with openings at three places which represent clear spaces, other places are opaque; whereas the diffraction screen in fig. (b) has three metallic disc portions (hence opaque) occupying the positions of the clear spaces in screen (a), at other places the screen (b) is transparent. Such screens are said to be complementary screens. The diffraction pattern due to the diffraction screen in (a) is similar to the diffraction pattern due to the screen in (b).

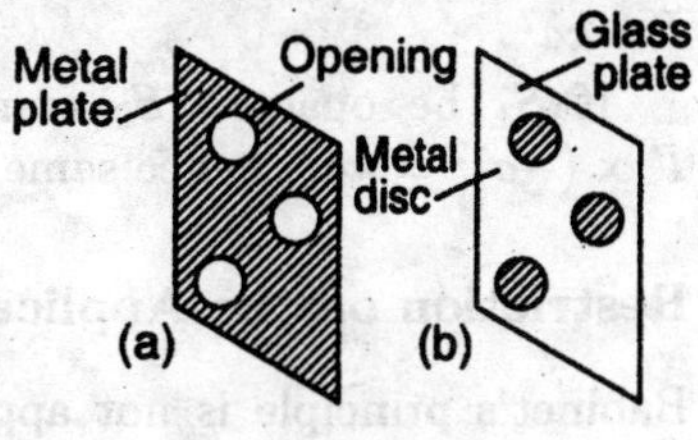

Fig 5.52

So the principle may be stated as—*the diffraction patterns in two complementary diffraction screens will be same.*

Illustration

In a plane transmission grating a is the width of the clear space and b is the width of the rulings (opaque space) so that the grating constant is $d = (a+b)$. Now, if the plane grating be constructed with b as the clear space and a as the width of the opaque space, the grating constant remains unaltered. With former specification of the grating, the IDF is given by,

$$I_p = I_0 \frac{\sin^2 \alpha}{\alpha^2} \cdot \frac{\sin^2 N\gamma}{\sin^2 \gamma}, \quad \text{where } \alpha = \frac{\pi a \sin\theta}{\lambda}, \gamma = \frac{\pi d \sin\theta}{\lambda}.$$

With the latter specification, it is given by

$$I'_p = I'_0 \frac{\sin^2 \alpha'}{\alpha'^2} \cdot \frac{\sin^2 N\gamma}{\sin^2 \gamma}.$$

Since γ remains unaltered, the following facts of the two gratings also remain the same—
(i) Positions of the principal maxima defined by $\sin\gamma = 0$ remain unchanged.
(ii) Positions of the secondary minima defined by $\sin N_\gamma = 0$ remain unchanged.

So the diffraction patterns due to diffraction in the two gratings are same. So the principle holds good.

Mathematical Formulation

When a small opening is there in an otherwise opaque screen, the light rays passing through the small hole are diffracted to produce a certain diffraction pattern on a receiving screen placed beyond the diffraction screen. The small opening may be divided into two parts S_1 and S_2 such that $S = S_1 + S_2$. At the point beyond the geometrical shadow where the intensity drops down to zero value, the amplitude due to one part is equal and opposite to the amplitude due to the other part. So if the displacement at the point P of zero intensity on the receiving screen due to the part S_1 of the opening be represented as

$$y_1 = R.P.C_1 e^{i(\omega t + \phi_1)}.$$

The displacement due to the part S_2 be

$$y_2 = R.P.C_2 e^{i(\omega t + \phi_2)},$$

then

$$y = y_1 + y_2 = 0 \quad \Rightarrow \quad C_1 = -C_2 = a, \phi_1 = \phi_2 = \phi$$

$$\therefore \quad y_1 = a\cos(\omega t + \phi), \quad y_2 = -a\cos(\omega t + \phi)$$

So if S_1 is transparent portion of the screen, S_2 opaque, then at P, the diffraction intensity $I = ka^2$.

If S_1 be opaque, S_2 transparent, then the diffraction intensity at the same point is $I' \propto (-a)^2 = ka^2$, hence same as I.

Restriction on the Applicability

Babinet's principle is not applicable to all cases of diffraction phenomena. Thus in case of Fresnel's diffraction by a circular opening and circular disc of the same size, there is patternwise difference and Babinet's principle is not applicable.

5.17 Analytical Approach to Fresnel's Diffraction

Without proceeding further with the consideration of Fresnel half-period elements, we now turn to the more rigorous, analytical approach to solve Fresnel's diffraction problems. It is through the formulation of Fresnel's integrals and explanation of the diffraction problems in terms of *Cornu spiral.*

Formulation of Fresnel's Integrals

Fresnel's integrals can be formulated from the first principle or through the use of Kirchhoff-Fresnel Diffraction equation, we discussed in the early part of our discussion.

To derive the Fresnel's integrals without any reference to any other equation, we consider fig. 5.53.

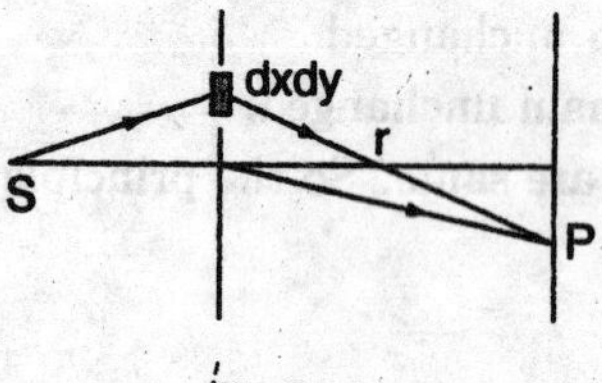

Fig 5.53

Consider an elementary area $dxdy$, the displacement of the source particles contained in the area is represented as

$$d\phi_s = Ae^{i\omega t}dxdy.$$

The displacement at P due to these source particles is written as

$$d\phi_p = \frac{Ae^{i(\omega t - kr)}}{r}dxdy.$$

Since in this case of diffraction, the screen is at a finite distance (unlike as in case of Fraunhofer type) this r is different for different source particles, but the observation screen being at large distance w.r.t. the diffracting aperture compared to the spread of the diffraction pattern on the receiving screen such that r in the denominator is replaced by d, the shortest distance. So for all the source particles, the effect can be written as

$$\phi_p = \iint \frac{A}{d}e^{-ikr}dxdy.$$

$$\therefore \quad r = \sqrt{d^2 + (x - X)^2 + (y - Y)^2} = d\sqrt{1 + \frac{(x - X)^2}{d^2} + \frac{(y - Y)^2}{d^2}}$$

$$= d\left[1 + \frac{1}{2}\frac{(x - X)^2}{d^2} + \frac{(y - Y)^2}{2d^2}\right] = d + \frac{(x - X)^2}{2d} + \frac{(y - Y)^2}{2d} \qquad (5.17.1)$$

$$\therefore \quad \phi_p = \iint \frac{A}{d}e^{-ik\left[d + \frac{(x-X)^2}{2d} + \frac{(y-Y)^2}{2d}\right]}dxdy = \bar{C}\iint e^{-ik\frac{(x-X)^2}{2d}}e^{-ik\frac{(y-Y)^2}{2d}}dxdy, \qquad (5.17.2)$$

where $\bar{C} = \frac{A}{d}e^{ikd}$, a constant.

Now, put

$$\frac{1}{2}\pi u^2 = k\frac{(x-X)^2}{2d} = \frac{\pi}{\lambda d}(x-X)^2$$

$$\frac{1}{2}\pi v^2 = k\frac{(y-Y)^2}{2d} = \frac{\pi}{\lambda d}(y-Y)^2$$

$$\Rightarrow \quad u = \sqrt{\frac{2}{\lambda d}(x-X)^2} = \sqrt{\frac{2}{\lambda}}(x-X) \tag{5.17.3}$$

$$v = \sqrt{\frac{2}{\lambda d}(y-Y)^2} = \sqrt{\frac{2}{\lambda d}}(y-Y) \tag{5.17.4}$$

$$\therefore \quad du = \sqrt{\frac{2}{\lambda d}}(dx),\ dv = \sqrt{\frac{2}{\lambda d}}(dy)$$

$$\therefore \quad \phi_p = \frac{\lambda d\bar{C}}{2}\iint e^{-i\frac{1}{2}\pi u^2}\cdot e^{-i\frac{1}{2}\pi v^2}dudv = \frac{1}{2}(\lambda d)\bar{C}\iint e^{-i\pi\frac{u^2}{2}}\cdot e^{-\pi\frac{v^2}{2}}dudv$$

$$= \frac{1}{2}\bar{C}(\lambda d)\int e^{-i\pi\frac{u^2}{2}}du\int e^{-i\pi\frac{v^2}{2}}dv \tag{5.17.5}$$

If the slit is a linear slit in x-direction, the transverse dimension being negligible, we can write

$$\phi_p = \bar{C}'\int e^{-i\pi\frac{u^2}{2}}du. \tag{5.17.6}$$

These equations (5.17.5) and (5.17.6) may be conveniently utilised according to the need of the problem.

Fresnel's Diffraction in a Straight Edge

In fig. 5.54 we have shown a straight edge illuminated by a light source S.

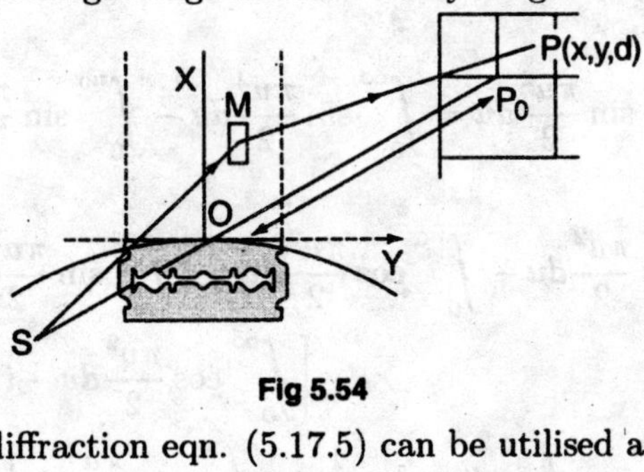

Fig 5.54

In this case of Fresnel's diffraction eqn. (5.17.5) can be utilised and we get,

$$\psi_p = \bar{C}'\int_{u_0}^{\infty}\int_{-\infty}^{\infty} e^{-i\pi\frac{u^2}{2}}\cdot e^{-i\pi\frac{v^2}{2}}dudv. \tag{5.17.7}$$

where

$$u_0 = -\sqrt{\frac{2}{\lambda d}}X \tag{5.17.8}$$

Now

$$\int_{-\infty}^{\infty} e^{-i\pi\frac{v^2}{2}}dv = \left[\int_{-\infty}^{\infty}\cos\pi\frac{v^2}{2}dv - i\int_{-\infty}^{\infty}\sin\pi\frac{v^2}{2}dv\right] = 2\left[\int_0^{\infty}\cos\frac{\pi v^2}{2}dv - i\int_0^{\infty}\sin\frac{\pi v^2}{2}dv\right]$$

since both $\sin \pi \frac{v^2}{2}$ and $\cos \pi \frac{v^2}{2}$ are even functions of v. If we set,

$$\int_0^{\tau} \cos \frac{\pi v^2}{2} dv = C(\tau) \tag{5.17.9}$$

$$\int_0^{\tau} \sin \frac{\pi v^2}{2} dv = S(\tau) \tag{5.17.10}$$

These two are the Fresnel's integrals having the following properties:
(i) $C(\infty) = \frac{1}{2} = S(\infty)$.
(ii) $C(-\tau) = -C(\tau), S(-\tau) = -S(\tau)$,
both $C(\tau)$ and $S(\tau)$ are odd functions.

$$\therefore \quad \phi_p = 2\bar{C}' \int_{u_0}^{\infty} e^{-i\pi \frac{u^2}{2}} du \int_0^{\infty} e^{-i\pi \frac{v^2}{2}} dv$$

$$= 2\bar{C}' \left[\int_{u_0}^{\infty} \cos\left(\frac{\pi u^2}{2}\right) du - i\int_{u_0}^{\infty} \sin\left(\frac{\pi u^2}{2}\right) du\right] \left[\int_0^{\infty} \cos\left(\pi \frac{v^2}{2}\right) dv - i\int_0^{\infty} \sin\left(\frac{\pi v^2}{2}\right) dv\right]$$

Now,

$$\int_0^{\infty} \cos \frac{\pi u^2}{2} du = \int_0^{u_0} \cos \frac{\pi u^2}{2} du + \int_{u_0}^{\infty} \cos \frac{\pi u^2}{2} du$$

$$\Rightarrow \int_{u_0}^{\infty} \cos \frac{\pi u^2}{2} du = \int_0^{\infty} \cos \frac{\pi u^2}{2} du - \int_0^{u_0} \cos \frac{\pi u^2}{2} du$$

Similarly,

$$\int_{u_0}^{\infty} \sin \frac{\pi u^2}{2} du = \int_0^{\infty} \sin \frac{\pi u^2}{2} du - \int_0^{u_0} \sin \frac{\pi u^2}{2} du.$$

$$\therefore \quad \phi_p = 2\bar{C}' \left[\int_0^{\infty} \cos \frac{\pi u^2}{2} du - \int_0^{u_0} \cos \frac{\pi u^2}{2} du - i\int_0^{\infty} \sin \frac{\pi u^2}{2} du + i\int_0^{u_0} \sin \frac{\pi u^2}{2} du\right] \times \left[\int_0^{\infty} \cos \frac{\pi v^2}{2} dv - i\int_0^{\infty} \sin \frac{\pi v^2}{2} dv\right]$$

$$= 2\bar{C}' \left[\frac{1}{2} - \int_0^{u_0} \cos \frac{\pi u^2}{2} du - i \times \frac{1}{2} + i\int_0^{u_0} \sin \frac{\pi u^2}{2} du\right] \left[\frac{1}{2} - i \times \frac{1}{2}\right]$$

$$= 2\bar{C}'[1-i]\frac{1}{2} \times \left[(1-i)\frac{1}{2} - \int_0^{u_0} \cos \frac{\pi u^2}{2} du + i\int_0^{u_0} \sin \frac{\pi u^2}{2} du\right]$$

$$= 2\bar{C}'[1-i]\frac{1}{2}\left[(1-i)\frac{1}{2} - C(u_0) + iS(u_0)\right]$$

$$= \bar{C}'[1-i]\left[\frac{1}{2} - C(u_0) + i\left\{S(u_0) - \frac{1}{2}\right\}\right]$$

$$\Rightarrow \quad \phi_p^* = \bar{C}'[1+i]\left[\frac{1}{2} - C(u_0) - i\left\{S(u_0) - \frac{1}{2}\right\}\right]$$

Table 1: Fresnel's Integrals

u_0	$C(u_0)$ (x_0)	$S(u_0)$ (y_0)	u_0	$C(u_0)$ (x_0)	$S(u_0)$ (y_0)	u_0	$C(u_0)$ (x_0)	$S(u_0)$ (y_0)
0.00	0.0000	0.0000	2.40	0.5550	0.6197	4.80	0.4338	0.4968
0.10	0.1000	0.0005	2.50	0.4574	0.6192	4.90	0.5002	0.4350
0.20	0.1999	0.0042	2.60	0.3890	0.5500	5.00	0.5637	0.4992
0.30	0.2994	0.0141	2.70	0.3925	0.4529	5.10	0.4998	0.5624
0.40	0.3975	0.0334	2.80	0.4675	0.3915	5.20	0.4389	0.4969
0.50	0.4923	0.0647	2.90	0.5624	0.4101	5.30	0.5078	0.4405
0.60	0.5811	0.1105	3.00	0.6058	0.4963	5.40	0.5573	0.5140
0.70	0.6597	0.1721	3.10	0.5616	0.5818	5.50	0.4784	0.5537
0.80	0.7230	0.2493	3.20	0.4664	0.5933	5.60	0.4517	0.4700
0.90	0.7648	0.3398	3.30	0.4058	0.5192	5.70	0.5385	0.4595
1.00	0.7799	0.4383	3.40	0.4385	0.4296	5.80	0.5298	0.5461
1.10	0.7638	0.5365	3.50	0.5326	0.4152	5.90	0.4486	0.5163
1.20	0.7154	0.6234	3.60	0.5880	0.4923	6.00	0.4995	0.4470
1.30	0.6386	0.6863	3.70	0.5420	0.5750	6.10	0.5495	0.5165
1.40	0.5431	0.7135	3.80	0.4481	0.5656	6.20	0.4676	0.5398
1.50	0.4453	0.6975	3.90	0.4223	0.4752	6.30	0.4760	0.4555
1.60	0.3655	0.6389	4.00	0.4984	0.4204	6.40	0.5496	0.4965
1.70	0.3238	0.5492	4.10	0.5738	0.4758	6.50	0.4816	0.5454
1.80	0.3336	0.4508	4.20	0.5418	0.5633	6.60	0.4690	0.4631
1.90	0.3944	0.3734	4.30	0.4494	0.5540	6.70	0.5467	0.4915
2.00	0.4882	0.3434	4.40	0.4383	0.4622	6.80	0.4831	0.5436
2.10	0.5815	0.3743	4.50	0.5261	0.4342	6.90	0.4732	0.4624
2.20	0.6363	0.4557	4.60	0.5673	0.5162	7.00	0.5207	0.4591
2.30	0.6266	0.5531	4.70	0.4914	0.5672			

$\therefore$ Intensity

$$I_p = \sqrt{\phi_p \phi_p^*} = \bar{C}'(1+1)\left[\left\{\frac{1}{2} - C(u_0)\right\}^2 + \left\{S(u_0) - \frac{1}{2}\right\}^2\right]$$

$$= 2\bar{C}'\left[\left\{\frac{1}{2} - C(u_0)\right\}^2 + \left\{\frac{1}{2} - S(u_0)\right\}^2\right] \tag{5.17.11}$$

When $u_0 = -\infty$, i.e., when the point is very much away from the geometrical shadow of the edge, $(X = \infty)$, then

$$I_p = 2\bar{C}'\left[\left\{\frac{1}{2} + \frac{1}{2}\right\}^2 + \left\{\frac{1}{2} + \frac{1}{2}\right\}^2\right] = 4\bar{C}' = I_0$$

$$\therefore \quad I_p = \frac{1}{2} I_0 \left[\left\{\frac{1}{2} - C(u_0)\right\}^2 + \left\{\frac{1}{2} - S(u_0)\right\}^2\right] \tag{5.17.12}$$

The Fresnel's integrals for different values of u_0 are given in the Table 1.

$$C(u_0) = \int_0^{u_0} \cos\frac{\pi u^2}{2} du = x_0 \tag{5.17.13}$$

$$S(u_0) = \int_0^{u_0} \sin\frac{\pi u^2}{2} du = y_0 \tag{5.17.14}$$

The values of u_0 determine the value of X, that is the x co-ordinate of the point of observation.

From the Table-1 when $u_0 = 0 \Rightarrow X = 0$, then intensity is given by,

$$I_p = \frac{1}{2} I_0 \left[\left\{ \frac{1}{2} - C(0) \right\}^2 + \left\{ \frac{1}{2} - S(0) \right\}^2 \right] = \frac{1}{2} I_0 \left[\left(\frac{1}{2} \right)^2 + \left(\frac{1}{2} \right)^2 \right] = \frac{1}{4} I_0.$$

Thus, at the edge of the geometrical shadow, the intensity is $\frac{1}{4}$ of the maximum intensity I_0.

For $u_0 = -$ve values, X become +ve, hence we get points on the x-direction upwards in the observation screen. When $u_0 = -\infty$, we get,

$$I_p = \frac{1}{2} I_0 \left[\left\{ \frac{1}{2} + \frac{1}{2} \right\}^2 + \left\{ \frac{1}{2} + \frac{1}{2} \right\}^2 \right] = I_0,$$

i.e., at such points the intensity would be same as that if the edge were absent.

But when $u_0 = \infty$, means a point at a large depth within the geometrical shadow boundary of the edge, we get,

$$I_p = \frac{1}{2} I_0 \left[\left\{ \frac{1}{2} - \frac{1}{2} \right\}^2 + \left\{ \frac{1}{2} - \frac{1}{2} \right\}^2 \right] = 0, \text{ as expected.}$$

Between these two extreme limits, the intensity variation will take place and the intensity can be calculated with reference to the table giving values of the Fresnel's integrals.

Let us write the IDF as

$$I_n = \frac{I_p}{I_0} = \frac{1}{2} \left[\left\{ \frac{1}{2} - C(u_0) \right\}^2 + \left\{ \frac{1}{2} - S(u_0) \right\}^2 \right]. \tag{5.17.15}$$

Table 2

u_0	$C(u_0)$	$S(u_0)$	I_n	
0.00	0	0	0.2500	For points below the edge of the geometrical shadow
0.20	0.1999	0.0042	0.1679	
0.40	0.3975	0.0334	0.1141	
0.60	0.5811	0.1105	0.0791	
0.80	0.7228	0.2493	0.0562	
1.00	0.7799	0.4383	0.0411	
1.20	0.7154	0.6234	0.0308	
1.40	0.5431	0.7135	0.0237	
1.60	0.3655	0.6389	0.0187	
1.80	0.3336	0.4509	0.0150	
2.00	0.4883	0.3434	0.01233	
−0.20			0.3720	For points above the geometrical shadow edge
−0.40			0.5450	
−0.60			0.7707	
−0.80			1.0283	
−1.00			1.2593	
−1.20			1.3696	
−1.40			1.2803	
−1.60			1.0231	
−1.80			0.7995	
−2.00			0.8440	
−2.20	−0.6363	−0.4557	1.1022	
−2.40	−0.6266	−0.5531	1.1891	
−2.60	−0.3890	−0.5500	0.9494	
−2.80	−0.4675	−0.3815	0.8654	
−3.00	−0.6057	−0.4963	1.1076	
−3.20	−0.4663	−0.5934	1.0646	
−3.40				
−3.60				

From the graphical plot of u_0 against I_n it appears that the intensity below the geometrical shadow edge decreases monotonically. I_n becomes 0.01233 at $u_0 = 2.00$.

If we take $\lambda = 6 \times 10^{-5}$ cm, $d = 120$ cm, then

$$u_0 = -\sqrt{\frac{2}{\lambda d}}X \quad \Rightarrow \quad X = -\sqrt{\frac{\lambda d}{2}}u_0 = -\sqrt{\frac{6 \times 10^{-5} \times 120}{2}} \times 2 = 0.12\,\text{cm}.$$

So the intensity falls off very sharply and tends to become zero within a very small depth below the geometrical shadow edge.

Above the geometrical shadow edge, the intensity fluctuates, becoming maximum equal to $1.37I_0, 1.20I_0$ and $1.15I_0$ at $u_0 = -1.22, -2.34, -3.08$, the first three maxima being referred to. The first three maxima occur at

$$u_0 = -1.22 \quad \Rightarrow \quad X_1 = \sqrt{\frac{6 \times 10^{-5} \times 120}{2}} \times 1.22 = 0.073\,\text{cm}$$

$$u_0 = -2.34 \quad \Rightarrow \quad X_2 = \sqrt{\frac{6 \times 10^{-5} \times 120}{2}} \times 2.34 = 0.14\,\text{cm}$$

$$u_0 = -3.08 \quad \Rightarrow \quad X_2 = \sqrt{\frac{6 \times 10^{-5} \times 120}{2}} \times 3.08 = 0.19\,\text{cm}$$

from the geometrical shadow edge.

This function I_n has been calculated for different values of u_0 (Table-2), $C(u_0)$ and $S(u_0)$ values being taken from the Table-1. From the Table-2 a few points become evident.

- **Below the edge of the geometrical shadow the intensity monotonically decreases and to a depth of 0.12 cm, it decreases down to $.012 I_0$, where I_0 is the intensity in absence of the edge for $\lambda = 6 \times 10^{-5}$ cm and $d = 120$ cm.**
- **Above the geometrical shadow edge, the intensity passes through several maxima and minima before finally attaining a steady intensity that would have been in absence of the edge. [see fig. 5.55(a)].**

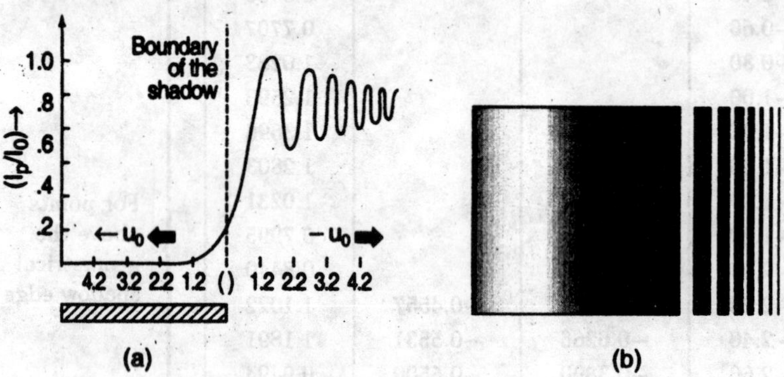

Fig 5.55

At the shadow edge, the intensity is $\frac{1}{4} I_0$ but this does not imply destruction of energy at this region, there is redistribution of energy near the edge due to diffraction adding more energy to a few maxima; intensity distribution in diffraction pattern is basically through interference.

The diffraction pattern observed from an edge is shown in fig. 5.55(b).

The variation of intensity due to diffraction at the sharp edge can be physically interpreted in a very simple way with the help of Cornu spiral.

Cornu Spiral and its a usefulness in Explaining Fresnel's Diffraction Pattern

The Cornu spiral is nothing but the curve connecting $C(u_0)$ and $S(u_0)$, these two functions are nothing but the Fresnel's integrals defined as

$$\int_0^{u_0} \cos \frac{\pi u^2}{2} du = C(u_0) = x_0 \text{ (say)} \qquad \int_0^{u_0} \sin \frac{\pi u^2}{2} du = S(u_0) = y_0.$$

When $u_0 = 0, C(u_0) = 0$, $S(u_0) = 0$, so the curve passes through the origin.

When u_0 is changed is sign, both $C(u_0)$ and $S(u_0)$ change in sign, the numerical values remaining same. It means, the curve is symmetric about the origin.

When $u_0 = \pm\infty$, $C(u_o) = \pm\frac{1}{2}$, $S(u_o) = \pm\frac{1}{2}$. The two points are called the *eye of the spiral*.

If ϕ be the angle made by the tangent drawn at some point on the curve, then

$$\tan\psi = \frac{dy_0}{dx_0} = \frac{\sin\left(\pi\frac{u_0^2}{2}\right)}{\cos\left(\pi\frac{u_0^2}{2}\right)} = \tan\left(\pi\frac{u_0^2}{2}\right) \Rightarrow \psi = \pi\frac{u_0^2}{2} \Rightarrow u_0^2 = \frac{2\psi}{\pi}.$$

Inferences

(i) $u_0 = 0$, $\psi = 0$, it means the curve is parallel to the x-axis at the origin.

(ii) $u_0 = 1$, $\psi = \frac{\pi}{2} \Rightarrow x_0(1) = 0.8$ and $y_0(1) = 0.4$; it means, the curve is parallel to the y-axis.

(iii) $u_0 = \sqrt{2}$, $\psi = \pi$, the curve again becomes parallel to the x-axis, $C(\sqrt{2}) = 0.54, S(\sqrt{2}) = 0.71$. This point on the curve defines the boundary of the first half-period element.

Now, $u_0 = \sqrt{2} \Rightarrow |X| = \sqrt{\frac{\lambda d}{2}}u_0 = 0.08$ cm.

Backed with these information we can draw the *Cornu spiral* as shown in fig. 5.56.

When a complete wavefront is operative, u_0 varies from $-\infty$ to ∞. As $u_0 \to \infty$, we get the lower spiral eye J_2 for which $C(u_0) = C(-\infty) = -\frac{1}{2}$ and when $u_0 \to \infty$, we get the upper spiral eye J_1 for which $C(u_0) = C(\infty) = \frac{1}{2}$. So for the whole wavefront unobstructed, the intensity at P_0 is given as proportional to the square of the length J_2J_1.

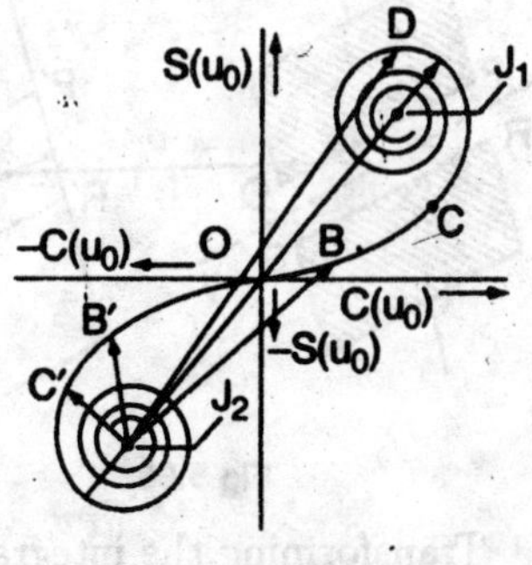

Fig 5.56

But in case of diffraction at a sharp edge, as shown in fig. (5.57) with respect to point P_0, A is the pole of the wave front PAQ, whose upper half is unobstructed, the lower half obstructed, so at P_0, the only the upper half of the wavefront is operative, so the effect is represented by the line joining the eye J_2 to O; thus J_2O represents the amplitude vector. But if we consider some point P_1 in the illuminated region, its pole is B, so at P_1 the upper half of the wavefront is operative and some portion of the lower half also becomes operative; hence now the amplitude vector may be obtained by considering one end of the eye J_2 and the tip moving over the upper spiral. Thus for points like P_1, P_2, etc., in the illuminated region the amplitude vectors are represented by lines like J_2B, J_2C, etc., which assume maxima and minima, and ultimately for a point of observation like P_n, practically the whole wavefront becomes unobstructed and the amplitude vector is given by J_2J_1.

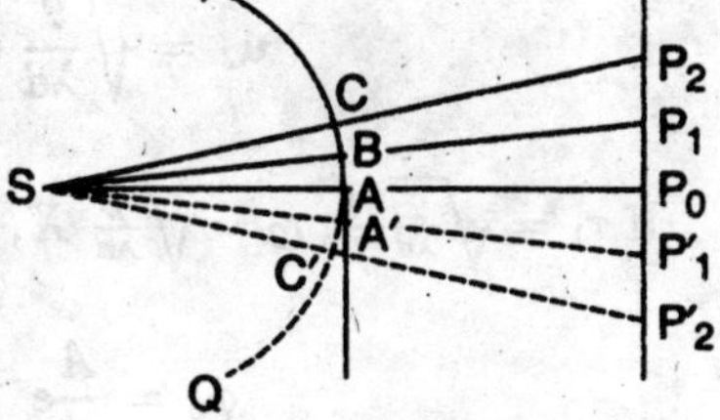

Fig 5.57

For a point like P_1' below the geometrical shadow edge P_0, the pole of the wavefront is A and so for such point on the observation screen, a portion of the upper half of the wavefront now becomes obstructed, so the amplitude vector is obtained by the line joining the eye J_2 to B'. Thus for points in the shadow region, the amplitude vectors are obtained by keeping one end of the line anchored at J_2 and moving the tip down the lower spiral. Thus in the shadow region the amplitude vector goes on shrinking in length, hence the intensity, monotonically decreases down to zero value within a very small depth inside the shadow region (c.f. fig. 5.56).

Fresnel's Diffraction by a Narrow Slit

Let us now consider the diffraction of a plane wave by a narrow slit parallel to the y-axis and the observation screen is at a finite distance from the diffracting slit $PQRS$. The centre of the slit is taken as the origin and edges of the slit are parallel to the y-axis. $P'Q'$ is the upper edge of the geometrical shadow and $R'S'$ is the lower edge.

The intensity distribution function can be derived in the same way as that in case of straight edge diffraction, so we would consider again eqn. (5.17.2) and write the displacement at P as

$$\phi_p = \bar{C}\int_{-\frac{b}{2}}^{\frac{b}{2}} e^{-ik(x-X)^2/2d}dx \int_{-\infty}^{\infty} e^{-ik(y-Y)^2/2d}dy$$

where $\bar{C} = \frac{A}{d}e^{-ikd}$.

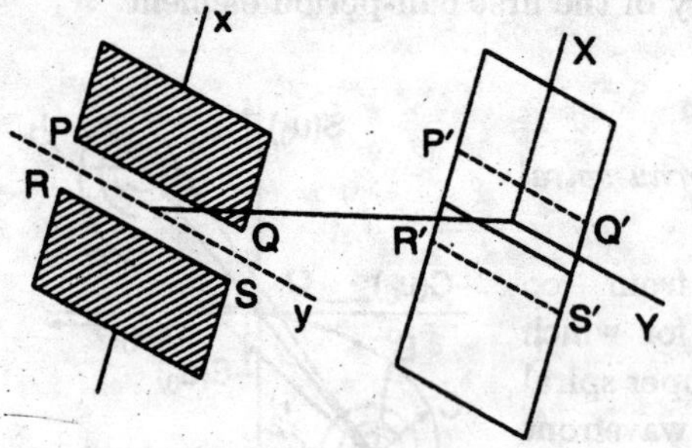

Fig 5.58

Put

$$\frac{\pi u^2}{2} = k\frac{(x-X)^2}{2d} = \frac{\pi}{\lambda d}(x-X)^2$$

$$\frac{\pi v^2}{2} = k\frac{(y-Y)^2}{2d} = \frac{\pi}{\lambda d}(y-Y)^2$$

$$\Rightarrow \quad u = \sqrt{\frac{2}{\lambda d}}(x-X), \quad v = \sqrt{\frac{2}{\lambda d}}(y-Y).$$

Transforming the integral into the variable u and v, we get,

$$\phi_p = \frac{A}{d}e^{-ikd}\int_{u_1}^{u_2}\sqrt{\frac{\lambda d}{2}}e^{\frac{-i\pi u^2}{2}}du\int_{-\infty}^{\infty}e^{\frac{-i\pi v^2}{2}}\sqrt{\frac{\lambda d}{2}}dv$$

where

$$u_1 = \sqrt{\frac{2}{\lambda d}}\left(-\frac{b}{2}-X\right) = -\sqrt{\frac{2}{\lambda d}}\left(\frac{b}{2}+X\right)$$

$$u_2 = \sqrt{\frac{2}{\lambda d}}\left(\frac{b}{2}-X\right) = -\sqrt{\frac{2}{\lambda d}}\left(X-\frac{b}{2}\right)$$

If $\tau_1 = \sqrt{\frac{2}{\lambda d}}\cdot\frac{b}{2}, \tau_2 = \sqrt{\frac{2}{\lambda d}}\,X$, then $u_1 = -(\tau_1+\tau_2)$, $u_2 = -(\tau_2-\tau_1)$.

$$\therefore \quad \phi_p = \frac{A}{d}e^{-ikd}\left(\frac{\lambda d}{2}\right)\int_{-(\tau_1+\tau_2)}^{-(\tau_2-\tau_1)}e^{\frac{-i\pi u^2}{2}}du\int_{-\infty}^{\infty}e^{-\pi\frac{v^2}{2}}dv$$

$$= A\lambda e^{ikd}\int_{-(\tau_1+\tau_2)}^{-(\tau_2-\tau_1)}e^{\frac{-i\pi u^2}{2}}du\int_{-\infty}^{\infty}e^{-\pi\frac{v^2}{2}}dv$$

$$= A\lambda e^{-ikd}\int_{-(\tau_1+\tau_2)}^{-(\tau_2-\tau_1)}e^{\frac{-i\pi u^2}{2}}du\left(\frac{1}{2}-i\times\frac{1}{2}\right)$$

$$= \frac{1}{2}(1-i)A\lambda e^{-ikd}\int_{-(\tau_1+\tau_2)}^{-(\tau_2-\tau_1)}e^{\frac{-i\pi u^2}{2}}du$$

Now,

$$\int_{-(\tau_1+\tau_2)}^{-(\tau_2-\tau_1)} e^{-i\pi\frac{u^2}{2}}du = \int_{-(\tau_1+\tau_2)}^{0} e^{-i\pi\frac{u^2}{2}}du + \int_{0}^{-(\tau_2-\tau_1)} e^{-i\pi\frac{u^2}{2}}du$$

$$= \int_{0}^{-(\tau_1-\tau_2)} e^{-i\pi\frac{u^2}{2}}du - \int_{0}^{-(\tau_1+\tau_2)} e^{-i\pi\frac{u^2}{2}}du$$

$$= \int_{0}^{-(\tau_2-\tau_1)} \left[\cos\left(\pi\frac{u^2}{2}\right) - i\sin\left(\pi\frac{u^2}{2}\right)\right]du$$

$$- \int_{0}^{-(\tau_1+\tau_2)} \left[\cos\left(\pi\frac{u^2}{2}\right) - i\sin\left(\pi\frac{u^2}{2}\right)\right]du$$

$$= -C(\tau_2-\tau_1) + iS(\tau_2-\tau_1) + C(\tau_1+\tau_2) - iS(\tau_1+\tau_2).$$

$$\therefore \quad \phi_p = \frac{(1-i)}{2}A\lambda e^{-ikd}\left[\{C(\tau_1+\tau_2) - C(\tau_2-\tau_1)\} + i\{S(\tau_2-\tau_1) - S(\tau_1+\tau_2)\}\right]$$

$$\phi_p^* = \frac{(1+i)}{2}A\lambda e^{ikd}\left[\{C(\tau_1+\tau_2) - C(\tau_2-\tau_1)\}^2 - i\{S(\tau_2-\tau_1) - S(\tau_1+\tau_2)\}\right]$$

$$\therefore \quad I_p = \phi_p\phi_p^* = \frac{1}{2}A^2\lambda^2\left[\{C(\tau_1+\tau_2) - C(\tau_2-\tau_1)\} + \{S(\tau_2-\tau_1) - S(\tau_1+\tau_2)\}^2\right] \tag{5.17.16}$$

This equation is the IDF for slit diffraction in case of Fresnel's diffraction. The value of τ_1 is determined by λ, d and b; so if we use some specifications $b = 0.1$ cm, $\lambda = 6\times10^{-5}$ cm, $d =$ 120 cm, then

$$\tau_1 = \sqrt{\frac{2}{\lambda d}}b = \sqrt{\frac{2}{6\times10^{-5}\times120}} \times 0.1 = 1.67.$$

As X changes, τ_2 changes, so the intensity varies.

Take $\tau_1 = 1.67$, take τ_2 different values and find I_p. Putting $A^2\lambda^2 = I_0$, we get,

$$I_p = \frac{1}{2}I_0\left[\{C(\tau_1+\tau_2) - C(\tau_2-\tau_1)\}^2 + \{S(\tau_1+\tau_2) - S(\tau_2-\tau_1)\}^2\right]. \tag{5.17.17}$$

The following table has been constructed for plotting the IDF for a single-slit Fresnel diffraction pattern.

Table 3: $\tau_1 = 1.7$

τ_2	$(\tau_2+\tau_1)$	$(\tau_2-\tau_1)$	$C(\tau_1+\tau_2)$	$C(\tau_2-\tau_1)$	$S(\tau_2+\tau_1)$	$S(\tau_2-\tau_1)$
0	1.7	−1.7	+0.3238	−.3238	.5492	−.5492
0.3	2.0	−1.4	0.4882	−.5431	.3434	−.7135
0.4	2.1	−1.3	0.5815	−.6386	.3743	−.6863
0.5	2.2	−1.2	0.6363	−.7154	.4557	−.6234
0.6	2.3	−1.1	0.6266	−.7638	.5531	−.5365
0.7	2.4	−1.0	0.5550	−.7799	.6197	−.4383
0.8	2.5	− .9	0.4574	−.7648	.6192	−.3398
0.9	2.6	− .8	0.3890	−.7230	.5500	−.2493

Table 3: $\tau_1 = 1.7$

τ_2	$(\tau_2+\tau_1)$	$(\tau_2-\tau_1)$	$C(\tau_1+\tau_2)$	$C(\tau_2-\tau_1)$	$S(\tau_2+\tau_1)$	$S(\tau_2-\tau_1)$
1.0	2.7	−.7	0.3925	−.6597	.4529	−.1721
1.1	2.8	− .6	0.4675	−.5811	.3915	−.2205
1.2	2.9	− .5	0.5624	−.4923	.4101	−.0647
1.3	3.0	− .4	0.6058	−.3975	.4963	−.0334
1.4	3.1	− .3	0.5616	−.2994	.5818	−.0141
1.5	3.2	− .2	0.4664	−.1994	.5933	−.0042
1.6	3.3	− .1	0.4058	−.1000	.5192	−.0005
1.7	3.4	−0.0	0.4385	−.0000	.4296	−.0000
1.8	3.5	+ .1	0.5326	.1000	.4152	.1000

$$\frac{I_p}{I_0} = \frac{1}{2}\left[\{C(\tau_1-\tau_2) - C(\tau_2-\tau_1)\}^2 + \{S(\tau_1+\tau_2) - S(\tau_2-\tau_1)\}^2\right]$$

τ_1	τ_2	I_p/I_0	τ_1	τ_2	I_p/I_0
1.7	0	.82		1.1	.68
	.3	1.09		1.2	.67
	.4	1.25		1.3	.64
	.5	1.50		1.4	.45
	.6	1.56		1.5	.40
	.7	1.45		1.6	.26
	.8	1.21		1.7	.19
	.9	.94		1.8	.14
	1.0	.75			

Transition from Fresnel to Fraunhofer

$\tau_2 = \sqrt{\frac{2}{\lambda d}}X = \sqrt{\frac{2d}{\lambda}}\cdot\frac{X}{d} \rightarrow \sqrt{\frac{2d}{\lambda}}\theta$, when d is very large compared to X which represents the condition of Fraunhofer. In this case, since d is large, $\tau_2 \rightarrow$ large value.

Fresnel's integral under this condition becomes

$$\begin{aligned}
C(u) &= \int_0^u \cos\frac{\pi u^2}{2}du = \int_0^\infty \cos\frac{\pi u^2}{2}du - \int_u^\infty \cos\frac{\pi u^2}{2}du \\
&= \frac{1}{2} - \int_u^\infty \cos\frac{\pi u^2}{2}du = \frac{1}{2} - \left[\int_u^\infty \frac{1}{\pi u}\cos\frac{\pi u^2}{2}d\left(\frac{\pi u^2}{2}\right)\right] \\
&= \frac{1}{2} - \left[\frac{1}{\pi u}\sin\left(\frac{\pi u^2}{2}\right)\bigg|_u^\infty - \int_u^\infty -\frac{1}{\pi u^2}\sin\frac{\pi u^2}{2}du\right. \\
&= \frac{1}{2} + \frac{1}{\pi u}\sin\frac{\pi u^2}{2} + \int_u^\infty \frac{1}{\pi u^2}\sin\frac{\pi u^2}{2}du \\
&\simeq \frac{1}{2} + \frac{1}{\pi u}\sin\frac{\pi u^2}{2}, \quad \text{neglecting small terms in } \frac{1}{u^3}, \text{ etc.}
\end{aligned}$$

Similarly,

$$S(u) = \int_0^u \sin\frac{\pi u^2}{2} du = \frac{1}{2} - \frac{1}{\pi u}\cos\frac{\pi u^2}{2}.$$

Since in Fraunhofer limit τ_2 is large and τ_1 is small,

$$\therefore \quad C(\tau_2 + \tau_1) = \frac{1}{2} + \frac{1}{\pi\tau_2}\sin\frac{\pi}{2}(\tau_1 + \tau_2)^2$$

$$C(\tau_2 - \tau_1) = \frac{1}{2} + \frac{1}{\pi\tau_2}\sin\frac{\pi}{2}(\tau_2 - \tau_1)^2.$$

$$\Rightarrow \quad C(\tau_2 + \tau_1) - C(\tau_2 - \tau_1) = \frac{1}{\pi\tau_2}\left[\sin\frac{\pi}{2}(\tau_1 + \tau_2)^2 - \sin\frac{\pi}{2}(\tau_2 - \tau_1)^2\right]$$

$$= \frac{2}{\pi\tau_2}\cos\frac{\pi}{2}(\tau_1^2 + \tau_2^2)\sin\pi\tau_1\tau_2$$

Similarly,

$$S(\tau_2 + \tau_1) - S(\tau_2 - \tau_1) = \frac{2}{\pi\tau_2}\sin\frac{\pi}{2}(\tau_1^2 + \tau_2^2)\sin\pi\tau_1\tau_2$$

$\therefore$ In the Fraunhofer region, the IDF eqn. (5.17.17) becomes

$$I_p = \frac{1}{2}I_0\left[\{C(\tau_2 + \tau_1) - C(\tau_2 - \tau_1)\}^2 + \{S(\tau_2 + \tau_1) - S(\tau_2 - \tau_1)\}^2\right]$$

$$= \frac{1}{2}I_0\left[\frac{4}{\pi^2\tau_2^2}\cos^2\frac{\pi}{2}(\tau_1^2 + \tau_2^2)\sin^2\pi\tau_1\tau_2 + \frac{4}{\pi^2\tau_2^2}\sin^2\frac{\pi}{2}(\tau_1^2 + \tau_2^2)\sin^2\pi\tau_1\tau_2\right]$$

$$= \frac{2}{\pi^2\tau_2^2}I_0\sin^2\pi\tau_1\tau_2 = 2I_0\tau_1^2\cdot\frac{\sin^2\pi\tau_1\tau_2}{(\pi\tau_1\tau_2)^2} = (I)_{\max}\frac{\sin^2\beta}{\beta^2} \qquad (5.17.18)$$

The equation is the familiar single-slit IDF in case of Fraunhofer diffraction.

Derivation of Fresnel's Integrals from Kirchhoff's Fresnel's Diffraction Equation

From the Kirchhoff-Fresnel diffraction equation, we can write, taking the amplitude factor and obliquity factor as constant,

$$\phi_p = \bar{C}\iint e^{ik(\omega b - \overline{r+r_0}) + i\frac{\pi}{2}}dA,$$

where r is the distance of the field-point and r_0 is the distance of the source-point w.r.t. an elementary area dA in the plane of the diffracting aperture.

If we set up a co-ordinate system as in fig. 5.59, then

$$\rho^2 = x^2 + y^2$$

$$\therefore \quad \phi_p = \bar{C}\iint e^{i(k\omega t + \frac{\pi}{2})}\cdot e^{-ik(r+r_0)}dA$$

$$= \bar{C}e^{i(k\omega t + \frac{\pi}{2})}\iint\limits_{\text{aperture}} e^{-ik(r+r_0)}dA$$

$$= \bar{C}e^{i(k\omega t + \frac{\pi}{2})}\iint\limits_{\text{aperture}} e^{-ik(r+r_0)}dA$$

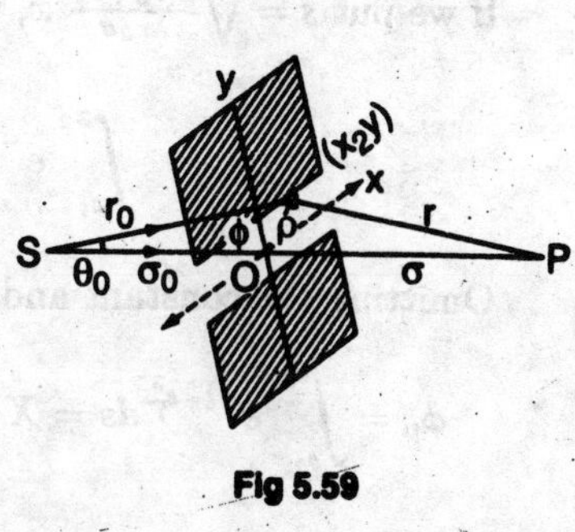

Fig 5.59

From fig. 5.59, we get,

$$\begin{aligned}
r^2 &= r_0^2 + (\sigma_0+\sigma)^2 - 2r_0(\sigma_0+\sigma)\cos\theta_0 \\
&= r_0^2 + (\sigma_0+\sigma)^2 - 2r_0(\sigma_0+\sigma)\left(1-\frac{\theta_0^2}{2}\right) \\
&= r_0^2 + (\sigma_0+\sigma)^2 - 2r_0(\sigma_0+\sigma)\left(1-\frac{\rho^2}{2r_0^2}\right)
\end{aligned}$$

Now, $r_0^2 = (x+\sigma_0)^2 + y^2$, since x, y are small compared to σ_0, we can write,

$$\begin{aligned}
r_0^2 &= \sigma_0^2 \\
\therefore \quad r^2 &= r_0^2 + (r_0+\sigma)^2 - 2r_0(r_0+\sigma)\left(1-\frac{\rho^2}{2r_0^2}\right) \\
&= 2r_0^2 + 2r_0\sigma + \sigma^2 - 2r_0(r_0+\sigma) + \frac{\rho^2}{r_0}(r_0+\sigma) \\
&= \sigma^2 + \frac{\rho^2}{r_0}(r_0+\sigma) = \sigma^2\left[1+\frac{\rho^2}{r_0\sigma^2}(r_0+\sigma)\right] \\
\Rightarrow \quad r &= \sigma\left[1+\frac{1}{2}\frac{\rho^2}{r_0\sigma^2}(r_0+\sigma)\right] = \sigma + \frac{1}{2}\cdot\frac{\rho^2}{r_0\sigma}(r_0+\sigma)
\end{aligned}$$

Now, $\rho^2 = x^2 + y^2$

$$\begin{aligned}
\therefore \quad \phi_p &= \bar{C}e^{i(k\omega t+\frac{\pi}{2})}\iint e^{-ik(\sigma_0+r)}dxdy \\
&= \bar{C}e^{i(k\omega t+\frac{\pi}{2}-k\sigma_0)}\iint e^{-ik[\sigma+\frac{1}{2}\frac{\rho^2}{r_0\sigma}(r_0+\sigma)]}dxdy \\
&= \bar{C}e^{i(k\omega t+\frac{\pi}{2}-k\sigma_0-k\sigma)}\iint e^{-ik(r_0+\sigma)\frac{\rho^2}{2r_0\sigma}}dxdy \\
&= \bar{C}e^{i[k\omega t+\frac{\pi}{2}-k(\sigma+\sigma_0)]}\iint e^{-\frac{ik(r_0+\sigma)}{2r_0\sigma}(x^2+y^2)}dxdy
\end{aligned}$$

If variation parallel to x-axis only considered, then

$$\phi_p = \bar{C}'e^{i[k\omega t+\frac{\pi}{2}-k(\sigma+\sigma_0)]}\int_{x_1}^{x_2} e^{-\frac{ik(r_0+\sigma)}{2r_0\sigma}x^2}dx.$$

If we put $s = \sqrt{\frac{2(r_0+\sigma)}{\lambda r_0\sigma}}\,x$, then we would get,

$$\int_{x_1}^{x_2} e^{-\frac{ik(r_0+\sigma)}{2r_0\sigma}x^2}dx = \int_{s_1}^{s_2} e^{-i\pi\frac{s^2}{2}}\sqrt{\frac{\lambda r_0\sigma}{2(r_0+\sigma)}}ds.$$

Omitting the constant and phase factor, we can write

$$\phi_p = \int_{s_1}^{s_2} e^{-i\pi\frac{s^2}{2}}ds = X+iY \Rightarrow \quad X = \int_{s_1}^{s_2}\cos\left(\frac{\pi s^2}{2}\right)ds, Y = \int_{s_1}^{s_2}\sin\left(\frac{\pi s^2}{2}\right)ds$$

The intensity is given by

$$I_p = X^2 + Y^2 = [\{C(s_2) - C(s_1)\}^2 + \{S(s_2) - S(s_1)\}^2] \quad (5.17.19)$$

Fresnel Diffraction in a Circular Aperture

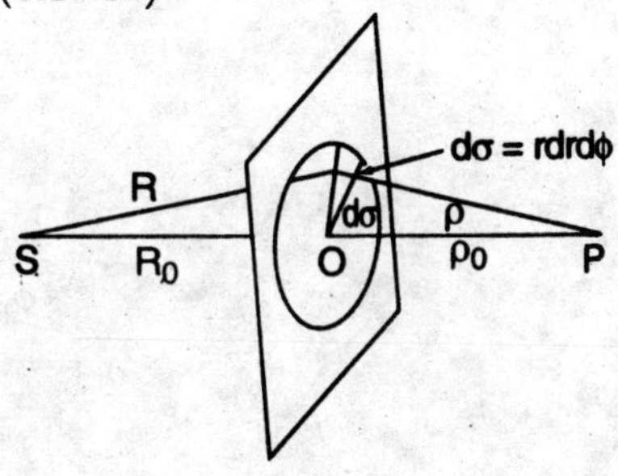

Fig 5.60

For convenience we will take $R_0 = \rho_0$, $R = \rho$; then if we consider an elementary area $d\sigma = r\ dr d\phi$, then displacement of the particles on $d\sigma$ due to the source S is given as

$$d\phi_1 = \frac{A}{R} e^{ik(\omega t - R)} d\sigma.$$

The displacement at P due to wavelets from $d\sigma$ is given by,

$$d\phi_p = \frac{d\phi_1}{\rho} e^{ik(\omega t - \rho)} = \frac{A}{\rho R} e^{ik(\omega t - R)\cdot ik(\omega t - \rho)} d\sigma = Ae^{ik\omega t}.\ \frac{e^{ik(R+\rho)}}{\rho R} \times d\sigma$$

$$\therefore \quad \phi_p = Ae^{ik\omega t} \iint_{\text{aperture}} \frac{e^{-ik(R+\rho)}}{\rho R} \times r\ dr d\phi = Ae^{ik\omega t} \int \frac{e^{-ik(R+\rho)}}{\rho R} r\ dr \int_0^{2\pi} d\phi$$

$$= 2\pi Ae^{ik\omega t} \int_{\text{aperture}} \frac{e^{-ik(R+\rho)}}{\rho R} r\ dr$$

Now, from fig. 5.60, $r^2 + \rho_0^2 = \rho^2 \Rightarrow r\ dr = \rho\ d\rho$

$$\therefore \quad \phi_p = 2\pi Ae^{ik\omega t} \int_0^a \frac{e^{-ik(\rho+R)}}{\rho^2} r\ dr = 2\pi Ae^{ik\omega t} \int_{\rho_0}^{\sqrt{a^2+\rho_0^2}} \frac{e^{-2ik\rho}}{\rho} d\rho.$$

Now,

$$\int_{\rho_0}^{\sqrt{a^2+\rho_0^2}} \frac{e^{-2ik\rho}}{\rho} d\rho = \frac{1}{\rho}\left(\frac{1}{-2ik} e^{-2ik\rho}\right)\Bigg|_{\rho_0}^{\sqrt{a^2+\rho_0^2}}$$

$$-\int_{\rho_0}^{\sqrt{a^2+\rho_0^2}} \frac{1}{\rho^2}\cdot\frac{1}{2ik} e^{-2ik\rho} d\rho = \frac{1}{-2ik}\frac{e^{-2ik\rho}}{\rho}\Bigg|_{\rho_0}^{\sqrt{a^2+\rho_0^2}} - \frac{1}{2ik}\int_{\rho_0}^{\sqrt{a^2\rho_0^2}} \frac{e^{-2ik\rho}}{\rho^2} d\rho$$

If we neglect terms containing $\frac{1}{\rho^2}$, since ρ is large, we get,

$$\int_{\rho_0}^{\sqrt{a^2+\rho_0^2}} e^{-2ik\rho} \rho\ d\rho = \frac{1}{-2ik}\left[\frac{e^{-2ik\sqrt{a^2+\rho_0^2}}}{\sqrt{a^2+\rho_0^2}} - \frac{e^{-2ik\rho_0}}{\rho_0}\right].$$

Since, $\rho_0 \gg a$, $\sqrt{a^2+\rho_0^2} = \rho_0\left(1 + \dfrac{a^2}{2\rho_0^2}\right)$

$$\int_{\rho_0}^{\sqrt{a^2+\rho_0^2}} \frac{e^{-2ik\rho}}{\rho} d\rho = \frac{1}{-2ik}\left[\frac{e^{-2ik\sqrt{a^2+\rho_0^2}}}{\sqrt{a^2+\rho_0^2}} - \frac{e^{-2ik\rho_0}}{\rho_0}\right]$$

$$= -\frac{1}{2ik\rho_0}\cdot e^{-2ik\rho_0}\left[\frac{e^{-ik(a^2/\rho_0)}}{(1+a^2/2\rho_0^2)} - 1\right]$$

$$\simeq -\frac{1}{2ik\rho_0}\cdot e^{-2ik\rho_0}\left[e^{-ika^2/\rho_0} - 1\right]$$

$$\phi_p = 2\pi A e^{ik\omega t}.\frac{1}{-2ik}.\frac{e^{-2ik\rho_0}}{\rho_0}.\left[e^{-ika^2/\rho_0} - 1\right]$$

$$= \frac{\pi A}{\rho_0} e^{ik(\omega t - 2\rho_0)} \frac{\left[e^{-ika^2/\rho_0} - 1\right]}{-ik}.$$

Put $(e^{-ika^2/\rho_0} - 1) = (e^{-ix} - 1)$, where $x = ka^2/\rho_0$.

$$\therefore \quad \phi_p^* = \frac{\pi A^{-ik(\omega t - 2\rho_0)}}{\rho_0}.\frac{\left[e^{+ika^2/\rho_0} - 1\right]}{+ik}$$

$$\therefore \quad \text{Intensity } I = \phi_p\phi_p^* = \frac{\pi^2 A^2}{\rho_0^2}\frac{1}{k^2}\left[e^{ix} - 1\right][e^{-ix} - 1]$$

$$= \frac{\pi^2 A^2}{(\rho_0 k)^2}\left[1 - (e^{ix} + e^{-ix}) + 1\right]$$

$$= \frac{\pi^2 A^2}{(\rho_0 k)^2}[2 - 2\cos x]$$

$$= \frac{4\pi^2 A^2}{(\rho_0 k)^2}\sin^2\frac{x}{2}$$

$$= \frac{4\pi^2 A^2}{(\rho_0 k)^2}\sin^2\left(\frac{ka^2}{2\rho_0}\right)$$

$$= \frac{\lambda^2 A^2}{\rho_0^2}\sin^2\left(\frac{\pi a^2}{\lambda\rho_0}\right)$$

$$= I_0\lambda^2\sin^2\left(\frac{\pi a^2}{\lambda\rho_0}\right) \tag{5.17.20}$$

where $I_0 = \frac{A^2}{\rho_0^2}$.

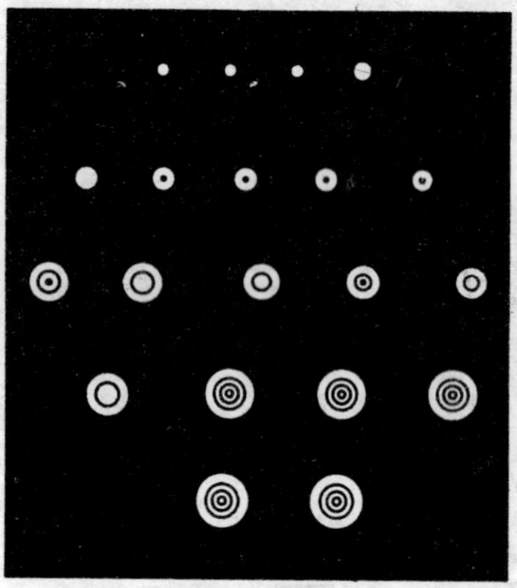

Fig 5.61

Thus the maxima or minima on the symmetry axis depend on the ratio (a^2/ρ_0). In fig. 5.61 the diffraction pattern determined by eqn. (5.17.20) for a circular aperture is shown with the radius of the aperture a increasing from figure to figure. It is, however, important to note that equation determines the intensity distribution along the symmetry axis of the circular aperture.

Fresnel Diffraction in a Circular Disc (Stop)

In this case will consider a circular disc (opaque) of the same radius as that of the circular aperture discussed in the earlier section. So this case may be treated as complementary to the circular aperture. So proceeding exactly in the same way, we would easily get,

$$\phi_p = 2\pi A e^{ik\omega t}\int_a^\infty \frac{e^{-ik(R+\rho)}}{\rho R} r\,dr, r > a$$

$$r^2+\rho_0^2 = \rho^2 \quad \Rightarrow \quad \rho\,d\rho = r\,dr$$

$$\therefore \quad \phi_p = 2\pi A e^{ik\omega t}\int_{\sqrt{a^2+\rho_0^2}}^\infty \frac{e^{-2ik\rho}}{\rho^2}\times \rho\,d\rho$$

$$= 2\pi A e^{ik\omega t}\int_{\sqrt{a^2+\rho_0^2}}^\infty \frac{e^{-2ik\rho}}{\rho} d\rho$$

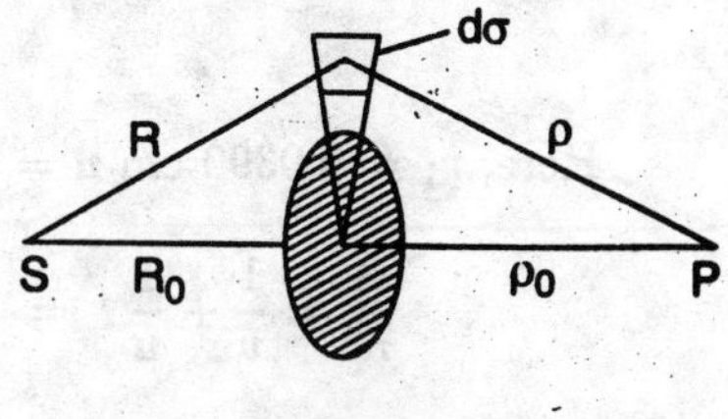

Fig 5.62

As done in case of circular aperture,

$$\int \frac{e^{-2ik\rho}}{\rho} = \frac{1}{-2ik}\cdot\frac{e^{-2ik\rho}}{\rho} - \int \frac{1}{2ik}\frac{e^{-2ik\rho}}{\rho^2} d\rho$$

neglecting terms containing $\frac{1}{\rho^2}$ and higher powers, we get,

$$\int \frac{e^{2ik\rho}}{\rho} = \frac{1}{-2ik}\cdot\frac{e^{-2ik\rho}}{\rho}$$

$$\therefore \phi_p = 2\pi A e^{ik\omega t}\int_{\sqrt{a^2+\rho_0^2}}^\infty \frac{e^{-2ik\rho}}{\rho} d\rho = \frac{\pi A e^{ik\omega t}}{-ik}\left[\frac{e^{-2ik\rho}}{\rho}\Bigg|_{\sqrt{a^2+\rho_0^2}}^\infty\right]$$

Now, $\sqrt{a^2+\rho_0^2} = \rho_0\left(1+\frac{1}{2}\frac{a^2}{\rho_0^2}\right)$.

The limiting value of the ratio at the upper limit can be obtained by applying L' Hospital's theorem

$$\lim_{\rho\to\infty}\frac{e^{-2ik\rho}}{\rho} = \lim_{\rho\to\infty}\frac{2ike^{-2ik\rho}}{1} = 0$$

$$\therefore \quad \phi_p = \frac{\pi A e^{ik\omega t}}{-ik}\left[0 - e^{-2ik\rho_0\left(1+\frac{1}{2}\frac{a^2}{\rho_0^2}\right)}\right]$$

$$= \pi A \frac{e^{ik(\omega t - 2\rho_0)}}{\rho_0}\left[\frac{e^{-ika^2/\rho_0}}{ik}\right]$$

$$\phi_p^* = \frac{\pi A}{\rho_0} e^{-ik(\omega t - 2\rho)}\frac{e^{ik\frac{a^2}{\rho_0}}}{-ik}$$

$$\therefore \quad I = \phi_p\phi_p^* = \frac{\pi^2 A^2}{k^2\rho_0^2} = \frac{\lambda^2 A^2}{4\rho_0^2} = \frac{1}{4}I_0\lambda^2 \qquad (5.17.21)$$

This equation represents the intensity distribution along the symmetry axis.

Solved Numerical Problems

(Fraunhofer and Fresnel)

1. A zone plate with radius 0.390 mm of the first zone is mounted on an optical bench 42 cm from a pin-hole illuminated by green light of wavelength 5461Å. Find the distance of the primary image.

Solution. From eqn. (5.16.1) we get,

$$r_1^2 = \lambda \frac{uv}{u+v}.$$

Here, $r_1 = 0.0390$ cm $u = 42$ cm, $\lambda = 5461 \times 10^{-8}$ cm.

$$\therefore \quad \frac{\lambda}{r_1^2} = \frac{1}{v} + \frac{1}{u} \quad \Rightarrow \quad \frac{1}{v} = \frac{\lambda}{r_1^2} - \frac{1}{u} = \frac{5461 \times 10^{-8}}{(.0390)^2} - \frac{1}{42} \quad \Rightarrow \quad v = 82.68 \text{ cm}$$

2. A zone plate with radius of 0.15 cm of the central zone is mounted on an optical bench 6 meter from a pin-hole illuminated by light of wavelength $\lambda = 6 \times 10^{-5}$ cm. Find the positions of the primary and secondary images.

Solution. From the zone plate equation, we can write [c.f. equation (5.16.4)]

$$\frac{1}{V} + \frac{1}{U} = \frac{1}{f_n} = (2n+1)\frac{\lambda}{r_n^2}.$$

For $n = 0$, primary image; $n = 1$, 1st secondary image, etc.

$$\therefore \quad \frac{1}{V_0} = \frac{\lambda}{r_0^2} - \frac{1}{U} = \frac{6 \times 10^{-5}}{(.15)^2} - \frac{1}{600} \quad \Rightarrow \quad V_0 = 10 \text{ meter}$$

Put $n = 1$.

$$\frac{1}{V_1} + \frac{1}{u} = \frac{1}{f'} = \frac{3\lambda}{r_n^2} \quad \Rightarrow \quad \frac{1}{V_1} = \frac{3\lambda}{r_n^2} - \frac{1}{U} = \frac{3 \times 6 \times 10^{-5}}{(.15)^2} - \frac{1}{600} \quad \Rightarrow \quad V_1 = 157.9 \text{ cm}$$

3. A strong parallel beam of monochromatic light is incident normally on a thin plate having a small circular hole of diameter 1 mm. If the screen be moved through a distance of 12.5 cm from the first position where the centre is black to the second similar position, find the wavelength of light used.

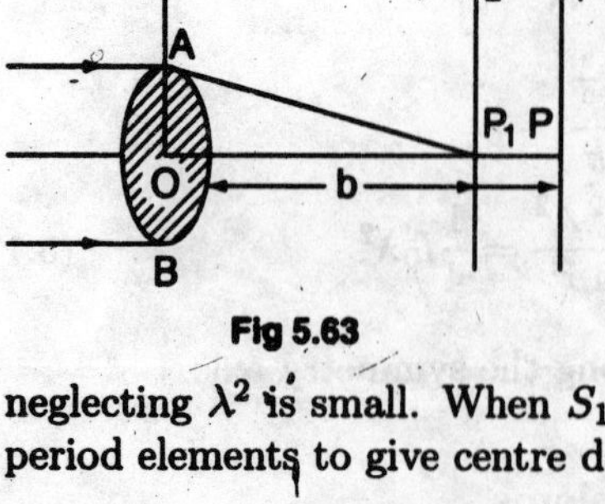

Fig 5.63

Solution. With respect to the point P, O is the pole. Depending on the size of the circular aperture and position P, the circular aperture when it transmits two half-period elements, then we would get darkness at P. So in this case,

$$PA = b + 2\frac{\lambda}{2} = (b + \lambda)$$

$$\therefore \quad (b+\lambda)^2 = r^2 + b^2 \quad \Rightarrow \quad 2b\lambda = r^2,$$

neglecting λ^2 is small. When S_1 shifts to S_2, the aperture w.r.t. P_1 transmits four half-period elements to give centre dark, so here

$$P_1A = \left(b - x + 4 \times \frac{\lambda}{2}\right) = (b + 2\lambda - x) = (b + 2\lambda - 12.5)$$

because in this case area of the half-period element is to be less, hence the screen must be nearer to the aperture.

$\therefore$ In the second position,

$$(b - 12.5 + 2\lambda)^2 = r^2 + (b - 12.5)^2 \quad \Rightarrow \quad 4b\lambda - 50\lambda = r^2$$

$$\therefore \; 2b\lambda = 4b\lambda - 50\lambda \quad \Rightarrow \quad 2b = 50 \quad \Rightarrow \quad b = 25\,\text{cm}$$

$$\therefore \; 2b\lambda = r^2 \quad \Rightarrow \quad 50 \times \lambda = (.05)^2 \quad \Rightarrow \quad \lambda = \frac{(.05)^2}{50} = 5 \times 10^{-5}\,\text{cm}.$$

4. (a) Discuss the nature of diffraction pattern due to a wire (or opaque strip).
   (b) Calculate the number of fringes observed in the geometrical shadow of a wire of diameter d at a distance a meter from the source. The screen is at a distance b meter from the wire.

Solution. (a) *Diffraction due to a wire:* In fig. 5.64, CD is a fine wire placed in the vertical position parallel to the vertical slit source. AB is the geometrical shadow into which fringes would appear near the boundary of the geometrical shadow both above the points A, B and also below.
The diffraction fringes above A and B are referred to as *exterior fringes.* The fringes below A and B on the screen are called *interior fringes.* The intensity distribution at points beyond the geometrical boundary of the shadow is due to the uncovered position of the wavefront.

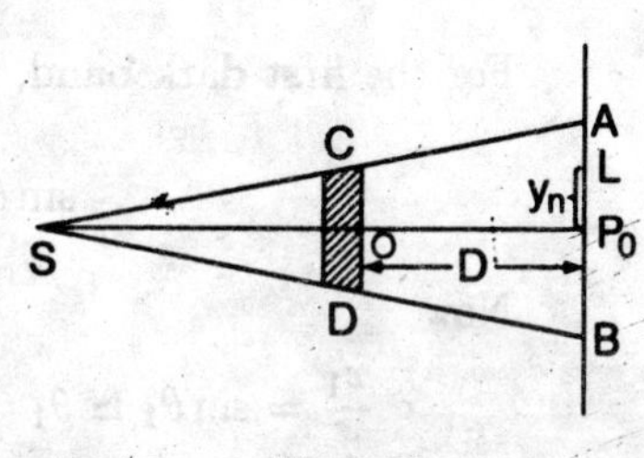

Fig 5.64

The interior fringes are formed through the interference of waves entering within the geometrical shadow by bending round the edges of the obstacle.
Thus at the point L, the displacement due to wavelets from the neighbourhood of the point C and D will depend on the phase condition at the time of union at L. The wavelets from C and D start out in same phase, and so the point L will be

$$\text{dark if } DL - CL = (2n+1)\frac{\lambda}{2}$$

$$\text{bright if } DL - CL = 2n \cdot \frac{\lambda}{2}$$

Now, let $OP_0 = D$, $P_0L = y_n$, $CD = d$, distance of the observation screen.
So

$$\text{for } n\text{th bright fringe at } L, y_n = \frac{D}{d} n\lambda$$

$$\text{for } n\text{th dark fringe at } L, y_n = \frac{D}{d}\left(n + \frac{1}{2}\right)\lambda$$

(b) The fringe width $\beta = y_{n+1} - y_n = \frac{D\lambda}{d}$ here $D = OP_0 = b$, $SO = a$.
From consideration of similar triangles, we immediately get,

$$\frac{CD}{AB} = \frac{SO}{SO + OP_0} = \frac{a}{(a+b)} \quad \Rightarrow \quad AB = \frac{(a+b)d}{a}.$$

$\therefore$ Number of fringes

$$N = \frac{(a+b)d}{a} \div \beta = \frac{(a+b)d}{a} \bigg/ \left(\frac{b\lambda}{d}\right) = \frac{(a+b)}{ab}\frac{d^2}{\lambda}.$$

5. What should be the diameter of a wire placed at 1 meter from Na-lamp ($\lambda = 5896$Å) so that there is no fringe in the geometrical shadow on a screen 1 meter away from the wire?
Solution. Here, $a = 100$ cm, $b = 100$ cm, $d = ?$
$N < 1$

$$\Rightarrow \quad \frac{(a+b)}{ab} \cdot \frac{d^2}{\lambda} < 1 \quad \Rightarrow \quad d^2 < \frac{\lambda ab}{(a+b)} \quad \Rightarrow \quad d^2 < 2.948 \times 10^{-3} \quad \Rightarrow \quad d < .054\,\text{cm}.$$

6. A lens whose focal length is 40 cm forms a Fraunhofer pattern of a slit of 0.3 mm width. Calculate distances of the first dark band and of the next bright band from the axis, $\lambda =$ 5890Å.
Solution. The minima for Fraunhofer single-slit diffraction are given by

$$a \sin\theta = n\lambda, \quad n = 1, 2, 3, \text{ etc.}$$

For the first dark band, $n = 1$

$$\therefore \quad \sin\theta_1 = \frac{\lambda}{a} = \frac{5890 \times 10^{-8}}{.03} \quad \Rightarrow \quad 1.963 \times 10^{-3}\,\text{rad}.$$

Now

$$\frac{x_1}{f} = \sin\theta_1 \simeq \theta_1 = 1.963 \times 10^{-3} \quad \Rightarrow \quad x_1 = 40 \times 1.963 \times 10^{-3} = .0785\,\text{cm}$$

The first secondary maxima occur at

$$\frac{\pi a \sin\theta_2}{\lambda} = 1.4303\pi \quad \Rightarrow \quad \sin\theta_2 = 1.4303\frac{\lambda}{a} \quad \Rightarrow \quad \theta_2 = 2.8077 \times 10^{-3}\,\text{rad}$$

$x_2 = 0.112$ cm. If on the otherhand, the first secondary maximum is taken approximately at $\frac{\pi a \sin\theta_2}{\lambda} = 1.5\pi, x_2 = .118\,\text{cm}$.

7. In a single-slit diffraction, the distance between the first minimum on either side is 2.6 mm (or equivalently, the distance between the first minimum on the right and the first minimum on the left). The screen on which the pattern is displayed is 80 cm from the slit and the wavelength used is 5460Å. Find the slit width.
Solution. The minima for single-slit diffraction are given by

$$a \sin\theta = n\lambda, n = 1, 2, 3, \text{ etc.} \quad \Rightarrow \quad \sin\theta_1 = \frac{\lambda}{a} \quad \Rightarrow \quad \theta_1 \simeq \frac{\lambda}{a}\,\text{rad}$$

$$\therefore \quad x_1 = 0.26 = D\theta_1 = 80 \times \frac{\lambda}{a} \quad \Rightarrow \quad a = \frac{80 \times \lambda}{0.26} = 0.0168\,\text{cm}.$$

8. Monochromatic light of wavelength 6.56×10^{-5} cm falls normally on a grating 2 cm wide. The first order spectrum is formed at angle $18°14'$ from the normal. What is the number of lines.
Solution. According to grating equation,

$$\sin\theta = mn\lambda.$$

Here $m = 1, n =$ number of rulings/cm, $\theta = 18°14'$.

$$\therefore \quad n = \left(\sin\frac{\theta}{\lambda}\right) = 4770, \text{ so total number} = 4770 \times 2 = 9540.$$

9. **A plane diffraction grating at normal incidence gives a green line $\lambda_1 = 5.40 \times 10^{-5}$ cm in a certain spectral order superposed on a violet line $\lambda_2 = 4.05 \times 10^{-5}$ cm of the next higher order. If the angle of diffraction be 30°, find the grating constant.**

Solution. According to the grating equation, we have

$$\sin\theta = mn\lambda.$$

So according to the given problem,

$$mn\lambda_1 = (m+1)n\lambda_2 \quad \Rightarrow \quad m \times 5.4 = (m+1)4.05 \quad \Rightarrow \quad m = 3$$

$$\therefore \ \sin 30° = 3 \times n \times 5.4 \times 10^{-5} \quad \Rightarrow \quad n = \frac{\sin 30°}{3 \times 5.4 \times 10^{-5}} = 3086 \text{ per cm}$$

$$\Rightarrow \ \text{grating constant } (a+b) = d = \frac{1}{n} = 3.24 \times 10^{-4} \text{ cm}$$

10. **Find the radii of the first three transparent zones of a zone plate whose first focal length is 1 meter for $\lambda = 6000$Å and the zone plate is illuminated by parallel rays.**

Solution. The zone plate formula gives us,

$$r_n^2 = n\lambda\frac{uv}{(u+v)} = n\lambda\frac{v}{1+v/u}.$$

As $u \to \infty, v = f$, so we get,

$$r_n = \sqrt{n\lambda f}$$

$$\therefore \quad r_1 = \sqrt{\lambda \times 1} = \sqrt{6000 \times 10^{-10} \times 1} = 0.775 \times 10^{-3} \text{ m}$$

$$r_3 = \sqrt{3 \times \lambda f} = \sqrt{3} \times .775 \times 10^{-3} = 1.34 \times 10^{-3}$$

$$r_5 = \sqrt{5\lambda f} = 1.73 \times 10^{-3} \text{ m}$$

11. **A circular aperture with a diameter 2 mm is illuminated by a plane wave. The most intense spot on the axis is at a distance of 200 cm from the aperture. Calculate the wavelength.**

Solution. According to the zone plate equation,

$$r_n^2 = n\lambda\frac{uv}{(u+v)} = n\lambda\frac{v}{1+v/u} \to n\lambda v \quad \text{as } u \to \infty$$

$$\therefore \quad r_n = \sqrt{n\lambda v} \quad \Rightarrow \quad 1 \times 10^{-1} = \sqrt{\lambda \times 200} \quad \Rightarrow \quad \lambda = 5 \times 10^{-5} \text{ cm}.$$

Theory Questions

1. **Describe in details the construction of a Fresnel's zone plate for a point source producing monochromatic radiation of wavelength λ and placed at a distance u from the plate. Find the areas of the successive half-period zones. Show that such a zone plate behaves as a multifoci converging lens.**
2. **Show how Fresnel explains the rectilinear propagation of light from wave theory. What is a zone plate? Define positive and negative zone plate. Show how does it resemble and differ from optical lenses.**
3. **Give the theory of Fresnel diffraction of a spherical wavefront at a circular aperture. What would happen if the wavefront is plane?**

4. Study theoretically the Fraunhofer diffraction pattern of a single-slit.
5. From Fresnel's theory of diffraction explain why the centre of the shadow of a small disc is bright.
6. Deduce an expression for the intensity of a plane diffraction grating spectrum. Explain the terms dispersive power and resolving power of a diffraction grating, and discuss their difference. In what respects does a grating spectrum differ from a prism spectrum?
7. Find the intensity expression for the Fraunhofer diffraction pattern of the double-slit. Deduce the conditions for maxima and minima. Discuss missing orders in a double-slit pattern.
8. Compare the diffraction effects produced by : (i) a narrow slit, (ii) a fine wire.
9. A circular aperture of radius a is illuminated by a plane wave of wavelength λ. Calculate the positions of the brightest and the darkest point on the axis. What would happen if the circular aperture is replaced by a circular disc of the same radius?
10. A zone plate is to have a principal focal length of 50 cm corresponding to wavelength $\lambda = 6 \times 10^{-5}$ cm, obtain values of the radii of different zones. [Apply $r_n = \sqrt{n\lambda f}$]
11. A straight edge is illuminated by a parallel beam of light with $\lambda = 6 \times 10^{-5}$ cm. Calculate the positions of the first few maxima and minima on a screen at a distance of 50 cm from the edge.
12. Consider a plane wave falling normally on a narrow slit. Study the diffraction pattern for various values of the distance between the slit and screen. What should be this distance for the pattern to be essentially Fraunhoferian?
13. Consider a rectangular aperture of dimension a and b. Obtain the positions of the first few maxima and minima in the Fraunhofer diffraction pattern.
14. A plane wave is incident on a convex lens of focal length 100 cm. If the diameter of the lens is 6 cm, calculate the radius of the first Ary disc. Assume $\lambda = 6 \times 10^{-5}$ cm.
[Treat the convex lens as a circular aperture of radius $a = 3$ cm. Hence, $R = 244 \times 10^{-3}$ cm.]
15. A zone plate has a diameter of 1 cm and has 50 zones. Calculate the distance of the remotest image for a source at a distance of 200 cm from the centre of the plate ($\lambda =$ 5000Å, use the formula $r_n = \sqrt{u\lambda f}$). **[Ans.** 2 meters]
16. A holder for a diffracting object is located 60 cm from the slit and fringes are observed 120 cm from the object. The diffracting object is a rod of diameter 1.8 mm, $\lambda = 5893$Å. Find the fringe width.
17. A very narrow vertical slit is illuminated with monochromatic light of wavelength 5461Å. A vertical wire of 0.1 cm diameter is placed in front of the slit, parallel to it. A screen is placed 3 meters from the wire. Calculate the fringe width within the geometrical shadow. **[Ans.** 0.1638 cm.]
18. If the diameter of the central zone is 3 mm and a point source of light of $\lambda = 6 \times 10^{-5}$ cm is placed 6 meters away from the zone plate, find the positions of the primary and secondary images. **[Ans.** 10 m and 157.9 cm.]
19. Light of wavelength $\lambda = 5000$Åfalls normally on a slit. The first minimum of the diffraction pattern is observed to lie at a distance of 5 mm from the central maximum on a screen placed at a distance of 2 meters from the slit. Calculate the width of the slit. **[Ans.** 0.02 cm.]

20. How many lines/cm are there in a plane diffraction grating which gives in the second order an angle of diffraction 30° for light of wavelength 5200Åincident normally on it?

[**Ans.** 4808 per cm.]

21. Na-light of wavelength 5890Åis incident normally upon a plane transmission diffraction grating of 10000 lines/inch. Calculate the angular separation of the Na-lines in the spectrum of the first order as observed in a telescope of which the objective has a focal length of 24 cm and eye-piece 2 cm. [**Ans.** 0.0028 rad.]

22. A parallel beam of monochromatic light is incident on a plane transmission grating normally. Write down an expression for the intensity pattern on to the focal plane of the telescope due to Fraunhofer diffraction at an angle θ.

    (a) Set up the condition for principal maxima.

    (b) Calculate chromatic dispersive power.

23. Distinguish between Fresnel type and Fraunhofer type of diffraction. Derive the expression for the intensity distribution of Fraunhofer diffraction pattern formed by a double slit.

References

1. **Born, M.** and **Wolf, E.** (1975), *Principles of Optics,* Pergamon Press, Oxford, England.

2. **Jenkins, F.A.** and **White, H.E.** (1976), *Fundamentals of Optics*, McGraw-Hill Book Co., New York, USA.

3. **Chakrabarti, P.K.** and **Kundu, S.N.** *Mathematical Physics*, New Central Book Agency (P) Ltd., Kolkata.

4. **Majumder, K.G.** *A Textbook of Light*, Modern Book Agency, Calcutta.

5. **Möller, K.D.** *Optics, University Science Books*, Mill Valley, California.

6. **Smith, C.J.** *A Degree Physics*, Part III, Optics, Radha Pub. House, Calcutta.

Chapter 6

Light as Electromagnetic Wave

6.1 Introduction

James Clerk Maxwell (1864) on the basis of Faraday's electrical researches was the first to formulate a mathematical theory of the electricity and magnetism. It was he who introduced the concept of *displacement current* in the dielectric media where the electric phenomena take place and also worked out the electromagnetic field equations—a set of differential equations representing the whole phenomenon of electromagnetism—assuming that the *displacement current* and the conduction current obey the same law.

The solutions to his differential equations were nothing but progressive waves, with electric and magnetic intensities at right angles to each other and contained in a plane perpendicular to the direction of propagation. Even in absence of solid matter, the displacement current occurs. It was also solved out that the electromagnetic wave propagates with a velocity $v = 1/\sqrt{\mu k}$, where μ = permeability and k = dielectric constant of the medium. Now, this quantity which has already been in our knowledge, it is equal to the velocity of light. This led Maxwell to conclude that light is also an electromagnetic phenomenon and thus the electromagnetic theory of light was born. This could well explain the different optical phenomena like reflection, refraction, dispersion, etc., and also the polarisation effect which is due to the transverse nature of light waves could satisfactorily be explained in terms of electromagnetic theory.

Maxwell's theory, however, does not exploit the microscopic nature of matter or, in other words, in this consideration, matter has no atomic structure, it is a continuum.

Electromagnetic Field vs Gravitational Field

We know that the space where gravitational forces act is called *gravitational field.* Here the value of the forces at any point is determined in Newtonian mechanics by the instantaneous positions of the gravitating bodies.

In electrodynamics, the picture is quite different. Interaction in electrodynamics is really an interaction of charged particles with an electromagnetic field. Now, elementary charges like protons, electrons and mesons very often do have velocities close to that of electromagnetic disturbance. So during the time an electromagnetic disturbance travels from one charge to another, the latter moves a very great distance. Though it is true that general theory of relativity attributes a finite velocity of propagation to the gravitational interaction, the effect is too small to cause any serious change in the laws of Newtonian mechanics. But in electrodynamics, this is not so.

Electromagnetic field is itself a real physical entity as much as the charged particles. The energy or momentum change of a charged particle under the action of electromagnetic field is imparted directly to the field because a finite time is needed for the change of energy and momentum of other particles. Gravitational field itself does not possess energy and momentum but it is the gravitating bodies that do contain them.

6.2 Basic Equation of Electrodynamics—Electromagnetic Field Equations

The field equations of Maxwell are just fundamental laws of electricity and magnetism suitably modified so as to apply to the general situation.

The four field equations due to Maxwell are just fundamental laws of electricity and magnetism suitably modified so as to apply to the general situation where the variations in the electric and magnetic fields are so rapid that there is considerable departure from the electrostatic and magnetostatic cases.

The four field equations due to Maxwell are

1. $\text{div}\vec{D} = \vec{\nabla}\cdot\vec{D} = 4\pi\rho$, or, $= \rho$ in S.I. unit
2. $\text{div}\vec{B} = \vec{\nabla}\cdot\vec{B} = 0$.
3. $\text{curl}\,\vec{E} = \vec{\nabla}\times\vec{E} = -\dfrac{1}{c}\dfrac{\partial\vec{B}}{\partial t}$, or, $= -\dfrac{\partial\vec{B}}{\partial t}$ where $\vec{B}$ is in Tesla in S.I.
4. $\text{curl}\,\vec{B} = \vec{\nabla}\times\vec{B} = \dfrac{4\pi\vec{J}}{c} + \dfrac{1}{c}\dfrac{\partial\vec{D}}{\partial t}$, or, $= \mu_0\left(\vec{J} + \dfrac{\partial\vec{D}}{\partial t}\right)$ in S.I.

Of the four equations above, the first pair is simply differential form of Gauss theorem in electrostatics and magnetostatics. The third equation is nothing but Faraday's law of electromagnetic induction and the last one, modified form of Ampere's circuital theorem for the magnetic field accompanying an electric current.

$\vec{D}$ and ρ are in e.s.u. and $\vec{B}$ in e.m.u. in c.g.s. system. Eqns. are also expressed in S.I. system (alternative expressions).

Equation of Continuity

This eqn. is the conservation of electric charges. The increase of total charge inside a volume dv must be at the cost of inflow of charge across the bounding surface.

So, we can write,

$$\frac{dq}{dt} = -\iint_s \vec{J}\cdot d\vec{s} \quad \text{or,} \quad \frac{d}{dt}\left[\iiint \rho dv\right] = -\iint \vec{J}\cdot d\vec{s}$$

$$\text{or,} \quad \iiint \frac{\partial\rho}{\partial t}dv = -\iiint \vec{\nabla}\cdot\vec{J}dv \quad \text{or,} \quad \iiint\left(\frac{\partial\rho}{dt} + \vec{\nabla}\cdot\vec{J}\right)dv = 0 \qquad (6.2.1)$$

Since eqn.(6.2.1) is true for any arbitrary value of dv, we must have $\vec{\nabla}\cdot\vec{J} = -\frac{\partial\rho}{\partial t}$, which is the *equation of continuity*. In the steady state, therefore, $\frac{\partial\rho}{\partial t} = 0$, and $\vec{\nabla}\cdot\vec{J} = 0$.

Setting up of Maxwell's Equations

(i) Eqn. $\vec{\nabla}\cdot\vec{D} = 4\pi\rho$, or, $= \rho$.

If we consider a dielectric medium and apply Gauss theorem, we must include the free charge as well as the polarisation charge density. The polarisation effect in a medium is to produce a volume density of polarisation charge equal to $-\vec{\nabla}\cdot\vec{P}$ where $\vec{P}$ = Polarisation vector in the medium,[1]

So applying Gauss's theorem, we get,

$$\iint \vec{E}\cdot d\vec{s} = 4\pi \iiint \left(\rho_{\text{true}} - \vec{\nabla}\cdot\vec{P}\right) dv \text{ or, } = \frac{1}{\epsilon_0}\iiint \left(\rho_{\text{true}} - \vec{\nabla}\cdot\vec{P}\right) dv \quad \text{(in S.I.)}$$

$$\Rightarrow \iiint \vec{\nabla}\cdot\vec{E} dv = 4\pi \iiint \left(\rho_{\text{true}} - \vec{\nabla}\cdot\vec{P}\right) dv \text{ or, } = \frac{1}{\epsilon_0}\iiint \left(\rho_{\text{true}} - \vec{\nabla}\cdot\vec{P}\right) dv \quad \text{(in S.I.)}$$

$$\Rightarrow \iiint \left[\vec{\nabla}\cdot\vec{E} - 4\pi\left(\rho_{\text{true}} - \vec{\nabla}\cdot\vec{P}\right)\right] dv = 0 \text{ or, } \iiint \left[\vec{\nabla}\cdot\vec{E} - \frac{1}{\epsilon_0}\left(\rho_{\text{true}} - \vec{\nabla}\cdot\vec{P}\right)\right] dv = 0 \tag{6.2.2}$$

Eqn.(6.2.2) is true for any value of dv, the integral must vanish.

$$\therefore \quad \vec{\nabla}\cdot\vec{E} = \pi\rho_{\text{true}} - 4\pi\vec{\nabla}\cdot\vec{P} \Rightarrow \vec{\nabla}\cdot\left(\vec{E} + 4\pi\vec{P}\right) = 4\pi\rho_{\text{true}}$$

$$\Rightarrow \vec{\nabla}\cdot\vec{D} = 4\pi\rho_{\text{true}} \tag{6.2.3}$$

where, $\vec{D} = \vec{E} + 4\pi\vec{P}$, called *electric displacement.*

In S.I. system,

$$\vec{\nabla}\cdot\left(\vec{E} + \frac{\vec{P}}{\epsilon_0}\right) = \frac{\rho_{\text{true}}}{\epsilon_0} \quad \Rightarrow \quad \vec{\nabla}\cdot\vec{D} = \rho_{\text{true}}$$

where $\vec{D} = \epsilon_0\vec{E} + \vec{P}$.

(ii) Eqn. div $\vec{B} = 0$.

The magnetic lines of force are either closed or go off to infinity. So across any arbitrary surface, the number of magnetic lines of force entering must also be leaving out of it.

Flux of magnetic induction across any surface, i.e., $\iint \vec{B}\cdot d\vec{s}$ must equate to zero.

$$\therefore \quad \iint \vec{B}\cdot d\vec{s} = 0 \quad \text{or,} \iiint \vec{\nabla}\cdot\vec{B} dv = 0. \quad \text{Hence, } \vec{\nabla}\cdot\vec{B} = 0 \tag{6.2.4}$$

because dv is arbitrary.

(iii) Eqn. $\vec{\nabla}\times\vec{E} = -\frac{1}{c}\frac{\partial\vec{B}}{\partial t}$ or, $= -\frac{\partial\vec{B}}{\partial t}$ (in S.I.)

According to Faraday—Newmann's law, the e.m.f. induced in a closed circuit due to variation of magnetic flux linked up with the circuit is given by,

$$e = -\frac{1}{c}\frac{\partial\psi}{\partial t} = -\frac{1}{c}\frac{\partial}{\partial t}\iint \vec{B}\cdot d\vec{s} = -\frac{1}{c}\iint \frac{\partial\vec{B}}{\partial t}\cdot d\vec{s}.$$

In S.I., $e = -\dfrac{\partial\psi}{\partial t} = -\displaystyle\iint \frac{\partial\vec{B}}{\partial t}\cdot d\vec{s}$.

[1]See *Advanced Textbook on Electricity & Magnetism* by the author.

Now, if the electric intensity $\vec{E}$ makes an angle θ with a line element $d\vec{l}$ of the closed path, then

$$e = \oint \vec{E} \cdot d\vec{l} \quad \therefore \quad \oint \vec{E} \cdot d\vec{l} = -\frac{1}{c} \iint \frac{\partial \vec{B}}{\partial t} \cdot d\vec{s}, \quad \text{or,} \quad = -\iint \frac{\partial \vec{B}}{\partial t} \cdot d\vec{s} \quad \text{(in S.I.)}$$

$$\Rightarrow \iint \left(\vec{\nabla} \times \vec{E}\right) \cdot d\vec{s} = -\frac{1}{c} \iint \frac{\partial \vec{B}}{\partial t} \cdot d\vec{s}, \quad \text{or,} \quad = -\iint \frac{\partial \vec{B}}{\partial t} \cdot d\vec{s} \quad \text{[Applying Stokes' theorem]}$$

$$\Rightarrow \iint \left[\left(\vec{\nabla} \times \vec{E}\right) + \frac{1}{c}\frac{\partial \vec{B}}{\partial t}\right] \cdot d\vec{s} = 0, \quad \text{or,} \quad = \iint \left[\vec{\nabla} \times \vec{E} + \frac{\partial \vec{B}}{\partial t}\right] \cdot d\vec{s} = 0 \quad \text{(in S.I.)}$$

This relation being true for any arbitrary value of $d\vec{s}$, we get from above

$$\vec{\nabla} \times \vec{E} = -\frac{1}{c}\frac{\partial \vec{B}}{\partial t} \quad \text{or,} \quad = -\frac{\partial \vec{B}}{\partial t} \quad \text{(in S.I.)} \tag{6.2.5}$$

(iv) Eqn. $\vec{\nabla} \times \vec{B} = \dfrac{4\pi \vec{J}}{c} + \dfrac{\partial \vec{D}}{c\partial t}$ or, $= \mu_0 \left(\vec{J} + \dfrac{\partial \vec{D}}{\partial t}\right)$ (in S.I.)

According to Ampere's circuital theorem, the work done in taking a unit magnetic pole once round a closed arbitrary circuit linking with current is given by,

$$\oint \vec{B} \cdot d\vec{l} = \frac{4\pi i}{c} \text{ in c.g.s. or } = \mu_0 i \text{ in S.I.}$$

$$\Rightarrow \int_s \left(\vec{\nabla} \times \vec{B}\right) \cdot d\vec{s} = \frac{4\pi}{e} \int \vec{J} \cdot d\vec{s} \text{ in c.g.s. or,} = \mu_0 \int \vec{J} \cdot d\vec{s} \text{ in S.I.}$$

Since, $d\vec{s}$ is arbitrary, so from the above relation we get,

$$\vec{\nabla} \times \vec{B} = \frac{4\pi \vec{J}}{c} \quad \text{or,} \quad = \mu_0 \vec{J} \text{ (in S.I.).} \tag{6.2.6}$$

Eqn.(6.2.6) is in direct conflict with the equation of continuity because this leads to $\vec{\nabla} \cdot \vec{J} = 0$.

Taking divergence of both sides of eqn.(6.2.6), we get,

$$\vec{\nabla} \cdot \left(\vec{\nabla} \times \vec{B}\right) = \frac{4\pi}{c} \vec{\nabla} \cdot \vec{J} \quad \text{or,} \quad = \mu_0 \left(\vec{\nabla} \cdot \vec{J}\right)$$

$$\Rightarrow \left(\vec{\nabla} \times \vec{\nabla}\right) \cdot \vec{B} = \frac{4\pi}{c} \vec{\nabla} \cdot \vec{J} \quad \therefore \quad \vec{\nabla} \cdot \vec{J} = 0.$$

This is possible only when $\rho =$ constant, i.e., only in the steady state. But when the electric field is changing, ρ is a function of time and so $\frac{\partial \rho}{\partial t} \neq 0$. So. $\vec{\nabla} \cdot \vec{J} \neq 0$ according to equation of continuity. At this point of crisis, Maxwell introduced the concept of displacement current. This produces some current density added to the current density produced by conduction current. So the total current density is different when the electric field is steady and when it is variable. We will come to this point again but before that let us discuss the concept of displacement current, which we may look upon as the kernel of Maxwell's electromagnetic theory.

Displacement Current

To resolve the crisis with Ampere's circuital theorem in the original form, Maxwell had to see other ways. The problem he understood as finding out a true current density in case of changing electric field. So the equation of Ampere is not enough in its form, we should put it as,

$$\vec{\nabla}\times\vec{B}=\frac{4\pi}{c}\vec{J}+? \quad \text{or,} \quad =\mu_0\vec{J}+? \text{ in c.g.s. and S.I. systems respectively,}$$

where? denotes something missing. So, if we write the missing term as $\frac{4\pi}{c}\vec{J}'$ or, $\epsilon_0\vec{J}'$ where $\vec{J}'$ is the added current density, then the eqn. becomes,

$$\vec{\nabla}\times\vec{B}=\frac{4\pi}{c}\left(\vec{J}+\vec{J}'\right)$$

In S.I.,

$$\vec{\nabla}\times\vec{B}=\mu_0\left(\vec{J}+\vec{J}'\right).$$

Taking divergence of both sides, we get,

$$\vec{\nabla}\cdot\left(\vec{\nabla}\times\vec{B}\right)=\frac{4\pi}{c}\vec{\nabla}\cdot\vec{J}+\frac{4\pi}{c}\vec{\nabla}\cdot\vec{J}', \quad \text{or,} \quad \vec{\nabla}\cdot\vec{J}=-\vec{\nabla}\cdot\vec{J}'$$

since, $\vec{\nabla}\cdot\left(\vec{\nabla}\times\vec{B}\right)=0.$

$$\therefore \quad \vec{\nabla}\cdot\vec{J}=-\vec{\nabla}\cdot\vec{J}'=-\frac{\partial\rho}{\partial t}.$$

Now, $\rho=\dfrac{\vec{\nabla}\cdot\vec{D}}{4\pi}$, or, $\vec{\nabla}\cdot$ (in S.I.)

$$\therefore \quad \vec{\nabla}\cdot\vec{J}'=\vec{\nabla}\cdot\left(\frac{1}{4\pi}\frac{\partial\vec{D}}{\partial t}\right) \quad \Rightarrow \quad \vec{J}'=\frac{1}{4\pi}\frac{\partial\vec{D}}{\partial t}.$$

Thus,

$$\vec{\nabla}\times\vec{B}=\frac{4\pi}{c}\frac{\partial\vec{D}}{\partial t}, \quad \text{or,} \quad \mu_0\left(\vec{J}+\frac{\partial\vec{D}}{\partial t}\right) \quad \text{in S.I.}$$

This added term $\frac{4\pi}{c}\vec{J}'$ or $\mu_0\vec{J}'$ which results in a solenoidal current is known as *displacement current density by Maxwell.* So the total current density is sum of the conduction current density $\vec{J}$ and displacement current density $\vec{J}'$. It is evident that as the electric field becomes non-changing, the displacement current ceases to exist. So when a changing electric field is there, there will be an attendant magnetic field due to the displacement current. The experimental confirmation of this magnetic field due to displacement current was made by Rowland by revolving a disc of ebonite containing static charge.

So the field equations of Maxwell are the following:

$$\left.\begin{array}{l}\vec{\nabla}\cdot\vec{D}=4\pi\rho \quad \text{or,} \quad =\rho \text{ (in S.I.)}\\ \vec{\nabla}\cdot\vec{B}=0\end{array}\right\} \text{ divergence equations and,}$$

$$\left.\begin{array}{l}\vec{\nabla}\times\vec{E}=-\dfrac{1}{c}\dfrac{\partial\vec{B}}{\partial t} \quad \text{or,} \quad =-\dfrac{\partial\vec{B}}{\partial t} \quad \text{(SI)}\\ \vec{\nabla}\times\vec{B}=\dfrac{4\pi}{c}\vec{J}+\dfrac{1}{c}\dfrac{\partial\vec{D}}{\partial t} \quad \text{or,}=\mu_0\left(\vec{J}+\dfrac{\partial\vec{D}}{\partial t}\right) \quad \text{(SI)}\end{array}\right\} \text{ circuital equations}$$

Remember that $\rho, \vec{J}, \vec{D}$ and $\vec{E}$ are measured in e.s.u. while $\vec{B}, \vec{H}$ in e.m.u. in c.g.s.

Now, these equations apply to case when the medium is at rest. But when the medium is a moving medium, both Faraday's law of induction and Maxwell's field equations need modification. In the development further, we will restrict ourselves to c.g.s. system. The reader may develop corresponding expressions in S.I. system.

6.3 Faraday's Law of Induction in Case of a Moving Medium

In case of stationary medium and changing magnetic field, Faraday's law of induction is given by equation (6.2.5). If now the medium is also moving, let us see how this modifies Faraday's induction law.

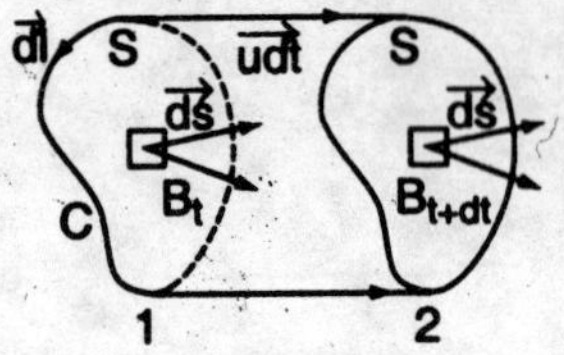

Fig 6.1

Let us consider certain contour C (fig.6.1) enclosing an area S in the position 1 where the induction $\vec{B}_t$. Now, let the medium be moving and so the induction $\vec{B}$ associated with the contour is a function of time also. Let the medium move along with a velocity $\vec{u}$ as shown and the second position 2 is associated with induction $\vec{B}_{t+dt}$. As a result of this displacement, the total flux of induction out is given by,

$$d\phi = \int_B \left(\vec{B}_{t+dt} - \vec{B}_t\right) \cdot d\vec{s} + \iint \left(\vec{u}dt \times d\vec{l}\right) \cdot \vec{B}_t = \int \frac{\partial \vec{B}_t}{\partial t} dt \cdot d\vec{s} - \oint dt \left(\vec{u} \times \vec{B}_t\right) \cdot d\vec{l}$$

$$\therefore \quad e = -\frac{1}{c}\frac{d\phi}{dt} = -\frac{1}{c}\int_S \frac{\partial \vec{B}_t}{\partial t} \cdot d\vec{s} + \frac{1}{c}\int \vec{\nabla} \times \left(\vec{u} \times \vec{B}_t\right) \cdot d\vec{s}$$

$$\text{or,} \quad \oint \vec{E}' \cdot d\vec{l} = -\frac{1}{c}\int \frac{\partial \vec{B}_t}{\partial t} \cdot d\vec{s} + \frac{1}{c}\int \vec{\nabla} \times \left(\vec{u} \times \vec{B}_t\right) \cdot d\vec{s}$$

$$\text{or,} \quad \int \left(\vec{\nabla} \times \vec{E}'\right) \cdot d\vec{s} = -\frac{1}{c}\int \frac{\partial \vec{B}_t}{\partial t} \cdot d\vec{s} + \frac{1}{c}\int \vec{\nabla} \times \left(\vec{u} \times \vec{B}_t\right) \cdot d\vec{s}$$

Since, $d\vec{s}$ is arbitrary,

$$\therefore \quad \vec{\nabla} \times \vec{E}' = -\frac{1}{c}\frac{\partial \vec{B}_t}{\partial t} - \frac{1}{c}\vec{\nabla} \times \left(\vec{B}_t \times \vec{u}\right),$$

where $\vec{E}'$ is the resulting electric field measured in a moving medium. The above equation can be written as,

$$\vec{\nabla} \times \left[\vec{E}' - \frac{1}{c}\left(\vec{u} \times \vec{B}_t\right)\right] = -\frac{1}{c}\frac{\partial \vec{B}_t}{\partial t} \tag{6.3.1}$$

Eqn.(6.3.1) shows that $\left[\vec{E}' - \frac{1}{c}\left(\vec{u} \times \vec{B}_t\right)\right]$ is the electric field as it appears to a stationary observer.

6.4 Maxwell's Field Equations for a Moving Medium

The field equations in case of a moving medium need modification, not all the equations but only the fourth of Maxwell's equations. The first two equations (i.e., divergence equations) are not affected by the motion of the medium in which the fields are measured. They can be written as,

(i) $\vec{\nabla}\cdot\vec{D}=4\pi\rho$; (ii) $\vec{\nabla}\cdot\vec{B}=0$;

(iii) $\vec{\nabla}\times\vec{E}=-\frac{1}{c}\frac{\partial\vec{B}}{\partial t}$; (iv) $\vec{\nabla}\times\vec{H}=\frac{4\pi}{c}\vec{J}+\frac{4\pi}{c}\rho\vec{u}+\frac{1}{c}\frac{\partial\vec{D}}{\partial t}$.

6.5 Maxwell's Field Equations Represent Propagation of Electromagnetic Waves

Maxwell's field equations admit of solution representing propagation of electromagnetic waves. In vacuum, they travel with the velocity of light ($c = 3\times10^{10}$ cms/sec) and this initiated the electromagnetic theory of light.

Solution of Maxwell's Field Equation in Isotropic Dielectric

In the case of isotropic dielectric, there is no free charge density, i.e., $\rho = 0$; $\vec{D}$ and $\vec{B}$ are in the direction of the electric field intensity $\vec{E}$ and magnetic field intensity $\vec{H}$.

So in such a case $\rho = 0$, $\vec{J} = 0$, $\vec{D} = k\vec{E}$, $\vec{B} = \mu\vec{H}$, where k and μ are respectively the dielectric constant and permeability of the medium.

So the field equations now become,

(i) $\vec{\nabla}\cdot\vec{D}=\vec{\nabla}\cdot\left(k\vec{E}\right)=0$ or, $\vec{\nabla}\cdot\vec{E}=0$; (ii) $\vec{\nabla}\cdot\vec{B}=0$ or, $\vec{\nabla}\cdot\vec{H}=0$;

(iii) $\vec{\nabla}\times\vec{E}=-\frac{\mu}{c}\frac{\partial\vec{H}}{\partial t}$; (iv) $\vec{\nabla}\times\vec{H}=\frac{k}{c}\frac{\partial\vec{E}}{\partial t}$.

Here, the displacement current is all the important.

Taking curl of equation (iii) we get,

$$\vec{\nabla}\times\left(\vec{\nabla}\times\vec{E}\right)=-\frac{\mu}{c}\frac{\partial}{\partial t}\left(\frac{k}{c}\frac{\partial\vec{E}}{\partial t}\right)\quad\text{using eqn.(iv)}$$

$$=-\frac{\mu k}{c^2}\frac{\partial^2\vec{E}}{\partial t^2}.$$

Expanding the left-hand expression, we have,

$$\vec{\nabla}\left(\vec{\nabla}\cdot\vec{E}\right)-\vec{\nabla}^2\vec{E}=-\frac{\mu k}{c^2}\frac{\partial^2\vec{E}}{\partial t^2}\quad\left[\text{cf. }\vec{\nabla}\times\left(\vec{\nabla}\times\vec{A}\right)=\vec{\nabla}\left(\vec{\nabla}\cdot\vec{A}\right)-\nabla^2\vec{A}\right]$$

$$\text{or,}\quad-\nabla^2\vec{E}=-\frac{\mu k}{c^2}\frac{\partial^2\vec{E}}{\partial t^2}\quad\text{or,}\quad\nabla^2\vec{E}=\frac{\mu k}{c^2}\frac{\partial^2\vec{E}}{\partial t^2}\quad\text{since }\vec{\nabla}\cdot\vec{E}=0 \tag{6.5.1}$$

where, ∇^2 = Laplacian operator.

Similarly, taking curl of eqn.(iv) and proceeding in the same way, we can have,

$$\nabla^2\vec{H}=\frac{\mu k}{c^2}\frac{\partial^2\vec{H}}{\partial t^2}=\frac{1}{v^2}\frac{\partial^2\vec{H}}{\partial t^2} \tag{6.5.2}$$

Eqns.(6.5.1) and (6.5.2) represent three-dimensional wave equation propagating with a velocity $v=\frac{c}{\sqrt{\mu k}}$. Thus, the electric field $\vec{E}$ and the magnetic field $\vec{H}$ satisfying Maxwell's equations are really propagating vectors with velocity v. In vacuum, $\mu = k = 1$, hence, the electromagnetic disturbance travels with velocity $v = c$ (i.e., the light velocity). This suggests light to be electromagnetic waves.

Plane Electromagnetic Waves in an Isotropic Dielectric Medium—Electromagnetic Waves are Transverse by Nature

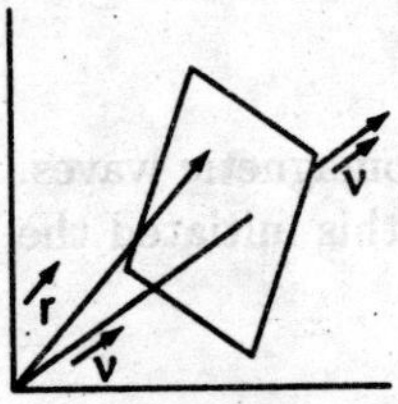

Fig 6.2

A plane wave is a wave such that the phases at all points in a plane drawn at right angles to the direction of propagation $\vec{\nu}$ (= unit vector in the propagation direction) is same. Such a plane is known as the *wavefront* and in an isotropic medium the wavefront is always normal to $\vec{\nu}$. The equation of the plane representing the wavefront at any instant is given by (cf. fig. 6.2),

$$\vec{r}\cdot\vec{\nu} = \text{const.}$$

Any disturbance which is to remain constant at all points in the wavefront which travels with a velocity v can be written as,

$$\psi = F(vt - \vec{r}\cdot\vec{\nu}).$$

We are to investigate the nature of such function such as F.

Let us take eqn.(6.5.1),

$$\nabla^2\vec{E} = \frac{\mu k}{c^2}\frac{\partial^2\vec{E}}{\partial t^2} = \frac{1}{v^2}\frac{\partial^2\vec{E}}{\partial t^2}.$$

Let $\vec{E}$ be a harmonic function of time, then an expression of the form, $\vec{E} = \vec{E}_0 e^{i\sigma(vt-\vec{r}\cdot\vec{\nu})}$ satisfies the wave equation as given above, $\sigma = \frac{2\pi}{\lambda}$ = propagation constant. This can be shown by actual calculation.

$$\frac{\partial\vec{E}}{\partial t} = i\sigma v\vec{E}_0 e^{i\sigma(vt-\vec{r}\cdot\vec{\nu})};\ \frac{\partial^2\vec{E}}{\partial t^2} = i^2\sigma^2 v^2\vec{E},\ \nabla^2\vec{E} = \vec{E}_0\nabla^2\left[e^{i\sigma(vt-\vec{r}\cdot\vec{\nu})}\right] = i\sigma v\vec{E}.$$

Now, $\vec{r}\cdot\vec{\nu} = lx + my + nz$.

$$\therefore\ \frac{\partial^2}{\partial x^2}\left[e^{i\sigma\{vt-(lx+my+nz)\}}\right] = i^2\sigma^2 l^2 e^{i\sigma(vt-\vec{r}\cdot\vec{\nu})}$$

$$\therefore\ \nabla^2\left[e^{i\sigma(vt-\vec{r}\cdot\vec{\nu})}\right] = i^2\sigma^2(l^2+m^2+n^2)e^{i\sigma(vt-\vec{r}\cdot\vec{\nu})} = i^2\sigma^2 e^{i\sigma(vt-\vec{r}\cdot\vec{\nu})}$$

$$\therefore\ \nabla^2\vec{E} = i^2\sigma^2\vec{E}_0 e^{i\sigma(vt-\vec{r}\cdot\vec{\nu})} = i^2\sigma^2\vec{E}.$$

Putting these values in the wave eqn., we get,

$$\nabla^2\vec{E} - \frac{\mu k}{c^2}\frac{\partial^2\vec{E}}{\partial t^2} = i^2\sigma^2\vec{E} - \frac{\mu k}{c^2}i^2\sigma^2 v^2\vec{E} = i^2\sigma^2\vec{E} - \frac{\mu k}{c^2}i^2\sigma^2\frac{c^2}{\mu k}\vec{E} = 0.$$

So, $\vec{E} = \vec{E}_0 e^{i\sigma(vt-\vec{r}\cdot\vec{\nu})}$ represents the solution of the wave equation. Similarly, it can be shown that, $\vec{H} = \vec{H}_0 e^{i\sigma(vt-\vec{r}\cdot\vec{\nu})}$ also represents the solution of the wave equations.

E_0 = amplitude of varying electric field

and H_0 = amplitude of varying magnetic field.

Taking divergence of eqn. $\vec{E} = \vec{E}_0 e^{i\sigma(vt - \vec{r}\cdot\vec{\nu})}$, we get,

$$\vec{\nabla}\cdot\vec{E} = \vec{\nabla}\cdot\left[\vec{E}_0 e^{i\sigma(vt-\vec{r}\cdot\vec{\nu})}\right] \tag{6.5.3}$$

Now, $\vec{\nabla}\cdot\left(\alpha\vec{A}\right) = \alpha\vec{\nabla}\cdot\vec{A} + \left(\vec{\nabla}\alpha\right)\cdot\vec{A} = \alpha\left(\vec{\nabla}\cdot\vec{A}\right) + \vec{A}\cdot\vec{\nabla}\alpha$, α = scalar quantity.

$$\begin{aligned}\therefore\quad \vec{\nabla}\cdot\vec{E} &= \vec{\nabla}\cdot(\alpha\vec{E}_0), \text{ here, } \alpha = e^{i\sigma(vt-\vec{r}\cdot\vec{\nu})}\\ &= e^{i\sigma(vt-\vec{r}\cdot\vec{\nu})}\left(\vec{\nabla}\cdot\vec{E}_0\right) + \vec{E}_0\cdot\vec{\nabla}\left(e^{i\sigma(vt-\vec{r}\cdot\vec{\nu})}\right)\\ &= \vec{E}_0\cdot\vec{\nabla}\left\{e^{i\sigma(vt-\vec{r}\cdot\vec{\nu})}\right\} \qquad \left[\because\ \vec{\nabla}\cdot\vec{E}_0 = 0\right]\end{aligned}$$

We know that $\vec{\nabla}\cdot\vec{E} = 0$ in the present problem, so eqn. (6.5.3) reduces to,

$$\vec{E}_0\cdot\vec{\nabla}\left\{e^{i\sigma(vt-\vec{r}\cdot\vec{\nu})}\right\} = 0 \quad \text{or,} \quad \vec{E}_0\cdot\left[-i\sigma e^{i\sigma(vt-\vec{r}\cdot\vec{\nu})}\vec{\nabla}(\vec{r}\cdot\vec{\nu})\right] = 0.$$

Now, $\vec{\nabla}(\vec{r}\cdot\vec{\nu}) = \left(\hat{i}\frac{\partial}{\partial x} + \hat{j}\frac{\partial}{\partial y} + \hat{k}\frac{\partial}{\partial z}\right)(lx + my + nz) = \hat{i}l + \hat{j}m + \hat{k}n = \vec{\nu}$.

$$\therefore\quad \vec{E}_0\cdot\left[-i\sigma e^{i\sigma(vt-\vec{r}\cdot\vec{\nu})}\vec{\nu}\right] = 0 \quad \text{or,} \quad \vec{E}_0\left[e^{i\sigma(vt-\vec{r}\cdot\vec{\nu})}\right]\cdot(-i\sigma\vec{\nu}) = 0 \quad \text{or,} \quad \vec{E}\cdot\vec{\nu} = 0. \tag{6.5.4}$$

Eqn.(6.5.4) shows that the electric field $\vec{E}$ is perpendicular to the wave-normal. So $\vec{E}$ will also be normal to the direction of wave propagation which takes place along the wave-normal in an isotropic medium.

Now, from Maxwell's equation, we have,

$$\vec{\nabla}\times\vec{E} = -\frac{\mu}{c}\frac{\partial\vec{H}}{\partial t} \quad \text{or,} \quad \vec{\nabla}\times\left[\vec{E}_0 e^{i\sigma(vt-\vec{r}\cdot\vec{\nu})}\right] = -\frac{\mu}{c}\frac{\partial\vec{H}}{\partial t}.$$

From the standard relation in vector calculus, we can write,

$$\vec{\nabla}\times\left(\alpha\vec{A}\right) = \alpha\left(\vec{\nabla}\times\vec{A}\right) - \vec{A}\times\left(\vec{\nabla}\alpha\right).$$

$$\begin{aligned}\therefore\quad \vec{\nabla}\times\left[\vec{E}_0 e^{i\sigma(vt-\vec{r}\cdot\vec{\nu})}\right] &= e^{i\sigma(vt-\vec{r}\cdot\vec{\nu})}\left(\vec{\nabla}\times\vec{E}_0\right) - \vec{E}_0\times\vec{\nabla}\left[e^{i\sigma(vt-\vec{r}\cdot\vec{\nu})}\right] \quad \left[\because\ \left(\vec{\nabla}\times\vec{E}_0\right) = 0\right]\\ &= -\vec{E}_0\times(-i\sigma)e^{i\sigma(vt-\vec{r}\cdot\vec{\nu})}\vec{\nu} = \left(\vec{E}\times\vec{\nu}\right)i\sigma.\end{aligned}$$

$$\therefore\quad -\frac{\mu}{c}\frac{\partial\vec{H}}{\partial t} = \left(\vec{E}_0\times\vec{\nu}\right)i\sigma e^{i\sigma(vt-\vec{r}\cdot\vec{\nu})} = i\sigma\vec{E}\times\vec{\nu}.$$

Integrating the above relation w.r.t. time,

$$-\frac{\mu}{c}\vec{H} = \frac{1}{v}\left(\vec{E}_0\times\vec{\nu}\right)e^{i\sigma(vt-\vec{r}\cdot\vec{\nu})} = \frac{1}{v}\left(\vec{E}\times\vec{\nu}\right)$$

$$\therefore\quad \vec{H} = -\frac{c}{\mu}\cdot\frac{\sqrt{k\mu}}{c}\left(\vec{E}\times\vec{\nu}\right) = -\sqrt{\frac{k}{\mu}}\left(\vec{E}\times\vec{\nu}\right), \quad \text{or,} \quad \vec{H} = \sqrt{\frac{k}{\mu}}\left(\vec{\nu}\times\vec{E}\right)$$

Thus, $\vec{H}$ is $\perp$ to $\vec{E}$ and $\vec{\nu}$. Thus, $\vec{E}, \vec{\nu}$ and $\vec{H}$ form an orthogonal triad. It is now put beyond any mistake that $\vec{E}$ and $\vec{H}$ whose variations give rise to electromagnetic disturbances travelling along the normal $\vec{\nu}$ are both perpendicular to the direction of propagation. It proves that the electromagnetic waves are transverse in character.

Refractive Index, Permeability and Dielectric Constant of a Medium

$$n = \frac{\text{velocity of light in vacuum}}{\text{velocity of light in the medium}} = \frac{c}{v} = \frac{c}{c/\sqrt{k\mu}} = \sqrt{k\mu} \quad \text{or,} \quad n^2 = k\mu,$$

i.e., $(r\cdot i)^2$ = permeability of the medium × dielectric const.

For a transparent material, $\mu \sim 1$, so $n^2 \simeq k$, known as *Maxwell's law*.

This relation is valid only for non-polar medium and for longer wavelength.

Hertz's Experiment on Production and Detection of Electromagnetic Waves

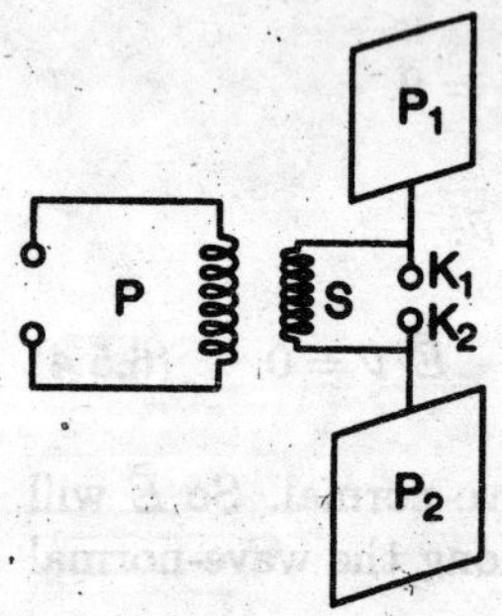

Fig 6.3

By a very simple device Hertz in 1888 could demonstrate the existence of electromagnetic waves as advanced by Maxwell in his electromagnetic theory.

The apparatus consists of two square sheets of brass of side 40 cm placed 60 cms apart, and connected to the two terminals of the secondary. These two plates along with the inductance of the secondary really form what is known as the *tank circuit* (*L-C* circuit) of an usual oscillator. As the primary circuit is made and broken, oscillations are set up in the secondary at each make and break of the primary. As a result rapidly varying alternating p.d. is developed across the gap *S* and electromagnetic waves are generated.

Hertz's Detector

In order to prove that the arrangement in fig.(6.3) produces electromagnetic waves, Hertz also devised a detector as shown in fig.(6.4) consisting of a loop (35 cm in radius) of thick wire having a small air gap. When this detector was placed at a distance (the range of detection of such detector was not more than a few yards), spark appeared at the gap of the detector when its air gap was parallel to the air gap of the source, but no such spark occurred when they were perpendicular to each other.

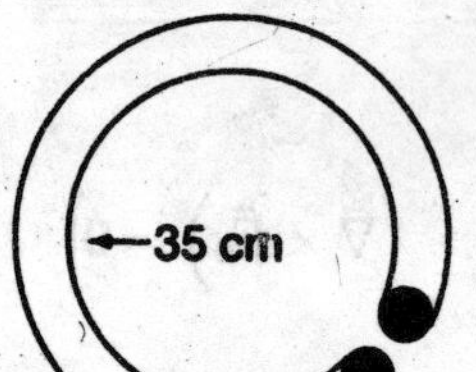

Fig 6.4

The transmitting as well as receiving devices have undergone enough of changes after Hertz devised his arrangement. The existence of electromagnetic waves is proved beyond doubt because nowadays transmission of signals either in radio or T.V. is done through electromagnetic waves; the reception also is done by different kinds of antenna systems.

6.6 Energy Flow in Electromagnetic Field—Poynting's Energy Theorem—Poynting's Vector

We have stated at the beginning that unlike the gravitational field where the gravitating bodies do possess energy, the electromagnetic field itself possesses energy and hence, as the wave propagates, there is energy flow. The energy flow due to passage of electromagnetic wave through a certain region can be calculated from the time variation of the density of the electromagnetic energy in that region. This has been expressed in the form of a theorem known as *Poynting's theorem* which can be stated as:

The flux of the vector $\vec{P} = \frac{c}{4\pi}\left(\vec{E}\times\vec{H}\right)$ across a closed surface enclosing a certain volume electromagnetic field represents the total energy flow across the bounding surface per unit time.

The vector $\vec{P}$ is known as *Poynting's vector* and the above statement is known as *Poyting's theorem.* Maxwell's circuital equations are,

$$\vec{\nabla}\times\vec{E} = -\frac{1}{c}\frac{\partial\vec{B}}{\partial t};\vec{\nabla}\times\vec{H} = \frac{4\pi\vec{J}}{c}+\frac{1}{c}\frac{\partial\vec{D}}{\partial t}.$$

Taking dot product of the first eqn. with $\vec{H}$ and the second with $\vec{E}$ and subtracting, we get,

$$\vec{E}\cdot\vec{\nabla}\times\vec{H}-\vec{H}\cdot\vec{\nabla}\times\vec{E} = \frac{4\pi}{c}\vec{J}.\vec{E}+\frac{1}{c}\frac{\partial\vec{D}}{\partial t}\cdot\vec{E}+\frac{1}{c}\vec{H}\cdot\frac{\partial\vec{B}}{\partial t}$$

$$\text{or, } \frac{c}{4\pi}\left(\vec{E}\cdot\vec{\nabla}\times\vec{H}-\vec{H}\cdot\vec{\nabla}\times\vec{E}\right) = \vec{J}\cdot\vec{E}+\frac{1}{4\pi}\left(\vec{E}\cdot\frac{\partial\vec{D}}{\partial t}+\vec{H}\cdot\frac{\partial\vec{B}}{\partial t}\right)$$

$$\text{or, } \frac{1}{4\pi}\left[\vec{E}\cdot\frac{\partial\vec{D}}{\partial t}+\vec{H}\cdot\frac{\partial\vec{B}}{\partial t}\right] = \frac{c}{4\pi}\left(\vec{E}\cdot\vec{\nabla}\times\vec{H}-\vec{H}\cdot\vec{\nabla}\times\vec{E}\right)-\vec{J}\cdot\vec{E} \qquad (6.6.1)$$

Now, $\vec{\nabla}\cdot\left(\vec{A}\times\vec{B}\right) = \vec{B}\cdot\vec{\nabla}\times\vec{A}-\vec{A}\cdot\vec{\nabla}\times\vec{B}$.

So using this relation the first term on the right of eqn.(6.6.1) can be written as,

$$\frac{c}{4\pi}\left(\vec{E}\cdot\vec{\nabla}\times\vec{H}-\vec{H}\cdot\vec{\nabla}\times\vec{E}\right) = \frac{c}{4\pi}\vec{\nabla}\cdot\left(\vec{H}\times\vec{E}\right)$$

$$\therefore\ \frac{1}{4\pi}\left[\vec{E}\cdot\frac{\partial\vec{D}}{\partial t}+\vec{H}\cdot\frac{\partial\vec{B}}{\partial t}\right] = -\frac{c}{4\pi}\vec{\nabla}\cdot\left(\vec{E}\times\vec{H}\right)-\vec{J}\cdot\vec{E}$$

Integrating the above relation over the volume V enclosed by the surface S, we get,

$$\frac{1}{4\pi}\iiint_V\left(\vec{E}\cdot\frac{\partial\vec{D}}{\partial t}+\vec{H}\cdot\frac{\partial\vec{B}}{\partial t}\right)dV = -\frac{c}{4\pi}\iiint_V\vec{\nabla}\cdot\left(\vec{E}\times\vec{H}\right)dV-\iiint_V\vec{J}\cdot\vec{E}dv$$

$$= -\frac{c}{4\pi}\iint_S\left(\vec{E}\times\vec{H}\right)\cdot d\vec{S}-\iiint_V\vec{J}\cdot\vec{E}dV$$

Now, in case of linear medium $\vec{D} = k\vec{E}$ and $\vec{B} = \mu\vec{H}$.

So, we can write, $\vec{E}\cdot\frac{\partial\vec{D}}{\partial t} = \frac{1}{2}\frac{\partial}{\partial t}\left(\vec{E}\cdot\vec{D}\right)$.

Even if the medium be anisotropic, the above relation is true. In such a case, we can write,

$$D_j = \sum k_{ij}E_j \quad \therefore\ \frac{1}{2}\frac{\partial}{\partial t}\left(\vec{E}\cdot\vec{D}\right) = \frac{1}{2}\sum_{i=1}^{S}\sum_{j=1}^{S}k_{ij}\left(E_i\frac{\partial E_j}{\partial t}+E_j\frac{\partial E_i}{\partial t}\right).$$

Now, $k_{ij} = k_{ji}$ from conservation of energy.

$$\therefore\ \frac{1}{2}\cdot\frac{\partial}{\partial t}\left(\vec{E}\cdot\vec{D}\right) = \sum_{i=1}^{S}\sum_{j=1}^{S}k_{ij}E_i\frac{\partial E_j}{\partial t} = \sum_{i=1}^{S}E_i\frac{\partial}{\partial t}\sum k_{ij}E_j$$

$$= \sum E_i\frac{\partial D_i}{\partial t} = \vec{E}\cdot\frac{\partial\vec{D}}{\partial t}$$

Thus we get,

$$\frac{1}{4\pi}\left(\vec{E}\cdot\frac{\partial\vec{D}}{\partial t}\right)=\frac{1}{8\pi}\frac{\partial}{\partial t}\left(\vec{E}\cdot\vec{D}\right)=\frac{\partial}{\partial t}\left(\frac{\vec{E}\cdot\vec{D}}{8\pi}\right).$$

Similarly,

$$\frac{1}{4\pi}\left(\vec{H}\cdot\frac{\partial\vec{B}}{\partial t}\right)=\frac{\partial}{\partial t}\left(\frac{\vec{H}\cdot\vec{B}}{8\pi}\right)$$

$$\therefore\ \frac{\partial}{\partial t}\iiint\limits_V\left(\frac{\vec{E}\cdot\vec{D}+\vec{H}\cdot\vec{B}}{8\pi}\right)dV=-\frac{c}{4\pi}\iint\left(\vec{E}\times\vec{H}\right)\cdot d\vec{s}-\iiint\limits_V\left(\vec{J}\cdot\vec{E}\right)dV$$

or, $$-\frac{\partial}{\partial t}\iiint\frac{\vec{E}\cdot\vec{D}+\vec{H}\cdot\vec{B}}{8\pi}dV=\frac{c}{4\pi}\iint\left(\vec{E}\times\vec{H}\right)\cdot d\vec{s}+\iiint\vec{J}\cdot\vec{E}dV \tag{6.6.2}$$

Now,

$$\frac{\vec{E}\cdot\vec{D}}{8\pi}=\text{electrostatic energy density}$$

$$\frac{\vec{H}\cdot\vec{B}}{8\pi}=\text{magnetostatic energy density}$$

So the right-hand side of eqn.(6.6.2) may be looked upon as the rate of decrease of electromagnetic field energy contained within certain volume V.

The last expression on the right-hand side of the same eqn. can now be taken for consideration.

$$\vec{E}\cdot\vec{J}dV=\rho dV\vec{v}\cdot\vec{E}=dq\vec{E}\cdot\frac{d\vec{r}}{dt},\quad dq=\rho dV$$

$\vec{E}dq$ represents the electric force experienced by the charge element and so $dq\vec{E}\cdot\frac{d\vec{r}}{dt}$ is the rate of work done by the electromagnetic forces on the charge element in moving it through $d\vec{r}$.

$\therefore$ Total rate of work done $=\iiint\vec{J}\cdot\vec{E}dV=$ total rate of change of kinetic energy of the charged particles $=\frac{\partial T}{\partial t}$.

$$\therefore\ -\frac{\partial}{\partial t}\iiint\frac{\vec{E}\cdot\vec{D}+\vec{H}\cdot\vec{B}}{8\pi}dV=\frac{\partial T}{\partial t}+\frac{c}{4\pi}\iint\left(\vec{E}\times\vec{H}\right)\cdot d\vec{s}.$$

Thus, the diminuation of electromagnetic energy stored in the finite volume V is partly accounted for by the increase of the kinetic energy of the charged particle and partly by the energy flow which is represented by the last term of the above equation. This term represents the flux of the vector $\vec{P}=\frac{c}{4\pi}\left(\vec{E}\times\vec{H}\right)$ across the surface bounding the volume V.

So the Poynting's vector is such that the flux of this vector across a closed surface enclosing a certain volume represents the total energy flow across the bounding surface per unit time. This does not in any way mean that $\vec{P}$ is the energy flow per unit time per unit area. The flux of $\vec{P}$ across a closed surface gives the total energy flow and so it is not logical to conclude therefrom the distribution of energy. In SI unit $\vec{P}=\left(\vec{E}\times\vec{H}\right)$.

Application of Poynting's Theorem

Let a battery of e.m.f. e feed a current of constant magnitude i to a resistor of conductivity σ (fig.6.5).

Here, $H = \frac{1}{4\pi}\frac{2i}{r}$ in SI, r = radius of the wire.

The electric intensity is in the direction of the current

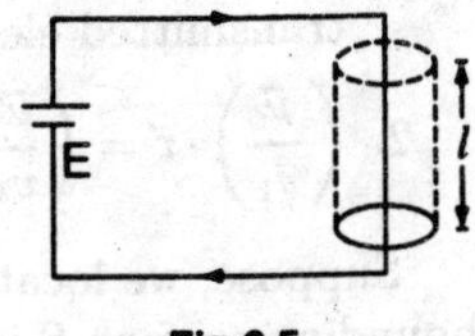

Fig 6.5

$$\therefore \quad P = \left(E\frac{1}{4\pi}\frac{2i}{r}\right).$$

$\therefore$ Total rate of flow of energy from outside through the surface of the conductor of length l is,

$$2\pi r l \cdot \frac{1}{4\pi} \cdot \frac{2iE}{r} = Eil = Vi.$$

$El = V$, Joule's heat for the same length of the conductor is also Vi. Thus, energy does not flow along the conductor from the source but it enters in the form of Poynting's vector from outside through the dielectric across the surface of the conductor.

6.7 Reflection and Refraction of Electromagnetic Waves

Let a plane electromagnetic wave travel from one homogeneous isotropic dielectric medium characterised by μ_1, k_1 to the second medium having the characteristic constants μ_2, k_2 across the surface of separation $S(X-Y$ plane). The plane of incidence is the $X-Z$ plane and the electric field and the magnetic field are shown in the wavefront. As the wave meets the surface of separation, reflection and refraction occur, the reflected ray returns to the first medium, the refracted ray advances through the second medium. If $\vec{r}$ represents the position vector of any point on the wavefront and $\vec{v}$, the direction of wave-normal, then, the solution of Maxwell's electromagnetic vector-wave equation is

$$\vec{E} = \vec{E}_0 e^{i\omega(t-\vec{r}\cdot\vec{v}/v)}; \quad \vec{H} = \sqrt{k/\mu}\left(\vec{v}\times\vec{E}\right).$$

A single progressive wave cannot satisfy the boundary conditions:

(i) tangential components of both $\vec{E}$ and $\vec{H}$ must be continuous across the boundary.

$$\left.\begin{aligned} \therefore \quad &\text{For the incident wave, } \vec{E}_i = \vec{E}_{0i} e^{i\omega_1(t-\vec{r}\cdot\vec{v}/v_1)} \\ &\text{For the reflected wave, } \vec{E}_r = \vec{E}_{0r} e^{i\omega_2(t-\vec{r}\cdot\vec{v}'/v_1)} \\ &\text{For the transmitted wave, } \vec{E}_t = \vec{E}_{0t} e^{i\omega_3(t-\vec{r}\cdot\vec{v}''/v_2)} \end{aligned}\right\} \tag{6.7.1}$$

Similarly, magnetic fields are given by,

$$\vec{H}_i = \sqrt{\frac{k_1}{\mu_1}}\left(\vec{v}\times\vec{E}_i\right)$$

$$\vec{H}_r = \sqrt{\frac{k_1}{\mu_1}}\left(\vec{v}\times\vec{E}_r\right)$$

$$\vec{H}_t = \sqrt{\frac{k_1}{\mu_1}}\left(\vec{v}\times\vec{E}_t\right)$$

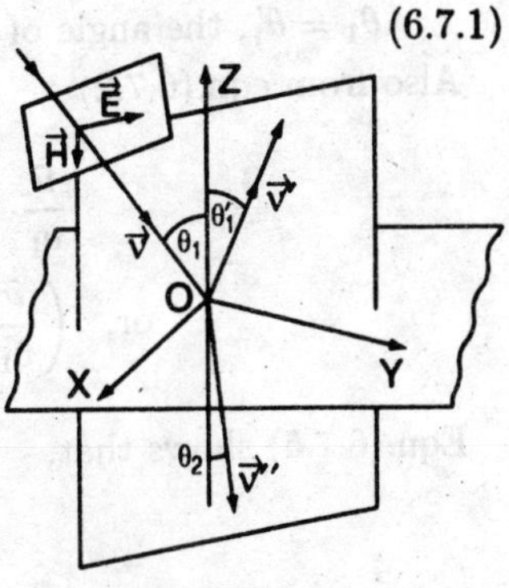

Fig 6.6

The tangential components are continuous across the surface of separation. This is possible only when the exponential terms of the three expressions in eqn.(6.7.1) are same. It means,

1. $\omega_1 = \omega_2 = \omega_3 = \omega$, i.e., frequency remains unchanged in incident, reflected and transmitted electromagnetic wave,
2. $\left(\frac{\vec{v}}{v_1}\right)\cdot\vec{r} = \left(\frac{\vec{v}'}{v_1}\right)\cdot\vec{r} = \left(\frac{\vec{v}''}{v_2}\right)\cdot\vec{r}$ at the bounding surface. (6.7.2)

Suppose, we locate our origin at O on the surface of separation S, then the equation of the bounding surface S is, $\vec{r}\cdot\vec{n} = 0$, $\vec{n}$ is a unit vector normal to the plane.

Thus, for any point on the plane,

$$\begin{aligned}\vec{n}\times(\vec{n}\times\vec{r}) &= (\vec{n}\cdot\vec{r})\,\vec{n} - \vec{r}\,(\vec{n}\cdot\vec{n})\\ &= -\vec{r} \qquad \left[\because \vec{A}\times\left(\vec{B}\times\vec{C}\right) = \left(\vec{A}\cdot\vec{C}\right)\vec{B} - \left(\vec{A}\cdot\vec{B}\right)\vec{C}\right]\end{aligned}$$

This is also evident if we remember that $(\vec{n}\times\vec{r})$ is a vector of magnitude r (length of the point on the plane w.r.t. O as origin) and perpendicular to the plane of incidence. This vector crossed with $\vec{n}$ produces a vector in the plane but opposite to $\vec{r}$ (position vector of the point).

From requirements (ii), eqn. (6.7.2), we get,

$$\left(\frac{\vec{v}}{v_1}\right)\cdot\vec{r} = \left(\frac{\vec{v}'}{v_1}\right)\cdot\vec{r} \text{ on the plane or, } \frac{\vec{v}}{v_1}\cdot\vec{n}\times(\vec{n}\times\vec{r}) = \frac{\vec{v}'}{v_1}\cdot\vec{n}\times(\vec{n}\times\vec{r})$$

$$\text{or, } (\vec{v}-\vec{v}')\cdot\vec{n}\times(\vec{n}\times\vec{r}) = 0 \quad \text{or,} \quad [(\vec{v}-\vec{v}')\times\vec{n}]\cdot(\vec{n}\times\vec{r}) = 0 \tag{6.7.3}$$

Now, $(\vec{v},\vec{n})$ define the plane of incidence, $(\vec{v}',\vec{n})$ define the plane of reflection. From eqn.(6.7.3), $(\vec{r}\times\vec{n})$ is a vector normal to the plane $\vec{r}\cdot\vec{n} = 0$. The first term of the dot product is a vector normal to the plane containing $\vec{n}$ and so the two vectors whose dot product is zero are not at right angles with each other. So the vanishing of the dot product means (since, $\vec{n}\times\vec{r}\neq 0$)

$$(\vec{v}-\vec{v}')\times\vec{n} = 0 \quad \text{or,} \quad \vec{v}\times\vec{n} = \vec{v}'\times\vec{n} \tag{6.7.4}$$

One information from equation (6.7.4) is that the plane of incidence is same as plane of reflection or in other words, the incident wave and reflected wave are contained in the same plane.

Second, $\sin(\pi-\theta_1) = \sin\theta_1'$ or, $\sin\theta_1 = \sin\theta_1'$

$\therefore\ \theta_1 = \theta_1'$, the angle of incidence is equal to the angle of reflection.

Also from eqn.(6.7.2)

$$\frac{\vec{v}}{v_1}\cdot\vec{n}\times(\vec{n}\times\vec{r}) = \frac{\vec{v}''}{v_2}\cdot\vec{n}\times(\vec{n}\times\vec{r}) \text{ at the plane } S$$

$$\text{or, } \left(\frac{\vec{v}}{v_1} - \frac{\vec{v}''}{v_2}\right)\times\vec{n}\cdot(\vec{n}\times\vec{r}) = 0 \tag{6.7.5}$$

Eqn.(6.7.5) shows that,

$$\frac{\vec{v}\times\vec{n}}{v_1} = \frac{\vec{v}''\times\vec{n}}{v_2}.$$

So the plane of incidence must be same as the plane of refraction.

Also,

$$\frac{\sin\theta_1}{v_1} = \frac{\sin\theta_2}{v_2} \tag{6.7.6}$$

$$\text{or,}\quad \frac{\sin\theta_1}{\sin\theta_2} = \frac{v_1}{v_2} = \frac{c/v_2}{c/v_1} = \frac{n_2}{n_1}$$

or, $n_1 \sin\theta_1 = n_2 \sin\theta_2$ — Snell's law of refraction.

6.8 Relation between Amplitudes of $\vec{E}_i$, $\vec{E}_r$ and $\vec{E}_t$: Fresnel's Laws

Applying boundary condition (about the continuity of tangential components), we can write,

$$\vec{n}\times\vec{E}_{0i} + \vec{n}\times\vec{E}_{0r} = \vec{n}\times\vec{E}_{0t} \tag{6.8.1}$$

$$\text{and,}\quad \vec{n}\times\vec{H}_{oi} + \vec{n}\times\vec{H}_{0r} = \vec{n}\times\vec{H}_{0t} \tag{6.8.2}$$

$$\text{or,}\quad \vec{n}\times\sqrt{\frac{k_1}{\mu_1}}\left(\vec{v}\times\vec{E}_{0i}\right) + \vec{n}\times\sqrt{\frac{k_1}{\mu_1}}\left(\vec{v}'\times\vec{E}_{0r}\right) = \vec{n}\times\sqrt{\frac{k_2}{\mu_2}}\left(\vec{v}''\times\vec{E}_{0t}\right) \tag{6.8.3}$$

From, eqn.(6.8.3)

$$\sqrt{\frac{k_1}{\mu_1}}\left[\vec{n}\times\left(\vec{v}\times\vec{E}_{0i}\right) + \left(\vec{n}\times\vec{v}'\times\vec{E}_{0r}\right)\right] = \sqrt{\frac{k_2}{\mu_2}}\vec{n}\times\left(\vec{v}''\times\vec{E}_{0t}\right)$$

$$\text{or,}\quad \sqrt{\frac{k_1}{\mu_1}}\left[\left\{\left(\vec{n}\cdot\vec{E}_{0i}\right)\vec{v} - (\vec{n}\cdot\vec{v})E_{0i}\right\} + \left\{\left(\vec{n}\cdot\vec{E}_{0r}\right)\vec{v}' - (\vec{n}\cdot\vec{v}')\vec{E}_{0r}\right\}\right]$$

$$= \sqrt{\frac{k_2}{\mu_2}}\left[\left(\vec{n}\cdot\vec{E}_{0t}\right)\vec{v}'' - (\vec{n}\cdot\vec{v}'')\vec{E}_{0t}\right] \tag{6.8.4}$$

$\vec{E}_{0i}$ may be quite arbitrary but it can be resolved into two components, one normal to the plane of incidence and hence tangential to the surface of separation AB and the second in the plane of incidence.

Case I. $\vec{E}_{0i}$ normal to the plane of incidence.

In this case, $\vec{E}_{0i}\cdot\vec{n} = \vec{E}_{0i}\cdot\vec{v} = 0$.

Since the medium is isotropic, the induced electric vectors in the transmitted and reflected waves must be parallel to $\vec{E}_{0i}$ also. Hence,

$$\vec{n}\cdot\vec{E}_{0r} = \vec{n}\cdot\vec{E}_{0t} = 0.$$

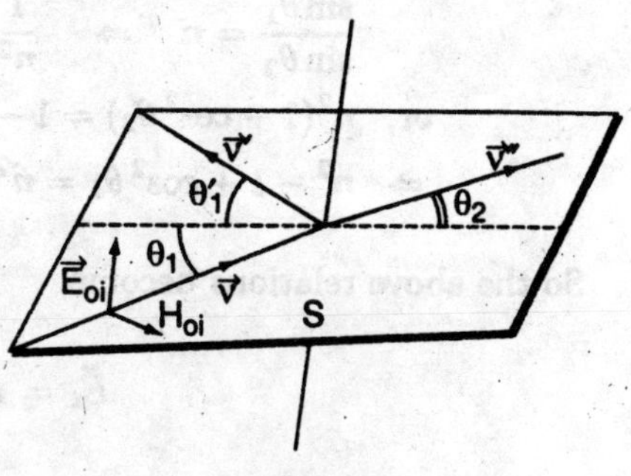

Fig 6.7

Now,

$$\vec{n}\cdot\vec{v} = \cos(\pi-\theta_1) = -\cos\theta_1$$

$$\vec{n}\cdot\vec{v}' = \cos\theta_1',$$

$$\vec{n}\cdot\vec{v}'' = \cos(\pi-\theta_2) = -\cos\theta_2$$

From (6.8.7) crossed with $\vec{n}$, we get,

$$\vec{n} \times \left[\vec{n} \times \left(\vec{E}_{0i} + \vec{E}_{0r}\right)\right] = \vec{n} \times \left(\vec{n} \times \vec{E}_{0t}\right)$$

$$\text{or,} \quad \left[\vec{n} \cdot \left(\vec{E}_{0i} + \vec{E}_{0r}\right)\right] \vec{n} - (\vec{n} \cdot \vec{n}) \left(\vec{E}_{0i} + \vec{E}_{0r}\right) = \left(\vec{n} \cdot \vec{E}_{0t}\right) \vec{n} - (\vec{n} \cdot \vec{n}) \vec{E}_{0t}$$

$$\text{or,} \quad \left[\left(\vec{n} \cdot \vec{E}_{0i} + \vec{n} \cdot \vec{E}_{0r}\right) \vec{n} - \left(\vec{E}_{0i} + \vec{E}_{0r}\right)\right] = \left(\vec{\nabla} \cdot \vec{E}_{0t}\right) \vec{\nabla} - \vec{E}_{0t}$$

$$\left[\because \ \vec{\nabla} \cdot \vec{E}_{0i} = 0, \quad \vec{\nabla} \cdot \vec{E}_{0r} = 0\right]$$

$$\text{or,} \quad \vec{E}_{0i} + \vec{E}_{0r} = \vec{E}_{0t}$$

From eqn.(6.8.4),

$$\sqrt{\frac{k_1}{\mu_1}} \cos\theta_1 \vec{E}_{0i} - \sqrt{\frac{k_1}{\mu_1}} \cos\theta_1' \vec{E}_{0r} = \sqrt{\frac{k_2}{\mu_2}} \cos\theta_2 \vec{E}_{0t}$$

$$\text{or,} \quad \frac{c}{v_1\mu_1}\left(\vec{E}_{0i}\cos\theta_1 - \vec{E}_{0r}\cos\theta_1'\right) = \frac{c}{v_2\mu_2}\cos\theta_2 \vec{E}_{0t}$$

$$\text{or,} \quad \vec{E}_{0i}\cos\theta_1 - \vec{E}_{0r}\cos\theta_1' = \frac{v_1}{v_2}\frac{\mu_1}{\mu_2}\cos\theta_2\vec{E}_{0t} = \frac{v_1}{v_2}\cos\theta_2\vec{E}_{0t} \tag{6.8.5}$$

taking $\mu_1 = \mu_2$.

Solving (6.8.4) and (6.8.5), we get,

$$\vec{E}_{0t} = \frac{2\vec{E}_{0i}\cos\theta_1}{\frac{v_1}{v_2}\cos\theta_2 + \cos\theta_1} \quad [\because \ \theta_1' = \theta_1]$$

$$\therefore \quad \vec{E}_t = \frac{2\vec{E}_i\cos\theta_1}{\frac{v_1}{v_2}\cos\theta_2 + \cos\theta_1} = \frac{2\vec{E}_i\cos\theta_1}{n_2\cos\theta_2 + \cos\theta_1} = \vec{E}_i \cdot \frac{2\cos\theta_1}{n\cos\theta_2 + \cos\theta_1}$$

[n = r.i. of medium 2 w.r.t. 1]

Similarly,

$$\vec{E}_r = -\vec{E}_i\frac{(v_1/v_2)\cos\theta_2 - \cos\theta_1}{(v_1/v_2)\cos\theta_2 + \cos\theta_1} = -\vec{E}_i\frac{n\cos\theta_2 - \cos\theta_1}{n\cos\theta_2 + \cos\theta_1}.$$

Again,

$$\frac{\sin\theta_1}{\sin\theta_2} = n \quad \Rightarrow \quad \frac{1}{n^2}\sin^2\theta_1 = \sin^2\theta_2$$

$$\text{or,} \quad n^2(1 - \cos^2\theta_2) = 1 - \cos^2\theta_1 \quad \Rightarrow \quad n^2 - n^2\cos^2\theta_2 = 1 - \cos^2\theta_1$$

$$\Rightarrow \quad n^2 - 1 + \cos^2\theta_1 = n^2\cos^2\theta_2 \quad \therefore \ n\cos\theta_2 = \sqrt{n^2 - 1 + \cos^2\theta_1}$$

So the above relations become,

$$\left.\begin{aligned} \vec{E}_t &= \vec{E}_i \cdot \frac{2\cos\theta_1}{\sqrt{n^2 - 1 + \cos^2\theta_1} + \cos\theta_1} \\ \text{and} \quad \vec{E}_r &= \vec{E}_i \cdot \frac{\cos\theta_1 - \sqrt{n^2 - 1 + \cos^2\theta_1}}{\cos\theta_1 + \sqrt{n^2 - 1 + \cos^2\theta_1}} \end{aligned}\right\} \tag{6.8.6}$$

Then, eqn.(6.8.2) crossed with $\vec{n}$,

$$\vec{n}\times\left(\vec{n}\times\vec{H}_{0i}\right)+\vec{n}\times\left(\vec{n}\times\vec{H}_{0r}\right)=\vec{n}\times\left(\vec{n}\times\vec{H}_{0t}\right)$$

$$\text{or,}\quad \left(\vec{n}\cdot\vec{H}_{0i}\right)\vec{n}-(\vec{n}\cdot\vec{n})\vec{H}_{0i}+\left(\vec{n}\cdot\vec{H}_{0r}\right)\vec{n}-(\vec{n}\cdot\vec{n})\vec{H}_{0r}=\left(\vec{n}\cdot\vec{H}_{0t}\right)\vec{n}-(\vec{n}\cdot n)\vec{H}_{0t}$$

$$\text{or,}\quad \vec{H}_{0i}+\vec{H}_{0r}=\vec{H}_{0t}\quad\Rightarrow\quad\vec{H}_i+\vec{H}_r=\vec{H}_t$$

$$\left[\because\left(\vec{n}\cdot\vec{H}_{0i}\right)=0,\ \left(\vec{n}\cdot\vec{H}_{0r}\right)=0,\ \left(\vec{n}\cdot\vec{H}_{0i}\right)=0\right]\tag{6.8.7}$$

since exponential terms same.

Again, we know that,

$$\vec{H}=\sqrt{\frac{k}{\mu}}\left(\vec{\nu}\times\vec{E}\right)$$

$$\therefore\quad\left(\vec{\nu}\times\vec{H}\right)=\sqrt{\frac{k}{\mu}}\vec{\nu}\times\left(\vec{\nu}\times\vec{E}\right)=\sqrt{\frac{k}{\mu}}\left[\left(\vec{\nu}\cdot\vec{E}\right)\vec{\nu}-(\vec{\nu}\cdot\vec{\nu})\vec{E}\right]$$

$$=-\sqrt{\frac{k}{\mu}}\vec{E}\quad[\text{since }\vec{\nu}\cdot\vec{E}=0\text{ because }\vec{E}\perp\vec{\nu}]$$

$$\therefore\quad\vec{E}=-\sqrt{\frac{\mu}{k}}\left(\vec{\nu}\times\vec{H}\right)$$

With the help of eqn.(6.8.1), we can write,

$$\vec{n}\times\sqrt{\frac{\mu_1}{k_1}}\left(\vec{\nu}\times\vec{H}_i\right)+\vec{n}\times\sqrt{\frac{\mu_1}{k_1}}\left(\vec{\nu}'\times\vec{H}_r\right)=\vec{n}\times\sqrt{\frac{\mu_2}{k_2}}\left(\vec{\nu}''\times\vec{H}_t\right)$$

$$\text{or,}\quad\sqrt{\frac{\mu_1}{k_1}}\left[\left(\vec{n}\cdot\vec{H}_i\right)\vec{\nu}-(\vec{n}\cdot\vec{\nu})\vec{H}_i\right]+\sqrt{\frac{\mu_1}{k_1}}\left[\left(\vec{n}\cdot\vec{H}_r\right)\vec{\nu}'-(\vec{n}\cdot\vec{\nu}')\vec{H}_r\right]$$

$$=\sqrt{\frac{\mu_2}{k_2}}\left[\left(\vec{n}\cdot\vec{H}_t\right)\vec{\nu}''-(\vec{n}\cdot\vec{\nu}'')\vec{H}_t\right]$$

$$\text{or,}\quad\sqrt{\frac{\mu_1}{k_1}}\vec{H}_i\cos\theta_1-\sqrt{\frac{\mu_1}{k_1}}\vec{H}_r\cos\theta_1=\sqrt{\frac{\mu_2}{k_2}}\vec{H}_t\cos\theta_1$$

$$\left(\vec{n}\cdot\vec{H}_i\right)=0,\left(\vec{n}\cdot\vec{H}_r\right)=0,\left(\vec{n}\cdot\vec{H}_t\right)=0$$

$$\text{or,}\quad\vec{H}_i\cos\theta_1-\vec{H}_r\cos\theta_1=\sqrt{\frac{\mu_2k_1}{\mu_1k_2}}\vec{H}_t\cos\theta_2=\frac{v_2}{v_1}\vec{H}_t\cos\theta_2\tag{6.8.8}$$

(taking $\mu_1=\mu_2$)

Solving eqns.(6.8.7) and (6.8.8),

$$\vec{H}_r=-\vec{H}_i\frac{\left(\frac{v_2}{v_1}\right)\cos\theta_2-\cos\theta_1}{\left(\frac{v_2}{v_1}\right)\cos\theta_2+\cos\theta_1}\quad\text{and}\quad\vec{H}_t=\vec{H}_i\frac{2\cos\theta_1}{\left(\frac{v_2}{v_1}\right)\cos\theta_2+\cos\theta_1}\tag{6.8.9}$$

Eqns.(6.8.6) and (6.8.9) are known as *Fresnel's formulae*.

Eqn.(6.8.6) can be put in another form. We know that,

$$\frac{\sin\theta_1}{\sin\theta_2} = \frac{v_1}{v_2}.$$

$$\therefore \quad \vec{E}_t = \frac{2\vec{E}_i \cos\theta_1}{\cos\theta_1 + \frac{v_1}{v_2}\cos\theta_2} = \frac{2\vec{E}_i \cos\theta_1}{\cos\theta_1 + \frac{\sin\theta_1\cos\theta_2}{\sin\theta_2}} = 2\vec{E}_i \frac{\sin\theta_2\cos\theta_1}{\sin(\theta_1+\theta_2)} \qquad (6.8.10)$$

$$\text{and} \quad \vec{E}_r = -\vec{E}_i \frac{\left(\frac{v_1}{v_2}\right)\cos\theta_2 - \cos\theta_1}{\left(\frac{v_1}{v_2}\right)\cos\theta_2 + \cos\theta_1} = \vec{E}_i \frac{\sin\theta_2\cos\theta_1 - \sin\theta_1\cos\theta_2}{\sin\theta_2\cos\theta_1 + \sin\theta_1\cos\theta_2}$$

$$= \frac{\sin(\theta_2 - \theta_1)}{\sin(\theta_2 + \theta_1)}\vec{E}_i \qquad (6.8.11)$$

6.9 Reflection and Transmission Co-efficients

The reflected energy is proportional to the reflected Poynting's vector and the transmitted energy is proportional to the transmitted Poynting's vector. So the reflection co-efficient R_n which is nothing but the ratio of reflected energy to the incident energy of the electromagnetic disturbance can be written as,

$$R_n = \frac{|\vec{E}_r \times \vec{H}_r|}{|\vec{E}_i \times \vec{H}_i|}$$

Similarly, the transmission co-efficient

$$R_n = \frac{|\vec{E}_t \times \vec{H}_t|}{|\vec{E}_i \times \vec{H}_i|}$$

These expressions may be determined when the electric field is in or perpendicular to the plane of incidence. If we take the case of $\vec{E}_i$ in the plane of incidence, then we have shown that [eqn. 6.8.9],

$$\vec{H}_r = \frac{\cos\theta_1 - \frac{v_2}{v_1}\cos\theta_2}{\cos\theta_1 + \frac{v_2}{v_1}\cos\theta_2}\vec{H}_i \quad \text{and} \quad \vec{H}_t = \frac{2\cos\theta_1}{\cos\theta_1 + \frac{v_2}{v_1}\cos\theta_2}\vec{H}_i$$

The corresponding electric field components can be written as,

$$\vec{E}_r = -\sqrt{\frac{\mu_1}{k_1}}\left(\vec{v} \times \vec{H}_r\right), \vec{E}_t = -\sqrt{\frac{\mu_2}{k_2}}\left(\vec{v}'' \times \vec{H}_t\right)$$

$$\therefore \quad \vec{E}_r \times \vec{H}_r = -\sqrt{\frac{\mu_1}{k_1}}\left(\vec{v} \times \vec{H}_r\right) \times \vec{H}_r = -\sqrt{\frac{\mu_1}{k_1}}\left[\left(\vec{v}\cdot\vec{H}_r\right)\vec{H}_r - \left(\vec{H}_r\cdot\vec{H}_r\right)\vec{v}\right]$$

$$= \sqrt{\frac{\mu_1}{k_1}} H_r^2 \vec{v}, \text{ since } \vec{v}\cdot\vec{H}_r = 0$$

$$\therefore \quad \left|\vec{E}_r \times \vec{H}_r\right| = \sqrt{\frac{\mu_1}{k_1}} H_r^2,$$

Similarly,

$$\left|\vec{E}_i \times \vec{H}_i\right| = \sqrt{\frac{\mu_1}{k_1}} H_i^2$$

$$\therefore \quad R_n = \frac{H_r}{H_i} = \left(\frac{\cos\theta_1 - \frac{v_2}{v_1}\cos\theta_2}{\cos\theta_1 + \frac{v_2}{v_1}\cos\theta_2}\right)^2 = \left(\frac{n_2\cos\theta_1 - n_1\cos\theta_2}{n_2\cos\theta_1 + n_1\cos\theta_2}\right)^2.$$

Similarly,

$$T_n = \frac{n_2}{n_1}\left(\frac{2n_1\cos\theta_1}{n_2\cos\theta_1 + n_1\cos\theta_2}\right)^2.$$

For normal incidence, $\theta_1 = 0$, $\theta_2 = 0$.

$$\therefore \quad R_n = \left(\frac{n_2 - n_1}{n_2 + n_1}\right)^2; \quad T_n = \frac{n_2}{n_1}\cdot\left(\frac{2n_1}{n_1 + n_2}\right)^2.$$

In case of air-glass interface, $n_2 = 1.5$ and $n_1 = 1$, so $R_n = 0.04$ and $T_n = 0.96$. It means that all the incident energy is either reflected or transmitted, there is no energy to be stored at the interface.

6.10 Some Interesting Results from Eqns. (6.8.10) and (6.8.11)

From eqn.(6.8.11), we see that,

$\vec{E}_r = 0$ when $\theta_1 = \theta_2$, i.e., there is no reflection if the two media have the same optical properties. Hence, reflection takes place only when $\theta_1 \neq \theta_2$.

From eqn.(6.8.9),

$$\vec{H}_r = \vec{H}_i \frac{\cos\theta_1 - \left(\frac{v_2}{v_1}\right)\cos\theta_2}{\cos\theta_1 + \left(\frac{v_2}{v_1}\right)\cos\theta_2} = \vec{H}_i \frac{\cos\theta_1 - \frac{\sin\theta_2\cos\theta_2}{\sin\theta_1}}{\cos\theta_1 + \frac{\sin\theta_2\cos\theta_2}{\sin\theta_1}} = \vec{H}_i \frac{\tan(\theta_1 - \theta_2)}{\tan(\theta_1 + \theta_2)}$$

So, $\vec{H}_r = 0$ when $\theta_1 + \theta_2 = \frac{\pi}{2}$. The angle of incidence for which this happens is known as *Brewster's angle* or *polarising angle* and so,

$$\frac{\sin\theta_1}{\sin\theta_2} = n \quad \text{or,} \quad \tan\theta_1 = n \quad \Rightarrow \quad \text{Brewster's law.}$$

Total Reflection

In deriving Fresnel's formulae, we have considered only real angles θ_1, θ_2. But let us see what happens when electromagnetic wave passes from a denser medium to the rarer medium.

$$\frac{\sin\theta_1}{\sin\theta_2} = \frac{v_1}{v_2} = n < 1 \text{ means,} \quad \sin\theta_2 > \sin\theta_1.$$

Then eqn.(6.8.11) can be written as,

$$\vec{E}_r = \frac{\cos\theta_1\sin\theta_2 - \cos\theta_2\sin\theta_1}{\cos\theta_1\sin\theta_2 + \cos\theta_2\sin\theta_1}\vec{E}_i$$

From the condition indicated above, if $v_2 > v_1$, the law of refraction will lead to values of $\sin\theta_2 > 1$ as a result of which $\cos\theta_2$ becomes a complex quantity which means a phase shift in the incident wave and there is also attenuation.

If we eliminate θ_2 with the help of Snell's law and remember that $n < 1$, then,

$$\vec{E}_r = \vec{E}_i \frac{\cos\theta_1 - i\sqrt{\sin^2\theta_1 - n^2}}{\cos\theta_1 + i\sqrt{\sin^2\theta_1 - n^2}} = \vec{E}_i \frac{\sqrt{a^2+b^2}e^{-i\phi}}{\sqrt{a^2+b^2}e^{i\phi}} = \vec{E}_i e^{-2i\phi}$$

where, $a = \cos\theta_1, b = (\sin^2\theta_1 - n^2)^{1/2}$ and $\tan\phi = \frac{b}{a}$.

So, the reflected amplitude is equal to that of incident wave but there is a phase change.

Again, for electric component in the plane of incidence, we have,

$$\vec{H}_r = \frac{\cos\theta_1 \sin\theta_1 - \cos\theta_2 \sin\theta_2}{\cos\theta_1 \sin\theta_1 + \cos\theta_2 \sin\theta_2} = \frac{\cos\theta_1 \sin\theta_1 - \frac{\sin\theta_1}{n}\sqrt{1-\sin^2\theta_2}}{\cos\theta_1 \sin\theta_1 + \frac{\sin\theta_1}{n}\sqrt{1-\sin^2\theta_2}}$$

$$= \frac{\cos\theta_1 \sin\theta_1 - \frac{\sin\theta_1}{n}\sqrt{1-\frac{\sin^2\theta_1}{n^2}}}{\cos\theta_1 \sin\theta_1 + \frac{\sin\theta_1}{n}\sqrt{1-\frac{\sin^2\theta_1}{n^2}}} = \frac{\cos\theta_1 - \frac{i}{n^2}\sqrt{\sin^2\theta_1 - n^2}}{\cos\theta_1 + \frac{i}{n^2}\sqrt{\sin^2\theta_1 - n^2}}\vec{H}_i$$

$$= \vec{H}_i \frac{c - id}{c + id} = \frac{\sqrt{c^2+d^2}e^{-i\psi}}{\sqrt{c^2+d^2}e^{i\psi}}\vec{H}_i = \vec{H}_i e^{-2i\psi}, \text{ where } \tan\psi = \frac{\sqrt{\sin^2\theta_1 - n^2}}{n^2\cos\theta_1}.$$

So, for an incident wave linearly polarised in a direction neither parallel nor normal to the plane of incidence, the electric vector can be resolved into two components, one in the plane of incidence, the other normal to the plane of incidence.

Such wave train under the present condition will be reflected with all the intensity and we call this phenomenon *total reflection*, but there will be phase change in both these components and the phase change is different. Thus, the reflected wave will be elliptically polarised wave and the difference of phase between the two components is given by,

$$\tan(\phi - \psi) = \frac{\tan\phi - \tan\psi}{1 + \tan\phi\tan\psi} = \frac{\frac{(\sin^2\theta_1 - n^2)^{1/2}}{\cos\theta_1} - \frac{\sqrt{\sin^2\theta_1 - n^2}}{n^2\cos\theta_1}}{1 + \frac{(\sin^2\theta_1 - n^2)}{n^2\cos^2\theta_1}}$$

$$= -\frac{\cos\theta_1(\sin^2\theta_1 - n^2)^{1/2}}{\sin^2\theta_1}$$

If the phase difference is zero, i.e., $(\phi - \psi) = 0$, then, for $\theta_1 \neq \frac{\pi}{2}$, we get,

$$\sin^2\theta_1 = n^2 \quad \text{or,} \quad \sin\theta_1 = n$$

where n is the refractive index of the second medium w.r.t. the first medium. If the second medium be air, then n represents the refractive index of air w.r.t. the first medium.

$\therefore\ \sin\theta_1 = {}^1n_a = \frac{1}{n_1}$, where n_1 is the r.i. of the first medium w.r.t. air. This angle is known as the *critical angle* for the two media.

$$\therefore \sin\theta_c = \frac{1}{n_1}.$$

6.11 Plane Electromagnetic Waves in Isotropic Conducting Medium

In case of conducting medium,

$$\vec{J} = \sigma\vec{E}, \quad \vec{D} = k\vec{E}, \quad \vec{B} = \mu\vec{H}.$$

So the field equations now become,

$$\vec{\nabla}\times\vec{H} = \frac{k}{c}\cdot\frac{\partial\vec{E}}{\partial t} + \frac{4\pi\sigma}{c}\vec{E} \tag{6.11.1}$$

$$\vec{\nabla}\times\vec{E} = -\frac{\mu}{c}\cdot\frac{\partial\vec{H}}{\partial t} \tag{6.11.2}$$

$$\vec{\nabla}\cdot\vec{E} = \frac{4\pi\rho}{k} \quad \text{and} \quad \vec{\nabla}\cdot\vec{H} = 0 \tag{6.11.3}$$

From this set of equations, we can get,

$$\vec{\nabla}\cdot\left(\vec{\nabla}\times\vec{H}\right) = \frac{k}{c}\cdot\frac{\partial}{\partial t}\left(\vec{\nabla}\cdot\vec{E}\right) + \frac{4\pi\sigma}{c}\left(\vec{\nabla}\cdot\vec{E}\right)$$

$$\text{or,} \quad \frac{k}{c}\,\frac{4\pi}{k}\cdot\frac{\partial\rho}{\partial t} + \frac{4\pi\sigma}{c}\cdot\frac{4\pi\rho}{k} = 0 \quad \text{or,} \quad \frac{\partial\rho}{\partial t} + \frac{4\pi\sigma}{k}\rho = 0 \tag{6.11.4}$$

Solution of (6.11.4) gives $\rho = \rho_0 e^{-t/\tau}$, where $\tau = \frac{k}{4\pi\sigma}$. This shows that any initial distribution of charge in a conductor is quickly transferred to the surface and so, $\rho = 0$ inside the conductor for all cases.

So, $\vec{\nabla}\cdot\vec{E} = 0$, from eqn.(6.11.3) (since $\rho = 0$).

Let us eliminate $\vec{H}$ in between eqns.(6.11.1) and (6.11.2). Taking curl of eqn.(6.11.2), we get,

$$\vec{\nabla}\times\left(\vec{\nabla}\times\vec{E}\right) = -\frac{\mu}{c}\cdot\frac{\partial}{\partial t}\left(\vec{\nabla}\times\vec{H}\right) = -\frac{\mu}{c}\frac{\partial}{\partial t}\left(\frac{k}{c}\cdot\frac{\partial\vec{E}}{\partial t} + \frac{4\pi\sigma}{c}\vec{E}\right)$$

$$= -\frac{\mu k}{c^2}\cdot\frac{\partial^2\vec{E}}{\partial t^2} - \frac{4\pi\sigma\mu}{c^2}\cdot\frac{\partial\vec{E}}{\partial t}$$

Now,

$$\vec{\nabla}\times\left(\vec{\nabla}\times\vec{E}\right) = \vec{\nabla}\left(\vec{\nabla}\cdot\vec{E}\right) - \vec{\nabla}^2\vec{E} = -\vec{\nabla}^2\vec{E}\left(\vec{\nabla}\cdot\vec{E}\right) = 0$$

$$\therefore \quad \vec{\nabla}^2\vec{E} = \frac{\mu k}{c^2}\cdot\frac{\partial^2\vec{E}}{\partial t^2} + \frac{4\pi\sigma\mu}{c^2}\cdot\frac{\partial\vec{E}}{\partial t}. \tag{6.11.5}$$

Similarly, eliminating $\vec{E}$, we could form similar expression for $\vec{H}$, i.e.,

$$\vec{\nabla}^2\vec{H} = \frac{\mu k}{c^2}\cdot\frac{\partial^2\vec{H}}{\partial t^2} + \frac{4\pi\sigma\mu}{c^2}\cdot\frac{\partial\vec{H}}{\partial t}. \tag{6.11.6}$$

The equations (6.11.5) and (6.11.6) are known as *equations of telegraphy*.

Take the solution of the vector wave equation (6.11.5) in the general form,

$$\vec{E} = \vec{E}_0 e^{i\omega(t-\vec{r}\cdot\vec{v}/v)}.$$

Now,

$$\frac{\partial \vec{E}}{\partial t} = i\omega \vec{E}_0 e^{i\omega(t-\vec{r}\cdot\vec{\nu}/v)} = i\omega \vec{E}$$

$$\therefore \quad \frac{\partial^2 \vec{E}}{\partial t^2} = i^2\omega^2 \vec{E} = i\omega \frac{\partial \vec{E}}{\partial t}$$

$$\therefore \quad \vec{\nabla}^2 \vec{E} = \frac{\mu k}{c^2}\cdot\frac{\partial^2 \vec{E}}{\partial t^2} + \frac{4\pi\sigma\mu}{c^2}\cdot\frac{1}{i\omega}\cdot\frac{\partial^2 \vec{E}}{\partial t^2} = \frac{\mu}{c^2}\left(k - \frac{4\pi\sigma i}{\omega}\right)\frac{\partial^2 \vec{E}}{\partial t^2} = \frac{\mu}{c^2}k^*\frac{\partial^2 \vec{E}}{\partial t^2} \quad (6.11.7)$$

Eqn.(6.11.7) shows that the electric field propagates with a velocity $v^* = \frac{c}{\sqrt{\mu k^*}}$ in a medium of dielectric constant $k^* = k - \frac{4\pi\sigma i}{\omega}$.

Both the velocity of propagation and the dielectric constant of the medium are complex quantities so far as the propagation of electromagnetic waves through conducting medium is concerned.

Now,

$$v^* = \frac{c}{\sqrt{\mu k^*}} = \frac{c}{n^*}.$$

The refractive index of the medium is complex. Because of the complex nature of the dielectric constant as regards the electromagnetic wave propagation through conducting medium, the electric field and magnetic field differ in phase.

$$n^* = \sqrt{\mu k^*} = \sqrt{\mu\left(k - \frac{4\pi\sigma i}{\omega}\right)} = n(1 - ic); \quad n^2(1-\alpha^2) = \mu k \quad (6.11.8)$$

$$n^2\alpha = \frac{2\pi\sigma\mu}{\omega} = \mu\sigma/f \quad (6.11.9)$$

$$\therefore \quad \vec{E} = \vec{E}_0 \exp\left\{i\omega\left(t - \vec{r}\cdot\frac{\vec{\nu}}{v^*}\right)\right\} = \vec{E}_0 \exp\left[i\omega\left\{t - \frac{(r\cdot\vec{\nu})}{c}n^*\right\}\right]$$

$$= \vec{E}_0 \exp\left[i\omega\left\{t - \frac{(\vec{r}\cdot\vec{\nu})}{c}n(1-i\alpha)\right\}\right] = \vec{E}_0 \exp\left\{-\frac{n\omega\alpha}{c}(\vec{r}\cdot\vec{\nu})\right\}\exp\left[i\omega\left\{t - (\vec{r}\cdot\vec{\nu})\frac{n}{c}\right\}\right]$$

$$\vec{H} = \sqrt{\frac{k^*}{\mu}}\left(\vec{\nu}\times\vec{E}\right) = \frac{1}{\mu}n(1-i\alpha)\left(\vec{\nu}\times\vec{E}\right) = \frac{n}{\mu}\sqrt{(1+\alpha^2)}\left(\vec{\nu}\times\vec{E}\right)e^{-i\phi}$$

$$= \frac{n}{\mu}\sqrt{(1+\alpha^2)}\vec{\nu}\times\vec{E}_0 \exp\left\{-\frac{n\omega\alpha(\vec{r}\cdot\vec{\nu})}{c}\right\}\exp\left\{i\omega\left(t - n\frac{(\vec{r}\cdot\vec{\nu})}{c} - \frac{\phi}{\omega}\right)\right\}$$

Thus, here also, $\vec{E}$ and $\vec{H}$ are $\perp$ to each other and also to $\vec{\nu}$ but unlike in dielectric medium, here there is a phase difference between them. Moreover, the amplitude $E_0 \exp\left\{-\frac{n\omega\alpha}{c}(\vec{r}\cdot\vec{\nu})\right\}$ diminishes exponentially as the wave advances through the conducting medium.

Now, $\frac{n\omega}{c} = \frac{2\pi}{\lambda}$, where λ is the wavelength of electromagnetic wave in the medium of r.i. n.

$\therefore$ The amplitude of the electric wave in the conducting medium $= E_0 \exp\left\{-\frac{2\pi\alpha}{\lambda}(\vec{r}\cdot\vec{\nu})\right\}$.

Since $I \propto (\text{amplitude})^2$

$$\therefore \quad I = I_0 \exp\left\{-\frac{4\pi\alpha}{\lambda}(\vec{r}\cdot\vec{\nu})\right\}.$$

If intensity falls to $\frac{1}{e}$ of its value after traversing a distance $\vec{r}\cdot\vec{\nu} = d$, then,

$$\frac{1}{e}I_0 = I_0 \exp\left(-\frac{4\pi\alpha}{\lambda}d\right) \text{ or, } \exp\left(\frac{4\pi\alpha d}{\lambda}\right) = e \text{ or, } \frac{4\pi\alpha d}{\lambda} = 1. \quad \therefore d = \frac{\lambda}{4\pi\alpha} \quad (6.11.10)$$

Greater the value of α for a given λ, the smaller will be the depth of penetration within the metallic conductor, meaning greater absorption or attenuation of intensity. That is why α is usually referred to as *attenuation index* or *extinction co-efficient.* The relation of α with n and k is given by eqns.(6.11.7), (6.11.8) and (6.11.9).

The co-efficient of $(\vec{r}\cdot\vec{v})$, i.e., $\frac{4\pi\alpha}{\lambda}$ is called *absorption co-efficient.*

6.12 Metallic Reflection

When an electromagnetic wave is incident on the plane surface of a conducting medium, the amplitudes of the reflected and transmitted waves can be calculated in exactly the same way as we have done for dielectric medium, only difference being that in this case the dielectric constant k_2 is complex and so also is the refractive index n of the metallic medium. Thus, we can write,

$$\left.\begin{aligned}\vec{E}_i &= \vec{E}_{0i}\exp\left[i\omega\left(t-\vec{r}\cdot\frac{\vec{v}}{v_1}\right)\right]\\ \vec{E}_r &= \vec{E}_{0r}\exp\left[i\omega\left(t-\vec{r}\cdot\frac{\vec{v}'}{v_1}\right)\right]\\ \vec{E}_t &= \vec{E}_{0t}\exp\left[i\omega\left(t-\vec{r}\cdot\frac{\vec{v}''}{v_2^*}\right)\right]\end{aligned}\right\} \tag{6.12.1}$$

where, $v_1 = \frac{c}{\sqrt{\mu_1 k_1}}, v_2^* = \frac{c}{\sqrt{\mu_2 k_2^*}}$.

The corresponding magnetic waves are given by,

$$\vec{H}_i = \sqrt{\frac{k_1}{\mu_1}}\left(\vec{v}\times\vec{E}_1\right),\quad \vec{H}_r = \sqrt{\frac{k_1}{\mu_1}}\left(\vec{v}'\times\vec{E}_r\right),\quad \vec{H}_t = \sqrt{\frac{k_2^*}{\mu_2}}\left(\vec{v}''\times\vec{E}_t\right). \tag{6.12.2}$$

Thus from previous calculation we can write for

Case I. When $\vec{E}_i$ is normal to the plane of incidence

$$\vec{E}_{0i}+\vec{E}_{0r} = \vec{E}_{0t} \tag{6.12.3}$$

for the tangential components of electric field to be continuous. Also the condition for the continuity of the tangential components of magnetic field, we can write,

$$\sqrt{\frac{k_1}{\mu_1}}\left(\vec{E}_{0i}-\vec{E}_{0r}\right)\cos\theta_1 = \sqrt{\frac{k_2^*}{\mu_2}}\vec{E}_{0t}\cos\theta_2$$

where, $k_2^* = k - i\frac{4\pi\sigma}{\omega}$.

The above eqn. can be written as,

$$\frac{\sqrt{\mu_1 k_1}}{\mu_1}\left(\vec{E}_{0i}-\vec{E}_{0r}\right)\cos\theta_1 = \frac{\sqrt{\mu_2 k_2^*}}{\mu_2}\vec{E}_{0t}\cos\theta_2.$$

Now, $\sqrt{\mu_1 k_1} = n_1$ (refractive index of first medium)

$$\sqrt{\mu_2 k_2^*} = n_2^* = \sqrt{\mu_2\left(k_2 - \frac{4\pi\sigma}{\omega}\right)} = n_2(1-i\alpha)$$

$$\therefore\quad n_2^2(1-\alpha^2) = \mu_2 k_2;\quad n_2^2\alpha = \frac{4\pi\sigma\mu}{\omega}.$$

$\therefore$ We get,

$$\frac{n_1}{\mu_1}\left(\vec{E}_{0i}-\vec{E}_{0r}\right)\cos\theta_1=\frac{n_2^*}{\mu_2}\vec{E}_{0t}\cos\theta_2. \qquad (6.12.4)$$

Since n_2^* is complex, $\vec{E}_{0r}$ and $\vec{E}_{0t}$ cannot both be real, thus showing that the phase shifts different from zero and π are possible for the reflected and transmitted waves.

Solving eqns. (6.12.3) and (6.12.4), we get,

$$\vec{E}_{0r}=\frac{\cos\theta_1-\frac{\mu_1}{\mu_2}.\frac{n_2^*}{n_1}\cos\theta_2}{\cos\theta_1+\frac{\mu_1}{\mu_2}.\frac{n_2^*}{n_1}\cos\theta_2}\vec{E}_{0i};\qquad \vec{E}_{0t}=\frac{2\cos\theta_1}{\cos\theta_1+\frac{\mu_1}{\mu_2}.\frac{n_2^*}{n_1}\cos\theta_2}\vec{E}_{0i}.$$

For normal incidence, these two expressions simplify to

$$\vec{E}_{0r}=\frac{1-\frac{\mu_1}{\mu_2}.\frac{n_2^*}{n_1}}{1+\frac{\mu_1}{\mu_2}.\frac{n_2^*}{n_1}}\vec{E}_{0i};\qquad \vec{E}_{0t}=\frac{2}{1+\frac{\mu_1}{\mu_2}.\frac{n_2^*}{n_1}}\vec{E}_{0i}.$$

For a perfect conductor, n_2^* is infinite and so,

$$\vec{E}_{0r}=-\vec{E}_{0i}\quad\text{and}\quad\vec{E}_{0t}=0.$$

It means all the incident energy is reflected and no penetration into the metal.

6.13 Waveguides

Waveguides are devices (tubes or parallel planes) which guide electromagnetic waves along a certain direction without attenuation of amplitude. Such waves are called *guided waves.*

When an electromagnetic wave is sent from an aerial into free space (or a non-conducting medium), its amplitude falls off inversely with distance as the wave spreads out in the form of spherical wavefront. Energy is not dissipated but the power flowing through unit area perpendicular to the wavefront inversely falls off as the square of distance. But a wave sent along a coaxial line suffers no diminuation in amplitude. This wave as referred to is the guided wave. We have shown in case of perfect conductors that at the surface of a conductor $\vec{E}_{0r}=-\vec{E}_{0i}$ and $\vec{E}_{0t}=0$, meaning the incident wave is wholly reflected from a metallic surface. So, if we consider a pair of conducting planes, there will be multiple reflections in between and again another pair of planes at right angles to the earlier pair, thus forming a rectangular waveguide guiding the electromagnetic wave without loss in amplitude.

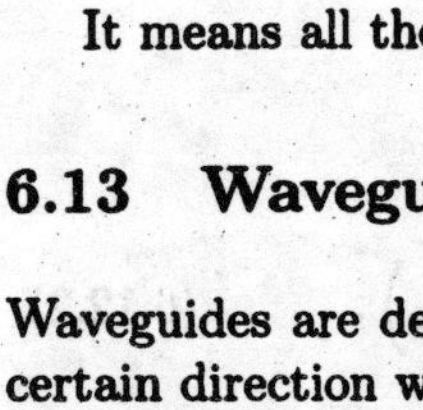

Fig 6.8.

Let us take such a rectangular waveguide as shown in the fig.(6.8). The co-ordinate axes are so set that Z-axis is along the axis of the waveguide and the walls are parallel to the planes YZ and ZX, and let a, b be the sides of the normal section, a being parallel to x-axis. We take Maxwell's equations

$$\vec{\nabla}\times\vec{H}=\frac{k}{c}.\frac{\partial\vec{E}}{\partial t}\quad\text{and}\quad\vec{\nabla}\times\vec{E}=-\frac{\mu}{c}.\frac{\partial\vec{H}}{\partial t}.$$

The solutions are taken as $\vec{E}=\vec{E}_0e^{-\alpha z}.e^{i\omega t}$, $\vec{H}=\vec{H}_0e^{-\alpha z}.e^{i\omega t}$ where resolutes $\vec{E}_0$ and $\vec{H}_0$ are independent of z. Putting these solutions in Maxwell's equations we get,

$$\frac{ik\omega}{c}\vec{E}_0=\alpha\vec{H}_0\times\hat{k}+\vec{\nabla}\times\vec{H}_0,\ \frac{i\mu\omega}{c}\vec{H}_0=\alpha\hat{k}\times\vec{E}_0-\vec{\nabla}\times\vec{E}_0$$

where $\hat{k}$ is a unit vector along OZ. These show that the electric and magnetic fields are not perpendicular to the direction of propagation, unless either or both of $\vec{\nabla} \times \vec{H}_0$ and $\vec{\nabla} \times \vec{E}_0$ vanish.

If $\vec{\nabla} \times \vec{H}_0 = 0$, the electric field is transverse to the direction of propagation, but $\vec{H}_0$ has a resolute in the direction of propagation, such a wave is known as *transverse electric* (TE) or H-wave. On the other hand, if $\vec{\nabla} \times \vec{E}_0 = 0$, then the magnetic field is transverse to the direction of propagation but $\vec{E}_0$ has resolutes in the direction of propagation, such waves are referred to as the *transverse magnetic* (TM) or E-wave.

6.14 Electromagnetic Waves through Antisotropic Medium : Fresnel's Eqns.

If the medium under consideration be antisotropic, then $\vec{D} = k\vec{E}$ relationship has to tell us a different story than when the medium is isotropic. In case of an isotropic medium, k is a scalar quantity, known as the *dielectric constant* of the medium and so, $\vec{D}$ is in the direction of $\vec{E}$. In such a case, in order to preserve linear relationship between $\vec{D}$ and $\vec{E}$, it is assumed that each component of $\vec{D}$ must be linearly related to the components of $\vec{E}$, i.e.,

$$\left.\begin{aligned} D_x &= k_{11}E_x - k_{12}E_y + k_{13}E_z \\ D_y &= k_{21}E_x - k_{22}E_y + k_{23}E_z \\ D_z &= k_{31}E_x - k_{32}E_y + k_{33}E_z \end{aligned}\right\} \tag{6.14.1}$$

Thus, k is an operator which operating on $\vec{E}$ not only changes the magnitude but also the direction. k has nine components k_{ij} which are scalars and dependent on the nature of the crystals. Hence, dielectric constant in such case is the tensor of rank two. The matrix equivalent of the operation in eqn.(6.14.1) can be written as,

$$\begin{bmatrix} D_x \\ D_y \\ D_z \end{bmatrix} = \begin{bmatrix} k_{11} & k_{12} & k_{13} \\ k_{21} & k_{22} & k_{23} \\ k_{31} & k_{32} & k_{33} \end{bmatrix} \begin{bmatrix} E_x \\ E_y \\ E_z \end{bmatrix} \tag{6.14.2}$$

These are axes, called *electric axes,* along which $k_{ij} = 0 (i \neq j)$ but k_{ij} exists only for $i = j$. If we set up such axes within the crystal called *principal axes,* the above matrix equation (6.14.2) becomes,

$$\begin{bmatrix} D_x \\ D_y \\ D_z \end{bmatrix} = \begin{bmatrix} k_{11} & 0 & 0 \\ 0 & k_{22} & 0 \\ 0 & 0 & k_{33} \end{bmatrix} \begin{bmatrix} E_x \\ E_y \\ E_z \end{bmatrix} \tag{6.14.3}$$

$\therefore\ D_x = k_{11}E_x = K_1E_x$, where K_1 is written for k_{11} falled principal dielectric constant.

$$D_y = k_{22}E_y = K_2E_y, D_z = k_{33}E_z = K_3E_z.$$

Let us now consider the propagation of electromagnetic waves through such a medium. From Maxwell's equations we have here,

$$\vec{\nabla}\cdot\vec{D} = 0, \quad \text{because the medium is free of charge,}$$

$$\vec{\nabla}\cdot\vec{H} = 0, \quad \mu \text{ is taken to be unity as is usually the case,}$$

$$\vec{\nabla} \times \vec{E} = -\frac{1}{c}\cdot\frac{\partial \vec{H}}{\partial t}$$

$$\vec{\nabla} \times \vec{H} = \frac{1}{c}\cdot\frac{\partial \vec{D}}{\partial t}, \quad \vec{J} = 0, \text{ since the medium is non-conducting.}$$

A plane harmonic wave advances with a phase velocity v along the direction of wave-normal $\vec{\nu}$. The electric vector in the electromagnetic wave is given by, $\vec{E} = \vec{E}_0 e^{i\sigma(vt-\vec{r}\cdot\vec{\nu})}$, where $\vec{r}$ is the radius vector drawn from the origin to the point (x, y, z).

Now,

$$\begin{aligned}\vec{\nabla}\times\vec{E} &= \vec{\nabla}\times\vec{E}_0 e^{i\sigma(vt-\vec{r}\cdot\vec{\nu})} = \vec{\nabla}\times\alpha\vec{A} = \alpha\left(\vec{\nabla}\times\vec{A}\right) - \vec{A}\times\left(\vec{\nabla}\alpha\right)\\ &= \alpha\left(\vec{\nabla}\times\vec{E}_0\right) - \vec{E}_0\times\vec{\nabla}\left\{e^{i\sigma(vt-\vec{r}\cdot\vec{\nu})}\right\}\\ &= -\vec{E}_0\times(-i\sigma)e^{i\sigma(vt-\vec{r}\cdot\vec{\nu})}\vec{\nu}, \text{ since } \left(\vec{\nabla}\times\vec{E}_0\right) = 0\\ &= i\sigma\left(\vec{E}\times\vec{\nu}\right)\end{aligned}$$

$$\therefore\quad -\frac{1}{c}\cdot\frac{\partial\vec{H}}{\partial t} = i\sigma\vec{E}_0 e^{i\sigma(vt-\vec{r}\cdot\vec{\nu})}\times\vec{\nu}$$

Integrating w.r.t. time,

$$-\frac{1}{c}\vec{H} = \frac{\vec{E}_0}{v}e^{i\sigma(vt-\vec{r}\cdot\vec{\nu})}\times\vec{\nu} \quad\text{or,}\quad \vec{H} = -\frac{c}{v}\left(\vec{E}\times\vec{\nu}\right) = \frac{c}{v}\left(\vec{\nu}\times\vec{E}\right).$$

Thus, $\vec{H}$ is $\perp$ to the plane containing $\vec{E}$ and $\vec{\nu}$.

Again, $\vec{\nabla}\times\vec{H} = \frac{1}{c}\cdot\frac{\partial\vec{D}}{\partial t}$.

Now,

$$\begin{aligned}\vec{H} &= e^{i\sigma(vt-\vec{r}\cdot\vec{\nu})}\times\frac{c}{v}\times\left(\vec{\nu}\times\vec{E}_0\right) = \vec{\nu}\times\frac{c}{v}\vec{E}_0 e^{i\sigma(vt-\vec{r}\cdot\vec{\nu})}\\ &= \vec{H}_0 e^{i\sigma(vt-\vec{r}\cdot\vec{\nu})},\quad \vec{H}_0 = \frac{c}{v}\left(\vec{\nu}\times\vec{E}_0\right)\end{aligned}$$

$$\therefore\quad \vec{\nabla}\times\vec{H} = \vec{\nabla}\times\vec{H}_0 e^{i\sigma(vt-\vec{r}\cdot\vec{\nu})} = -\vec{H}_0\times(-i\sigma)e^{i\sigma(vt-\vec{r}\cdot\vec{\nu})}\vec{\nu} = i\sigma\vec{H}_0 e^{i\sigma(vt-\vec{r}\cdot\vec{\nu})}\times\vec{\nu}$$

$$\therefore\quad \frac{\partial\vec{D}}{\partial t} = ci\sigma\vec{H}_0 e^{i\sigma(vt-\vec{r}\cdot\vec{\nu})}\times\vec{\nu}$$

$$\vec{D} = \frac{c}{v}\vec{H}_0 e^{i\sigma(vt-\vec{r}\cdot\vec{\nu})}\times\vec{\nu} = -\frac{c^2}{v^2}\vec{\nu}\times\left(\vec{\nu}\times\vec{E}\right) \tag{6.14.4}$$

$$= -\frac{c}{v}\left(\vec{\nu}\times\vec{H}\right) \tag{6.14.5}$$

Thus $\vec{D}$ is $\perp$ to the plane containing $\vec{\nu}$ and $\vec{H}$. So both $\vec{H}$ and $\vec{D}$ are normal to $\vec{\nu}$, i.e., $\vec{H}$ and $\vec{D}$ lie in the wavefront. The electromagnetic waves are thus transverse w.r.t. $\vec{H}$ and $\vec{D}$. $\vec{D}, \vec{E}$ and $\vec{\nu}$ are coplanar since they are normal to $\vec{H}$ (fig.6.9). From eqn.(6.14.4),

$$\vec{D} = -\frac{c^2}{v^2}\left[\left(\vec{\nu}\cdot\vec{E}\right)\vec{\nu} - \vec{E}(\vec{\nu}\cdot\vec{\nu})\right] = \frac{c^2}{v^2}\left[\vec{E} - \left(\vec{\nu}\cdot\vec{E}\right)\vec{\nu}\right].$$

Again, Poynting's vector $\vec{P}$ is, $\vec{P} = \frac{c}{4\pi}\left(\vec{E}\times\vec{H}\right)$ which is a vector representing the direction of energy flow and its direction is $\perp$ to the plane of $\vec{E}$ and $\vec{H}$. It is also evident that $\vec{P}$ is coplanar with $\vec{D}, \vec{E}$ and $\vec{\nu}$. Also the angle between $\vec{D}$ and $\vec{E}$ is equal to the angle between their normals, i.e., between $\vec{P}$ and $\vec{\nu}$.

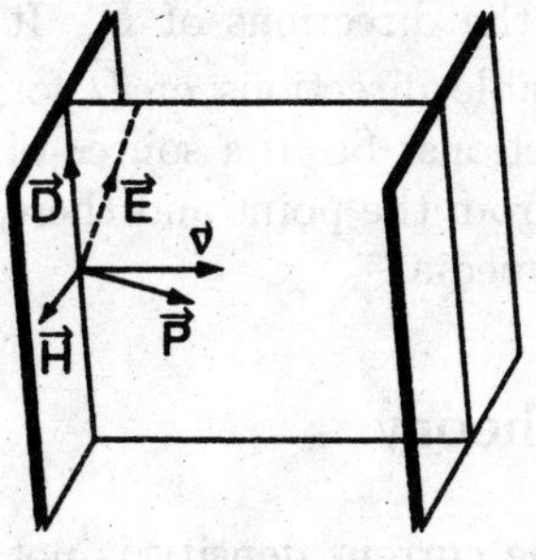

Fig 6.9

Thus, $\vec{P}, \vec{E}$ and $\vec{H}$ form one set of orthogonal triplets, so also do the three other vectors $\vec{D}$, $\vec{v}$, $\vec{H}$. Hence, $\vec{P}$ has direction different from that of $\vec{v}$. The vector $\vec{v}$ represents direction of wave-normal (i.e., direction of advancing wave) while $\vec{P}$ gives the direction of energy flow, which we designate as *ray velocity.*

Thus, wave velocity v is along the direction $\vec{v}$, but the ray travels with velocity V along the direction of $\vec{P}$. So we can write, $V\cos\theta = v$. It means, the wave velocity is the component of ray velocity in the direction of wave-normal.

Suppose, with respect to the principal axes, the vector $\vec{v}$ makes angles α, β, γ such that $\cos\alpha = l$, $\cos\beta = m$ and $\cos\gamma = n$.

Then,

$$D_x = \frac{c^2}{v^2}\left[E_x - \left(\vec{v}\cdot\vec{E}\right)l\right], D_y = \frac{c^2}{v^2}\left[E_y - \left(\vec{v}\cdot\vec{E}\right)m\right], D_z = \frac{c^2}{v^2}\left[E_z - \left(\vec{v}\cdot\vec{E}\right)n\right].$$

But,

$$D_x = k_1 E_x = \frac{c^2}{v_x^2}E_x \qquad \left[\because\ v_x = \frac{c}{\sqrt{\mu k_1}} = \frac{c}{\sqrt{k_1}}, \mu = 1\right]$$

$$\therefore\ D_x = \frac{c^2}{v^2}\left[\frac{v_x^2 D_x}{c^2} - \left(\vec{v}\cdot\vec{E}\right)l\right] \quad \text{or,}\quad D_x\left(1 - \frac{v_x^2}{v^2}\right) = -\frac{c^2}{v^2}\left(\vec{v}\cdot\vec{E}\right)l$$

$$\text{or,}\quad D_x\left(\frac{v_x^2 - v^2}{v^2}\right) = \frac{c^2}{v^2}\left(\vec{v}\cdot\vec{E}\right)l$$

$$D_x = \frac{c^2}{(v_x^2 - v^2)}\left(\vec{v}\cdot\vec{E}\right)l,\ D_y = \frac{c^2}{(v_y^2 - v^2)}\left(\vec{v}\cdot\vec{E}\right)m,\ D_z = \frac{c^2}{(v_z^2 - v^2)}\left(\vec{v}\cdot\vec{E}\right)n$$

Now, since $\vec{D}$ is $\perp$ to $\vec{v}$, we have,

$$\vec{D}\cdot\vec{v} = 0$$

$$\text{or,}\quad lD_x + mD_y + nD_z = 0$$

$$\text{or,}\quad \vec{v}\cdot\vec{E}\left[\frac{l^2}{(v_x^2 - v^2)} + \frac{m^2}{(v_y^2 - v^2)} + \frac{n^2}{(v_z^2 - v^2)}\right] = 0.$$

Since, $\vec{v}\cdot\vec{E} \neq 0$

$$\frac{l^2}{(v_x^2 - v^2)} + \frac{m^2}{(v_y^2 - v^2)} + \frac{n^2}{(v_z^2 - v^2)} = 0. \tag{6.14.6}$$

Eqn.(6.14.6) is known as *Fresnel's wave equation.* This equation is biquadratic in v or quadratic in v^2. So, if the values of v^2 be s and r, then $v = \pm s, \pm r$. Thus, the phase velocity in a given direction determined by the direction cosines (l, m, n) may have two values $(+s, +r)$, we are considering the $+ve$ signs, because $-ve$ signs denote wave propagation in the opposite direction to the direction denoted by (l, m, n). If the direction be taken along a particular direction called *optic axis,* the two values of the phase velocities become identical. Such a crystal is a uniaxial crystal. But there are also many anisotropic media where two directions (optic axes) along which there is one phase velocity. Such crystals are known as *biaxial crystal.*

In a direction other than optic axis, the values of v can be found out from eqn.(6.14.6) and these values on substitution in the eqns. for D_x, D_y, D_z will give the directions of $\vec{D}$. It can be shown that $\vec{D}$ is $\perp$ to v. In general, there will be two possible directions of $\vec{D}$ for any given direction of the wave-normal. Thus, there are two ray directions. So, if a source of electromagnetic disturbance be placed at a point the rays spread out from the point and there will be two expanding spherical waves unlike that in case of isotropic media.

6.15 Skin Effect in the Light of Electromagnetic Theory

We know that when an alternating current passes through a wire, the current density is not uniform across the cross-section of the wire. The current density is confined mainly within a very thin shell, which effect is known as *skin effect,* as a result, the resistance of the wire is greater with an a.c. flowing through the wire than with d.c. An explanation of the effect, derivation of the ratio of $\frac{R_{a.c.}}{R_{d.c.}}$ of the wire has been done elsewhere[2]. We are here to solve out the distribution of current density inside the wire from electromagnetic field theory.

We have seen that when a current flows through a conducting medium, the electric field produced satisfies the eqn.

$$\vec{\nabla}^2\vec{E} = \frac{\mu k}{c^2}\cdot\frac{\partial^2\vec{E}}{\partial t^2} + \frac{4\pi\sigma\mu}{c^2}\cdot\frac{\partial\vec{E}}{\partial t}.$$

Now, the first term arises from the displacement current which in case of electrical conduction through metals can be neglected. This can be shown to be the case if we consider the field variables to vary harmonically with time as $e^{+i\omega t}$, then

$$\vec{\nabla}^2\vec{E} = \left(\frac{4\pi\sigma\mu}{c^2}i\omega - \frac{\omega^2 k\mu}{c^2}\right)\vec{E}.$$

Thus, if $k\omega \ll \sigma$, then the term arising out of the displacement current can be neglected. Thus for conduction of electric current through metals, the error in neglecting the displacement current term is not appreciable unless ω is comparable with the frequency of light waves. So we have the eqn. of electric field associated with current flow of frequency $\omega = 2\pi f$ as,

$$\vec{\nabla}^2\vec{E} = \frac{4\pi\sigma\mu}{c^2}\cdot\frac{\partial\vec{E}}{\partial t}.$$

Now, $\vec{J} = \sigma\vec{E}$, putting this value in the above equation, we get,

$$\vec{\nabla}^2\vec{J} = \frac{4\pi\sigma\mu}{c^2}\cdot\frac{\partial\vec{J}}{\partial t} \tag{6.15.1}$$

Suppose, $\vec{J}$ is along the axis of the wire supposed to be in the direction of x. At a depth z below the surface of the wire, let $\vec{J} = \vec{A}(z)e^{i\omega t}$, corresponding to any time t.

Then the equation (6.15.1) known as skin effect equation becomes,

$$\frac{d^2\vec{A}}{dz^2} = \frac{4\pi\sigma\mu i\omega}{c^2}\vec{A} = \frac{4\pi\sigma\mu\omega}{c^2}\vec{A}e^{i(\pi/2)}.$$

[2]See *Advanced Textbook on Electricity and Magnetism,* Vol.II by the author.

A solution of this equation is,

$$\vec{A} = R\cdot P\cdot [\vec{a}\exp]\left\{\frac{p(1+i)}{\sqrt{2}}\cdot z\right\} + \vec{b}\exp\left\{-\frac{p(1-i)}{\sqrt{2}}\cdot z\right\}$$

where $p^2 = \frac{4\pi\sigma\mu\omega}{c^2}$.

Now, $\vec{A} \neq \infty$ for $z = \infty$, so, $\vec{a} = 0$.

$$\therefore \quad \vec{A} = \vec{b}e^{-pz/\sqrt{2}}e^{ipz/\sqrt{2}}$$

$$\therefore \quad \vec{J} = R\cdot P\cdot \vec{b}e^{-pz/\sqrt{2}}\exp\left\{i\left(\frac{pz}{\sqrt{2}} - \omega t\right)\right\}$$

$$= \vec{b}\exp\left\{-\sqrt{\frac{2\pi\sigma'\mu\omega}{c}}z\right\}\cos\left(\sqrt{\frac{2\pi\sigma\mu\omega}{c}}z - \omega t\right),$$

where, σ is in e.s.u. and $\sigma' = \frac{\sigma}{c}$ is in e.m.u.

It shows that the current density decreases exponentially with depth below the surface. If the conductivity is high ($\sigma'\omega$ is large), the current is confined to a thin layer below the surface.

Now, for copper $\sigma' = 6.2 \times 10^{-4}$ e.m.u. and so $\sigma = 6.2 \times 10^{-4} \times c$ (e.s.u.).

$$\therefore \quad \sqrt{\frac{2\pi\sigma'\mu\omega}{c}} = \sqrt{2\pi \times 6.2 \times 10^{-4} \times 2\pi \times 1000} \simeq 5, \quad \text{taking } \mu = 1.$$

Thus at a depth 1 cm below in surface of the conductor, the current density falls off to e^{-5} times its value at the surface. If at a depth $z = d$, the current density falls to e^{-1} times its value at the surface, then,

$$\exp\left\{\sqrt{\frac{2\pi\sigma'\pi\mu\omega}{c}}d\right\} \quad \therefore \quad d = \sqrt{\frac{c}{2\pi\sigma'\mu\omega}} \quad \text{is known as } \textit{skin depth}.$$

Example 1 *Assuming that all the energy from a 1000 watt lamp is radiated uniformly, calculate the average value of electric field of radiation at a distance of* 2×100 *cms from the lamp.*

Solution. 1000 watts = 1000 joules/sec = 1000×10^7 ergs/sec.

At a distance 2×100 cms from the source of radiation, the total radiation flows across the surface of a sphere of radius 200 cms.

$\therefore$ Total area $= 4\pi \times (2 \times 100)^2$

$$\therefore \quad \frac{c}{4\pi}\iint\left(\vec{E} \times \vec{H}\right)\cdot\vec{ds} = \text{total flux out of the bounding surface.}$$

$$\therefore \quad \frac{c}{4\pi}EH\sin\frac{\pi}{2}\cdot 4\pi \times (2 \times 100)^2 = 1000 \times 10^7, \quad c\left(\frac{EH}{4\pi}\right) = \frac{1000 \times 10^7}{16\pi \times (100)^2} = \frac{10^6}{16\pi}.$$

Now, $\vec{H} = \sqrt{\frac{k}{\mu}}\left(\vec{\nu} \times \vec{E}\right)$.

Since, the medium is air, $k = 1 = \mu$.

$$\therefore \quad |\vec{H}| = |\vec{E}|$$

$$\therefore \quad \frac{cE^2}{4\pi} = \frac{10^6}{16\pi}$$

$$\therefore \quad E = \sqrt{\frac{10^6}{4 \times 3 \times 10^{10}}} = \frac{10^{-2}}{\sqrt{12}} \text{ e.s.u. of p.d.} = \frac{10^{-2} \times 300}{\sqrt{12}} \text{ volt/cm} = .866 \text{ volt/cm}.$$

Example 2 *Determine the mean value of the intensity of the electric field in air at a distance of 5 metres from a one kilo-watt radiating point source.*

Solution. 1 kilo-watt = 10^3 watts = $10^3 \times 10^7$ ergs/sec.

$$(EH)\frac{c}{4\pi} = \frac{10^3 \times 10^7}{4\pi \times (5 \times 10^2)^2}$$

Again, $E = H$,

$$\therefore \quad \frac{c}{4\pi}E^2 = \frac{10^3 \times 10^7}{4\pi \times 25 \times 10^4} \quad \text{or,} \quad E^2 = \frac{10^6}{25 \times 3 \times 10^{10}} = \frac{10^{-4}}{5^2 \times 3}$$

$$\therefore \quad E = \frac{10^{-2}}{5 \times \sqrt{3}} \text{ e.s.u. of p.d.} = \frac{10^{-2}}{5 \times \sqrt{3}} \times 300 \text{ volts/cm} = .2 \times \sqrt{3} = .346 \text{ volts/cm.}$$

Example 3 *A large sphere of radius b is made of material of conductivity σ and dielectric constant k. At time t = 0, a charge is uniformly distributed over the surface of a small concentric sphere of radius a. Show that at time t, the charge Q on the inner is given by $Q = Q_0 e^{-\alpha t}$, where $\alpha = \frac{4\pi\sigma}{k}$. Show further that the total Joule heat loss during the discharge is $\frac{Q_0^2(b-a)}{2abk}$.*

Solution. From the standard eqn. we find that $\rho = \rho_0 e^{-t/\tau} = \rho_0 e^{-\alpha t}$ where $\tau = \frac{k}{4\pi\sigma} = \alpha^{-1}$

$$\therefore \quad Q = \int_v \rho dv = e^{-\alpha t}\int_v \rho_0 dv = Q_0 e^{-\alpha t}$$

$$\therefore \quad \text{Current } i = \frac{dQ}{dt} = -\alpha Q_0 e^{-\alpha t}.$$

The potential drop ϕ between the two spheres $= \frac{Q}{k}\cdot\left(\frac{1}{a} - \frac{1}{b}\right)$.

$\therefore$ Joule's heat loss

$$\int_0^\infty i\phi dt = -\frac{\alpha}{k}\int_0^\infty Q_0 e^{-\alpha t}\cdot Q_0\left(\frac{1}{a} - \frac{1}{b}\right)e^{-\alpha t} = \frac{Q_0^2}{2k}\cdot\left(\frac{1}{a} - \frac{1}{b}\right). \quad (6.15.2)$$

Again, capacity of two concentric spheres is given by,

$$C = \frac{kab}{b-a}.$$

$\therefore$ Energy of the condenser initially

$$E_i = \frac{1}{2}\cdot\frac{Q_0^2}{C} = \frac{Q_0^2}{2abk}(b-a). \quad (6.15.3)$$

Eqns. (6.15.2) and (6.15.3) represent the same amount of energy.

Example 4 *Current density depends on the distance r from the axis and satisfies the scalar equation of the following form in cylindrical coordinate,*

$$\frac{\partial^2 J}{\partial r^2} + \frac{1}{r}\cdot\frac{\partial J}{\partial r} = \frac{4\pi\sigma\mu}{c^2}\cdot\frac{\partial J}{\partial t},$$

J = current density. Find J.

Solution. Since J is a function of time, let $J = Ae^{iet}$.
Then,

$$\frac{d^2A}{dr^2}e^{i\omega t} + \frac{1}{r}\cdot\frac{dA}{dr}e^{i\omega t} = \frac{4\pi\sigma\mu}{c^2}i\omega Ae^{i\omega t}$$

$$\text{or,}\quad \frac{d^2A}{dr^2} + \frac{1}{r}\cdot\frac{dA}{dr} = -p^2A, p^2 = -\frac{4\pi\sigma\mu}{c^2}i\omega$$

Putting $x = pr, \frac{dA}{dr} = p\frac{dA}{dx}, \frac{d^2A}{dr^2} = p^2\frac{d^2A}{dx^2}$

$$\therefore\quad p^2\frac{d^2A}{dx^2} + \frac{p^2}{x}\cdot\frac{dA}{dx} + p^2A = 0 \quad\text{or,}\quad \frac{d^2A}{dx^2} + \frac{1}{x}\cdot\frac{dA}{dx} + A = 0$$

This is a Bessel's differential equation of zeroth order and its solution will be,

$$A = a_0J_0(x)$$

$$\therefore\quad J = Ae^{i\omega t} = a_0J_0(x)e^{i\omega t} = a_0J_0(pr)e^{i\omega t} = a_0J_0\left[\sqrt{-i}\sqrt{4\pi\sigma\mu\omega}\frac{r}{c}\right]e^{i\omega t}$$

$$= a_0e^{i\omega t}\left[\text{ber}\left(\sqrt{\frac{4\pi\sigma\mu\omega}{c^2}}\cdot r\right) + i\,\text{bei}\left(\sqrt{\frac{4\pi\sigma\mu\omega}{c^2}}r\right)\right]$$

$$= a_0e^{i\omega t}\left[\text{ber}\left(\sqrt{\frac{4\pi\sigma'\mu\omega}{c}}\cdot r\right) + i\,\text{bei}\left(\sqrt{\frac{4\pi\sigma'\mu\omega}{c}}r\right)\right]$$

[Here, σ is in e.s.u. and $\frac{\sigma}{c} = \sigma'$ is in e.m.u.]

Note. ber = Bessel real, bei = Bessel imaginary.
These are the real and imaginary parts of the function $J_0(x\sqrt{-i})$, i.e.,

$$\text{ber}\,x = R\left[J_0\left(x\sqrt{-i}\right)\right],\ \text{bei}\,x = I\left[J_0\left(x\sqrt{-i}\right)\right].$$

Example 5 *Solve the eqn.*

$$\vec{\nabla}^2\vec{E} = \left(\frac{\mu k}{c^2}\right)\frac{\partial^2\vec{E}}{\partial t^2} + \left(\frac{4\pi\sigma'\mu}{c}\right)\frac{\partial\vec{E}}{\partial t}$$

for waves of frequency $f = \frac{\omega}{2\pi}$ *in a conducting medium by taking* $E = a\exp\{i\omega(t - qz), 0, 0\}$, *where* q *is given in terms of* ω *by the equation,*

$$q^2 = \frac{k\mu}{c^2}\left(1 - \frac{i4\pi\sigma'c}{k\omega}\right).$$

By taking the real part of E_x, *show that without absorption a harmonic wave cannot propagate in such a medium.*

Solution. If $E = a\exp\{i\omega(t - qz), 0, 0\}$, it means,

$$E_x = a\exp\{i\omega(t - qz)\},\quad E_y = 0,\quad E_z = 0,$$

i.e., electric field is along the x-direction.

∴ The eqn. of wave propagation in a conducting medium now stands as,

$$\vec{\nabla}^2 E_x = \frac{k\mu}{c^2}\cdot\frac{\partial^2 E_x}{\partial t^2} + \frac{4\pi\sigma'\mu}{c}\cdot\frac{\partial E_x}{\partial t}.$$

Now, $E_x = ae^{i\omega(t-qz)}$.

$$\therefore \quad \dot{E}_x = ai\omega e^{i\omega(t-qz)}, \ddot{E}_x = -a\omega^2 e^{i\omega(t-qz)}, \nabla^2 E_x = \frac{\partial^2 E_x}{\partial z^2} = -\omega^2 q^2 a e^{i\omega(t-qz)}$$

Substituting these values in the eqn. of wave propagation, we will get,

$$q^2 = \frac{k\mu}{c^2} - \frac{4\pi\sigma'\mu}{\omega c}i = \frac{\mu k}{c^2}\left\{1 - \frac{4\pi\sigma' c}{\omega k}i\right\} = -\frac{\mu k}{c^2}\cdot\frac{4\pi\sigma' c}{k\omega}i\left(1 - \frac{k\omega}{4\pi\sigma c}i\right).$$

If $k\omega \ll 4\pi\sigma' c$, then

$$q^2 = -\frac{4\pi\sigma'\mu}{\omega c}i = \frac{2\pi\sigma'\mu}{\omega c}(-2i) = \frac{2\pi\sigma'\mu}{\omega c}(1-i)^2 \quad \therefore \quad q = \sqrt{\frac{2\pi\sigma'\mu}{\omega c}}(1-i).$$

The reciprocal of this q actually represents the velocity of the wave in the conducting medium. So, the refractive index which is a complex quantity (since q is complex) is given by,

$$n^* = cq.$$

When $k\omega \ll 4\pi\sigma' c$, then

$$q = \sqrt{\frac{2\pi\sigma'\mu}{\omega c}}(1-i).$$

$$\therefore \quad E_x = ae^{i\omega(t-qz)} = ae^{i\omega t}.e^{-i\omega qz} = ae^{i\omega t}.\exp\left\{-i\omega\sqrt{\frac{2\pi\sigma'\mu}{\omega c}}(1-i)z\right\}$$

$$= ae^{i\omega t}\cdot\exp\left\{-i\omega\sqrt{\frac{2\pi\sigma'\mu}{\omega c}}z\right\}\exp\left\{-\sqrt{\frac{2\pi\sigma'\mu}{\omega c}}\omega z\right\}$$

$$= a\exp\left\{-\sqrt{\frac{2\pi\sigma'\mu\omega}{c}}z\right\}\exp\left\{i\left(\omega t - \sqrt{\frac{2\pi\sigma'\mu\omega}{c}}z\right)\right\}$$

So the electric field within a conductor propagates with absorption. It decreases exponentially with depth from the surface.

6.16 Scalar and Vector Potentials in Maxwell's Equations: Inhomogeneous Wave Equation

We know that for a changing magnetic field, $\vec{\nabla}\times\vec{E}$ is not zero and hence, $\vec{E}$ cannot be derived solely from a scalar function, called *potential*. But the divergence of induction vector $\vec{B}$ is always zero, i.e.,

$$\vec{\nabla}\cdot\vec{B} = 0 \tag{6.16.1}$$

and so we can put, $\vec{B} = \vec{\nabla}\times\vec{A}$, it means that $\vec{B}$ can be derived from the vector potential A.

If $\vec{B} = \vec{\nabla} \times \vec{A}$, then from Maxwell's circuital equation, we get,

$$\vec{\nabla} \times \vec{E} = -\frac{1}{c} \cdot \frac{\partial \vec{B}}{\partial t} = -\frac{1}{c} \cdot \frac{\partial}{\partial t} \left(\vec{\nabla} \times \vec{A} \right) = \vec{\nabla} \times \left(-\frac{1}{c} \cdot \frac{\partial \vec{A}}{\partial t} \right)$$

$$\text{or,} \quad \vec{\nabla} \times \left(\vec{E} + \frac{1}{c} \cdot \frac{\partial \vec{A}}{\partial t} \right) = 0. \tag{6.16.2}$$

Expression (6.16.2) suggests that,

$$\vec{E} + \frac{1}{c} \cdot \frac{\partial \vec{A}}{\partial t} = -\vec{\nabla}\phi, \quad \text{where, } \phi = \text{scalar potential.} \quad \therefore \ \vec{E} = -\vec{\nabla}\phi - \frac{1}{c} \cdot \frac{\partial \vec{A}}{\partial t}. \tag{6.16.3}$$

So the electric field may be derived from the scalar potential function ϕ and from vector potential $\vec{A}$.

Now, $\vec{\nabla} \cdot \vec{E} = 4\pi\rho$, Maxwell's first field equation, or, $\vec{\nabla} \cdot \vec{E} = \frac{4\pi\rho}{k}$, for an isotropic dielectric $\vec{D} = k\vec{E}$.

$$\because \ \vec{E} = -\vec{\nabla}\phi - \frac{1}{c} \cdot \frac{\partial \vec{A}}{\partial t}$$

$$\therefore \ \vec{\nabla} \cdot \left[-\vec{\nabla}\phi - \frac{1}{c} \cdot \frac{\partial \vec{A}}{\partial t} \right] = \frac{4\pi\rho}{k}$$

$$\text{or,} \quad \nabla^2\phi + \frac{1}{c}\vec{\nabla} \cdot \left(\frac{\partial \vec{A}}{\partial t} \right) = -\frac{4\pi\rho}{k} \tag{6.16.4}$$

Maxwell's fourth relation is $\vec{\nabla} \times \vec{H} = \frac{4\pi}{c}\vec{J} + \frac{1}{c} \cdot \frac{\partial \vec{D}}{\partial t}$.

Now, $\vec{B} = \mu\vec{H}$.

$$\therefore \ \vec{\nabla} \times \vec{H} = \vec{\nabla} \times \left(\frac{\vec{B}}{\mu} \right) = \frac{1}{\mu}\vec{\nabla} \times \left(\vec{\nabla} \times \vec{A} \right)$$

$$\therefore \ \vec{\nabla} \times \left(\vec{\nabla} \times \vec{A} \right) = \frac{4\pi\mu}{c}\vec{J} + \frac{\mu k}{c} \cdot \frac{\partial \vec{E}}{\partial t}$$

$$\text{or,} \quad \vec{\nabla}\left(\vec{\nabla} \cdot \vec{A} \right) - \vec{\nabla}^2\vec{A} = \frac{4\pi\mu}{c}\vec{J} + \frac{\mu k}{c} \frac{\partial}{\partial t} \left[-\vec{\nabla}\phi - \frac{1}{c} \cdot \frac{\partial \vec{A}}{\partial t} \right]$$

$$= \frac{4\pi\mu}{c}\vec{J} - \frac{\mu k}{c}\vec{\nabla}\left(\frac{\partial \phi}{\partial t} \right) - \frac{\mu k}{c^2} \cdot \frac{\partial^2 \vec{A}}{\partial t^2} \tag{6.16.5}$$

Uptil now we have not in any way defined the vector potential $\vec{A}$. Let us define $\vec{A}$ such that,

$$\vec{\nabla} \cdot \vec{A} + \frac{\mu k}{c} \cdot \frac{\partial \phi}{\partial t} = 0. \tag{6.16.6}$$

Then from the eqn. (6.16.5), we have,

$$\nabla^2\vec{A} = -\frac{4\pi\mu}{c}\vec{J} + \frac{\mu k}{c^2} \cdot \frac{\partial^2 \vec{A}}{\partial t^2} \quad \text{or,} \quad \vec{\nabla}^2\vec{A} - \frac{\mu k}{c^2} \cdot \frac{\partial^2 \vec{A}}{\partial t^2} = -\frac{4\pi\mu}{c}\vec{J}. \tag{6.16.7}$$

From eqn.(6.16.4), we can have, with the help of eqn.(6.16.6),

$$\nabla^2\phi + \frac{1}{c}\cdot\frac{\partial}{\partial t}\left(\vec{\nabla}\cdot\vec{A}\right) = -\frac{4\pi\rho}{k} \quad \text{or,} \quad \nabla^2\phi - \frac{\mu k}{c^2}\cdot\frac{\partial^2\phi}{\partial t^2} = -\frac{4\pi\rho}{k}. \tag{6.16.8}$$

So, we have,

$$\nabla^2\phi - \frac{\mu k}{c^2}\cdot\frac{\partial^2\phi}{\partial t^2} = -\frac{4\pi\rho}{k}, \quad \nabla^2\vec{A} - \frac{\mu k}{c^2}\cdot\frac{\partial^2\vec{A}}{\partial t^2} = -\frac{4\pi\mu}{c}\vec{J}. \tag{6.16.9}$$

The quantities ϕ and $\vec{A}$ are known as *electromagnetic potentials* and eqn. (6.16.9) is known as *inhomogeneous wave equation.* If the right-hand side of eqn.(6.16.9) becomes zero, then we get wave equation which is homogeneous. Introducing a new operator $\Box$ which goes by the name of d'Alembertian operator defined as,

$$\Box = \nabla^2 - \frac{k\mu}{c^2}\cdot\frac{\partial^2}{\partial t^2},$$

the inhomogeneous wave eqn. (6.16.9) becomes,

$$\Box^2\phi = -\frac{4\pi\rho}{k}, \quad \Box^2\vec{A} = -\frac{4\pi\mu}{c}\vec{J}. \tag{6.16.10}$$

Method 1 : Retarded Potential (Solution of Equation 6.16.10)

Let us take the first of the eqn. (6.16.10), so we have,

$$\Box^2\phi = -4\pi\rho/k.$$

If the potential ϕ be due to static distribution of charges, then $\phi = \phi(x,y,z)$ and $\rho = \rho(x,y,z)$ and the above eqn. then simply reduces to $\nabla^2\phi = -4\pi\rho/k$, which is nothing but Poisson's equation and the method of solving this equation has been shown elsewhere[3]. We will be here interested in solving the wave equation under prescribed charge distribution $\rho(\vec{r},t)$ and current distribution $\vec{J}(\vec{r},t)$. Unlike in the statical case, where ϕ at any point at any instant of time t is due to charge at the origin at that instant, here, the potential at any point $\vec{r}$ and at any time w.r.t. origin where the charge is located is not related to the instantaneous charge at the origin but it is due to charge distribution $\left(t - \frac{r}{v}\right)$ earlier because any change in the distribution of charge at the origin will have its effect felt at the field-point distant r in time $\frac{r}{v}$ later. Thus the potential at the field point is said to be retarded potential and the time of retardation is given by $\frac{r}{v}$. In case of static field or slowly varying field or if the field point is very near to the origin, then the retardation is practically nil.

To solve the inhomogeneous wave equation we first find the solution for a point charge located at the origin of co-ordinates and then sum over all the charge elements ρdV in the prescribed charge distribution. Thus, $\Box^2\phi = 0$ must be satisfied everywhere except at the origin where in a small volume δV surrounding the origin,

$$\int_{\delta V} \Box^2\phi dV = -4\pi q(t) \quad \text{must be satisfied.}$$

[3]See *Advanced Textbook of Electricity and Magnetism* by the Author.

Now,

$$\Box^2\phi = 0 \quad \text{or,} \quad \nabla^2\phi - \frac{k\mu}{c^2}\cdot\frac{\partial^2\phi}{\partial t^2} = 0$$

ϕ is symmetric and hence, the spatial dependence of ϕ be only on r.

Now,

$$\nabla^2\phi = \frac{\partial^2\phi}{\partial x^2} + \frac{\partial^2\phi}{\partial y^2} + \frac{\partial^2\phi}{\partial z^2}, \quad r = \sqrt{x^2+y^2+z^2}$$

$$\therefore \quad \frac{\partial\phi}{\partial x} = \frac{\partial\phi}{\partial x}\cdot\frac{\partial r}{\partial x} = \frac{x}{\sqrt{x^2+y^2+z^2}}\cdot\frac{\partial\phi}{\partial r}$$

$$\therefore \quad \frac{\partial^2\phi}{\partial x^2} = \frac{x^2}{x^2+y^2+z^2}\cdot\frac{\partial^2\phi}{\partial r^2} + \frac{1}{\sqrt{x^2+y^2+z^2}}\cdot\frac{\partial\phi}{\partial r} - \frac{x^2}{(x^2+y^2+z^2)^{3/2}}\cdot\frac{\partial\phi}{\partial r}.$$

Similarly, expressions for $\frac{\partial^2\phi}{\partial y^2}$ and $\frac{\partial^2\phi}{\partial z^2}$ can be formed.

$$\therefore \quad \nabla^2\phi = \frac{\partial^2\phi}{\partial r^2} + \frac{2}{r}\cdot\frac{\partial\phi}{\partial r} \quad \therefore \quad \frac{1}{r^2}\cdot\frac{\partial}{\partial r}\left(r^2\frac{\partial\phi}{\partial r}\right) - \frac{k\mu}{c^2}\cdot\frac{\partial^2\phi}{\partial t^2} = 0.$$

Put $\phi = \frac{\xi(r,t)}{r}$

$$\therefore \quad \frac{\partial^2\xi}{\partial r^2} - \frac{k\mu}{c^2}\cdot\frac{\partial^2\xi}{\partial t^2} = 0 \quad \text{or,} \quad \frac{\partial^2\xi}{\partial r^2} - \frac{1}{v^2}\cdot\frac{\partial^2\xi}{\partial t^2} = 0.$$

It means a one-dimensional wave equation which has for its solution,

$$\xi = f_1\left(t - \frac{r}{v}\right) + f_2\left(t + \frac{r}{v}\right)$$

So, we can write for the potential ϕ,

$$\phi = \frac{f_1\left(t - \frac{r}{v}\right)}{r} + \frac{f_2\left(t + \frac{r}{v}\right)}{r} \tag{6.16.11}$$

Now, we know that,

(i) $\phi(t \to -\infty, r \to \infty) = 0$

(ii) $\vec{A}(t \to -\infty, r \to \infty) = 0$

The meaning of the condition (i) is obvious, it means that there is no field at an infinitely distant region at an infinitely large interval of time before the initiation of radiation by the radiator. The solution $f_1\left(t - \frac{r}{v}\right)$ depends on the argument $\left(t - \frac{r}{v}\right)$ and the other solution depends on the argument $\left(t + \frac{r}{v}\right)$. The second argument for $t \to \infty, r \to \infty$ has completely indeterminate from $\infty - \infty$, i.e., it is equal to any quantity, but we know that under this condition $\phi \to 0$ and so $f_2\left(t + \frac{r}{v}\right) \to 0$ for any value of the argument. But for $t \to \infty$ and $r \to \infty$, $f_1(-\infty)$ is to be zero, which means this solution becomes zero only at $-\infty$, elsewhere, otherwise.

$$\therefore \quad \phi = \frac{f_1\left(t - \frac{r}{v}\right)}{r}.$$

The function f_1 is yet not defined. The choice of the function can be made from the fact that in the static case,

$$\phi = \frac{q}{r}, \quad \therefore \quad \phi(r,t) = \frac{q\left(t - \frac{r}{v}\right)}{r}$$

So, the first eqn.(6.16.10) is satisfied by,

$$\phi(\vec{r},t) = \frac{1}{k} \cdot \frac{\int \rho\left(\vec{r}_s, t - \frac{|\vec{r}-\vec{r}_s|}{v}\right) dV}{|\vec{r}-\vec{r}_s|} \tag{6.16.12}$$

where, $\vec{r}_s$ is the position vector of the source and $\vec{r}$, the position vector of the field point. In a similar way we can find the vector potential $\vec{A}$ that satisfies the second expression of eqn.(6.16.10). It is given by,

$$\vec{A}(\vec{r},t) = \frac{u}{c} \frac{\int \vec{J}\left(\vec{r}_s, t - \frac{|\vec{r}-\vec{r}_s|}{v}\right) dV}{|\vec{r}-\vec{r}_s|} \tag{6.16.13}$$

Meaning of Equations (6.6.12) and (6.6.13)

The values of ϕ and $\vec{A}$ at a time t and at a point distant $|\vec{r}-\vec{r}_s|$ from the element of volume containing ρ and $\vec{J}$ are related not to the value of ρ and $\vec{J}$ at the origin at the same time but to those values obtained at a time $\left(t - \frac{r}{v}\right)$. To put it in another way, the disturbance set up by the values of ρ and $\vec{J}$ at the origin is propagated with velocity v and reaches the field point $\frac{r}{v}$ units of time later. Thus, the disturbance at the field point is related to what happened at the origin $\frac{r}{v}$ secs. earlier. If the charge at the origin be suddenly changed, the resulting effect is not at once felt at the field point. The cause and effect are linked through the time of travel by the wave to field point. This is almost similar to the phenomenon of light reaching us from distant stellar bodies. The information it conveys to us at any time is really something of the past history of the source when it had left that body and not an information up to date and time. These values of ϕ and $\vec{A}$ are, therefore, called *retarded potentials.*

Method 2 : Kirchhoff's Solution

Take the inhomogeneous scalar wave equation

$$\nabla^2\phi - \frac{1}{v^2} \cdot \frac{\partial^2 \phi}{\partial t^2} = -\frac{4\pi\rho(x,y,z,t)}{k}.$$

We want to find the potential ϕ at any pint due to the given charge distribution. This problem mathematically speaking, becomes one of relating a field to its source which can be done by integrating an inhomogeneous differential equation as above.

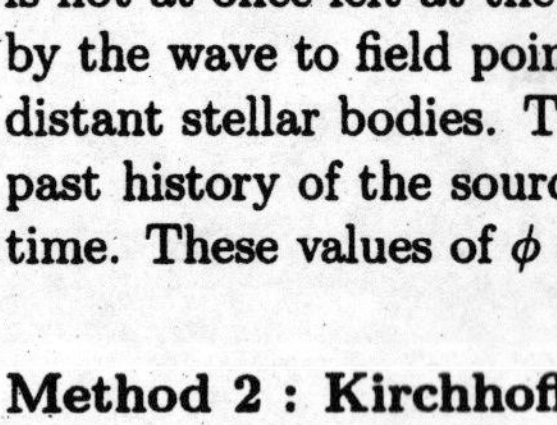

Fig 6.10

In the equation, ϕ represents scalar potential and $\rho(x,y,z,t)$, the source density. To integrate the inhomogeneous wave equation, a closed domain of volume V bounded by the surface is considered (as Kirchhoff did) and let ϕ and ψ be any two scalar functions such that the first and second derivatives are continuous throughout V and also on S.

The function ϕ satisfies the inhomogeneous wave equation and ψ satisfies the homogeneous wave equation $\Box^2\psi = 0$.

The sperically symmetric solution of the homogeneous wave equation is,

$$\psi = \frac{1}{r} f\left(t + \frac{r}{v}\right).$$

Thus, ψ has a singularity at $r = 0$ and so, this point is to be excluded from the somain of integration. Consider a spherical surface S' about the point $P(x, y, z)$ of radius r_1. The volume V is bounded externally by S and internally by S' and the unit vectors outward from S and S' are $\vec{n}$ and $\vec{n}'(\vec{n} = -\vec{n}')$ [cf. fig.(6.10)]. Let us consider a small volume element dV at a distance r from the point P. then by Green's theorem,

$$\iiint_R (\phi\nabla^2\psi - \psi\nabla^2\phi)\, dV = \iint_S \left(\phi\frac{\partial\psi}{\partial n} - \psi\frac{\partial\psi}{\partial n}\right) dS + \iint_{S'} \left(\phi\frac{\partial\psi}{\partial n'} - \psi\frac{\partial\phi}{\partial n'}\right) dS'$$

where the volume integration extends over the space between S and S'. Now, $\frac{\partial}{\partial n} = -\frac{\partial}{\partial n'}$, because the direction of the unit vectors outward are in opposite sense

$$\iiint_R (\phi\nabla^2\psi - \psi\nabla^2\phi)\, dV = \iint_S \left(\phi\frac{\partial\psi}{\partial n} - \psi\frac{\partial\psi}{\partial n}\right) dS - \iint_{S'} \left(\phi\frac{\partial\psi}{\partial n} - \psi\frac{\partial\phi}{\partial n}\right) dS'.$$

Now, since

$$\nabla^2\phi - \frac{1}{v^2}\cdot\frac{\partial^2\phi}{\partial t^2} = -\frac{4\pi\rho}{k} \quad \text{and} \quad \nabla^2\psi - \frac{1}{v^2}\cdot\frac{\partial^2\psi}{\partial t^2} = 0$$

$$\therefore \iiint_R \left[\phi\cdot\frac{1}{v^2}\cdot\frac{\partial^2\psi}{\partial t^2} - \psi\left(\frac{1}{v^2}\cdot\frac{\partial^2\phi}{\partial t^2} - \frac{4\pi\rho}{k}\right)\right] dV$$

$$= \iint_S \left(\phi\frac{\partial\psi}{\partial n} - \psi\frac{\partial\psi}{\partial n}\right) dS - \iint_S \left(\phi\frac{\partial\psi}{\partial n} - \psi\frac{\partial\psi}{\partial n}\right) dS'$$

$$\text{or,}\quad \frac{1}{v^2}\iiint_R \left(\phi\frac{\partial^2\psi}{\partial t^2} - \psi\frac{\partial^2\phi}{\partial t^2}\right) dV + \frac{4\pi}{k}\iiint_R \psi\rho dV$$

$$= \iint_S \left(\phi\frac{\partial\psi}{\partial n} - \psi\frac{\partial\phi}{\partial n}\right) dS - \iint_S \left(\phi\frac{\partial\psi}{\partial n} - \psi\frac{\partial\phi}{\partial n}\right) dS'$$

$$\text{or,}\quad -\iint_S \left(\phi\frac{\partial\psi}{\partial n} - \psi\frac{\partial\phi}{\partial n}\right) dS' = \frac{1}{v^2}\iiint_R \left(\phi\frac{\partial^2\psi}{\partial t^2} - \psi\frac{\partial^2\phi}{\partial t^2}\right) dV$$

$$+\frac{4\pi}{k}\iiint_R \rho\psi dV - \iint_S \left(\phi\frac{\partial\psi}{\partial n} - \psi\frac{\partial\phi}{\partial n}\right) dS.$$

Now,

$$\left(\phi\frac{\partial^2\psi}{\partial t^2} - \psi\frac{\partial^2\phi}{\partial t^2}\right) = \frac{\partial}{\partial t}\left(\phi\frac{\partial\psi}{\partial t} - \psi\frac{\partial\phi}{\partial t}\right)$$

$$\therefore -\iint_{S'} \left(\phi\frac{\partial\psi}{\partial n} - \psi\frac{\partial\phi}{\partial n}\right) dS' = \frac{1}{v^2}\cdot\frac{\partial}{\partial t}\iiint_R \left(\phi\frac{\partial\psi}{\partial t} - \psi\frac{\partial\phi}{\partial t}\right) dV$$

$$+\frac{4\pi}{k}\iiint_R \psi\rho dV - \iint_{S'} \left(\phi\frac{\partial\psi}{\partial n} - \psi\frac{\partial\phi}{\partial n}\right) dS$$

Integrating the above equation between $t=\infty$ to $t=-\infty$, we get,

$$-\int_{-\infty}^{\infty} dt \iint_{S'} \left(\phi\frac{\partial\psi}{\partial n} - \psi\frac{\partial\phi}{\partial n}\right) dS'$$

$$= \frac{1}{v^2}\iiint_R \left[\left(\phi\frac{\partial\psi}{\partial n} - \psi\frac{\partial\phi}{\partial n}\right)\right]_{-\infty}^{\infty} dV + \frac{4\pi}{k}\int_{-\infty}^{\infty} dt \iiint_R \rho\psi dV$$

$$-\int_{-\infty}^{\infty} dt \iint_{S'} \left(\phi\frac{\partial\psi}{\partial n} - \psi\frac{\partial\phi}{\partial n}\right) dS \qquad (6.16.14)$$

Now, $\psi = \frac{1}{r}f\left(t+\frac{r}{v}\right)$, where f is an arbitrary function. We now choose the function such that it and all its differential coefficients vanish for all values of time except $t=-\frac{r}{v}$. Such a function may be taken as Dirac δ-function, $\delta(x)$ which has the following important properties:

1. At $x=0$, it is so large that the integral of the function over an interval containing the point $x=0$ is equal to unity, i.e.,

$$\int \delta(x)dx = 1.$$

2. $\int_a^b F(x)\delta(x)dx = F(0)$, for any function $F(x)$, provided the interval a, b contains the point $x=0$. This property means, the product of $\delta(x)$ with any continuous function of x reduces simply to replace the argument of the function by zero.
3. By shifting the origin, property (2) gives

$$\int F(x)\delta(x-a)dx = F(a).$$

4. $\int \delta'(x)F(x)dx = -F'(0)$,where $\delta'(x)$ is the derivative of δ function w.r.t. x.

So with such choice of the function, the first term in the right-hand side of eqn. (6.16.14) vanishes for $t=-\infty$ to ∞, and we get,

$$\frac{4\pi}{k}\int_{-\infty}^{\infty} dt \iiint_R \rho\psi dV = \int_{-\infty}^{\infty} dt \iint_S \left(\phi\frac{\partial\psi}{\partial n} - \psi\frac{\partial\phi}{\partial n}\right) dS + \int_{-\infty}^{\infty} dt \iint_{S'} \left(\psi\frac{\partial\phi}{\partial n} - \phi\frac{\partial\psi}{\partial n}\right) dS'. \qquad (6.16.15)$$

Now,

$$\int_{-\infty}^{\infty} dt \iiint_R \rho\psi dV = \iiint_R \int_{-\infty}^{\infty} \frac{1}{r}\rho\delta\left(t+\frac{r}{v}\right) dtdV = \iiint_R \frac{1}{r}\rho\left(t=-\frac{r}{v}\right) dV,$$

utilising the property (3) of δ function, since ρ is a function of time.

If the value of ρ at $t=-\frac{r}{v}$ is denoted as $[\rho]$, then we get,

$$\int_{-\infty}^{\infty} dt \iiint_R \rho\psi dv = \iiint_R \frac{[\rho]}{r} dV$$

Again,

$$\int_{-\infty}^{\infty} dt \iint_{S'} \psi \frac{\partial \phi}{\partial n} dS' = \iint_{S'} \left[\int_{-\infty}^{\infty} \psi \frac{\partial \psi}{\partial n} dt\right] dS'$$

$$= \iint_{S'} \frac{dS'}{r} \int_{-\infty}^{\infty} \frac{\partial \phi}{\partial n} \delta\left(t + \frac{r}{v}\right) dt = \iint_{S'} \frac{\left[\frac{\partial \phi}{\partial n}\right]}{r} dS'$$

Also, since,

$$\frac{\partial \psi}{\partial n} = \frac{\partial}{\partial n}\left(\frac{1}{r}\right) \delta\left(t + \frac{r}{v}\right) + \frac{1}{v} \cdot \frac{1}{r} \cdot \frac{\partial r}{\partial n} \delta'\left(t + \frac{r}{v}\right).$$

$$\therefore \int_{-\infty}^{\infty} dt \iint_{S'} \phi \frac{\partial \psi}{\partial n} dS' = \int_{-\infty}^{\infty} dt \iint_{S'} \left[\phi \frac{\partial}{\partial n}\left(\frac{1}{r}\right) \delta\left(t + \frac{r}{v}\right) + \frac{1}{v} \cdot \frac{1}{r} \cdot \frac{\partial r}{\partial n} \delta'\left(t + \frac{1}{v}\right)\right] dS'$$

$$= \iint_{S'} \frac{\partial}{\partial n}\left(\frac{1}{r}\right) dS' \int_{-\infty}^{\infty} \phi \delta\left(t + \frac{r}{v}\right) dt + \frac{1}{v} \iint_{S'} \frac{1}{r} \frac{\partial r}{\partial n} dS' \int_{-\infty}^{\infty} \phi \delta'\left(t + \frac{r}{v}\right) dt$$

$$= \iint \frac{\partial}{\partial n}\left(\frac{1}{r}\right) [\phi] dS' - \frac{1}{v} \iint_{S'} \frac{1}{r} \cdot \frac{\partial r}{\partial n} [\phi'] dS'$$

Utilising properties (3) and (4) of the δ function as outlined above.

Similarly,

$$\int_{-\infty}^{\infty} dt \iint_{S} \phi \frac{\partial \psi}{\partial n} dS = \iint_{S} \frac{\partial}{\partial n}\left(\frac{1}{r}\right) [\phi] dS - \frac{1}{v} \iint_{S} \frac{1}{r} \cdot \frac{\partial r}{\partial n} [\phi'] dS$$

and $$\int_{-\infty}^{\infty} dt \iint_{S} \psi \frac{\partial \phi}{\partial n} dS = \iint_{S} \frac{\left[\frac{\partial \phi}{\partial n}\right]}{r} dS$$

$$\therefore \int_{-\infty}^{\infty} dt \iint_{S} \left(\phi \frac{\partial \psi}{\partial n} - \psi \frac{\partial \phi}{\partial n}\right) dS = \iint \left\{\frac{\partial}{\partial n}\left(\frac{1}{r}\right) [\phi] - \frac{1}{vr} \cdot \frac{\partial r}{\partial n} \left[\frac{\partial \phi}{\partial t}\right] - \frac{\left[\frac{\partial \phi}{\partial n}\right]}{r}\right\} dS$$

and $$\int_{-\infty}^{+\infty} dt \iint_{S'} \left(\psi \frac{\partial \phi}{\partial n} - \phi \frac{\partial \psi}{\partial n}\right) dS' = \iint_{S'} \left\{\frac{1}{r}\left[\frac{\partial \phi}{\partial n}\right] - \frac{\partial}{\partial n}\left(\frac{1}{r}\right) [\phi] + \frac{1}{vr}\left[\frac{\partial \phi}{\partial t}\right] \frac{\partial r}{\partial n}\right\} dS'$$

$$= \iint_{S'} \left\{\frac{1}{r}\left[\frac{\partial \phi}{\partial n}\right] + \frac{1}{r^2}[\phi] + \frac{1}{vr}\left[\frac{\partial \phi}{\partial t}\right]\right\} dS' \qquad (6.16.16)$$

since on the sphere S', $\frac{\partial}{\partial r} = \frac{\partial}{\partial n}$.

Again, since on the surface of the sphere of radius r_1, an elementary solid angle with vertex at the centre of the sphere is given by, $dw = \frac{dS'}{r_1^2}$ so eqn. (6.16.16) in case of spherical surface of radius r_1 becomes in the limiting case $r_1 \to 0$,

$$\iint [\phi]_{t=0} \frac{dS'}{r_1^2} = 4\pi (\phi_p)_{t=0}$$

Now, if the position vector to the point P be $\vec{R}$ and that to the field point $\vec{r}$, then,

$$|\vec{r}| = |\vec{r}' - \vec{R}|.$$

$$\therefore\ \phi\left(\vec{R},0\right) = \frac{1}{4\pi}\iint\limits_{S}\left\{\frac{1}{v}\cdot\frac{1}{r}\left[\frac{\partial\phi}{\partial t}\right] + \frac{1}{r}\left[\frac{\partial\phi}{\partial n}\right] - \frac{\partial}{\partial n}\left(\frac{1}{r}\right)[\phi]\right\}dS + \frac{1}{k}\iiint\frac{[\rho]}{r}dV \quad (6.16.17)$$

The values of the expressions within [ ] are all related to time $t = -\frac{r}{v}$. Now, if we choose the surface S infinitely distant such that the electromagnetic disturbance propagating with finite velocity cannot reach it, then the first term on the right-hand side of eqn. (6.16.17) vanishes, i.e., ϕ and $\frac{\partial\phi}{\partial t} = 0$ on S.

$$\therefore\quad \phi\left(\vec{R},0\right) = \frac{1}{k}\iiint\frac{[\rho]}{r}dV,$$

this equation means, the potential at the point $P\left(\vec{R}\right)$ at time $t = 0$ is not determined by the charge distribution at a distance r from the field point P but by the distribution $\frac{r}{v}$ sec. earlier. So the potential at P at any time t is related to the charge distribution at r at time $t' = \left(t - \frac{r}{v}\right)$.

$$\therefore\ \phi\left(\vec{R},t\right) = \iiint\frac{[\rho']}{k|\vec{r}' - \vec{R}|}dV.$$

Similar expression for $\vec{A}$,

$$\vec{A}\left(\vec{R},t\right) = \frac{\mu}{c}\iiint\frac{[\vec{J}]'}{|\vec{r}' - \vec{R}|}dV$$

where, $[\rho]'$ and $[\vec{J}]'$ denote values calculated at time $t' = t - \frac{r}{v}$. So the above two expressions written explicitly stand out as,

$$\phi = \phi\left(\vec{R},t\right) = \iiint\frac{1}{k}\cdot\frac{\rho\left(\vec{r}',t - \frac{|\vec{r}'-\vec{R}|}{v}\right)dV}{|\vec{r}' - \vec{R}|} \quad (6.16.18)$$

$$\vec{A} = \vec{A}\left(\vec{R},t\right) = \iiint\frac{\mu}{c}\cdot\frac{\vec{J}\left(\vec{r}',t - \frac{|\vec{r}'-\vec{R}|}{v}\right)dV}{|\vec{r}' - \vec{R}|} \quad (6.16.19)$$

These expressions are known as *retarded potentials* and the significance of the expression has already been elaborated earlier.

6.17 Application of the Retarded Potentials in the Computation of Radiation Fields due to an Oscillating Dipole : Electromagnetic Oscillator of Hertz

The origin of electromagnetic waves lies in the oscillation of charge. The most convenient way of producing the necessary vibratory motion of the charge is by Hertzian oscillator, which is essentially an oscillating dipole, with equal positive and negative charges vibrating along the wire $A'B$ with simple harmonic motion (Fig. 6.11). Such as oscillating dipole is the source of electromagnetic radiation and we shall assume that oscillations are maintained with constant amplitude.

Two opposite charges $\pm q_0 \sin \omega t$ at $\frac{l}{2}$ and $-\frac{l}{2}$ constitute an oscillating dipole of moment $\vec{p}_0 \sin \omega t = \vec{l} q_0 \sin \omega t$ and the corresponding current is,

$$i = \frac{dq}{dt} = \omega q_0 \cos \omega t = I_0 \cos \omega t.$$

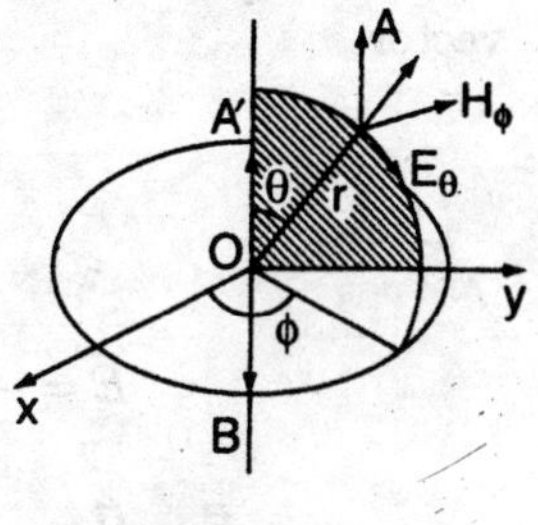

Fig 6.11

Then, since $\vec{J} dV = I d\vec{l}$, the retarded value

$$\vec{J}\left(\vec{r}', t - \frac{r}{v}\right) = I\left(\vec{r}', t - \frac{r}{v}\right) d\vec{l}.$$

If l is small compared with $r = |\vec{r}' - \vec{R}|$, then r does not change significantly during the integration in eqn. (6.16.19) and we may write,

$$\vec{A} = \frac{\mu}{cr} \int I\left(\vec{r}', t - \frac{r}{v}\right) d\vec{l} = \frac{\mu}{cr} I\left(\vec{r}', t - \frac{r}{v}\right) \vec{l} = \frac{\mu}{cr} [I'] \vec{l}. \tag{6.17.1}$$

Equation (6.17.1) shows that $\vec{A}$ is parallel to $\vec{l}$. Now,

$$\vec{B} = \text{curl}\, \vec{A} = \frac{\mu}{c} \text{curl} \frac{[I']}{r} \vec{l} = \frac{\mu}{c} \left\{ \frac{[I']}{r} \text{curl}\, \vec{l} - \vec{l} \times \text{grad} \frac{[I']}{r} \right\} = -\frac{\mu}{c} \vec{l} \times \text{grad} \frac{[I']}{r} \tag{6.17.2}$$

because, $\vec{l}$ being a constant $\vec{\nabla} \times \vec{l} = 0 \cdot$ grad $\frac{[I']}{r} = \frac{\partial}{\partial r}\left[\frac{[I']}{r}\right] \hat{r}$, where $\hat{r}$ is a unit vector along $\vec{r}$.

$$\therefore \quad \vec{H} = -\frac{1}{c}(\vec{l} \times \hat{r}) \frac{\partial}{\partial r}\left[\frac{I'(r)}{r}\right] \tag{6.17.3}$$

Eqn. (6.17.3) shows that $\vec{H}$ is $\perp$ to $\vec{l}$ and $\hat{r}$, thus having only one component H_ϕ in spherical polar co-ordinate.

Now,

$$[I'] = I_0 \cos \omega \left(t - \frac{r}{v}\right)$$

$$\therefore \quad \frac{\partial}{\partial r}\left[\frac{I'(r)}{r}\right] = -\frac{I_0}{r^2} \cos \omega \left(t - \frac{r}{v}\right) + \frac{I_0 \omega}{vr} \sin \omega \left(t - \frac{r}{v}\right)$$

$$= -\frac{I_0}{r^2} \cos \omega \left(t - \frac{r}{v}\right) + \frac{2\pi I_0}{r\lambda} \sin \omega \left(t - \frac{r}{v}\right)$$

where λ is the wavelength of radiation.

So the magnetic field H_ϕ is given by,

$$H_\phi = \frac{I_0}{cr^2} \cos \omega \left(t - \frac{r}{v}\right) \sin \theta - \frac{2\pi I_0 l \sin \theta}{cr\lambda} \sin \omega \left(t - \frac{r}{v}\right) \tag{6.17.4}$$

The first term in equation (6.17.4) predominates when $r \ll \lambda$ and the magnetic field is given by the well-known *Biot-Savart's law.* The second term which is predominant at a few wavelengths from the origin represents the radiation field falling off as $(\lambda r)^{-1}$ and it constitutes the radiation field. The computation of electric field at large distance from the origin where the spherical wavefront transforms into a plane wavefront is not very difficult, because in that case $\vec{E}, \vec{H}$ are normal to the direction of propagation.

To compute the rate at which the dipole radiates energy, we take recourse to Poynting's vector

$$\vec{P} = \frac{c}{4\pi}\left(\vec{E} \times \vec{H}\right).$$

Now,

$$\vec{E} = -\sqrt{\frac{\mu}{k}}\left(\vec{v} \times \vec{H}\right)$$

$$\therefore \quad \vec{E} \times \vec{H} = -\sqrt{\frac{\mu}{k}}\left(\vec{v} \times \vec{H}\right) \times \vec{H} = -\sqrt{\frac{\mu}{k}}\left[\left(\vec{v} \cdot \vec{H}\right)\vec{H} - \left(\vec{H} \cdot \vec{H}\right)\vec{v}\right] = \sqrt{\frac{\mu}{k}} H^2 \vec{v}.$$

The total energy radiated in all directions/sec

$$W = \int_0^{\pi} 2\pi r^2 \sin\theta d\theta < N >$$

where $< N >$ = mean value of energy flow per unit area averaged over a cycle of oscillation.

$$\therefore \quad < N > = \frac{1}{T}\int_0^T \frac{c}{4\pi}\left[\sqrt{\frac{\mu}{k}} H^2\right] dt = \frac{c}{4\pi}\sqrt{\frac{\mu}{k}}\frac{1}{T}\int_0^T \frac{4\pi^2 l^2 I_0^2 \sin^2\theta}{c^2 r^2 \lambda^2} \sin^2 \omega \left(t - \frac{r}{v}\right) dt,$$

taking the long range component

$$= \frac{c}{4\pi} \times \sqrt{\frac{\mu}{k}} \cdot \frac{4\pi^2 l^2 I_0^2 \sin^2\theta}{2c^2 r^2 \lambda^2}$$

$$\therefore \quad W = \frac{c}{4\pi}\sqrt{\frac{\mu}{k}} \cdot \frac{8\pi^2 (I_0 l)^2}{2c^2\lambda^2}\int_0^{\pi} \sin^3\theta d\theta = \frac{c}{4\pi}\sqrt{\frac{\mu}{k}} \cdot \frac{16\pi^2}{3}\left(\frac{l I_0}{c\lambda}\right)^2 \qquad (6.17.5)$$

The power radiated may be expressed in terms of a resistance called *damping resistance* (or *radiation resistance*) written as,

$$W = \frac{1}{2} R I_0^2 \quad \therefore \quad R = \frac{c}{6}\sqrt{\frac{\mu}{k}}\frac{l^2}{\lambda^2} \qquad (6.17.6)$$

6.18 Dispersion—Normal and Anomalous and their Interpretations in terms of Electromagnetic Theory

It is found that when light of composite nature passes through a dispersive medium, the component colours are separated out and spectrum is obtained on a screen. The phenomenon of separation of light into its constituents is known as *dispersion*. This occurs because refractive index changes with wavelength. The variation of refractive index with wavelength is empirically proposed by Cauchy in the form expressed as,

$$n = A + \frac{B}{\lambda^2} + \frac{C}{\lambda^4} \qquad (6.18.1)$$

where A, B, C are constants, known as *Cauchy's constants.* For all practical purposes only first two terms are retained in the expression (6.18.1) and so we write,

$$n = A + \frac{AB}{\lambda^2}. \tag{6.18.2}$$

Differentiating (6.18.2) w.r.t. λ, we get,

$$\frac{dn}{d\lambda} = -\frac{2B}{\lambda^3} \tag{6.18.3}$$

Eqn. (6.18.1) shows that refractive index decreases with increase in wavelength and $\frac{dn}{d\lambda}$ which represents dispersion is inversely proportional to the cube of the wavelength. Such dispersion where the refractive index decreases with wavelength is known as *normal dispersion.*

Every transparent substance selectively absorbs some wavelenghs giving rise to absorption band. In the neighbourhood of the absorption band, the variation of the refractive index with wavelength departs from normal behaviour, it exhibits abnormality in the sense that the refractive index increases with wavelength. This type of dispersion is known as *anomalous dispersion.* Na-vapour, iodine vapour, fuchine (an organic dye) etc., show anomalous dispersion in the visible range, so such a prism would show a spectrum where the red will be deviated most, violet the least.

Experiment done to explore the change of refractive index with wavelength has yielded the results shown in fig.(6.12).

The shaded region represents the absorption band and the variation of n with wavelength on the shorter wavelength side and longer wavelength side of the absorption band are shown. If we consider the curve EFG, which shows normal dispersion behaviour, it more rapidly falls off beyond the point G in the neighbourhood of the absorption band. On the other side of the band, the refractive index is very high and then falls off (JK portion) as in normal dispersion case. Corresponding to the absorption wavelength light is refused transmission. The presence of the discontinuity in the n vs. λ curve and the behaviour in the neighbourhood of the absorption band showing increase of n with wavelength is known as the *anomalous dispersion.*

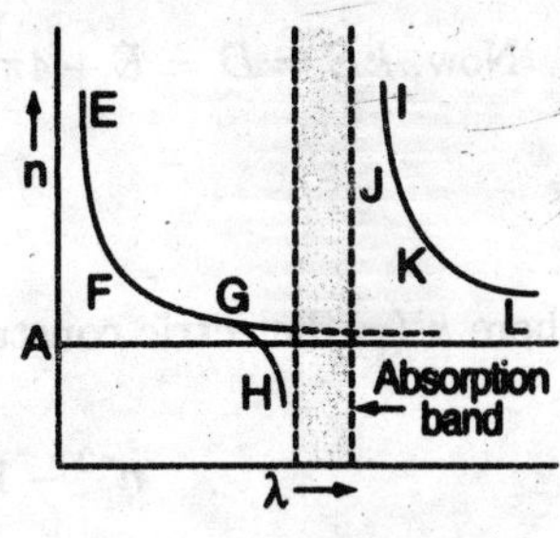

Fig 6.12

Theoretical background

Let us consider an electromagnetic wave train incident on the substance. Under the influence of the fluctuating electric field associated with the wave, there will be creation of dipoles which will be treated as harmonic oscillator of natural frequency ω_1. The equation of motion of the harmonic oscillator can be written as,

$$\frac{d^2x}{dt^2} + \omega\frac{dx}{dt} + \omega_1^2 x = \left(E + \frac{4\pi}{3}P\right)\frac{e}{m}. \tag{6.18.4}$$

The term $\frac{4\pi}{3}P$ accounts for the effect of polarisation. If we consider that there are N of such oscillators per unit volume, then polarisation P is given by,

$$P = Nex \tag{6.18.5}$$

Using equation (6.18.5) in (6.18.4) we get,

$$\ddot{P}+\omega\dot{P}+\omega_1^2P=\frac{Ne^2}{m}\left(E+\frac{4\pi}{3}P\right). \tag{6.18.6}$$

Let the harmonic wave train be represented by a suitable rotating vector $\vec{\xi}$, such that

$$E=\hat{E}_0e^{jpt}\vec{\nu}\cdot\vec{\nu}=\vec{\xi}\cdot\vec{\nu}.$$

So the instantaneous polarisation can be written as,

$$P=\vec{\eta}\cdot\vec{\nu},$$

where $\vec{\eta}$ is the rotating vector representing polarisation of the medium.

$$\dot{P}=jP\vec{\eta}\cdot\vec{\nu},\quad \ddot{P}=j^2p^2\vec{\eta}\cdot\vec{\nu}$$

So from eqn. (6.18.6) we get,

$$[(\omega_1^2-p^2)+j\omega p]\,\vec{\eta}\cdot\vec{\nu}=\frac{Ne^2}{m}\left(\vec{\xi}\cdot\vec{\nu}+\frac{4\pi}{3}\vec{\eta}\cdot\vec{\nu}\right)$$

$$\text{or, }\vec{\eta}\cdot\vec{\nu}=\frac{\frac{Ne^2}{m}\vec{\xi}\cdot\vec{\nu}}{[(\omega_1^2-p^2)+j\omega p-\frac{4\pi}{3}\cdot\frac{Ne^2}{m}]} \tag{6.18.7}$$

Now, $kE=D=E+4\pi P$.

$$\therefore\quad 4\pi\frac{P}{E}=k-1\simeq n^{*2}-1$$

where k (= dielectric constant) (here it is complex) is approximately given as $k\simeq n^{*2}$.

$$\therefore\quad n^{*2}-1=\frac{(4\pi Ne^2)/m}{[(\omega_1^2-p^2)+j\omega p-\frac{4\pi}{3}\cdot\frac{Ne^2}{m}]}$$

$$=\frac{4\pi Ne^2/m}{(\omega_1^2-\frac{4\pi}{3}\cdot\frac{Ne^2}{m})-p^2+j\omega p}=\frac{4\pi Ne^2/m}{(\omega_1'^2-p^2)+j\omega p} \tag{6.18.8}$$

Eqn. (6.18.8) is derived on the basis of the existence of same frequency oscillators. This would give rise to one absorption band. But, if the medium possesses more than one absorption band we would generalise eqn. (6.18.8).

Suppose that there are N molecules/c.c. and there are f_1 electrons per molecule belonging to type 1 with frequency ω_1, f_2 molecules belonging to type 2 with frequency ω_2 and so on

$$\therefore\ P=Ne\sum f_kx_k$$

$$\therefore\ \omega_k'^2=\omega_k^2-\frac{4\pi}{3}\cdot\frac{Ne^2}{m}f_k$$

$$\therefore\ n^{*2}-1=\frac{4\pi Ne^2}{m}\sum\frac{f_k}{(\omega_k'^2-p^2)+j\omega p} \tag{6.18.9}$$

Normal Dispersion

If the frequency of the incident light is remote from the natural frequencies such that

$$\omega_k'^2 - p^2 \gg \omega p,$$

then absorption can be neglected and the refractive index is real, we get,

$$n^2 = 1 + \frac{4\pi N e^2}{m} \sum \frac{f_k}{\omega_k'^2 - p^2}.$$

Now,

$$\lambda = \frac{2\pi c}{p}, \quad \lambda_k' = \frac{2\pi c}{\omega_k'}$$

$$\therefore n^2 = 1 + \frac{4\pi N e^2}{m} \sum \frac{f_k}{4\pi^2 c^2 \left(\frac{1}{\lambda_k'^2} - \frac{1}{\lambda^2}\right)} = 1 + \sum \frac{A_k \lambda^2}{(\lambda^2 - \lambda_k^2)} \tag{6.18.10}$$

where $A_k = \frac{Ne^2}{\pi c^2 m} f_k \lambda_k^2$.

Eqn. (6.18.10) is known as *Sellmeier's formula.* This formula holds good for wavelength remote from the absorption wavelength. Now, when λ remaining away from the absorption wavelength tends to it from greater wavelength side, n value increases which explains normal dispersion phenomenon.

Anomalous Dispersion

Taking absorption into account, we get,

$$n^{*2} - 1 = \frac{4\pi N e^2}{m} \sum \frac{f_k}{(\omega_k'^2 - p^2) + j\omega p}$$

This equation holds for all frequencies or wavelengths

$$n^{*2} = n^2(1 - jx)^2 = 1 + \frac{4\pi N e^2}{m} \sum \frac{f_k}{(\omega_k'^2 - p^2) + j\omega p}$$

Equating real and imaginary parts,

$$n^2(1 - x^2) = 1 + \frac{4\pi N e^2}{m} \sum \frac{f_k(\omega_k'^2 - p^2)}{(\omega_k'^2 - p^2)^2 + \omega^2 p^2}$$

$$\text{and} \quad n^2 x = \frac{2\pi N e^2}{m} \sum \frac{\omega p f_k}{(\omega_k'^2 - p^2)^2 + \omega^2 p^2}$$

When x is small (i.e., for weak absorption)

$$n^2 = 1 + \frac{4\pi N e^2}{m} \sum \frac{f_k(\omega_k'^2 - p^2)}{(\omega_k'^2 - p^2)^2 + \omega^2 p^2}$$

Introducing wavelengths, the above equation becomes

$$n^2 = 1 + \frac{4\pi N e^2}{m} \sum \frac{f_k \lambda_k^2 \lambda^2 (\lambda^2 - \lambda_k^2)}{4\pi^2 c^2 (\lambda^2 - \lambda_k^2)^2 + \omega^2 \lambda^2 \lambda_k^4}$$

In the neighbourhood of the absorption wavelength $\lambda \simeq \lambda_k$, we can make the approximation $\lambda_k^2\lambda^2 \simeq \lambda_k^4$, $\lambda_k^4\lambda^2 \simeq \lambda_k^6$, $\lambda + \lambda_k \simeq 2\lambda_k$.

$$\therefore \quad n^2 = 1 + \frac{4\pi Ne^2}{m}\sum \frac{f_k\lambda_k^4(\lambda^2-\lambda_k^2)}{4\pi^2c^2(\lambda^2-\lambda_k^2)^2+\omega^2\lambda_k^6} = 1 + \frac{4\pi Ne^2}{m}\sum \frac{2f_k\lambda_k^5(\lambda-\lambda_k)}{4\pi^2c^2(\lambda^2-\lambda_k^2)^2+\omega^2\lambda_k^6}$$

$\because$ n is approximately equal to unity, we can write,

$$\therefore \quad n = 1 + \frac{4\pi Ne^2}{m}\sum \frac{f_k\lambda_k^5(\lambda-\lambda_k)}{4\pi^2c^2(\lambda^2-\lambda_k^2)^2+\omega^2\lambda_k^6} = 1 + \frac{4\pi Ne^2}{m}\sum \frac{A_k(\lambda-\lambda_k)}{B(\lambda^2-\lambda_k^2)^2+D} \quad (6.18.11)$$

When λ approaches λ_k from greater wavelength side, the denominator decreases and n value increases and in the immediate neighbourhood, the second factor tends to a large value, thus giving n a large magnitude. This accounts for the behaviour of the curve after the absorption band. On the other hand, from lower wavelength side as $\lambda \to \lambda_k$, the value of n decreases, thus explaining the normal dispersion phenomenon. On the other side of the absorption band, the n value starts from a high value, then it decreases. The behaviour of n vs. wavelength in the neighbourhood of the absorption band on the longer wavelength side represents the anomalous dispersion phenomenon.

When equation (6.18.11) is plotted, it produces the curve as shown in fig. (6.13) when the absorption is weak.

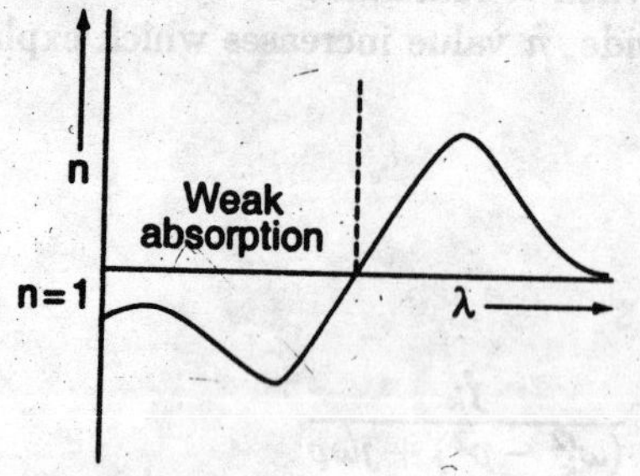

Fig 6.13

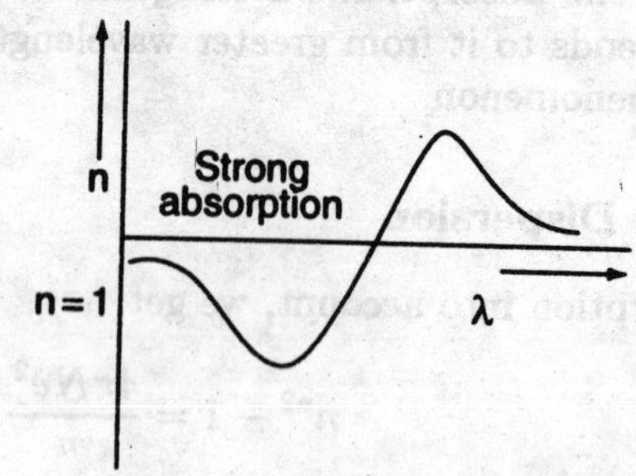

Fig 6.14

The fig. (6.14) represents behaviour when the absorption is strong.

Experimental Demonstration

As already remarked Na-vapour, fuchine dye, idone vapour exhibit anomalous dispersion in the visible range. In 1904 R.W. Wood demonstrated the dispersion phenomenon in case of Na-vapour by the arrangement as shown in fig. (6.15).

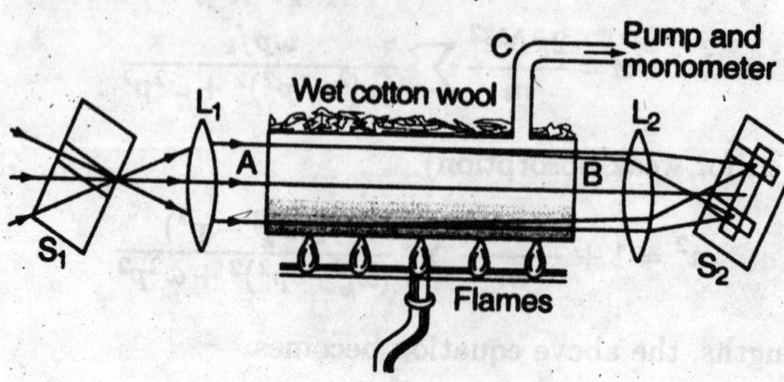

Fig 6.15

AB is a steel tube 40 cm long, 3 cm in diameter, the two ends being sealed airtight by glass plates. On the top of the tube there is placed wet cotton wool for a special purpose. There is arrangement in the form of a side tube for connection to a pump required to maintain the pressure inside to the desired amount and also connection to a manometer for the measurement of pressure.

Keeping the steel tube horizontal a few pieces of clean and dry sodium are placed at the bottom inside the tube along its length and by means of a series of burners, the tube is gradually heated. The absorbed hydrogen in sodium pieces will be liberated and by means of the pump most of the hydrogen is expelled from the tube leaving only a trace of hydrogen inside. When the pressure inside is about 1-2 cm of Hg, Na-vapour will be emitted. The wet cotton wool at the top of the tube will cause vapour to diffuse upwards and the trace of hydrogen present in the tube opposing the diffusion creates a density gradient, density gradually decreasing upwards. Thus the Na-vapour in the tube behaves as if a prism with its base towards the bottom, its apex upwards.

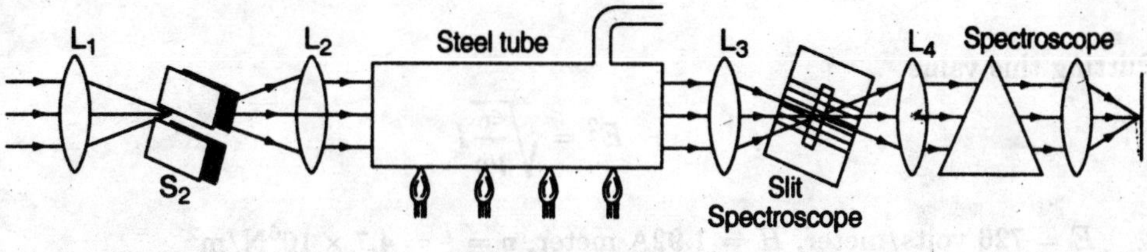

Fig 6.16

In this experiment (fig.6.16) light from the sun or arc is allowed to pass through the horizontal slit in the screen S_1, the diverging rays from S_1 are rendered into a parallel beam by means of a lens L_2 and allowed to pass through the steel tube. The lens L_3 produces a real image of the horizontal slit onto the slit of the spectrometer. The slit in S_1 is horizontal but that of the spectrometer vertical. Thus the image formed on the slit of the spectrometer is at right angles to the slit.

Initially the steel tube is not heated and, therefore, the white light passing through the slit S_1 and then through the spectrometer forms a spectrum in the field of view consisting of known seven colours VIBGYOR (fig. 6.17a). Now the steel tube is heated gently and the pump run at the same time such that the pressure reduces to 1-2 cm of Hg, driving the air and the absorbed H_2 in Na. The steel tube is then strongly heated.

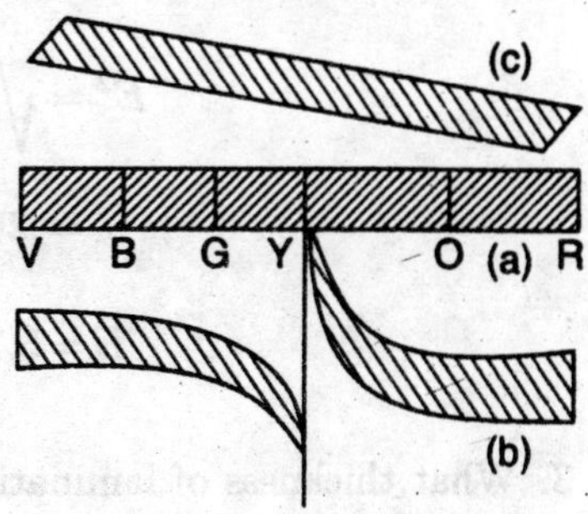

Fig 6.17

Na-vapour will be formed and formed into a prism. Thus there will be dispersion in the steel tube and there will be a number of images formed onto the spectrometer slit. Due to dispersion in Na-vapour the different colours will be relatively displaced. Due to absorption of sodium D yellow line, there would appear discontinuity. From yellow to green, the spectrum is displaced downwards which means, the rays passing through Na-vapour prism have bent upwards (because there is an inversion by the spectrometer). It shows, therefore, that due to passage through the vapour the phase velocity is greater than the velocity in vacuum thus turning the rays opposite to the base, yellow and red colours will show opposite phenomenon. The spectrum will be as shown in the fig. (6.17b). Had it been normal dispersion all rays should bend towards the base of the prism, thus the violet would appear in the field of view to be displaced relatively upwards fig. (6.17c).

Solved Numerical Problems

1. The rate at which solar energy falls on the earth's surface is approximately 2 cal/cm²/min. Calculate the r.m.s. values of the electric and magnetic fields at the earth's surface and the pressure exerted on it, assuming it to behave as perfect absorber.

Solution. The energy received per m² per sec.

$$I = \frac{2}{60} \times 10^4 \times 4.2 \text{ joules/m}^2\text{/s}$$

Now, EH (in R.M.K.S) $= I$

$$E = \sqrt{\frac{\epsilon_0}{\mu_0}}H, \epsilon_0 = 8.8547 \times 10^{-12} \text{ F/m} \quad \mu_0 = 4\mu \times 10^{-7}.$$

Putting this value,

$$E^2 = \sqrt{\frac{\epsilon_0}{\mu_0}}I$$

$\therefore \quad E = 726$ volts/meter, $H = 1.92$A meter, $p = \frac{I}{c} = 4.7 \times 10^6 \text{N/m}^2$

2. Find the mean value of the intensities of the electric and magnetic field in air at a distance r cm from a radiating point source of P watts.

Solution. Poynting energy flow is given by,

$$EH = \frac{P \times 10^4}{r^2} \text{ joules/s/m}^2 \quad \text{or,} \quad \sqrt{\frac{\mu_0}{\epsilon_0}}E^2 = \frac{P \times 10^4}{r^2}$$

$$\therefore \quad E^2 = \sqrt{\frac{\epsilon_0}{\mu_0}} \times \frac{P \times 10^4}{r^2} = \sqrt{\frac{8.8547 \times 10^{-12}}{4\pi \times 10^{-7}}} \times \frac{P \times 10^4}{r^2}$$

$$= .265 \times 10^{-2} \times \frac{P \times 10^4}{r^2} = 26.5\frac{P}{r^2}$$

$$\therefore \quad E = 5.148\frac{\sqrt{P}}{r} \text{ volt/m.}$$

3. What thickness of lamination is required to make the skin depth equal to the thickness at a frequency of 50 cps?
Given $\mu = 10^5$, $\rho = 5 \times 10^{-7} W$-meter.

Solution. The skin depth δ is given by,

$$\delta = (\pi\sigma f\mu\mu_0)^{\frac{1}{2}}.$$

Here, $\sigma = \frac{1}{5 \times 10^{-7}} = .2 \times 10^7$, $f = 50$.

$$\therefore \quad \delta = (\pi \times .2 \times 10^7 \times 50 \times 10^5 \times 4\pi \times 10^{-7})^{-\frac{1}{2}} = .16 \text{ mm.}$$

Exercises

1. Set up Maxwell's field equations. Explain clearly the displacement current.
2. Show that Maxwell's field equations represent waves. Also show that in the electromagnetic waves, the electric and magnetic waves are transverse in character.
3. What do you mean by skin depth? Find an expression for the skin depth at a frequency f cps.
4. Explain the phenomena of reflection and refraction of electromagnetic waves.
5. Derive Fresnel's formula for electromagnetic waves reflected and refracted in dielectric media.
6. The introduction of the concept of displacement current was inevitable in the field theory by Maxwell. Discuss the necessity of the introduction and derive the field equation involving the displacement current.
7. The electromagnetic energy flow per sec. across a closed surface is given by the flux of a vector across the surface. Derive an expression for the said vector and state the theorem in this connection.
8. For glass-air interface it is found that the relfection co-efficient $R_n = 4\%$ and the transmission co-efficient $T_n = 96\%$. Can you justify this result theoretically? Derive the necessary formula in support of the above result.
9. What do you mean by dispersion? Classify the phenomenon as normal and anomalous. Explain theoretically the normal and anomalous dispersion.
10. Give the experimental arrangement to demonstrate the anomalous dispersion in case of Na-vapour.

References

1. **Reitz** and **Milford** (1962) : *Foundation of Electromagnetic Theory,* Addison Wesley Publishing Co. (USA).
2. **Panofsky** and **Phillips** (1962) : *Classical Electricity and Magnetism,* Addison Wesley Publishing Co. (USA).
3. **Feynman, Leighton** and **Sands** (1965) : *The Feynman Lectures on Physics,* Vol. I, Addison Wesley Publishing Co. (USA).

Chapter 7

Resolving Power

7.1 Introduction

Resolving power is an important property of an optical device, related to the capability of the device to separate closely spaced spectral lines or source images. The separation of the spectral lines is called *resolution*. Different spectroscopic devices have different resolving powers depending on the availability of illumination or on the limitations set by diffraction. As a result, the criterion for resolution is also different in different cases. The criterion which applies to the consideration of the resolving power of a diffraction grating may not apply to the case of spectroscopic devices like Michelson Interferometer or Fabry-Perot Interferometer. A very important criterion for resolution of spectral lines is, however, that of Rayleigh, known as *Rayleigh's criterion of resolution.*

In this chapter we will consider the resolving powers of a number of spectroscopic devices like grating, prisms, microscope and telescope, etc. The resolving powers of Michelson interferometer and Fabry-Perot interferometer are discussed with respective interferometers.

7.2 Rayleigh's Criterion

Though there is a number of criteria about the resolution of spectral lines, Rayleigh's criterion may be looked upon as *tablets of gold written by an angel* because it may be applied with slight modification to almost all problems of resolution. This criterion is easy to conceive and it is also easily reproducible.

Whatever may be the criterion put forward it is to remember that ultimate receptor is our eye, so to recognise two spectral lines or two sources of light as separate we must not lose sight of the fact that our eye is not able to differentiate between two regions if the contrast between them is less than roughly 20%. So, if two spectral lines through interference or diffraction produce fringe systems such that overlap of the two patterns occurs to an extent that in between the positions of the peak of intensity of the two spectral lines, the resultant intensity drops at least by 20%, then the two peaks can be distinguished, hence the spectral lines giving the peaks can be distinguished or resolved. On the other hand, if the difference of the two wavelengths giving rise to the peaks of the intensity distribution is so small that variation of intensity between the two peaks is less than 20% of the peak intensity, the two peaks cannot be identified as different. Based on this consideration Rayleigh has framed a criterion which has been accepted universally as the criterion of resolution.

Formally stated, this criterion says that (in case of diffraction fringe pattern) *two spectral lines of wavelengths* λ *and* $(\lambda \pm d\lambda)$ *will be just resolved if the pth principal maximum of wavelength* $(\lambda \pm d\lambda)$ *just falls on the first minimum after the corresponding principal maximum of wavelength* λ.

Let us now consider the criterion as stated above elaborately and see what it means physically. If the condition or criterion of Rayleigh is fulfilled, then at the mid-way between the centres of maxima of wavelengths λ and $(\lambda \pm d\lambda)$, the intensity is reduced to about 81% of that existing at the centre of each equally bright maximum. In case of two sources for the same wavelength, the closeness must not exceed a certain limit such that the overlap of the diffraction pattern produces resultant intensity more than 80% of the central intensity maximum. In this connection we should note that depending on whether the optical device under consideration is used to resolve two intimate spectral lines or closely neighboured spectral sources, resolving power may be formed as chromatic resolving power or positional resolving power. The chromatic resolving power of an instrument is given by $\frac{\lambda}{d\lambda}$, which means smaller the difference of wavelength $\delta\lambda$, greater is the resolution of the instrument. In the second case, smaller is the limiting distance of separation between two sources visualised as separate sources by an instrument, greater is the positional resolving power of the instrument. We will first discuss in details the chromatic resolution by an instrument and then the positional resolution. In this chapter we will also develope expressions for the resolving power (R.P.) of some important instruments.

7.3 Spectral Resolution : Chromatic Resolving Power

When overlapping diffraction patterns are produced by a diffracting aperture (single slit) due to two wavelengths λ_1 and λ_2, the diffraction patterns to be identifiable, there is a certain limiting difference in wavelength such that there is a dip of 20% in the resultant intensity distribution in between the central maxima in the patterns by the two spectral lines.

Moreover, spectral resolution also depends on the relative intensities of the two patterns.

In fig. (7.1) we have shown superposition of two spectral patterns with relative intensity ratio 1 : 1 and 5 : 1. In figures (a), (b), (c) and (d), the spectral angular separations $\Delta\theta$ are taken to be $0.6\Delta\lambda_R, 1.0\Delta\lambda_R, 1.2\Delta\lambda_R$ and $1.5\Delta\lambda_R$ respectively where $\Delta\lambda_R$ is the spectral wavelength difference as required in Rayleigh criterion.

Fig. 7.1(a) shows that in both cases of intensity ratio, the resultant intensity pattern is such that it does not show any dip in between the central maxima, hence the spectral lines cannot be resolved.

In fig. 7.1(b) where the intensity ratio is unity, the resultant intensity curve shows an identification dip of about 20% between the two central maxima but at the same angular separation for intensity ratio 5, the identification dip does not occur. In fig. 7.1(c), where $\Delta\theta = 1.2\Delta\lambda_R$, there is a dip of about 40% in between the central maxima and it is said to be complete resolution. But for the same value of $\Delta\theta$ in case of intensity ratio 5, such dip does not appear, but, however, presence of two spectral patterns can be detected. For complete resolution with intensity ratio 5, the angular separation required is $\Delta\theta = 1.5\Delta\lambda_R$. So it appears that, in case of equal intensity pattern, Rayleigh's criterion holds.

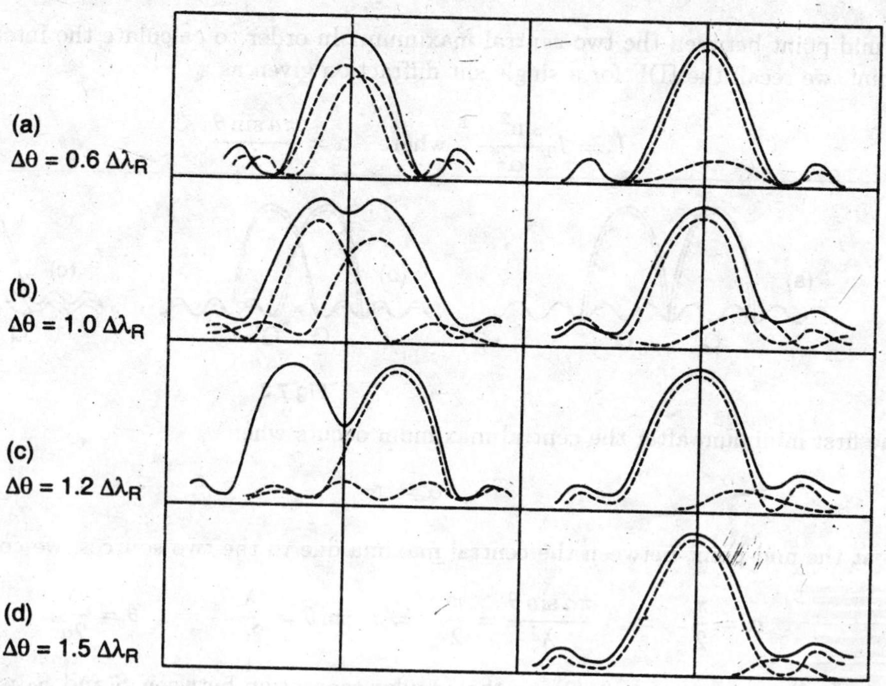

Fig 7.1

Positional Resolution and Positional Resolving Power

In fig. (7.2) two sources S and S_2, independent and hence incoherent, illuminate the slit; the angular separation of the two sources being θ. The waves coming from these two sources and getting diffracted by the slits will produce two diffraction patterns of Fraunhofer type by the two lenses.

The diffraction pattern due to the source S has its central maximum at Q surrounded by its characteristic secondary maxima and minima; the pattern due to the other source S_2 has its central maximum at Q'; the angular separation between the positions of the central maxima in the two patterns is also equal to θ. The resultant intensity will be obtained by addition of the individual intensities. So the resultant intensity pattern will depend on the angular separation between the two sources S and S_2.

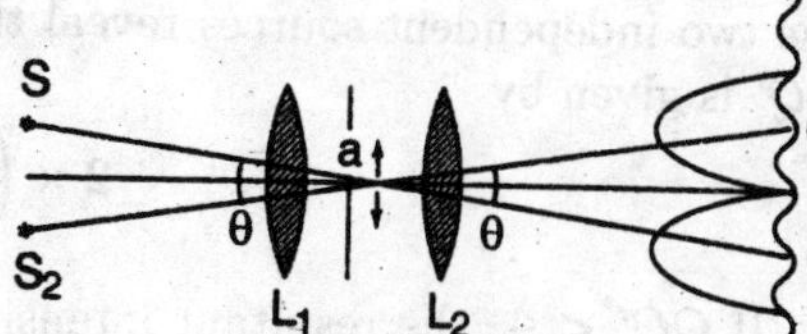

Fig 7.2

Thus in fig. 7.3(a), the position of the two sources is such that the central maximum of the pattern due to S_2 falls on the second minimum of the pattern due to source S. The continuous curve obtained by addition of two intensities gives the resultant intensity curve. In this case at the mid-point between the central maxima due to the sources S and S_2, the intensity drops almost to zero value, thus the two patterns could easily be identified.

In fig. 7.3(b), the angular separation of the two sources is such that the central maximum due to source S_2 coincides in position with the first minimum after the central maximum due to the other source S. In this case the resultant intensity does not drop down to zero value at the mid-point between the two central maximum. In order to calculate the intensity at the mid-point, we recall the IDF for a single slit diffraction given as

$$I_p = I_0 \frac{\sin^2 \alpha}{\alpha^2} \quad \text{where} \quad \alpha = \frac{\pi a \sin\theta}{\lambda}.$$

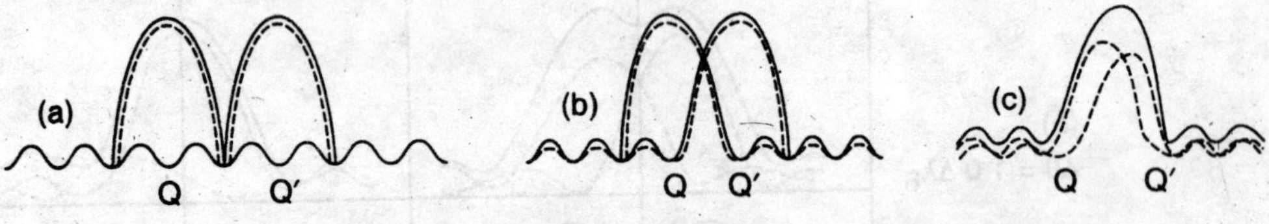

Fig 7.3

The first minimum after the central maximum occurs when

$$\alpha = \pi$$

So at the mid-point between the central maxima due to the two sources, we could write

$$\alpha = \frac{\pi}{2} \quad \Rightarrow \quad \frac{\pi a \sin\theta}{\lambda} = \frac{\pi}{2} \quad \Rightarrow \quad \sin\theta = \frac{\lambda}{2a} \quad \Rightarrow \quad \theta = \frac{\lambda}{2a}.$$

In fig.(7.3) it is shown as $\theta_0/2$), so the angular separation between S and S_2 is given by

$$2\left(\frac{\theta_0}{2}\right) = \frac{\lambda}{a} \tag{7.3.1}$$

At this position,

$$I_p = I_0 \times \frac{\sin^2 \pi/2}{(\pi^2/4)} = \frac{4}{\pi^2} I_0 = 0.4053 I_0 \tag{7.3.2}$$

Adding the contributions by both, the resultant intensity at this mid-point is $0.8053 I_0$, thus at this separation between the two sources $S$and S_2, there is a 20% dip in the resultant intensity curve in between the central maxima due to the two sources, so the central maxima and hence the two independent sources reveal their identity. In this case the angular separation between QQ' is given by

$$2 \times \left(\frac{\theta_0}{2}\right) = \frac{\lambda}{a} \quad \Rightarrow \quad \theta_0 = \frac{\lambda}{a}$$

If $QQ' < \frac{\lambda}{a}$, the resultant intensity curve is shown in fig.(7.3c), at this separation the two patterns overlap so much that at the mid-point between the central maxima by the two series there is no dip in the resultant intensity curve, rather the resultant intensity curve has its peak at the mid-point. Thus the two patterns appear to be fused into one and cannot be identified as separate.

Resolving Power of a Prism Spectroscope

The resolving power of a prism spectroscope can be found by applying Rayleigh's criterion and reference may be made to the fig. (7.4).

A prism ABC throws spectrum on to a screen by means of two lenses L_1 and L_2. The first lens L_1 renders the rays from S into a parallel beam. Suppose there are two wavelengths

λ_1 and λ_2 in the light. After emergence each wavelength will form the image on the screen. After emergence from the face AC, the wavelength λ_1 forms a parallel beam of width CE and that of λ_2 forms a parallel beam of width CD. Due to dispersion in the prism, the angular separation between the two emerging beams is $\Delta\theta_0$ which is also equal to the angle between the two wavefronts CE and CD.

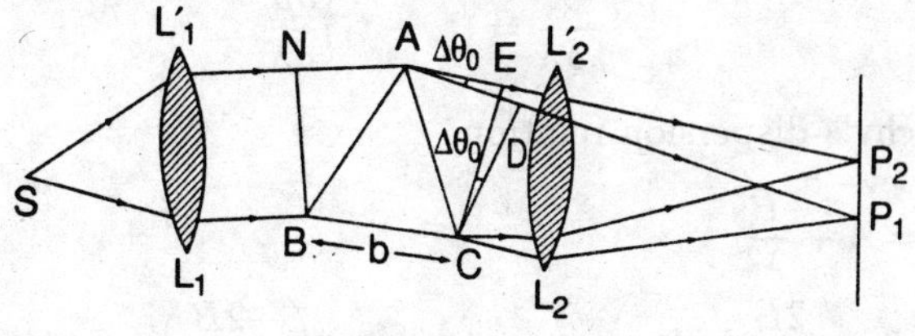

Fig 7.4

According to the principle due to Fermat, the optical paths (O.P.) connecting S and P_1 or P_2 must be same.

For wavelength λ_1,

$$SL_1' + L_1'N + NA + AE + EP_2 = SL_1 + L_1B + \mu_1 BC + CP_2$$

Again, $SL_1' + L_1'N = SL_1 + L_1B, EP_2 = CP_2$

$\therefore\ NA + AE = \mu_1 BC$, where, μ_1 is the r.i. of the prism for wavelength λ_1.

Similarly, for the wavelength λ_2, we can write,

$$NA + AD = \mu_2 BC, \quad \mu_2 \text{ being the r.i. for wavelength } \lambda_2$$

$$\therefore \quad AD - AE = (\mu_2 - \mu_1)BC = l\delta\mu \tag{7.3.3}$$

$$\Rightarrow \quad AD - AE = l\frac{\delta\mu}{\delta\lambda}\delta\lambda$$

where l = base width of the prism

$$\delta\mu = (\mu_2 - \mu_1) \tag{7.3.4}$$

$$\delta\lambda = (\lambda_1 - \lambda_2) \tag{7.3.5}$$

The emergent face of the prism acts as a rectangular aperture to the emerging beam, the width of this aperture for wavelength λ_1 and λ_2 is respectively given by CE and CD; now since $\Delta\theta_0$ is very small $CE \simeq CD$. As shown in case of a single slit resolution, the resolution limit is given by,

$$\Delta\theta_0 = \frac{\lambda}{a} = \frac{\lambda}{CE}$$

Now,

$$\Delta\theta_0 = \frac{AD - AE}{CE} \quad \Rightarrow \quad \Delta\theta_0 \cdot CE = AD - AE = l\frac{\delta\mu}{\delta\lambda}\delta\lambda$$

$$\Rightarrow \quad \Delta\theta_0 = \frac{l}{CE}\cdot\frac{\delta\mu}{\delta\lambda}\delta\lambda$$

$$\Rightarrow \quad \frac{\lambda}{CE} = \frac{l}{CE}\cdot\frac{\delta\lambda}{\delta\lambda}\delta\lambda$$

$$\Rightarrow \quad \frac{\lambda}{\Delta\lambda} = l\frac{\delta\mu}{\delta\lambda} \tag{7.3.6}$$

Thus eqn. (7.3.6) defines the chromatic resolving power of a prism spectroscope.

If the width of the incident parallel beam does not cover the full width of the prism but the extreme rays traverse t_1 and t_2 thickness of the prism, then l in eqn. (7.3.6) should be replaced by $(t_2 - t_1)$, hence the R.P. is then given by,

$$\frac{\lambda}{\delta\lambda} = (t_2 - t_1)\frac{\delta\mu}{\delta\lambda} \tag{7.3.7}$$

According to Cauchy's dispersion relation,

$$\mu = A + \frac{B}{\lambda^2}$$

$$\therefore \quad \frac{\delta\mu}{\delta\lambda} = -\frac{2B}{\lambda^3} \quad \Rightarrow \quad \frac{\lambda}{\delta\lambda} = (t_2 - t_1)\left(\frac{-2B}{\lambda^3}\right) = -2(t_2 - t_1)\frac{B}{\lambda^3}$$

$$\therefore \quad \text{C.R.P.} = \frac{\lambda}{\delta\lambda} = l\left|\frac{\delta\mu}{\delta\lambda}\right| = 2(t_2 - t_1)\frac{B}{\lambda^3} \tag{7.3.8}$$

In order to find out the chromatic resolving power of a prism, the prism is set in the position of minimum deviation. In this case then, the paths of the rays within the prism will be parallel to the base and the angle of incidence would be equal to the angle of emergence.

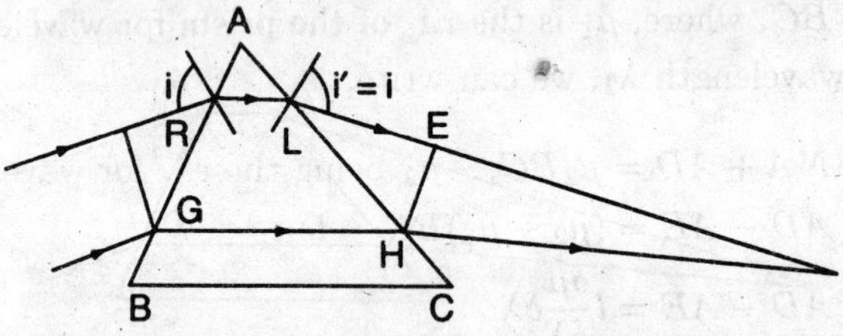

Fig 7.5

Since the prism is placed in the position of minimum deviation, the path of the rays corresponding to the wavelength λ inside the prism will be parallel to the base and the incident wavefront and emerging wavefront will make the same angle i with the faces AB and AC. Then from fig. (7.5) we can write,

$$GH = t_2 = 2AH \sin\frac{A}{2} \tag{7.3.9}$$

$$RL = t_1 = 2AL \sin\frac{A}{2}, \quad \text{where } \angle A \text{ is the angle of the prism} \tag{7.3.10}$$

$$\therefore \quad \text{C.R.P.} = 4(AH - AL)\sin\frac{A}{2} \times \frac{B}{\lambda^3} = 4LH \sin\frac{A}{2} \times \frac{B}{\lambda^3}$$

The width of the beam a is represented by $HE = LH\cos i \quad \Rightarrow \quad LH = \left(\frac{a}{\cos i}\right)$

$$\therefore \quad \text{C.R.P.} = 4\left(\frac{a}{\cos i}\right)\sin\frac{A}{2} \times \frac{B}{\lambda^3}. \tag{7.3.11}$$

Now, r.i. values μ_1 and μ_2 for the wavelengths λ_1 and λ_2 are found out by spectrometric method and hence the value of B found out.

We know,

$$\mu = A + \frac{B}{\lambda^2}; \quad \therefore \quad \mu_1 = A + \frac{B}{\lambda_1^2}, \quad \mu_2 = A + \frac{B}{\lambda_2^2}.$$

Taking the difference, we get,

$$(\mu_1 - \mu_2) = \frac{B(\lambda_2^2 - \lambda_1^2)}{\lambda_1^2\lambda_2^2} \quad \Rightarrow \quad B = \frac{\lambda_1^2\lambda_2^2(\mu_1 - \mu_2)}{\lambda_2^2 - \lambda_1^2)}$$

Thus finding B and measuring the width of the beam, the chromatic resolving power (C.R.P.) can be immediately found using eqn. (7.3.11).

Also from eqn.(7.3.8) it is evident that the chromatic resolving power depends on $(t_2 - t_1)$ which ultimately determines the width of the emerging beam, hence the width of the equivalent slit.

Example. A prism has $\frac{d\mu}{d\lambda} = 10^3$, find the base length of the prism required to resolve the D_1 and D_2 lines of Na.

Solution. D_1 line has wavelength $\lambda_1 = 5896$Å, D_2 has $\lambda_2 = 5890$Å

$$\therefore \quad \frac{\lambda}{\delta\lambda} = \frac{5893}{6} = l\left(\frac{\delta\mu}{\delta\lambda}\right) = l \times 10^3 \quad \Rightarrow \quad l \simeq 1\,\text{cm}.$$

Resolving Power of a Grating

The resolving power of a grating has been derived earlier in connection with our discussion on grating spectrum. We have seen there that the chromatic resolving power is given by $\frac{\lambda}{\delta\lambda} = mN$, where m is the order number and N is the total number of lines in the grating. So, from this expression it is evident that for greater resolution with a grating, the number of lines must be greater. But from the consideration of grating equation the order of spectrum available will be less if N is greater. So, for the purpose of determining unknown wavelength by a grating, the number of line should not be so high that we would only get first or second order of the spectrum. But, for the study of fine structure of lines, resolution of the grating should be made high which requires greater number of lines.

7.4 Spatial Resolution : Resolving Power of Microscope and Telescope

In the consideration of resolution of spectral lines, we referred to two wavelengths λ_1 and λ_2, and discussed criterion of resolution of these two wavelengths. Such resolution of spectral lines is referred to as *chromatic resolutions*.

In the present discussion we would introduce another type of resolution concerned with the same wavelength but emitted from differently positioned point sources. In this case, therefore, the question is not what is the minimum wavelength difference such that the diffraction patterns due to two wavelengths could be recognised but here the question could be—what is the minimum separation between two point sources such that the diffraction patterns due to them could be identified as different. This consideration is usually referred to as *spatial resolution* and may be applied to find the resolving power in case of instruments like telescope, microscope, Michelson's stellar interferometer, etc. We will consider the spatial resolution of a telescope and microscope here but that of Michelson's stellar interferometer in the chapter on *interferometers*.

The objective of a telescope or a microscope behaves as a circular aperture and so even if the lens be corrected for all types of aberrations, we won't get a point image of a point object. It is because of the diffraction effect. When light from a bright point source is incident on the objective, we would get a central bright disc, called *Airy's disc* surrounded by alternate dark and bright circular rings of rapidly decreasing intensity.

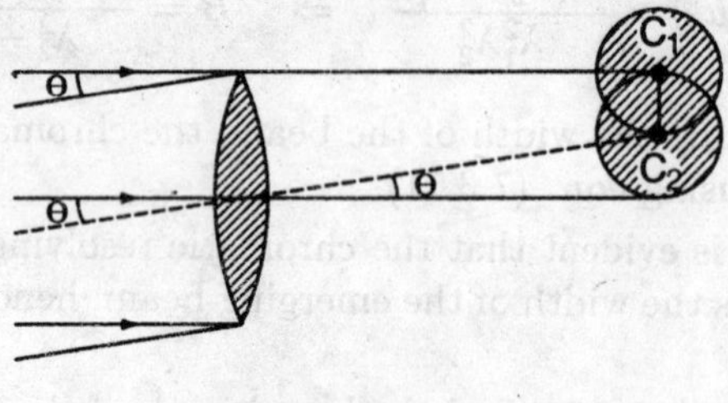

Fig 7.6

Two such bright point sources like a double-star would form diffraction patterns in the focal plane of the objective. Closeness of the sources S_1 and S_2 determines the extent of overlap. According to Rayleigh, the two sources are just resolved if the central maximum of one pattern just falls on the first minimum after the central maximum of the other. In figures (7.7) we have shown the diffraction intensity curves due to circular aperture corresponding to two neighbouring sources. The resultant intensity curve in fig. 7.7(b) shows a dip of about 20% in between the two central maxima due to a certain spacing of the two sources. In fig. (7.8a) and (7.8b) actual diffraction patterns due to the two sources are shown. The fig.(7.8a) shows the pattern due to two sources well separated, hence fully resolved and in fig.(7.8b) the closeness of the two sources is such that they are just resolved.

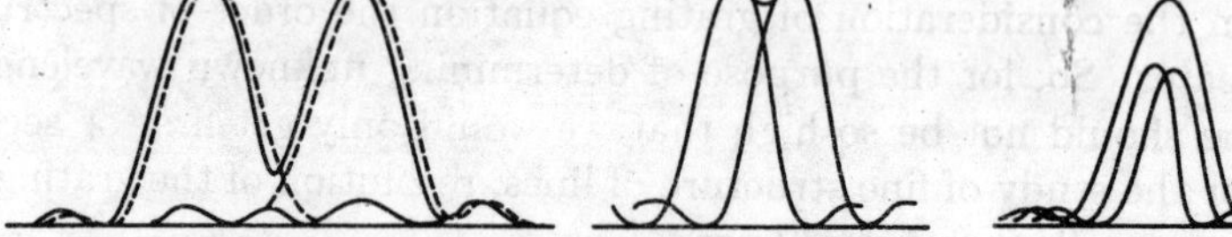

(a) (b) (c)

Fig 7.7

Smaller is the distance of separation between the two sources for which they can be resolved by an instrument, greater is said to be the positional resolving power of the instrument. But the resolvable closeness is ultimately determined by the Rayleigh's criterion.

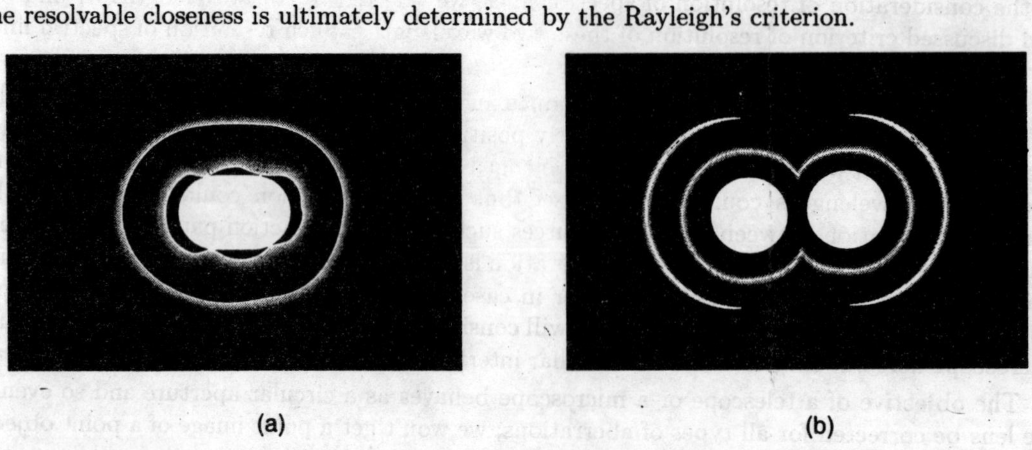

(a) (b)

Fig 7.8

Resolving Power of a Telescope

Telescope is an instrument by which we can see distant objects like stars and planets. When a telescope is directed to closely spaced stars (a double star), each star will produce its characteristic diffraction pattern, that of a circular aperture, consisting of Airy's disc surrounded by alternate dark and bright circular rings. Whether the two stars be resolved or not depends on the positional separation of the centres of the two Airy's discs.

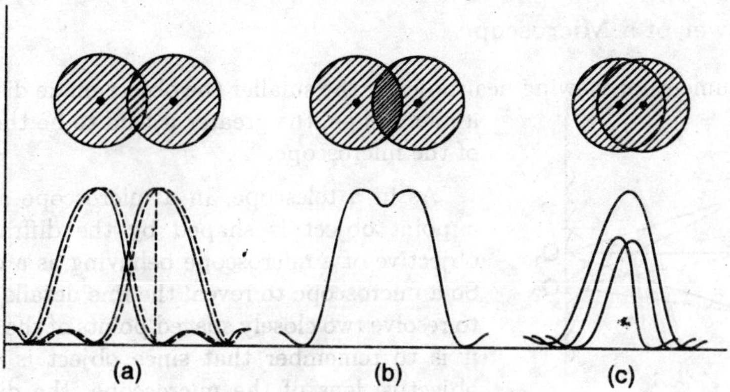

Fig 7.9

In figs.(7.9 a,b,c) we have shown the centres of the Airy's discs depending on the separation of the two sources. In fig. 7.9(a), the sources are so positioned that the centres of the two Airy's discs are separated more than the radius of the Airy's disc and the central maxima are distinguishably evident. In fig. 7.9(b), the separation is equal to the radius where there occurs at the mid-point between the two central maxima a dip of intensity by about 20%, hence it denotes the limiting separation set by Rayleigh's criterion. In the third case, fig. 7.9(c), the separation is less than the radius and at this separation, the two central maxima fuse into one peak and hence the instrument fails to produce the two patterns to be recognised as distinctly different. So we say that at this separation the two sources have gone beyond the resolution limit of the telescope.

From the theory of diffraction by a circular aperture, we know that the angular radius of the Airy's disc is given by

$$\theta = (1.22)\frac{\lambda}{D},$$

where D is the diameter of the aperture, here it is the diameter of the objective lens. In case of rectangular shape of the objective, the multiplying factor would be unity in place of 1.22. From figure (7.6), we get the separation in the limit of resolution between the centres C_1 and C_2 as

$$C_1C_2 = \frac{1.22\lambda f}{D},$$

where f is the focal length of the objective of the telescope.

Thus two point sources subtending an angle θ at the objective of a telescope given by

$$\theta = \frac{1.22\lambda}{D}$$

will be just resolved. If two sources are closer such that the angle subtended is less, then the two sources cannot be resolved. So the resolving power of a telescope will be written as

$$\text{R.P.} = \frac{1}{\theta} = \frac{D}{1.22\lambda}. \tag{7.4.1}$$

So, the resolving power will increase if D is greater and λ is smaller.

Resolving Power of a Microscope

This is an instrument for viewing near objects; the smaller the object made distinctly visible by a microscope, the greater is said to be the resolving power of the microscope.

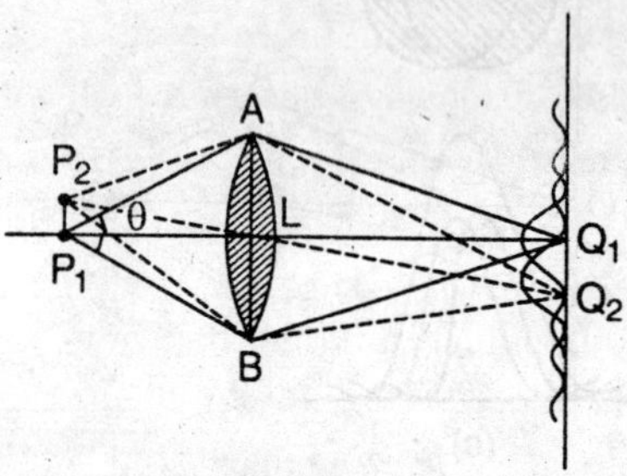

Fig 7.10

As in a telescope, in a microscope also the image of a point object is shaped by the diffraction effect, the objective of a microscope behaving as a circular aperture. So a microscope to reveal the fine details of the object has to resolve two closely spaced points of the object. However, it is to remember that since object is very near to the objective lens of the microscope, the diffraction pattern is not Fraunhoferian but it is Fresnelian. In case of a telescope, the two distant sources subtend very small angle at a point on the focal plane of the objective, but in microscope, the points of a small object placed very near to the objective lens, each subtends a large angle at the objective (see fig. 7.10). The point P_1 forms its characteristic Airy's pattern with centre at Q_1; the Airy's disc due to the other point P_2 has its centre at Q_2. If Q_2 happens to be the position of the first minimum for the source point P_2, the radius of the first dark ring is given by,

$$R = \frac{1.22\lambda}{D}v,$$

where v is the distance between the lens and the screen or image plane.

$$\therefore \quad Q_1Q_2 = R = \frac{1.22\lambda}{D}v \tag{7.4.2}$$

Again from the fig. (7.10), we get,

$$\frac{Q_1Q_2}{P_1P_2} = \frac{v}{u} \quad \Rightarrow \quad P_1P_2 = \frac{u}{v}\cdot Q_1Q_2 = \frac{1.22\lambda u}{D}$$

Again, $\frac{D/2}{u} = \sin\theta \quad \Rightarrow \quad \frac{u}{D} = \frac{1}{2\sin\theta}$

$$\therefore \quad P_1P_2 = \frac{1.22\lambda}{2\sin\theta} \tag{7.4.3}$$

Thus the minimum distance of separation that can be resolved by the microscope is given by,

$$P_1P_2 = \frac{1.22\lambda}{2\sin\theta}.$$

If μ be the r.i. of the objective space, then we can write, $\mu = \frac{c}{v} = (\lambda_0/\lambda)$, where λ_0 is the wavelength of light in vacuum, λ is the wavelength in the object space.

$$\lambda = \frac{\lambda_0}{\mu}, \quad \therefore \quad P_1P_2 = \frac{1.22\lambda_0}{2\mu \sin\theta}. \tag{7.4.4}$$

Eqn. (7.4.4) defines the limit of resolution by the microscope and the resolving power is defined as

$$\text{R.P.} = \frac{1}{\text{limit of resolution}} = \frac{2\mu\sin\theta}{1.22\lambda_0} = \frac{2}{1.22\lambda_0}(\mu\sin\theta)$$

$\mu \sin\theta$ is termed as *numerical aperture* (N.A.) of the objective. Hence,

$$\text{R.P. of a microscope} = \frac{2}{1.22\lambda_0} \times (\text{N.A.}) \tag{7.4.5}$$

So to increase the resolving power

(i) λ_0 should be low. In this regard an electron microscope has greater resolving power than an optical microscope. The matter-wave associated with electrons have wavelength given by,

$$\lambda = \frac{h}{mv}.$$

If the electron beam be accelerated by a p.d. V, then

$$\frac{1}{2}mv^2 = Ve \quad \Rightarrow \quad v^2 = \frac{2Ve}{m} \quad \Rightarrow \quad v = \sqrt{\frac{2Ve}{m}}$$

$$\therefore \quad \lambda = \frac{h}{\sqrt{2mVe}} = \frac{h}{\sqrt{2me}\sqrt{V}} = \frac{12}{\sqrt{V}}\text{Å} \quad \text{(app.)}$$

In most of the electron microscope, $V = 60$ kV.

$\therefore$ $\lambda = 0.05$Å hence in electron microscope the resolving power is greater.

In the same way a proton microscope has greater resolving power than that of electron microscope.

(ii) Resolving power also depends on the value of the numerical aperture of the microscope. Now, N.A. $= \mu\sin\theta$, so if the r.i. of the object space is increased as in the case of oil immersion objective, the resolving power of the microscope is also going to increase.

Example. If $\mu = 1.6, \theta = 60°$ and $\lambda_0 = 6000$Å, we get,

$$\text{R.P.} = \frac{2}{1.22\lambda_0}(\mu\sin\theta) = \frac{2}{1.22 \times 6000 \times 10^{-8}} \times \left(1.6 \times \frac{\sqrt{3}}{2}\right) = 37859.$$

Example. In an electron microscope, $V = 60{,}000$ volts, $\theta = 30°, \mu = 1$, then the resolving power is given by,

$$\text{R.P.} = \frac{2\sin\theta}{1.22\lambda_0} = \frac{2 \times \sin 30°}{1.22 \times \frac{12}{\sqrt{V}} \times 10^{-8}}$$

$$= \frac{2 \times \frac{1}{2}}{1.22 \times \frac{12}{\sqrt{60000}} \times 10^{-8}} = 16.7 \times 10^8$$

The minimum distance of separation that can be resolved is given by,

$$P_1P_2 = \frac{1}{\text{R.P.}} = .059 \times 10^{-8}\text{ cm} = .0059\text{ mm}$$

The resolving powers of other optical devices like Fabry-Perot, Michelson's interferometer, Lummer-Gehrcke plate, etc., are discussed in the chapter on *interferometers.*

Exercises

1. What should be the length of a side of a 60° glass prism in order to resolve the sodium D lines at minimum deviation?
   (μ values at wavelengths 5461, 5893 and 6438Å are respectively 1.6546, 1.6499 and 1.6434)
   [**Hints.** $\frac{\lambda}{\delta\lambda} = l\frac{\delta\mu}{\delta\lambda}$, find $\frac{\delta\mu}{\delta\lambda}$ from the given data.]
2. An equilateral prism is to be made of a material whose μ is given by $\mu = 1.47 + 1.8 \times 10^{-10}/\lambda^2$, so that it will have the resolving power as that of a plane diffraction grating of width 15 cm and having 1000 lines per cm in the first order. Find the length of the base of the prism $\lambda = 6000$Å. [$l = 9.0$ cm]
3. The μ value of a certain glass is given by $\mu_c = 1.6545, \mu_E = 1.6635$ corresponding to wavelengths 6563 and 5270Å respectively, calculate the length of the base of a 60° prism made of this glass which is just capable of resolving the D-lines of wavelengths 5890 and 5896Å. [1.41 cm]
4. In the second order spectrum of a plane diffraction grating, a certain spectral line appears at an angle of 10°, while another line of wavelength 5×10^{-9} cm greater appears at an angle of 3″ greater. Find the wavelengths of the lines and the minimum grating width required to resolve them, sin 10° = 0.1736 and cos 10° = 0.9848. [4.2 cm]
5. What do you mean by resolving power of an instrument? What is Rayleigh's criterion of resolution? Explain the meaning of chromatic resolving power and positional resolving power.
6. Derive an expression for the resolving power of a prism spectrometer. Can it have the same resolving power as that of a grating?
7. In case of a microscope, which term will be needed—chromatic resolving power or spatial resolving power? Derive an expression for the same. How can you increase the resolving power of a microscope?
8. An electron microscope has greater resolving power than an optical microscope. Why?
9. In Rayleigh's criterion of resolution 20% variation of intensity at the mid-point w.r.t. the central maxima is referred to. Explain, why.
10. Light is incident normally on a grating 0.5 cm wide with 2500 lines. Find the angular separation of the two Na lines in the first order spectrum. Can they be seen distinctly? Wavelengths are 5890 and 5896Å. [$\theta = 17°8'$, yes]
11. Why is the resolving power of a proton microscope greater than that of an electron microscope?
12. Which phenomenon sets the limit of resolution by a telescope or a microscope? Derive an expression for the resolving power of a telescope.

Chapter 8

Interferometry

8.1 Introduction

Instruments based on the phenomenon of interference are called *interferometers.* Some of these instruments utilise the interference of the primary beams from two coherent sources, some of them works on the interference of the diffracted beams. To the first class belong the Michelson's interferometer, the Fabry-Perot interferometer, Jamin's interferometer and Lummer-Gehrcke plate interferometer. To the second class belongs the Michelson stellar interferometer. However, the purpose of any interferometer is to make spectroscopic study, analysis of hyperfine structure of spectral lines; to study the variation of refractive index of gases or salt solutions with density and temperature or to determine the linear diameter of stars. These instruments are usually of high resolving power and this makes the spectral analysis more convincing.

8.2 Michelson's Interferometer

This is a very important piece of apparatus designed in 1880 by Albert Michelson with a specific purpose of testing the validity of the either hypothesis and now used as a powerful tool for study of atomic spectra for which he was awarded the Nobel prize in 1907.

Leaving apart the details of the instrument, the basic construction necessary for developing the theory is represented as in fig. (8.1). The formation of fringe pattern in Michelson's interferometer is essentially the fringe pattern obtained in a parallel-sided plate or wedge-shaped film, of course, with a slight difference.

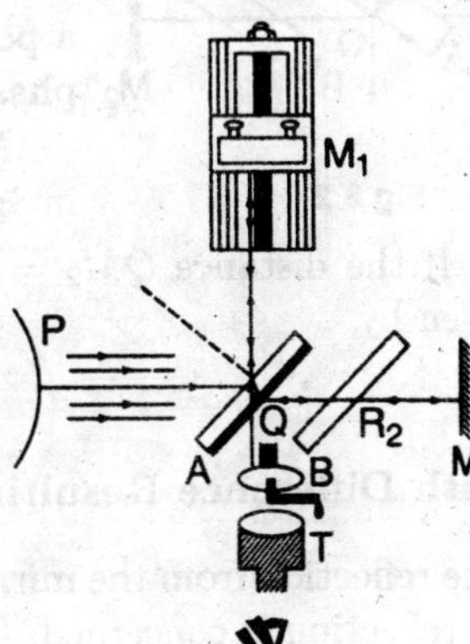

Fig 8.1

Consider the figure, where we have shown rays essential for the formation of interference fringes. The glass-plate A partially silvered is held inclined at an angle of 45° w.r.t. the mirrors M_1 and M_2 whose reflecting faces are at right angles fig. 8.1. The glass-plate A is silvered such that a ray PQ, one of a parallel beam of rays, incident at the point Q half reflects and half transmits. So it acts as an ideal beam-splitter. The reflected portion moves vertically up to be incident on the mirror M_1 normally, it reflects back also normally. The portion transmitted through the glass-plate A passes through a compensating plate B and finally falls normally on M_2 and turns back. This plate B is similar to A in all respect except that it is unsilvered. The ray reflected from M_1 and the ray reflected from M_2 superpose as they reach the point Q of the glass-plate A. These two rays, one from M_2, the other from M_1, are derived from the same ray, hence they are coherent disturbances capable of producing interference fringes. So in the field of view of a telescope T, we see circular fringes as obtained in case of parallel air film.

Functions of Different Components of the Instrument

Source P

It is an extended source such that a large number of rays incident on the glass-plate A at a particular angle of incidence is available. In the diagram a parallel beam of axial rays is shown, which make an angle of 45° with the normal to the glass-plate A.

Glass-plate A

It is a partially silvered glass-plate at which the incident light is split into two portions of almost equal intensity (amplitude division) one part is the reflected part, the other is the transmitted part. So this plate may be looked upon as an ideal beam-splitter.

Glass-plate B

It is an unsilvered glass-plate similar to the plate A in shape and orientation and acts as a compensator. At the point Q the incident ray is divided into two parts, the part going to the mirror M_1 and coming back again to Q after reflection traverses the glass-plate twice; to compensate this path in glass, the compensator plate B is used.

Formation of Interference Fringes

At the point Q the incident ray is amplitude divided into (i) a transmitted ray to the mirror M_2 and (ii) reflected to the mirror M_1. The difference in path traversed in glass medium by the component travelling towards M_1 is compensated by the compensator plate B. The two components uniting at Q after reflection from M_1 and M_2 are derived from the same ray, hence are in a position to interfere constructively or destructively depending on the phase difference.

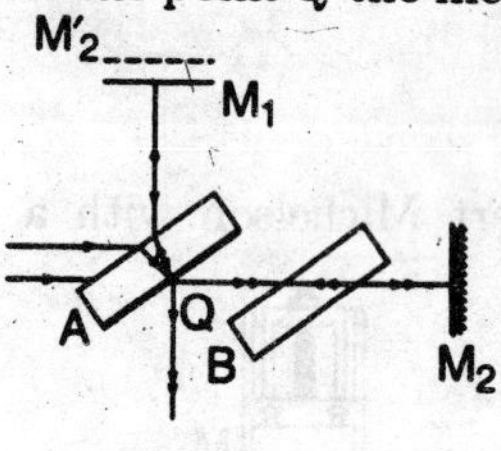

Fig 8.2

Now, the mirror M_2 can be taken as being replaced by the mirror image M_2' of M_2 formed due to reflection in the silvered face of A.

If the distance $QM_2 = x_2$, $QM_1 = x_1$, the path difference between the interfering rays is given by,

$$\Delta_1 = 2(x_2 \sim x_1). \tag{8.2.1}$$

Path Difference Resulting from Phase Change due to Reflection

The reflection from the mirrors M_1 and M_2 affects the two rays equally so far as phase variation on reflection is concerned. There is an additional phase difference of π introduced into the ray component reflected from M_2 on its reflection at Q in its return path. The other component going to the mirror M_1 undergoes a phase change of π on its reflection at Q in its forward journey, if silvering is not thin.

$\therefore$ Thus the total path difference[1]

$$\Delta = 2(x_2 \sim x_1) \tag{8.2.2}$$

[1] If the silvering is thin, the total path difference $= 2(x_2 \sim x_1) \pm \frac{\lambda}{2}$

So the formation of fringe system in this case may be looked upon as in a parallel air-film enclosed between M_1 and the mirror-image M_2' of M_2 formed due to reflection in the silvered plate A. In case the glass-plate is heavily silvered, the total path difference would be

$$\Delta = 2(x_2 \sim x_1) \tag{8.2.3}$$

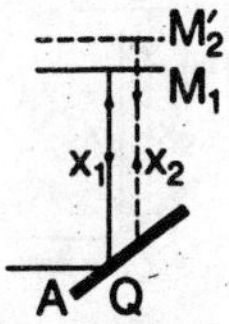

Fig 8.3

so according to the condition of interference, if,

$$2(x_2 \sim x_1) = 2m \cdot \frac{\lambda}{2} \quad \text{BRIGHTNESS}$$

$$\Rightarrow 2d = m\lambda \tag{8.2.4}$$

$$\text{and} \quad 2(x_2 \sim x_1) = (2m+1)\frac{\lambda}{2} \quad \text{DARKNESS} \tag{8.2.5}$$

So, if we consider axial rays from a point of the source for the formation of interference fringes, we get a bright point or a dark point according to the condition satisfied for brightness or darkness.

Formation of Circular Fringes

Fringes of different nature may be obtained in a Michelson's apparatus but most commonly used form of the fringes is circular. The formation of these circular fringes is explained with reference to the fig. 8.4. The rays from any point S_1 of the source after refraction through the lens are rendered parallel and these rays interfere constructively or destructively to give a bright or dark point. Thus all the point sources like S_1, S_2, S_3 which happen to lie on the circumference of a circle will form a series of bright points lying on a circle. The source used necessarily being broad will give rise to an interference pattern consisting bright and dark rings fig. 8.5.

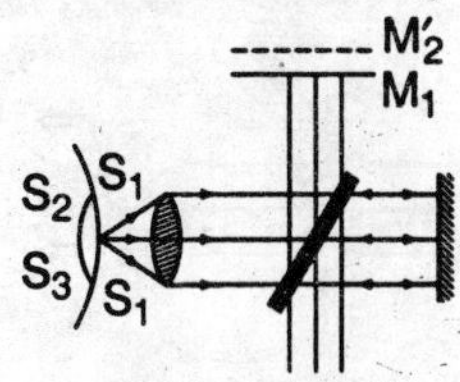

Fig 8.4

Analytical Formulation

The light wave divided by the beam-splitter into two components, one going to M_2 and reflected back by the mirror M_2 at a distance x_2 and the displacement of a particle at Q due to this component can be written as,

$$y_1 = A_1 e^{i\{wt - k(2x_2 + \lambda/2)\}}. \tag{8.2.6}$$

The displacement due to the other component is;

$$y_2 = A_2 e^{i\{wt - k(2x_1 + \lambda/2)\}}. \tag{8.2.7}$$

$$\therefore \quad y = y_1 + y_2 = A_1 e^{i\{wt - k(2x_2 + \lambda/2)\}} + A_2 e^{i\{wt - k(2x_1 + \lambda/2)\}}$$

If we take $A_1 = A_2 = A$, then

$$y = A e^{iwt}\left\{e^{-ik(2x_2+\lambda/2)} + e^{-ik(2x_1+\lambda/2)}\right\} \tag{8.2.8}$$

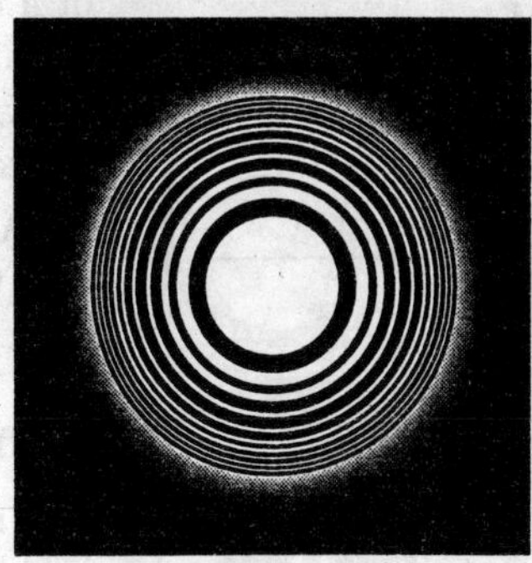
Fig 8.5

Let,
$$k\left(2x_2 + \frac{\lambda}{2}\right) = \alpha_1, k\left(2x_1 + \frac{\lambda}{2}\right) = (\alpha_1 - \phi)$$

$$\Rightarrow \phi = k\left(2x_2 + \frac{\lambda}{2}\right) - k\left(2x_1 + \frac{\lambda}{2}\right) \tag{8.2.9}$$

$$\therefore\ y = Ae^{iwt}\left[e^{-i\alpha_1} + e^{-i(\alpha_1-\phi)}\right] = Ae^{iwt}\cdot e^{-i\alpha_1}\left[1 + e^{+i\phi}\right]$$
$$= Ae^{i(wt-\alpha_1)}\left[e^{-i\phi/2} + e^{+i\phi/2}\right]e^{+i\phi/2} = 2Ae^{i(wt-\alpha_1)}\cos(\phi/2)e^{+i\phi/2}$$
$$= 2A\cos(\phi/2)e^{i(wt-\alpha_1+\phi/2)}.$$

$$|y| = \sqrt{yy^*} = 2A\cos\frac{\phi}{2}. \tag{8.2.10}$$

$$\therefore\quad I = 4A^2\cos^2\frac{\phi}{2}. \tag{8.2.11}$$

So I will be max, when $\cos\frac{\phi}{2} = \pm 1$.

$$\Rightarrow\quad \frac{\phi}{2} = m\pi \quad\Rightarrow\quad \frac{1}{2}\left[k\left(2x_2 + \frac{\lambda}{2}\right) - k\left(2x_1 + \frac{\lambda}{2}\right)\right] = m\pi$$

$$\Rightarrow\quad \frac{2\pi}{\lambda}(2x_2 - 2x_1) = 2m\pi \quad\Rightarrow\quad 2(x_2 - x_1) = m\lambda \tag{8.2.12}$$

For minimum, $\cos\frac{\phi}{2} = 0$

$$\Rightarrow\quad \frac{\phi}{2} = (2m+1)\frac{\pi}{2} \quad\Rightarrow\quad \frac{2\pi}{\lambda}(2x_2 - 2x_1) = (2m+1)\pi$$

$$\Rightarrow\quad 2(x_2 - x_1) = (2m+1)\frac{\lambda}{2}. \tag{8.2.13}$$

Formation of Fringes due to Rays at an Angle θ w.r.t. the Axial Rays

In this case the path difference between the two components would be given by,

$$\Delta_1 = 2(x_2 \sim x_1)\cos\theta = 2d\cos\theta. \tag{8.2.14}$$

The rays which make the same angle θ w.r.t. the axial ray will be lying on the surface of a cone having its apex at the source point. So depending on the condition whether constructive or destructive interference is satisfied, we will get bright or dark points lying on the circumference of a circle, hence we get alternate bright or dark ring systems.

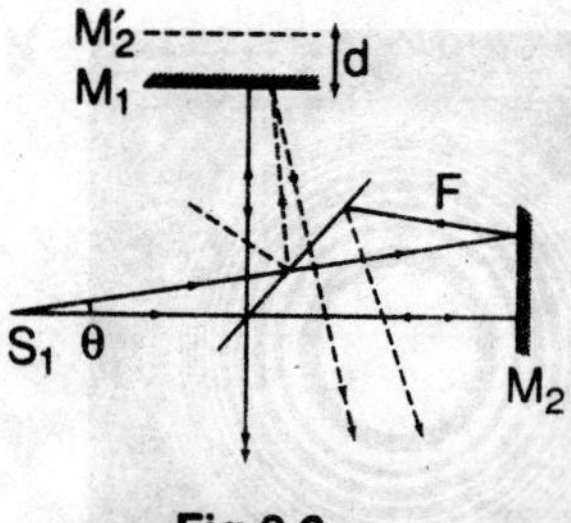

Fig 8.6

As we have remarked that the beam splitter splits the incident ray into two components—one getting reflected from M_1 and the other from M_2 which may be taken to be replaced by its image M_2' behind M_1. These two components interfere to produce fringes provided the condition of coherency continues to be satisfied.

The coherence referred to in this case is the temporal coherence or time coherence. The concept of temporal coherence has been discussed at length in the chapter on interference but here again we refer it back.

In case of atomic emitters, the wavetrain consists of pulses, hence it is not a continuous wavetrain but there are sudden breaks. This is because of the fact that the transit time for an excited atom to go to the ground state is of the order of 10^{10} sec, called *coherence time* τ_c, the distance travelled by light in this time is the *coherence length* l_c. So, if two observation points differ in path less than the coherence length, then the disturbances at these two points must be coming from the same atomic emitter. We can write,

$$l_c = c\tau_c. \tag{8.2.15}$$

Now, in Michelson's interferometer, interference occurs between two components of light—one light that goes to the mirror M_1 and reflects back to Q; the other that goes to M_2, reflects and gets back to the point Q or equivalently the other that gets reflected from M_2' which is the image of M_2 in the silvered plate A. If the distance between M_1 and M_2' be written as d and if

$$2d \ll l_c \quad \text{or,} \quad 2d/c \ll \tau_c$$

the two interfering beams continue to be originated from the same atomic emitter, hence they continue to remain coherent. Under this condition fringes are available. But, on the other hand, keeping one of the mirrors, say M_1, fixed, if M_2 is varied such that $\frac{2d}{c} \ll \tau_c$, there is no fixed phase relationship between the two interfering beams and no interference pattern is observed. Thus from the condition $\frac{2d}{c} \ll \tau_c$ giving well-defined fringes, as we go on increasing the value of $\frac{2d}{c}$ by shifting the position of M_2, the fringes go on losing distinctiveness and ultimately for $\frac{2d}{c} \gg \tau_c$, the fringe pattern completely disappears.

From the appearance of well-defined fringes to no appearance of fringes at all, there is gradual transition, the two interfering beams gradually lose in degree of coherence, ultimately they completely lose the correlation in phase and degree of coherence is equal to zero. In case of well-defined fringes the degree of coherence is unity and in case of non-visibility of fringes, it is zero. So we have introduced a term, called *visibility of the fringes* and it is related to the degree of coherence.

Visibility of Fringes

Michelson has introduced this term and it is intimately related to the degree of coherence. Since the appearance of the fringes or no appearance is not sharp, but it is gradual, we have introduced the term *visibility of the fringes*.

The visibility of the fringes is defined as

$$V = \frac{I_{max} - I_{min}}{I_{max} + I_{min}} \tag{8.2.16}$$

I_{max} represents maximum intensity of a fringe and I_{min} is the minimum intensity of its neighbour.

To clarify the meaning of eqn. (8.2.16) we consider the formation of well-defined circular fringes with Michelson's interferometer. If the light used is strictly monochromatic, the fringe pattern would consist of circular fringes alternately bright and dark. If one of the mirrors, say M_1, is displaced by $\frac{\lambda}{2}$ towards A or away from it, the position of one bright fringe will be occupied by its next one and the field remains unaltered.

In this case $I_{min} = 0$ and therefore, $V = 1$, meaning degree of monochromacity is unity. But, if the source consists of two wavelengths closely spaced, say λ_1 and λ_2 (as in case of Na-lamp), then there would be fringe patterns due to both and for certain value of separation between M_1 and M_2', the intensity minimum of wavelength λ_2 coincides with the intensity maximum of wavelength λ_1, so $I_{min} \neq 0$ in this case and $V \neq 1$. If the wave-amplitude of the two waves be same, then at every point $I_{max} = I_{min}$ and $V = 0$.

The fact that $V \neq 1$ in all cases shows that no source is strictly monochromatic. Thus V may be taken as a measure of degree of monochromacity of the source.

Now, the wavelengths λ_1 and λ_2 may be such and the displacement of one mirror M_1 be such that the bright rings due to one coincide with the bright rings due to the other; the fringes, then, are said to be in consonance, the field is distinct. The field becomes indistinct and the fringes are said to be in dissonance if the bright rings due to one coincide with dark rings of the other.

White Light Fringes in Michelson's Interferometer

In Michelson's interferometer a monochromatic source is usually used for experiment. Circular fringes cannot be seen with white light. The reason is not very difficult to see: The width of the fringes for different wavelengths would be different and this difference becomes more prominent as the order number of the fringes increases. It might so happen, therefore, that a some place the maximum of one falls on the minimum due to the other. If instead of two wavelengths a number of wavelengths is present, then at all points maximum intensity due to some wavelengths would be present which will give rise to an average intensity and average colour and this colour will be white. Only for the central fringe all the colours will be of equal intensity and hence it will be achromatic. As one proceeds outward from the centre, differences will more and more creep in and the fringes will become chromatic.

In order to realise white light fringes the path difference should not exceed a few wavelengths. The mirrors should first be adjusted to get straight fringes with monochromatic light. Then a slight adjustment of the mirror M_1 will bring fringes having a central dark fringe bordered on either side by 8 or 10 coloured fringes.

Adjustments for Michelson Interferometer

The two mirrors M_1 and M_2 are adjusted to be at equal distances w.r.t. Q. A pin is placed between the lens L and the glass-plate A, we would get two pairs of distinct images—one pair due to reflection at the silvered surface, the other pair due to reflection at the unsilvered surface of A. By adjusting the inclination of M_2, the first pair of images are made to coincide on placing a Na-flame at the focal plane of the lens and removing the pin, interference bands are seen. By adjustment of the inclination, the fringes are made well-defined. Also any form of the fringes will be available by adjustment of the mirrors.

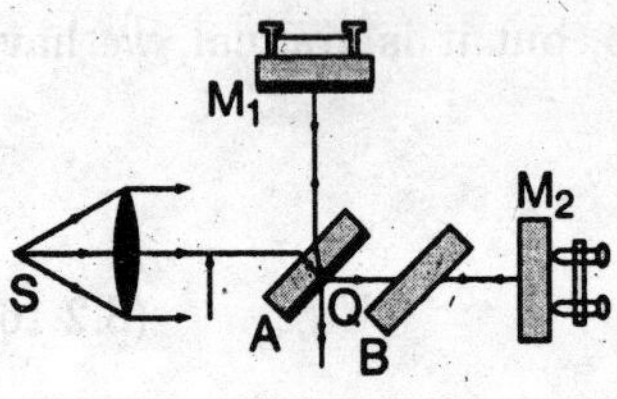

Fig 8.7

Resolving Power of Michelson's Interferometer

According to eqn. (8.2.11), the intensity distribution function is written as

$$I = 4A^2 \cos^2 \frac{\phi}{2}, \tag{8.2.17}$$

where, $\phi = 2k(x_2 - x_1)$.

Now, intensity maximum or bright fringes are available when

$$\cos \frac{\phi}{2} = \pm 1 \Rightarrow \frac{\phi}{2} = m\pi$$

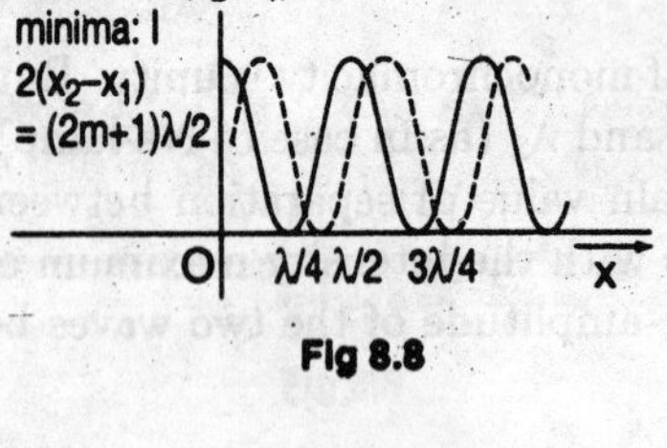

Fig 8.8

$$\Rightarrow \frac{1}{2}k\{2(x_2 - x_1)\} = m\pi \quad \Rightarrow, \quad \frac{1}{2} \times \frac{2\pi}{\lambda}\{2(x_2 - x_1)\} = m\pi$$

$$\Rightarrow 2(x_2 - x_1) = m\lambda \quad \Rightarrow \quad 2d = m\lambda \tag{8.2.18}$$

If we get $x = x_2 - x_1$, i.e., the difference in the distance of the two mirrors and take it as a parameter and plot the IDF vs. x curve, we get the maxima and minima as shown in fig. 8.8, the variation in x is brought about by varying the position of the movable mirror M_2; but, of course, satisfying the condition of temporal coherence $\frac{2d}{c} \ll \tau_c$.

Let us consider two wavelengths λ_1 and λ_2, the fringe patterns due to them being represented by the continuous and broken curves respectively. The maximum of the mth order in wavelength λ is given by

$$2x(\text{for } \lambda_1) = m\lambda_1 = m\lambda, \quad \lambda_1 = \lambda \tag{8.2.19}$$

and for the other wavelength

$$2x \text{ (for } \lambda_2) = m\lambda_2 = m(\lambda + d\lambda), \quad \lambda_2 = \lambda + d\lambda. \tag{8.2.20}$$

For the resolution of two wavelengths λ_1 and λ_2, maximum due to one falls on the minimum due to the other (no fringes observed), then we would have,

$$m(\lambda_2 - \lambda_1) = md\lambda = \left(m + \frac{1}{2}\right)\lambda - m\lambda = \frac{\lambda}{2} \quad \Rightarrow \quad \frac{\lambda}{d\lambda} = 2m. \tag{8.2.21}$$

This eqn. (8.2.21) defines the resolving power of Michelson's interferometer. It is to note that the resolving power is not dependent on the characteristic parameter of the device unlike as in grating, where the resolving power depends on the order m as well as on the total number of lines.

A few facts of Practical Importance in Michelson's Interferometer

From the interferometer equation given as

$$\begin{aligned} 2d &= m\lambda \quad \text{for bright circular fringe} \\ \text{or,} \quad 2d\cos\theta &= m\lambda \quad \text{for bright circular fringe} \end{aligned}$$

as the case may be, we can draw a few important conclusions.

(i) The centre of the Michelson's interferometer pattern may be dark or bright depending on whether or not a phase difference π is introduced due to reflection between the two interfering beams. In case of heavy silvering, the extra-phase difference due to reflections is balanced out, hence in this case the centre of the pattern corresponding to $d = 0$ is bright. But in case of light silvering, there is a phase-difference of π introduced because of reflection and in this case, for $d = 0$, the centre of the pattern would be dark.

(ii) When the two mirrors M_1 and M_2 are equidistant w.r.t. the point Q, they are said to be in symmetric position and $d = 0$. As they are moved out of the symmetric position, d value increases but for a particular order m and fixed wavelength λ, since $m\lambda$ is to remain constant, it means rings expand and new rings appear at the centre. When one mirror is moved through $\frac{\lambda}{2}$, a new fringe appears at the centre fig. 8.9.

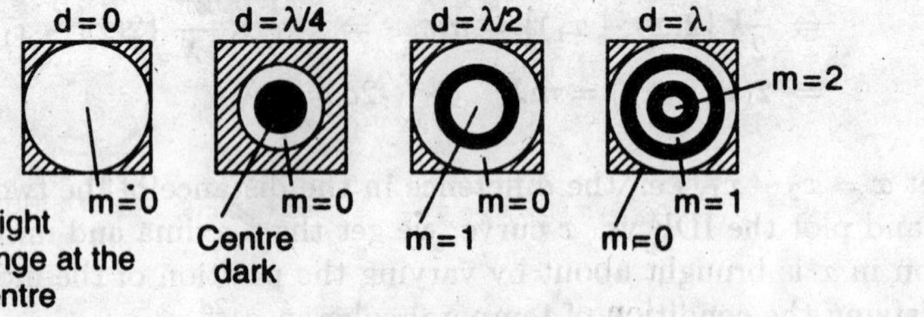

Fig 8.9

So as d increases, the fringes move out and disappear and how many fringes move out across the cross-wire due to variation of d may be counted. For each fringe crossing, $d = \frac{\lambda}{2}$; so, if n fringes move across the cross-wire, then

$$d_s = n\frac{\lambda}{2}.$$

(iii) If the mirrors are tilted and are not perpendicular to each other, the fringes will appear to be straight.

Applications of the Interferometer

Michelson's interferometer, though originally designed for detecting and hence verifying either hypothesis, is a versatile piece of apparatus which can be applied for a number of purposes. Of the different purposes to which it can be put, it does not claim to be best suited, for example, in the study of fine structure of spectral line. Fabry-Perot's interferometer is more suitable than Michelson's apparatus for the same purpose. Ignoring this fact, the specific purposes to which Michelson's interferometer may be put are—

(i) Determination of unknown wavelength

(ii) Measurement of distance or length—standardisation of Meter

(iii) Study of fine structure of spectral line

(iv) Measurement of refractive index.

We will now discuss the different applications of the instrument.

Determination of Wavelength

The circular fringes with the given light of wavelength λ is formed in Michelson's interferometer with a certain value of $d = (x_2 - x_1)$. If the movable mirror M_2 is moved through a distance $\frac{\lambda}{2}$, then the path difference between the interfering beams becomes

$$\Delta = (x_2 - x_1) + \frac{\lambda}{2}.$$

In the first pattern, the mth order maximum occurs at a point defined by

$$2(x_2 - x_1)i = m\lambda.$$

When $(x_2 - x_1)$ is increased by $\frac{\lambda}{2}$, we get, $2(x_2 - x_1)$ is increased by λ.

$$\therefore \quad 2(x_2 - x_1)f = m\lambda + \lambda = (m+1)\lambda,$$

so the position of one bright fringe is occupied by the next bright fringe. Let the mirror be moved through a distance x, the number of fringes crossing the cross-wire of the receiving telescope is n, suppose; then

$$x = n\frac{\lambda}{2}. \tag{8.2.22}$$

From eqn. (8.2.22), noting down the number n and measuring the distance x, which can be done with the help of linear scale S' and vernier scale V', we can determine the value of λ.

Standardisation of Meter in terms of Wavelength

Fundamental unit of length, *meter* is standardised with reference to wavelength of spectral line since it is invariable and easily reproducible at any time. It is very difficult to count the number of fringes passing across the cross-wire as the movable mirror M_2 is shifted from one end of the meter to the other. Actually it will be a fantastically large number. To get an idea about the largencess, consider Na-light of wavelength $\lambda = 5893 \times 10^{-8}$ cm. If x is such that there is shift by one fringe, then $n = 1$, and

$$x = \frac{\lambda}{2} = \frac{1}{2} \times 5893 \times 10^{-8} = 2.947 \times 10^{-5} \text{ cm} \tag{8.2.23}$$

With this instrument measurement of the shift by $\frac{1}{10}$th of a fringe can be made, hence, measurement of length up to $\frac{1}{10} \times 2.947 \times 10^{-5}$ cm $\simeq 2 \times 10^{-6}$ cm can be possible. So the number of fringes crossing the cross-wire when the mirror M_1 (say) moves through 1 meter is given by

$$N = \frac{2x}{\lambda} = \frac{2 \times 100}{5893 \times 10^{-8}} = 3 \times 10^6. \tag{8.2.24}$$

As this is a huge number, it is not convenient to count the number. So Michelson and Benoit in 1892 introduced intermediate standards, ten in number, called *etalon* with two basic difficulties:

(i) difficulty with counting number of fringes of the order of 10^6 across the cross-wire,

(ii) when x is 1 meter, the width of the fringes will become so low that they cannot be seen at all. Michelson and Benoit followed a procedure for the realisation of standard meter and the procedure consists of the following operations:

(1) Construction of the etalon
(2) Determination of the number of waves in the smallest etalon
(3) Comparison of the etalons
(4) Comparison of the longest etalon with the standard meter.

Construction of Etalons

An etalon consists of two plane mirrors A_1 and B_1 separated by a distance apart as shown in fig. 8.10.

Of the ten etalons, the longest one is roughly 10 cm long and the shortest one roughly .02 cm. These ten etalons are so constructed that the second one is roughly half of the first, the third one is roughly half of the second and so on and ultimately the tenth etalon becomes 0.02 cm approximately.

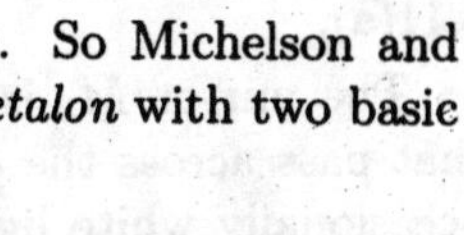

Fig 8.10

Determination of the Number of Waves in the Smallest Etalon

After the construction of the etalons, the shortest etalon was first taken and the number of waves contained in it was determined with help of the arrangement shown in fig. 8.11, a modified version of Michelson's interferometer.

In this modified arrangement, A_1 and B_1 are the mirrors in the smallest etalon and A_1', B_1' in the second. The field of view is made much wider in this case such that the fringe pattern in the four mirrors (A_1, B_1) and (A_1', B_1') in the shortest etalon and the next could be seen.

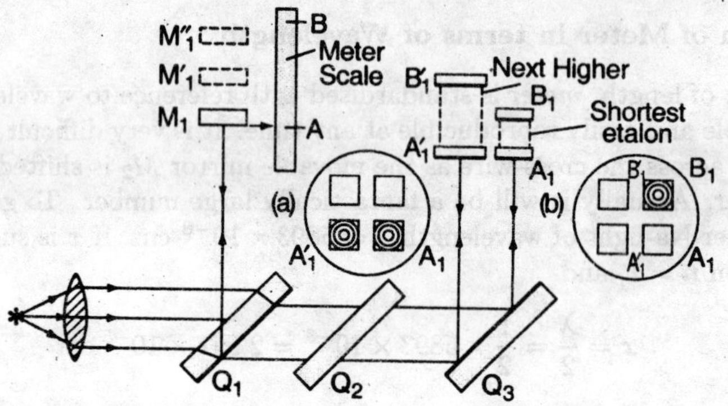

Fig 8.11

The distances of M_1 and the front mirrors A_1 and A_1' of the first two etalons are at first made nearly equal by measurements and then by white light fringes they are made equal. In the field of view, all the four mirrors are visible fig. (8.11). The white light source is replaced by the red line of cadmium. Circular fringes would now be seen both in mirrors A_1 and A_1' fig. 8.11(a).

The mirror M_1 is now moved from M_1 towards M_1' and the number of cadmium fringes that pass across the cross-wire noted carefully. While this movement is allowed to continue, occasionally white light is employed to see whether white light fringes appear in the field or not. When white light fringes appear once again at position M_1', it ensures equality of distances between M_1' and B_1 w.r.t. the point Q_1. This time circular fringes would appear in the mirror B_1. The integral number of fringes as well as any fraction crossing across the cross-wire is carefully noted during this movement of the mirror M_1 to the position M_1'. The etalon A_1B_1 is moved through its own length such that A_1 now occupies the position previously occupied by B_1 when fringe appears again in mirror A_1. During this movement of the etalon, the fringes are not counted. The mirror M_1 from the position M_1' is shifted to M_1'' when fringes would appear in the new position of B_1 again so that in this position the distances of M_1'' and B_1 are equal. During this shift also there is no need of counting the fringes.

The mirror is slightly shifted from the position M_1'' until fringes appear in the mirror B_1'. The number of fringes in this movement crossing the cross-wire is carefully noted.

If N_1 be the integral number of fringes crossing the cross-wire and f_1 be the fraction in the movement of M_1 from the position M_1 to the position M_1', the shortest etalon is equal to $(N_1 + f_1)$ fringes. If for the movement of the mirror M_1 from M_1'' until fringes are found in the mirror B_1' of the second etalon, then the second etalon is equivalent to a number of fringes given by

$$N = 2(N_1 + f_1) + (N_2 + f_2) \tag{8.2.25}$$

in this manner the number of fringes in the other longer etalon can be found.

Take the longest etalon (roughly 10 cm) and in ten steps the whole length of the standard meter can be expressed in terms of the fringes.

The mirror M_1 is made to coincide with the mark A of the meter and its distance is made equal to the distance of the front mirror of the longest etalon. This time fringes would appear in the mirror A_1' of the longest etalon. The mirror M_1 is then shifted to the position when fringes would appear in the mirror B_1' of the longest etalon. In this way through a process of ten steps, the other end of the meter is reached. In all these steps, there is no need of counting the fringes, because the number of fringes contained in the longest etalon has already been found. After the tenth step, the additional number of fringes crossing the cross-wire when the mirror M_1 is ultimately made to coincide with the other end-mark of the meter is carefully noted.

The red line of cadmium has been accepted in spectroscopy as primary standard. The average value of the wavelength has been evaluated from four experimental results—

Michelson and *Benoit* (1895)

$$\lambda_R = 6438.4691\text{Å}$$

Benoit, Fabry and *Perot* (1906)

$$\lambda_R = 6438.4703\text{Å}$$

Watenabe and *Imaizumi* (1928)

$$\lambda_R = 6438.4682\text{Å}$$

Sears and *Bayrel* (1934)

$$\lambda_R = 6438.4708\text{Å}$$

Average of these four results is given as

$$\lambda_R = 6438.4696\text{Å}$$

From the measurement of the fringe shift of cadmium red line, the meter standard is written as

$$1 \text{ meter} \equiv 1553, 163.5\lambda_R.$$

Study of Fine Structure of Spectral Line

In this analysis Fabry-Perot interferometer is more suitable but with the help of the Michelson's interferometer also two wavelengths having a small difference can be determined.

If the source has two wavelengths λ_1 and $\lambda_2 (\lambda_1 > \lambda_2)$,then each will produce its characteristic interference fringes. Now, if the mirror M_1 be shifted, the fringes will be alternately visible and invisible, referred to as the condition of consonance and dissonance respectively. When in consonance, the bright rings of one wave coincide with the bright rings of the other. When dissonance occurs, the bright rings of one coincide with the dark rings of the other.

If the mirror M_1 be shifted through a distance d so that fringes pass from one consonance to the next, then

$$d = \frac{n\lambda_1}{2} = (n+1)\frac{\lambda_2}{2}, \tag{8.2.26}$$

where n and $(n+1)$ represent the number of fringe shifts for lights of wavelengths λ_1 and λ_2.

$$\therefore \quad \frac{2d}{\lambda_2} - \frac{2d}{\lambda_1} = 1 \tag{8.2.27}$$

So if λ_1 is known, the wavelength λ_2 very close to λ_1 can be determined.

Use of the Interferometer as a Refractometer

With the help of Michelson's interferometer r.i. of a thin film or a gaseous medium can be determined.

To determine the refractive index of a thin film, it is to be introduced in the path of one of the interfering beams. The central fringe will be displaced from the cross-wire. The mirror M_1 is adjusted such that the central fringe again coincides with the wire. If the displacement of the mirror is d, then we can write,

$$d = (\mu - 1)t,$$

where t is the thickness of the film. So, from a knowledge of one of μ or t, the other can be found.

To determine the refractive index of a gas, two identical glass tubes whose two ends are closed air-tight are introduced, one in the reflected beam path to M_1 and the other in the transmitted beam path to M_2. At first the two tubes are empty and the central white light fringe which is achromatic is made to coincide with the wire. Then the experimental gas at a suitable pressure is introduced into one tube. The central position. If the displacement of the mirror M_1 is d, then $d = (\mu - 1)l$, where l is the length of the tube.

Example 1 *Using red cadmium light, λ = 643.8 nm, Michelson in his original experiment could still see interference fringes after he had moved one of the mirrors 25 cm away from the coincidence position. How many fringes did he count?*

Solution.

$$d = \frac{n\lambda}{2}, \lambda = 643.8\,\text{nm} = 6438 \times 10^{-8}\,\text{cm} \quad \Rightarrow \quad 25 = n \times 6438 \times \frac{10^{-8}}{2} \quad \Rightarrow \quad n = 776638.$$

Example 2 *How far must one of the mirrors of a Michelson interferometer be moved for* 400 *fringes of light of wavelength* 500 mm *to cross the centre of the field of view?*

[Similar to C.U. Hons. 1982]

Solution.

$$d = \frac{n\lambda}{2}, \lambda = 5000 \times 10^{-8}\,\text{cm}, n = 400. \quad \therefore \quad d = \frac{400 \times 5000 \times 10^{-8}}{2} = 0.01\,\text{cm}.$$

Example 3 *Calculate the distance between two successive positions of the movable mirror of Michelson's interferometer giving best fringes in case of Na having lines of wavelengths* 5893Å *and* 5896Å.

Solution. Using eqn. (8.2.26), we get,

$$\frac{2d}{\lambda_2} - \frac{2d}{\lambda_1} = 1, d = ?, \ \lambda_1 = 5896 \times 10^{-8}, \ \ \lambda_2 = 5893 \times 10^{-8}$$

$$\Rightarrow \ 2d \times \frac{(\lambda_1 - \lambda_2)}{\lambda_1\lambda_2} = 1 \quad \Rightarrow \quad d = \frac{\lambda_1\lambda_2}{2(\lambda_1 - \lambda_2)} = 0.058\,\text{cm}.$$

Example 4 *If after obtaining fringes in Michelson interferometer with white light, the white light is replaced by sodium light and if fringes are found to disappear when one mirror is moved through* 0.289 mm, *find the wavelength of the longer wavelength component when the shorter one has the value* 5890Å.

Solution. The fringes pass from the position of consonance to the condition of dissonance, hence,

$$2d = n\lambda_1 = \left(n + \frac{1}{2}\right)\lambda_2$$

$$\Rightarrow \quad \frac{\lambda_1}{\lambda_2} = \frac{(n+\frac{1}{2})}{n} \quad \Rightarrow \quad \frac{\lambda_1 - \lambda_2}{\lambda_2} = \frac{1}{2n} = \frac{\lambda_1}{4d}$$

$$\Rightarrow \quad \Delta\lambda = \frac{\lambda_1\lambda_2}{4d} \simeq \frac{\lambda_1^2}{4d} = \frac{(5890 \times 10^{-8})^2}{4 \times 0.0289} = 3 \times 10^{-8}$$

$$\therefore \quad \lambda_2 = \lambda_2 + \Delta\lambda = 5893 \times 10^{-8}\ \text{cm.} = 5893\text{Å}.$$

Example 5. *The distance traversed by the mirror in Michelson's interferometer between two successive disappearance is* 0.289 mm. *Calculate the difference in wavelengths of D_1 and D_2 lines,* $\lambda = 5890$Å.

Solution. As in the previous example, in this case we would have,

$$2d = (2n+1)\lambda_1 = (2n+2)\lambda_2$$

$$\Rightarrow \quad \frac{\lambda_1}{\lambda_2} = \frac{2n+2}{2n+1} \quad \Rightarrow \quad \frac{\lambda_1 - \lambda_2}{\lambda_2} = \frac{1}{n} = \frac{\lambda_1}{2d}$$

$$\Rightarrow \quad \Delta\lambda = (\lambda_1 - \lambda_2) = \frac{\lambda_1\lambda_2}{2d} = \frac{\lambda_1^2}{2d} = \frac{(5890 \times 10^{-8})^2}{2 \times 0.0289} = 6 \times 10^{-8} = 6\text{Å}.$$

Example 6. *If one arm of a Michelson interferometer contains a tube* 2.5 cm *long which is first evacuated and then slowly filled with air* ($\mu = 1.0003$), *how many fringes will cross the centre? Assume* $\lambda = 600$ nm.

Solution. The path difference introduced = $(\mu - 1)t$.
For the shift of n fringes $d = \frac{n\lambda}{2}$.

$$\therefore \quad (\mu - 1)t = n\frac{\lambda}{2} \quad \Rightarrow \quad n = \frac{2(\mu-1)t}{\lambda} = 25.$$

Example 7. *If the contrast in an interference pattern is 50% and if the maxima receive 15 units of light, how much do the minima receive?*

Solution. Contrast or visibility

$$V = \frac{I_{\max} - I_{\min}}{I_{\max} + I_{\min}} \quad \Rightarrow \quad \frac{1+V}{1-V} = \frac{I_{\max}}{I_{\min}} \quad \Rightarrow \quad I_{\min} = 5.$$

8.3 Fabry-Perot Interferometer

Fabry-Perot interferometer is a very important and sensitive interferometer used for various purposes, specially suited for the determination of wavelength with sufficient accuracy, more so for the study of hyperfine structure of spectral line; has a commendable resolving power and it can be also used for realisation of standard meter like the Michelson's interferometer.

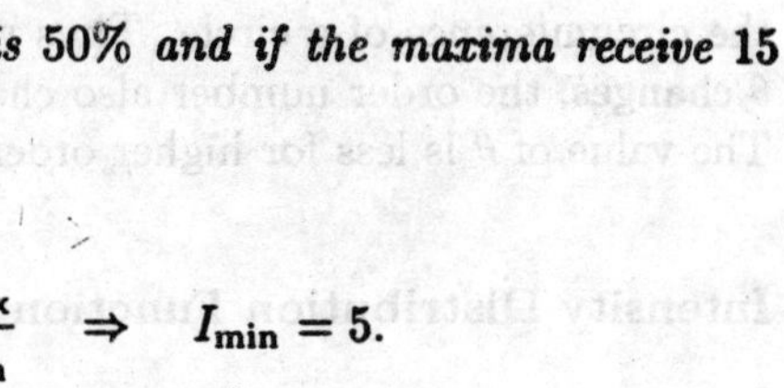

Fig 8.12

Construction. This interferometer consists of two plane parallel glass-plates or quartz plates P_1 and P_2 whose inside faces, thinly silvered, are perfectly parallel while the front face AB of P_1 and outside face CD of P_2 are themselves parallel but they are inclined to the silvered faces to avoid interference effects that would occur due to multiple reflections and refractions. As AB and CD faces are parallel, the emergent ray from CD will be parallel to the incident ray on AB.

Formation of Fringes and their Nature

This is a case of fringe formation in a parallel air-film enclosed between the silvered faces of the interferometer plates and fringes are observed in the transmitted light.

In fig. 8.13 we have shown two transmitted rays emerging out at an angle θ, the different angles equal to θ are shown in the diagram, the optical path difference between these two rays is

$$\Delta = \mu(BC + CD) - BE. \tag{8.3.1}$$

Since the film enclosed between the silvered faces is air, here $\mu = 1$

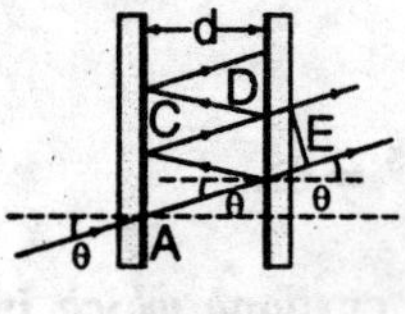

Fig 8.13

$$\therefore \quad \Delta = (BC + CD) - BE$$
$$= \frac{2d}{\cos\theta} - BD\sin\theta$$
$$= \frac{2d}{\cos\theta} - 2d\tan\theta\sin\theta = 2d\cos\theta. \tag{8.3.2}$$

So the phase difference $\delta = \frac{2\pi}{\lambda} \times 2d\cos\theta$.

So the condition for brightness is

$$\frac{2\pi}{\lambda} \times 2d\cos\theta = 2m\pi \quad \Rightarrow \quad 2d\cos\theta = m\lambda. \tag{8.3.3}$$

The points satisfying the equation (8.3.3) will be all bright points and they will be lying on the circumference of a circle. Thus we get circular fringes, alternately bright and dark. When θ changes, the order number also changes. Such fringes are called *fringes of equal inclination.* The value of θ is less for higher order number.

Intensity Distribution Function (IDF)

The intensity distribution function has been evaluated in Chapter 4, considering the superposition of the transmitted components. From this calculation we get,

$$I_T = \frac{a^2}{1 + \frac{4r^2}{(1-r)^2} sin^2\frac{\delta}{2}} = \frac{a^2}{1 + g^2\sin^2\frac{\delta}{2}}, \tag{8.3.4}$$

where a is the amplitude of the incident light and $g^2 = \frac{4r^2}{(1-r)^2}$ is called the *Finesse co-efficient.* This co-efficient depends on the value of r, called the *reflection co-efficient.* The expression for I_T as given above is true for an infinite number of rays, but for finite area of the interferometer plates, the number of transmitted rays is also finite. So the expression for I_T is also to be modified for this finiteness and can be written as

$$I_T = a^2\left(1-r^2\right)^2 \times \frac{\left(1-r^{2N}\right)^2 + 4r^2\sin^2\frac{N\delta}{2}}{\left(1-r^2\right)^2 + 4r^2\sin^2\frac{\delta}{2}}$$

$$= \frac{a^2}{1+g^2\sin^2\frac{\delta}{2}}\left[\left(1-r^{2N}\right)^2 + 4r^2\sin^2\frac{N\delta}{2}\right] \qquad (8.3.5)$$

Fringe Visibility and Sharpness of Fabry-Perot Fringes

The visibility of the fringes, according to Michelson, is determined in terms of the visibility factor V given as

$$V = \frac{I_{\max} - I_{\min}}{I_{\max} + I_{\min}}. \qquad (8.3.6)$$

Now, the IDF for the Fabry-Perot interferometer, in case the transmitted rays are of very large number, is given by,

$$I_T = \frac{a^2}{1+g^2\sin^2\frac{\delta}{2}}$$

For a particular value of g^2, when $\sin\frac{\delta}{2} = 0 \;\Rightarrow\; \frac{\delta}{2} = m\pi, I_T \to \max = a^2 = I_0$ and when $\sin\frac{\delta}{2} = 1, I_T \to \min = \frac{a^2}{(1+g^2)} = I_{\min}$.

$\therefore$ From equation (8.3.6), we get,

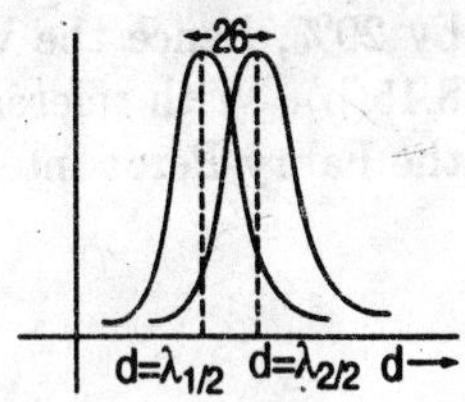

Fig 8.14

$$V = \frac{a^2 - \frac{a^2}{(1+g^2)}}{a^2 + \frac{a^2}{(1+g^2)}} = \frac{g^2}{(2+g^2)} = \frac{\frac{4r^2}{(1-r^2)^2}}{2+\frac{4r^2}{(1-r^2)^2}}$$

$$= \frac{4r^2}{2(1-r^2)^2 + 4r^2} = \frac{2r^2}{1+r^4} \qquad (8.3.7)$$

It has been remarked earlier and it is also evident from the actual photographs of the fringes that the fringe-system in transmitted light is sharper than the fringe-systems observed in reflected light. This fact can be explained in terms of the visibility of the fringes.

A light silvering on the interferometer plate increases the value of r and it approaches the value of 1. So the *Finesse co-efficient* $g^2 = \frac{4r^2}{(1-r^2)^2} \to$ large value. As g becomes large, the peakiness of the intensity maxima becomes much pronounced. So bright fringes are very narrow and they are separated by broad patches of almost zero intensity. From the expression for visibility

$$V = \frac{g^2}{2+g^2} = \frac{1}{1+2/g^2},$$

we could also infer that as g increases, the denominator decreases and hence V increases. Hence we conclude that in Fabry-Perot interferometer sharp circular fringes are obtained when the interferometer plates are lightly silvered. If silvering is heavily done, r becomes equal to unity but in this case, there will be practically no transmission.) So r value is chosen to lie between 0.8 to 0.9.

Resolving Power

Fabry-Perot interferometer is a high resolving power instrument and hence it is of great practical importance in the study of hyperfine structure of spectral line. In this respect this instrument is better than Michelson's interferometer. The resolving power depends on the number of transmitted rays available and this number greatly increases even when the plates are lightly silvered. In case of diffraction grating, the resolving power also depends on the number of rulings on the grating surface. This number is limited by the precision of the instrument by which the rulings are made but in case of Fabry-Perot interferometer, the number of transmitted components can be very simply increased by silvering.

To find an expression for the resolving power, we again consider Rayleigh's criterion. We would consider rings near the centre for which $\cos\theta \simeq 1$.

$$\therefore \quad 2d = m\lambda$$

gives the mth order bright fringe for the wavelength λ. But application of Rayleigh's criterion in Fabry-Perot instrument is different from what has been said in case of a grating. In case of Fabry-Perot pattern, two maxima are widely separated as shown in fig. 8.15(a), the minimum occurs mid-way between two successive orders of maxima. So in this case, if two wavelengths are such that their difference $\delta\lambda$ makes the same order maxima for the two wavelengths cross at a point, where the intensity is half the maximum of either, then at that point the resultant intensity would be equal to the maximum intensity of any one of them. But in this condition, the intensity maximum of individual fringes would increase by about 15% to 20%. Thus, at the mid-point between the two increased maxima (of the same order), there is a dip of intensity by 20%, hence the wavelengths will be said to be resolved. These facts are indicated in fig. 8.15(b). With reference to this discussion it is now possible to work out the resolving power of the Fabry-Perot interferometer.

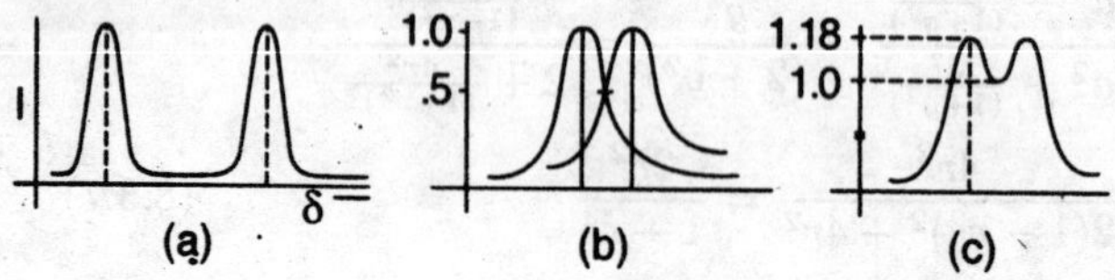

Fig 8.15

Consider two spectral lines λ_1 and λ_2, if they intersect at half-height, we would have fig.8.15(b).

$$\frac{1}{2} = \frac{1}{1 + g^2 \sin^2\left\{\frac{2\pi}{\lambda_1}(d+\epsilon)\right\}} \tag{8.3.8}$$

$$\text{and} \quad \frac{1}{2} = \frac{1}{1 + g^2 \sin^2\left\{\frac{2\pi}{\lambda_2}(d-\epsilon)\right\}} \tag{8.3.9}$$

Equating the above two equations, we get,

$$1 + g^2 \sin\left\{\frac{2\pi}{\lambda_1}(d+\epsilon)\right\} = 1 + g^2 \sin^2\left\{\frac{2\pi}{\lambda_2}(d-\epsilon)\right\}$$

$$\Rightarrow \quad \frac{2\pi}{\lambda_1}(d+\epsilon) = \frac{2\pi}{\lambda_2}(d-\epsilon) \quad \Rightarrow \quad \frac{d+\epsilon}{\lambda_1} = \frac{d-\epsilon}{\lambda_2} \tag{8.3.10}$$

Now, if $\lambda_1 = \lambda_2 - \Delta\lambda$, we get,

$$\frac{d+\epsilon}{\lambda_2 - \Delta\lambda} = \frac{d-\epsilon}{\lambda_2}$$
$$\Rightarrow \lambda_2 d + \lambda_2\epsilon = \lambda_2 d - \lambda_2\epsilon - d\times\Delta\lambda + \epsilon\Delta\lambda$$
$$\Rightarrow (\epsilon - d)\Delta\lambda = 2\lambda_2\epsilon \quad\Rightarrow\quad \frac{\lambda_2}{\delta\lambda} \equiv \frac{\lambda}{\Delta\lambda} = \frac{(\epsilon-d)}{2\epsilon} \tag{8.3.11}$$

To find the value of ϵ, we consider the half-height of one of the spectral lines

$$\frac{1}{2} = \frac{1}{1+g^2\sin^2\{\frac{2\pi}{\lambda}(d+\epsilon)\}} = \frac{1}{1+g^2\left[\sin\frac{2\pi}{\lambda}d\cos\frac{2\pi}{\lambda}\epsilon + \cos\frac{2\pi}{\lambda}d\sin\frac{2\pi}{\lambda}\epsilon\right]^2} \tag{8.3.12}$$

Now, for maximum we would have,

$$d = m\pi \quad\Rightarrow\quad \frac{2\pi}{\lambda}d\cos\theta = m\pi.$$

If we observe fringes near the centre, $\theta \simeq 0$.

$$\therefore \quad \frac{2\pi}{\lambda}\times d = m\pi.$$

So, $\left|\sin\frac{2\pi}{\lambda}d\right| = \sin m\pi = 0, \quad \left|\cos\frac{2\pi}{\lambda}d\right| = 1$

$$\therefore \quad \frac{1}{2} = \frac{1}{1+g^2\sin^2\frac{2\pi\epsilon}{\lambda}} = \frac{1}{1+g^2\times\frac{4\pi^2\epsilon^2}{\lambda^2}}$$
$$\Rightarrow g^2\times\frac{4\pi^2\epsilon^2}{\lambda^2} = 1 \quad\Rightarrow\quad 2\epsilon = \frac{\lambda}{\pi g}$$
$$\therefore \quad \left|\frac{\lambda}{\Delta\lambda}\right| = \left|\frac{\epsilon-d}{2\epsilon}\right| = \frac{d}{2\epsilon}, \tag{8.3.13}$$

neglecting ϵ compared to d in the numerator.

Putting the value of 2ϵ, we get,

$$\left|\frac{\lambda}{\Delta\lambda}\right| = \frac{d}{(\lambda/\pi g)} = \frac{\pi g d}{\lambda} = \frac{\pi g}{\lambda}\times\frac{m\lambda}{2} = \frac{m\pi g}{2} = \frac{m\pi}{2}\cdot\frac{2r}{(1-r^2)} = \frac{m\pi r}{(1-r^2)}. \tag{8.3.14}$$

Thus eqn. (8.3.14) shows that the R.P. of the Fabry-Perot interferometer depends on the order number m and on r, called the *reflection co-efficient*.

To get an idea of the shortest difference of wavelength $\Delta\lambda$, we consider specific values in eqn. (8.3.14), take $r = 0.9, \lambda = 5000\times10^{-8}$ cm, $d = 1$ cm.

$$\frac{\lambda}{\Delta\lambda} = \frac{m\pi r}{(1-r^2)} = \frac{\pi r}{(1-r^2)}\times\frac{2d}{\lambda} = \frac{3.14\times0.9}{(1-0.9^2)}\times\frac{2\times1}{5000\times10^{-8}} = 594947$$
$$\Rightarrow \Delta\lambda = \frac{\lambda}{594947} = \frac{5000\times10^{-8}}{594947} = 8.404\times10^{-11} = 0.0084\text{Å}. \tag{8.3.15}$$

Example 1 *Determine the least separation of two spectral lines, near* 500 nm, *that can be resolved by an etalon that is* 10 mm *long and has a reflecting coefficient* 0.95.

Solution. Here,

$$m = \frac{2d}{\lambda} = \frac{2 \times 10 \times 10^{-1}}{5000 \times 10^{-8}} = 40,000$$

$$\frac{\lambda}{\Delta\lambda} = \frac{m\pi r}{(1-r^2)} = \frac{40 \times 10^3 \times \pi \times 0.95}{(1-0.95^2)} \quad \Rightarrow \quad \Delta\lambda = 0.0041\text{Å}.$$

Example 2 *What plate separation is needed in a Fabry-Perot interferometer to resolve two spectral lines* 0.05Å *apart if the average wavelength is* 633 nm *and reflection coefficient* $r = 0.89$?

Solution. Here, $\Delta\lambda = 0.05\text{Å} = 0.05\text{Å} \times 10^{-8}$ cm, $\lambda = 6330 \times 10^{-8}$

$$\frac{\lambda}{\Delta\lambda} = \frac{m\pi r}{(1-r^2)} = \frac{2d}{\lambda}\cdot\frac{\pi r}{(1-r^2)} \quad \Rightarrow \quad d = \frac{\lambda^2(1-r^2)}{2\pi r\Delta\lambda} = 0.298 \text{ cm}.$$

Applications of Fabry-Perot Interferometer

This interferometer can be applied like Michelson's one to make the following determinations but in some of these determinations, Fabry-Perot's instrument is better than Michelson's.

1. To determine small wavelength difference between spectral lines,
2. To determine wavelength accurately,
3. To study the hyperfine structure of a spectral line.

Fabry-Perot Etalon

Etalon is an arrangement in which the two interferometer silvered plates are kept separated at fixed distances apart.

Determination of Small Difference of Wavelength

Very small difference of wavelength can be determined with the help of Fabry-Perot interferometer. There are two methods for this purpose, one is known as (i) the Method of Coincidence, the other (ii) Method of Exact Fractions.

Method of coincidence. If the source emits two wavelengths λ_1 and λ_2 (as in case of Na-yellow line or yellow lines of Hg), the difference between them can be determined with the help of Fabry-Perot interferometer. The method of coincidence and discordance of the fringe systems due to the wavelengths is utilised for this purpose.

When the separation between two plates is very small, the two fringe systems nearly coincide, but as one of the plates moved out, there will appear relative motion between the fringe systems and at a particular separation there will be discordance between fringe systems when the bright fringes due to one would coincide with dark fringes due to the other. Passing through this discordance again there will occur at further separation coincidence of the fringe systems.

At the first coincidence position, we can write if consideration is made near the centre,

$$2d_1 = m_1\lambda_1 = m_2\lambda_2.$$

In the second position of coincidence,

$$2d_2 = (m_1 + p)\lambda_1 = (m_2 + p + 1)\lambda_2.$$

From these two eqns., we get,

$$2(d_2 - d_1) = p\lambda_1 = (p+1)\lambda_2 \quad (8.3.16)$$

$$\Rightarrow \frac{1}{\lambda_2} - \frac{1}{\lambda_2} = \frac{p+1}{2(d_2 - d_1)} - \frac{p}{2(d_2 - d_1)} = \frac{1}{2(d_2 - d_1)}$$

$$\Rightarrow (\lambda_1 - \lambda_2) = \frac{\lambda_1 \lambda_2}{2(d_2 - d_1)} \quad (8.3.17)$$

If $(d_2 - d_1)$ be determined, then from eqn. (8.3.16) we can get the value of p when one of the wavelengths λ_1 or λ_2 is known. Hence from eqn. (8.3.17), $(\lambda_1 - \lambda_2)$ can be determined. Alternatively by counting the number of fringe shift of one wavelength, say λ_1, the value of p is determined and hence $(\lambda_1 - \lambda_2)$ is determined. This method, though simple, is not very accurate, whereas the method of exact fractions is very accurate but at the same time very complicated.

8.4 Investigation of Hyperfine Structure

A spectral line may consist of a number closely spaced lines differing in wavelengths ranging between 0.1Å to 0.001Å. One of them is called the *main line*, the other lines are called its *satellites*. The spectral line is then said to have hyperfine structure. A Fabry-Perot interferometer because of its high resolution can be used to investigate hyperfine structure of spectral lines.

We might enquire into the cause of the hyperfine structure. Two reasons individually or conjointly may be responsible for the origin of the hyperfine structure.

One reason may be tbe isotopic constitution of the element. Isotope of an element possesses different nuclear masses and as a result, the lines from each isotope are slightly displaced w.r.t. each other. To put it in other way, according to Bohr model, the wave-number $\bar{\nu}$ of a spectral line is given by,

$$\bar{\nu} = \frac{2\pi^2 \mu e^4 z^2}{ch^3}\left(\frac{1}{n_1^2} - \frac{1}{n_2^2}\right),$$

where n_1, n_2 are the quantum numbers; h is the Planck's constant and $\mu = \frac{mM}{m+M}$, called the *reduced mass of the atom*, m being the mass of the electron and M that of the nucleus. So, for isotopic constitution, since M is different for the isotopes, a spectral line may consist of closely spaced group of lines or in other words, it is said to have hyperfine structure.

Apart from what has been stated above, hyperfine structure may also occur due to the existence of nuclear magnetic moment of an atom. That there is a magnetic moment possessed by the nucleus of an atom is beyond any dispute and Pauli in 1924 and Russell in 1927 concluded from independent premises that this nuclear magnetic moment gives rise to hyperfine structure of spectral lines.

The first effect is referred to as isotopic hyperfine structure effect, demonstrated and analysed by Urey, Schüler and others to study not only the isotopic constitution of elements but also more important nuclear properties like nuclear volume and nuclear symmetry. Soon it was found that an element which has only one isotope also produces hyperfine structure. Bismuth, for example, has one isotope but it has a hyperfine structure. So it was thought that there could be other reason other than the isotope effect to have hyperfine structure. It was Pauli who first proposed nuclear spin tobe another alternative. According to him, the nuclear spin with its attending magnetic moment interacts with the magnetic field due to the motion of the extranuclear electrons to produce hyperfine structure. In other words, the action of the orbital electrons in the field of the nuclear magnetic dipole will cause the different atomic states to have slightly different energies, hence hyperfine separation of levels.

The order of separation in the two cases is same and so it is not always very easy to say which effect is responsible for the hyperfine structure.

the splitting of spectral lines due to either of them is so small that only high resolving power instrument like Fabry-Perot interferometer can only show it. From such study it has been found that the hyperfine structure of tantalum consists of eight components. The isotopic separation in case of tungsten produces three components corresponding to isotopic masses 182, 184 and 186 which have greater relative abundances.

To study the hyperfine structure of the spectral line with the help of Fabry-Perot interferometer, the plate separation is to be adjusted such that at a particular separation there will appear fringe systems due to different components. Thus we would get for any order not a fringe but a fringe system.

For a particular order m for a wavelength λ_1 (main line) in the neighbourhood of the centre,

$$2d\cos\theta_1 \simeq 2d = m\lambda_1$$

For the next order outwards,

$$2d = (m-1)\lambda_1$$

If there be a line λ_2 very close to λ_1, then applying the method of coincidence, the fringe of order m in λ_2 is made to coincide with $(m-1)$th fringe of wavelength λ_1, then

$$2d = m\lambda_2 = (m-1)\lambda_1 \tag{8.4.1}$$

Now, if $\lambda_1 = \lambda_2 + \Delta\lambda$, then from eqn. (8.4.1) we get,

$$m(\lambda_1 - \Delta\lambda) = (m-1)\lambda_1 \quad \Rightarrow \quad \Delta\lambda = \frac{\lambda_1}{m} \tag{8.4.2}$$

From eqn. (8.4.1), $m = \frac{2d}{\lambda_2}$

$$\therefore \quad \Delta\lambda = \frac{\lambda_1\lambda_2}{2d} \simeq \frac{\lambda_1^2}{2d} \tag{8.4.3}$$

To see how small $\Delta\lambda$ can be determined by Fabry-Perot interferometer, we consider the following parameters of the instrument—

Take $d = 10$ cm, $\lambda = 5000$Å, then

$$\Delta\lambda = \frac{(5000\times10^{-8})^2}{2\times10} = \frac{25\times10^{-10}}{2\times10} = 12.5\times10^{-3}\text{Å} = 0.0125\text{Å}$$

If the relative shift could be measured to $\frac{1}{10}$th of a fringe, then

$$\Delta\lambda = \frac{1}{10}\times 0.0125\text{Å} \simeq 0.00125\text{Å}$$

Thus we can very easily study the hyperfine structure of spectral line with the help of Fabry-Perot interferometer.

Short Theory Questions with Answers

1. Under what condition straight and circular fringes are produced in a Michelson's interferometer?

Ans. Fringes of different nature, straight, circular or curved convex outwards or inwards will be produced depending on the orientation of the two mirrors.

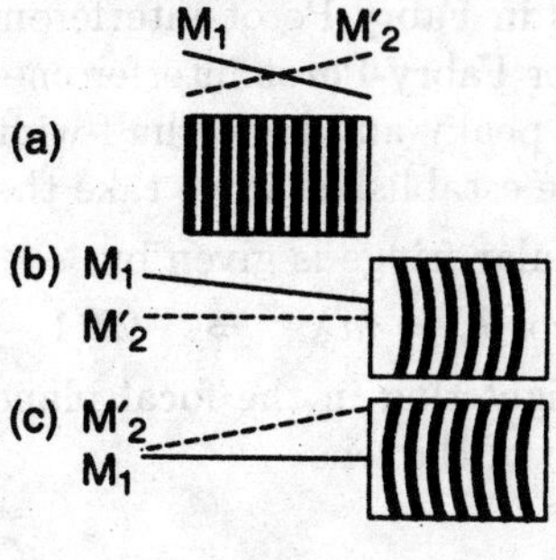

Fig 8.16

When the two mirrors are inclined such that the mirror M_1 and M'_2 which is the image of M_2 by reflection intersect each other as shown in fig. 8.16(a), we would get straight fringes (localised) parallel to the line of intersection. These fringes are called *fringes of equal width.* If the inclination be such as in fig. 8.16(b), fringes would be convex outwards, towards the line of intersection, whereas if the inclination be as in fig. 8.16(c), the fringes would again be convex inwards again towards the line of intersection.

But when the mirrors M_1 and M_2 are exactly perpendicular to each other so that M_1 and M'_2 are exactly parallel, we would get circular fringes which would be fringes of equal inclination. The straight fringes are localised fringes and to see them eye must be focussed at finite distance, whereas the circular fringes are non-localised in the sense that they are formed at infinity and to see them eye must be focussed at infinity.

2. Compare and contrast the resolving power and shape of fringes of Michelson's interferometer with those of Fabry-Perot interferometer.

Ans. Both are high resolving power instruments but Fabry-Perot interferometer is of much higher resolving power than that of Michelson's interferometer. The difference of wavelength measurable with a Michelson's interferometer is of an order of an angstrom ($\simeq 10^{-8}$ cm), whereas the difference of wavelength measurable with a Fabry-Perot interferometer is of the order of 10^{-3} Å. It is because of this high power, Fabry-Perot interferometer is used for the study of hyperfine structure.

The resolving power of a Michelson's interferometer is given as

$$(\text{R.P.})_{\text{Michelson}} = 2m \quad \text{(c.f. eqn. 8.2.21)}$$

The resolving power of a Fabry-Perot apparatus is given as

$$(\text{R.P.})_{\text{Fabry-Perot}} = \frac{m\pi r}{(1-r^2)} = \frac{2d}{\lambda} \cdot \frac{\pi r}{(1-r^2)}.$$

So the resolving power of Michelson's interferometer depends only on the order number, whereas the resolving power of a Fabry-Perot apparatus depends only on r, the reflection co-efficient and d, the separation between the plates in addition to the order number.

The fringes in Michelson's interferometer may be straight or curved and circular depending on the orientation of the two mirrors, hence may be localised fringes or not, but the fringes in Fabry-Perot are always circular or in other words, the fringes in Michelson's interferometer may be of equal width or inclination but in Fabry-Perot interferometer, the fringes are always of equal inclination.

3. Explain why the fringes in Fabry-Perot interferometer are very narrow and resolving power very high.

Ans. The circular fringes in Fabry-Perot interferometer are very sharp and narrow. It is evident from the IDF for Fabry-Perot interferometer which shows that as g increases, the maxima become more peaky and hence half-width becomes narrow. The narrowness of the fringes could also be established if we take the problem analytically.

The mth order bright circular fringe is given by,

$$2d\cos\theta_m = m\lambda \quad \Rightarrow \quad 2d(1-\theta_m^2/2) = m\lambda.$$

The radius of the mth bright ring in the focal plane of the lens is given as $R_m = f\theta_m$, where f is the focal length of the lens

$$\Rightarrow \quad R_m^2 = f^2\theta_m^2 = 2f^2\left(1-\frac{m\lambda}{2d}\right)$$

The next order outwards is lesser than m, hence,

$$R_{m-1}^2 = 2f^2\left(1-\overline{m-1}\frac{\lambda}{2d}\right)$$

$$\therefore \quad R_{m-1}^2 - R_m^2 = 2f^2\left(1-\overline{m-1}\frac{\lambda}{2d}\right) - 2f^2\left(1-\frac{m\lambda}{2d}\right) = 2f^2\left(\frac{\lambda}{2d}\right) = \frac{\lambda f^2}{d}$$

$$\Rightarrow (R_{m-1} - R_m) = \frac{\lambda f^2}{d}\cdot\frac{1}{(R_{m-1}+R_m)} = \frac{\lambda f^2}{2R_m d} \tag{8.4.4}$$

Eqn. (8.4.4) shows that the fringe-width is extremely small and since the fringe-width is very small, the fringes are very sharp.

The resolving power of Fabry-Perot interferometer is given by

$$\frac{\lambda}{d\lambda} = \frac{m\pi r}{(1-r^2)}$$

Usually $r \simeq 0.95$, hence the resolving power of interferometer is very high because the denominator is a small quantity.

Physically, the spacing between the silvered plates at some distances act as resonator for certain wavelength and the separation space is filled with air, where damping is small, hence resonance peak becomes sharply defined. There is a considerable separation between two sharply defined peaks and hence the visibility of the fringes is also high.

4. Compare Fabry-Perot fringes with Newton's rings.

Ans. The fringes in both cases are circular, hence they are fringes of equal inclination; but Fabry-Perot fringes are narrower and hence sharper than the Newton's rings. Also fringe visibility in case of Fabry-fringes is greater than that of Newton's rings.

5. If the separation between the plates of a Fabry-Perot etalon is slowly increased, do the radii increase or decrease?

Ans. The radius of the mth bright fringe is given by

$$R_m^2 = 2f^2\left(1 - \frac{m\lambda}{2d}\right).$$

So if d increases, the second term within the bracket decrease, it means R_m increase.

6. What is the intensity of the dark fringes in F-P interferometer?

Ans. The minimum intensity is given by

$$I_{\text{min}} = \frac{a^2}{1+g^2},$$

where g^2 is called the *Finesse co-efficient* and is given by

$$g^2 = \frac{4r^2}{(1-r^2)^2},$$

g value is quite high depending on the value of r. So as g value is very high, I_{min} is very small; when $g \to \infty, I_{\text{min}} \to 0$.

7. What do you mean by spectral range? Derive expressions for the spectral range of Michelson's interferometer and Fabry-Perot interferometer.

Ans. By spectral range of an interferometer we mean the change in wavelength $\delta\lambda$ corresponding to a displacement by one order. If the spectral range is greater, it would physically mean that overlapping of orders is smaller.
In Michelson or in Fabry-Perot interferometer, the basic interferometer equation is

$$2d\cos\theta = m\lambda$$

If $\delta\lambda$ be the difference between the two wavelengths such that mth order of one coincides with $(m+1)$th order of the other, then

$$2d\cos\theta = m(\lambda + \delta\lambda) = (m+1)\lambda \quad \Rightarrow \quad \delta\lambda = \frac{\lambda}{m} = \frac{\lambda(\lambda+\delta\lambda)}{2d\cos\theta} \simeq \frac{\lambda^2}{2d},$$

where we have neglected $\lambda\delta\lambda$ and $\cos\theta = 1$ has been assumed. Thus $\delta\lambda$ is inversely proportional to the separation d of the plates and also the spectral range depends on the wavelength.

Solved Numerical Problems

1. A transparent film of glass of $\mu = \frac{3}{2}$ is introduced normally in the path of one of the interfering beams of a Michelson's interferometer with light of $\lambda = 4800$Å. This causes 500 dark fringes to sweep across the field. Determine the thickness of the film.

Solution. $2(\mu - 1)t = m\lambda$; here $\mu = \frac{3}{2}$, $m = 500$, $\lambda = 4800 \times 10^{-8}$ cm.

$$\therefore \quad t = 0.24 \text{ mm}.$$

2. How small one must make the plate separation of a Fabry-Perot interferometer in order to obtain a free spectral range of 0.05Å at a wavelength of 5000Å?

Solution.

$$\delta\lambda = \frac{\lambda^2}{2d} \quad \Rightarrow \quad d = \frac{\lambda^2}{2\delta\lambda} = \frac{(5000 \times 10^{-8})^2}{2 \times 0.05 \times 10^{-8}} = 2.5 \text{ cm}.$$

3. **Show that in the system of fringes formed in a Fabry-Perot interferometer, the ratio of the intensity of the maxima to that mid-way between two maxima is $\frac{(1+r^2)^2}{(1-r^2)^2}$.**

Solution. The intensity maximum is given by

$$(I_T)_{\text{max}} = I_0 = a^2.$$

Now,

$$I_T = \frac{I_0}{1 + g^2 \sin^2 \delta/2}$$

The maxima are defined by

$$\frac{\delta}{2} = n\pi$$

So the value of δ mid-way between two maxima is given by

$$\delta_1 = \frac{\delta}{4} = \frac{1}{2}n\pi$$

$$\therefore \quad (I_T)_{\text{Half}} = \frac{a^2}{1 + g^2 \sin^2\left(\frac{1}{2}n\pi\right)} = \frac{a^2}{1+g^2}$$

$$\therefore \quad (I_T)_{\text{max}}/(I_T)_{\text{Half}} = \frac{a^2}{\left(\frac{a^2}{1+g^2}\right)} = (1+g^2) = \frac{(1+r^2)^2}{(1-r^2)^2}$$

4. **White light fringes are formed in a Michelson's interferometer. Now, Na-light replaces the white light source. If for the displacement of one mirror by 0.0145 cm, the visibility of the fringes becomes minimum, find the greater wavelength present in Na-light. Take shorter wavelength component as 5890Å.**

Solution. For this displacement, bright fringes due to one wavelength coincide with dark fringes due to the other component.

$$2d = n\lambda_1 = \left(n + \frac{1}{2}\right)\lambda_2 \quad \Rightarrow \quad \frac{2d}{\lambda_2} - \frac{2d}{\lambda_1} = \frac{1}{2}$$

$$\Rightarrow \quad \frac{2d}{\lambda_2} - \frac{1}{2} = \frac{2d}{\lambda_1} \quad \Rightarrow \quad \lambda_1 = 5895.98\text{Å}.$$

5. **Shorter wavelength component is to be determined after comparing with the standard wavelength of 5000Å. The plates of a Fabry-Perot interferometer are to be kept separated by distances 1.4 mm, 2.8 mm and 4.2 mm for coincidence. Find the required wavelength.**

Solution. Using eqn. (8.3.16), we get

$$\lambda_1 - \lambda_2 = \frac{\lambda_1\lambda_2}{2(d_2 - d_1)}$$

Here, $\lambda_1 = 5000\text{Å}, (d_2 - d_1) = 1.4 \text{ mm} = 0.14 \text{ cm}.$

$$\therefore \quad \lambda_1 - \lambda_2 = \frac{\lambda_1^2}{2(d_2 - d_1)}, \text{ putting } \lambda_1 = \lambda_2 \text{ in } \lambda_1\lambda_2$$

$$\Rightarrow \quad \lambda_1 - \lambda_2 = 0.893\text{Å} \quad \Rightarrow \quad \lambda_2 = 4999.1072\text{Å}.$$

8.5 Lummer-Gehrcke Plate

Lummer and Gehrcke introduced this interferometer which consists of a plane parallel glass or quartz plate 10 cm to 20 cm long and 1 to 2 cm wide, and high reflecting power was available by utilising the property of total internal reflection. In Fabry-Perot interferometer, multiple reflections occurred in the thin air-film between the silvered faces or faces coated with reflecting film. The loss of light due to absorption in the silvered film of Fabry-Perot interferometer is avoided to great extent in this plate by making the angle θ almost equal to the critical angle for glass-air interface, hence the angle of refraction is almost $\frac{\pi}{2}$ fig. (8.17).

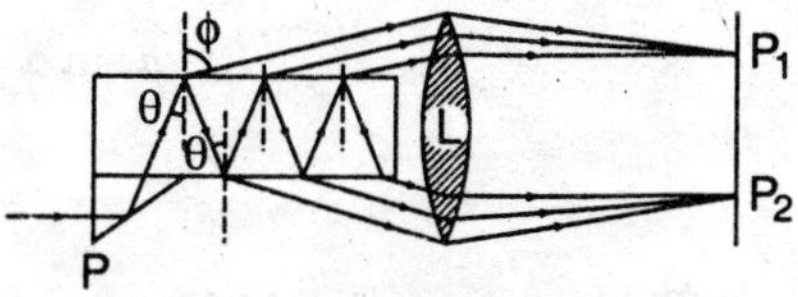

Fig 8.17

To avoid loss of light by reflection at the end-face of the plate, a prism P is placed in optical contact with one face of the plate such that the angle of incidence becomes equal to θ, very near to the critical angle. We would then get a series of rays emerging from the top surface and another series from the bottom. Interference fringes are formed both above and below the plate, the pattern consists of narrow sharp maxima with broad minima between them. The two sets of fringe pattern are separated by a dark zone.

Resolving Power of the Lummer-Gehrcke Plate

The criterion due to Rayleigh can be applied in this case also to find an expression for the resolving power of the instrument, but the difficulty that arises in this case is that the fringes are not of equal intensity and of infinite number because of the finite length of the plate and the restriction of the angle of incidence to be in the neighbourhood of the critical angle. The analysis that follows and hence the expression for the resolving power thus derived holds in case the angle of emergence is approximately 90°.

AB represents the emerging wavefront at some instant, it is of width

$$w = AC\cos\phi = l\cos\phi, \quad \text{where } l = AC. \tag{8.5.1}$$

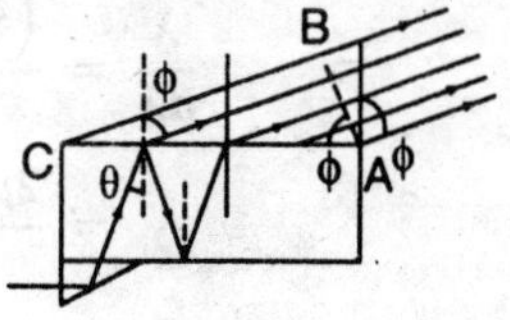

Fig 8.18

Positions of Maxima and Minima

The parallel beam of rays emerging from the top or bottom will give rise to maxima or minima at the focal plane of the lens depending on the path difference.

The path difference between two successive rays is given by,

$$\Delta = 2\mu d\cos\theta.$$

So for maxima, we would have,

$$2\mu d\cos\theta = m\lambda \quad\Rightarrow\quad 4\mu^2 d^2(1-\sin^2\theta) = m^2\lambda^2$$

$$\Rightarrow\quad 4\mu^2 d^2\left(1-\frac{\sin^2\phi}{\mu^2}\right) = m^2\lambda^2 \quad\Rightarrow\quad 4d^2(\mu^2-\sin^2\phi) = m^2\lambda^2 \tag{8.5.2}$$

For a given λ, the angular separation between two successive maxima is obtained by differentiating eqn. (8.5.2), we get,

$$d\phi = -\frac{\lambda^2 m\,dm}{2d^2\sin 2\phi} = -\frac{\lambda^2 m}{2d^2\sin 2\phi}, \quad \text{taking } dm = 1 \tag{8.5.3}$$

Now, according to eqn. (8.5.1), w may be taken as the width of a single slit, hence the angular separation between the central maximum and the next minimum is given by,

$$a \sin\alpha = \lambda \Rightarrow a\alpha = \lambda, \quad \text{since } \alpha \text{ is small}$$

$$\Rightarrow \alpha = \frac{\lambda}{a} = \frac{\lambda}{l\cos\phi} \tag{8.5.4}$$

This angle α should be the angular separation between the centres of the maxima of two wavelengths λ and $\lambda + d\lambda$ for resolution of these two wavelengths λ and $\lambda + d\lambda$ according to Rayleigh's criterion. The angular separation between the maxima of wavelengths λ and $\lambda + d\lambda$ is obtained by differentiating eqn. (8.5.2) w.r.t. λ, we get,

$$8d^2\mu\frac{d\mu}{d\lambda} - 8d^2 \sin\phi\cos\phi\frac{d\phi}{d\lambda} = 2m^2\lambda$$

$$\Rightarrow 8d^2\sin\phi\cos\phi d\phi = 8\mu d^2 d\mu - 2m^2\lambda d\lambda$$

$$\Rightarrow d\phi = (4\mu d^2 d\mu - m^2\lambda d\lambda)/2d^2\sin 2\phi \tag{8.5.5}$$

The angular separation $d\phi$ given by eqn. (8.5.5) and the angle α given by eqn. (8.5.3) must be equal.

$$\therefore \quad \frac{\lambda}{l\cos\phi} = \frac{4\mu d^2 d\mu - m^2\lambda d\lambda}{2d^2\sin 2\phi}$$

$$\Rightarrow \quad \frac{\lambda}{d\lambda} = \frac{\left(4\mu d^2\frac{d\mu}{d\lambda} - m^2\lambda\right) l\cos\phi}{2d^2\sin 2\phi} = \frac{4\mu d^2\frac{d\mu}{d\lambda} - \frac{4d^2}{\lambda}(\mu^2 - \sin^2\phi)}{2d^2\sin 2\phi} \times l\cos\phi$$

$$= \frac{2\mu\lambda d^2\frac{d\mu}{d\lambda} - 2d^2(\mu^2 - \sin^2\phi)}{\lambda d^2\sin 2\phi} \times l\cos\phi = \frac{l}{\lambda\sin\phi}\left(\mu\lambda\frac{d\mu}{d\lambda} - \mu^2 + \sin^2\phi\right)$$

$$= -\frac{l}{\lambda\sin\phi}\left(\mu^2 - \mu\lambda\frac{d\mu}{d\lambda} - \sin^2\phi\right) \tag{8.5.6}$$

Thus the chromatic resolving power is given by eqn. (8.5.6). The negative sign expresses the fact that as λ increases ϕ decreases and vice versa.

$\therefore$ Chromatic resolving power

$$\left|\frac{\lambda}{d\lambda}\right| = \frac{l}{\lambda\sin\phi}\left(\mu^2 - \mu\lambda\frac{d\mu}{d\lambda} - \sin^2\phi\right) \tag{8.5.7}$$

If $\phi \simeq 90°$ and $\frac{d\mu}{d\lambda}$ neglected, then,

$$\left|\frac{\lambda}{d\lambda}\right| = \frac{l}{\lambda}(\mu^2 - 1). \tag{8.5.8}$$

Suppose, $l = 10$ cm, $\lambda = 600$ nm, $\mu = \frac{3}{2}$, then from eqn. (8.5.8), we get,

$$\left|\frac{\lambda}{d\lambda}\right| = \frac{10}{6000 \times 10^{-8}}\left(\frac{9}{4} - 1\right) = \frac{50}{4 \times 6 \times 10^{-5}}$$

$$\Rightarrow \quad d\lambda = \frac{6000 \times 10^{-8} \times 24 \times 10^{-5}}{50} = 0.029\text{Å} \tag{8.5.9}$$

But remember, this resolution by the plate is subject to $\phi \simeq 90°$ and equality of intensity of all the emergent rays. So eqn. (8.5.8) may be taken as the highest limit of resolving power.

Actually, however, a small length of the plate is excluded for reflection. The effective length for getting the emergent beam is $\frac{2}{3}l$.

So,

$$\left|\frac{\lambda}{d\lambda}\right| = \frac{2}{3}\frac{l}{\lambda}(\mu^2 - 1) \tag{8.5.10}$$

Dispersion in L-G Plate

From eqn. (8.5.5) we get,

$$\frac{d\phi}{d\lambda} = \frac{2\mu\lambda\frac{d\mu}{d\lambda} - 2(\mu^2 - \sin^2\phi)}{\lambda \sin 2\phi}. \tag{8.5.11}$$

So the angular dispersion is independent of the dimension of the plate, it depends on the angle of emergence ϕ, $\frac{d\mu}{d\lambda}$ of the plate and also on the wavelength.

This interferometer requires high degree of optical quality for the plates in the sense that they must be perfectly plane and parallel. The material of the plate must be highly uniform, otherwise ghost images will come up. The fringes available in this interferometer and straight fringes.

8.6 Jamin's Interferometer

It is based on the formation of interference fringes in a thick plate, called *Brewster's fringes.*

Two thick glass plates, heavily silvered at the back, are positioned parallel to each other and in the vertical plane such that rays from an extended source fall on them at Brewster's angle which is nearly 45°. One such ray is shown as PQ. This ray PQ on incidence at Q splits into two parts, one part moves as $QRST$, the other part as $QUVT$ on their travel towards the receiving telescope T_1. As these two rays are derived from the same ray, they are in a position to interfere with each other to produce interference fringes. If the plates are perfectly parallel, the two systems of rays travel equal paths and are in phase. So, if one of the plates be turned thro' an angle, path difference will occur.

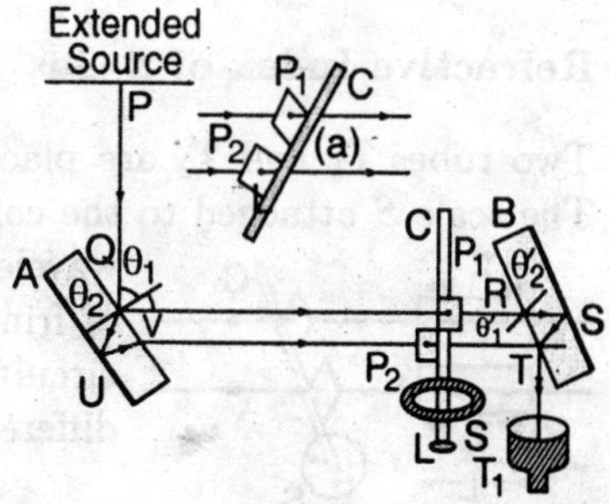

Fig 8.19

The Path Difference

The two plates A and B are identical plates with thickness d. The path difference produced by the first plate is given as

$$\Delta_1 = 2\mu d \cos\theta_2 - \frac{\lambda}{2},$$

and the path difference introduced by the second plate is

$$\Delta_2 = 2\mu d \cos\theta_2' - \frac{\lambda}{2}$$

$\therefore$ The total path difference introduced by the two plates is

$$\Delta = \Delta_1 - \Delta_2 = 2\mu d(\cos\theta_2 - \cos\theta_2') = 2\mu d\left(\sqrt{1-\sin^2\theta_2} - \sqrt{1-\sin^2\theta_2'}\right)$$

Now, $\sin\theta_1 = \mu\sin\theta_2, \quad \sin\theta_1' = \mu\sin\theta_2'$

$$\therefore \quad \Delta = 2d\left(\sqrt{\mu^2 - \sin^2\theta_1} - \sqrt{\mu^2 - \sin^2\theta_1'}\right), \tag{8.6.1}$$

where, θ_1, θ_2, θ_1' and θ_2' are the angles as indicated in fig. 8.19. In case of exact parallelism, $\theta_1 = \theta_1'$, hence, $\Delta = 0$ so the two systems of rays will be in phase everywhere and hence no interference bands but if the plate B be rotated about a horizontal axis, the line of intersection of the plates will be horizontal and we would get horizontal fringes. If the path difference given by eqn. (8.6.1) becomes equal to $m\lambda$, we would get bright fringes.

$$\therefore \quad 2d\left(\sqrt{\mu^2 - \sin^2\theta_1} - \sqrt{\mu^2 - \sin^2\theta_1'}\right) = m\lambda \quad \Rightarrow \quad \text{Bright fringe} \tag{8.6.2}$$

Use of Compensator Plates P_1 and P_2

Two plates P_1, P_2 fixed on to an arm CL at adjustable inclination are called *compensator plates.* They are placed one in each of the paths of the interfering beams so as to produce path difference between the rays. The circular scale attached to the compensator arm is calibrated with reference to fringe shift of a monochromatic light.

8.7 Applications

Refractive Index of a Gas

Two tubes T_1 and T_2 are placed in the path of the interfering beams as shown in fig. 8.20. The scale S attached to the compensator arm CL is calibrated with reference to the standard wavelength λ. Interference fringes are produced with λ and the number of fringes moving across the cross-wire is noted. If x divisions of the circular scale need be rotated to cause a shift of n fringes, then path difference for one division is

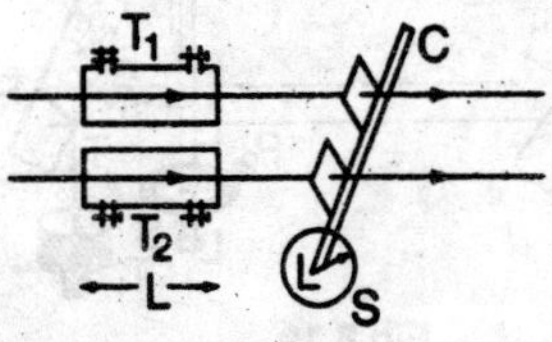

Fig 8.20

$$p = \frac{n\lambda}{x} \tag{8.7.1}$$

The calibration of the scale being over, the two tubes T_1 and T_2 are placed in position. First both tubes are made empty and central bright fringe is taken on the cross-wire. Then one tube is filled with gas under certain pressure. There will be displacement of the central fringe. If the compensator requires to be rotated through y divs. of the circular scale to bring the central fringe back on the cross-wire, then

$$(\mu - 1)l = py = \frac{n\lambda}{x}\cdot y, \tag{8.7.2}$$

where l is the length of the tube T_1 or T_2.

Thickness of a Thin Film

Fringes are first formed in this interferometer with white light, central fringe is taken on the cross-wire. The thin film is then placed in the path of one of the interfering beams. The central fringe shifts. The compensator is rotated through m divs. to bring the central fringe back on to cross-wire. Then,

$$(\mu - 1)t = mx\left(\frac{n\lambda}{x}\right) \tag{8.7.3}$$

8.8 Michelson's Stellar Interferometer

This interferometer designed by Michelson aims to determine the angular separation of two stars and also the angular diameter of a single star.

Two circular sources, such as two stars, are separated by a certain distance and their separation can be determined by placing a double slit in front of the objective of a telescope. Each source will produce Fraunhoffer double-slit diffraction pattern of its own centered on its geometrical image. But as the slit width is very narrow, the fringes are practically the interference fringes whose intensity is controlled by the $\cos^2$-factor. So, if the condition of spatial coherence is satisfied we would get clear fringes.

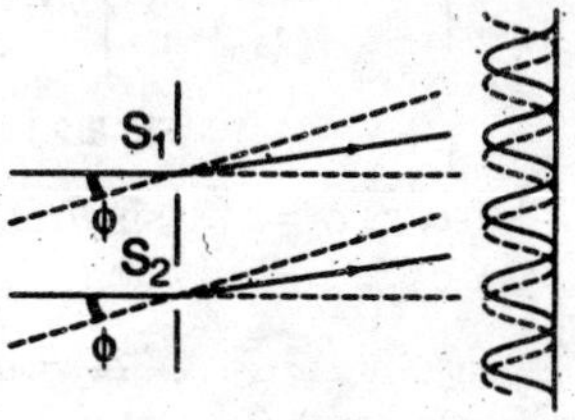

Fig 8.21

Consider the arrangement as shown in fig. 8.21 P, Q are the two sources having angular separation equal to ϕ, the two slits of width a and having separation equal to b are placed in the path of rays from the sources. P, Q may be taken to be two distant stars or two points at the ends of a diameter of a distant star. In the former case, ϕ would be called the *angular separation* of two stars and in the latter case ϕ would mean angular diameter of a distant star.

When the two sources P, Q are very close, the two patterns practically coincide. As the angular separation increases, spatial coherence begins to fail and at a particular separation it so happens that central maximum of one coincides with the minimum of the other, spatial coherence completely fails and interference fringes disappear.

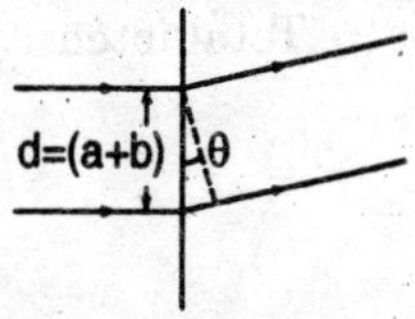

Fig 8.22

The minimum of interference fringes due to one source (say Q) is given by

$$d\sin\theta = (2n+1)\frac{\lambda}{2} \quad \Rightarrow \quad \theta = (2n+1)\frac{\lambda}{2d} \tag{8.8.1}$$

when the angular separation ϕ becomes equal to θ, the disappearance of fringes would occur. So for disappearance of fringes

$$\phi = \theta = (2n+1)\frac{\lambda}{2\phi} \tag{8.8.2}$$

The first disappearance would occur when

$$\phi = \frac{\lambda}{2d} \tag{8.8.3}$$

Now, $\frac{\lambda}{\phi}$ is known as the *lateral coherence width* and

$$d = \frac{\lambda}{2d}$$

is the lateral distance over which the beam may be supposed to be spatially coherent. Thus getting the position of the first disappearance of fringes, where the condition satisfied is given by eqn. (8.8.3) and knowing λ and d, the angular separation of two stars can be calculated.

Angular Diameter of a Single Star

To determine the angular diameter of a star, we will assume the star to be a uniformly illuminated circular disc. Divide the disc into large number of strips parallel to the y-axis.

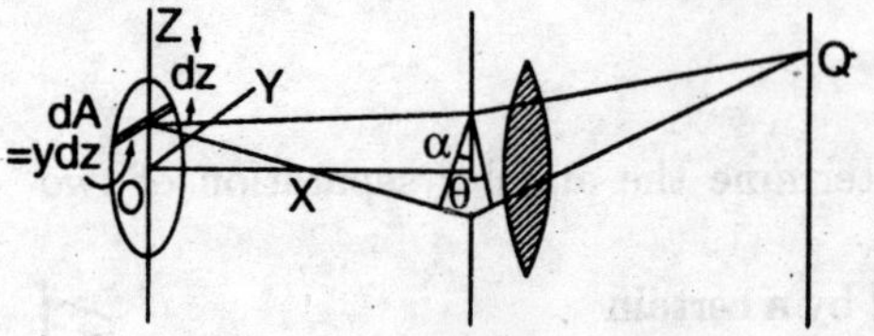

Fig 8.23

The strip shown in fig. 8.23 has the area

$$dA = 2ydZ = 2(R^2 - Z^2)^2 dZ.$$

The point Q in the focal plane of the objective of a telescope has path difference for two rays from the point P given by,

$$\Delta = d(\sin\alpha + \sin\theta) \qquad (8.8.4)$$

$\sin\alpha \simeq \alpha = \frac{Z}{X}$, $\sin\theta \simeq \theta$.

$$\therefore\ \Delta = d\left(\frac{Z}{X} + \theta\right)$$

$$\therefore\ \text{Phase diff. } \delta = \frac{2\pi}{\lambda}\times\Delta = \frac{2\pi d}{\lambda}\left(\frac{Z}{X} + \theta\right) = m\left(\frac{Z}{X} + \theta\right)$$

Due to the elementary strip shown in fig. 8.21, the intensity dI is given as,

$$dI = dA(1+\cos\delta) = 2(R^2 - Z^2)^{1/2} dZ(1+\cos\delta).$$

$\therefore$ Total intensity at Q is given as,

$$I = \int_{-R}^{R} 2(R^2 - Z^2)^{1/2}\left[1 + \cos m\left(\frac{Z}{X} + \theta\right)\right] dZ$$

$$= 2R\int_{-R}^{R}\left(1 - \frac{Z^2}{R^2}\right)^{1/2}\left[1 + \cos m\left(\frac{Z}{X} + \theta\right)\right] dZ$$

Set $\frac{Z}{R} = u \quad \Rightarrow \quad \frac{1}{R}dZ = du \quad \Rightarrow \quad dz = Rdu$

$$\therefore\quad I = 2R^2\int_{-1}^{1}(1-u^2)^{1/2}\left[1 + \cos m\left(\frac{uR}{X} + \theta\right)\right] du$$

$$= 2R^2\int_{-1}^{1}(1-u^2)^{1/2}du + 2R^2\int_{-1}^{1}\cos\left(\frac{muR}{X} + m\theta\right) du \qquad (8.8.5)$$

$$\int_{-1}^{1}(1-u^2)^{1/2}du, \text{ let } u = \sin\theta, \quad du = \cos\theta d\theta$$

$$= \int_{\frac{-\pi}{2}}^{\frac{\pi}{2}}\cos^2\theta d\theta = \int_{\frac{-\pi}{2}}^{\frac{\pi}{2}}\frac{1+\cos 2\theta}{2}d\theta = \frac{\pi}{2}$$

$$\therefore\quad I = \pi R^2 + 2R^2\int_{-1}^{1}\left(\cos\frac{muR}{X}\cos m\theta - \sin\frac{muR}{X}\sin m\theta\right)(1-u)^{1/2}du$$

Now, $\int_{-1}^{1} \sin \frac{muR}{X} du = 0$

$$\therefore \quad I = \pi R^2 + 2R^2 \int_{-1}^{1} \cos \frac{muR}{X} \cos m\theta (1-u^2)^{1/2} du$$

$$= \pi R^2 + 2R^2 \cos m\theta \int_{-1}^{1} \cos \frac{muR}{X} (1-u^2)^{1/2} du$$

Using the standard result for Bessel function, we can write,

$$\int (1-u^2)^{1/2} \cos\left(\frac{mR}{X} u\right) du = \frac{\pi J_1(mR/X)}{(mR/X)} \tag{8.8.6}$$

using relation (8.8.6) we get,

$$I = \pi R^2 + 2R^2 \cos m\theta \times \frac{\pi J_1(mR/X)}{(mR/X)} \tag{8.8.7}$$

Eqn. (8.8.7) gives the resultant intensity at any point Q due to an illuminating disc source (like a star).

The second term represents the intensity distribution in the fringes and so fringes would disappear when the $\frac{J_1(x)}{x} = 0$. This happens when $x = 1.22\pi$.

$\therefore$ When the fringes disappear, we would get,

$$\frac{mR}{X} = 1.22\pi \Rightarrow \frac{2\pi d}{\lambda} \cdot \frac{R}{X} = 1.22\pi$$

$$\Rightarrow \frac{2R}{X} \cdot \frac{d}{\lambda} = 1.22 \quad \Rightarrow \quad \phi \cdot \frac{d}{\lambda} = 1.22$$

or, angular diameter

$$\phi = 1.22 \frac{\lambda}{d} \tag{8.8.8}$$

Thus the angular diameter of a star can be determined with the help of eqn. (8.8.8) observing the disappearance of the fringes. But to get the disappearance of the fringes, we would get large value of d. To appreciate the statement numerically, we take

$$\lambda = 5000\text{Å}, \quad \phi = 0.01 \text{ sec} = \frac{0.01}{57.4 \times 60 \times 60} \text{ rad}.$$

$$\therefore \quad d = \frac{1.22\lambda}{\phi} = \frac{1.22 \times 5500 \times 10^{-8} \times 57.4 \times 60 \times 60}{0.01} = 1390 \text{ cm} = 13.9 \text{ meters}$$

To realise this distance, the objective of the telescope must have diameter as large as this value. It is not practically possible to get it. Moreover, for this value of d, the fringe width would become so small that it cannot be observed.

To resolve these difficulties Michelson in his stellar interferometer made an ingenious modification as shown in fig. 8.24.

In fig. 8.24 M_1 and M_2 are fixed mirrors, whereas M_3 and M_4 movable mirrors symmetrically w.r.t. the axis. Parallel rays from the ends of the diameter of the star fall on M_3 and M_4, the pair of parallel rays marked (I, I) come from one end while the other pair of parallel rays marked

(II, II) come from the other end. After reflection by the two mirrors, they fall on the double slit and produce interference fringes on a screen and its theory has been worked out in details through eqns. (8.8.5) to (8.8.7). The angle between the two systems of rays is given by ϕ such that when from a small distance apart when the mirrors M_3 and M_4 are symmetrically moved, then at a distance h between them first disappearance of fringes occurs. Then we can write from eqn. (8.8.8), that

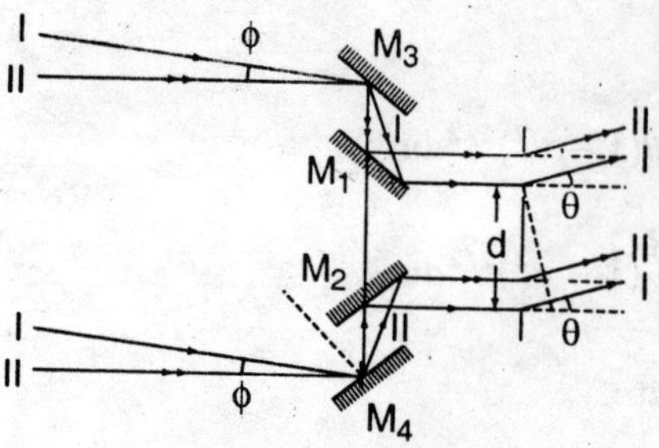

Fig 8.24

$$\phi = \frac{1.22\lambda}{h}.$$

If the mirrors M_3, M_4 are not used, for the disappearance of the fringes, we would get,

$$\phi = \frac{1.22\lambda}{d}$$

Suppose in an arrangement using Michelson's stellar interferometer,

$$\lambda = 5700\text{Å}, \quad \phi = 22.7 \times 10^{-8}\text{ rad}$$

$$\therefore \quad h = \frac{1.22\lambda}{\phi} = 306.8\text{ cm} \simeq 121\text{ inches.}$$

Now, in case of circular disc type source, the coherence width is given by

$$l_w = \frac{1.22\lambda}{\phi}.$$

So, without M_3, M_4, the coherence width

$$l_w = \frac{1.22\lambda}{\phi} = d \quad \text{(double-slit separation)}$$

In presence of M_3, M_4 introduced by Michelson,

$$l_w = \frac{1.22\lambda}{\phi} = h \quad (M_3 - M_4\text{—mirror separation})$$

Thus, in presence of M_3, M_4, the coherence width increases.

Solved Numerical Problems

1. Using stellar interferometer Michelson found that for star the fringes disappear when the distance between the movable mirrors is 25″. Assume $\lambda = 6 \times 10^{-5}$ cm, calculate the angular diameter of the star.

Solution. Here, $h = 25'' = 63.5\,\text{cm}$, $\lambda = 6 \times 10^{-5}$ cm

$$\therefore \quad \text{Angular diameter } \phi = \frac{1.22\lambda}{h} = 0.238\,\text{sec}.$$

2. Write down the expression for the lateral coherence length in case of a circular source. Use this expression to find the double slit constant for clear fringes to be available for a source of linear diameter 0.2 cm and the distance of the slits from the source being 1,000 cm. Assume $\lambda = 5 \times 10^{-5}$ cm.

Solution. For a circular source, the lateral coherence width is given by,

$$l_w = \frac{1.22\lambda}{\phi}, \text{ here, } \phi = \frac{0.2}{1000} = 2 \times 10^{-4} \text{ rad}$$

$$\therefore \quad l_w = \frac{1.22\lambda}{2 \times 10^{-4}} = 0.31 \text{ cm}$$

So the slit constant must be less than this value for clear fringes to be seen.

3. For the star Betelgeuse the mirror separation for the first disappearance of the fringes was 3 meters. If the distance of the star from the earth is 1.7×10^{15} km, calculate its diameter, assuming effective wavelength emitted as 5750Å.

Solution. $2R/X = \phi$.

Also,

$$\phi = \frac{1.22\lambda}{h}, \text{ here, } h = 3 \times 10^2 \text{ cm} = 2.34 \times 10^{-7} \text{ rad}$$

$$2R = \text{diameter} = 1.7 \times 10^{15} \times 2.34 \times 10^{-7} = 3.98 \times 10^8 \text{ km.}$$

4. In a Jamin's refractometer, two evacuated tubes each of length 20 cm are placed in the two beams. A gas at some temperature and pressure is slowly introduced and 200 fringes cross the centre of the field. Calculate the refractive index and refractivity of the gas, $\lambda = 5890$Å.

Solution. $(\mu - 1)l = n\lambda$, here $l = 20$ cm, $n = 200$, $\lambda = 5890 \times 10^{-8}$ cm

$$\therefore \quad \mu = \frac{n\lambda}{l} + 1 = 1.00059$$

$$\text{Refractivity} = (\mu - 1) = \frac{n\lambda}{l} = 59 \times 10^{-5}.$$

Exercises A

1. How are the interference fringes formed in a Fabry-Perot interferometer? Derive an expression for the chromatic resolving power in such an interferometer.
A Fabry-Perot interferometer with a mirror separation of 0.5 cm and having a reflecting power of 50% is placed to receive a radiation whose mean wavelength is 5000Å. Find the resolving power. $\left[\text{Use the formula, } \frac{\lambda}{\Delta\lambda} = \frac{m\pi r}{(1-r^2)} = \frac{\pi r}{(1-r^2)} \times \frac{2d\cos\theta}{\lambda}\right]$
2. Derive an expression for the intensity of interference fringes formed by transmission in a Fabry-Perot interferometer and its chromatic resolving power. Explain why the fringes are very narrow. How can you study the hyperfine structure with this interferometer?
3. The plates of a Fabry-Perot interferometer have reflectance of 0.9. Find the separation between the plates when its resolving power is equal to that of a 5 inch grating of 15,000 lines/inch in the first order. Assume $\lambda = 5000$Å.

4. Explain the formation of fringes in a Michelson's interferometer. Compare and contrast the resolving power and shape of fringes of this interferometer with those of Fabry-Perot interferometer.

5. Explain the formation of fringes in Fabry-Perot interferometer. What is the shape of these fringes? How would you use it for the measurement of wavelength of light?

6. Explain the formation of sharp maxima by a Fabry-Perot interferometer. Describe an arrangement in which the interferometer can be used to compare the values of two nearly equal wavelengths.

7. By means of Fabry-Perot interferometer an unknown wavelength less than 6000Å is to be compared with a standard wavelength of 6000Å. Coincidence occurs when plates are separated by 0.0015 m, 0.003 m. and 0.0045 m. Find the unknown wavelength. [5998.8Å]

8. Fringes of equal inclination are formed in a Michelson's interferometer. As one of the mirrors is moved back 1 mm, 3663 fringes move out from the centre of the pattern. Calculate the wavelength of light. What will be observed if one then changes to white light? [5460Å]

9. A Michelson interferometer is adjusted so that fringes can be observed with white light and the source is changed to a quasi-monochromatic source. Describe the appearance of the fringes. Fringes are photographed using black and white film with a quasi-monochromatic source emitting (a) blue light, (b) green light. How can one distinguish between the photographs?

10. Explain the formation of fringes in a Michelson's interferometer illuminated by (a) monochromatic light, (b) white light. If one mirror is slowly moved back with velocity v, calculate directly the number of monochromatic fringes crossing a point in the field of view in one second. [$2v/\lambda$ fringes]

11. Describe the construction and principle of Michelson's interferometer. In what circumstances is the compensating plate essential?

    Describe and explain the procedure for determining the wavelength separation of the two *D*-lines with the help of the interferometer.

12. Two mirrors of a Fabry-Perot interferometer have reflecting power 0.80. Estimate in each instance the separation of the mirrors when the resolving power is equal to that in the first order of a 5 inch grating of 15,000 lines/inch. $\lambda = 5000$Å. [0.176 cm]

Exercises B

1. What is Jamin's interferometer? What is the role of the compensator? What type of fringe do you get in this interferometer?

   The two tubes of a Jamin's interferometer are each 40 cm long. One is evacuated, the other contains gaseous Argon at atmospheric pressure. The refractive index of the gas is 1.00028. How many fringes of green light ($\lambda = 5500$Å) would be counted when the gas is pumped out? [203.6]

2. Describe a suitable arrangement for measuring accurately the refractive index of (a) a liquid, (b) a gas.

3. Describe a suitable arrangement for measuring accurately the refractive index of (a) a liquid, (b) a gas.

4. Discuss the importance of Lummer-Gehrcke plate. derive an expression for its chromatic resolving power.
5. How did Michelson determine the angular diameter of a star with his stellar interferometer?
6. What is the lateral coherence width for a circular source? Derive an expression for it.
7. How can you determine the angular separation between two stars in a double-star system?
8. Compare the interference fringes produced by Fabry-Perot interferometer and Lummer plate.
9. The length of a gas-filled tube in Janin's interferometer is 20 cm. When the pressure of the gas is changed by 50 cm of Hg, 50 fringes pass across the cross-wire of the telescope. Find the value of μ of the gas at normal pressure. Take $\lambda = 5893$Å and use the standard relation $(\mu - 1) = A\rho$, where A is some constant and ρ is the density of the gas. [1.000224]

References

1. **S. Tolansky,** *An Introduction to Interferometry*, New York, John Wiley & Sons, Inc., 1973.
2. **J. Dyson,** *Interferometry as a Measuring Tool*, Brighton, England : The Machinery Publishing Co. Ltd., 1970.

Chapter 9

Polarised Light

9.1 Introduction

We have considered earlier two optical phenomena, viz., *Interference* and *Diffraction.* We need not look into the nature of the wave; it was sufficient to explain the phenomena and the associated problems by considering light simply as a wave having three characteristics:

1. It propagates in a medium with a certain speed and at every point, there is a fixed and measurable physical quantity.
2. This physical quantity changes periodically.
3. The state of the particle of the medium repeats on an immediate neighbour in a succeeding moment.

These information about the wave are sufficient for explaining the interference and diffraction phenomena, no need of informing about the nature of the wave.

But subsequently an optical phenomenon was discovered which for its explanation required specific information about the nature of the wave. We know that as the wave propagates, the particles of the medium do vibrate either in the direction or at right angles to the direction of propagation. The first type is referred to as longitudinal disturbance, the second as transverse disturbance. The phenomena of reflection, refraction, interference and diffraction—all these optical facts could be simply explained if we assume wave nature of light, without any specific prescription about the nature of the wave. But the phenomenon of polarisation, which we are at present going to discuss, requires specific information about the nature of the wave. This is the phenomenon which for its explanation requires that light be considered as a transverse disturbance. We have seen in the chapter on Electromagnetic Theory (Chapter Six) that light which is an electro-magnetic disturbance consists of electric wave and magnetic wave at right angles to each other and also at right angles to the direction of propagation of light. So we conclude that light is a transverse disturbance, a fact so much needed for the explanation of the phenomenon of *polarisation.*

What is Polarisation? Polarised Light and Unpolarised Light

Unpolarised light is light that exhibits no long-term preference to vibration pattern whereas polarised light shows sectional pattern either linear, circular or elliptical. Unpolarised or ordinary light consists of oscillations in any plane about and at right angles to the direction of propagation because such light is emitted by an atom or a group of atoms of the source vibrating independently. The coherence time is of the order of 10^{-8}s, so while the individual wavetrain oscillates in some constant mode and direction, the modes and directions of oscillations vary from wavetrain to wavetrain, the result being the sequence of oscillations oriented at random. If by some means, these oscillations in an unpolarised light are so cut off that only oscillation of a constant mode and direction remain, the light so obtained will be called *polarised light*, the phenomenon is called *polarisation*. The plane which contains the vibration of the polarised light is called the *plane of vibration* and a plane normal to the plane of vibration is called the *plane of polarisation.*

In fig.(9.1) we have represented by ↕ the direction of vibration of a linearly polarised light and the plane containing these vibrations is termed as the plane of vibration; the shaded plane normal to the plane of vibration is the *plane of polarisation.*

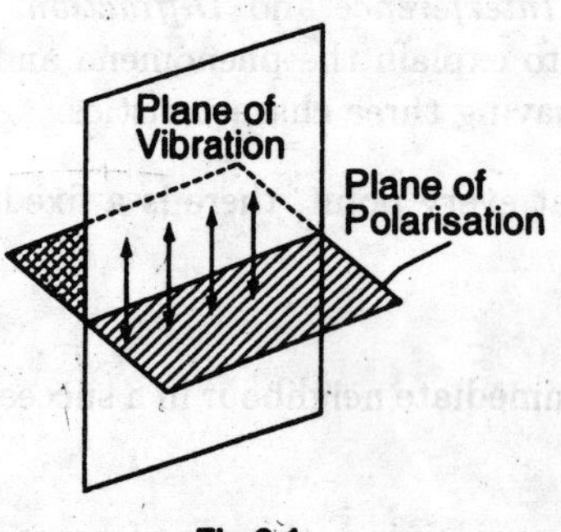

Fig 9.1

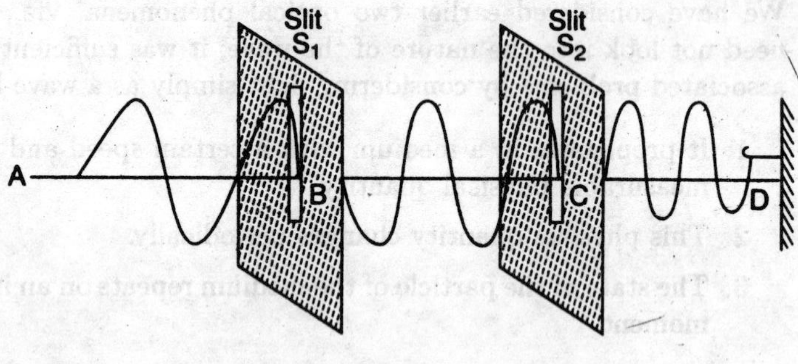

Fig 9.2

A mechanical example, I suppose, will make the matter clear [see fig.(9.2)]. Two card-boards containing narrow slits S_1 and S_2 are placed parallel to each other at a certain distance apart. The width of the slit is just sufficient to allow a wire to pass through it. One end of the wire (left-hand end say) is held by hand and the other end after taking the wire through the slits is fixed to a hook. The free end is vibrated transversely. The portion of the wire on the left of the first slit can be thrown into vibration in any plane, but when the wire is set to vibration parallel to the slit, wavy undulations will be observed over the entire length, right up to the end fixed to the hook. The vibration can be set up in the portion AB in any plane, hence unqualified vibration but vibration in the portion BC will be confined to the direction parallel to the linear dimension of the slit S_1; when the slit S_2 is also parallel to S_1, vibration is observed in the portion CD also. When the slit S_2 is rotated with the wire as axis, the vibration gradually dies away and it completely fades out when the slit S_2 is at right angles to S_1 (called *crossed position of the slits*). Since vibration in the portion BC is confined only to a particular plane, it is called *polarised vibration* and the first slit acts as a polariser. The second slit whose rotation from parallel orientation to crossed orientation fades out the vibration in the portion CD acts as an analyser.

It is to note that only in case of transverse vibration set up in the wire such creation of polarised disturbance is possible. If longitudinal disturbance is set up in the wire by rubbing along the length by resined cloth or chamois leather, the disturbance always spreads over the entire length irrespective of the orientation of the slits.

9.2 Types of Polarised Light

Polarised light may be primarily of three varieties—(1) Linearly polarised light, (2) Circularly polarised light, (3) Elliptically polarised light.

Linearly polarised light

Light in which oscillation is contained in a plane or in a constant orientation is referred to as linearly polarised light. Thus in fig.(9.3a) the oscillation is in the horizontal plane and propagation direction is along the X-axis. So it is said to be horizontally linearly polarised, the representation of such polarised light is shown to the right.

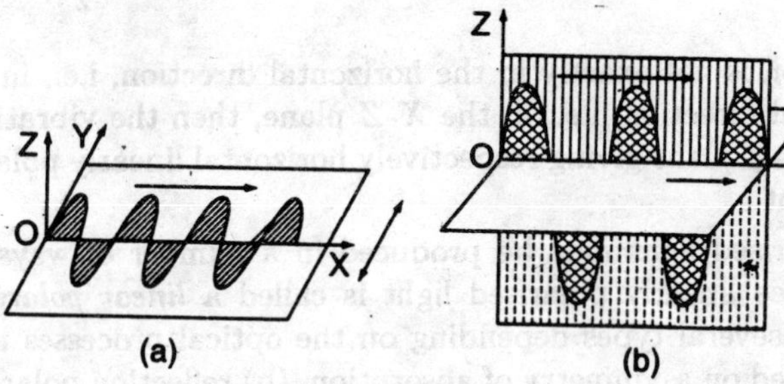

Fig 9.3

In fig.(9.3b) we have shown another type, linearly polarised but it is in vertical plane, the propagation direction being same, its representation is shown to the right.

9.3 Linear Polariser

The device or mechanism to produce linearly polarised light is called a *linear polariser*. There are several types of linear polarisers due to different optical processes utilised for its production.

Linear polarisers may be divided into (a) Dichroic polariser, (b) Reflection polariser, (c) Refraction polariser, (d) Scattering polariser.

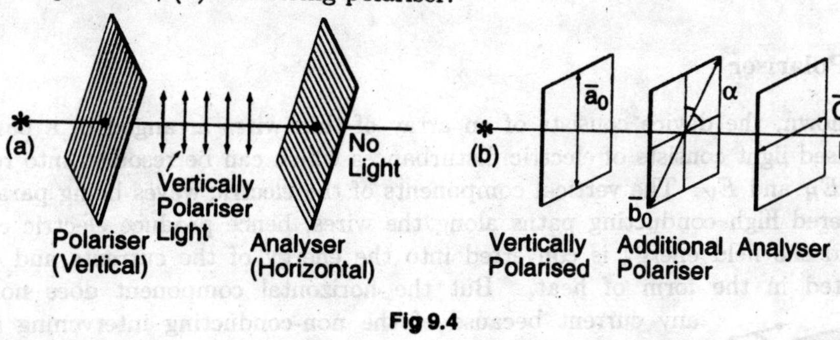

Fig 9.4

The function of a linear polariser may be understood with reference to the figure (9.4). In first figure ordinary light is made to be incident on a linear polariser producing polarised light in the vertical direction.

Consider the arrangement shown in fig.(9.4b) where the first polariser is a vertical polariser, this vertically polarised light is then allowed to pass through the additional polariser whose transmission axis makes an angle of β with the vertical direction represented by unit vector $\bar{a}_0$;

the directional unit vector along the transmission axis of the additional polariser is represented by $\bar{b}_0$ and the analyser transmission axis is at right angles to the first polariser. Let the amplitude of the polarised light by the vertical polariser be A, then the amplitude passing through the polariser can be written as $A\bar{a}_0$, the amplitude through the additional polariser is given by

$$A_c = A\,(\bar{a}_0 \cdot \bar{b}_0)$$

The amplitude of the emergent light through the analyser is written as

$$\begin{aligned} A_0 &= A_c\bar{b}_0 \cdot \bar{c}_0 = A\,(\bar{a}_0 \cdot \bar{b}_0)\,(\bar{b}_0 \cdot \bar{c}_0) \\ &= A\cos\beta\cos\alpha = A\cos(90-\alpha)\cos\alpha = \frac{1}{2}A\sin 2\alpha \end{aligned}$$

So, if the vibration is not exactly in the horizontal direction, i.e., in the X-Y plane or not exactly in the vertical direction, i.e., in the X-Z plane, then the vibration may be resolved in the X-Y plane and X-Z plane giving respectively horizontal linearly polarised light and vertical linearly polarised light.

The linearly polarised light may be produced in a number of ways and such a device or mechanism to produce linearly polarised light is called a *linear polariser*. Linear polarisers may be divided into several types depending on the optical processes involved. They are (a) dichroic polariser based on asymmetry of absorption, (b) reflection polariser based on reflection phenomenon, (c) refraction polariser based on the phenomenon of refraction, (d) scattering polariser utilising the phenomenon of scattering.

9.4 Dichroic Polariser

The action of this type of polariser is related to the symmetry of absorption of the polarising element. The simplest member of this family is the *wire-grid polariser* whose chemical version came in the form of H-sheet polariser invented by Land in 1938, the world's most popular polariser; J-sheet polariser, the first sheet type polariser; K-sheet polariser, HR-sheet polariser, etc. Let us begin our discussion with wire-grid polariser.

Wire-grid Polariser

In the fig. shown, the device consists of an array of thin wires arranged in a parallel grid. The unpolarised light consists of electric disturbances which can be resolved into rectangular components E_H and E_V. The vertical components of the electric waves being parallel to the wires are offered high conducting paths along the wires, hence produce electric currents in the wires and the field energy is converted into the energy of the currents and ultimately it is dissipated in the form of heat. But the horizontal component does not produce any current because of the non-conducting intervening space and no energy is dissipated corresponding to this horizontal component, thus such a grid-wire arrangement transmits the horizontal electric field component through without any attenuation of energy but the component parallel to the wires is preferentially absorbed or in other words dissipated due to current flow in the wires. Thus we get a linearly polarised light, polarised in the horizontal direction.

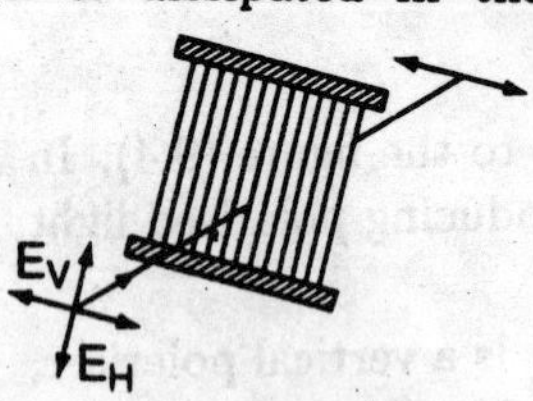

Fig 9.5

Though the basic principle is very easy to understand, it is very difficult to make a satisfactory wire-grid polariser. The difficulty is very easy to understand because in slightly larger space than a postage stamp, we have to incorporate about 30,000 wires. Definitely, it is a fantastic task. But this task was accomplished by G.R. Bird and M. Parish who formed a grid of wires by evaporating in vacuum stream of gold atoms onto the ridges of a transparent diffraction grating of 50,000 lines/inch. Each ridge became coated with gold atoms along one side. The coatings, a few hundred atoms thick provide good conductivity in one direction.

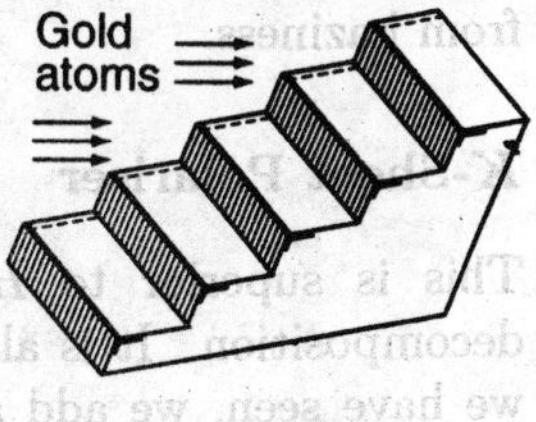

Fig 9.6

The performance of a wire-grid is very bad if the wires are thick and the spaces between them are large. A thick wire permits some current flow in transverse direction across individual wire and thus absorbs some of the wanted component of light. If the wires are randomly oriented, both the components are absorbed and the device would not transmit any light at all.

H-Sheet Polariser

This is the most popular sheet polariser, to be considered as the chemical version of the wire-grid polariser as described above, invented by E.H. Land in 1938.

Instead of the wires as in grid-wire polariser, here we employ long, thin molecules—long-chain polymeric molecules that contain many iodine atoms. These long, straight molecules are aligned almost perfectly parallel to one another. These iodine atoms provide conductivity for electric field component parallel to the aligment, thus absorbing it. The component perpendicular to it passes through, with very little absorption.

Alignment of the conducting molecules is done in a very ingenious way. A large transparent sheet of easily stretched and chemically reactive plastic—usually polyvinyl alcohol is taken. The sheet is warmed and very quickly stretched to many times its original length. With increase in linear dimension it becomes also narrower and thinner.

During stretching by hand or preferably by machine, most of the long polymeric molecules, initially at random orientation, come to align in the direction of the stretching force. Thus we can realise an alignment of billions of molecules. As soon as stretching is complete, the sheet is cemented to a rigid sheet of cellulose acetate, so that no unstretching can occur. It is then dipped into a liquid solution that is iodine-rich. Very quickly iodine diffuses into the polyvinyl alcohol layer and gets attached to the molecules. Thus the iodine atoms tend to form long, thin chains too small to be seen under microscope but only can be examined by X-ray diffraction techniques.

The sheet thus prepared is washed, dried and cut into pieces of any desired shape and size. Cover plates of glass are then put to protect the plastic layers from scratches or fingerprints.

The long, thin iodine chains act like wires. They absorb electric vibrations parallel to the aligment axis, but freely transmit transverse vibrations. *H*-sheet polarisers are designated as *HN*-32 meaning 32% transmission of the incident light; *HN*-22 transmits 22%.

J-Sheet Polariser

Although *J*-sheet polariser is no longer produced, it is to be noted that it was the first of its kind invented by B.H. Land, employing actual crystals, millions of them of dichroic variety. Land used crystals of *herapathite*, a strongly dichroic mineral which absorbs light to different extent depending on the vibration direction.

These crystals have a needle-shape structure and hence can be aligned relatively simply. When aligned, they all agree to absorb one vibration component, transmitting the orthogonal component. Thus, a large sheet of plastic containing alignment of millions of herapathite crystals acts like one huge crystal.

The main difficulty with this sheet polariser was its haziness. In this respect *H*-sheet polariser being a molecular polariser rather than a micro-crystalline polariser does not suffer from haziness.

K-Sheet Polariser

This is superior to *H*-sheet polariser since it can withstand high temperature without decomposition. It is also made from polyvinyl alcohol like the *H*-sheet. In making *H*-sheet, we have seen, we add atoms to the sheet, but here instead of adding atoms, we take atoms away. Employing hydrogen chloride catalyst and a high temperature oven Land and Ragers removed $2N$ hydrogen atoms and N oxygen atoms, thus producing a different type of polymeric molecule, called *polyvinylene.* Polyvinylene molecules are aligned by stretching the sheet in one particular direction.

Other Types of Dichroic Polariser

Dichroic polariser may be made of dyes consisting long and thin molecules. A large crystal of appropriate type, as for the case with tourmaline, may serve the purpose of asymmetric absorption of the incident light.

Reflection Polariser

This type of polariser depends for its action on the asymmetry in reflection from a glass-plate or any polished surface, advantage of this type is that it can be used in any portion of the specimen.

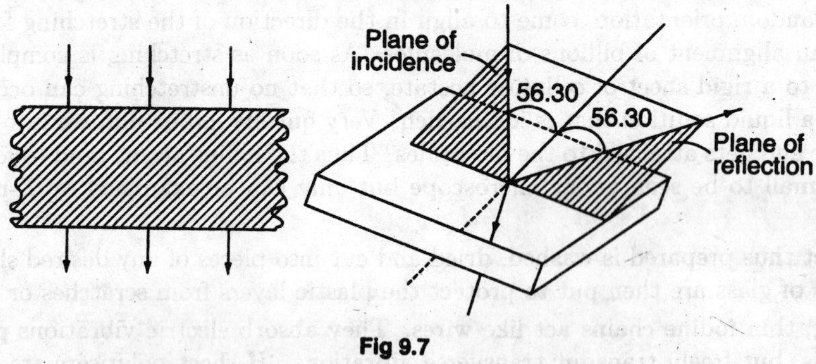

Fig 9.7

When an unpolarised beam (ordinary light) is incident on a glass-plate normally, the transmitted light is also unpolarised. But when the glass-plate is titled, an interesting change appears as observed by Malus in 1808. The transmitted beam is found to be partially polarised. The reflected beam is even more polarised.

The vibrations of the two parts are orthogonal. There is a certain angle of incidence, called *Brewster's angle* or also called *angle of polarisation* such that the reflected light gets almost completely polarised. The vibration continued in the reflected beam is perpendicular to the plane of incidence.

9.5 Refraction Polariser : Double-refraction Phenomenon (or Bi-refringence)

This mechanism of getting linearly polarised light is based on the asymmetry in the refraction of light in a crystal. Actually the phenomenon of polarisation was discovered in connection with this type of refraction.

There are two types of crystals, uni-axial and bi-axial, where, we find double-refraction of the incident beam. We have already referred to the essence of crystallography when we discussed X-ray diffraction in the chapter on diffraction. We have come across seven classes to which all the crystals will belong. These seven classes are triclinic, monoclinic, trigonal, hexagonal, orthorhombic, tetragonal and cubic. We have also stated the fact that the cubic crystals have got maximum number of symmetry elements which include (i) axis of symmetry, (ii) plane of symmetry and (iii) centre of symmetry.

In cubic crystals like sodium and potassium chloride, diamond, etc. many properties such as electrical, thermal, optical are same in all directions and, therefore, such crystals are also termed as isotropic crystals. In such a crystal, double-refraction phenomenon does not take place. But the remaining six classes of crystals show the phenomenon of double refraction. Of these six, trigonal, tetragonal and hexagonal crystals are distinctly different from orthorhombic, monoclinic and triclinic crystals in the sense that in the former trio, there is one particular direction such that if light is allowed along that direction, double refraction does not occur. This direction is identified with the optic axis of the crystal. But in the latter trio, there are two such directions along which double refraction does not take place. So a crystal belonging to either of trigonal, tetragonal or hexagonal class is called a *uni-axial crystal* while a crystal belonging to either of orthorhombic, monoclinic or triclinic class is called a *bi-axial crystal.* Whether uni-axial or bi-axial does not matter, a crystal exhibiting double refraction is said to bi-refringent.

What is the phenomenon of Double Refraction?

We know that when ordinary light (or unpolarised light) is incident on the surface of glass, there is a refracted ray. But it was observed by Erasmus Bertholinus when he passed light through a crystal of iceland spar (calcite crystal) belonging to rhombohedral class that instead of one refracted ray, there were two refracted rays. Of these two refracted rays, one ray is found to obey laws of refraction due to Snell and hence, it is termed as *Ordinary ray* (or *O-ray*) and the other ray does not obey Snell's law and hence, it is called *Extraordinary ray* (or *E-ray*). It was further observed that both these rays are plane polarised and the vibrations in them are orthogonal. This phenomenon in which one incident ray gives rise to two refracted rays due to refraction in an anisotropic crystal is known as *double refraction* (or *bi-refringence*). The state of polarisation of these two refracted rays is orthogonal.

A Few Common Terms

Optic Axis of a Crystal

Consider a calcite crystal, viz., the transparent variety, called *iceland spar.* Like any other crystal, this is also bounded by plane faces inclined to one another at fixed angles. The repetitive unit out of which a big calcite crystal is formed is shown in fig.(9.8), whose two faces are designated as $ABCD$, $A_1B_1C_1D_1$. It is a rhombohedral body having six parallelogram faces, the angles are 78°5′ and 101°55′. Each of its eight corners is formed as a meeting of three plane faces. The pair of corners such as A and C_1 are formed by meeting of three plane faces with obtuse angles adjacent to each other. If we consider a line within the crystal either at A or C_1 such that it is equally inclined to all the three faces, it will be called an *optic axis*. Actually, however, optic axis is not a line but it is a direction, so the line drawn at A within the crystal and equally inclined with the three plane faces is definitely an optic axis but this is not the only optic axis, any line parallel to this optic axis at A is also an optic axis. This is such a direction within the crystal that light directed along that direction will not show any double refraction phenomenon. In a uni-axial bi-refringent crystal, there is one such direction but in a bi-axial crystal there are two such directions.

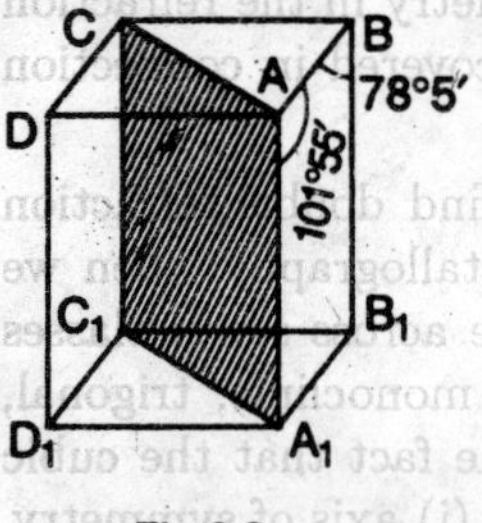

Fig 9.8

Principal Plane and Principal Section

Principal section of a crystal is the section perpendicular to its two opposite refracting faces and through the optic axis. The plane obtained through principal section is a principal plane.

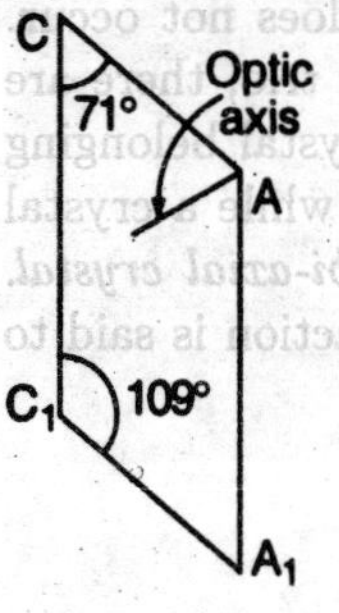

Fig 9.9

In fig.(9.9), the section cut along the shorter diagonal AC and also containing the optic axis, is the principal section and the plane thus obtained as ACC_1A_1 is the principal plane. Since the optic axis is not unique and any direction parallel to the direction shown as optic axis in fig.(9.9) is also an optic axis, it means that we can have a number of principal sections and consequently principal planes. The end-on-view representation of the principal section and principal planes are shown in fig.(9.10).

Take a principal section ACC_1A_1 of the crystal; a narrow beam of light from a narrow pin-hole is allowed to fall normally or obliquely on the face AC. Looking from the opposite face A_1C_1, we will get two images of the slit separated from each other. These two images correspond to two refracted rays. In fig.(9.11), the incident ray is normal, we get one image in line and the other image laterally deviated but parallel to the incident direction. If the crystal be rotated about the incident ray as axis, one image that which corresponds to the direction in line with the incident direction in fig.(9.11) or that which occurs nearer to the normal at the point of incidence in fig.(9.12) remains stationary, the other image rotates about the stationary image. Now, rotation of the crystal about the incident ray leaves the angle of incidence unchanged, so the image which remains stationary, meaning constancy of the angle of refraction, is due to refracted ray obeying Snell's law and hence this refracted ray is called *ordinary ray*. The other ray corresponding to which the image rotates and is not restricted by Snell's law is called *extraordinary ray* or *E-ray*.

Both the O-ray and E-ray are polarised and on examination by an analyser (say a tourmaline crystal, the mode of analysis discussed later), the state of polarisation of these two rays is found to be orthogonal. The vibration of the ordinary ray is perpendicular to the plane of the crystal and indicated by a dot (·) and the vibration of the extraordinary ray is parallel to the plane and indicated by (↔). Thus we can produce linearly polarised light through double-refraction. The crystals like calcite, tourmaline, etc., which exhibit double refraction when unpolarised light is allowed through them are called *bi-refringent crystals*.

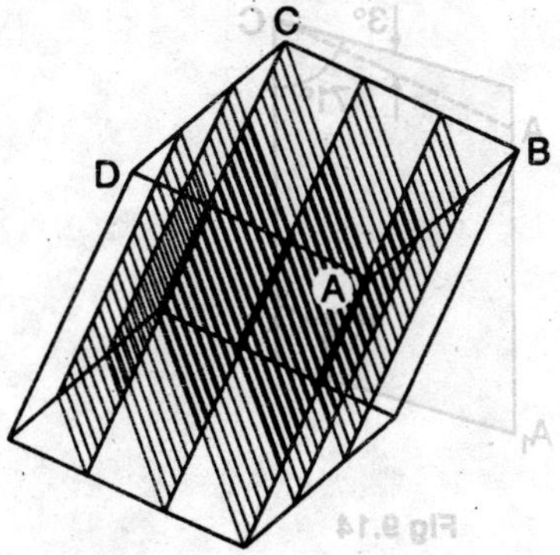

Fig 9.10

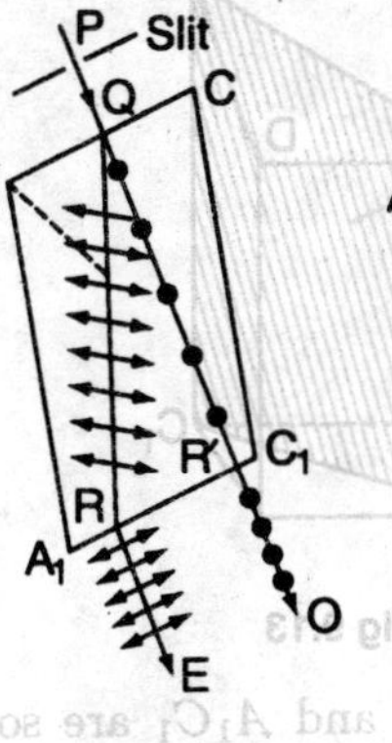

Fig 9.11

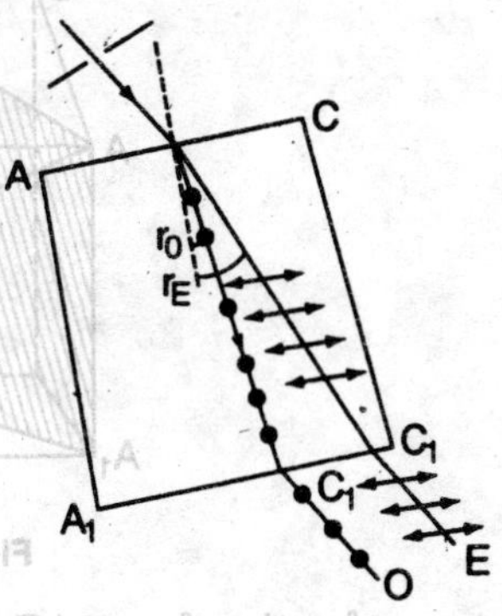

Fig 9.12

Now, when light is allowed to pass through a bi-refringent crystal, along the optic axis, both ordinary and extraordinary ray travel with the same speed. But when travelling takes place in a different direction, they travel with unequal speeds. As already remarked, ordinary ray satisfied Snell's law but the extraordinary does not.

So the r.i. value of the crystal for ordinary ray is a constant and is defined as,

$$\mu_o = \frac{\text{velocity of light in vacuum}}{\text{velocity of O-ray in the crystal}}.$$

But, the r.i. for E-ray is not a constant, it is dependent on the direction of traverse. So for +ve crystal,

$$\mu_e = \frac{\text{velocity of light in vacuum}}{\text{minimum velocity of E-wave}}.$$

For a negative crystal,

$$\mu_e = \frac{\text{velocity of light in vacuum}}{\text{maximum velocity of E-wave}}.$$

Linearly Polarised Light by Nicol Prism

The most important of double refraction type of polariser is the *Nicol prism polariser*.

Nicol, a Scottish physicist, in the year 1828 invented the most useful polariser of double refraction type. It was extensively used in laboratory work previously but because of its complexity, it is not much in vogue nowadays.

It consists of a rhombohedron whose length is roughly three times the breadth. Take the principal section of the rhombohedral crystal which is shown as shaded in fig.(9.13) and it is denoted as ACC_1A_1. It is a parallelogram whose angles are 71° and 109° as shown separately in fig.(9.14).

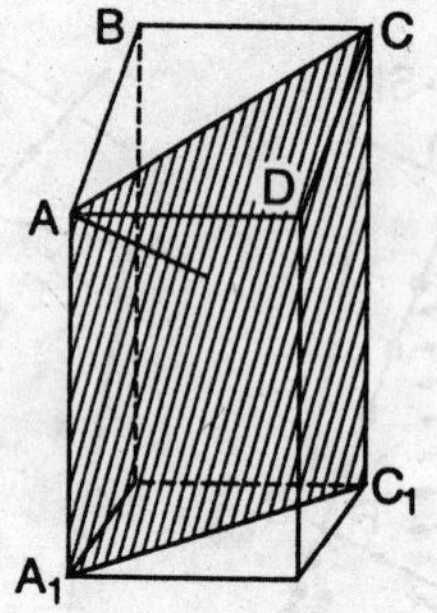

Fig 9.13

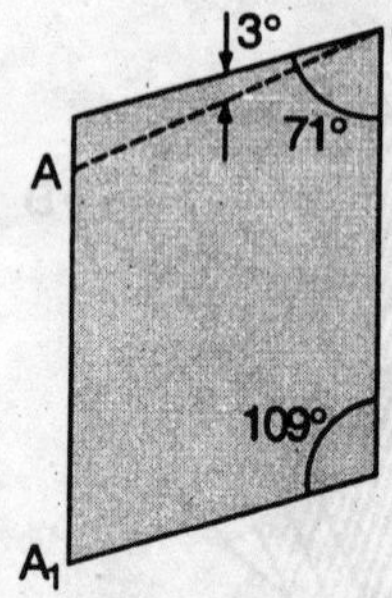

Fig 9.14

The two refracting faces AC and A_1C_1 are so polished and ground that the angle 71° is reduced to 68°, i.e., 3° cut out. After this adjustment, the principal section would look like what has been shown in fig.(9.15).

It is now cut along $A'C'$ by a plane ⊥ to both the principal section and the end faces of the rhomb such that $\angle CA'C' = 90°$. The two cut-out halves are again cemented together by an interposed layer of Canada balsam. The complete outfit is now what is shown in fig.(9.16).

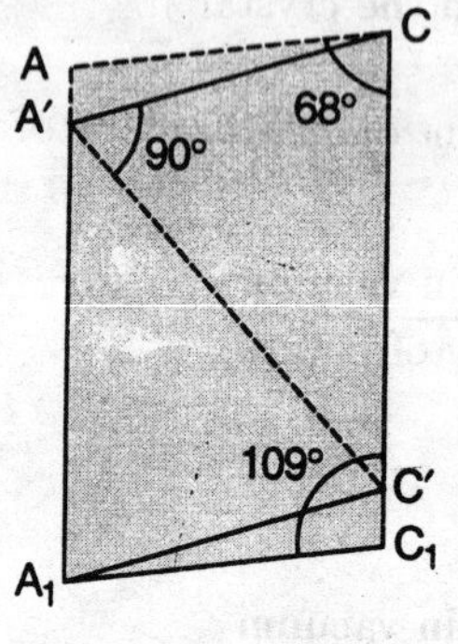

Fig 9.15

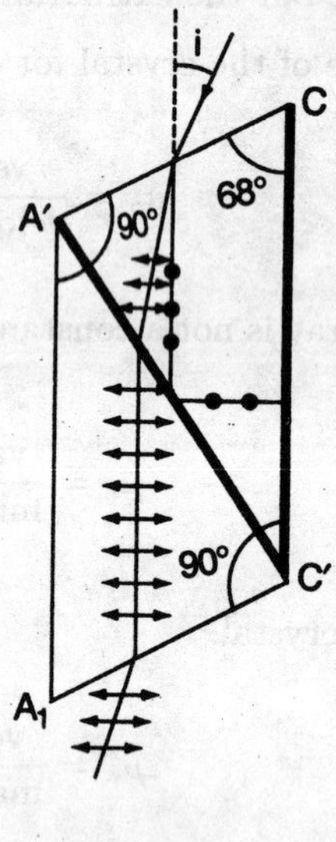

Fig 9.16

How does it act?

A ray incident nearly parallel to the longer dimension of the Nicol is split into two components—one O-ray and the other E-ray respectively. Both these components are linearly polarised, O-ray polarised ⊥ to the plane of incidence (principal section of the Nicol) and E-ray parallel to the plane of incidence. They are represented by · and ↔ respectively.

The O-ray component gets internally reflected at the Canada balsam layer and gets absorbed in the black coating on the side CC' but the extraordinary ray component passes through and emerges out as a linearly polarised light.

Why does the O-ray suffer total reflection and not the E-ray?

The refractive index of a material depends on the wavelength. Calcite is a negative crystal which means, the extraordinary ray travels faster than the O-ray in it.

Now, $\mu = \frac{c}{v}$, where c is the velocity of light in air and v is the velocity in a medium. So,

$$\mu_e = \frac{c}{v_e} \quad \text{and} \quad \mu_e = \frac{c}{v_o}.$$

since in a negative crystal $v_e > v_o$ we get, $\mu_e < \mu_O$.

Now, the r.i. of calcite for O-ray $\mu_O = 1.66$ and the r.i. of calcite for E-ray $\mu_e = 1.49$. The refractive index of Canada balsam is 1.55. Thus for O-ray as it tries to pass across the dividing surface laid with Canada balsam, it is a passage of light from denser medium to rarer but for E-ray it is just opposite, from rarer to denser. Thus the O-ray has the possibility of suffering internal reflection but the E-ray has not.

The critical angle for calcite Canada balsam interface is given by,

$$\sin C = \frac{\mu_{\text{balsam}}}{\mu_{\text{calcite}}} = \frac{1.55}{1.66} \quad \Rightarrow \quad C = 69^\circ.$$

To get an angle of incidence greater than the critical angle for O-ray, the length of the Nicol is made three times its breadth so that the balsam layer is sufficiently oblique which makes the angle of incidence exceed the required value of 69°. For the angle of incidence greater at the Canada balsam layer, the angle i required to be within 14°. Thus the incoming beam must be kept confined within a cone of semi-vertical angle 14°.

Functions of a Nicol Prism

A Nicol prism can be used as a polariser and also as analyser of linearly polarised light.

As a Polariser

When unpolarised light is made incident on the first face of the prism within an angle of 14°, it will break up on entry into the prism into two components, ordinary and extraordinary ray, polarised respectively perpendicular and parallel to the principal section of the prism. As explained earlier, the O-ray will be reflected by the Canada balsam layer and absorbed in the black coating on the side face whereas the E-ray will be freely transmitted and it will emerge out of the opposite end face. Thus we would get linearly polaried light with orientation axis parallel to the principal section of the prism.

As an Analyser

A Nicol prism can also be used as an analyser for linearly polarised light. To analyse a linearly polarised light, we would require to get information about the plane of vibration or orientation direction of the plane polarised light.

Consider a plane polarised light with orientation making an angle θ with the principal section of the Nicol. Let the amplitude be denoted as a, then it can be resolved into two components, one being $a\cos\theta$ parallel to the principal section and it will be the extraordinary ray, the other, $a\sin\theta \perp$ to the principal section and be the ordinary ray w.r.t. the analysing Nicol. If the Nicol be rotated about the incident light, θ value changes. When θ is zero, the extraordinary ray which is freely transmitted by the Nicol becomes most intense attaining the amplitude equal to a. When $\theta = \frac{\pi}{2}$, the E-ray has its amplitude zero and so the field becomes completely dark. At intermediate values, the intensity of the field of view varies.

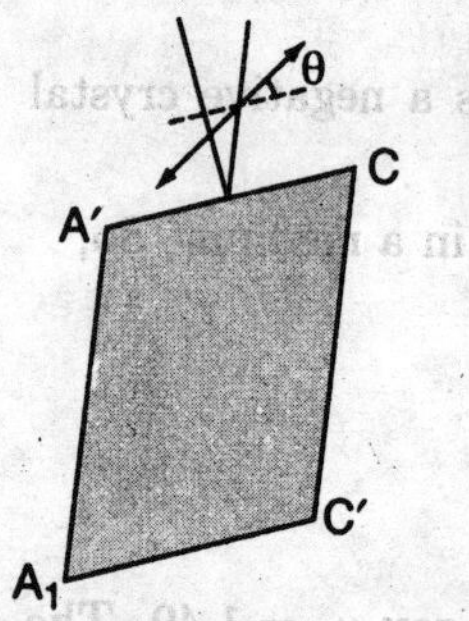

Fig 9.17

9.6 Methods to Separate O-ray and E-ray

Rochon's Prism

In Nicol prism, the ordinary ray is separated from the E-ray by its total reflection at the balsam layer but it is absorbed in the Nicol prism. So, if the black coating is not given, the ordinary and extraordinary linearly polarised light may be available from Nicol prism.

To separate the ordinary ray and extraordinary ray special types of prisms have been constructed—Rochon's prism, Walloston's prism are such examples. Rochon's prism is shown in fig.(9.18), it consists of two right-angled prisms ABC and ADC of iceland spar (of which calcite is one variety) cemented together along their hypotenuse sides so as to form a rectangular slab. The optic axis in the prism ABC is perpendicular to the face AB of the prism and also to the edge of A and the optic axis in the prism ADC is parallel to the refracting face CD.

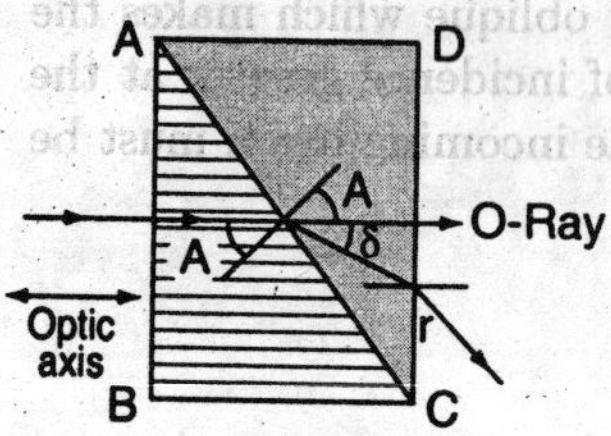

Fig 9.18

A ray incident normally on AB moves along the optic axis and the ordinary and extraordinary ray move with the same speed. So there is no separation of the ordinary and extraordinary ray as light travels through the first prism. When it enters the second prism the light is no longer along the optic axis and hence it now separates into ordinary and extraordinary ray. The ordinary ray will pass out undeviated but the extraordinary ray will follow a different path. Thus the ordinary ray and the extraordinary ray get separated.

Wolloston's Prism

This prism also consists of two right-angled prism ABC and ADC cemented together along the hypotenuse side. The optic axis in ABC is parallel to the face AB but perpendicular to the edge A. The optic axis in the other prism is parallel to the face DC and also to the edge D.

A ray is incident normal to the optic axis, so though the two rays travel with different speeds along the normal direction, there will be no angular separation between them. On entry into the second prism, the ordinary ray in the first prism would become the extraordinary ray and the extraordinary ray would behave as the ordinary ray. So there will be angular deviation produced in both of them and it will be further increased at the face CD.

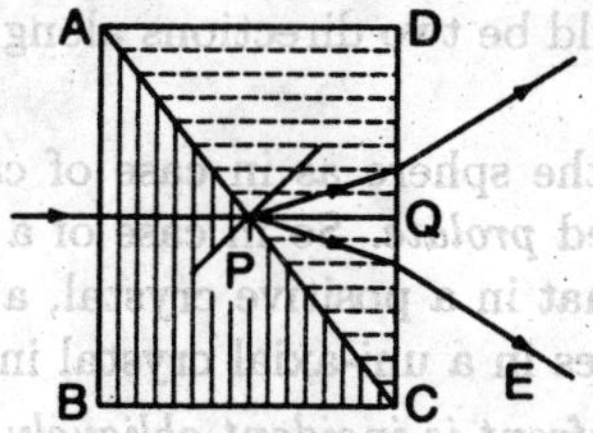

Fig 9.19

9.7 Wave Surfaces in Uni-axial Crystals : Huygens' Construction

There are crystals in which an unpolarised light is broken up into two components—one is called *Ordinary ray,* the other, *extraordinary ray.* This phenomenon is already referred to and is known as double refraction. Such a crystal is called *bi-refringent crystal.* There are two types of bi-refringent crystals, uni-axial and bi-axial. In a uni-axial crystal, there is one and only one direction along which ordinary and extraordinary waves travel with the same speed. Such a direction is called the *optic axis.* But in a bi-axial crystal, there are two directions in which the two velocities are equal.

After the discovery of the double refraction phenomenon, Huygens was very eager to explain the propagation of the two waves in an anisotropic crystal in the same way as he could explain the propagation of refracted wave in an isotropic medium.

There would be two wave surfaces in a uni-axial crystal, one corresponding to ordinary wave, the other to the extraordinary wave. Huygens' construction of wave surfaces is based on the fact that ordinary wave which obeys Snell's law propagates in the form of spheres while the extraordinary wave surface is assumed to be spheroidal (ellipsoid of revolution). Since in a uni-axial crystal there is one direction, called the *optic axis* along which the two waves travel with equal speed, it was argued that the two wave surfaces would touch at two points whose join would represent the optic axis.

In a negative crystal like calcite, the extraordinary wave travels faster than the ordinary wave, the spherical wave surface was supposed to be included wihin the spheroidal wave surface (fig. 9.20a) while in a positive crystal like quartz, mica, etc., the spheroidal wave surface was taken to be included within the sphere. The ordinary ray would be normal to the sperical wave surface but the extraordinary ray won't be normal to the spheroidal wave surface.

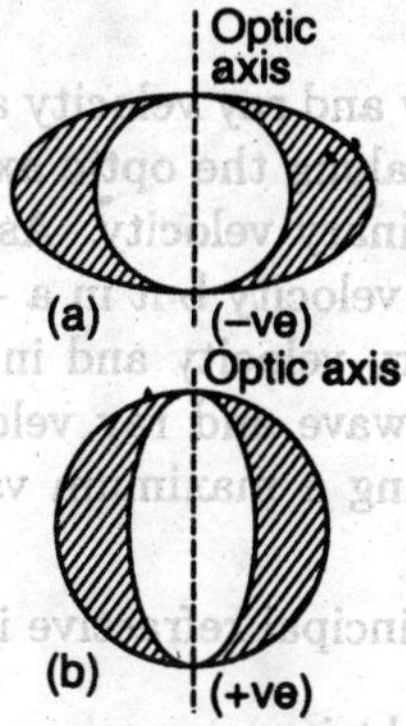

Fig 9.20

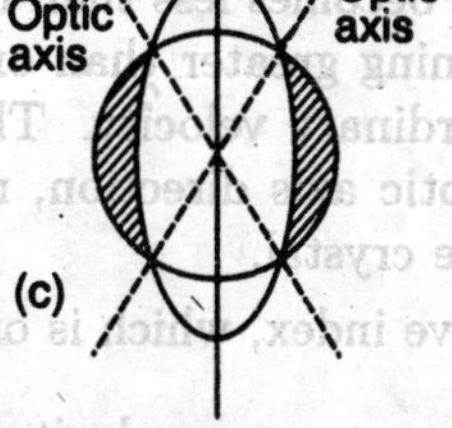

Fig 9.21

In a bi-axial crystal there would be two directions along which the two wave surfaces would intersect each other (fig.9.21).

When the spheroid includes the sphere as in case of calcite, the spheroid is called *oblate* while that in case of quartz, called *prolate*. So in case of a negative crystal, the extraordinary wave surface is an *oblate* while that in a positive crystal, a *prolate*. Based on this concept, we would now construct wave surfaces in a uni-axial crystal in different cases.

Case I. *When the plane wavefront is incident obliquely and the optic axis is inclined to the surface?*

Consider parallel rays to be incident on the surface of a uni-axial crystal at A and A', the optic axis direction is shown by the broken line. The incident wavefront is a plane passing through AN, where N is the foot of the perpendicular dropped from A. On incidence at A, there would be two refracted waves, one a sphere and another a spheroid included within it, touching each other at P on the optic axis. The sections of the sphere and the spheroid by the plane of the diagram are shown as a circle and an ellipse about A.

Let the disturbance from N take time t to reach the point A' on the surface. Draw tangents from A' on to the circle and ellipse, they are respectively represented by $A'N_1$ and $A'N_2$. $A'N_1$ and $A'N_2$ represent the refracted wavefront corresponding to the ordinary and extraordinary disturbances and AN_1 and AN_2 represent the ordinary ray and the extraordinary ray. It is to note that the ordinary ray is along the wave normal but not the extraordinary ray. The wave velocity is the rate of increase of wave normal, so if t be the time after which the wavefront traces are as shown in fig.(9.22), the ordinary wave velocity v_o and extraordinary wave velocity v_e are given by,

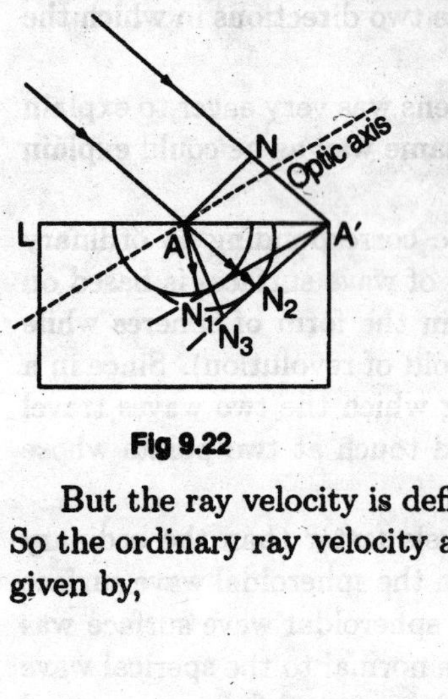

Fig 9.22

$$v_o = \frac{AN_1}{t} \quad (9.7.1)$$

$$v_e = \frac{AN_3}{t}. \quad (9.7.2)$$

But the ray velocity is defined as the rate of description of distance along the ray direction. So the ordinary ray velocity and extraordinary ray velocity with reference to the fig.(9.22) are given by,

$$v_o = AN_1/t \quad (9.7.3)$$

$$v_e = AN_2/t \quad (9.7.4)$$

Thus it appears that the ordinary wave velocity and ray velocity are equal but extraordinary wave velocity and ray velocity are different. But along the optic axis, the extraordinary wave and ray velocities are equal and equal to the ordinary velocity. As one moves away from the axis, the E-wave velocity becomes less than E-ray velocity but in a –ve crystal like that shown in fig.(9.23), both remaining greater than ordinary velocity and in a +ve crystal like quartz, both remain less than ordinary velocity. The E-wave and ray velocities again become equal at right angles to the optic axis direction, reaching a maximum value for a –ve crystal and minimum value for a +ve crystal.

The ordinary refractive index, which is one principal refractive index, is defined as,

$$\mu_o = \frac{\text{velocity of light in vacuum}}{\text{ordinary wave (or ray) velocity}}$$

It is a constant quantity. The extraordinary refractive, also called the *second principal refractive index*, is defined as,

$$\mu_o = \frac{\text{velocity of light in vacuum}}{\text{maximum velocity of the E-wave}} \quad \text{for a } -\text{ve crystal}$$

Also

$$\mu_e = \frac{\text{velocity of light in vacuum}}{\text{minimum velocity of the E-wave}} \quad \text{for } +\text{ve crystal}$$

Case II. *The optic axis is parallel to the surface of the crystal.*

When the optic axis is parallel to the crystal surface the ordinary and extraordinary wave surfaces are respectively spheres and spheroids, the intersection by the plane of incidence being circle and ellipse as shown in fig.(9.23) touching at two points along the optic axis. The extraordinary and ordinary waves travel along the same direction, wave and ray velocity becoming equal; but the speed of extraordinary wave is greater than ordinary wave as it is a negative crystal. If we draw common tangents to the ordinary wave surfaces generated at the points P and Q, we would get ordinary wavefront at some instant such as $ABCD$; similarly we would construct extraordinary wavefront at the same instant as $A'B'C'D'$.

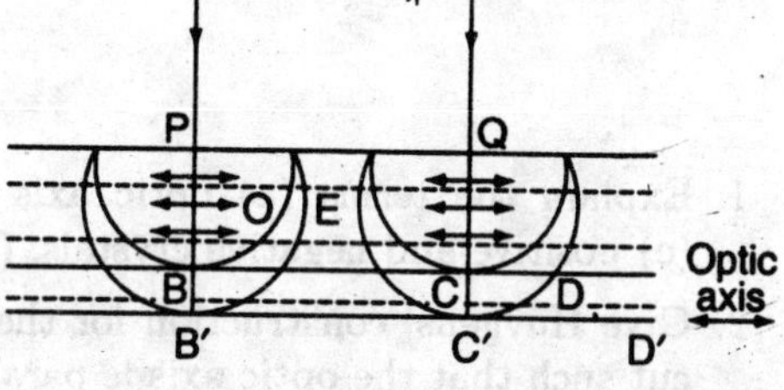

Fig 9.23

In case of oblique incidence we would refer to the fig.(9.24) from A dkrop a $\perp$ on to the second ray BC at B.

Let the light disturbance take t sec to travel from B to C after the first ray strikes the crystal surface at A. From A both ordinary and extraordinary wave travel into the crystal and from C draw tangents CD and CD' to the two wave surfaces. Then, ray velocity for the ordinary disturbance, also equal to the wave velocity is given by $v_o = V_o = \frac{AD}{t}$. The extraordinary wavefront is CD' and the extraordinary wave velocity V_e is proportional to the normal from A on to the wavefront while the extraordinary ray velocity is proportional to AD'.

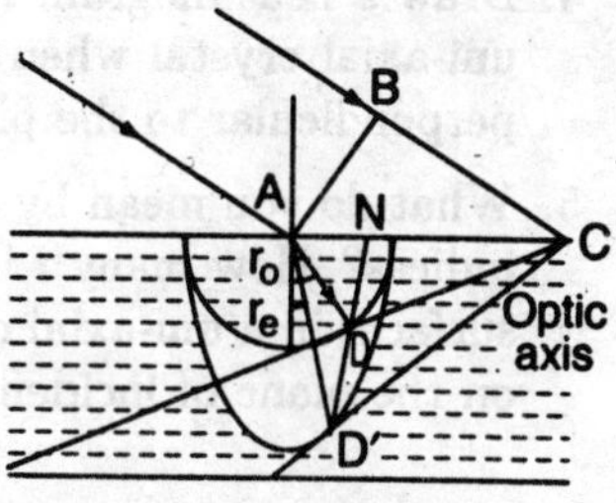

Fig 9.24

If r_o and r_e be the angles as indicated in the fig.(9.24), we would at once get,

$$\tan r_o = \frac{AN}{ND}, \quad \tan r_e = \frac{AN}{ND'} \quad \therefore \quad \frac{\tan r_o}{\tan r_e} = \frac{ND'}{ND}.$$

Also,

$$\mu_e \propto \frac{1}{ND}, \quad \mu_o \propto \frac{1}{ND'} \quad \therefore \quad \frac{\mu_o}{\mu_e} = \frac{ND'}{ND} = \frac{\tan r_o}{\tan r_e}.$$

This relation was verified by Malus, thus justifying Huygens' construction of wave surfaces.

Case III. *When the plane of incidence is $\perp$ to the optic axis of the crystal.*

In this case, the optic axis lies on the surface of the crystal but it is perpendicular to the plane of incidence (see fig.9.25).

The plane of the crystal face contains the optic axis which is $\perp$ to the plane of incidence. After the incident ray strikes the point A on the crystal face, ordinary wave and extraordinary wave propagate into the crystal as sphere and spheroid touching each other on the optic axis, so the section of these wave surfaces by the plane of the crystal is represented as PQR for the ordinary wave and as $P'Q'R'$ for the extraordinary wave. The intersection of the wave surfaces by the incident plane would be both circles shown in the figure by the broken line. From the point C, if we draw tangents to the circles intercepted by the incident plane, we would get the refracted wavefront corresponding to the ordinary and extraordinary component of light. It is also to note in this case that the wave velocity and ray velocity for the extraordinary components are identical.

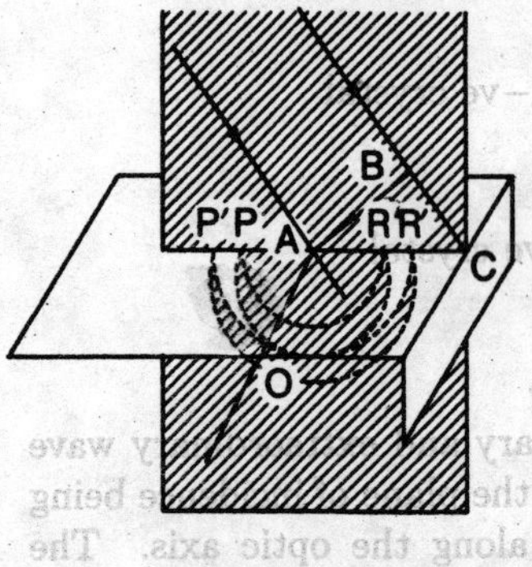

Fig 9.25

Exercises

1. Explain the terms (a) optic axis of a crystal, (b) ordinary and extraordinary rays, (c) positive and negative crystals, (d) principal section of a crystal.
2. Give Huygens' construction for the ordinary and extraordinary rays in a quartz crystal cut such that the optic axis is parallel to the refracting surface.
3. Discuss the nature of wave surfaces in a uni-axial crystal according to Huygens' construction. What are positive and negative crystals?
4. Draw a neat diagram to show the wave surfaces on the plane of incidence for a negative uni-axial crystal when the optic axis of the crystal lies on the surface of incidence but perpendicular to the plane of incidence.
5. What do you mean by a bi-axial crystal? What is the meaning of the principal refractive indices? How many r.i.-s are there? Construct according to Huygens' concept the wave surfaces in a uni-axial crystal when the optic axis lies on the surface of incidence and also on the plane of incidence?

Chapter 10

Matrix Formulation of Linearly Polarised Light: Jones' Vector and Jones' Matrix

10.1 Introduction

There are many modern mathematical formalism developed to deal with the problems of polarisation, one is due to Stokes, another is developed by Jones, still another by Wolf. Of these methods having relative merits and demerits, the formalism due to Jones (1941) is simpler to understand and that is why we are here making use of this method to deal with the problems of polarisation.

In this method, a polarised light is represented by a column vector. The first element represents the amplitude and phase of the Y-component, the second element does it for Z component of the light wave (according to Electromagnetic theory, Y-component of electric vibration), when the wave is supposed to be travelling along the X-direction.

Thus, in fig. (10.1) the oscillation is contained in horizontal plane, hence it is linearly polarised, horizontally. Mathematically, such a disturbance can be represented as

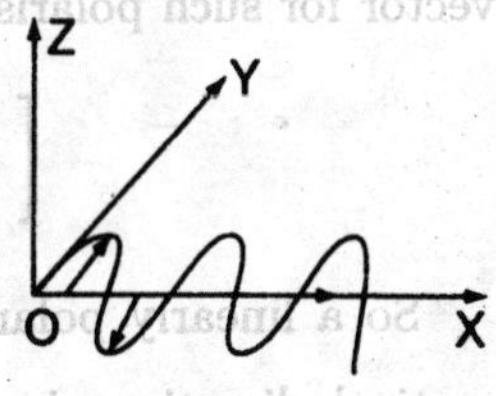

Fig 10.1

$$\vec{Y} = \hat{j} a_2 e^{i(k_1 x - \omega t)},$$

where a_2 is the amplitude of oscillation and $k_1 = \frac{2\pi}{\lambda}$. This disturbance has no Z-component. So the horizontally polarised light disturbance can be written as

$$y = a_2 e^{i\phi_x}, \quad \phi_x = kx - \omega t, \quad z = 0 \quad \therefore \quad \begin{bmatrix} y \\ z \end{bmatrix} = \begin{bmatrix} a_2 e^{i\phi_x} \\ 0 \end{bmatrix} \tag{10.1.1}$$

Thus, in (10.1.1), the first element of the column matrix reresents amplitude and phase of horizontal oscillation. If the intensity is normalised and the phase indicating factor omitted, the relation (10.1.1) becomes

$$\begin{bmatrix} y \\ z \end{bmatrix} = \begin{bmatrix} 1 \\ 0 \end{bmatrix} \tag{10.1.2}$$

So, in Jones' formalism, a horizontally polarised light is represented by Jones' vector J_1 given as

$$J_1 = \begin{bmatrix} 1 \\ 0 \end{bmatrix} \longleftrightarrow \quad \text{for horizontally polarised light.}$$

Similarly, the oscillation in the vertical plane is shown in fig. (10.2).

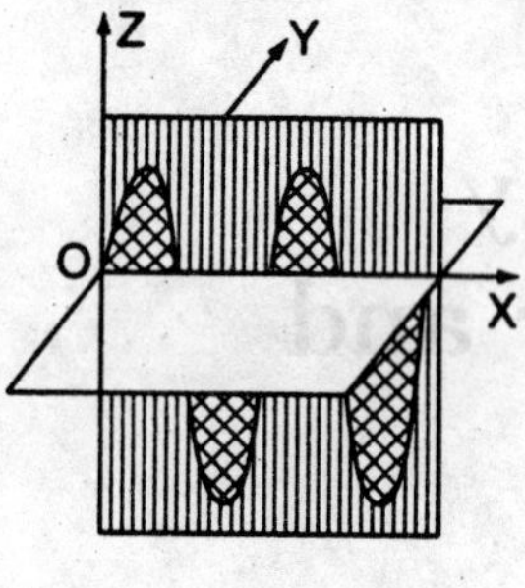

Fig 10.2

In fig. (10.2) is shown a disturbance in the vertical plane, represented by an equation

$$\bar{Z} = \hat{k} a_3 e^{i(kx-\omega t+\phi)}$$

where a_3 is the amplitude, ϕ is the phase difference w.r.t. the horizontal disturbance (taken as phase reference) and $\hat{k}$ is the unit vector along Z-direction. This polarised wave can be written as

$$y = 0, \quad z = a_3 e^{i\phi_x} e^{i\phi} \qquad \therefore \quad \begin{bmatrix} y \\ z \end{bmatrix} = \begin{bmatrix} 0 \\ a_3 e^{i\phi_x} e^{i\phi} \end{bmatrix}$$

For normalised intensity and if the phase factor be omitted, the vertically polarised disturbance can be represented as Jones' vector J_2, given as

$$J_2 = \begin{bmatrix} 0 \\ e^{i\phi} \end{bmatrix} \longrightarrow \begin{bmatrix} 0 \\ 1 \end{bmatrix} \quad \text{if} \quad \phi \to 0.$$

Thus,

$$J_2 = \begin{bmatrix} 0 \\ 1 \end{bmatrix} \quad \text{for vertically polarised light} \quad \updownarrow$$

If the light disturbance is polarised at 45° w.r.t. horizontal or vertical direction, then Jones' vector for such polarised light can be wirtten as

$$\begin{bmatrix} y \\ z \end{bmatrix} = \begin{bmatrix} ae^{i\phi_x} \cos \pi/4 \\ ae^{i\phi_x} \sin \pi/4 \end{bmatrix} \longrightarrow \begin{bmatrix} 1/\sqrt{2} \\ 1/\sqrt{2} \end{bmatrix} = \frac{1}{\sqrt{2}} \begin{bmatrix} 1 \\ 1 \end{bmatrix}$$

So a linearly polarised light polarised at an angle of 45° with either of the horizontal or vertical directions is represented by Jones' vector $\frac{1}{\sqrt{2}} \begin{bmatrix} 1 \\ 1 \end{bmatrix}$. This polarised light may be obtained by addition of a horizontally polarised light and vertically polarised light of equal intensity. Thus,

$$\underset{\text{Horizontal}}{\frac{1}{\sqrt{2}} \begin{bmatrix} 1 \\ 0 \end{bmatrix}} + \underset{\text{Vertical}}{\frac{1}{\sqrt{2}} \begin{bmatrix} 0 \\ 1 \end{bmatrix}} = \underset{\text{Linearly polarised light at 45°}}{\frac{1}{\sqrt{2}} \begin{bmatrix} 1 \\ 1 \end{bmatrix}}$$

The normalised linearly polarised light at 45° is represented by $J_3 = \frac{1}{\sqrt{2}} \begin{bmatrix} 1 \\ 1 \end{bmatrix}$.

So, if we have a light disturbance represented as

$$\xi = \begin{bmatrix} a \\ b \end{bmatrix}$$

it will mean a disturbance polarised along the direction making an angle θ w.r.t. the Y-direction given by $\tan\theta = (b/a)$, so if $a = b, \tan\theta = 1 \Rightarrow \theta = \frac{\pi}{4}$.

We note that $\begin{bmatrix} 1 \\ 0 \end{bmatrix}$ and $\begin{bmatrix} 0 \\ 1 \end{bmatrix}$ are mutually orthogonal and individually normalised. For orthogonality of two vectors represented by matrices $\begin{bmatrix} a \\ b \end{bmatrix}$ and $\begin{bmatrix} a' \\ b' \end{bmatrix}$, we will have[1]

$$[a^* b^*] \begin{bmatrix} a' \\ b' \end{bmatrix} = a^*a' + b^*b' = 0,$$

here $[a^*b^*]$ is the transpose of the matrix $\begin{bmatrix} a \\ b \end{bmatrix}$ and a^*, b^* are the complex conjugates.

For the condition of normalisation,

$$[a^* b^*] \begin{bmatrix} a \\ b \end{bmatrix} = a^*a + b^*b = 1$$

With this idea in the background, let us consider the two Jones' vectors as stated above.

$$J_1 = \begin{bmatrix} 1 \\ 0 \end{bmatrix}, \quad J_2 = \begin{bmatrix} 0 \\ 1 \end{bmatrix}, \quad [1 \quad 0] \begin{bmatrix} 1 \\ 0 \end{bmatrix} = 1 + 0 = 1,$$

hence, $\begin{bmatrix} 1 \\ 0 \end{bmatrix}$ is normalised, so also is $\begin{bmatrix} 0 \\ 1 \end{bmatrix}$.

Again, $[1 \quad 0] \begin{bmatrix} 0 \\ 1 \end{bmatrix} = 0$, so it satisfies the condition of orthogonality.

Thus the two Jones' vectors J_1 and J_2 are mutually orthogonal and individually normalised, but a combination J_3 of these two linearly polarised light is not normalised, because

$$J_3 = \begin{bmatrix} 1 \\ 1 \end{bmatrix} \quad \therefore \quad [1 \quad 1] \begin{bmatrix} 1 \\ 1 \end{bmatrix} = 2, \quad \text{hence} \quad \frac{1}{\sqrt{2}} \begin{bmatrix} 1 \\ 1 \end{bmatrix}$$

is a normalised vector.

A device which transforms unpolarised light or ordinary light into a linearly polarised light is referred to as a linear polariser. Since the function of the device is to effect transformation, matrix formalism may be conveniently employed. So let us introduce Jones' matrix for linear polariser. Let us consider a matrix

$$[M_t] = \begin{bmatrix} 1 & 0 \\ 0 & 0 \end{bmatrix}$$

Let it operate on another matrix $\begin{bmatrix} a \\ b \end{bmatrix}$, thus

$$\begin{bmatrix} 1 & 0 \\ 0 & 0 \end{bmatrix} \begin{bmatrix} a \\ b \end{bmatrix} = \begin{bmatrix} a \\ 0 \end{bmatrix} = a \begin{bmatrix} 1 \\ 0 \end{bmatrix}$$

[1] For clear understanding see 'A Text-book of Mathematical Physics' Chakrabarti and Kundu, New Central Book Agency (P) Ltd., Kolkata.

So operating the matrix $[M_t]$ representing the transformation matrix on the matrix $\begin{bmatrix} a \\ b \end{bmatrix}$ representing the incident light, we get light linearly polarised in the horizontal direction with amplitude a times as great. Such a matrix is called Jones' linear polariser matrix.

So a linear polariser with its transmission axis horizontal is given as

$$[L_h] = \begin{bmatrix} 1 & 0 \\ 0 & 0 \end{bmatrix}$$

With this mathematical formalism, the matrix transformation may be utilised to represent the functioning of a device producing linear polarisation. The formalism may be schematically developed as given below—

Let $\begin{bmatrix} A \\ B \end{bmatrix}$ be the incident matrix $[M_i]$ representing the incident light; the transformation matrix $[M_t]$, describing the behaviour of the polarising device that operates on it resulting in the product matrix $[M_f]$ representing the nature of the emergent light.

Thus,

$$\underset{\text{Transformation matrix}}{\begin{bmatrix} a & b \\ c & d \end{bmatrix}} \quad \underset{\text{Incident light matrix}}{\begin{bmatrix} A \\ B \end{bmatrix}} = \underset{\text{Emergent light matrix}}{\begin{bmatrix} A' \\ B' \end{bmatrix}}$$

$$\Rightarrow \quad [M_t]\begin{bmatrix} A \\ B \end{bmatrix} = [M_f]$$

A simple example related to linear polarisation will make the formalism aptly clear.

Example 1. *Linearly polarised light of unit intensity oscillating in the horizontal direction is added to a 16-times as intense vertically polarised beam. Find the nature of the emergent beam.*

Solution: The two beams are represented by Jones' vectors $\begin{bmatrix} 1 \\ 0 \end{bmatrix}$ and $\begin{bmatrix} 0 \\ 4 \end{bmatrix}$. So the emergent light has Jones' vector given as

$$J = \begin{bmatrix} 1 \\ 0 \end{bmatrix} + \begin{bmatrix} 0 \\ 4 \end{bmatrix} = \begin{bmatrix} 1 \\ 4 \end{bmatrix}$$

which represents a linearly polarised light at an angle θ given by $\tan\theta = \left(\frac{4}{1}\right) \Rightarrow \theta = 76°$, the intensity is given by $(1^2 + 4^2) = 17$. So we remember that Jones' vectors represent either the incident or emergent beam but Jones' matrix represents the functioning device or the polariser.

Matrix for linear polariser

Since a polariser transforms state of polarisation, any polarising device can be represented by a matrix, called transformation matrix or polarising element matrix.

Thus, the matrix $\begin{bmatrix} 1 & 0 \\ 0 & 0 \end{bmatrix}$ represents a linear polariser with transmission axis horizontal; the matrix $\begin{bmatrix} 0 & 0 \\ 0 & 1 \end{bmatrix}$ represents a linear polariser with transmission axis vertical; the matrix

$\frac{1}{2}\begin{bmatrix} 1 & 1 \\ 1 & 1 \end{bmatrix}$ represents a linear polariser with transmission axis at 45°, because the first matrix operates on an incident hight transforming it into a horizontally polarised light; the second into a vertically polarised light, etc.

Illustration: The following Jones matrices represent horizontal, vertical, +45° polariser

$$\underset{\text{Horizontal}}{\begin{bmatrix} 1 & 0 \\ 0 & 0 \end{bmatrix}}, \quad \underset{\text{Vertical}}{\begin{bmatrix} 0 & 0 \\ 0 & 1 \end{bmatrix}}, \quad \underset{+45^\circ}{\frac{1}{2}\begin{bmatrix} 1 & 1 \\ 1 & 1 \end{bmatrix}}$$

Give interpretation of the matrices.

The first matrix represents a polarising element, called a horizontal polariser, because it transforms the incident light into a horizontally linearly polarised light.

Thus,

$$\begin{bmatrix} 1 & 0 \\ 0 & 0 \end{bmatrix}\begin{bmatrix} ae^{i\phi_1} \\ be^{i\phi_2} \end{bmatrix} = \begin{bmatrix} ae^{i\phi_1} \\ 0 \end{bmatrix} \rightarrow \begin{bmatrix} 1 \\ 0 \end{bmatrix}, \quad \text{i.e., horizontal polarisation.}$$

Similarly,

$$\begin{bmatrix} 0 & 0 \\ 0 & 1 \end{bmatrix}\begin{bmatrix} ae^{i\phi_1} \\ be^{i\phi_2} \end{bmatrix} = \begin{bmatrix} 0 \\ be^{i\phi_2} \end{bmatrix} \rightarrow \begin{bmatrix} 0 \\ 1 \end{bmatrix}, \quad \text{i.e., vertically polarised light.}$$

Similarly,

$$\frac{1}{2}\begin{bmatrix} 1 & 1 \\ 1 & 1 \end{bmatrix}\begin{bmatrix} ae^{i\phi_1} \\ be^{i\phi_2} \end{bmatrix} = \frac{1}{2}\begin{bmatrix} (ae^{i\phi_1} + be^{i\phi_2}) \\ (ae^{i\phi_1} + be^{i\phi_2}) \end{bmatrix}$$

If $\phi_1 = \phi_2 = 0$, $a = b = 1$, then

$$\frac{1}{2}\begin{bmatrix} 1 & 1 \\ 1 & 1 \end{bmatrix}\begin{bmatrix} ae^{i\phi_1} \\ be^{i\phi_2} \end{bmatrix} = \frac{1}{2}\begin{bmatrix} 2 \\ 2 \end{bmatrix} = \begin{bmatrix} 1 \\ 1 \end{bmatrix} \quad \text{linearly polarised light at } +45^\circ.$$

Example 2. *Show that the matrix* $\frac{1}{2}\begin{bmatrix} 1 & -1 \\ -1 & 1 \end{bmatrix}$ *represents a polariser of* $-45°$.

Solution:

$$\frac{1}{2}\begin{bmatrix} 1 & -1 \\ -1 & 1 \end{bmatrix}\begin{bmatrix} ae^{i\phi_1} \\ be^{i\phi_2} \end{bmatrix} \text{ if } \phi_1 = \phi_2 = 0, \quad a = b = 1, \quad \frac{1}{2}\begin{bmatrix} 1 & -1 \\ -1 & 1 \end{bmatrix}\begin{bmatrix} 1 \\ 1 \end{bmatrix} = \frac{1}{2}\begin{bmatrix} 0 \\ 0 \end{bmatrix}$$

The matrix operates on a polarised light at +45° producing the emergent light of zero intensity, i.e., light is cut off. So the matrix represents a polariser of −45°.

Example 3. *Operate* $\frac{1}{2}\begin{bmatrix} 1 & -1 \\ -1 & 1 \end{bmatrix}$ *on* −45° *polarised light. Interpret the emergent light.*

Solution:

$$\frac{1}{2}\begin{bmatrix} 1 & -1 \\ -1 & 1 \end{bmatrix}\begin{bmatrix} -1 \\ +1 \end{bmatrix} = \frac{1}{2}\begin{bmatrix} -2 \\ +2 \end{bmatrix} = \begin{bmatrix} -1 \\ +1 \end{bmatrix}$$

The emergent light is linearly polarised at −45°.

Example 4. *Let light first be horizontally polarised, then pass through a +45° polariser, then through a −45° polariser. What is the polarisation of emergent light?*

Solution: The complete step by step process is represented by a sequence of matrix operations given as below:

$$\frac{1}{2}\begin{bmatrix} 1 & -1 \\ -1 & 1 \end{bmatrix}\frac{1}{2}\begin{bmatrix} 1 & 1 \\ 1 & 1 \end{bmatrix}\begin{bmatrix} 1 \\ 0 \end{bmatrix} = \frac{1}{2}\begin{bmatrix} 1 & -1 \\ -1 & 1 \end{bmatrix}\frac{1}{2}\begin{bmatrix} 1 \\ 1 \end{bmatrix} = \frac{1}{4}\begin{bmatrix} 0 \\ 0 \end{bmatrix} = \begin{bmatrix} 0 \\ 0 \end{bmatrix}.$$

No light is transmitted.

In general terms, we can represent polarising action of each of the optical elements by one matrix and operate with the product on the incident light giving the polarisation of the emergent light. Thus the schematic representation of the action of a set of polarisers is given as

$$\begin{bmatrix} 1 & 0 \\ 0 & e^{i\phi_1} \end{bmatrix}\begin{bmatrix} 1 & 0 \\ 0 & e^{i\phi_2} \end{bmatrix}\cdots\begin{bmatrix} 1 & 0 \\ 0 & e^{i\phi_n} \end{bmatrix}\begin{bmatrix} \text{Incident} \\ \text{polarised} \\ \text{light} \end{bmatrix} = \begin{bmatrix} \text{Emergent} \\ \text{polarised} \\ \text{light} \end{bmatrix}$$

Example 5. *A linear polariser with its transmission axis horizontal is taken. Describe what would happen when*

(i) *linearly polarised light of horizontal orientation is incident on it.*

(ii) *The same light but now with vertical orientation is incident on it.*

Solution: (i) A linear polariser with its transmission axis horizontal is represented by a matrix $\begin{bmatrix} 1 & 0 \\ 0 & 0 \end{bmatrix}$. It operates on the matrix $\begin{bmatrix} 1 \\ 0 \end{bmatrix}$ representing linearly polarised light in the horizontal direction to give the state of polarisation of the emergent light.

$$\therefore \begin{bmatrix} 1 & 0 \\ 0 & 0 \end{bmatrix}\begin{bmatrix} 1 \\ 0 \end{bmatrix} = \begin{bmatrix} 1 \\ 0 \end{bmatrix}$$

↓
Linear horizontally polarised

(ii)

$$\begin{bmatrix} 1 & 0 \\ 0 & 0 \end{bmatrix}\begin{bmatrix} 0 \\ 1 \end{bmatrix} = \begin{bmatrix} 0 \\ 0 \end{bmatrix}$$

↓
No transmission

Example 6. *A linearly polarised light of horizontal orientation is incident on a linear polariser oriented at +45°. Then*

(i) *the light enters a polariser whose transmission axis is vertical.*

(ii) *the vertical polariser is placed ahead of the 45° polariser. Describe the emergent ray.*

Solution: (i) $\frac{1}{2}\begin{bmatrix} 1 & 1 \\ 1 & 1 \end{bmatrix}\begin{bmatrix} 1 \\ 0 \end{bmatrix} = \frac{1}{2}\begin{bmatrix} 1 \\ 1 \end{bmatrix}$.

So the emergent light is linearly polarised at 45°, and of the same intensity as that of the incident light. The polariser with axis vertical is represented by a matrix $\begin{bmatrix} 0 & 0 \\ 0 & 1 \end{bmatrix}$.

$$\therefore \begin{bmatrix} 0 & 0 \\ 0 & 1 \end{bmatrix}\begin{bmatrix} 1 \\ 1 \end{bmatrix} = \begin{bmatrix} 0 \\ 1 \end{bmatrix}$$

Thus the emergent light is linearly polarised of vertical orientation, the intensity of the emergent light is reduced to half of the original value.

(ii) The step by step functioning here is represented by a sequence of matrix operation as shown

$$\begin{bmatrix} 0 & 0 \\ 0 & 1 \end{bmatrix}\begin{bmatrix} 1 \\ 0 \end{bmatrix} = \begin{bmatrix} 0 \\ 0 \end{bmatrix}, \text{ then } \frac{1}{2}\begin{bmatrix} 1 & 1 \\ 1 & 1 \end{bmatrix}\begin{bmatrix} 0 \\ 0 \end{bmatrix} = \frac{1}{2}\begin{bmatrix} 0 \\ 0 \end{bmatrix}$$

There is no transmission of light.

Remember the following Jones' vectors and Jones' matrix for solution of polarisation problems:

	Jones' Vector	Jones' Matrix
Linearly polarised light with horizontal orientation	$\begin{bmatrix} 1 \\ 0 \end{bmatrix}$	—
Linearly polarised light with vertical orientation	$\begin{bmatrix} 0 \\ 1 \end{bmatrix}$	—
Linearly polarised light at +45° orientation	$\frac{1}{\sqrt{2}}\begin{bmatrix} 1 \\ 1 \end{bmatrix}$	—
Linearly polarised light at −45° orientation	$\frac{1}{\sqrt{2}}\begin{bmatrix} -1 \\ 1 \end{bmatrix}$	—
Linear polariser with axis horizontal	—	$\begin{bmatrix} 1 & 0 \\ 0 & 0 \end{bmatrix}$
Linear polariser with axis vertical	—	$\begin{bmatrix} 0 & 0 \\ 0 & 1 \end{bmatrix}$
Linear polariser with axis at +45°	—	$\frac{1}{2}\begin{bmatrix} 1 & 1 \\ 1 & 1 \end{bmatrix}$
Linear polariser with axis at −45°	—	$\frac{1}{2}\begin{bmatrix} 1 & -1 \\ -1 & 1 \end{bmatrix}$

If a light disturbance of amplitude a is incident with orientation angle θ with the y-direction and $(90 - \theta) = \beta$ with the z-direction, then the disturbance is represented as

$$y = a\cos\theta e^{i(kx-\omega t)} = a\cos\theta e^{i\phi_x}$$

$$z = a\sin\theta e^{i(kx-\omega t)} = a\sin\theta e^{i\phi_x}$$

$$\therefore \begin{bmatrix} y \\ z \end{bmatrix} = \begin{bmatrix} a\cos\theta e^{i\phi_x} \\ a\cos\beta e^{i\phi_x} \end{bmatrix} \longrightarrow \begin{bmatrix} a\cos\theta \\ a\cos\beta \end{bmatrix} \longrightarrow \begin{bmatrix} A_1 \\ A_2 \end{bmatrix}$$

Thus we get a linearly polarised light with the orientation axis making an angle ϕ, given by,

$$\tan\phi = \frac{A_2}{A_1} = \frac{a\cos\beta}{a\cos\theta} = \frac{a\sin\theta}{a\cos\theta} = \tan\theta \quad \Rightarrow \quad \phi = \theta.$$

So, the linear polariser matrix transforming an incident linearly polarised light into another linearly polarised light with orientation θ with y-direction is written as

$$\begin{bmatrix} A_1 & A_1 \\ A_2 & A_2 \end{bmatrix}.$$

Thus,

$$\begin{bmatrix} A_1 & A_1 \\ A_2 & A_2 \end{bmatrix} \begin{bmatrix} 1 \\ 0 \end{bmatrix} = \begin{bmatrix} A_1 \\ A_2 \end{bmatrix}$$

$\begin{bmatrix} 1 \\ 0 \end{bmatrix}$ ↓ Horizontally polarised; $\begin{bmatrix} A_1 \\ A_2 \end{bmatrix}$ ↓ Linearly polarised

$$\begin{bmatrix} A_1 & A_1 \\ A_2 & A_2 \end{bmatrix} \begin{bmatrix} 0 \\ 1 \end{bmatrix} = \begin{bmatrix} A_1 \\ A_2 \end{bmatrix}$$

$\begin{bmatrix} 0 \\ 1 \end{bmatrix}$ ↓ Vertically polarised; $\begin{bmatrix} A_1 \\ A_2 \end{bmatrix}$ ↓ Linearly polarised

$$\begin{bmatrix} A_1 & A_1 \\ A_2 & A_2 \end{bmatrix} \frac{1}{\sqrt{2}} \begin{bmatrix} 1 \\ 1 \end{bmatrix} = \frac{1}{\sqrt{2}} \begin{bmatrix} 2A_1 \\ 2A_2 \end{bmatrix} = \sqrt{2} \begin{bmatrix} A_1 \\ A_2 \end{bmatrix}$$

Writing out the full form in the last case we get

$$y = \sqrt{2}A_1\cos(kx - \omega t), \quad z = \sqrt{2}A_2\cos(kx - \omega t). \quad \therefore \quad \tan\theta = \frac{z}{y} = \frac{A_2}{A_1},$$

so here also the emergent light is plane polarised with orientation in the direction making an angle θ with y-direction.

10.2 Circular and Elliptical Polarisation

We have seen earlier how to produce linearly polarised light, polarised either vertically or horizontally or polarised at arbitrary orientation w.r.t. vertical or horizontal direction. Now, two such polarised components under certain conditions on superposition produces circularly polarised light where the tip of the displacement vector lies on a circle or on an ellipse.

In this section we will be concerned with circular polarisation and elliptical polarisation.

Circularly polarised light

In a linearly polarised light with orientation in the horizontal direction, vibration will be in a horizontal plane, if such light disturbance combines with an orthogonal disturbance at some place, the resultant vibration may be linear, circular or elliptical depending on the phase and amplitude relationship of these two orthogonal linear vibrations under consideration. Thus, let the y-direction vibration be represented as

$$\bar{Y} = \hat{j}a_1 e^{i(kx-\omega t)} = \hat{j}a_1 e^{i\phi_x},$$

where $k = \frac{2\pi}{\lambda}$, called the propagation constant and $\phi_x = kx - \omega t$, which definitely depends on time. This is represented in fig. [10.3(a)].

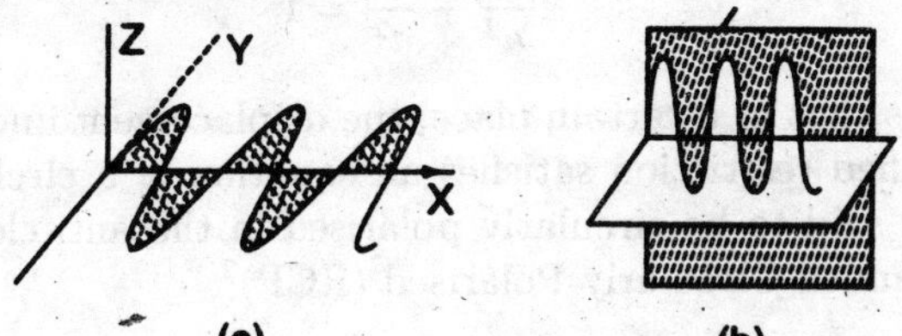

Fig 10.3

The orthogonal vibration where displacement occurs in the Z-direction while propagation direction is along X-axis is shown in fig. [10.3(b)] and is written as

$$\bar{Z} = \hat{k} a_2 e^{i(\phi_x + \phi)}$$

where, ϕ represents the phase difference between the two linear vibrations. If ϕ is positive, Z-vibration is in phase lead and if ϕ is negative, Z-vibration is phase lag w.r.t. Y-vibration. The resultant can be written and shown as in fig (10.4).

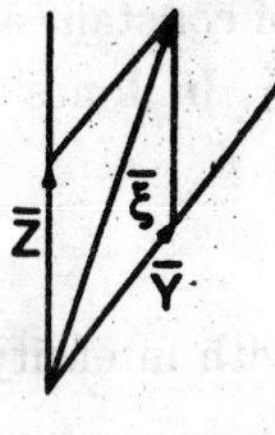

Fig 10.4

$$\bar{\xi} = \bar{Y} + \bar{Z} = (\hat{j} a_1 + \hat{k} a_2 e^{i\phi}) e^{i\phi_x}.$$

Now, if $a_1 = a_2 = a$, $\phi = \frac{\pi}{2}$, then $\bar{\xi} = a(\hat{j} + \hat{k} e^{i\pi/2}) e^{i\phi_x}$.

The two components fluctuate in different phases, the two vibrations here are said to be in quadrature; at any place, defined by a fixed value of x, the displacement due to one is maximum when that due to the other is minimum but the resultant amplitude is always constant, hence the tip of the amplitude vector lies on a circle.

When Z-component leads in phase by $\frac{\pi}{2}$

Under this case, the two components are represented as

$$y = \text{R.P. } a e^{i\phi_x} = a\cos\phi_x \tag{10.2.1}$$

$$z = \text{R.P. } a e^{i(\phi_x + \phi)} = a\cos(\phi_x + \phi) = a\cos(\phi_x + \pi/2) = -a\sin\phi_x \tag{10.2.2}$$

where $\phi_x = kx - \omega t$.

Now, at $t = t_1$, let $\phi_x = 0$, then, we get

$$\left.\begin{array}{c} y = a \\ z = 0 \end{array}\right\} \text{ corresponding to } t_1 = \frac{kx}{w}$$

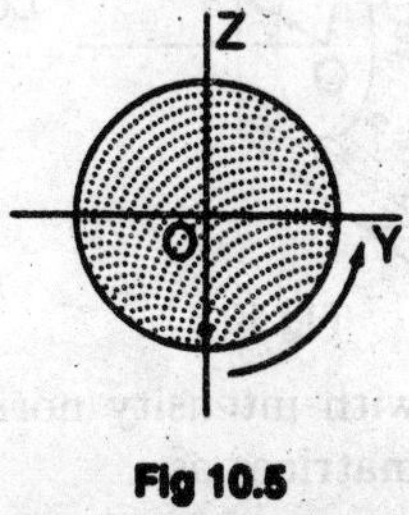

Also, let at $t = t_2$, $\phi_x = \frac{\pi}{2}$, then

$$\left.\begin{array}{c} y = 0 \\ z = -a \end{array}\right\} \text{ corresponding to } t_2 = \frac{kx}{w} - \frac{\pi}{2w}$$

Fig 10.5

Since $t_1 > t_2$, we can say that as time passes on, the resultant displacement vector passes from the configuration defined by $y = 0$, $z = -a$ to the configuration $y = a$, $z = 0$, as shown in fig. (10.5). So it represents that the resultant displacement vector rotates in a circle in the anti-clockwise sense.

Also, from eqns. (10.2.1) and (10.2.2), we easily get

$$\frac{y^2}{a^2} + \frac{z^2}{a^2} = 1 \tag{10.2.3}$$

Eqn. (10.2.3) shows that at a certain place, the displacement under the influence of the two vibrations under specified restriction satisfies an equation of a circle traversed anti-clockwise. So light in this case is said to be circularly polarised in the anti-clockwise sense. This is also referred to as Right-handed Circularly Polarised (RCP)[2]

So first conclusion

Of the two orthogonal vibrations, one along Y-direction, the other along Z-direction, if Z-vibration leads in phase by $\left(2p+\frac{1}{2}\right)\pi$, where $p = 0, 1, 2$, etc., the resultant vibration would be of constant amplitude, not fixed in space but rotating anti-clockwise in a circle.

In Jones' matrix formalism, this type of circularly polarised light is represented as

$$[J]_{\text{RCP}} = \begin{bmatrix} a \\ ae^{i\pi/2} \end{bmatrix} e^{i\phi_x} \quad \rightarrow \quad \frac{1}{\sqrt{2}}\begin{bmatrix} 1 \\ +i \end{bmatrix}, \text{ RCP}$$

with intensity normalised. It is the right circularly polarised light.

When the Z-component lags in phase by $\frac{\pi}{2}$

In this case, the two component vibrations are represented as

$$y = a\cos\phi_x, \quad z = a\cos\left(\phi_x - \frac{\pi}{2}\right) = a\sin\phi_x$$

When $\phi_x = 0 \quad \Rightarrow \quad t_1 = \frac{kx}{\omega}$, we have,

$$y = a, \quad z = 0.$$

When $\phi_x = \frac{\pi}{2} \quad \Rightarrow \quad t_2 = \frac{kx}{\omega} - \frac{\pi}{2\omega}$, we have,

$$y = 0, \quad z = a.$$

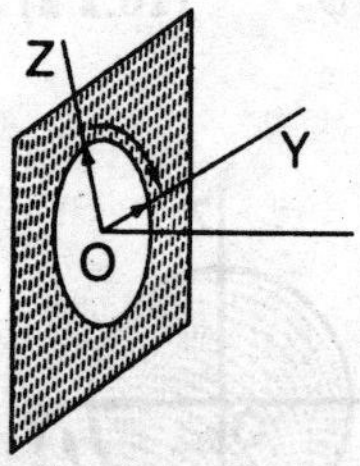

Fig 10.6

So, as time passes on, the vector passes from the configuration defined by $y = 0$, $z = a$, to the configuration defined by $y = a$, $z = 0$, as shown in fig. (10.6). It is said to be circularly polarised vibration in the clockwise sense or Left-handed Circularly Polarised (LCP).

So, in Jones' formalism,

$$[J]_{\text{LCP}} = \begin{bmatrix} a \\ ae^{-i\pi/2} \end{bmatrix} e^{i\phi_x} \longrightarrow \frac{a}{\sqrt{2}}\begin{bmatrix} 1 \\ -i \end{bmatrix} \longrightarrow \underbrace{\frac{1}{\sqrt{2}}\begin{bmatrix} 1 \\ -i \end{bmatrix}}_{\text{LCP}}$$

with intensity normalised. Thus the two circularly polarised lights are represented by Jones' matrices as

$$[J]_{\text{RCP}} = \frac{1}{\sqrt{2}}\begin{bmatrix} 1 \\ i \end{bmatrix}, \quad [J]_{\text{LCP}} = \frac{1}{\sqrt{2}}\begin{bmatrix} 1 \\ -i \end{bmatrix}$$

[2]In naming anti-clockwise circularly polarised light as Right-handed one, we have followed the convention due to Feynman. In some books, however, opposite convention is followed.

When the Y-component leads by $\frac{\pi}{2}$.

In this case, we can write,

$$y = \text{R.P. } ae^{i(\phi_x+\pi/2)} = a\cos\left(\frac{\pi}{2}+\phi_x\right) = -a\sin\phi_x$$

$$z = \text{R.P. } ae^{i\phi_x} = a\cos\phi_x.$$

When $\phi_x = 0 \Rightarrow t_1 = \frac{kx}{\omega}$, we get,

$$y = 0, \quad z = a$$

When $\phi_x = \frac{\pi}{2} \Rightarrow t_2 = \frac{kx}{\omega} - \frac{\pi}{2\omega}$, we get,

$$y = -a, \quad z = 0$$

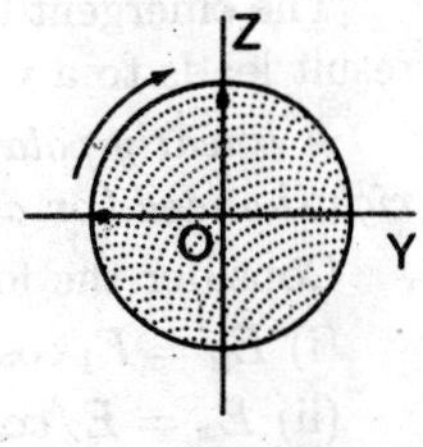

Fig 10.7

So, as time passes on, the configuration changes from one defined by $y = -a$, $z = 0$ to $y = 0$, $z = a$ as shown in fig. (10.7). It is a clockwise circularly polarised vibration (LCP).

In Jones' formalism,

$$[J]_{\text{LCP}} = \frac{1}{\sqrt{2}}\begin{bmatrix} i \\ 1 \end{bmatrix} \quad \text{— LCP,}$$

the Y-component leads. Similarly, if Y-component lags, then,

$$[J]_{\text{RCP}} = \frac{1}{\sqrt{2}}\begin{bmatrix} -i \\ 1 \end{bmatrix} \quad \text{— RCP.}$$

Thus,

$$[J]_{\text{RCP}} = \frac{1}{\sqrt{2}}\begin{bmatrix} 1 \\ i \end{bmatrix} \quad \text{for RCP when } Z\text{-component leads}$$

$$= \frac{1}{\sqrt{2}}\begin{bmatrix} -i \\ 1 \end{bmatrix} \quad \text{for RCP when } Y\text{-component lags}$$

$$[J]_{\text{LCP}} = \frac{1}{\sqrt{2}}\begin{bmatrix} 1 \\ -i \end{bmatrix} \quad \text{for LCP when } Z\text{-component lags}$$

$$\text{or,} \quad \frac{1}{\sqrt{2}}\begin{bmatrix} i \\ 1 \end{bmatrix} \quad \text{for LCP when } Y\text{-component leads}$$

Remember the following table:

Right circularly polarised light (RCP)	$\frac{1}{\sqrt{2}}\begin{bmatrix} 1 \\ i \end{bmatrix}$	or,	$\frac{1}{\sqrt{2}}\begin{bmatrix} -i \\ 1 \end{bmatrix}$	Anti-clockwise
Left circularly polarised light (LCP)	$\frac{1}{\sqrt{2}}\begin{bmatrix} 1 \\ -i \end{bmatrix}$	or,	$\frac{1}{\sqrt{2}}\begin{bmatrix} i \\ 1 \end{bmatrix}$	Clockwise

If an RCP and an LCP are superposed, we would get the emergent light as

$$\frac{1}{\sqrt{2}}\begin{bmatrix} 1 \\ i \end{bmatrix} + \frac{1}{\sqrt{2}}\begin{bmatrix} 1 \\ -i \end{bmatrix} = \frac{1}{\sqrt{2}}\begin{bmatrix} 2 \\ 0 \end{bmatrix} = \begin{bmatrix} \sqrt{2} \\ 0 \end{bmatrix}$$

So the emergent light is lineary polarised, with horizontal orientation and intensity $[(\sqrt{2})^2 + 0^2]$, i.e., 2 (sum of the two incident intensities). Similarly,

$$\frac{1}{\sqrt{2}}\begin{bmatrix} -i \\ 1 \end{bmatrix} + \frac{1}{\sqrt{2}}\begin{bmatrix} i \\ 1 \end{bmatrix} = \frac{1}{\sqrt{2}}\begin{bmatrix} 0 \\ 2 \end{bmatrix} = \begin{bmatrix} 0 \\ \sqrt{2} \end{bmatrix}$$

The emergent light is linearly polarised with orientation in the vertical direction. The above result leads to a very important conclusion—

A linearly polarised light may be obtained by superposing two circularly polarised light, one right-handed (or anti-clockwise) and another, left-handed (or clockwise).

Consider the following waves:

(i) $E_x = E_1 \cos(\omega t - kz), E_y = 0$;

(ii) $E_x = E_2 \cos(\omega t - kz), E_y = E_2 \cos(\omega t - kz - \pi/2)$;

(iii) $E_x = E_3 \cos(\omega t + kz), E_y = E_3 \cos(\omega t + kz + \pi/2)$;

(iv) $E_x = E_4 \cos(\omega t - kz), E_y = E_5 \cos(\omega t - kz)$.

Identify the state of polarisation of each of the waves. **[B.U. '99]**

Ans. (i) $E_x = \text{R.P.} E_1 e^{i\phi_z}$, $\phi_z = \omega t - kz$, $E_y = 0$

$$\Rightarrow \begin{bmatrix} E_x \\ E_y \end{bmatrix} = \begin{bmatrix} E_1 \\ 0 \end{bmatrix} e^{i\phi_z} \rightarrow \begin{bmatrix} 1 \\ 0 \end{bmatrix},$$

so it is a horizontally polarised with orientation axis along the X-direction.

(ii) $E_x = \text{R.P.} E_2 e^{i\phi_z}, E_y = \text{R.P.} E_2 e^{i(\phi_z - \pi/2)}$

$$\therefore \begin{bmatrix} E_x \\ E_y \end{bmatrix} = \begin{bmatrix} E_2 \\ E_2 e^{-i\pi/2} \end{bmatrix} e^{i\phi_z} \rightarrow \frac{1}{\sqrt{2}}\begin{bmatrix} 1 \\ -i \end{bmatrix}$$

So it represents an LCP.

(iii) It is represented as $\frac{1}{\sqrt{2}}\begin{bmatrix} 1 \\ i \end{bmatrix}$, hence as RCP.

(iv) $E_x = \text{R.P.} E_4 e^{i\phi_z}, \quad E_y = \text{R.P.} E_5 e^{i\phi_z}$

$$\begin{bmatrix} E_x \\ E_y \end{bmatrix} = \begin{bmatrix} E_4 \\ E_5 \end{bmatrix} e^{i\phi_z} \rightarrow \begin{bmatrix} E_4 \\ E_5 \end{bmatrix}$$

So it is also a plane polarised light with orientation axis making an angle θ with the X-direction given by $\tan\theta = (E_5/E_4)$.

Retarder plates: Production of circularly polarised light

From the above discussion it is clear that to obtain a circularly polarised light, we have to take two linearly polarised lights orthogonal to each other and produce in one a phase difference of $\frac{\pi}{2}$. A phase difference of $\frac{\pi}{2}$ is equivalent to a path difference of

$$\Delta = \frac{\lambda}{2\pi} \times \frac{\pi}{2} = \frac{\lambda}{4} \quad \text{(cf. the relation phase diff.} = \tfrac{2\pi}{\lambda} \times \text{path diff.)}.$$

A device which introduces a phase difference of $\frac{\pi}{2}$ is called a quartz-wave plate or a $\lambda/4$-plate or a $\frac{\pi}{2}$-retarder. Such a retarder can be used to produce and analyse circularly polarised and elliptically polarised light.

Construction of a Quarter-wave plate

Wave plates, also called retarders are converters of polarisation forms. With the help of appropriate retarder a given form of polarisation can be converted into any other form.

A quarter-wave plate introduces a phase-change of $\frac{\pi}{2}$. It may be made of a thin plate of calcite or quartz whose optic axis lies on the surface. Both of these materials are bi-refringent but calcite is said to be a −ve crystal because in it *the extraordinary ray* (E-ray) travels faster than the ordinary ray (O-ray), whereas quartz is a +ve crystal where the ordinary ray travels faster than the extraordinary ray.

We know from double refraction phenomenon that when light, ordinary or polarised, is incident on a crystal, in a direction other than the optic axis direction, it breaks up into two waves—ordinary and extraordinary. Both are linearly polarised and they are orthogonal to each other. Depending on the nature of the crystal (negative or positive), extraordinary or ordinary ray will travel faster. The direction along which speed is greater, is called 'fast axis', the other direction is the 'slow axis' of the crystal.

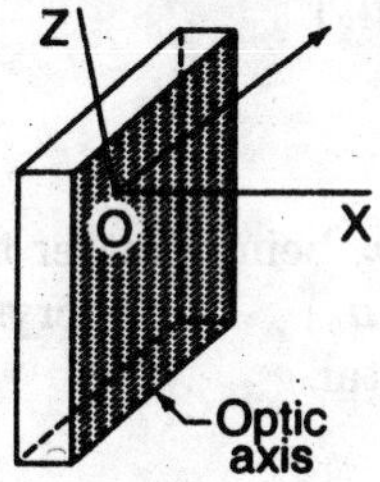

Fig 10.8

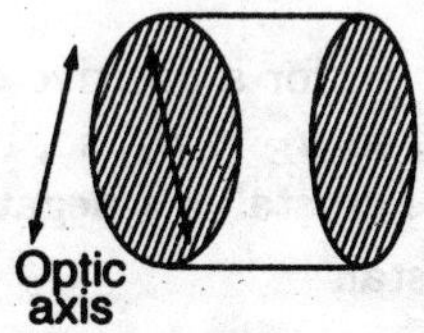

Fig 10.9

In case of a retarder it is true that both of ordinary ray and extraordinary ray get retarded, but it is the relative phase difference that is of concern to us.

Consider a calcite crystal whose plane faces are parallel to the optic axis and let light fall on the crystal face from left; also suppose that light is linearly polarised with horizontal orientation. The crystal is so oriented that the vibration makes an angle of 45° with the optic axis. The amplitude of vibration is, suppose, a. On incidence it breaks up into two vibrations, one along the optic axis or parallel to the optic axis, called E-ray; the other, orthogonal to it, called O-ray. Since in a calcite crystal, extraordinary travels faster, the optic axis direction is the fast axis, and direction $\perp$ to the optic axis, along which ordinary travels, is the slow axis. The calcite crystal retarder retards both these components, but due to difference in speeds of the ordinary and extraordinary rays and the thickness of the plate is so chosen that there is a relative change of phase by an amount equal to $\frac{\pi}{2}$ on emergence.

Thickness calculation: for a $\frac{\lambda}{4}$-plate (90°-retarder)

Let us take up once again the eqns.,

$$\vec{Y} = \hat{j}ae^{i(k_1x-\omega t)}, \quad \vec{Z} = \hat{k}ae^{i(k_2x-\omega t)}$$

Let after emergence ($x = L$), the relative phase change introduced by the plate is $\frac{\pi}{2}$, then,

$$(k_2 - k_1)L = \frac{\pi}{2} \quad \Rightarrow \quad L = \frac{\pi}{2(k_2 - k_1)} \tag{10.2.4}$$

Now, $k = \frac{2\pi}{\lambda}$, hence,

$$k_2 = \text{propagation constant for } E\text{-ray} = \frac{2\pi}{\lambda_E} = k_e$$

Again,

$$\mu_E = \frac{c}{v_e} = \frac{\lambda_{\text{air}}}{\lambda_E} = \frac{\lambda}{\lambda_E}.$$

Similarly, for ordinary ray,

$$k_1 = k_o = \frac{2\pi}{\lambda_o}, \quad \mu_o = \frac{c}{v_o} = \frac{\lambda_{\text{air}}}{\lambda_o} = \frac{\lambda}{\lambda_o}$$

$$k_2 - k_1 = k_e - k_o = 2\pi\left(\frac{1}{\lambda_E} - \frac{1}{\lambda_o}\right) = 2\pi\left(\frac{\mu_E}{\lambda} - \frac{\mu_o}{\lambda}\right) = \frac{2\pi}{\lambda}(\mu_E - \mu_o)$$

$$\therefore \quad L = \frac{\pi}{2} \cdot \frac{\lambda}{2\pi(\mu_E - \mu_o)} = \frac{\lambda}{4} \cdot \frac{1}{(\mu_E - \mu_o)} \tag{10.2.5}$$

$$\Rightarrow \quad L(\mu_E - \mu_o) = \frac{\lambda}{4}$$

Since $\mu_E = \frac{c}{v_e}$, $\mu_o = \frac{c}{v_o}$, for a positive crystal v_o being greater than v_e, $\mu_E > \mu_o$, and for a negative crystal $\nu_e > v_o$, so $\mu_E < \mu_o$, so, if $(\mu_E - \mu_o) > 0$, the crystal is +ve; if, on the other hand, $(\mu_E - \mu_o) < 0$, the crystal is a negative crystal.

In case of calcite crystal,

$$\mu_o = 1.6584, \qquad \mu_E = 1.4864$$

If we take Na-light, $\lambda = 5893 \times 10^{-8}$ cm.

$$\therefore \quad L|(\mu_E - \mu_o)| = \frac{\lambda}{4}$$

$$\Rightarrow L = \frac{5893 \times 10^{-8}}{4} \times \frac{1}{|1.4864 - 1.6584|} = 8565 \times 10^{-8} = .00086 \,\text{mm}.$$

So it is extremely thin, hence calcite is not suitable quarter-wave plate material, because the difference of $(\mu_E - \mu_o)$, called *bi-refringence* is large for calcite. If $\mu_E > \mu_o$ in a crystal, it is said to be +ve crystal; if $\mu_E < \mu_o$, it is a −ve crystal.

For mica, the bi-refringence is,

$$\mu_E - \mu_o = 1.5977 - 1.5936 = 4.1 \times 10^{-3},$$

hence, $\mu_E > \mu_o$, therefore, mica is a +ve bi-axial crystal and since its bi-refringence is small, it is suitable for making the wave plate. The thickness required is,

$$L = \frac{\lambda}{4(\mu_E - \mu_o)} = \frac{5893 \times 10^{-8}}{4 \times 4.1 \times 10^{-3}} = 3.59 \times 10^{-3}\,\text{cm} = .036\,\text{mm}.$$

Thus in Jones' matrix formalism, a matrix represented as

$$[M] = \begin{bmatrix} 1 & 0 \\ 0 & e^{i\delta} \end{bmatrix},$$

where δ is the phase difference introduced, would represent a linear retarder, either right-handed or left-handed depending on the value of δ.

Jones' matrix written as

$$[J_c] = \begin{bmatrix} 1 & 0 \\ 0 & e^{\pm i\pi/2} \end{bmatrix}$$

will be called quarter-wave plate matrix producing right circularly polarised light with the positive sign; with negative sign, it will be called a quarter-wave plate matrix producing left circularly polarised light. Thus,

$$[J_c] = \begin{bmatrix} 1 & 0 \\ 0 & i \end{bmatrix} \longrightarrow \text{Quarter-wave plate matrix producing RCP}$$

$$[J_c] = \begin{bmatrix} 1 & 0 \\ 0 & -i \end{bmatrix} \longrightarrow \text{Quarter-wave plate matrix producing LCP}$$

Consider the matrix $\begin{bmatrix} 1 & 0 \\ 0 & i \end{bmatrix}$ and let it operate on a linearly polarised light with orientation 45°, then

$$\begin{bmatrix} 1 & 0 \\ 0 & i \end{bmatrix} \frac{1}{\sqrt{2}} \begin{bmatrix} 1 \\ 1 \end{bmatrix} = \frac{1}{\sqrt{2}} \begin{bmatrix} 1 \\ i \end{bmatrix} \quad \text{RCP}$$

$$\begin{bmatrix} 1 & 0 \\ 0 & -i \end{bmatrix} \frac{1}{\sqrt{2}} \begin{bmatrix} 1 \\ 1 \end{bmatrix} = \frac{1}{\sqrt{2}} \begin{bmatrix} 1 \\ -i \end{bmatrix} \quad \text{LCP.}$$

In general, if the linearly polarised light be represented as $\begin{bmatrix} a \\ b \end{bmatrix}$, then

$$\begin{bmatrix} 1 & 0 \\ 0 & i \end{bmatrix} \begin{bmatrix} a \\ b \end{bmatrix} = \begin{bmatrix} a \\ ib \end{bmatrix} = \begin{bmatrix} a \\ be^{i\pi/2} \end{bmatrix}$$

Writing out the y and z components we get,

$$y = \text{R.P.} \quad ae^{i\phi_x} \quad \Rightarrow \quad y = a\cos\phi_x$$

$$z = \text{R.P.} \quad be^{(i\phi_x + \pi/2)} \quad \Rightarrow \quad z = -b\sin\phi_x$$

Thus, when $\phi_x = kx - \omega t_1 = 0 \quad \Rightarrow \quad t_1 = \frac{kx}{\omega}$

$$y = a \qquad z = 0$$

When, $\phi_x = kx - \omega t_2 = \frac{\pi}{2} \quad \Rightarrow \quad t_2 = \frac{kx}{\omega} - \frac{\pi}{2\omega}$,

$$y = 0, \quad z = -b = -a \quad \text{when} \quad a = b$$

Thus we get the emergent light as circularly polarised in the anti-clockwise direction (or RCP).

Example 7. *A plane polarised light at +45° to the Y-axis enters a $\frac{\lambda}{4}$ plate, whose fast axis is vertical. Determine nature of emergent light.*

Solution: A plane polarised light at +45° is represented by Jones' vector $\frac{1}{\sqrt{2}}\begin{bmatrix}1\\1\end{bmatrix}$.

A $\lambda/4$ plate with fast axis vertical or horizontal is represented by Jones' matrix

$$\begin{bmatrix}1 & 0\\0 & i\end{bmatrix} \text{ or, } \begin{bmatrix}1 & 0\\0 & -i\end{bmatrix} \quad \text{respectively}$$

$$\therefore \quad \begin{bmatrix}1 & 0\\0 & i\end{bmatrix}\frac{1}{\sqrt{2}}\begin{bmatrix}1\\1\end{bmatrix} = \underset{\text{RCP}}{\frac{1}{\sqrt{2}}\begin{bmatrix}1\\-i\end{bmatrix}} \quad \text{Right-handed circularly polarised}$$

$$\text{or,} \quad \begin{bmatrix}1 & 0\\0 & -i\end{bmatrix}\frac{1}{\sqrt{2}}\begin{bmatrix}1\\1\end{bmatrix} = \underset{\text{LCP}}{\frac{1}{\sqrt{2}}\begin{bmatrix}1\\-i\end{bmatrix}} \quad \text{Left-handed circularly polarised light.}$$

10.3 Half-wave Plate or 180°-Retarder

If the relative phase difference introduced by a plate be equal to π or path retardation be $\frac{\lambda}{2}$, then the plate will be termed as a half-wave plate.

Suppose, the incident beam is linearly polarised at +45°. The incident beam is equivalent to two simple harmonic vibrations, vibrations being parallel to Y and Z axes and are in same phase on incidence. But, if one component be retarded within the plate by an amount equal to 180°, (say, Z-component), it must be represented by a line OB' which is opposite to OB. The resultant lies on a line at −45°, thus a 180°-retarder converts linearly polarised light at +45° to a linearly polarised light at −45°. This transformation may be carried through a matrix transformation and thus a half-wave retarder matrix can be realised. Thus,

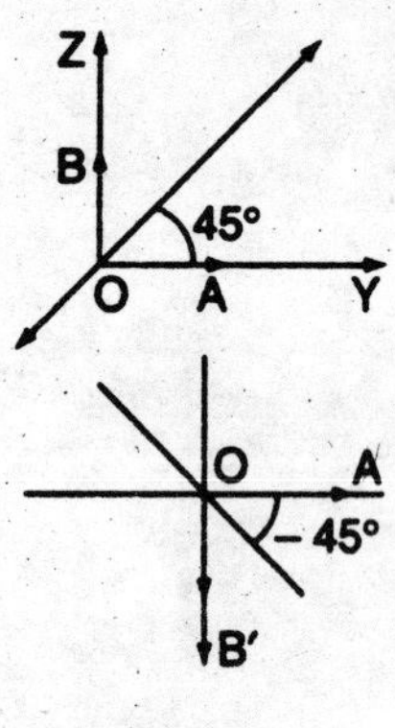

Fig 10.10

$$\begin{bmatrix}1 & 0\\0 & -1\end{bmatrix}\frac{1}{\sqrt{2}}\begin{bmatrix}1\\1\end{bmatrix} = \frac{1}{\sqrt{2}}\begin{bmatrix}1\\-1\end{bmatrix},$$

hence for a linear half-wave retarder with fast axis vertical, the Jones' matrix is

$$\begin{bmatrix}1 & 0\\0 & -1\end{bmatrix}.$$

It is as if a 180°-retarder had reflected the incident beam's vibration direction, the mirror being supposed to be parallel to the fast or slow axis.

The half-wave plates are in practice never made of calcite, because its bi-refringence is large, hence its thickness is extremely small and, therefore, it becomes fragile; usually quartz or mica is used.

An important observation

Two retarders, one right-handed and one left-handed circularly polarised light producer, acting in succession, produce no overall change. Thus,

$$\underset{\left(\frac{\lambda}{4}\text{-plate matrix for RCP}\right)}{\begin{bmatrix}1 & 0\\0 & i\end{bmatrix}} \quad \underset{\left(\text{linearly polarised light}\right)}{\frac{1}{\sqrt{2}}\begin{bmatrix}1\\1\end{bmatrix}} = \underset{\text{(RCP)}}{\frac{1}{\sqrt{2}}\begin{bmatrix}1\\i\end{bmatrix}}$$

$$\begin{bmatrix} 1 & 0 \\ 0 & -i \end{bmatrix} \qquad s\frac{1}{\sqrt{2}}\begin{bmatrix} 1 \\ i \end{bmatrix} = \qquad \frac{1}{\sqrt{2}}\begin{bmatrix} 1 \\ 1 \end{bmatrix}$$

$\left(\begin{matrix}\frac{\lambda}{4}\text{-plate matrix} \\ \text{for LCP}\end{matrix}\right)$ $\qquad$ $\left(\begin{matrix}\text{linearly polarised} \\ \text{light given}\end{matrix}\right)$

Thus,

$$\begin{bmatrix} 1 & 0 \\ 0 & -i \end{bmatrix}\begin{bmatrix} 1 & 0 \\ 0 & i \end{bmatrix}\frac{1}{\sqrt{2}}\begin{bmatrix} 1 \\ 1 \end{bmatrix} = \frac{1}{\sqrt{2}}\begin{bmatrix} 1 \\ 1 \end{bmatrix}$$

operation of the two elements as taken above really yields a unitary matrix, thus producing ultimately no overall change.

Thus,

$$\begin{bmatrix} 1 & 0 \\ 0 & i \end{bmatrix}\begin{bmatrix} 1 & 0 \\ 0 & -i \end{bmatrix} = \begin{bmatrix} 1 & 0 \\ 0 & 1 \end{bmatrix}$$

But, if each element acts twice, for example, if two left circularly polarising quarter-wave plates act one after the other, producing a total phase difference of π, then we get the action of a half-wave plate, thus

$$\begin{bmatrix} 1 & 0 \\ 0 & i \end{bmatrix}\begin{bmatrix} 1 & 0 \\ 0 & i \end{bmatrix} = \begin{bmatrix} 1 & 0 \\ 0 & -1 \end{bmatrix}$$

↓

half-wave plate polarising element matrix

Also,

$$\begin{bmatrix} 1 & 0 \\ 0 & -i \end{bmatrix}\begin{bmatrix} 1 & 0 \\ 0 & -i \end{bmatrix} = \begin{bmatrix} 1 & 0 \\ 0 & -1 \end{bmatrix}$$

A few more Illustrative Examples

Example 8. *A quarter-wave plate, fast axis horizontal is followed by a linear polariser, transmission axis at +45° and then by another quarter-wave plate, fast axis vertical. Describe the system matrix.*

Solution: A quarter-wave plate with fast axis vertical is represented by a matrix

$$[J_1] = \begin{bmatrix} 1 & 0 \\ 0 & i \end{bmatrix}.$$

A quarter-wave plate with fast axis horizontal is represented by a matrix

$$[J_2] = \begin{bmatrix} 1 & 0 \\ 0 & -i \end{bmatrix}.$$

A linear polariser whose transmission axis is at +45° is

$$[J_3] = \frac{1}{2}\begin{bmatrix} 1 & 1 \\ 1 & 1 \end{bmatrix}$$

$\therefore$ The system matrix is represented by,

$$[S] = \begin{bmatrix} 1 & 0 \\ 0 & i \end{bmatrix} \begin{bmatrix} 1 & 0 \\ 0 & -i \end{bmatrix} \frac{1}{2} \begin{bmatrix} 1 & 1 \\ 1 & 1 \end{bmatrix} = \begin{bmatrix} 1 & 0 \\ 0 & +1 \end{bmatrix} \frac{1}{2} \begin{bmatrix} 1 & 1 \\ 1 & 1 \end{bmatrix} = \frac{1}{2} \begin{bmatrix} 1 & 1 \\ 1 & 1 \end{bmatrix}$$

Thus the system matrix is the same as that of a linear polariser with axis at +45°.

Example 9. *Ice is a +ve uniaxial crystal with indices of refraction* 1.309 *and* 1.310. *How thick must the ice be to act as a quarter-wave plate for light of 600 nm wavelength?*

Solution:

$$\mu_E - \mu_o < 0, \quad -\text{ve crystal}, \quad \mu_E - \mu_o > 0, \quad +\text{ve crystal}$$

$$\therefore \quad \mu_o = 1.309, \quad \mu_E = 1.310$$

$$\therefore \quad L = \frac{\lambda_{\text{air}}}{4(\mu_E - \mu_o)} = \frac{600 \times 10^{-9}}{41 \times 10^{-3}} = 1.5 \times 10^{-4} \text{ m} = 1.5 \times 10^{-1} \text{ mm} = .15 \text{ mm}.$$

Example 10. *A $\frac{\lambda}{4}$-plate with its fast axis horizontal is inserted into a beam of light linearly polarised and oscillating at 45°. Determine the polarisation of the emergent light.*

Solution:

$$\begin{bmatrix} 1 & 0 \\ 0 & -i \end{bmatrix} \frac{1}{\sqrt{2}} \begin{bmatrix} 1 \\ 1 \end{bmatrix} = \frac{1}{\sqrt{2}} \begin{bmatrix} 1 \\ -i \end{bmatrix}$$

LCP with the same normalised intensity.

The following table in the next page shows the Jones' vector and matrices for circularly polarised light and $\frac{\lambda}{4}$-retarder.

Example 11. *What is the action of First a quarter-wave plate, then a half-wave plate; First a half-wave plate, then a quarter-wave plate?*

Solution:

$$\underset{\text{Half-wave plate}}{\begin{bmatrix} 1 & 0 \\ 0 & -1 \end{bmatrix}} \quad \underset{\text{RCP } \frac{\lambda}{4}\text{-plate}}{\begin{bmatrix} 1 & 0 \\ 0 & i \end{bmatrix}} = \underset{\text{LCP } \frac{\lambda}{4}\text{-plate}}{\begin{bmatrix} 1 & 0 \\ 0 & -i \end{bmatrix}}$$

$$\underset{\text{RCP } \frac{\lambda}{4}\text{-plate}}{\begin{bmatrix} 1 & 0 \\ 0 & i \end{bmatrix}} \quad \underset{\text{Half-wave plate}}{\begin{bmatrix} 1 & 0 \\ 0 & -1 \end{bmatrix}} = \underset{\text{LCP } \frac{\lambda}{4}\text{-plate}}{\begin{bmatrix} 1 & 0 \\ 0 & -i \end{bmatrix}}$$

So the result is same in the two cases.

Example 12. *What is the action of first a half-wave retarder, followed by a $\frac{\lambda}{4}$-plate, then again followed by a $\frac{\lambda}{2}$-plate?*

Solution:

$$\begin{bmatrix} 1 & 0 \\ 0 & -1 \end{bmatrix} \quad \begin{bmatrix} 1 & 0 \\ 0 & i \end{bmatrix} \quad \begin{bmatrix} 1 & 0 \\ 0 & -1 \end{bmatrix}$$

$$= \underset{\frac{\lambda}{2}\text{-plate}}{\begin{bmatrix} 1 & 0 \\ 0 & -1 \end{bmatrix}} \quad \underset{\text{LCP } \frac{\lambda}{4}\text{-plate}}{\begin{bmatrix} 1 & 0 \\ 0 & -i \end{bmatrix}} = \underset{\text{RCP } \frac{\lambda}{4}\text{-plate}}{\begin{bmatrix} 1 & 0 \\ 0 & i \end{bmatrix}}$$

Example 13. *A left circularly polarised beam of $\lambda = 5893Å$ is incident normally on a calcite crystal of thickness 0.005141 mm. What will be the state of polarisation of the emergent beam?*

Solution. Phase difference introduced between extraordinary ray and ordinary ray is given by

$$L = \phi \cdot \frac{\lambda}{2\pi|\mu_E - \mu_o|}, \quad \phi = \frac{2\pi L|\mu_E - \mu_o|}{\lambda}$$

$$\mu_E = 1.4864, \quad \mu_o = 1.6584$$

$$\therefore \quad \phi = \frac{2\pi L \times 0.172}{5893 \times 10^{-8}} = \frac{2 \times \pi \times .0005141 \times .172}{5893 \times 10^{-8}}$$

$$= 9.921 \times 10^{-6} \times 10^{8} = 3\pi$$

According to Jones' formalism,

$$\begin{bmatrix} 1 & 0 \\ 0 & e^{i3\pi} \end{bmatrix} \frac{1}{\sqrt{2}} \begin{bmatrix} 1 \\ i \end{bmatrix} = \frac{1}{\sqrt{2}} \begin{bmatrix} 1 \\ ie^{i3\pi} \end{bmatrix}$$

Writing in full, the emergent light is represented by

$$\begin{bmatrix} y \\ z \end{bmatrix} = \frac{1}{\sqrt{2}} \begin{bmatrix} 1 \\ e^{i\pi/2} . e^{i3\pi} \end{bmatrix} ae^{i\phi_x}$$

$$y = \frac{a}{\sqrt{2}} \cos\phi_x = \frac{a}{\sqrt{2}} \cos(kx - \omega t)$$

$$z = \frac{a}{\sqrt{2}} \cos\left(3\pi + \frac{\pi}{2} + \phi_x\right)$$

$$= \frac{a}{\sqrt{2}} \cos\left(\frac{\pi}{2} + \phi_x\right) = -\frac{a}{\sqrt{2}} \sin\phi_x$$

$$= -\frac{a}{\sqrt{2}} \sin(kx - \omega t)$$

$\frac{y^2}{a^2} + \frac{z^2}{a^2} = 1$, so the emergent ray is circularly polarised. It is right-handed circularly polarised.

Table for Retarder Matrix and Circular Polarisation Matrix

Right circularly polarised light (RCP)	$\frac{1}{\sqrt{2}}\begin{bmatrix} 1 \\ i \end{bmatrix}$ or, $\frac{1}{\sqrt{2}}\begin{bmatrix} -i \\ 1 \end{bmatrix}$
Left circularly polarised light (LCP)	$\frac{1}{\sqrt{2}}\begin{bmatrix} 1 \\ -i \end{bmatrix}$ or, $\frac{1}{\sqrt{2}}\begin{bmatrix} i \\ 1 \end{bmatrix}$
$\frac{\lambda}{4}$-plate retarder with fast axis vertical	$\begin{bmatrix} 1 & 0 \\ 0 & i \end{bmatrix}$
$\frac{\lambda}{4}$-plate retarder with fast axis horizontal	$\begin{bmatrix} 1 & 0 \\ 0 & -i \end{bmatrix}$

Elliptical polarisation

It is a general case of polarisation under the influence of two linearly polarised light orthogonal to each other and having a phase difference. Circularly polarised light is a special case. Thus suppose, the horizontally polarised light is represented as

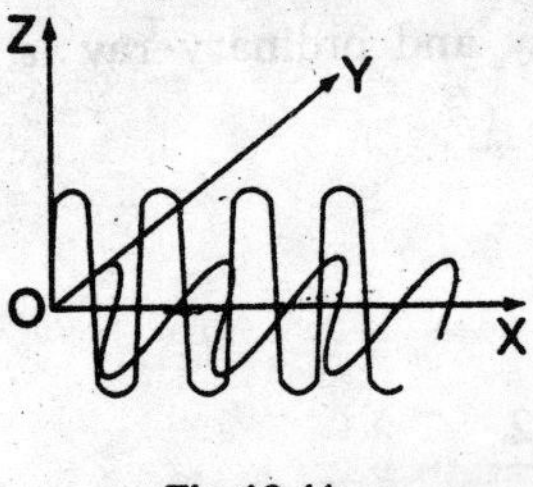

Fig 10.11

$$\bar{Y} = \hat{j}A_1 e^{i(k_1 x - \omega t)} = \hat{j}A_1 e^{i\phi_x}.$$

The vertically polarised light is represented as

$$\bar{Z} = \hat{k}A_2 e^{i(k_1 x - \omega t + \phi)} = \hat{k}A_2 e^{i\phi_x} \cdot e^{i\phi}$$

where, the disturbance is supposed to move along X-direction, the propagation constant being $k_1 = \frac{2\pi}{\lambda}$, and ϕ is the phase difference between the two orthogonal polarised disturbances. The combined displacement is written as

$$\bar{\xi} = \bar{Y} + \bar{Z} = \hat{j}A_1 e^{i\phi_x} + \hat{k}A_2 e^{i(\phi_x + \phi)} = (\hat{j}A_1 + \hat{k}A_2 e^{i\phi})e^{i\phi_x}$$

$\therefore$ y-component of the resultant vibration is given by,

$$y = \text{R.P. } A_1 e^{i\phi_x} = A_1 \cos\phi_x.$$

Similarly, z-component is given by,

$$z = \text{R.P. } A_2 e^{i(\phi_x + \phi)} = A_2 \cos(\phi_x + \phi)$$

Since ϕ_x is a function of time, as time passes on, both these components change, but there is a phase difference ϕ between these two components.

From the two relations, we get,

$$\frac{y^2}{A_1^2} + \frac{z^2}{A_2^2} - \frac{2yz}{A_1 A_2}\cos\phi = \sin^2\phi \tag{10.3.1}$$

Eqn. (10.3.1) is a general equation of an ellipse, so light under this condition is said to be elliptically polarised. For different values of ϕ, we would get either linearly polarised or elliptically polarised light, right-handed or left-handed.

The general eqn. (10.3.1) can be thrown into the form

$$Ay^2 + Bz^2 + 2hyz = 1 \tag{10.3.2}$$

where we have written,

$$A = \frac{1}{A_1^2 \sin^2\phi}, \quad B = \frac{1}{A_2^2 \sin^2\phi}, \quad h = \frac{-\cos\phi}{A_1 A_2 \sin^2\phi}$$

Rotate the axes of co-ordinates through an angle θ such that the new co-ordinates are given by,

$$y = Y\cos\theta - Z\sin\theta, \quad z = Y\sin\theta + Z\cos\theta$$

Putting the values of y, z in eqn. (10.3.2), we get,

$$A(Y\cos\theta - Z\sin\theta)^2 + B(Y\sin\theta + Z\cos\theta)^2$$
$$+2h(Y\cos\theta - Z\sin\theta)(Y\sin\theta + Z\cos\theta) = 1$$
$$\Rightarrow \quad Y^2(A\cos^2\theta + 2h\cos\theta\sin\theta + B\sin^2\theta) + Z^2(A\sin^2\theta - 2h\sin\theta\cos\theta + B\cos^2\theta)$$
$$+2YZ\{h(\cos^2\theta - \sin^2\theta) - (A-B)\sin\theta\cos\theta\} = 1$$

Choose the value of θ such that the term containing the product YZ vanishes.
So the choice is given by,

$$h(\cos^2\theta - \sin^2\theta) = (A-B)\sin\theta\cos\theta \quad \Rightarrow \quad \tan 2\theta = \frac{2h}{(A-B)} \qquad (10.3.3)$$

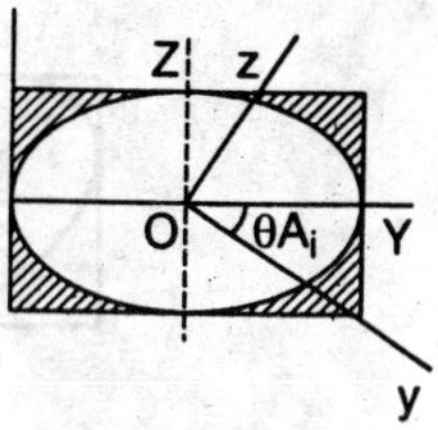

Fig 10.12

The eqn. (10.3.2) when plotted results in an elliptic curve as shown in fig. (10.12), the angle between the co-ordinate axes and those of the ellipse being θ defined by eqn. (10.3.3).

Depending on the value of θ, we would get different types of polarised light.

Case I: Linearly polarised light

When $\phi = s\pi$, the eqn. (10.3.1) can be written as

$$\left[\frac{y}{A_1} - (-1)^s \frac{z}{A_2}\right]^2 = 0 \quad \Rightarrow \quad \frac{y}{A_1} = (-1)^s \frac{z}{A_2} \qquad (10.3.4)$$

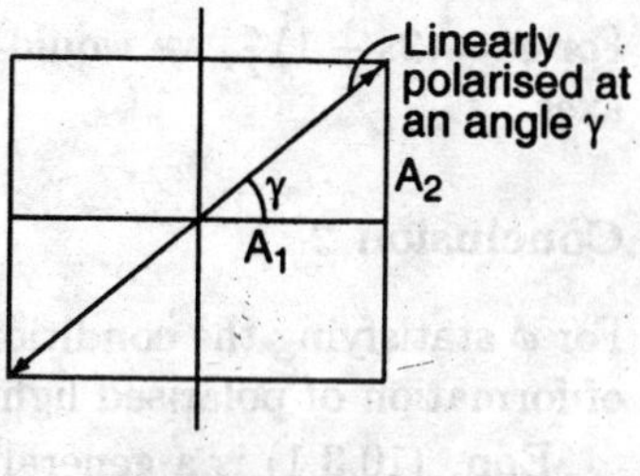

Fig 10.13

So, when $s = 0, 2, 4 \quad \Rightarrow \quad \phi = 0, 2\pi, 4\pi$, etc.,

$$z = \frac{A_2}{A_1} y \qquad (10.3.5)$$

which is a linearly polarised light with orientation γ with respect y-direction given as

$$\tan\gamma = \frac{A_2}{A_1} \qquad (10.3.6)$$

This is represented as in fig. (10.13)

If on the other hand, $s = 1, 3, 5$, etc. $\Rightarrow \quad \phi = \pi, 3\pi, 5\pi$, etc., we get,

$$z = -\frac{A_2}{A_1} y$$

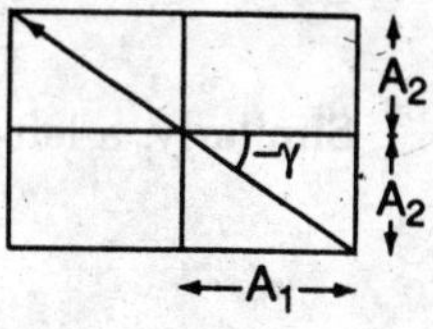

Fig 10.14

a linearly polarised light at an angle $-\gamma$ w.r.t. the y-axis and it is shown in fig. (10.4).

Case II

When $\phi = (2s+1)\frac{\pi}{2}$, $s = 0, 1, 2$, etc., we would get, eqn. (10.3.1) as

$$\frac{y}{A_1^2} + \frac{z}{A_2^2} = 1 \qquad (10.3.7)$$

Thus we would get elliptically polarised light whose axes would coincide with the co-ordinate axes. So we draw a sequence of conclusions diagrammatically represented by a series of figures as in fig. [10.15 (a), (b), (c), (d), (e), (f), (g), (h)].

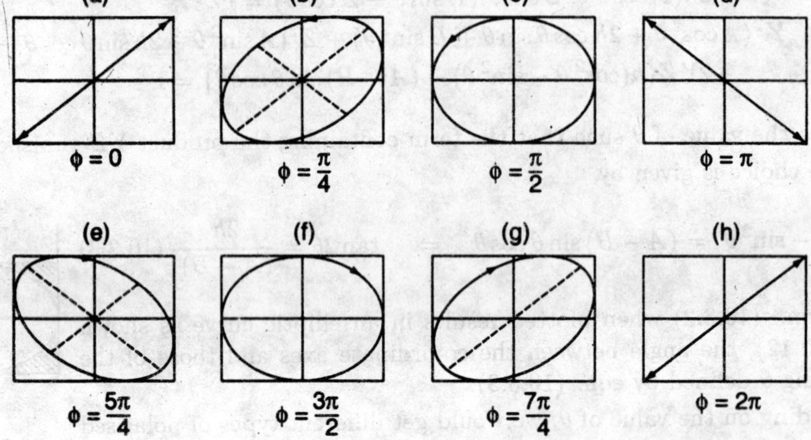

Fig 10.15

Conclusion 1

For $\phi = (2s+1)\frac{\pi}{2}$, we would get elliptically polarised light with its axes along the co-ordinate axes.

Conclusion 2

For ϕ statisfying the condition $\phi = 2s \times \frac{\pi}{2}$, we would get linearly polarised light. The sequence of formation of polarised light is shown.

Eqn. (10.3.1) is a general eqn. for an ellipse, which reduces to standard form when $\phi = \frac{\pi}{2}$ as shown in case I above.

In Jones' formalism, the elliptically polarised light is represented as

$$[J_{\text{REP}}] = \begin{bmatrix} A_1 \\ A_2 e^{i\phi} \end{bmatrix} \rightarrow \begin{bmatrix} A_1 \\ iA_2 \end{bmatrix}$$

Right elliptically polarised (REP)

Similarly, a left-handed elliptically polarised light is represented as

$$[J_{\text{LEP}}] = \begin{bmatrix} A_1 \\ -iA_2 \end{bmatrix}$$

Table for Jones' Matrix for Elliptically Polarised Light

Right-handed elliptically polarised light (REP)	$\begin{bmatrix} A_1 \\ iA_2 \end{bmatrix}$
Left-handed elliptically polarised (LEP)	$\begin{bmatrix} A_1 \\ -iA_2 \end{bmatrix}$

Production of elliptically polarised light by quarter-wave plate

When linearly polarised light (either of horizontal or vertical orientation or any orientation) is incident on a quarter-wave plate such that the vibration direction is at an angle θ with the optic axis (which lies on the left-hand face and is parallel to the z-direction), the vibration at the entry-end is resolved into two components along and $\perp$ to the optic axis as E-ray and O-ray respectively. The ordinary ray (O-ray) is represented as

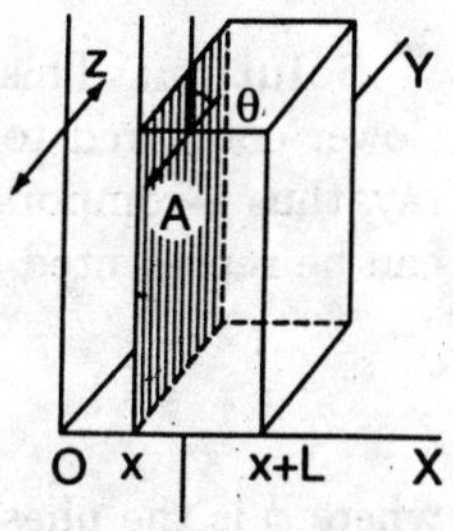

$$y = A\sin\theta e^{i\phi_x} \quad (O\text{-ray})$$

and the extraordinary ray (E-ray) as

$$z = A\cos\theta e^{i\phi_x} \quad (E\text{-ray})$$

Fig 10.16

If the $\frac{\lambda}{4}$-plate be made of a negative crystal (e.g., calcite), the E-ray travels faster than the O-ray and in case of a +ve crystal (e.g., quartz), the reverse is true. So, if the E-ray travels faster, at emergence the two components are represented as

$$y = A\sin\theta e^{i\phi_{x+L}} = A_1 e^{i\phi_{x+L}}, \quad z = A\cos\theta e^{i\phi_{x+L}} e^{i\phi} = A_2 e^{i\phi_{x+L}} e^{i\phi}$$

where L is the thickness of the plate.

In Jones' vector notation, the elliptically polarised light is represented as

$$\begin{bmatrix} A\sin\theta \\ A\cos\theta e^{i\phi} \end{bmatrix} \to \begin{bmatrix} A_1 \\ A_2 e^{i\phi} \end{bmatrix}$$

The elliptic polariser can be represented as

$$[J_e] = \begin{bmatrix} A_1 & 0 \\ 0 & A_2 e^{i\phi} \end{bmatrix}$$

Thus, when a circularly polarised light is incident on an elliptic polariser, the emergent light nature is available through matrix operation as

$$\underset{\substack{\downarrow \\ \text{Elliptic} \\ \text{polariser}}}{\begin{bmatrix} A_1 & 0 \\ 0 & A_2 e^{i\phi} \end{bmatrix}} \quad \underset{\substack{\downarrow \\ \text{Circularly} \\ \text{polarised light}}}{\begin{bmatrix} 1 \\ i \end{bmatrix}} = \underset{\substack{\downarrow \\ \text{Elliptically} \\ \text{polarised light} \\ \text{(right-handed)}}}{\begin{bmatrix} A_1 \\ iA_2 e^{i\phi} \end{bmatrix}} = \begin{bmatrix} A_1 \\ A_2 e^{i(\phi + \frac{\pi}{2})} \end{bmatrix}$$

The quarter-wave plate introduces a phase difference of $\frac{\pi}{2}$ between the orthogonal components, hence quarter-wave matrix producing elliptically polarised light is represented as

$$\begin{bmatrix} A_1 & 0 \\ 0 & A_2 e^{i\pi/2} \end{bmatrix} \to \begin{bmatrix} A_1 & 0 \\ 0 & iA_2 \end{bmatrix} \quad \text{(Right-handed)}$$

Similarly,

$$\begin{bmatrix} A_1 & 0 \\ 0 & iA_2 \end{bmatrix} \begin{bmatrix} a \\ b \end{bmatrix} = \begin{bmatrix} A_1 a \\ ibA_2 \end{bmatrix} = \underset{\substack{\downarrow \\ \text{Elliptically} \\ \text{polarised}}}{\begin{bmatrix} B_1 \\ iB_2 \end{bmatrix}}$$

Example 14. *A left circularly polarised light of $\lambda = 5893 \text{Å}$ is incident on a quartz crystal (with its optic axis parallel to the surface) of thickness 0.025 mm. Determine the state of polarisation of the emergent beam. Take $\mu_o = 1.54425$ and $\mu_E = 1.55336$.*

Solution: This type of retarder is a +ve retarder, where the extraordinary ray travels slower compared to the O-ray. So after emergence, the extraordinary ray lags behind the O-ray, thus z-component (parallel to the optic axis) will lag in phase. The retarder matrix here can be represented as

$$\begin{bmatrix} 1 & 0 \\ 0 & e^{-i\phi} \end{bmatrix}$$

where ϕ is the phase difference introduced in the path of traverse.

$$\therefore \underbrace{\begin{bmatrix} 1 & 0 \\ 0 & e^{-i\phi} \end{bmatrix}}_{\text{Retarder matrix}} \underbrace{\frac{1}{\sqrt{2}} \begin{bmatrix} 1 \\ -i \end{bmatrix}}_{\text{Left circular polarised light}} = \underbrace{\frac{1}{\sqrt{2}} \begin{bmatrix} 1 \\ -ie^{-i\phi} \end{bmatrix}}_{\text{Emergent polarised light}} = \frac{1}{\sqrt{2}} \begin{bmatrix} 1 \\ e^{-i(\pi/2+\phi)} \end{bmatrix}$$

To study the nature of the light, we take the full form of the emergent polarised light matrix written as

$$\begin{bmatrix} y \\ z \end{bmatrix} = \frac{1}{\sqrt{2}} \begin{bmatrix} 1 \\ -ie^{-i\phi} \end{bmatrix} e^{i\phi_{x+L}} = \frac{1}{\sqrt{2}} \begin{bmatrix} 1 \\ e^{-i(\pi/2+\phi)} \end{bmatrix} e^{i\phi_{x+L}}$$

$$\Rightarrow \quad y = \frac{A}{\sqrt{2}} \cos\phi_{x+L} = \frac{A}{\sqrt{2}} \cos\{k_1(x+L) - \omega t\} \tag{10.3.8}$$

$$z = \frac{A}{\sqrt{2}} \cos\left\{k_1(x+L) - \omega t - \frac{\pi}{2} - \phi\right\} = \frac{A}{\sqrt{2}} \sin\{k_1(x+L) - \omega t - \phi\} \tag{10.3.9}$$

Let us now see how do y and z change with time.

At the instant when $k_1(x+L) - \omega t = \frac{\pi}{2}$, we get

$$y = 0, \quad z = \frac{A}{\sqrt{2}} \sin\left(\frac{\pi}{2} + \phi\right) = \frac{A}{\sqrt{2}} \cos\phi.$$

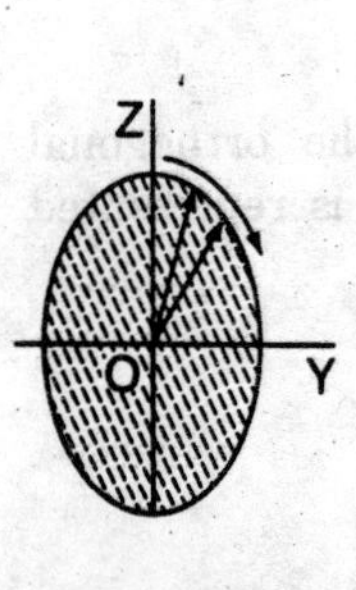

Fig 10.17

These two quantities are represented as in fig. (10.17). As time passes on and at the instant when

$$k_1(x+L) - \omega t + \phi = \frac{\pi}{2} \quad \Rightarrow \quad \omega t = k_1(x+L) + \phi - \frac{\pi}{2},$$

we get

$$y = \frac{A}{\sqrt{2}} \cos\left(\frac{\pi}{2} - \phi\right) = \frac{A}{\sqrt{2}} \sin\phi, \quad z = 0$$

Thus, the tip of the displacement vector lies on an ellipse, left-handed. So such a wave plate acts so as to produce a left-handed elliptically polarised light.

10.4 Babinet's Compensator: Its use in the Production of Polarised Light

It is a very important and useful piece of apparatus for production and analysis of different types of polarised light.

To make a Babinet's compensator, two thin right-angled prisms ABD and BDC made of quartz are joined along the hypotenuse side so as to form a rectangular block. The optic axis in the first prism lies parallel to the face of the prism and the optic axis in the second lies ⊥ to the face of the second. Thus the two optic axes are mutually at right angles. There is only one direction such as EF along which a ray traverses equal thickness of the prisms. This may be taken as the symmetry axis of the compensator and for any other ray either towards the left or towards the right of the symmetry axis, path in one increases as the path in the other decreases (or vice versa), see rays like GH or IJ. Since the optic axes of the two prisms are orthogonal so the vibrations parallel and ⊥ to the optic axis of the first prism become respectively ⊥ and parallel to the optic axis of the second.

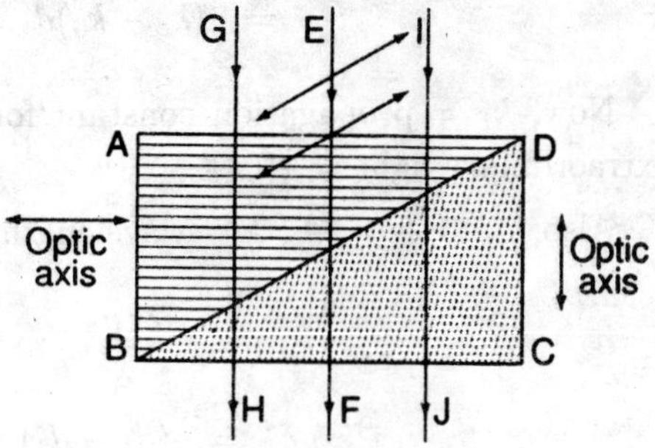

Fig 10.18

Physical Basis of the Action of the Compensator

Plane polarised light with the axis of orientation making an angle θ with the optic axis of the first prism is allowed to fall and pass through the compensator. On incidence this vibration is broken up into two components—one, parallel to the optic axis, called extraordinary ray (E-ray) and the other, perpendicular to the optic axis, called the ordinary ray (O-ray).

Consider the ray IJ not along the symmetry axis such that the thickness traversed in the first prism is d_1 and that in the second prism, d_2. Let the plane polarised light be incident normally on the compensator with the direction of vibration making an angle θ with the optic axis of the prism ABD. The vibration is taken to have amplitude equal to a (say). Then resolving along and ⊥ to the optic axis (taken along the Y-axis), we can write,

$$(y_0)_{\text{ex}} = a\cos\theta = A_1 \quad \text{— } E\text{-ray}, \qquad (z_0)_{\text{or}} = a\sin\theta = A_2 \quad \text{— } O\text{-ray}$$

After its passage through the first prism of thickness d_1, the extraordinary ray (whose vibration is parallel to the Y-direction) can be represented as

$$y = \text{R.P. } A_1 e^{i\phi_e} \qquad \text{E-ray}$$

and, the ordinary ray can be represented as

$$z = \text{R.P. } A_2 e^{i\phi_0} \qquad \text{O-ray}$$

Since the compensator is made of quartz, it is a +ve crystal and hence O-ray travels faster and so it will lead in phase.

∴ phase difference introduced is

$$\delta_1 = \phi_0 - \phi_e = (k_0 - k_e)d_1 \tag{10.4.1}$$

k_0 for O-ray and k_e for E-ray.

In the second prism, the ordinary ray and the extraordinary interchange their character and so after emergence the y-component and the z-component can be represented as

$$y_o = \text{R.P. } ye^{i\alpha_o} = \text{R.P. } A_1 e^{i(\phi_e+\alpha_o)}, \quad z_e = \text{R.P. } ze^{i\alpha_e} = \text{R.P. } A_2 e^{i(\phi_o+\alpha_e)} \tag{10.4.2}$$

where $(\alpha_e - \alpha_o) = (k_o - k_e)d_2$.

So the phase difference between emerging components is given by,

$$\begin{aligned}\delta &= (\phi_e + \alpha_o) - (\phi_o + \alpha_e) = (\phi_e - \phi_o) - (\alpha_2 - \alpha_o) \\ &= (k_e - k_o)d_1 - (k_e - k_o)d_2 = (d_1 - d_2)(k_e - k_o)\end{aligned} \tag{10.4.3}$$

Now, k_1 = propagation constant for ordinary light $= \frac{2\pi}{\lambda_o}$, k_2 = propagation constant for extraordinary light $= \frac{2\pi}{\lambda_E}$.

Also, $\mu_o = \frac{c}{v_o} = \frac{\lambda}{\lambda_o}$, λ = wavelength in air.

$$\therefore \quad k_o = \frac{2\pi\mu_o}{\lambda}, \quad \text{similarly,} \quad k_e = \frac{2\pi\mu_e}{\lambda}$$

$$\therefore \quad \delta = (d_1 - d_2)\cdot\frac{2\pi}{\lambda}(\mu_o - \mu_e) = \frac{2\pi}{\lambda}(d_1 - d_2)(\mu_o - \mu_e) \tag{10.4.4}$$

Jones' matrix formalism

The above optical process which ultimately leads to emergent light from the compensator may be very simply obtained through Jones' matrix for retarder.

The first prism may be looked upon as a retarder plate with fast axis horizontal whose matrix representation according to Jones is

$$J_1 = \begin{bmatrix} 1 & 0 \\ 0 & e^{-i\delta_1} \end{bmatrix}$$

The second prism where the fast axis is vertical is represented by Jones' matrix

$$J_2 = \begin{bmatrix} 1 & 0 \\ 0 & e^{i\delta_2} \end{bmatrix}$$

The plane polarised light whose vibration or orientation is at an angle θ with the horizontal is represented as $\begin{bmatrix} A \\ B \end{bmatrix}$ where $\frac{B}{A} = \tan\theta$.

$$\therefore \quad \begin{bmatrix} 1 & 0 \\ 0 & e^{i\delta_2} \end{bmatrix}\begin{bmatrix} 1 & 0 \\ 0 & e^{-i\delta_1} \end{bmatrix}\underset{\substack{\downarrow \\ \text{Incident light}}}{\begin{bmatrix} A \\ B \end{bmatrix}} = \begin{bmatrix} 1 & 0 \\ 0 & e^{i\delta_2} \end{bmatrix}\begin{bmatrix} A \\ Be^{-i\delta_1} \end{bmatrix} = \begin{bmatrix} A \\ Be^{i(\delta_2-\delta_1)} \end{bmatrix} \tag{10.4.5}$$

In the present case, $A = a\cos\theta$, $B = a\sin\theta$.

Thus from eqn. (10.4.5) a few important conclusions can be arrived at.

Conclusion 1. When $\delta = (\delta_2 - \delta_1) = 0$, this will be true only for symmetric passage where $d_2 = d_1$.

Then the emergent light matrix is given by,

$$J_e = \begin{bmatrix} A \\ B \end{bmatrix} = \begin{bmatrix} a\cos\theta \\ a\sin\theta \end{bmatrix}$$

It represents linearly polarised light with orientation parallel to the incident orientation (see fig. 10.19).

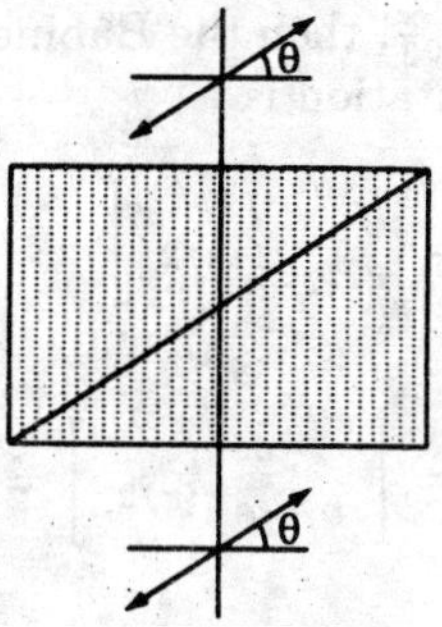

Fig 10.19

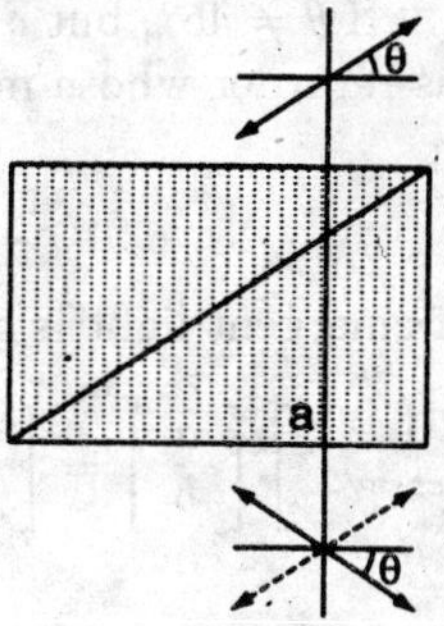

Fig 10.20

Conclusion 2. If we move out of the symmetric line, the thicknesses traversed in the two prisms differ and hence path difference and as a result phase difference is produced. For the departure of a certain distance a from the symmetric line, if the phase difference introduced between the two components be π, then

$$\delta = (\delta_2 - \delta_1) = \pi.$$

In this case the Babinet's compensator will act as a $\frac{\lambda}{2}$-plate whose Jones' matrix is represented as

$$J_{\lambda/2} = \begin{bmatrix} 1 & 0 \\ 0 & -1 \end{bmatrix}.$$

So the transformation of the incident light into emergent light is carried through

$$\begin{bmatrix} 1 & 0 \\ 0 & -1 \end{bmatrix} \begin{bmatrix} a\cos\theta \\ a\sin\theta \end{bmatrix} = \underbrace{\begin{bmatrix} a\cos\theta \\ -a\sin\theta \end{bmatrix}}_{\text{Emergent light}} \tag{10.4.6}$$

Eqn. (10.4.6) shows that in this case, the emergent vibration is also plane polarised with the orientation given by,

$$\tan\phi = \frac{-a\sin\theta}{a\cos\theta} = -\tan\theta \quad \Rightarrow \quad \phi = (\pi - \theta) \tag{10.4.7}$$

Eqn. (10.4.7) suggests that the Babinet's compensator in this case produces linear vibration but direction of vibration rotates through 2θ w.r.t. the incident vibration direction (fig. 10.21).

Conclusion 3. *Circularly polarised light*

If $\theta = 45°$, $\delta_2 - \delta_1 = \frac{\pi}{2}$, the Babinet's compensator acts as a $\frac{\lambda}{4}$-plate whose matrix representation is given by,

$$J_{\lambda/4} = \begin{bmatrix} 1 & 0 \\ 0 & e^{\pm i\pi/2} \end{bmatrix} = \begin{bmatrix} 1 & 0 \\ 0 & \pm i \end{bmatrix}$$

The nature of emergent light is obtained through matrix transformation as

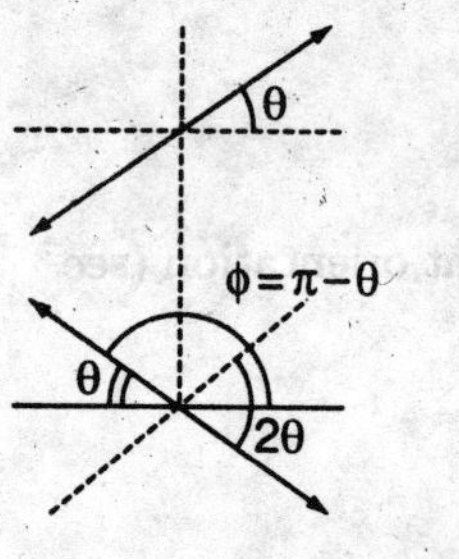

Fig 10.21

$$\begin{bmatrix} 1 & 0 \\ 0 & \pm i \end{bmatrix} \frac{1}{\sqrt{2}} \begin{bmatrix} 1 \\ 1 \end{bmatrix} = \frac{1}{\sqrt{2}} \begin{bmatrix} 1 \\ \pm i \end{bmatrix},$$

either a RCP or LCP emergent light.

Conclusion 4. *Elliptically polarised light*

If $\theta \neq 45°$, but $\delta = (\delta_2 - \delta_1) = \frac{\pi}{2}$, then the Babinet's compensator acts as retarder whose matrix representation is

$$J = \begin{bmatrix} A_1 & 0 \\ 0 & A_2 e^{\pm i\pi/2} \end{bmatrix}$$

∴ The nature of emergent light is given as

$$\begin{bmatrix} A_1 & 0 \\ 0 & A_2 e^{\pm i\pi/2} \end{bmatrix} \begin{bmatrix} a \\ b \end{bmatrix} = \begin{bmatrix} aA_1 \\ bA_2 e^{\pm i\pi/2} \end{bmatrix} = \begin{bmatrix} aA_1 \\ bA_2 e^{\pm i\pi/2} \end{bmatrix} \doteq \begin{bmatrix} A \\ \pm iB \end{bmatrix}$$

↓ Emergent light matrix

Thus in this case the emergent light is elliptically polarised, either right-handed (corresponding to +ve sign) or left-handed (corresponding to −ve sign). This will be more evident if we write out the full expression of the components on emergence

$$\begin{bmatrix} y \\ z \end{bmatrix} = \begin{bmatrix} A \\ \pm iB \end{bmatrix} e^{i\phi_x}$$

$$\Rightarrow \quad y = A\cos\phi_x$$

$$z = B\cos\left(\phi_x \pm \frac{\pi}{2}\right) = B\cos\left(\frac{\pi}{2} \pm \phi_x\right) = \pm B\sin\phi_x$$

$$\therefore \quad \frac{y^2}{A^2} + \frac{z^2}{B^2} = 1, \quad \text{an ellipse.}$$

Thus using Babinet's compensator under specified condition, linearly polarised light, circularly polarised light or elliptically polarised light may be produced.

Jamin's modification of Babinet's compensator

In Jamin's modified form, one of the two prisms is shifted by a micrometer screw instead of shifting the eyepiece. Advantage is two-fold. One is that for one particular setting of the eyepiece, the different amount of path difference $(d_2 - d_1)$ may be made, hence different phase difference producing linearly polarised, circularly polarised or elliptically polarised light. The second advantage is that in Jamin's modification, one prism is to be shifted twice as great as when the eye-piece is to be shifted for introducing the same amount of phase difference. So this arrangement is more sensitive. The Babinet's compensator with Jamin's modification is shown in fig. (10.22).

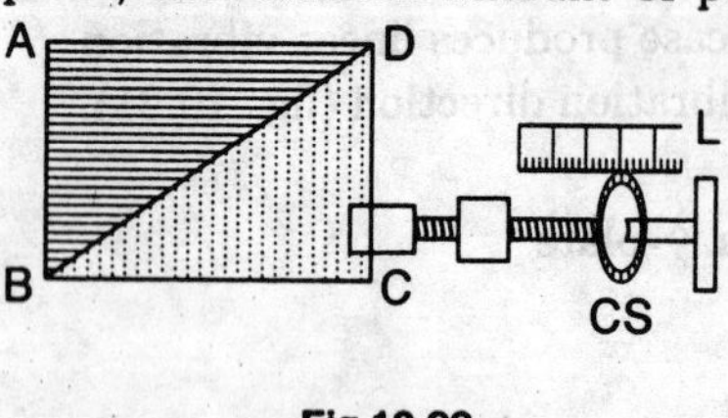

Fig 10.22

So, in Jamin's modification, if the prism BCD is to be shifted by a distance $2b$ to have consecutive dark band appear across the fixed cross-wire, then the path difference between the emergent rectangular components would be λ. So, for a shift of x of the prism, the path difference would be $\frac{\lambda}{2b} \cdot x$.

$$\therefore \quad (d_2 - d_1)(\mu_o - \mu_E) = \frac{\lambda}{2b} \cdot x$$

Application of the compensator

The Babinet's compensator serves dual purpose—production as well as analysis of linear, circular or elliptic polarisation.

The theoretical discussion has been made w.r.t. the production of the different types of polarised light. We again here summarise what we have said earlier.

How to produce linearly polarised light?

At some distances with respect to the symmetric axis, we would get path difference equal to λ. At the symmetric point, the path difference is zero and suppose for a linear distance $2a$ from the symmetric axis (or $2b$ displacement of one prism), the path difference first becomes equal to λ; then for all points linearly separated by distances equal to $2a$ from each other, the path difference would become equal to λ. At all these points, we would get linearly polarised light whose vibration direction is parallel to the incident vibration direction (see fig. 10.23). Theoretical discussion is given in conclusion 1.

Also, at those points where we would have path difference $\Delta = \frac{\lambda}{2}$, the phase difference would be π, hence the compensator would act as a half-wave retarder. At these points also, the emergent light will be plane polarised but with vibration direction rotated through twice the angle of the incident vibration direction [see fig (10.23)].

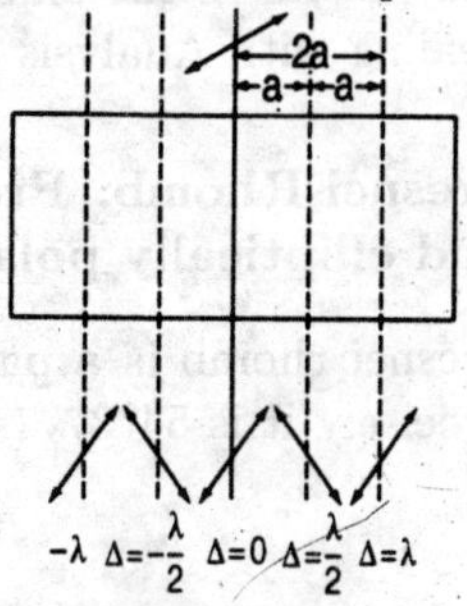

Fig 10.23

How to produce elliptically polarised light?

At those points for which Δ has value

$$0 < \Delta < \frac{\lambda}{4}$$

and θ is different from 45°, we would get elliptically polarised light of different orientation and shape see fig. (10.24). Since, at a separation equal to $2a$, the path difference is λ, the separation to get path difference equal to $\frac{\lambda}{4}$ is $\frac{a}{2}$. So, within the distance $\frac{a}{2}$ we will get elliptically polarised light; for separation greater than $\frac{a}{2}$ but less than a, we would again get elliptically polarised light of different orientation and shape. At a separation a, we would get linearly polarised light with orientation rotated through twice the angle of incident orientation.

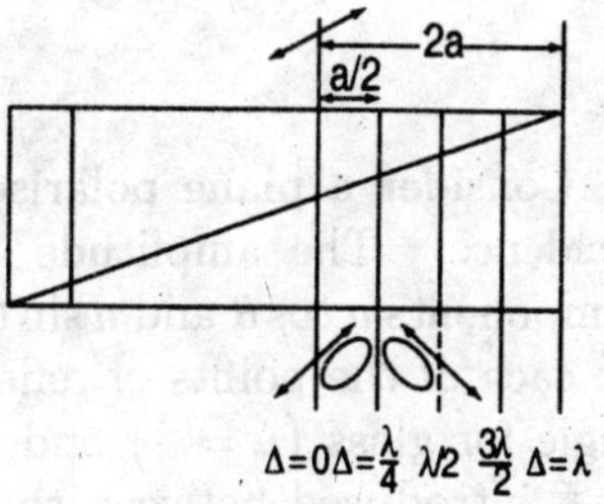

Fig 10.24

How to get circularly polarised light?

If the incident orientation is 45° w.r.t. the optic axis of the first prism, then elliptically polarised light degenerates into circularly polarised light.

Experimental set-up

The experimental set-up to demonstrate the formation of different types of polarised light is shown in fig. (10.25). S is the source, L_1 is the lens rendering the rays parallel; P is the polarising Nicol transforming the ordinary light into a linearly polarised light with direction of vibration specified by the transmission axis of the polariser. The polarised light then passes through the Babinet's compensator transforming the incident polarised light into emergent polarised light of different nature, linear, circular or elliptic as discussed earlier. For particular setting of the analyser we may get dark fringes corresponding to cross-position of the linearly polarised light.

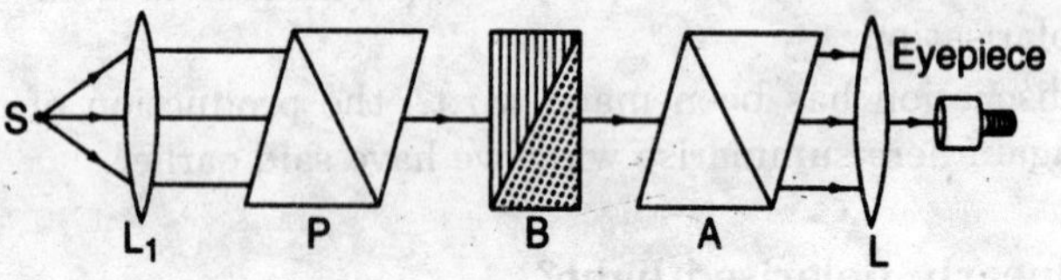

Fig 10.25

Analysis of Polarised light by Babinet's Compensator: Babinet's compensator is conveniently used for the analysis of given polarised light of unknown nature. It is particularly very useful in the analysis of elliptically polarised light, the details being given in the section dealing with Analysis of Polarised Light.

Fresnel Rhomb: Production of linearly polarised light, circularly polarised light and elliptically polarised light

Fresnel rhomb is a parallelopiped of glass whose opposite acute angles are nearly 55° (more precisely it is 54°37′ (see fig. 10.26).

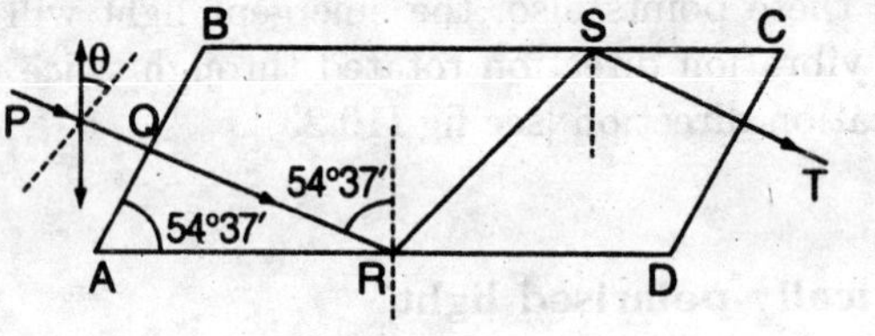

Fig 10.26

Consider a plane polarised light with transmission axis at an angle θ with the plane of incidence. The amplitude of incident vibration being taken as a, it is resolved into two components $a\cos\theta$ and $a\sin\theta$ respectively parallel and perpendicular to the plane of incidence. At each of the points of reflection R and S, the angle of incidence is greater than the critical angle for glass ($\mu = \frac{3}{2}$) and air interface. At each reflection, there will be a phase difference of $\frac{\pi}{4}$ introduced between the two rectangular components. So Jones' matrix operation for a Fresnel rhomb for two reflections can be represented as

$$\begin{bmatrix} 1 & 0 \\ 0 & e^{i\pi/4} \end{bmatrix} \begin{bmatrix} 1 & 0 \\ 0 & e^{i\pi/4} \end{bmatrix}$$

So the emergent light from a Fresnel rhomb, considering two internal reflections can be obtained as

$$\begin{bmatrix} 1 & 0 \\ 0 & e^{i\pi/4} \end{bmatrix} \begin{bmatrix} 1 & 0 \\ 0 & e^{i\pi/4} \end{bmatrix} \begin{bmatrix} A_1 \\ A_2 \end{bmatrix}$$

where $A_1 = a\cos\theta, A_2 = a\sin\theta$.

Analysis of polarised light

According to Stokes, there may be seven possible states of polarisation of a beam of light and they are—(a) Unpolarised, (b) Linearly polarised (or plane polarised), (c) Circularly polarised, (d) Elliptically polarised, (e) Mixture of unpolarised and linearly polarised, (f) Mixture of unpolarised and circularly polarised, (g) Mixture of unpolarised and elliptically polarised.

The device by which analysis of a given light can be done is called an analyser. Such an analyser may be a Nicol prism (for linearly polarised light), Quarter-wave plate, Babinet's compensator or Fresnel rhomb. The scheme of analysis is given below.

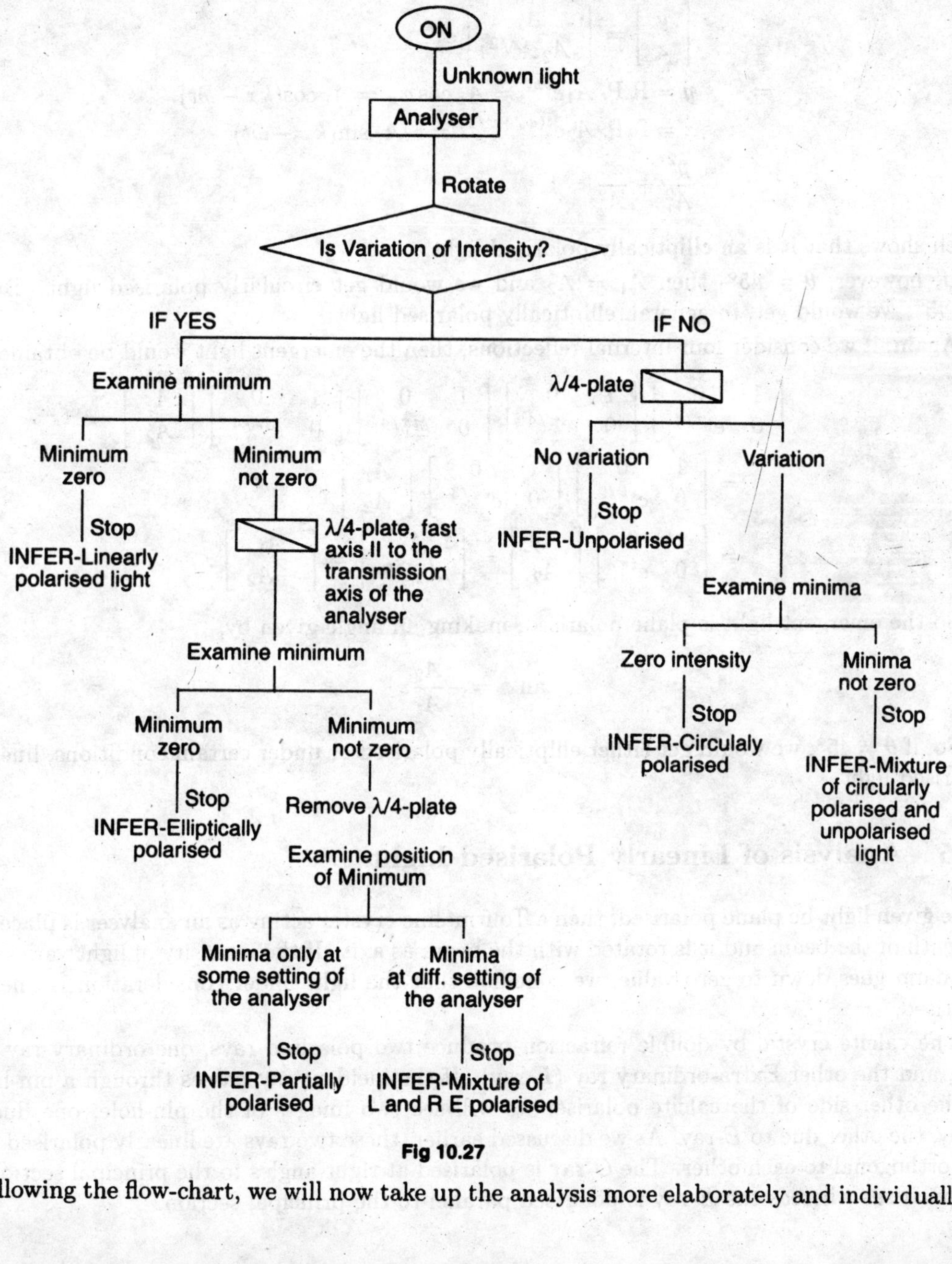

Fig 10.27

Following the flow-chart, we will now take up the analysis more elaborately and individually.

So,

$$\begin{bmatrix} 1 & 0 \\ 0 & e^{i\pi/4} \end{bmatrix}\begin{bmatrix} 1 & 0 \\ 0 & e^{i\pi/4} \end{bmatrix}\begin{bmatrix} A_1 \\ A_2 \end{bmatrix} = \begin{bmatrix} 1 & 0 \\ 0 & e^{i\pi/4} \end{bmatrix}\begin{bmatrix} A_1 \\ A_2 e^{i\pi/4} \end{bmatrix} = \underbrace{\begin{bmatrix} A_1 \\ A_2 e^{i\pi/2} \end{bmatrix}}_{\text{Emergent light}}$$

If the parallel component be taken along y-axis and perpendicular component be taken along the z-direction, writing out the full form,

$$\begin{bmatrix} y \\ z \end{bmatrix} = \begin{bmatrix} A_1 \\ A_2 e^{i\pi/2} \end{bmatrix} e^{i\phi_x}$$

$$\Rightarrow \quad y = \text{R.P. } A_1 e^{i\phi_x} = A_1 \cos\phi_x = A_1 \cos(kx - \omega t)$$

$$z = \text{R.P. } A_2 e^{i(\phi_x + \pi/2)} = -A_2 \sin(kx - \omega t)$$

$$\therefore \quad \frac{y^2}{A_1^2} + \frac{z^2}{A_2^2} = 1,$$

which shows that it is an elliptically polarised light.

If, however, $\theta = 45°$, then $A_1 = A_2$ and we would get circularly polarised light. But if $\theta \neq 45°$, we would get, in general, elliptically polarised light.

Again, if we consider four internal reflections, then the emergent light would be obtained as

$$\begin{bmatrix} 1 & 0 \\ 0 & e^{i\pi/4} \end{bmatrix}\begin{bmatrix} 1 & 0 \\ 0 & e^{i\pi/4} \end{bmatrix}\begin{bmatrix} 1 & 0 \\ 0 & e^{i\pi/4} \end{bmatrix}\begin{bmatrix} 1 & 0 \\ 0 & e^{i\pi/4} \end{bmatrix}\begin{bmatrix} A_1 \\ A_2 \end{bmatrix}$$

$$= \begin{bmatrix} 1 & 0 \\ 0 & e^{i\pi/2} \end{bmatrix}\begin{bmatrix} 1 & 0 \\ 0 & e^{i\pi/4} \end{bmatrix}\begin{bmatrix} A_1 \\ A_2 \end{bmatrix}$$

$$= \begin{bmatrix} 1 & 0 \\ 0 & e^{i\pi} \end{bmatrix}\begin{bmatrix} A_1 \\ A_2 \end{bmatrix} = \begin{bmatrix} A_1 \\ A_2 e^{i\pi} \end{bmatrix} = \begin{bmatrix} A_1 \\ -A_2 \end{bmatrix}.$$

So the emergent light is plane polarised, making an angle given by,

$$\tan\phi = -\frac{A_2}{A_1}.$$

So, if $\theta \neq 45°$, we would get either elliptically polarised or under certain conditions, linearly polarised light.

10.5 Analysis of Linearly Polarised Light

If the given light be plane polarised, then a Tourmaline crystal acting as an analyser is placed in the path of the beam and it is rotated with the beam as axis. If the intensity of light varies and minimum goes down to zero value, we conclude that the light under consideration is linearly polarised.

The calcite crystal by double refraction produce two polarised rays, one ordinary ray (O-ray), and the other Extra-ordinary ray (E-ray). If the incident ray passes through a pin-hole, on the other side of the calcite polariser we will see two images of the pin-hole; one due to O-ray, the other due to E-ray. As we discussed earlier, these two rays are linearly polarised and also orthogonal to each other. The O-ray is polarised at right angles to the principal section of the polariser whereas the E-ray is polarised parallel to the principal section.

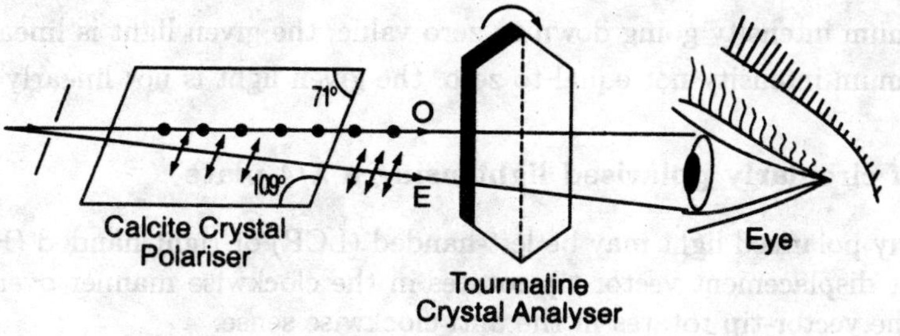

Fig 10.28

Now, to test the state of polarisation, the emergent beam is allowed to pass through an analyser, say a tourmaline crystal. An eye placed behind the tourmaline crystal sees two images, but as the tourmaline crystal is rotated about the incident ray as axis, one of the two images, say the rotating image disappears or reduces to zero intensity while the stationary image appears in its brightest at one position. The rotating image is due to E-ray, which does not obey Snell's law, while the stationary image is due to O-ray which obeys Snell's law. In this position, the optic axis of the tourmaline analyser is perpendicular to the shorter diagonal of the calcite polariser. Since the E-ray is polarised in the plane containing the shorter diagonal and the optic axis of the analyser is set at right angles, the extraordinary ray is refused transmission; the ordinary ray under this condition is fully transmitted producing stationary image as brightest. When the shorter diagonal of the calcite polariser in another setting is parallel to the optic axis of the analyser, the stationary image disappears and the rotating image appears in its brightest. Thus the state of polarisation of the emergent beam consisting of O-ray and E-ray is orthogonal.

Thus, any linearly polarised light examined by a tourmaline crystal as analyser will show gradual fading in brightness and when the optic axis of the analyser is at right angles to the direction of vibration of the polarised light, the intensity minimum goes down to zero value.

In Jones' language the interpretation comes in the following way:

A linearly polarised light with orientation horizontal is represented by $\begin{bmatrix} 1 \\ 0 \end{bmatrix}$, an analyser whose fast axis is vertical is represented by Jones' matrix $\begin{bmatrix} 1 & 0 \\ 0 & 0 \end{bmatrix}$, so

$$\begin{bmatrix} 1 & 0 \\ 0 & 0 \end{bmatrix} \begin{bmatrix} 1 \\ 0 \end{bmatrix} = \begin{bmatrix} 0 \\ 0 \end{bmatrix}$$

↓

Emergent light matrix

So the intensity of the emergent light is zero. But if the intensity does not go down to zero value, the given light is not definitely linearly polarised. Thus,

$$\begin{bmatrix} 1 & 0 \\ 0 & 0 \end{bmatrix} \frac{1}{\sqrt{2}} \begin{bmatrix} 1 \\ e^{+i(\pi/2+\theta)} \end{bmatrix} = \frac{1}{\sqrt{2}} \begin{bmatrix} 1 \\ 0 \end{bmatrix}.$$

So the intensity of the emergent light is not zero under the same condition of the analyser.

So our conclusion is—

If the given light is found to show variation of intensity when an analyser (say a tourmaline crystal) is rotated in its path with the incident direction as axis, then for

(i) minimum intensity going down to zero value, the given light is linearly polarised,

(ii) minimum intensity not equal to zero, the given light is not linearly polarised.

Analysis of circularly polarised light using a $\lambda/4$-plate

The circularly polarised light may be left-handed (LCP) or right-handed (RCP); in the former, the resultant displacement vector-tip rotates in the clockwise manner over a circle whereas in the latter, the vector-tip rotates in the anti-clockwise sense.

If the light disturbance be assumed to travel in the X-direction, and two orthogonal vibrations polarised in the Y and Z directions differ in phase by $\frac{\pi}{2}$, then the resultant is a circularly polarised light. If the Z-component leads in phase over the Y-component by the said phase, the circularly polarised light is clockwise or left circularly polarised (LCP) if it lags, it is anti-clockwise or right circularly polarised (RCP). Thus in the Jones' formalism the two orthogonal polarised lights are represented as

$$\begin{bmatrix} 1 \\ 0 \end{bmatrix} \quad \text{— horizontally polarised } (Y\text{-vibration})$$

$$\begin{bmatrix} 0 \\ e^{\pm i\pi/2} \end{bmatrix} \quad \text{— vertically polarised, but with a phase diff. } \pi/2 \ (Z\text{-vibration})$$

Thus,

$$\frac{1}{\sqrt{2}}\begin{bmatrix} 1 \\ 0 \end{bmatrix} + \frac{1}{\sqrt{2}}\begin{bmatrix} 0 \\ e^{i\pi/2} \end{bmatrix} = \frac{1}{\sqrt{2}}\begin{bmatrix} 1 \\ i \end{bmatrix} = \boxed{\frac{1}{\sqrt{2}}\begin{bmatrix} 1 \\ i \end{bmatrix} - \text{RCP}}$$

$$\frac{1}{\sqrt{2}}\begin{bmatrix} 1 \\ 0 \end{bmatrix} + \frac{1}{\sqrt{2}}\begin{bmatrix} 0 \\ e^{-i\pi/2} \end{bmatrix} = \frac{1}{\sqrt{2}}\begin{bmatrix} 1 \\ -i \end{bmatrix} = \boxed{\frac{1}{\sqrt{2}}\begin{bmatrix} 1 \\ -i \end{bmatrix} - \text{LCP}}$$

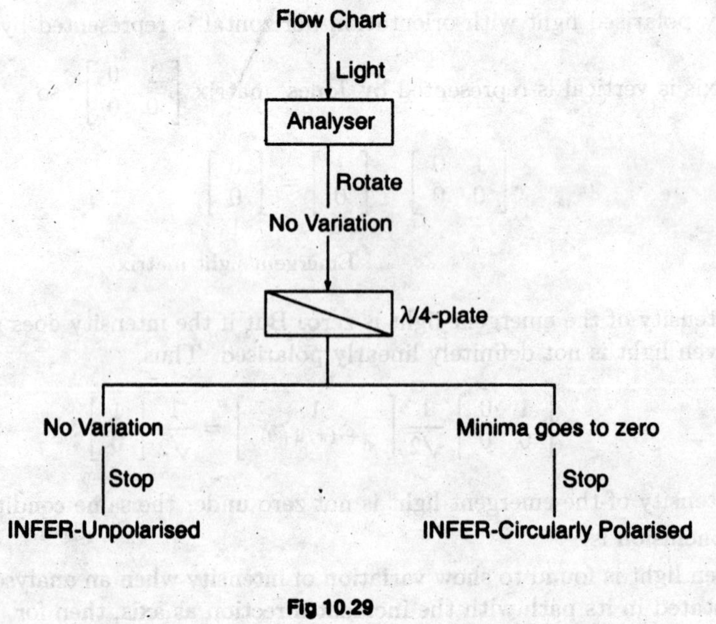

Fig 10.29

The reverse of this process of synthesis is also true; in other words, it means, an RCP or LCP can be analysed into two orthogonal linearly polarised light of equal intensity differing in phase by $\frac{\pi}{2}$. Thus, if a circularly polarised light is passed through an analyser (say, a tourmaline crystal), and the analyser is rotated, intensity of the emergent light is never changing. Conclusion is two-fold, either the light is unpolarised or the light is circularly polarised.

To confirm the nature, introduce a $\frac{\lambda}{4}$-plate after the analyser; if now also no variation of intensity is observed, the given light is confirmed as unpolarised; but if variation is observed and minimum of intensity reduces to zero, the given light is confirmed as circularly polarised.

Explanation

When a circularly polarised light passes through an analyser, on rotation there is no intensity variation, a fact not very difficult to understand from the discussion we had till now.

Then, when it is passed through a $\frac{\lambda}{4}$-plate, depending on the orientation of the fast axis and slow axis of the retarder, there is a phase difference of $\frac{\pi}{2}$ between the two orthogonal components, hence they combine to form a linearly polarised light which may be cut off by another analyser at proper orientation. But, if after passing through a $\frac{\lambda}{4}$-plate, there is no variation of intensity still, the given light is confirmed to be unpolarised.

This information is conveyed through Jones' formalism in the following way:

A quarter-wave plate with the fast axis vertical is represented as $\begin{bmatrix} 1 & 0 \\ 0 & i \end{bmatrix}$, a RCP is represented as $\frac{1}{\sqrt{2}}\begin{bmatrix} 1 \\ i \end{bmatrix}$. So the action of the $\frac{\lambda}{4}$-plate on the RCP is expressed through the matrix product

$$\underbrace{\begin{bmatrix} 1 & 0 \\ 0 & i \end{bmatrix}}_{\text{Polariser matrix}} \frac{1}{\sqrt{2}} \underbrace{\begin{bmatrix} 1 \\ i \end{bmatrix}}_{\text{RCP matrix}} = \frac{1}{\sqrt{2}} \underbrace{\begin{bmatrix} 1 \\ -1 \end{bmatrix}}_{\text{Emergent light matrix}}$$

So the emergent light is plane polarised at $-45°$ with the slow axis of the $\frac{\lambda}{4}$-plate. Similarly, when the incident light is LCP, the $\frac{\lambda}{4}$-plate will transform it into a linearly polarised light as

$$\begin{bmatrix} 1 & 0 \\ 0 & i \end{bmatrix} \frac{1}{\sqrt{2}} \underbrace{\begin{bmatrix} 1 \\ -i \end{bmatrix}}_{\text{LCP}} = \frac{1}{\sqrt{2}} \underbrace{\begin{bmatrix} 1 \\ 1 \end{bmatrix}}_{\text{Linearly polarised at } 45° \text{ with the slow axis}}$$

So this process of conversion and analysis of the circularly polarised light by using a $\frac{\lambda}{4}$-plate is diagrammatically represented as in fig. (10.30).

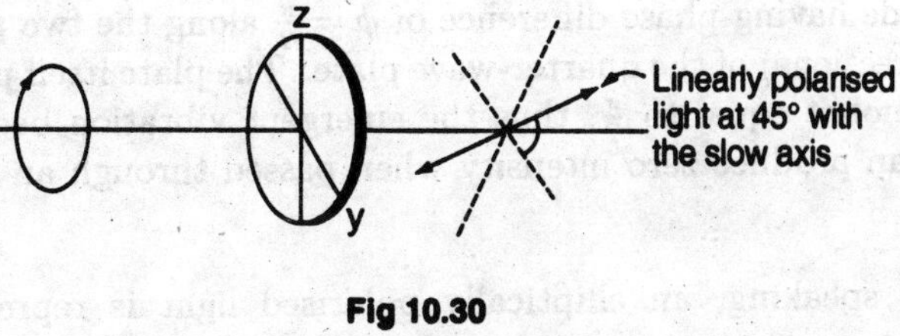

Fig 10.30

So the action of a $\frac{\lambda}{4}$-plate on a circularly polarised light is to convert it into a linearly polarised light which may be cut off by an analyser suitably oriented.

Analysis of elliptically polarised light using a $\frac{\lambda}{4}$-plate

We have seen that an elliptically polarised light consists of two rectangular vibrations of unequal amplitudes having a phase difference $\phi\left(0 \leq \phi \leq \frac{\pi}{2}\right)$. So, if an elliptically polarised light is passed through an analyser and the analyser rotated, there will occur variation of intensity. Conclusion is two-fold, the incident light is either linearly polarised or elliptically polarised. Confirmation is arrived at by a $\frac{\lambda}{4}$-plate. The scheme of analysis is shown as below.

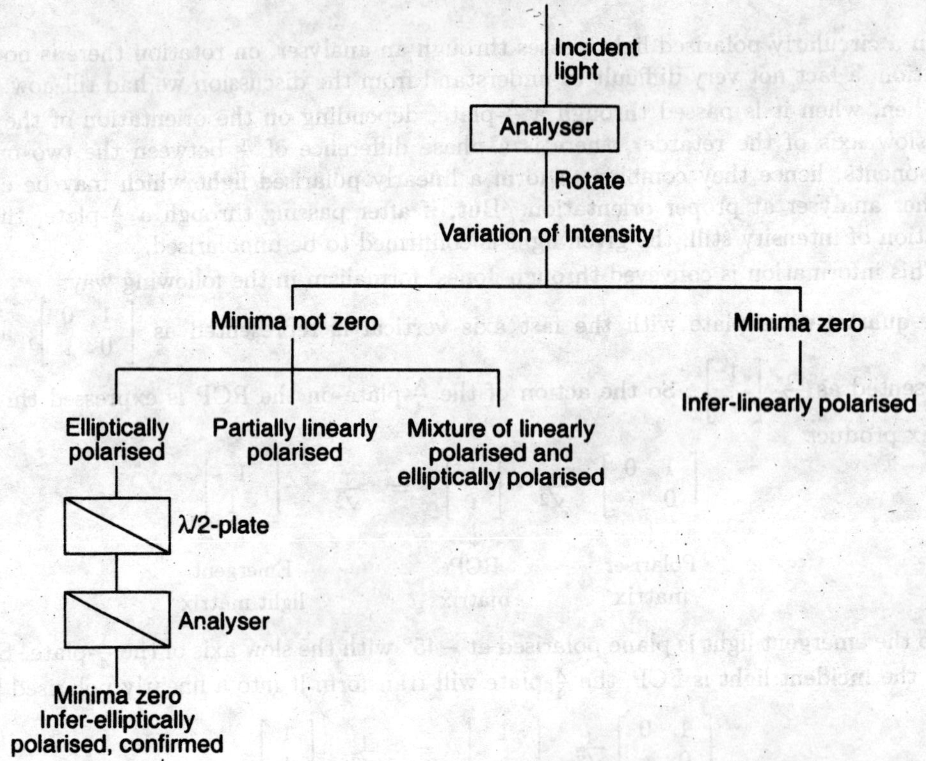

Fig 10.31

Explanation

Elliptically polarised light on incidence on the $\frac{\lambda}{4}$-plate is resolved into two orthogonal vibrations of unequal amplitude having phase difference of $\phi = \frac{\pi}{2}$ along the two principal directions (also called privileged directions) of the quarter-wave plate. The plate itself produces an overall phase difference by an amount equal to $\frac{\pi}{2}$, thus the emergent vibration becomes linearly polarised, which, therefore, can produce zero intensity when passed through an analyser (a Nicol prism, say).

Mathematically speaking, an elliptically polarised light is represented as $\begin{bmatrix} A_1 \\ iA_2 \end{bmatrix}$. A quarter-wave plate whose fast axis is vertical is represented as

$$\begin{bmatrix} 1 & 0 \\ 0 & i \end{bmatrix}$$

So the action of $\frac{\lambda}{4}$-retarder on the elliptically polarised light is expressed as

$$\begin{bmatrix} 1 & 0 \\ 0 & i \end{bmatrix} \underset{\text{REP}}{\underset{\downarrow}{\begin{bmatrix} A_1 \\ iA_2 \end{bmatrix}}} = \underset{\text{Linearly polarised}}{\underset{\downarrow}{\begin{bmatrix} A_1 \\ -A_2 \end{bmatrix}}}$$

So the emergent light is linearly polarised making an angle θ with the slow axis given by,

$$\tan\theta = -\frac{A_2}{A_1}.$$

Hence the minimum would go to zero value when the transmission axis of the analyser placed after the retarder-plate is at right angles to the emergent linear vibration.

Writing out once again the eqns. representing the emergent vibration, we get in full form

$$\begin{bmatrix} y \\ z \end{bmatrix} = \begin{bmatrix} A_1 \\ -A_2 \end{bmatrix} e^{i\phi_x}$$

$$\Rightarrow \quad \begin{aligned} y &= \text{R.P. } A_1 e^{i\phi_x} = A_1 \cos\phi_x \\ z &= \text{R.P.} - A_2 e^{i\phi_x} = -\text{R.P. } A_2 e^{i\phi_x} = -A_2 \cos\phi_x \end{aligned}$$

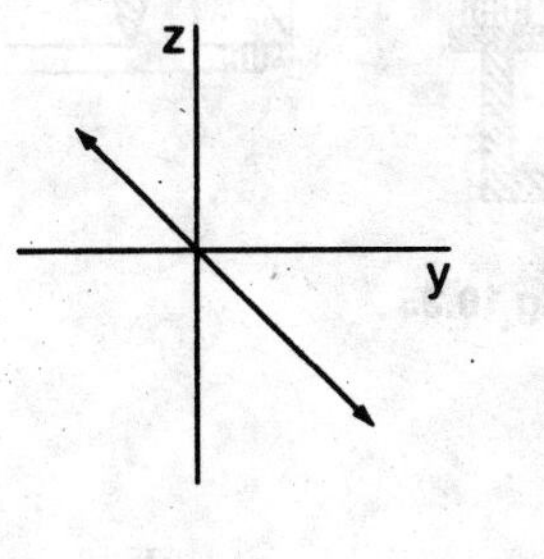

Fig 10.32

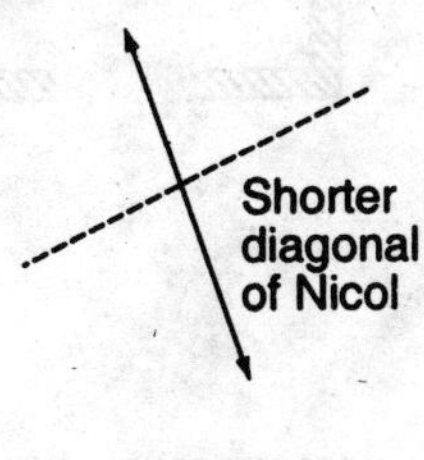

Fig 10.33

Thus, we get,

$$y = -\frac{A_1}{A_2} z = -mz$$

It is a linear vibration with orientation given by, $\tan\phi = -\frac{A_1}{A_2}$ (note: if $A_1 = A_2, \tan\phi = -1 \quad \Rightarrow \quad \phi = 135°$ or, $-45°$ w.r.t. y-direction). So such a linear vibration will be completely cut-off by the analysing Nicol when its transmission axis makes an angle of 45° w.r.t. the y-direction.

10.6 Experimental Procedure

Circularly polarised light: Using $\lambda/4$-plate as analyser

In the analysis we have seen that if the circularly polarised light be RCP, then on incidence on a $\frac{\lambda}{4}$-plate (of a +ve crystal) whose fast axis is vertical and slow axis horizontal, the circularly polarised light will be resolved into two rectangular components whose z-component (along fast axis) leads by $\frac{\pi}{2}$ over its y-component (along slow axis). Due to the action of the $\frac{\lambda}{4}$-plate, there will be additional advancement of phase by $\frac{\pi}{2}$, thus the emergent vibration will be a linearly polarised light with orientation at $-45°$ with the slow axis.

Mathematically, the fact is stated as—the $\frac{\lambda}{4}$-retarder matrix operates on an RCP producing a linearly polarised light at $-45°$ with the slow axis of the plate. Thus,

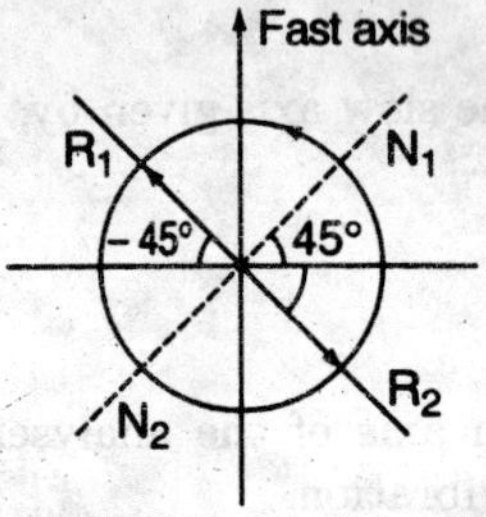

$$\underset{\text{Retarder matrix}}{\begin{bmatrix} 1 & 0 \\ 0 & i \end{bmatrix}} \frac{1}{\sqrt{2}} \underset{\text{RCP}}{\begin{bmatrix} 1 \\ i \end{bmatrix}} = \frac{1}{\sqrt{2}} \underset{\text{Linearly polarised at } -45°}{\begin{bmatrix} 1 \\ -1 \end{bmatrix}}$$

Fig 10.34

In the fig. (10.34) R_1R_2 represents the direction of vibration of the linearly polarised light. So an analysing Nicol whose shorter diagonal N_1N_2 is set making an angle of $+45°$ with the slow axis will cut off the linearly polarised light producing zero intensity.

The above scheme of analysis is carried out w.r.t. the experimental set-up as shown in fig. (10.35). Over the collimator of an ordinary spectrometer is slipped a tube T_1 containing a polariser P; another tube T_2 contains an analyser A and telescope T. In between the polariser and analyser, there is provision for placing another tube T_3 where a $\frac{\lambda}{4}$-plate may be put.

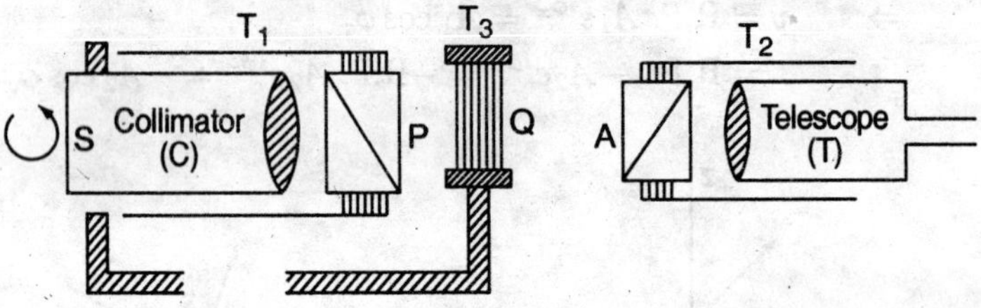

Fig 10.35

1. Initial setting

The collimator and telescope are adjusted for parallel rays. Without placing P, Q and A, a Wollaston's prism is placed in the position of P in the tube T_1. Two images of the slit are observed and the tube T_1 is rotated to bring its two images in one vertical line. The vibrations of the two images are now horizontal and vertical.

2. Alignment of the analyser

The analysing Nicol is now placed in the tube T_2 with its shorter diagonal approximately vertical and rotated until one of the images disappears. The principal section of A is now vertical. The circular scale attached to T_2 is kept at zero of the scale in this position.

3. Alignment of the polariser

The Wollaston's prism is removed, the polariser P is placed in its position with its shorter diagonal horizontal and adjusted by rotation on either side until the field is dark. The principal section of the polariser is now exactly horizontal. The reading of the circular scale attached with the tube T_1 is adjusted for zero.

4. Alignment of the quarter-wave plate

The slit is illuminated by light suitable for the $\frac{\lambda}{4}$-plate. The quarter-wave plate will produce elliptically polarised light, so the field of the analyser will appear bright. The $\frac{\lambda}{4}$-plate is now rotated until the field is dark. This time the two principal directions of the $\frac{\lambda}{4}$-plate are horizontal and vertical, knowing the fast axis of the plate, it is adjusted to be vertical (hence, slow axis horizontal). The circular scale attached the tube T_3 is adjusted to read zero in this setting.

The complete adjustment now is represented in fig. (10.35) where we have shown only the orientational configurations.

Analysis procedure: Circularly polarised light

The polariser P is removed. The slit is illuminated by the given circularly polarised light. The analyser is rotated to make the field dark. If the shorter diagonal makes an angle of 45° with the slow axis of $\frac{\lambda}{4}$-plate in this position, the given light is a right circularly polarised (RCP).

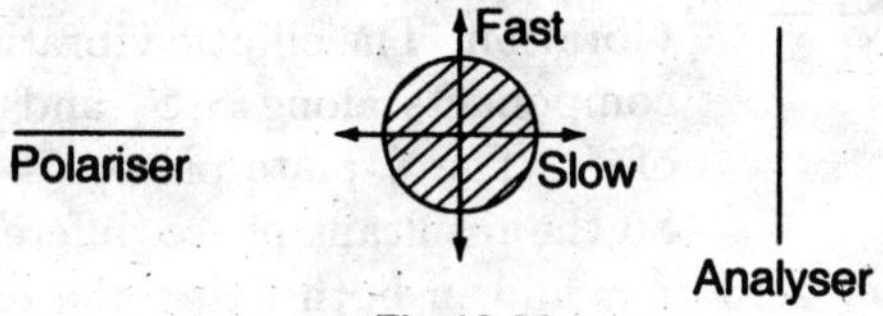

Fig 10.36

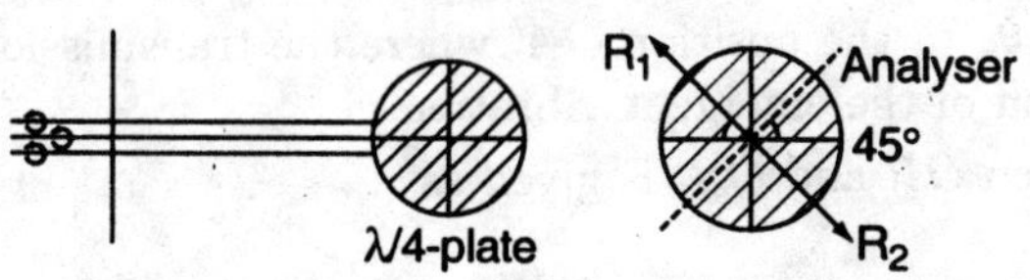

Fig 10.37

But, if the shorter diagonal of the analyser is to be set at $-45°$ with the slow axis of the $\frac{\lambda}{4}$-plate, the given circularly polarised light is left-handed (LCP).

The mathematical representation of the process is given earlier for an RCP light. If the light is an LCP, the mathematical representation of the process is

$$\underbrace{\begin{bmatrix} 1 & 0 \\ 0 & i \end{bmatrix}}_{\text{Retarder matrix}} \frac{1}{\sqrt{2}} \cdot \underbrace{\begin{bmatrix} 1 \\ -i \end{bmatrix}}_{\text{LCP}} = \frac{1}{\sqrt{2}} \underbrace{\begin{bmatrix} 1 \\ 1 \end{bmatrix}}_{\text{Linearly polarised at 45°}}$$

So the shorter diagonal of the analyser is set at $-45°$ w.r.t. slow axis to cut off light.

Elliptically polarised light

Initial alignment up to step (4) as done in circularly polarised light is made here also. Then the polariser is removed, the slit is illuminated by the given elliptically polarised light. The field of the telescope will now be bright. The tube T_3 containing the $\frac{\lambda}{4}$-plate is rotated in steps, at the same time the analyser A also rotated to get minimum intensity; this process of simultaneous rotation of the $\frac{\lambda}{4}$-plate and the analyser is continued till we get the field completely dark. This time the principal directions of the $\frac{\lambda}{4}$-plate will give the axes of the elliptic vibration under analysis.

To study an elliptically polarised light, we require to get information on

(i) The phase difference between the two rectangular components

(ii) Orientation of the major and minor axes

(iii) Ratio of the magnitude of the minor and major axes of the ellipse.

In fig. (10.38) we have shown the essence of the analysis.

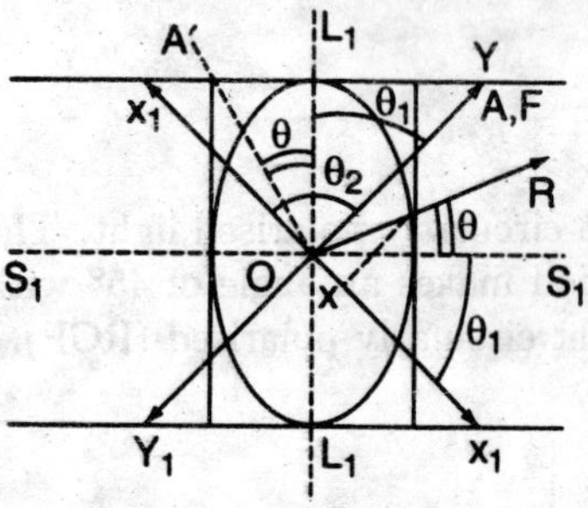

Fig 10.38

Let A, F represent the initial direction of the transmission axis of the analyser and fast axis of the $\frac{\lambda}{4}$-plate; P, S represent the initial direction of the transmission axis of the polariser and slow axis of the $\frac{\lambda}{4}$-plate (cf. plan shown in fig. 10.36). These two directions are respectively vertical and horizontal directions.

The $\frac{\lambda}{4}$-plate is rotated through an angle θ_1 so as to make the fast axis and slow axis coincide with the axes of the elliptic vibration. The elliptic vibration then will be resolved into two components along S_1S_1 and L_1L_1 having a phase difference of $\frac{\pi}{2}$. The $\frac{\lambda}{4}$-plate produces a further phase difference of $\frac{\pi}{2}$. So the resultant phase difference of the emerging rectangular components will be either zero or π and in both cases, the emergent vibration will be plane polarised along way OR. This may be cut-off by rotating the analyser from its initial position OA through an angle θ_2 to the position OA' where the transmission axis of the analyser is in cross with the direction of the emergent vibration.

The angle θ between OR and OS_1 is given as

$$\theta = (\theta_2 - \theta_1)$$

which, therefore, can be obtained from the difference between the angles of rotation of the polariser and analyser from their initial position to get the field of view completely dark with the $\frac{\lambda}{4}$-plate kept in the position in the tube T_3.

Analysis of elliptically polarised light by babinet's compensator

As already remarked analysis of elliptically polarised light would mean determination of several characteristics of elliptic polarisation.

1. The phase difference between the two rectangular components of elliptic polarisation.
2. Orientation of the axes of the ellipse.
3. Ratio of the axes of the ellipse.

1. Determination of the phase difference between the components

As the elliptically polarised light enters the compensator, it is broken up into two rectangular vibrations coincident with the optic axes of the compensator. They can be represented as

$$[J_{\text{REP}}] = \begin{bmatrix} A_1 \\ iA_2 \end{bmatrix}$$

As this passes through the compensator, there will be additional path difference, hence phase difference introduced. The phase difference δ at the point where the thicknesses become d and d' is given by

$$\delta = \frac{2\pi}{\lambda}(d - d')(\mu_o - \mu_E).$$

In the elliptical polarisation, the z-component had a lead of phase by $\frac{\pi}{2}$. As a result of the passage through the compensator if the phase of z-component lags by δ, then the Jones' matrix for the emergent light is written as

$$[J_e] = \begin{bmatrix} A_1 \\ A_2 e^{i\pi/2 - i\delta} \end{bmatrix} = \begin{bmatrix} A_1 \\ A_2 e^{i(\pi/2 - \delta)} \end{bmatrix}$$

So, if $\delta = \frac{\pi}{2}$, then

$$[J_e] = \begin{bmatrix} A_1 \\ A_2 \end{bmatrix},$$

a linearly polarised light. In general, an elliptically polarised.

$$[J_i] = \begin{bmatrix} A_1 \\ A_2 e^{i\delta} \end{bmatrix}.$$

If δ' is the phase lag introduced,

$$[J_e] = \begin{bmatrix} A_1 \\ A_2 e^{i(\delta - \delta')} \end{bmatrix}.$$

So a linearly polarised light is obtained when

$$\delta = \delta' = \frac{2\pi}{\lambda}(d - d')(\mu_o - \mu_E).$$

This linearly polarised light can be extinguished by an analysing Nicol.

So darkness in the field of view of the eyepiece will be obtained at different points of the compensator corresponding to the path difference $\Delta = 0, \pm\lambda, \pm 2\lambda, \pm 3\lambda$, etc.

So at first, fringes are formed in the field of view with plane polarised light. The cross-wire of the eyepiece is made coincident with the central (dark) fringe. Now, elliptically polarised light is allowed to pass through the compensator, the phase difference does not become equal to zero at the central point. So the central fringe shifts towards one side. At this position, the phase difference is reduced to zero value. The eyepiece is shifted by x to bring the cross-wire in coincidence with the shifted central fringe (see experimental arrangement fig. 10.25).

If for a displacement $2s$ w.r.t. the central line, the path difference becomes λ, then

$$\frac{x}{2s} = \frac{\delta'}{2\pi} \quad \Rightarrow \quad \delta' = \pi\frac{x}{s} = \delta.$$

Hence the phase difference can be determined.

2. Determination of the orientation of the axes of the elliptically polarised light

When the axes of the ellipse coincide with the optic axes of the compensator, the rectangular components of the elliptic vibration have a phase difference equal to $\frac{\pi}{2}$. This fact can be utilised for the determination of the orientation of the axes of the ellipse.

First, fringes are made available in the field of view of the eye-piece with plane polarised light. The cross-wire is made coincident with the central fringe (corresponding to the point where $d = d'$). Then one half of the compensator is shifted linearly through a distance such that the fringes suffer a displacement by $\frac{s}{2}$. We have known previously that $\delta' = \frac{\pi x}{s} = \frac{\pi}{s} \cdot \frac{s}{2} = \frac{\pi}{2}$. It means that at the position of the corss-wire, a phase of difference of $\frac{\pi}{2}$ exists.

Now, the plane polarised light is replaced by the experimental elliptically polarised light. The central fringe will no longer remain in coincidence with the cross-wire. The compensator is rotated in its own plane about a line normal to the plane of AD of the compensator. With this rotation, the fringes will undergo displacement and at one time, the central fringe will coincide again with the cross-wire. At the position of the cross-wire there existed a phase difference o. $\frac{\pi}{2}$ and when the compensator is rotated, at certain orientation, the directions of optic axes of the compensator coincide with the axes of the elliptic vibration. When this occurs, as the elliptically polarised light falls on the compensator, it will be resolved into two rectangular components differing in phase by $\frac{\pi}{2}$. The cross-wire position is already at a phase difference of $\frac{\pi}{2}$. The compensator with its axes coinciding with the axes of the elliptic vibration will resolve into components having a phase difference equal to $\frac{\pi}{2}$, equal and opposite to that which existed at the position of the cross-wire. Thus at the cross-wire position, the phase difference is again reduced to zero, and hence central fringe again comes in coincidence with the cross-wire. The orientation of the optic axes of the compensator gives us the orientation of the axes of the elliptically polarised light.

3. Determination of the ratio of the axes of the ellipse

When the axes of the compensator coincide with the axes of the ellipse, then the central fringe would correspond to zero phase difference and so in that condition the emergent light will be plane polarised, which may be cut off by the analysing Nicol.

In the position of complete extinction, the transmission axis of the analysing Nicol would be perpendicular to the direction of vibration of the polarised light. For further details, see $Q2$ given below.

Short Theory Questions with Answers

1. What is meant by elliptically polarised light? Show that plane and circularly polarised light are special cases of elliptically polarised light.

   Ans. Elliptically polarised light is the result of superposition of two rectangular vibrations having a phase difference $0 \le \phi \le \frac{\pi}{2}$. In Jones' matrix formalism,

$$[J_{\text{REP}}] = \begin{bmatrix} A_1 \\ A_2^{i\phi} \end{bmatrix} \quad \text{Right elliptically polarised light}$$

$$[J_{\text{LEP}}] = \begin{bmatrix} A_1 \\ A_2^{-i\phi} \end{bmatrix} \quad \text{Left elliptically polarised light}$$

If $A_1 = A_2, \phi = 0$, then

$$[J] = \begin{bmatrix} A \\ A \end{bmatrix} \rightarrow \begin{bmatrix} 1 \\ 1 \end{bmatrix}, \quad \text{linearly polarised light.}$$

If $\phi = \frac{\pi}{4}, A_1 = A_2$, then

$$[J_{\text{RCP}}] = \begin{bmatrix} 1 \\ +i \end{bmatrix}, \quad \text{Right circularly polarised.}$$

2. Elliptically polarised light is passed through a $\frac{\lambda}{4}$-plate and then through a Nicol. Extinction occurs when the optic axis of the plate and the principal section of the Nicol are inclined to the horizontal at angles of 30° and 60° respectively, the angles being measured in the same direction. Find the orientation of the ellipse and the ratio of its axis.

Ans. The elliptically polarised light on incidence on the $\frac{\lambda}{4}$ has its resolved components in the form of linearly polarised light parallel to the privileged axes of the $\frac{\lambda}{4}$-plate. On the first face, the two rectangular vibrations will have a phase difference of ϕ. We have seen that when $\phi = \frac{\pi}{2}$, the $\frac{\lambda}{4}$-plate produces an additional phase difference of $\frac{\pi}{2}$ and so the beam on exit out of the plate will be plane polarised having a certain orientation and it will be thrown to extinction when the principal section of the analysing Nicol is at right angle w.r.t. the emerging linearly polarised light.

Mathematically,

$$\begin{bmatrix} 1 & 0 \\ 0 & i \end{bmatrix} \begin{bmatrix} A_1 \\ iA_2 \end{bmatrix} = \begin{bmatrix} A_1 \\ -A_2 \end{bmatrix}$$

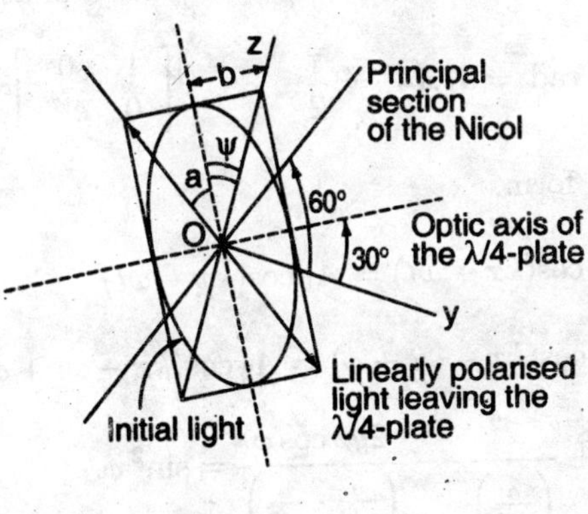

Fig 10.39

This mathematical transformation is diagrammatically represented as in fig. (10.39). From fig. (10.39) showing complete exinction, we get, the ratio of the axes of the ellipse given as $\frac{b}{a} = \tan\psi = \tan 30° = \frac{1}{\sqrt{3}}$. The minor axis is at an angle of 30° w.r.t. the horizontal.

3. A quartz quarter-wave plate which is designed for light of wavelength 5893Å is used in conjunction with a Nicol prism to produce circularly polarised light. Show that if there is also present some light of wavelength 5460Å, this latter will be elliptically polarised with axes at 45° to the axes of the plate. Find the ratio of the major and minor axes of the ellipse.

Ans. Let the slow axis of the plate be along the y-direction. If the incident wave have amplitude a, then the Nicol acting as a polariser at 45° will transform the incident beam into polarised light at 45°, hence the incident light will be represented by

$$[J_i] = \frac{1}{\sqrt{2}}\begin{bmatrix}1\\1\end{bmatrix}.$$

The quarter-wave plate producing a phase difference of $\frac{\pi}{2}$ transforms the incident ray into a circularly polarised light. The transformation is carried through matrix product of the Jones' incident ray matrix and $\frac{\lambda}{4}$-plate matrix which is represented as

$$[J_{\lambda/4}] = \begin{bmatrix}1 & 0\\0 & i\end{bmatrix}. \quad \therefore \quad \begin{bmatrix}1 & 0\\0 & i\end{bmatrix}\frac{1}{\sqrt{2}}\begin{bmatrix}1\\1\end{bmatrix} = \frac{1}{\sqrt{2}}\begin{bmatrix}1\\i\end{bmatrix}$$

Writing out in full form, the emergent light can be represented as

$$\begin{bmatrix}y\\z\end{bmatrix} = \frac{a}{\sqrt{2}}\begin{bmatrix}1\\i\end{bmatrix}e^{i\phi_x}$$

$$\Rightarrow \quad y = \text{R.P.}\frac{a}{\sqrt{2}}e^{i\phi_x} = \frac{a}{\sqrt{2}}\cos\phi_x = \frac{a}{\sqrt{2}}\cos(kx-\omega t)$$

$$z = \text{R.P.}\frac{a}{\sqrt{2}}\cos\left(\phi_x + \frac{\pi}{2}\right) = \frac{a}{\sqrt{2}}\sin(kx-\omega t)$$

$$\therefore \quad y^2 + z^2 = \frac{a^2}{2}, \quad \text{a circularly polarised light.}$$

If the wavelength is not 5893, the plate will produce a phase difference

$$\phi = \frac{2\pi}{5460}\times\frac{5893}{4}\text{ rad} = 97.14° \neq \frac{\pi}{2}. \quad \therefore \quad \begin{bmatrix}1 & 0\\0 & e^{i\phi}\end{bmatrix}\frac{1}{\sqrt{2}}\begin{bmatrix}1\\1\end{bmatrix} = \frac{1}{\sqrt{2}}\begin{bmatrix}1\\e^{i\phi}\end{bmatrix}$$

Or, writing out in full form,

$$y = \frac{a}{\sqrt{2}}\cos(kx-\omega t) = A_1\cos(kx-\omega t)$$

$$z = \frac{a}{\sqrt{2}}\cos(kx-\omega t+\phi) = A_2\cos(kx-\omega t+\phi) \quad \text{(hence, } A_1 = A_2)$$

$$\Rightarrow \quad \frac{y^2}{\left(\frac{a}{\sqrt{2}}\right)^2} + \frac{z^2}{\left(\frac{a}{\sqrt{2}}\right)^2} - \frac{2yz\cos\phi}{\left(\frac{a}{\sqrt{2}}\cdot\frac{a}{\sqrt{2}}\right)} = \sin^2\phi.$$

The angle between the axes of the ellipse and co-ordinate axes is given by,

$$\tan 2\theta = \frac{2A_1A_2\cos\phi}{(A_1^2 - A_2^2)}.$$

Here, $A_1 = \frac{a}{\sqrt{2}}, A_2 = \frac{a}{\sqrt{2}}$.

$$\therefore \quad \tan 2\theta = \frac{2\cdot\frac{a^2}{2}\cos\phi}{\left(\frac{a^2}{2} - \frac{a^2}{2}\right)} = \infty \quad \Rightarrow \quad 2\theta = \frac{\pi}{2} \quad \Rightarrow \quad \theta = \frac{\pi}{4}.$$

Rotate the axes through 45°, then we would get,

$$z = \frac{(Z-Y)}{\sqrt{2}}, \quad y = \frac{(Z+Y)}{\sqrt{2}}$$

$$\therefore \quad \frac{(Z+Y)^2}{a^2} + \frac{(Z-Y)^2}{a^2} - \frac{4}{a^2} \cdot \frac{(Z-Y)}{\sqrt{2}} \cdot \frac{(Z+Y)}{\sqrt{2}} \cos\phi = \sin^2\phi$$

$$\Rightarrow \quad (Z+Y)^2 + (Z-Y)^2 - 2(Z^2 - Y^2)\cos\phi = a^2 \sin^2\phi$$

$$\Rightarrow \quad 2(Z^2 + Y^2) - 2(Z^2 - Y^2)\cos\phi = a^2 \sin^2\phi$$

$$\Rightarrow \quad Z^2(1-\cos\phi) + Y^2(1+\cos\phi) = \frac{a^2\sin^2\phi}{2}$$

$$\Rightarrow \quad \frac{Y^2}{\frac{a^2\sin^2\phi}{2(1+\cos\phi)}} + \frac{Z^2}{\frac{a^2\sin^2\phi}{2(1-\cos\phi)}} = 1$$

$$\therefore \quad \text{Ratio of the axes} = \sqrt{\frac{1-\cos\phi}{1+\cos\phi}} = \tan\frac{\phi}{2} = \tan\frac{97.14}{2} = 1.13.$$

Numerical Problems

1. Find the thickness of a quarter-wave plate made of quartz for light of wavelength $\lambda =$ 5893Å. The r.i. values for O- and E-ray are respectively 1.541 and 1.551.

Solution: According to eqn. (10.2.5), we can write

$$L = \frac{\lambda}{4} \cdot \frac{1}{(\mu_E - \mu_o)}$$

Putting the values of μ_E and μ_o, the thickness L can be calculated.

2. Through a calcite plate plane polarised Na-light (λ = 5893Å) is passed so as to produce circularly polarised light. If μ_o = 1.660 and μ_E = 1.495, find the thickness.

Solution:

$$\underset{\text{Retarder matrix}}{\underset{\downarrow}{\begin{bmatrix} 1 & 0 \\ 0 & i \end{bmatrix}}} \frac{1}{\sqrt{2}} \begin{bmatrix} 1 \\ 1 \end{bmatrix} = \frac{1}{\sqrt{2}} \underset{\text{RCP}}{\underset{\downarrow}{\begin{bmatrix} 1 \\ i \end{bmatrix}}}$$

So the incident plane polarised light represented as

$$y' = \frac{a}{\sqrt{2}}\cos(kx - \omega t), \quad z' = \frac{a}{\sqrt{2}}\cos(kx - \omega t)$$

Through the calcite plate there is a phase difference $\phi = \frac{\pi}{2}$ so as to get a circularly polarised light. If z axis be the fast axis, then the emergent light is represented as

$$y = \frac{a}{\sqrt{2}}\cos(k_1 L - \omega t), \quad z = \frac{a}{\sqrt{2}}\cos(k_2 L - \omega t)$$

where $(k_1 - k_2) = \frac{\pi}{2}$

$$\Rightarrow \quad \frac{2\pi}{\lambda}(\mu_o - \mu_E)L = \frac{\pi}{2}$$

$$\Rightarrow \quad L = \frac{\lambda}{4} \Big/ (\mu_o - \mu_E) = \left(\frac{5893 \times 10^{-8}}{4}\right) \Big/ (1.660 - 1.495)$$

$$= \left(\frac{5893 \times 10^{-8}}{4}\right) \Big/ 0.165 = 8.929 \times 10^{-5} \text{ cm}.$$

3. A left circularly polarised light ($\lambda = 5893$Å) is incident on a quartz crystal (with its optic axis parallel to the surface) of thickness 0.025 mm. Determine the state of polarisation of the emergent light. Assume $\mu_o = 1.54425$ and $\mu_E = 1.55336$ respectively.

Solution:

$$\delta = (k_2 - k_1)L = \frac{2\pi}{\lambda}(\mu_E - \mu_o)L = \frac{2\pi}{\lambda}(1.55336 - 1.54425)L = 0.76\pi.$$

4. A left circularly polarised beam ($\lambda = 5893$Å) is incident normally on a calcite crystal (with its optic axis cut parallel to the surface) of thickness 0.005141 mm. What will be the state of polarisation of the emergent beam?

Solution: The phase difference introduced by the plate is given by,

$$L = \phi \frac{\lambda}{2\pi(\mu_E - \mu_o)}$$

Here, $L = .005141$ mm $= .005141$ cm., $\lambda = 5893 \times 10^{-8}$ cm., $\mu_E - \mu_o = 0.17195$.

$$\therefore \quad \phi \simeq 3\pi$$

$$\therefore \quad \begin{bmatrix} 1 & 0 \\ 0 & e^{i\phi} \end{bmatrix} \frac{1}{\sqrt{2}} \begin{bmatrix} 1 \\ -i \end{bmatrix} = \frac{1}{\sqrt{2}} \begin{bmatrix} 1 \\ -ie^{i\phi} \end{bmatrix}$$

Here, $\phi = 3\pi \Rightarrow e^{i\phi} = \cos 3\pi + i \sin 3\pi = \cos(2\pi + \pi) = -1$

$$\therefore \quad \underset{\text{Retarder matrix}}{\begin{bmatrix} 1 & 0 \\ 0 & e^{i\phi} \end{bmatrix}} \frac{1}{\sqrt{2}} \underset{\text{Left circularly polarised light (incident)}}{\begin{bmatrix} 1 \\ -i \end{bmatrix}} = \frac{1}{\sqrt{2}} \underset{\text{Emergent light (right circularly polarised light)}}{\begin{bmatrix} 1 \\ i \end{bmatrix}}$$

Assuming the horizontal direction as the direction of the fast axis, Jones' matrix transformation gives as

$$\underset{\text{Retarder matrix with fast axis horizontal}}{\begin{bmatrix} 1 & 0 \\ 0 & e^{-i\delta} \end{bmatrix}} \frac{1}{\sqrt{2}} \underset{\text{Left circularly polarised light matrix (LCP)}}{\begin{bmatrix} 1 \\ -i \end{bmatrix}} = \frac{1}{\sqrt{2}} \underset{\text{Emergent light}}{\begin{bmatrix} 1 \\ -ie^{-i\delta} \end{bmatrix}}$$

Writing out in full form, the emergent light components can be written as

$$\begin{aligned} y &= \text{R.P. } \frac{a}{\sqrt{2}} e^{i\phi_x} = \frac{a}{\sqrt{2}} \cos \phi_x. \\ z &= \text{R.P. } \frac{a}{\sqrt{2}} e^{i(\phi_x - \pi/2 + \delta)} = -\frac{a}{\sqrt{2}} \sin(\phi_x - \delta) \end{aligned}$$

This would represent a clockwise elliptic vibration, so we would conclude that the emergent light is a right-handed elliptically polarised light.

The conclusion is drawn from the consideration given below when

$$\phi_x = \frac{\pi}{2} \quad \Rightarrow \quad t_1 = \frac{kx}{\omega} - \frac{\pi}{2\omega}, \quad \text{we get,} \quad y = 0, \quad z = -\frac{a}{\sqrt{2}} \cos\delta$$

This is represented by a line OA in fig (10.40).

When

$$\phi_x - \delta = 0 \quad \Rightarrow \quad \phi_x = \delta \quad \Rightarrow \quad t_2 = \frac{kx}{\omega} - \frac{\delta}{\omega}, \quad y = \frac{a}{\sqrt{2}} \cos\delta, \quad z = 0.$$

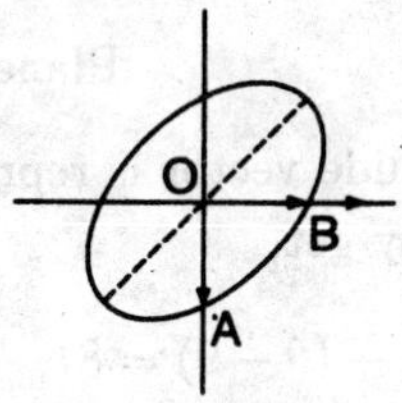

Fig 10.40

This is represented by a line OB. Now, if $\delta < \frac{\pi}{2}, t_2 > t_1$. If $\delta > \frac{\pi}{2}, t_2 < t_1$. So the displacement vector remaining constant in magnitude rotates in the clockwise direction, since δ here is greater than $\frac{\pi}{2}$.

5. A horizontal elliptically polarised parallel beam of light is passed through first a quarter-wave plate and then a Nicol prism. It is found that for a certain orientation of the quarter-wave plate, the emergent light is plane polarised at 23° to the vertical and that when the quarter-wave plate is turned through a right angle, it is plane polarised at 81° to the vertical. Determine the ratio.

Solution: The quarter-wave matrix with fast axis vertical is given by $\begin{bmatrix} 1 & 0 \\ 0 & i \end{bmatrix}$ and when the fast axis is horizontal it is given by $\begin{bmatrix} 1 & 0 \\ 0 & -i \end{bmatrix}$. The elliptically polarised light is represented by the matrix $\begin{bmatrix} A_1 \\ -iA_2 \end{bmatrix}$, the left-handed one or by $\begin{bmatrix} A_1 \\ iA_2 \end{bmatrix}$, the right-handed one.

$$\therefore \quad \begin{bmatrix} 1 & 0 \\ 0 & i \end{bmatrix} \begin{bmatrix} A_1 \\ -iA_2 \end{bmatrix} = \underbrace{\begin{bmatrix} A_1 \\ A_2 \end{bmatrix}}_{\text{Plane polarised}}$$

So considering a left-handed elliptically polarised light, the emergent light is plane polarised at an angle ϕ_o given by,

$$\tan\phi_o = \frac{A_2}{A_1},$$

and this is shown in fig (10.41). The ellipse is included within the rectangle of sides $2A_1$ and $2A_2$.

The resultant amplitude vector is represented by OB. From fig. (10.41), we can write,

$$23° + \phi_o + \theta = \frac{\pi}{2}$$

$$\Rightarrow \quad \phi_o = \left(\frac{\pi}{2} - 23 - \theta\right) = (67 - \theta)$$

$$\therefore \quad \tan\phi_o = \frac{A_2}{A_1} = \tan(67 - \theta)$$

When the quarter-wave plate is rotated through 90°, the fast axis becomes now horizontal and so

$$\begin{bmatrix} 1 & 0 \\ 0 & -i \end{bmatrix} \begin{bmatrix} A_1 \\ -iA_2 \end{bmatrix} = \begin{bmatrix} A_1 \\ -A_2 \end{bmatrix}$$

↓
Plane polarised light

In this case, the resultant amplitude vector is represented by OA.
From fig. (10.41), we immediately get,

$$23° + \phi_o + (\theta - 9) = 81, \quad \phi_o + \theta = 67°$$

Now, $\phi_o = (\theta - 9)$

$$\therefore \quad 2\phi_o + 9 = 67° \quad \Rightarrow \quad 2\phi_o = 58° \quad \Rightarrow \quad \phi_o = 29°$$

Since, $\tan\phi_o = \frac{A_2}{A_1} = 0.5543 \quad \Rightarrow \quad \frac{A_1}{A_2} = 1.804.$

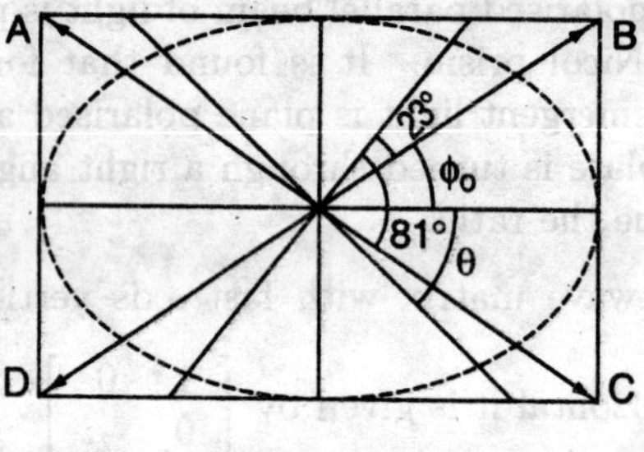

Fig 10.41

Also, $\phi_o + \theta = 67°$

$$\therefore \quad \theta = 67° - \phi_o = 67 - 29 = 38°.$$

So the major axis makes an angle (90° – 38°) w.r.t the vertical.

6. Fig. (10.42) shows a section of a wave surface for extraordinary ray. The optic axis being along OX. Show that refractive index of the medium along the direction making an angle θ is given by

$$\mu_\theta = \sqrt{\mu_o^2 \cos^2\theta + \mu_e^2 \sin^2\theta}$$

Solution: Fresnel's wave equation for propagation in an anisotropic crystal along a direction having direction cosines l, m, n is given by (see chapter on EM Theory)

$$\frac{l^2}{(v_x^2 - v^2)} + \frac{m^2}{(v_y^2 - v^2)} + \frac{n^2}{(v_z^2 - v^2)} = 0$$

In $X-Y$ plane we have,

$$\frac{l^2}{(v_x^2 - v^2)} + \frac{m^2}{(v_y^2 - v^2)} = 0.$$

Here, $l = \cos\theta, m = \sin\theta, v_x = v_e$ (extraordinary velocity) $v_y = v_o$ (ordinary ray velocity), v is the velocity along the direction θ.

Now, $\frac{c}{v} = \mu_\theta, \frac{c}{v_e} = \mu_e, \frac{c}{v_o} = \mu_o$. Putting these values in the wave surface equation, we get

$$\mu_\theta^2 = \mu_o^2 \cos^2\theta + \mu_e^2 \sin^2\theta \quad \Rightarrow \quad \mu_\theta = \sqrt{\mu_o^2 \cos^2\theta + \mu_e^2 \sin^2\theta}.$$

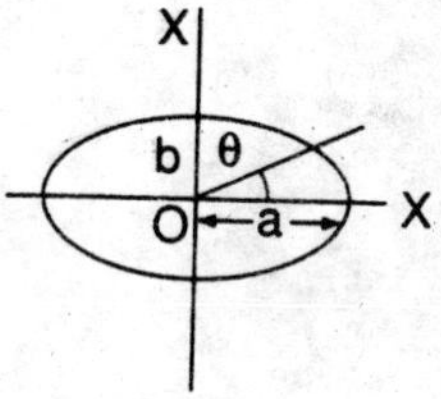

Fig 10.42

Theory Questions Exercise

1. What happens when an RCP and an LCP are superposed?
2. Two $\frac{\lambda}{4}$-plates with their fast axes horizontal are placed in succession. What happens when an incident light is allowed to pass through the combination?
3. Formulate the Jones' matrix for a linear polariser with axis horizontal.
4. What is Jones vector for linearly polarised light at $\pm 45°$ orientation?
5. Formulate the Jones matrix for a linear polariser with vertical orientation.
6. Work out the nature of the emergent representation below:

$$\frac{1}{\sqrt{2}}\begin{bmatrix} 1 \\ 11 \end{bmatrix}\begin{bmatrix} 1 & 0 \\ 0 & \pm i \end{bmatrix} = \frac{1}{\sqrt{2}}\begin{bmatrix} 1 \\ \pm i \end{bmatrix}$$

7. Show by Jones' method that two left or right circularly polarising quarter wave plates acting in succession are in effect equivalent to a half-wave plate.
8. A $\lambda/4$-plate with its fast axis horizontal is followed by a linear polariser with its transmission axis at 45°, followed by another $\lambda/4$-plate with its fast axis vertical. Find the system matrix.
9. Show by matrix method that elliptically polarised light is the general form of polarised light because it leads to both plane polarised or circularly polarised light as in special cases.
10. Determine the nature of the emergent light which passes initially through the process depicted by matrix operation as below:

(i) $\frac{1}{2}\begin{bmatrix} 1 & -1 \\ -1 & 1 \end{bmatrix}\frac{1}{2}\begin{bmatrix} 1 & 1 \\ 1 & 1 \end{bmatrix}\begin{bmatrix} 1 \\ 0 \end{bmatrix} = ?$

(ii) $\begin{bmatrix} 1 & 0 \\ 0 & i \end{bmatrix} \frac{1}{\sqrt{2}} \begin{bmatrix} 1 \\ 1 \end{bmatrix} = ?$

(iiii) $\begin{bmatrix} 1 & 0 \\ 0 & -i \end{bmatrix} \frac{1}{\sqrt{2}} \begin{bmatrix} 1 \\ 1 \end{bmatrix} = ?$

11. Explain the action of a Babinet's compensator by Jones matrix formalism.

Chapter 11

Optical Activity, Rotatory Dispersion and Faraday Effect

11.1 Introduction

There are some substances, solid, liquid or gas (under the influence of magnetic field) which rotate the plane of vibration of a linearly polarised light, either towards the right or towards the left. The substance which rotates the plane of vibration of the incident light is called an *active* substance and if it rotates towards the right, the substance will be called *dextro-rotatory* but if the rotation is towards the left, the substance is referred to as *laevo-rotatory* substance. The extent of rotation depends on the nature of the active substance, concentration, pathlength within the active substance. A very common active substance is the solution of common sugar, *dextrose.* The solution of sugar is dextro-rotatory and the angle of rotation is given by,

$$\theta = cls, \tag{11.1.1}$$

where s is the specific rotation of the substance, l is the length of the solution, c is the concentration. The equation is very important in sugar industry, because with the help of this formula one can find out the concentration of sugar in a solution.

In a liquid, there is no axis as in the case of a solid, so here the amount of rotation per unit pathlength is the same whatever may be the direction of the incident beam.

Many crystals show optical activity and this activity depends on the arrangement of the atoms in the crystal lattice. A quartz crystal exhibits optical activity only if light is travelling approximately parallel to the optic axis. A crystal of quartz that shows optical activity when in crystal form loses it completely when the crystal is dissolved. The crystal Cinnabar is noted for its large value of rotation, about 5800° per centimeter of pathlength. A few liquid crystals like cholesteryl benzoate, etc., produce even more optical rotation per centimeter of pathlength.

In case of gases, most of them do not exhibit optical activity under ordinary condition, but they do it when under a strong magnetic field. Such an effect has a special name, usually referred to as *Faraday effect.*

For a given optically active substance, the amount of rotation depends on the wavelength, so when white light passes through an optically active medium different colours will be rotated through different angles, separating the different colours, thus producing the phenomenon of what is known as *Rotatory Dispersion.*

11.2 Optical Rotation—Optical Activity

We have already remarked that some substances, solid like quartz, liquid like sugar solution behave w.r.t. incident linearly polarised light in such a way that after emergence through the substance or medium, the plane of vibration rotates. This rotation of the plane of vibration is referred to as optical activity of the medium.

Experimental demonstration

From the source light is allowed to pass through the polarising Nicol P, the active medium is not now placed and the analysing Nicol (A) is rotated such that no light is received on the screen. In this position, the direction of vibration of linearly polarised light and the optic axis of the analyser are at right angles. Now, the active medium (say, a quartz crystal or sugar solution) is introduced in between the polariser and the analyser when light will now be received on the screen. The analyser is adjusted so that once again the field beyond the analyser may be rendered dark. The angle through which the analyser needs be rotated is the angle of rotation produced. Looking at the source from the analyser side, if the rotation of the plane of vibration be anti-clock, the active medium is laevo-rotatory; if it is clockwise, it is dextro-rotatory.

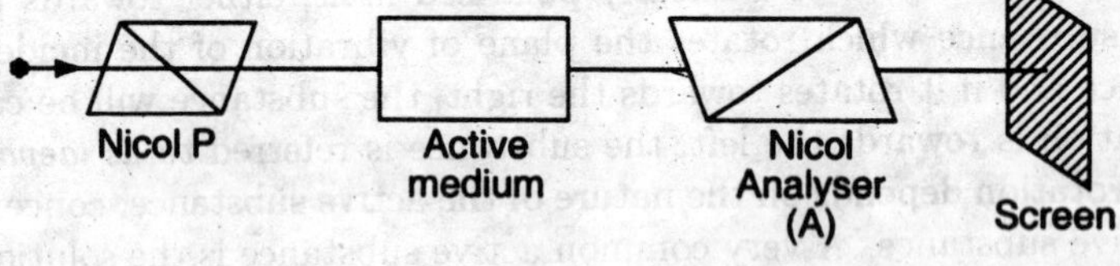

Fig 11.1

Physical basis of rotation

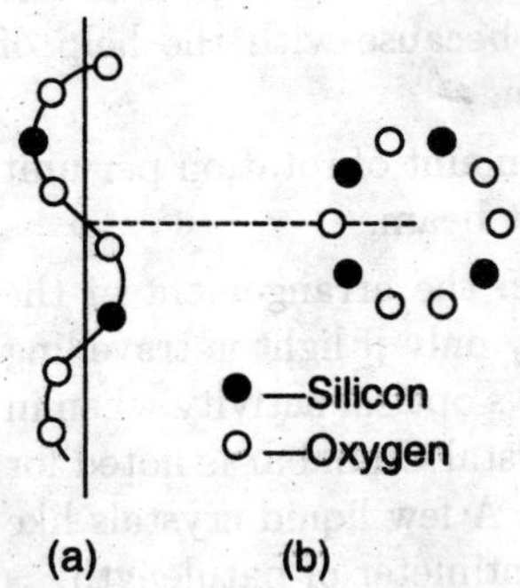

Fig 11.2

Why some substances produce rotation of the plane of vibration, while others not the reason is to be found in the structure of the substance? The rotation of the linearly polarised light by the active substance is associated with the helicity of the structure of the subtance.

If we consider the structure of quartz which is an active medium, Si and Oxygen atoms are so arranged that it gives crystal of hexagonal class, shown in fig.(11.2) we have shown a helical arrangement which one encounters as one follows down the arrangement of atoms. As light travels down the helical path, the direction of vibration gets changed. It is to note that the helical path may be right-handed or left-handed.

Fresnel's explanation : Theoretical analysis

Fresnel has given an explanation about the rotation of the plane of vibration by the optically active medium.

Any linearly polarised light may be thought of being composed of two circularly polarised lights, one left circularly polarised (LCP), the other right circularly polarised (RCP). Thus in

Jones' formalism,

$$\frac{1}{\sqrt{2}}\underset{\text{LCP}}{\begin{bmatrix}1\\-i\end{bmatrix}}+\frac{1}{\sqrt{2}}\underset{\text{RCP}}{\begin{bmatrix}1\\i\end{bmatrix}}=\frac{1}{\sqrt{2}}\begin{bmatrix}2\\0\end{bmatrix}=\begin{bmatrix}\sqrt{2}\\0\end{bmatrix}=\sqrt{2}\begin{bmatrix}1\\0\end{bmatrix}$$

Dividing throughout by $\sqrt{2}$, we get,

$$\frac{1}{2}\begin{bmatrix}1\\-i\end{bmatrix}+\frac{1}{2}\begin{bmatrix}1\\i\end{bmatrix}=\begin{bmatrix}1\\0\end{bmatrix} \tag{11.2.1}$$

If the medium be dextro-rotatory, there will be a phase difference introduced between the two circularly polarised lights such that the RCP advances in phase over the LCP.

$\therefore$ at emergence, the RCP is represented as $\frac{1}{\sqrt{2}}\begin{bmatrix}e^{i\phi_1}\\+ie^{i\phi_1}\end{bmatrix}$ where $\phi_1(=k_R l)$ is the phase introduced due to travel throough the active medium of length l.

The LCP is represented as, $\frac{1}{\sqrt{2}}\begin{bmatrix}e^{i\phi_2}\\-ie^{i\phi_2}\end{bmatrix}$, where $\phi_2 = k_L l$.

So at emergence, the light is represented by,

$$\begin{aligned}\frac{1}{\sqrt{2}}\begin{bmatrix}e^{i\phi_1}\\+ie^{i\phi_1}\end{bmatrix}+\frac{1}{\sqrt{2}}\begin{bmatrix}e^{i\phi_2}\\-ie^{i\phi_2}\end{bmatrix} &= \frac{1}{\sqrt{2}}\begin{bmatrix}e^{i\phi_1}\\e^{i(\phi_1+\pi/2)}\end{bmatrix}+\frac{1}{\sqrt{2}}\begin{bmatrix}e^{i\phi_2}\\e^{i(\phi_2-\pi/2)}\end{bmatrix}\\ &= \frac{1}{\sqrt{2}}\begin{bmatrix}e^{i\phi_1}+e^{i\phi_2}\\e^{i(\phi_1+\pi/2)}+e^{i(\phi_2-\pi/2)}\end{bmatrix}\end{aligned}$$

Now,

$$\begin{aligned}e^{i\phi_1}+e^{i\phi_2} &= (\cos\phi_1+\cos\phi_2)+i(\sin\phi_1+\sin\phi_2)\\ &= a+bi=\sqrt{a^2+b^2}\,e^{i\theta_1}\end{aligned}$$

$$\begin{aligned}\therefore\quad a^2+b^2 &= (\cos\phi_1+\cos\phi_2)^2+(\sin\phi_1+\sin\phi_2)^2\\ &= 2+2\cos(\phi_1-\phi_2)=4\cos^2\frac{\phi_1-\phi_2}{2}.\end{aligned}$$

$$\therefore\quad \sqrt{a^2+b^2}=2\cos\frac{\phi_1-\phi_2}{2}$$

$$\begin{aligned}e^{i(\phi_1+\pi/2)}+e^{i(\phi_2-\pi/2)} &= \cos(\phi_1+\pi/2)+\cos(\phi_2-\pi/2)\\ &\quad +i\sin(\phi_1+\pi/2)+i\sin(\phi_2-\pi/2)\\ &= (-\sin\phi_1+\sin\phi_2)+i(\cos\phi_1-\cos\phi_2)\\ &= c+id=\sqrt{c^2+d^2}\,e^{+i\theta_2}.\end{aligned}$$

$$\begin{aligned}c^2+d^2 &= (-\sin\phi_1+\sin\phi_2)^2+(\cos\phi_1-\cos\phi_2)^2\\ &= 2-2\cos(\phi_1-\phi_2)=4\sin^2\frac{\phi_1-\phi_2}{2}.\end{aligned}$$

$$\therefore \quad \sqrt{c^2 + d^2} = 2 \sin \frac{\phi_1 - \phi_2}{2}$$

$\therefore$ The emergent light vibration is represented by Jones' vector given as

$$\frac{1}{\sqrt{2}} \begin{bmatrix} 2\cos \frac{\phi_1 - \phi_2}{2} e^{i\theta_1} \\ 2\sin \frac{\phi_1 - \phi_2}{2} e^{i\theta_2} \end{bmatrix} = \sqrt{2} \begin{bmatrix} A_1 e^{i\theta_1} \\ A_2 e^{i\theta_2} \end{bmatrix}.$$

This represents a linear vibration inclined at an angle α with the y-direction given by

$$\tan\alpha = \frac{A_2}{A_1} = \frac{\sin\left(\frac{\phi_1 - \phi_2}{2}\right)}{\cos\left(\frac{\phi_1 - \phi_2}{2}\right)} = \tan\frac{\phi_1 - \phi_2}{2} \tag{11.2.2}$$

$$\Rightarrow \quad \alpha = \frac{\phi_1 - \phi_2}{2}.$$

Thus a horizontally polarised light after passing through a length l of the active medium rotates through an angle α given by the above relation.

Now,

$$\phi_1 = k_R l, \quad \phi_2 = k_L l.$$

Again,

$$k = \frac{2\pi}{\lambda}, \quad \mu = \frac{c}{v} = \frac{\nu\lambda}{\nu\lambda_1} = \frac{\lambda}{\lambda_1}$$

$$\therefore \quad \mu_R = \frac{\lambda}{\lambda_R} \quad \Rightarrow \quad k_R = \frac{2\pi}{\lambda_R} = \frac{2\pi\mu_R}{\lambda}.$$

$$\therefore \quad \alpha = \frac{\phi_1 - \phi_2}{2} = \frac{(k_R - k_L)l}{2} = \frac{\pi}{\lambda}(\mu_R - \mu_L)l.$$

The specific rotatory power of the medium is given by $\frac{\alpha}{l}$ and denoted as s, thus,

$$s = \frac{\alpha}{l} = \frac{\pi}{\lambda}(\mu_R - \mu_L) \tag{11.2.3}$$

Example 1. *The specific rotation of quartz in light of* 589 mm *wavelength is* 21.7°/mm.. *What thickness of quartz, cut perpendicularly to its axis and inserted between parallel polarisers, will cause no light to be transmitted?*

Solution : Here, $\alpha = \frac{\pi}{2}$.

$$\therefore \quad \frac{\pi}{2} = \frac{\pi}{\lambda}(\mu_R - \mu_L)l = sl \quad \Rightarrow \quad l = \frac{\pi}{2s}.$$

$$s = 21.7°/\text{mm.} = 217°/\text{cm.} = \frac{\pi}{180} \times 217\,\text{rad}/\text{cm.}.$$

$$\therefore \quad l = \frac{\pi}{2 \times \frac{\pi}{180} \times 217} = \frac{180}{2 \times 217} = 0.415\,\text{cm.} = 4.15\,\text{mm.}$$

Example 2. *A glucose solution of unknown concentration is contained in a* 12 cm *long tube and then seen to rotate linearly polarised light by* 2.5°. *Specific rotation of glucose is* 52°, *find the concentration.*

Solution : $\theta = 2.5° = \frac{\pi}{180} \times 2.5, \quad s = 52 = \frac{52}{180} \times \pi.$

$$\therefore \quad \theta = cls \quad \Rightarrow \quad \frac{\pi}{180} \times 2.5 = c \times 1.2 \times \frac{52}{180} \times \pi \quad \Rightarrow \quad c = \frac{2.5}{1.2 \times 52} = .04.$$

It means the solution contains .4 gms/lit.

Example 3. *A quartz plate cut with its surface ⊥ to the optic axis is required to annul completely the rotation of the plane of polarisation of red light* ($\lambda = 7600$ Å) *by a* 26.7 cm *length of lactose solution containing the* 100 gms *of active substance/lit of the solution. What should be the thickness of the plate? (Sp. rotation* 52.5°, $\mu_L = 1.53920$, $\mu_R = 1.53914$).

Solution : $\theta = slm$, l = length in decimeter $= 52.5 \times 2.67 \times .1 = 14.02 \times \frac{\pi}{180}$ rad

$$\therefore \quad 14.02 \times \frac{\pi}{180} = \frac{\pi}{180} = \frac{\pi}{\lambda}(\mu_L - \mu_R)L \quad \Rightarrow L = \frac{14.02 \times 7600 \times 10^{-8}}{180(1.53920 - 1.53914)}$$

$$\therefore \quad L = .0987 \text{ cm.} = 0.987 \text{ mm.}$$

Rotatory dispersion

Since optical rotation is wavelength dependent and the rotation almost follows Cauchian law of dispersion, it tells us that if white light source is taken, then optical rotation suffered by violet will be more than that by the red.

The Cauchian law of dispersion gives, $\theta = A + \frac{B}{\lambda^2}$, hence θ is larger for smaller wavelength.

Thus, violet wavelength being smaller than red wavelength, $\theta_v > \theta_R$ which means that plane of vibration of different polarised light (i.e., light of different wavelength) will turn through different angles by the optically active medium and this phenomenon is referred to as *Rotatory Dispersion.*

The experimental arrangement for the realisation of the rotatory dispersion is shown in the fig. (11.3).

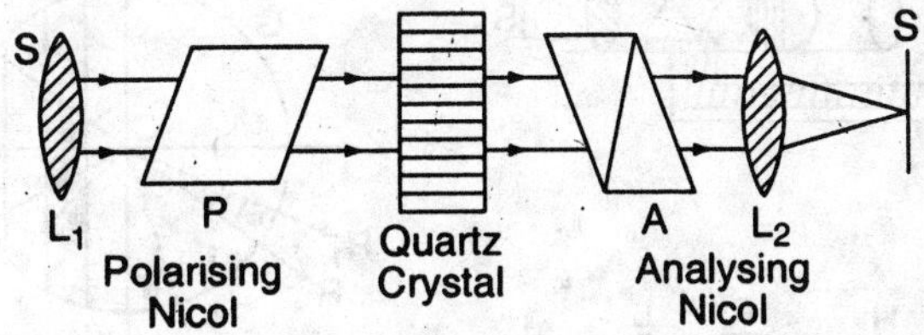

Fig 11.3

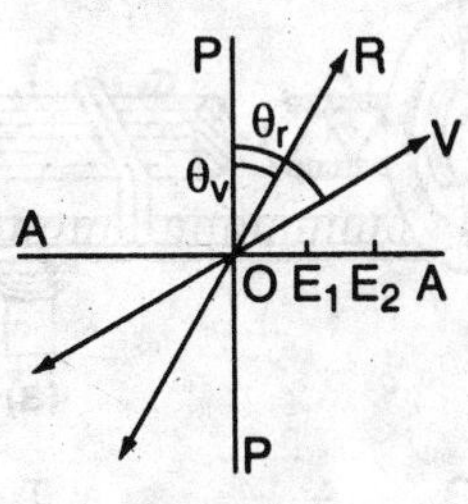

Fig 11.4

At first, the quartz crystal is not placed in position. Light rays, rendered parallel by the lens L_1 get polarised by the polarising Nicol P whose shorter diagonal is vertical, hence the direction of vibration of the emergent light is vertical; the shorter diagonal of the analysing Nicol when set horizontal will cut off light and the screen S will be dark. Now, the quartz crystal is introduced in position as indicated in the figure with its optic axis perpendicular to the surface. As light emerges out of the crystal, different colours get rotated through different angles depending on the wavelength as shown in fig. (11.4). The red colour having greater wavelength suffers optical rotation θ_r and the violet colour suffers greater amount of rotation θ_v. The components OE_1 for red and OE_2 for violet will be transmitted through the analyser and we will now get coloured image on the screen. This phenomenon is known as *Rotatory Dispersion.*

Now, if the principal plane of the analyser is set perpendicular to the greenish-yellow colour, then it will be cut off and the illumination of the field will be minimum due to absence of illumination producing part of the visible spectrum. Then the colour of the emergent beam becomes greyish-violet due to combination of less intense red and violet parts. This greyish-violet colour is called *tint of passage.* If the analyser is rotated to one side of this position, the field become red due to greater transmission of red light; if it is rotated to the other side, field becomes blue due to greater transmission of blue. The transition from red to blue is very quick and forms the basis of a very sensitive polarimeter, called a *biquartz polarimeter.*

11.3 Polarimeters

These are instruments by which we find out the unknown concentration of an active solution by measuring the plane of vibration of the linearly polarised light.

The polarimeter consists of two Nicol prisms A and P, respectively called analyser and polariser; the polarising Nicol polarises the light and the analysing Nicol detects the polarised light.

There are two types of polarimeters—(1) Laurent's half-shade polarimeter or Saccharimeter, (2) Biquartz polarimeter.

Laurent's polarimeter

Such a polarimeter is shown in fig. (11.5) where at one end of the tube T_1 there is a slit S' at the focal plane of the lens L, this tube also contains a Nicol P, serving as the polariser, the half-shade plate H. The tube T_2 contains Nicol A, serving as the analyser and a telescope consisting of an objective O and eye-piece E. This tube can be rotated about a horizontal axis but keeping always the axis of the two tubes T_1, T_2 in line. The middle tube T whose axis also in line contains the active solution.

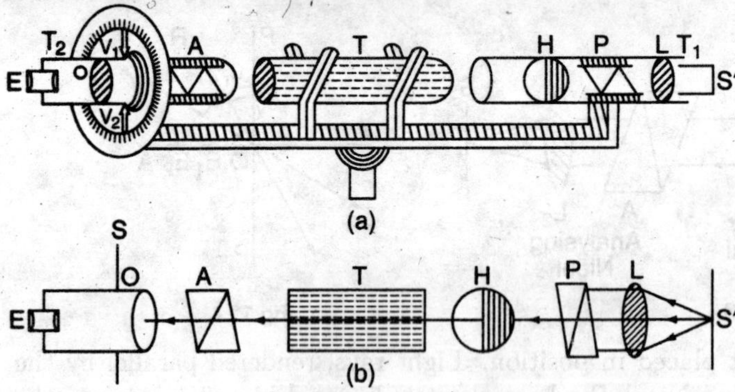

Fig 11.5

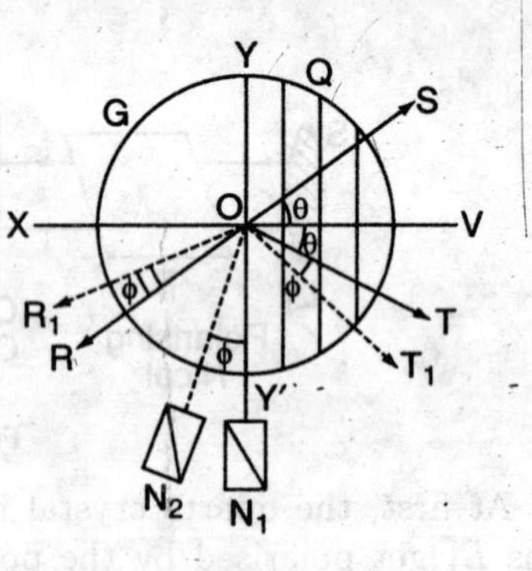

Fig 11.6

Half-shade plate H

A half-shade plate which is a circular plate, one half (G) of glass, the other half (Q) of quartz is shown in fig. (11.6). The thickness of both halves are same and the optic axis of quartz half Q is on its surface and parallel to the dividing the YY'.

Action

The slit S' is illuminated by suitable light for half-shade plate, the lens L renders it into a parallel beam which passes through the polarising Nicol, gets polarised and falls on the half-shade plate. Let θ be the angle which the direction of vibration OS makes with OX, a line $\perp$ to the axis of the quartz. This vibration will be resolved into two components, one parallel to the axis, called *extraordinary* ray, the other $\perp$ to the axis, called *ordinary ray.* These two rays travelling through the quartz with unequal speed will suffer a path difference $\lambda/2$ or phase difference π, the thickness of quartz is so chosen and so they will after emergence combine to form a linear vibration parallel to OT making the same angle θ with the OX-direction (cf. fig. 11.6). But the directions of vibration of the rays incident on the glass-half (G) will remain parallel to OR even on emergence. So if the principal section of the analyser be placed at N_1 along the bisector of the angle $\angle ROT$, both halves will appear equally bright. Now, if the tube T contains an active solution, then there will be rotation of OR and OT by the same amount ϕ (say) and so at the position of the analyser N_1, the two halves will be unequally bright, so when rotated through ϕ to the position at N_2, the two halves again become equally bright.

Biquartz polarimeter

This type of polarimeter differs from the half-shade type in the fact that here half-shade plate is replaced by a biquartz plate, having both halves made of quartz of equal thickness. One half is made of Left-handed quartz (L), the other Right-handed (R). A plane polarised white light is incident on the biquartz plate and the thickness is such that the greenish-yellow part of the spectrum is rotated by each half through 90°. When the analyser A is set with its principal section parallel to that of the polariser (P) the greenish-yellow part will be cut off and both halves will appear greyish-violet, knwon as *tint of passage.* After passing through the biquartz when the light further passes through an active solution, there will be further rotation and one half quartz will appear red, the other half blue. Rotate the tube T_2 when both halves again appear greyish-violet. The angle of rotation gives the angle through which plane of polarisation has been turned by the active substance.

Points to note

Biquartz polarimeter is more sensitive than Laurent's polarimeter; moreover, white light can be used with biquartz polarimeter, whereas a suitable wavelength of light is to be used for a half-shade polarimeter.

Interference of polarised light

In the chapter on *Interference* we remarked that for interference of polarised light, the two coherent sources must send out waves having same plane of vibration. The formation of interference fringes with polarised light can simply be demonstrated w.r.t. fig. (11.7).

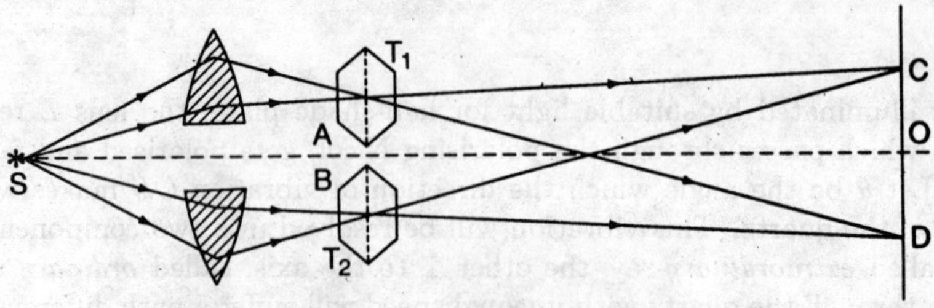

Fig 11.7

The arrangement consists of Billet's split lens which forms two coherent sources A, B out of the single sources as a result of refraction through the two halves of the lens. The rays from these two sources superpose on the portion COD producing permanent interference fringes on the screen.

Now, two tourmaline crystals T_1 and T_2 are placed at A and B with their optic axes parallel. Interference fringes are still available on the screen. Keeping one crystal fixed, the other is rotated about the incident ray direction as axis. It is found that the interference pattern gradually disappears and it disappears completely when the optic axes of the two tourmaline crystals are crossed with each other.

Thus, when the optic axes were parallel, the light from the two sources becomes polarised and the direction of vibration is parallel and contained in parallel planes, hence interference of polarised light takes place but when they are crossed, the light from the two sources are polarised in orthogonal planes, hence, no interference pattern available.

Interference of parallel plane polarised light

Between a polariser and an analyser if a crystal is placed, we observe different colour changes due to interference.

Light from the sources is rendered parallel by means of the lens L_1 and then is made incident on the polariser P. The emergent light gets plane polarised. This polarised light can be represented as $\begin{bmatrix} \cos\alpha \\ \sin\alpha \end{bmatrix}$. It then passes through the crystal whose axis is along OY-direction. On emergence, in case of a +ve crystal, the vibration along z-axis, gains in phase, hence the emergent light may be represented as $\begin{bmatrix} \cos\alpha \\ \sin\alpha e^{i\delta} \end{bmatrix}$.

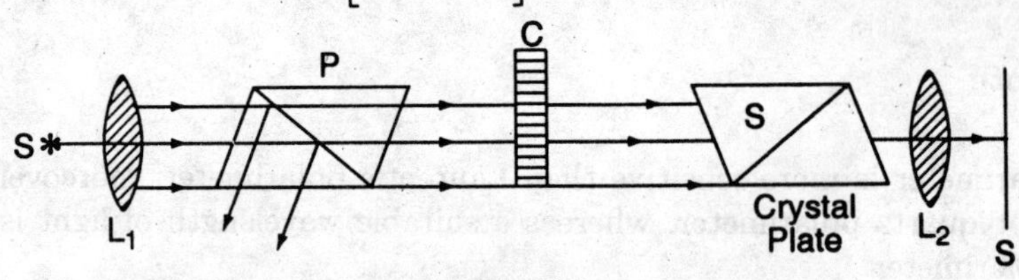

Fig 11.8

This vibration then passes through the analyser whose axis is making angle β with the OY-direction. So the emergent light from the analyser can be represented as

$$\begin{bmatrix} \cos\beta & 0 \\ 0 & \sin\beta \end{bmatrix} \begin{bmatrix} \cos\alpha \\ \sin\alpha e^{i\delta} \end{bmatrix} = \begin{bmatrix} \cos\beta\cos\alpha \\ \sin\beta\sin\alpha e^{i\delta} \end{bmatrix}.$$

So writing in full, we get the emergent light represented by,

$$y_e = a\cos\beta\cos\alpha\cos(k_1 x - \omega t) \quad (11.3.1)$$

$$z_e = a\sin\beta\sin\alpha\cos(k_1x - \omega t + \delta) \tag{11.3.2}$$

$$\begin{aligned} \therefore\quad \xi &= y_e + z_e = a\cos\beta\cos\alpha\cos(k_1x-\omega t) \\ &\quad + a\sin\beta\sin\alpha\cos(k_1x-\omega t)\cos\delta - a\sin\beta\sin\alpha\sin(k_1x-\omega t)\sin\delta \\ &= a\{[\cos\beta\cos\alpha + \sin\beta\sin\alpha\cos\delta]\cos(k_1x-\omega t) - \sin\beta\sin\alpha\sin\delta(\sin k_1x - wt)\} \end{aligned}$$

Putting,

$$a(\cos\beta\cos\alpha + \sin\beta\sin\alpha\cos\delta) = A\cos\phi, \quad a(\sin\beta\sin\alpha\sin\delta) = A\sin\phi$$

$$\therefore\quad \xi = A\cos\phi\cos(k_1x-\omega t) - A\sin\phi\sin(k_1x-\omega t) = A\cos(k_1x-\omega t+\phi)$$

So the intensity of the light transmitted through the analyser is

$$\begin{aligned} I &= A^2 = a^2\{(\cos\beta\cos\alpha + \sin\beta\sin\alpha\cos\delta)^2 + \sin^2\beta\sin^2\alpha\sin^2\delta\} \\ &= a^2\{\cos^2\beta\cos^2\alpha + \sin^2\beta\sin^2\alpha\cos^2\delta \\ &\quad + 2\cos\beta\cos\alpha\sin\beta\sin\alpha\cos\delta + \sin^2\beta\sin^2\alpha\sin^2\delta\} \\ &= a^2\left\{\cos^2\beta\cos^2\alpha + \sin 2\beta\sin 2\alpha + \frac{1}{2}\sin 2\beta\sin 2\alpha\cos\delta\right\} \\ &= a^2\left[\cos^2(\beta-\alpha) - \frac{1}{2}\sin 2\alpha\sin 2\beta + \frac{1}{2}\sin 2\beta\sin 2\alpha\cos\delta\right] \\ &= a^2[\cos^2(\beta-\alpha) - \sin 2\alpha\sin 2\beta\sin^2\delta/2]. \end{aligned}$$

For white light,

$$I = \cos^2(\beta-\alpha)\Sigma a^2 - \sin 2\alpha\sin 2\beta\Sigma a^2\sin^2\delta/2 \tag{11.3.3}$$

Thus from equation (11.3.3) we can discuss the different cases of interference of polarised light. The first term in equation (11.3.3) is the sum of intensities of all colours hence it gives white light. The second term accounts for the colour because here δ is different for different values of λ.

Discussion of equation (11.3.3) in case of a unixial crystal cut perpendicular to the optic axis

Refer to fig (11.9) where we have taken a unixial crystal cut perpendicular to the optic axis. Take a convergent beam of plane polarised light, obtained by a suitable polariser with its transmission axis at an angle of α with the optic axis of the crystal. The polarised light on incidence on the plate resolves into two rectangular components as O- and E-ray. But rays which are incident on the circumference of the circle have the same angle of incidence.

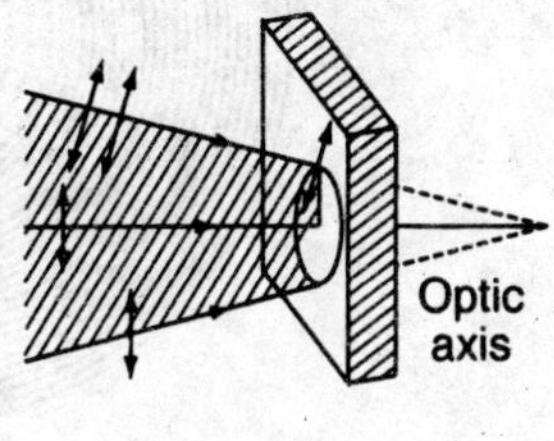

Fig 11.9

Achromatic or Neutral lines

When the angle $\alpha = 0$ or $\pi/2$, in both cases we would get from equation (11.3.3)

$$I = \cos^2(\beta-\alpha)\Sigma a^2 \tag{11.3.4}$$

So along and perpendicular to the transmission axis of the polariser we get a rectangular white cross.

When $\beta = 0$ or $\pi/2$, we would again get the intensity function as $I = \cos^2(\beta - \alpha)\Sigma a^2$.

Thus along and perpendicular to the transmission axis of the analyser, we would get another rectangular white cross.

Thus we get two systems of rectangular white crosses—one system will be parallel and perpendicular to the transmission axis of the polariser, the other system is parallel and perpendicular to the transmission axis of the analyser.

When the transmission axes of the polariser and the analyser are parallel

In this case, $(\beta - \alpha) = 0$, and so we would get,

$$I = \Sigma a^2 - \sin^2 2\alpha \Sigma a^2 \sin^2 \delta/2 = \Sigma a^2(1 - \sin^2 2\alpha \sin^2 \delta/2) \tag{11.3.5}$$

In addition if $\alpha = 0$, or $\pi/2$, we get from equation (11.3.3), $I = \Sigma a^2$. Thus, if the polariser and analyser transmission axes are parallel and set along perpendicular to the optic axis of the crystal, we would get one rectangular white cross having directions parallel and perpendicular to the transmission axis of the polariser or the analyser.

When the polariser and the analyser are crossed

In this case, $\beta - \alpha = \frac{\pi}{2}$, so we get from equation (11.3.3)

$$I = -\sin^2 2\alpha \sin(\pi + 2\alpha)\Sigma a^2 \sin^2 \delta/2 = \sin^2 2\alpha \Sigma a^2 \sin^2 \delta/2$$

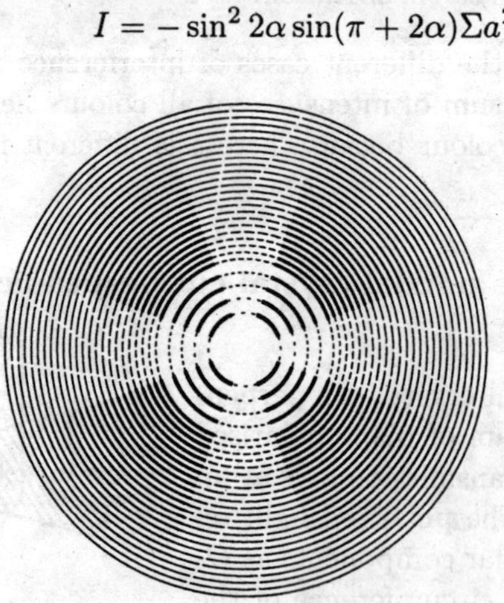

(a) Polarizer and Analyser crossed.

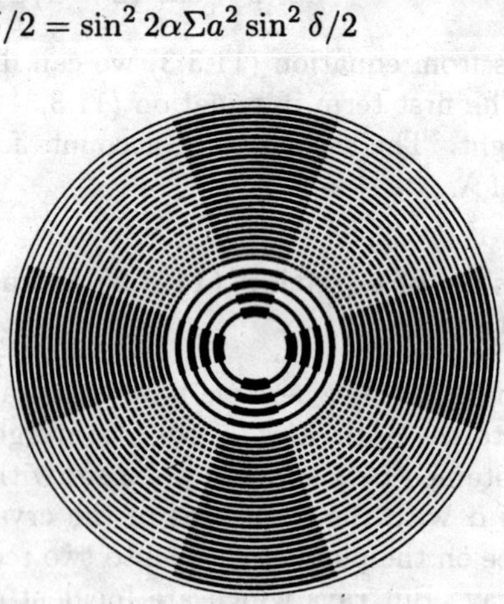

(b) Polarizer and Analyser parallel.

Fig 11.10

In addition, if the transmission axis of the polariser or the analyser is parallel to the optic axis, then α being equal to zero, we get $I = 0$. So we get a rectangular dark cross in the directions parallel to the transmission axes of the polariser and the analyser.

Isochromatic Lines

When the transmission axes of the analyser and polariser are at right angles and the phase difference introduced by the crystal plate is a multiple of 2π, i.e., $\delta = m \times 2\pi$, then from equation (11.3.6), we get, $I = 0$.

So there will be a series of concentric dark rings in the image plane and between these dark rings there will be maxima. The radii of the rings depend on the wavelength and coloured rings are observed with white light. Since there is symmetry about the axis of the system, the colour is constant round any circle centred on the axis. Such a line is termed as an *isochromatic line.*

When the analyser and polariser are mutually at right angles, we would get two perpendicular diameters of zero intensity. Since these dark lines will be the same for all wavelengths, they are referred to as *achromatic lines.* The intensities of the isochromatic maxima gradually build as we pass from the achromatic lines. The isochromatic lines are usually called the rings and the achromatic lines are called the brushes [see fig. 11.10(a) and 11.10(b)].

11.4 Faraday Effect

We have seen earlier that in case of a medium having asymmetrical molecular arrangement, the plane of polarisation of a polarised light rotates, the phenomenon is referred to as Optical Activity.

It was observed by Faraday that when a plane polarised light is made to pass through an isotropic substance of high refractive index value (e.g., a glass slab), the plane of polarisation rotates when placed in a strong magnetic field. This effect of rotation of the plane of polarisation under a magnetic field is called *Faraday effect.* This rotation θ is proportional to the length of the medium traversed, the strength of the magnetic field applied, so we can put $\theta = VLB$, where L is the length, B is the magnetic field and V is the constant of proportionality, called *Verdet's constant.* This rotation is $+ve$ or $-ve$ depending on the direction whether in the direction of or opposite to the direction of current producing the magnetic field. These substances showing Faraday rotation are ordinarily optically inactive.

Difference between Faraday Rotation and Optical Rotation

Faraday rotation is independent of the direction of light but natural rotation depends on the direction of light. The natural optical rotation can be annulled by reversal of the direction of light but Faraday rotation is doubled on reversal. Also the magnetic rotation is maximum when it travels in the direction of the magnetic field and minimum when at right angles.

Experimental Demonstration

When the rotation is of the order of 0.005°, Lippich's double or triple field polarimeter is very suitable but for larger rotation, biquartz polarimeter may be employed.

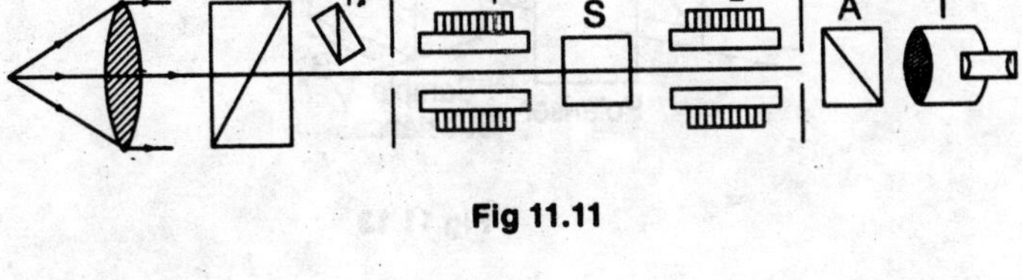

Fig 11.11

We have shown an arrangement for demonstrating the Faraday rotation P is the polariser, A is the analyser, M_1M_2 is an electromagnet that provides the magnetic field, S is the medium producing the rotation. P_1 is a small Nicol which covers half of the aperture of the polariser P. The principal section of P_1 is slightly inclined by an angle α (say) to the principal section of the Nicol P. This angle is called the *shadow angle*.

In the absence of the magnetic field supplied by M_1M_2, the two halves of the field of view will be equally bright when the analyser axis AA' is at right angles to the bisector A_1A_1' of the shadow angle. The polarised light can pass through the longitudinal holes bored in the electromagnet. The sample substance S is placed between the pole-pieces. When the magnetic field is switched on, the substance S rotates the plane of polarisation P_1 on one half and form P on the other by the same amount and in the same direction. As a result, the principal section of the analyser is not longer perpendicular to the bisector of the shadow angle. The analyser A and the telescope T should be rotated in the proper direction so as to get equal brightness or matching of both halves. This angle of rotation θ is found to depend on the strength of the magnetic field as well as on the length of the sample substance and can be related as $\theta = VLB$.

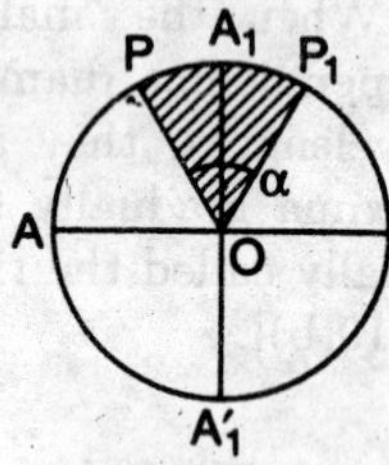

Fig 11.12

Some Characteristics of Faraday Rotation

1. The Faraday-effect or in otherwords the magnetically induced rotation is largest when the field is parallel to the direction of the light beam (longitudinal field) and it vanishes when the field is transverse.
2. The optical rotation is reversible, i.e., the rotation is completely annulled on reversal but Faraday rotation is cummulative.
3. The amount of Faraday rotation depends on the frequency of radiation. Thus the rotation in case of ultra-violet is larger.

11.5 Classical Theory

Let us consider the arrangement as given in fig.(11.13) where the magnetic field has been applied along the direction of propagation of light (i.e., along x-direction). According to the electromagnetic theory of Maxwell, light consists of oscillation of electric and magnetic vectors which are orthogonal to each other and also at right angles to the direction of propagation. Faraday effect is due to interaction between the electrons of the atoms of the medium and the electromagnetic field. The electric field causes the electron to oscillate, the magnetic field vector has practically no effect on the motion of the electron compared to the effect due to the electric component of the electromagnetic wave and that due to the applied magnetic field.

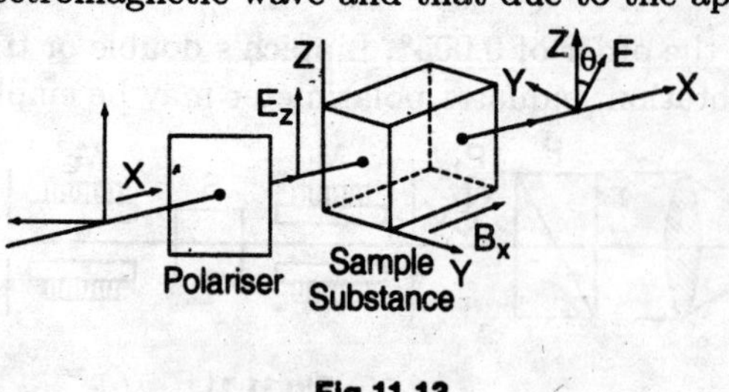

Fig 11.13

Consider the incident light to be on the face of the block in the Y-Z plane, the electric vector at incidence is supposed to be in the z-direction. As light travels through the mediums, the plane of polarisation rotates so that the electric vector now has components E_z and E_y.

Let us say that at some instant inside the medium, the position vector of an electron is represented by $\vec{r}$.

So the equation of motion can be written as,

$$m\ddot{\vec{r}} = -e\bar{E} - e\begin{vmatrix} \hat{i} & \hat{j} & \hat{k} \\ \dot{x} & \dot{y} & \dot{z} \\ B & 0 & 0 \end{vmatrix} - s(\hat{i}x + \hat{j}y + \hat{k}z)$$

$$= -e(\hat{j}E_y + \hat{k}E_z) - e\begin{vmatrix} \hat{i} & \hat{j} & \hat{k} \\ \dot{x} & \dot{y} & \dot{z} \\ B & 0 & 0 \end{vmatrix} - s(\hat{i}x + \hat{j}y + \hat{k}z)$$

The component equations of motion are

$$m\ddot{x} = -e \times 0 - sx \Rightarrow \ddot{x} + \frac{s}{m}x = 0 \Rightarrow \ddot{x} + w_0^2 x = 0 \tag{11.5.1}$$

$$m\ddot{y} = -eE_y - eB\dot{z} - sy \Rightarrow \ddot{y} + \frac{Be}{m}\dot{z} + \frac{e}{m}E_y + \frac{s}{m}y = 0$$

$$\Rightarrow \ddot{y} + g\dot{z} + hE_y + w_0^2 y = 0 \tag{11.5.2}$$

$$m\ddot{z} = -eE_z + Be\dot{y} - sz \Rightarrow \ddot{z} - \frac{Be}{m}\dot{y} + \frac{e}{m}E_z + \frac{s}{m}z = 0$$

$$\Rightarrow \ddot{z} + g\dot{y} + hE_z + w_0^2 z = 0 \tag{11.5.3}$$

Multiplying (11.5.2) by j where $j^2 = -1$ and then adding to equation (11.5.3), we get,

$$(\ddot{z} + j\ddot{y}) + g(j\dot{z} - \dot{y}) + h(E_z + jE_y) + w_0^2(z + jy) = 0.$$

Setting $\xi = z + jy, j\xi = jz - y \Rightarrow j\dot{\xi} = j\dot{z} - \dot{y}$, we have,

$$\ddot{\xi} + jg\dot{\xi} + h(E_z + jE_y) + w_0^2\xi = 0 \tag{11.5.4}$$

Equation (11.5.2) multiplied by j and subtracting from (11.5.3), we get,

$$(\ddot{z} - j\ddot{y}) - g(\dot{y} + j\dot{z}) + h(E_z - jE_y) + w_0^2(z - jy) = 0.$$

Setting $z - jy = \eta, j\eta = jz + y \Rightarrow j\eta = j\dot{z} + \dot{y}$.

$$\therefore \quad \eta - jg\eta + h(E_z - jE_y) + w_0^2\eta = 0 \tag{11.5.5}$$

Let $E_+ = Ae^{j(\omega t - k_+x)} = E_z + jE_y,\ E_- = Ae^{j(\omega t - k_-x)} = E_z - jE_y$

$$\Rightarrow \quad E_z = \frac{1}{2}(E_+ + E_-) = \frac{A}{2}[e^{j(\omega t - k_+x)} + e^{i(\omega t - k_-x)};$$

$$E_y = \frac{1}{2j}A[e^{i(\omega t - k_+x)} - e^{j(\omega t - k_-x)}]$$

E_z is the real part of E_+ or E_-, we write,

$$E_z = \frac{A}{2}[\cos(\omega t - k_+x) + \cos(\omega t - k_-x)] \tag{11.5.6}$$

E_y can be taken as

$$E_y = \frac{A}{2}[\sin(\omega t - k_+x) - \sin(\omega t - k_-x)] \tag{11.5.7}$$

Take equation (11.5.6) once again, we get

$$\begin{aligned} E_z &= \frac{A}{2}[\cos(\omega t - k_+x) + \cos(\omega t - k_-x)] \\ &= A\cos\left\{\omega t - \frac{(k_+ + k_-)}{2}x\right\}\cos\frac{(k_-k_+)}{2}x. \end{aligned} \tag{11.5.8}$$

Similarly,

$$E_y = A\cos\left\{\omega t - \frac{(k_+ + k_{-1})}{2}x\right\}\sin\frac{(k_- - k_+)x}{2} \tag{11.5.9}$$

These relations obviously satisfy the boundary condition that at the incident face,

$$x = 0,\ E_y = 0,\ E_z = A\cos\omega t. \quad \therefore \quad \frac{E_y}{E_z} = \tan\frac{(k_- - k_+)x}{2}.$$

So as light travels through the medium, the electric vector rotates through an angle δ w.r.t. the original electric vector direction after traversing a length x of the medium given by

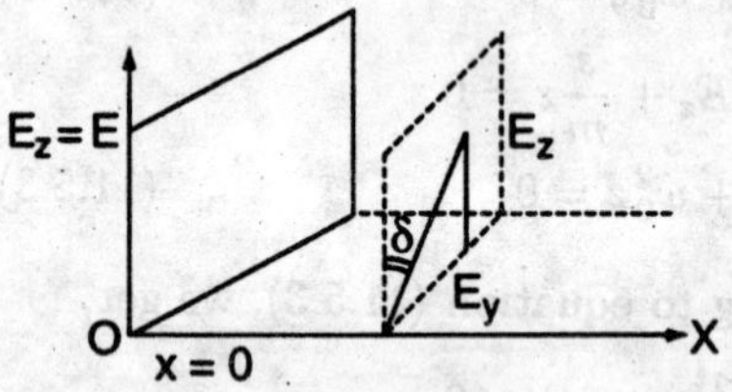

Fig 11.14

$$\delta = \left(\frac{k_- - k_+}{2}\right)x \tag{11.5.10}$$

Now, let us solve equation (11.5.4)

$$\ddot{\xi} + jg\dot{\xi} + h(E_z + jE_y) + w_0^2\xi = 0.$$

Try, $\xi = Be^{j(\omega t - k_+x)}$; $\dot{\xi} = jwBe^{j(\omega t - k_+x)}$; $\ddot{\xi} = j^2w^2Be^{j(\omega t - k_+x)}$

$$\therefore \quad (j^2w^2 + j^2gw + w_0^2)Be^{j(\omega t - k_+x)} = -hE_+ = -hAe^{j(\omega t - k_+x)}$$

$$\Rightarrow \quad \xi = \frac{-hE_+}{(w_0^2 - w^2 - gw)}.$$

Inside the medium, $D_+ = E_+ + 4\pi Ne\xi$, N being the number of electrons/c.c.
Also, $D_+ = KE_+$.

$$\therefore \quad KE_+ = E_+ + 4\pi Ne\xi$$

$$\Rightarrow \quad K = 1 + 4\pi Ne\frac{-h}{(w_0^2 - w^2 - gw)} = 1 - \frac{4\pi hNe}{(w_0^2 - w^2 - gw)} \tag{11.5.11}$$

Similarly, solving equation (11.5.5) taking $\eta = B'e^{j(\omega t - k_-x)}$

$$(j^2w^2 - jg(jw) + w_0^2)(B'e^{j(\omega t - k_-x)}) = -hE_-$$

$$\Rightarrow \quad \eta = \frac{-hE_-}{(w_0^2 - w^2 + gw)} \quad \therefore \quad KE_- = E_- + 4\pi Ne\eta$$

$$\Rightarrow \quad K = 1 - \frac{4\pi Neh}{(w_0^2 - w^2 + gw)} \tag{11.5.12}$$

Equations (11.5.10) and (11.5.11) can be put in one single equation as

$$K = 1 - \frac{4\pi Neh}{(w_0^2 - w^2 \mp gw)} \tag{11.5.13}$$

Now, $K = n_\pm^2(1 - i\beta)^2$ and for a transparent medium $\beta = 0$.

$$\therefore \quad K = n_\pm^2.$$

So we get,

$$K = n_\pm^2 = 1 - \frac{4\pi Neh}{(w_0^2 - w^2 \mp gw)}.$$

$$\therefore \quad n_+^2 = 1 - \frac{4\pi Neh}{(w_0^2 - w^2) - gw}, \quad n_-^2 = 1 - \frac{4\pi Neh}{(w_0^2 - w^2) + gw}.$$

$$\begin{aligned}
\therefore \quad n_+^2 - n_-^2 &= \frac{4\pi Neh}{(w_0^2 - w^2)^2 + gw} - \frac{4\pi Neh}{(w_0^2 - w^2)^2 - gw} \\
&= 4\pi Neh\left[\frac{(w_0^2 - w^2)^2 - gw - (w_0^2 - w^2)^2 - gw}{(w_0^2 - w^2)^2 - g^2w^2}\right] \\
&= -\frac{8\pi Nehgw}{(w_0^2 - w^2)^2 - g^2w^2}
\end{aligned}$$

$$\Rightarrow \quad (n_+ + n_-)(n_+ - n_-) = -\frac{8\pi Nehgw}{(w_0^2 - w^2)^2 - g^2w^2}.$$

Now, $n_+ + n_- = 2n$, where n is the refractive index of the medium in absence of the magnetic field.

$$\therefore \quad 2n(n_+ - n_-) = -\frac{8\pi Nehgw}{(w_0^2 - w^2)^2 - g^2w^2} \quad \Rightarrow \quad (n_- - n_+) = \frac{4\pi Nehgw}{n(w_0^2 - w^2)^2},$$

neglecting g^2w^2 in the denominator, compared to the first term in the denominator.

Now, $w_0 = \sqrt{s/m}$, $w = 2\pi\nu$, $= 2\pi\nu_0$, $g = \frac{Be}{m}$, $h = \frac{e}{m}$, then from above we get,

$$n_- - n_+ = \frac{4\pi Ne \times \frac{e}{m} \times \frac{Be}{m} \times 2\pi\nu}{4\pi^4 n(\nu_0^2 - \nu^2)^2} = \frac{BNe^3\nu}{2\pi^2 n(\nu_0^2 - \nu^2)^2 m^2} \tag{11.5.14}$$

Faraday rotation δ is given by (cf. equation 11.5.11)

$$\delta = \frac{(k_- - k_+)}{2} l,$$

where l is the length of the medium traversed.

$$n_+ = \frac{c}{v_+}, \quad n_- = \frac{c}{v_-},$$

where n_+ and n_- represent the r.i. values of the medium for the z and y-component of the electric wave.

$$v_+ = \nu\lambda_+ = \frac{2\pi\nu}{2\pi}\lambda_+ = \frac{w\lambda_+}{2\pi}; \quad v_- = \frac{w\lambda_-}{2\pi}.$$

$$\therefore \quad \frac{c}{n_+} = \frac{w\lambda_+}{2\pi} \quad \Rightarrow \quad n_+ = \frac{2\pi c}{w\lambda_+} \quad \Rightarrow \quad k_+ = \frac{wn_+}{c}.$$

Similarly,

$$n_- = \frac{2\pi c}{w\lambda_-} \quad \Rightarrow \quad k_- = \frac{wn_-}{c}$$

$$\Rightarrow \quad \delta = \frac{(k_- - k_+)}{2} l = \frac{w}{2c}(n_- - n_+)l = \frac{\nu}{c}(n_- - n_+)l.$$

Putting the value of $(n_- - n_+)$ from equation (11.5.14), we get

$$\delta = \frac{\nu}{c}\left\{\frac{BNe^3\nu}{2\pi^2 nm^2(\nu_0^2 - \nu^2)^2}\right\} = BlV, \quad \text{where } V = \text{Verdet's const.}$$

$$\therefore \quad V = \frac{Ne^3\nu^2}{2\pi^2 nm^2 c(\nu_0^2 - \nu^2)^2} \qquad (11.5.15)$$

References

1. **W.A. Shurcliff**—*Polarised Light Production and Use;* Cambridge, MA: Harvard University Press, 1962.
2. **P. S. Theocaris** and **E.E. Gdoutos**—*Matrix Theory of Photolasticity.*
3. **E. Charney**—*The Molecular Basis of Optical Activity: Optical Rotatory Dispersion and Circular Dichroism,* New York, John-Wiley and Sons, Inc., 1979.

Chapter 12

Laser Optics

12.1 Introduction

One gift of science that has found a good number of applications in the fields of medicine, biology, chemistry, metallurgy, radar and ranging; one device by which to produce an extraordinarily brilliant and sharply defined parallel beams of monochromatic and coherent light—is the LASER, an acronym for LIGHT AMPLIFICATION by STIMULATED EMISSION OF RADIATION.

We have come across different sources of light; sources giving radiation which is continuous over a large frequency range as in blackbody radiation, also sources of atomic emission confined to well-defined frequency intervals. So blackbody spectrum extends over a broad spectral range while atomic emission spectrum consists of emission lines in a narrow range. To get light of still narrower width, a suitable technique of amplification by stimulated emission of radiation is to be adopted giving rise to first MASER (microwave amplification by stimulated emission of radiation) and then LASER (light amplification by stimulated emission of radiation). MASER was introduced in 1954 by Gordon, Zeiger and Townes using NH_3 gas and the operating frequency was 2.4×10^{10} c/s. In 1958 A.L. Schawlow and C.H. Townes tried maser technique in the visible frequency range. It was Maiman who constructed the first optical Maser or Laser in 1960. It was a Ruby laser, made of Ruby doped with chromium. Ruby laser was followed by gas laser by Javan, Bennet and Herriot in 1961, semi-conductor laser in 1962. At present we have also constructed Dye laser.

Laser Principle

The idea of obtaining laser beam was hinted by Einstein first, with the introduction of *stimulated emission* in addition to the process of *spontaneous emission.*

The idea of spontaneous emission is not difficult. Every atom can remain in discrete energy levels or states without radiating energy. The minimum energy state, also called the ground state, is the stable state and the other states are referred to as *excited states.* The ground state is densely populated. An atom on absorption of energy is raised to one of the excited states and then falls spontaneously back to the ground state with emission of energy. In the lasing process, there is a third state, called *metatable state,* where an atom can stay longer than in the excited state. This is shown in fig. (12.1).

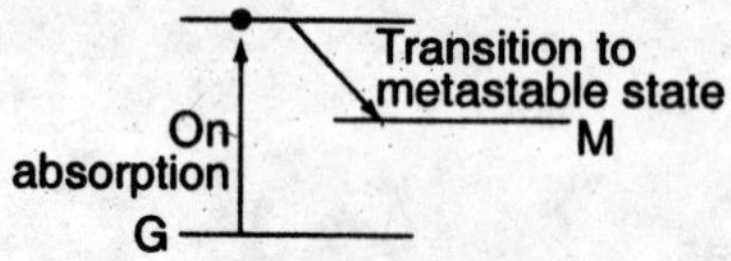

Fig 12.1

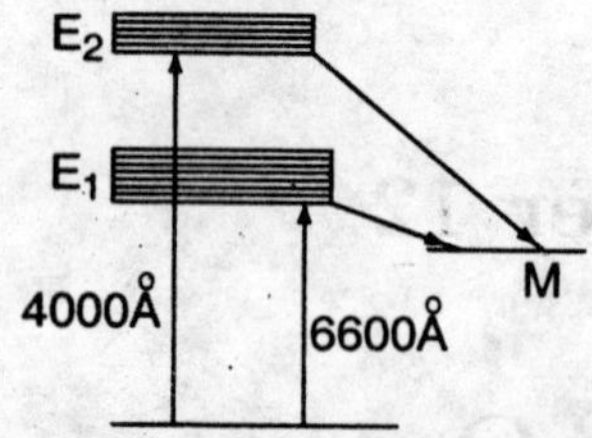

Fig 12.2

In ruby laser, chromium was doped. The energy states of chromium ions are as shown in fig. (12.1), gound state G, excited state E and the metastable state M. The life time of a chromium ion in E is of the order of 10^{-8} sec, while in M it is of the order of 10^{-3} sec. An ion of chromium absorbing a photon (having wavelength around 6600Å) makes a transition to one of states in E_1 band and on absorption of a photon of wavelength abound 4000Å, makes a transition to one of the states in E_2. In either case, there is immediate transition to the metastable state M in time $\simeq 10^{-9}$ sec. Since in the metastable state the lifetime is comparatively very long, by such non-radiative transition there is greater population in the metasble state than even in the ground state. This is referred to as *population inversion.* So the metastable state has more number of ions than that present in the gound state. All the ions in the metastable state have the same energy. If now a photon having the same energy is incident on the excited ions, the excited ions in their transition to the ground state emit radiation all of the same frequency, same phase and same direction as the incident photon. Thus we get light amplification due to transition of the excited ions in the metastable to the ground state in unison. The population inversion is carried through a process called optical pumping.

So for a laser, the essential components are:

(i) A laser material, a solid or a liquid or a gas;

(ii) Impurity atoms doped to provide metastable state;

(iii) An optical pumping device to produce population inversion;

(iv) An optical resonator to bring about stimulated emission in chain reaction;

(v) Cooling arrangement to prevent generation of heat.

A Solid Laser: Ruby Laser

The first Laser device was introduced by Maiman and it was a solid laser system. It consists of a long rod of synthetic ruby crystal doped with chromium. The ruby (Al_2O_3) crystal is the lasing material, the chromium atoms serving as the impurity atoms to provide metastable states. The rod ends are made precisely flat and parallel and partially silvered [fig. 12.3(a)].

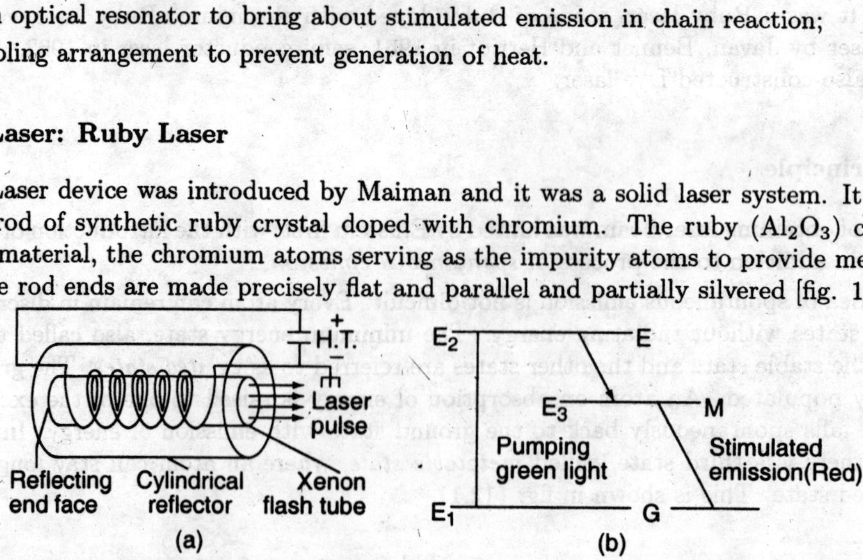

Fig 12.3

The rod is surrounded by a Xenon flash tube of spiral shape pumping green light. The whole arrangement is inside a cylindrical reflector and there is also a cooling device (not shown in the fig.) Due to the presence of chromium impurity, three energy states are provided, E_1, E_2 and E_3 (fig. 12.3b) respectively called the ground state G, excited state E and metable state M.

The Xenon flash lamp, the optical pump, produces green light exciting chromium atoms to the energy level E_2. The excited atoms by non-radiative transition transferring some of their energy to the crystal lattice pass on to the metastable state where they can relax for a comparatively longer time. Thus density of population in the metastable state is made greater than in the ground state and hence population inversion is achieved. If a photon obtained by spontaneous emission is made to strike an atom in the excited state, then, the excited atom is stimulated to emit a photon of the same energy as that of the stimulating photon. The energy of the stimulated photon is added to that of the stimulating photon. The stimulating and the stimulated photon are made to traverse the medium back and forth by reflection stimulating other excited atoms in the path to make them follow suit. It is some sort of chain reaction by which in a very short time energy density in the medium attains an enormous value when a giant pulse of short duration comes out. Thus output energy in pulses becomes greater than the input energy and thus light amplification by stimulated emission has been achieved.

Theory of a Laser

The idea of stimulated emission was first advanced by Einstein in 1917.

From an excited state an atom can make transition to the ground state in two ways, by spontaneous emission and by stimulated emission.

Consider fig. (12.4), where we have shown three states, ground state G, excited state E and a metastable state M. An atom on absorption of energy is raised to the state E and in its back journey emits radiation, this process being called *spontaneous emission.* In spontaneous emission no triggering is needed. It might so happen in some cases that the excited atom or ion imparts some energy to the lattice in which it is embeded and suffers non-radiative transition to the metastable state M. A triggering pulse initiates transition to the ground state emitting radiation of the same phase, same frequency and polarisation.

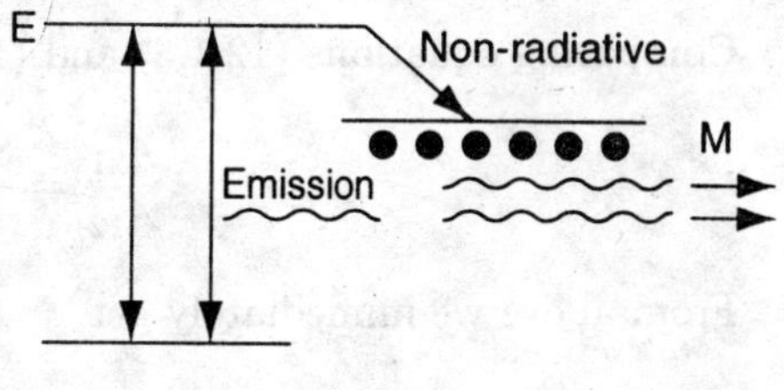

Fig 12.4

The rate of absorption of radiation of frequency ν effecting transition from G to E is proportional to the energy density of radiation $u(\nu)$. So the number of transitions per unit time per unit volume from G to E can be written as

$$T_1 = A_{12}N_1u(\nu) \tag{12.1.1}$$

where N_1 is the number of atoms/volume in state G.

The number of spontaneous emission per unit time per unit volume will be proportional to N_2 which represents the number of atoms per unit volume in state E.

$$\therefore \quad T_2 = A_{21}N_2 \tag{12.1.2}$$

This N_2 would also relax in the metasble state where from stimulated emission under the influence of stimulating photon would occur, so the number of stimulated emission per unit time per unit volume would depend on N_2 as well as energy density.

$$\therefore \quad T_3 = A_{31} N_2 u(\nu) \tag{12.1.3}$$

where the subscript 3 refers to the third metastable state.

$\therefore$ In thermal equilibrium,

$$N_1 A_{12} u(\nu) = A_{21} N_2 + A_{31} N_2 u(\nu) \tag{12.1.4}$$

$$\Rightarrow \quad (N_1 A_{12} - A_{31} N_2) u(\nu) = A_{21} N_2$$

$$\Rightarrow \quad u(\nu) = \frac{A_{21} N_2}{N_2 \left(A_{12} \frac{N_1}{N_2} - A_{31} \right)} = \frac{A_{21}}{\left(A_{12} \frac{N_1}{N_2} - A_{31} \right)} \tag{12.1.5}$$

These terms A_{11}, A_{21}, A_{31} represent Einstein's coefficients.

Also from Boltzmann's equation we get,

$$\frac{N_2}{N_1} = \exp\{-(E_2 - E_1)/kT\} \tag{12.1.6}$$

$$\therefore \quad u(\nu) = \frac{A_{21}}{A_{12} \exp\{(E_2 - E_1)/kT\} - A_{31}} \tag{12.1.7}$$

$$= \frac{A_{21}}{A_{12} e^{h\nu/kT} - A_{31}} = \frac{A_{21}}{A_{12}} \cdot \frac{1}{e^{h\nu/kT} - A_{31}/A_{12}} \tag{12.1.8}$$

From the radiation law of Planck, we get

$$u(\nu) = \frac{8\pi h \nu^3}{c^3} \cdot \frac{1}{e^{h\nu/kT} - 1} \tag{12.1.9}$$

Comparing equations (12.1.8) and (12.1.9), we definitely get,

$$\frac{A_{21}}{A_{12}} = \frac{8\pi h \nu^3}{c^3} \quad \text{and} \quad \frac{A_{31}}{A_{12}} = 1.$$

From above we immediately get,

$$A_{31} = A_{12} \tag{12.1.10}$$

$$\text{and} \quad \frac{8\pi h \nu^3}{c^3} A_{12} = A_{21} \tag{12.1.11}$$

These relations are called Einstein's relations.

Population Inversions Phenomenon

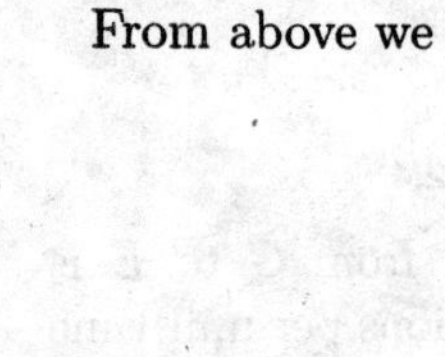

Fig 12.5

Depending on the value of the absorption coefficinet of the medium, the Laser source may be used as oscillator.

Consider the propagation of a collimated beam of light along X-direction. The change of light intensity as the collimated beam propagates through the layer of atoms bounded by the two planes of area S and distance of separation δx may be calculated.

Let $n_1(\nu)d\nu$ represent the number of atoms per unit volume capable of transition from the plane at $x = x$ to a higher energy plane at $x = x + \delta x$ on absorption of radiation having frequency between ν and $\nu + d\nu$ per unit time, such that $\int n_1(\nu)d\nu = N_1$.

Then the rate of transitions from 1 to 2, can be written as

$$n_1(\nu)d\nu \times A_{12} \times u(\nu)S\delta x \tag{12.1.12}$$

Each transition is through an absorption of a photon of energy $h\nu$, so the energy decrease per sec would be

$$h\nu \times n_1(\nu)A_{12}u(\nu)d\nu S\delta x \tag{12.1.13}$$

Let $n_2(\nu)d\nu$ represent the number density of atoms which are capable of emitting radiation of frequency between ν and $\nu + d\nu$. Then the number of stimulated emissions/sec would be

$$n_2(\nu)d\nu \times A_{31}u(\nu)S\delta x \tag{12.1.14}$$

Each emission is accompanied by a release of energy $h\nu$, thus gain of energy/sec, is given by

$$h\nu \times [n_2(\nu)A_{31}u(\nu)d\nu]S\delta x \tag{12.1.15}$$

$u(\nu)d\nu S\delta x$ being the energy of radiation contained within the volume $S\delta x$, then

$$\begin{aligned} \frac{d}{dt}[u(\nu)d\nu S\delta x] &= h\nu[n_2(\nu)A_{31}u(\nu)d\nu]S\delta x - h\nu n_1(\nu)A_{12}u(\nu)d\nu]S\delta x \\ \Rightarrow \quad -\frac{d}{dt}[u(\nu)d\nu S\delta x] &= -h\nu[n_2(\nu)A_{31}u(\nu)d\nu - n_1(\nu)A_{12}u(\nu)d\nu]S\delta x \end{aligned} \tag{12.1.16}$$

Now we would introduce the term intensity of the beam. $I_\nu(x)$ denotes the intensity at the section at $x = x$, it means then the rate of energy flow across it for radiation of frequency ν. So the rate of energy flow out of the volume under consideration is

$$[-I_\nu(x + \delta x) + I_\nu(x)]Sd\nu$$

which when Taylor expanded would land to

$$-S\frac{\partial I_\nu}{\partial x}\delta x d\nu \tag{12.1.17}$$

Equating (12.1.16) and (12.1.17), we at once get,

$$\begin{aligned} -S\frac{\partial I_\nu}{\partial x}\delta x d\nu &= -h\nu[n_2(\nu)A_{31}u(\nu)d\nu - n_1(\nu)A_{12}u(\nu)d\nu]S\delta x \\ \Rightarrow \quad \frac{\partial I_\nu}{\partial x} &= h\nu[n_2(\nu)A_{31}u(\nu) - n_1(\nu)A_{12}u(\nu)] \\ &= h\nu[n_2(\nu)A_{12} - n_1(\nu)A_{12}]u(\nu) = -h\nu A_{12}[n_1(\nu) - n_2(\nu)]u(\nu) \end{aligned} \tag{12.1.18}$$

where we have used equation (12.1.20).

Now, $u(\nu) \times c = I_\nu$, c respresenting velocity of propagation in the medium. Also, $\frac{c_0}{c}\mu$ (r.i.), so we get,

$$\begin{aligned} \frac{\partial I_\nu}{\partial x} &= -A_{12}h\nu[n_1(\nu) - n_2(\nu)]\frac{I_\nu}{c} = -I_\nu A_{12}\frac{h\nu\mu}{c_0}[n_1(\nu) - n_2(\nu)] \\ \Rightarrow \quad \frac{1}{I_\nu}\frac{\partial I_\nu}{\partial x} &= -\frac{h\nu A_{12}\mu}{c_0}[n_1(\nu) - n_2(\nu)] \end{aligned} \tag{12.1.19}$$

Also, we have,

$$I_\nu(x) = I_\nu(x=0)e^{-\alpha_\nu x} \quad \Rightarrow \quad \frac{\partial I_\nu}{\partial x} = -\alpha_\nu I_\nu(x=0)e^{-\alpha_\nu x} = -\alpha_\nu I_\nu \tag{12.1.20}$$

Comparing equations (12.1.19) and (12.1.20), we immediately get,

$$\alpha_\nu = \frac{h\nu A_{12}\mu}{c_0}[n_1(\nu) - n_2(\nu)] \tag{12.1.21}$$

Ordinarily, $n_1(\nu) > n_2(\nu) \Rightarrow a_\nu$ is positive which means absorption of the beam as it propagates thro' the medium.

If on the other hand, $n_1(\nu) < n_2(\nu)$, it would imply α_ν negative, it would mean that the beam would be amplified. Since under this condition population in the higher energy state is greater, this is sometimes referred to as *population inversion*.

Reference

1. **Allen, L.:** (1969) *Essentials of Laser*, Pergamon Press, Oxford.
2. **Ghatak, A.:** (1977) *Optics,* Tata McGraw-Hill Publishing Co., Ltd.
3. **Schawlow, A.:** (1969) *Lasers and Light,* A Collection of Scientific Americans Articles.

Chapter 13

Holography and Fibre Optics

13.1 Introduction

With the development of Laser, two important branches were gaining ground, one is the *holography*, the other is the *fibre communication optics*.

Three-dimensional recording of an object is usually referred to as the holography. Though originally thought of by Gabor (1949) to get the image by reconstruction of wavefronts, the technique of holography had a rapid development leading to 3-D photography with coherent Laser beam.

Laser initiated another important branch in communication engineering—fibre communication. Prior to the introduction of fibre optics, electromagnetic waves having a frequency spectrum down to microwave region were used for communication purpose. It is known that the amount of information carried by electromagnetic wave is proportional to the frequency of the wave. A wave having a frequency of 100 can have 100 pulses at best, hence can transmit 100 pulses per sec. But with Laser through optical fibre it was possible to transmit 44.7 million pulses per sec.

So these two branches had a rapid and radical progress with the development of Laser. In this chapter we will give a short account of these two aspects of modern optics.

13.2 Holography

Ordinary photograph gives the measure of light intensity from different parts of the object photographed. So it carries information about the mean square value of the amplitude of the original light wave from the object. So it loses information about the phase of the wave arriving at the film from different points of the object. But in holography both amplitude and phase of light waves are recorded on the film. Such a photograph will be called a *hologram,* greek word 'holos' means whole' a hologram, therefore, conveys whole information about the light wave, its phase as well as its amplitude. Originally the idea was advanced by Gabor in a paper entitled 'microscopy by reconstructed wavefronts', but practical realisation in a most satisfactory way was not possible before Leith and Upatnieks succeeded in devising holograms illuminated with coherent Laser light.

Basic Principle

The point object A is embeded in a parallel beam of coherent light which fall on the photo-plate PQ. The diffracted light from the object A surperposes on the coherent incident light producing a hologram on the photo-plate. (fig. 13.1)

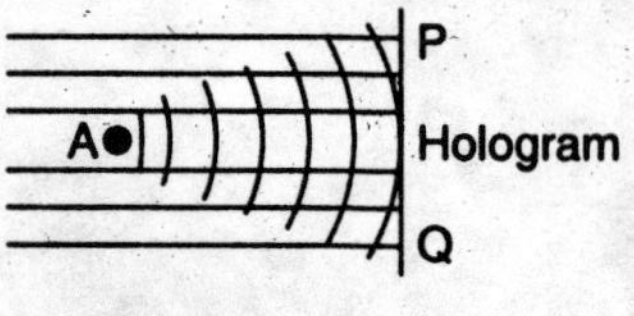

Fig 13.1

The hologram consists of a speckled pattern of fine dots; when such a photo-plate or a hologram is illuminated by Laser light, the object is reconstructed. The reconstruction is represented in fig. (13.2) where we have a real image as well as a virtual image.

But a slightly off-axial arrangement would make it possible to see virtual image without seeing the real image.

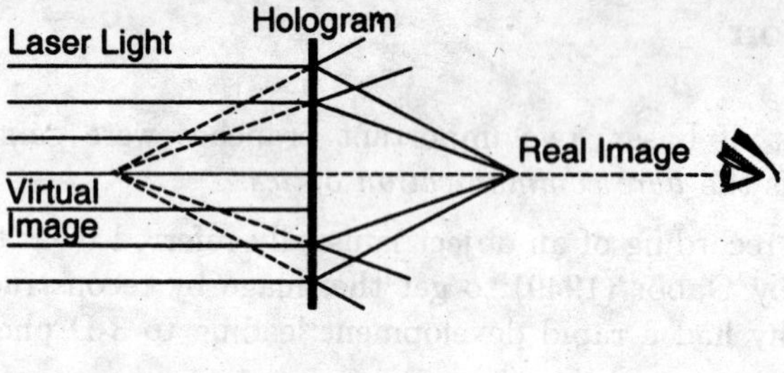

Fig 13.2

Making a Hologram

The technique of holography consists of two operations (i) optical analysis, i.e., making a hologram of the object, (ii) optical synthesis, i.e., reconstructing the image of the original object. Analysis and synthesis can be both performed by light of the same wavelength; sometimes, the synthesis is carried through by longer wavelength whereas analysis by smaller one.

In fig. (13.3) we have shown an arrangement for making a hologram.

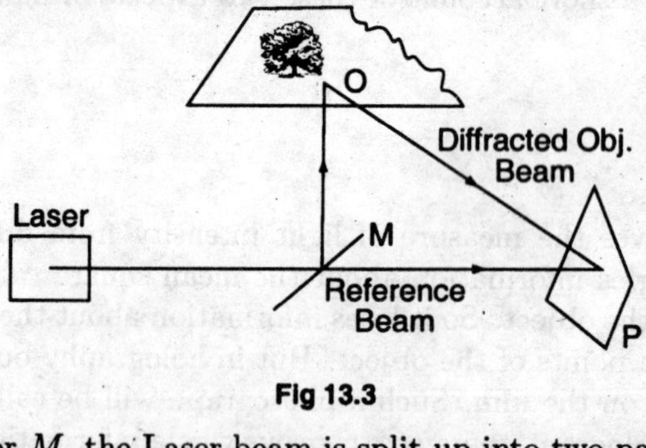

Fig 13.3

By a half-silvered mirror M, the Laser beam is split up into two parts (i) reflected towards the object O, (ii) transmitted through the mirror straight to the plate. The object diffracts the beam incident on to it and this diffracted beam then travels towards the plate P. Thus the photo-plate is illuminated by the 'object beam' diffracted or scattered by the object and the direct beam, called the reference beam. The reference beam is equivalent to 'carrier wave' in electrical communication. As in broadcasting the audio-frequency currents are used to modulate the carrier wave, here in making a hologram, the optical analogue, the reference beam is modulated by interference with the object beam at the photo-plate. The numerous points making up the image on P are formed due to interference between the coherent object beam and reference beam. The hologram would consist of numerous closely spaced points, invisible to the naked eye but can be made visible under powerful microscope.

Reconstruction of the Image from the Hologram

In fig. (13.4) we have shown an arrangement to reconstruct the image from the hologram. This is a reverse of making a hologram.

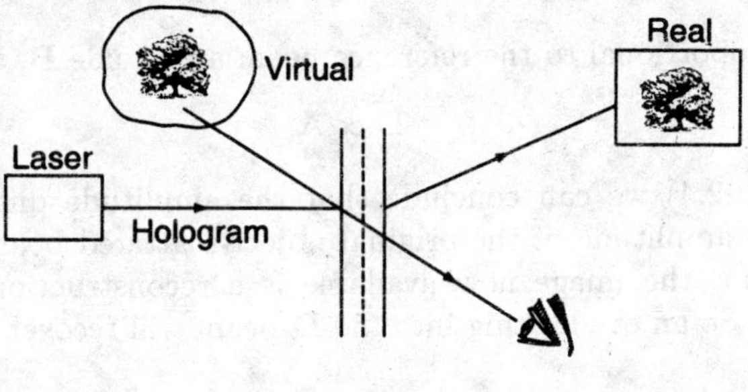

Fig 13.4

The hologram is illuminated by coherent light from a Laser. The waves diffracted through the hologram carry the phase and amplitude of the waves originally diffracted by the object when the hologram was made. The object wavefronts have thus been reconstructed. One of the diffracted beams forms real image, while another forms a virtual image as shown in fig. (13.4).

The remarkable feature of a hologram is that by moving the head while looking through it one can see more of the object. Thus a hologram may be taken as a record of a three-dimensional view on a two-dimensional photo-plate. Moreover, if a hologram is cut into small pieces, each piece will reproduce the whole image.

Mathematical Theory

Holography is a matter of interference. The interference occurs between the object beam, which is scattered by the object and the reference beam which directly reaches the emulsion.

The photographic emulsion responds to energies and so we can write $I(x, y) = (\mathbf{A} + \mathbf{B})^2$, where **A** and **B** are both complex. If $\mathbf{A}^*$ and $\mathbf{B}^*$ represent complex conjugates to A and B respectively, then

$$\mathbf{I(y, x)} = (\mathbf{A} + \mathbf{B})(\mathbf{A}^* + \mathbf{B}^*) = |\mathbf{A}|^2 + |\mathbf{B}|^2 + \mathbf{AB}^* + \mathbf{BA}^* \tag{13.2.1}$$

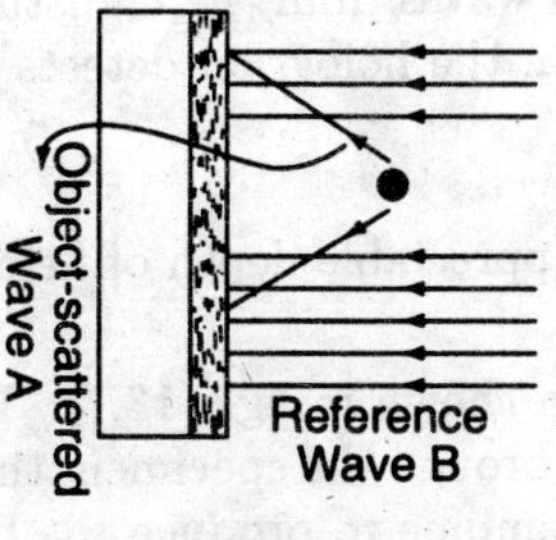

Fig 13.5

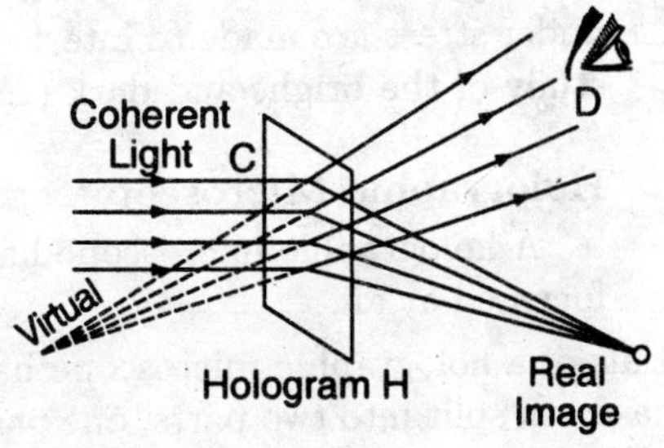

Fig 13.6

In the wave reconstruction from a hologram, light of amplitude **C** illuminates the hologram H, that contains the recorded information while making the hologram. The last two terms in equation (13.2.1) represents the interference related terms. So the transmittance function of the hologram is

$$\mathbf{T(y,z) = AB^* + BA^*} \tag{13.2.2}$$

The incident illuminating beam **C** is diffracted by the hologram into the image reconstruction beam **D** given by

$$\mathbf{D = CT(y,z)} \quad \Rightarrow \quad \mathbf{D \propto (CAB^* + CBA^*)} \tag{13.2.3}$$

If **C** is equal or proportional to the reference beam amplitude **B**, then

$$\mathbf{D \propto A} \tag{13.2.4}$$

From equation (13.2.4) we can conclude that the amplitude diffracted by the hologram is proportional to the amplitude of the original object-scattered beam utilised in the making of the hologram. Thus the image now available is a reconstruction of the object-beam in amplitude and phase. So an eye looking into the **D**-beam will recover the object with its three dimensionality.

Applications of Holography

As already remarked with the introduction of Lasers, the holographic process is gaining grounds. There is a long list of applications of holography, only a few of them will be mentioned here. For more details the readers are advised to consult the references given at the end of this chapter.

Holographic Interferometry

This is one of the important applications of holography. Holographic interferometry falls into three classes—*real time, time lapse* and *time averaged* interferometry.

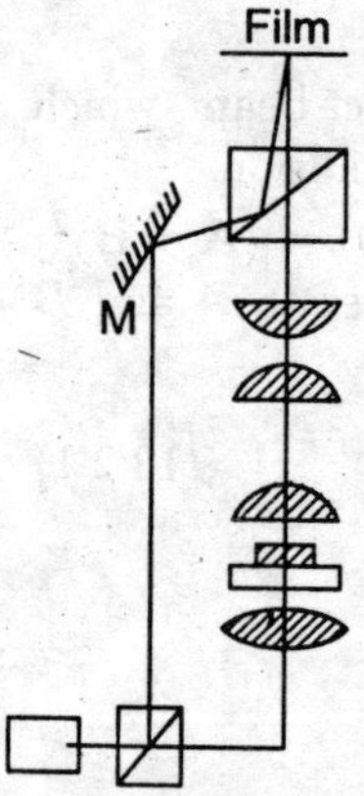

Fig 13.7

In real time holographic interferometry a hologram of an object is made in the usual way. The object is then changed or distorted in some way. This changed object is then placed in the position of the virtual image in the reconstruction arrangement. An observer looking through the hologram will see the original object superimposed on the modified object. Since the two are viewed in coherent light, difference fringes would appear that would indicate the changes the object has undergone since making the hologram. This interferometric method finds an application in detecting defects in air-craft types. In this case coherent waves from the tyre surface momentarity under stress are made to interfere with waves from the undisturbed surface. A study of the bright and dark regions on the hologram detects flaws.

Holographic Microscopy

A holographic microscope has an appreciable depth of field, it need not be focussed at all.

The essentials of a holographic microscope has been shown in fig. (13.7). The light from a Laser source has been split into two parts, one passing through the specimen, the other through a different path via the mirror M. The two beams recombine to produce the hologram on the film. The reconstructed image can be seen in any cross-section desired.

Fibre Optics

From ancient times we tried to design and improve communication system by which messages are sent over long distances. For every communication system there must be a transmitter to which message from a message source is put in; the transmitter then couples this message onto a transmission channel in the form of a signal and the channel bridges the gap between the transmitter and the receiver. This channel may be either a guided transmission line such as a wire or a waveguide. In such cases of transmission line signal gets progressively attenuated and distorted. So improvement in the communication process would mean motivation to improve the transmission fidelity and at the same time to improve the data rate of transmission. Starting from the ancient crude method of using signal lamp and passing through the discovery of Morse system of telegraphy and then through electromagnetic transmission, we have now reached the age of optical fibre communication system.

A wave that has a maximum of 100 pulses can only transmit a maximum of 100 pulses per sec. With the introduction of Laser light and fibre as the transmission medium, a single fibre can carry 44.7 million pulses per second. What an amazingly large data rate of transmission now possible! This communication mechanism has now been pssible through the works of Kao and Hockman and A. Werts and most recently by Kapron, Keck and Maurer.

The essential components of an optical fibre communication link is shown in fig. (13.8) by block diagram.

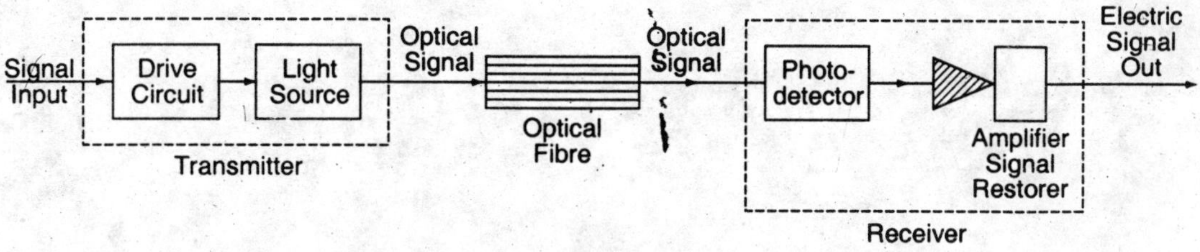

Fig 13.8

So in this transmission link one essential component is the optical fibre about which some discussion is needed.

Optical fibres are fibre guides which are of two types—one is the *step-index fibres* and the other the GRIN (graded refractive index) fibre. Step-index fibres consist of a transparent core of glass or plastic of certain r.i. surrounded by a cladding layer of lower r.i. Most of the light travels through the core and is confined there by total internal reflection.

Refer to fig. (13.9) where we have shown a step-index fibre; n_1 is the r.i. of the core, n_2 is that of the cladding material. Let n, be the r.i. of the outside medium. The angles α, β and γ are as shown. To contain light within the core, the angle of incidence at the core-clad interface must not be less than the value given by

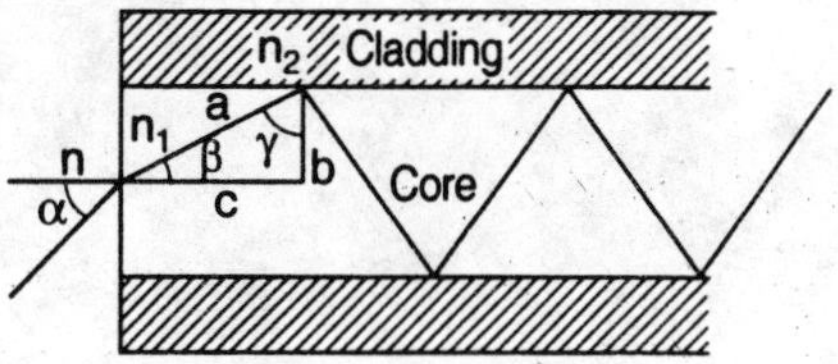

Fig 13.9

$$\sin\gamma = \frac{n_2}{n_1} = \frac{c}{a}.$$

Again,

$$a^2 = c^2 + b^2 \quad \Rightarrow \quad 1 = \frac{c^2}{a^2} + \frac{b^2}{a^2} = \frac{n_2^2}{n_1^2} - \sin^2\beta \quad \Rightarrow \quad \sin\beta = \frac{\sqrt{n_1^2 - n_2^2}}{n_1}.$$

From Snell's law, $n\sin\alpha = n_1\sin\beta = \sqrt{n_1^2 - n_2^2} n\sin\alpha$ is the numerical aperture of the fibre.

In a GRIN fibre, on the other hand, the r.i. is a function of the radial distance and decreases with it w.r.t. the axes. Light travels along the axis and if it deviates from the axis, it is returned by the gradient.

In case of a number of media of r.i. values n_1, n_2, n_3, etc., by Snell's law we would get, $n_1 \sin \alpha_1 = n_2 \sin \alpha_2 = n_3 \sin \alpha_3$, etc.

Thus, at the nth boundary, at a distance r from the axis, we can write, $n(r) \sin \alpha(r) =$ constant.

So, if n decreases, α increases and light is returned to the axis. In it all rays travelling through have the same optical path. The more detailed theory of refractive index gradient has been developed in Geometrical Optics (Part I) of our book on optics.

Mixed Theory Questions Exercises

1. What are cardinal points? Define principal points and nodal points? Deduce Newton's formula using cardinal points.

2. Show that the planar surface of a plano-convex or a plano-concave lens does not contribute to the system matrix.

3. (a) What is a system matrix? Obtain the system matrix for a thick lens.

   (b) Using the system matrix as in (a), prove the following relation for a thin lens:

$$\frac{1}{v} - \frac{1}{u} = \frac{1}{f}, \text{ where } \frac{1}{f} = (n-1)\left(\frac{1}{R_1} - \frac{1}{R_2}\right).$$

4. Two thin lenses, one convex of focal length 20 cm, the other concave of focal length 10 cm placed at a distance of 8 cm apart form a coaxial system. An object 1 cm tall is at a distance of 40 cm from the convex lens, the whole system is an air.

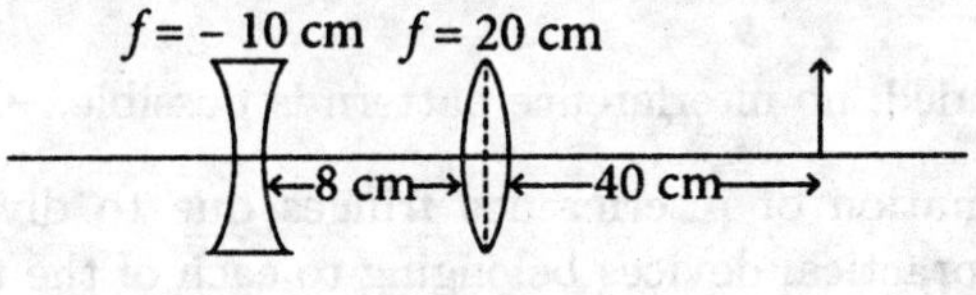

Fig. ME 1

   Find (i) the system matrix, (ii) the equivalent focal length of the system, (iii) the position and size of the image.

5. The magnitudes of the radii of curvature of the two surfaces of a thick equiconvex lens (r.i. = 1.5) are 4 cm each. The thickness of the lens is 1 cm and it is placed in air. Determine the focal length, the positions of unit planes and nodal planes.

6. What is aberration of a lens? How many types of aberration are there in case of a lens? What is Seidel aberration?

7. What do you mean by chromatic aberrations? Find the condition of achromatism of two thin lenses (a) in contact (b) separated by a small distance apart in air.

8. What are the nature and causes of different Seidal aberrations? State the various methods for reducing these aberrations?

9. Find the positions of cardinal points of (a) Ramsden's eyepiece (b) Huygens' eyepiece. Find also their equivalent focal lengths.

10. What kind of achromatism is required in (i) photographic camera? (ii) eyepiece? (iii) telescope objective? Which error of the image is most harmful in case of the objective of a high power microscope?

11. Describe Huygens' and Ramsden's eyepiece. Develop their system matrices and hence find their focal lengths.

12. Explain the terms (i) ray (ii) wavefront. What is the frequency range that is visible to an average human being?

13. What do you mean by 'optical path' in a certain medium? A ray of light traverses certain geometrical path in an optical medium of r.i. n. What is the equivalent air-path travelled?

14. State and explain Huygens' principle of wave propagation. Applying this principle explain the phenomena of reflection and refraction of plane waves at plane surface.

15. Explain the phenomenon of interference of light. State the conditions to be satisfied for the production of sustained interference fringe pattern.

16. Show that the dark and bright fringes produced in Young's experiment are equispaced. Find expression for fringe width.

17. For permanent interference fringe pattern, the sources must satisfy the conditions of spatial coherency as well as temporal coherency. Explain the meaning of the two types of coherency.

18. If source is extended, no interference pattern is possible.—Explain.

19. Discuss the formation of interference fringes due to division of (i) wavefront (ii) amplitude. Cite practical devices belonging to each of the two categories.

20. Find the intensity distribution function (IDF) in case of Young's Double-slit experiment and hence derive expression for the fringe width.

21. What is your argument for interferrometric confirmation of the fact that light travels with greater speed in air than in any medium?

22. Two coherent sources of different intensities send wave which interfere. In the fringe pattern, the ratio $\frac{I_{max}}{I_{min}} = 25$. What is the ratio of the intensities of the two sources?

23. The inclined faces of a biprism of r.i. 1.6 make an angle of 2° with its base. A slit illuminated by a monochromatic light is placed at a distance of 0.1 metre from the biprism. Fringe width of 0.24 mm is observed on a screen at a distance of 1.9 m from the biprism. Determine the wavelength of light used.

24. "When a drop of oil is spread on a water surface, it displays beautiful in daylight".— Explain.

25. In Young's double-slit experiment, a thin glass plate is introduced in one of the interfering beam-path—explain the changes you will observe.

26. Describe Newton's ring experiment. Why do we use a convex lens of large radius of curvature for producing Newton's rings?

27. Flatness of a surface can be tested by interferometric method—explain how?

28. Consider a plane which is normal to the line joining the two point coherent sources S_1 and S_2 in Young's experiment. Analyse the shape of the interference fringes?

29. How do you get, coherent sources are produced in (i) Newton's rings arrangement? (ii) biprism arrangement? In Young's double-slit experiment, show that half-width of the intensity distribution curve is (λ/d), where d is the separation distance between two virtual coherent sources and λ is the wavelength of light used.

30. Explain how interference fringes are formed by a thin wedge shaped film when examined by normally reflected light. Derive expression for fringe width. How would you estimate the difference of film thickness between two points?

31. How can you measure thickness of a thin film using Fresnel's biprism? Develop the necessary theoretical background. Hence, prove that the velocity of light is smaller in a denser medium.

32. Why are the outer surfaces of the plates of a Fabry-Perot interferometer kept inclined? What are Haidinger fringes? Prove that in the fringe system formed in a Fabry-Perot interferometer, the ratio of the intensity of maxima to the intensity midway between the minima is given by $\dfrac{(1+r^2)^2}{(1-r^2)^2}$, where r is the the reflection coefficient.

33. What is the meaning of the terms (i) finesse coefficient and (ii) finesse. Discuss the effect of the reflection coefficient on the sharpness of the rings.

34. What is diffraction? What is the significance of this phenomenon? Discuss the difference between the two phenomena: (i) Interference (ii) Diffraction.

35. How many types of diffraction phenomenon are there? Explain the differences between them.

36. Show that the areas of various half-period zones are nearly equal and are independent of the order no. of zones.

37. How do you compare the function of a zone plate with that of a converging lens? Show that the amplitude of a complete wavefront at a point is half of what would be caused by the first zone.

38. A plane wavefront of $\lambda = 6000$ Å advancing towards a point is divided into a number of half-period zones. The amplitude contribution of these half-period zones is given by 1.00, 0.98, 0.96 up to zero. Compare the intensities at the point when first 31 and 36 half-period zones only are exposed respectively.

39. Find the IDF of Fraunhofer diffraction pattern due to a single slit. Also find the condition for (i) principal maxima (ii) secondary maxima and minima. Plot the intensity distribution curve.

40. What do you mean by plane transmission grating? What is its difference from crystal grating?

A plane diffraction grating at normal incidence gives a green line $\lambda_1 = 5.4 \times 10^{-5}$ cm in a certain spectral order superposed on a violet line $\lambda_2 = 4.05 \times 10^{-5}$ cm of the next higher order. If the angle of diffraction $\theta = 30°$, find the grating constant.

41. In a multiple diffraction pattern produced by N slits, find the expression for width of the principal maxima.

42. What do you mean by absent spectra, and ghost line? In a multiple slits (N) diffraction pattern between two principal maxima, there are $(N - 1)$ minima—explain and analytically establish it.

43. What is resolving power of an optical instrument? State and explain Rayleigh criterion of resolution.

44. Derive an expression for the resolving power of a microscope. What sets the limit of resolution of a microscope? Explain why in microscope of high resolution, the space between the object and the objective is filled with cedar oil.

45. Derive an expression for the resolving power of a grating. Compare between grating spectra and prism spectra.

46. A transmission is 4 cm long and it has 4000 lines/cm. Compute the resolving power of the grating for $\lambda = 5900$ Å in the first order. Can this grating resolve the two lines in the Na-doublet of $\lambda_1 = 5890$ Å and $\lambda_2 = 5896$ Å?

47. Distinguish between the dispersive power and resolving power of a grating.

48. In the second order spectrum of a plane diffraction grating, a certain spectral line appears at an angle of 10° while another line of wavelength 0.5×10^{-10} m appears at an angle 3″ greater. Find the wavelength of the lines and the minimum grating width required to resolve them. Given, $\sin 10° = 0.1736$ and $\cos 10° = 0.9848$.

49. What do you mean by polarised light? What is plane of polarisation and plane of vibration for a polarised light?

50. How can you get polarised light by (i) reflection (ii) refraction? What is Brewster's law. State and explain Malus' law.

51. What is birefringence? How does it explain double refraction?

52. Define (a) Optic axis (b) Principal section of the crystal (iii) Principal plane.

Following Huygens' construction discuss the nature of wave surfaces in a uniaxial crystal.

53. How do you construct a Nicol prism and show how a plane polarised light can be obtained by using it.

54. Show how a Nicol prism can be used as a polariser or as an analyser.

55. What is a Babinet's compensator? How to use it to produce polarised light?

56. There are at least a few mathematical formalisms to deal with the problems of polarisation. Of them the simplest is the Jones formalism. What is the essence of this formalism?

57. What is a Jones vector and what is a Jones matrix? In this formalism a matrix equation is represented as $\begin{bmatrix} a & b \\ c & d \end{bmatrix}\begin{bmatrix} A \\ B \end{bmatrix} = \begin{bmatrix} C \\ D \end{bmatrix}$.

Interpret the above matrices.

58. Represent Jones matrix which acts as a linear polariser matrix $\frac{1}{2}\begin{bmatrix} 1 & 1 \\ 1 & 1 \end{bmatrix}, \frac{1}{2}\begin{bmatrix} 1 & -1 \\ -1 & 1 \end{bmatrix}$.

59. Develop Jones matrix for circular polariser, elliptic polariser. Show that these matrices operate on the plane polarised light at 45° and we get circularly polarised light and elliptically polarised light.

60. What happens when

(i) $\begin{bmatrix} 1 & 0 \\ 1 & i \end{bmatrix} \frac{1}{\sqrt{2}} \begin{bmatrix} 1 \\ i \end{bmatrix} = [?]$ (ii) $\begin{bmatrix} 0 & 0 \\ 0 & 1 \end{bmatrix}\begin{bmatrix} 1 \\ 0 \end{bmatrix} = [?]$.

61. Explain (i) $\begin{bmatrix} 1 & 0 \\ 0 & 0 \end{bmatrix}\begin{bmatrix} 1 \\ 0 \end{bmatrix} = \begin{bmatrix} 1 \\ 0 \end{bmatrix}$ (ii) $\begin{bmatrix} 1 & 0 \\ 0 & i \end{bmatrix} \frac{1}{\sqrt{2}} \begin{bmatrix} 1 \\ i \end{bmatrix} = \frac{1}{\sqrt{2}}\begin{bmatrix} 1 \\ -1 \end{bmatrix}$.

62. What conclusion can you draw about the state of polarisation when the x and y components of the electric fields are as below?

(i) $E_x = E_0 \sin(\omega t + kz), E_y = E_0 \cos(\omega t + kz)$

(ii) $E_x = E_0 \cos(\omega t + kz), E_y = \frac{E_0}{\sqrt{2}} \cos(\omega t + kz + \pi)$.

63. Two linearly polarised light waves are in phase but have different amplitudes. They are represented by

$$\vec{E}_1(z,t) = \hat{\imath} A_1 \cos(kz - \omega t) + \hat{\jmath} B_1 \cos(kz - \omega t)$$
$$\vec{E}_2(z,t) = \hat{\imath} A_2 \cos(kz - \omega t) + \hat{\jmath} B_2 \cos(kz - \omega t).$$

Show that $\vec{E} = \vec{E}_1 + \vec{E}_2$ is also linearly polarised. Find polarisation direction.

64. If n_O is the refractive index for O-rays, n_e is the principal refractive index for E-rays, then show that the refractive index n_θ for E-rays in a direction θ with optic axis is given by

$$\frac{1}{n_\theta^2} = \frac{\cos^2\theta}{n_\theta^2} + \frac{\sin^2\theta}{n_e^2}.$$

Bibliography

1. **A. Millikan (1868-1953):** Robert Andrews Milikan grew up in American Midwest. He was the only physics student at Columbia University during 1893-95. He joined (1896) Michelson in his pioneering experimental project in Chicago until he went to the California Institute of Technology in 1921.

   Millikan was an experimental physicist who was famous for his experiment on electronic charge determination popularly known as 'Oil drop experiment' performed during the period 1909-13. The accuracy of his experimental results on electronic charge measurement was unparalleled and remained unsurpassed till 1928. He extensively and assiduously studied photoelectric effect (1912-16). He himself wrote, 'I spent ten years of my life testing the 1905 equations of Einstein and contrary to all my expectations I was compelled to assent to its unambiguous verification in spite of its unreasonableness since it seems to violate everything we know about the interference of light'. For his experimental skill and accuracy of measurement of e and h, he was awarded Nobel prize in 1923 for Physics.

2. **Albert Einstein (1879-1947):** Here is a name which shines and shines brightly like a brilliant star in the sky of science. Albert Einstein was a German-Swiss Theoretical Physicist who was very famous for his Theory of Relativity of Space and time through his Special Theory and of Gravitation through his general theory.

   Einstein was born much ahead of his time. So, he was not properly recognised during his early academic career, rather I should say that 'his morning did not show his day'. His academic results did not show that glitter of a diamond as he really was. So after completion of studies in Swiss Institute, he had to join a petty post in a patent office and that too at the recommendation of someone. What an irony of fate it was! But brilliance cannot be shadowed by any means. Here also it came to be true.

   While working in the patent office, Einstein's brilliance first shot out in the form of his first trio-series of papers of theoretical reasoning which was published in 1905, one on Brownian motion, the other on Photoelectric effect and the third on the special theory of relativity. This first two phenomena, which stood out as unsolved scientific riddle so long, came to be solved by Einstein using the revolutionary idea of Planck's energy quantisation principle.

   The third paper contained the seed of a new era of scientific thought processing, it was the beginning of the era of Relativity. This special theory of relativity tried to advocate that space is not absolute, time is not absolute, even the sequence of events is not absolute. The emergence of this new concept of relativity of space and time was so fundamental and profound that it shook the foundation of Newtonian frame of classical mechanics. The change in ideas from Newtonian to Einsteinian is not just a mere change but Einsteinian concept brought an upheaval, an earthquake in the foundation of classical physics.

It is true that all the predictions of Special Theory of Einstein could not be verified experimentally during the lifetime of Einstein but with time as more and more fast moving ventricular space-craft came into existence, more and more predictions came into experimental confidence. The tremendous amount of energy released due to nuclear fission or fusion based on Einstein's relation $\Delta E = \Delta mc^2$ is now an experimental truth; the devastation of Nagasaki and Hiroshima due to release of huge amout of nuclear energy is already a historical fact of colossal wastage of power due to Einstein's mass-energy relation.

The general theory of relativity is really Einstein's theory of Gravitation which appeared in 1915, but he was awarded the Nobel Prize in 1921 for his services to Theoretical Physics and especially for his discovery of the law of photoelectric effect. Nobody knows why so much of delay in declaring Nobel prize for him. It might be due to the fact that his idea was much ahead of his time. Einstein himself observed, 'Discovery in the grand manner is for young people ... and hence for me a thing of the past'.

3. **Amedeo Avogadro (1776-1856):** Born in Italy before unification, Avogadro was a Lawyer at an early age of sixteen, turned into a physicist-cum-chemist; he proposed the law that equal volumes of all gases under the same condition of pressure and temperature contained the same number of molecules. It was Amedeo Avogadro who first introduced the difference between atoms and molecules. To honour him we have introduced a number, a universal constant called Avogadro number $N_A = 6.025 \times 10^{23}$ molecules/mole.

4. **Andrè-Marie Ampere (1775-1836):** French Physicist, famous for finding the relation between the direction of current and the magnetic field produced. Ampere went on to making quantitative studies of the force between two parallel current carrying wires. He made the mathematical formulation of the law known as Ampere's law on the relation between the magnetic field produced and the current that produces it.

5. **Anders Jonas Ångström (1814-74):** A Swedish spectroscopist who first detected the presence of hydrogen in the Sun. Ångström anticipated that good emitters are good absorbers too. In his honour the unit called ångström ($\mp 10^{-10}$ am $= 10^{-8}$ m) is in use.

6. **Arnold Johannes Wilhelm Sommerfeld (1868-1951):** German Physicist, spent most of his career in Munich making splendid contribution to quantum theory and relativity and writing excellent series of textbooks.

   Sommerfeld was a real man-maker giving the world a team of brilliant scientists in the name of Bethe, Heisenberg, Debye, Heitler, Pauli and many more; modified Bohr's theory of atomic spectra. During the formative period of quantum mechanics, he was the leading light of the third nodal centre at Munich with Max Born at Göttingen, Niels Bohr at Copenhagen.

7. **Arthur H. Compton (1892-1962):** American Physicist, a PhD from the University of Princeton, N.J., USA. A.H. Compton's experiment of X-ray scattering by the electrons of the scatterer is an example of interaction between matter (electron) and wave (X-ray).

   He started his research career under E. Rutherford who was them at Cambridge University doing scattering experiment of α-particles. At that time St. Louis became professor at Chicago in 1923 where he became very busy doing experiment with scattering of X-ray from paraffin. He reported an increase in wavelength. Compton was inspired by this scattering experiment. He and Debye together worked out the physical understanding of the elastic scattering of photon by the electron. Compton and CTR Wilson received 1927 Nobel Prize jointly.

8. **Bohr, Niels (1885-1962):** Danish physicist with excellent idea, Bohr's father was a well-known physiologist, his brother a gifted mathematician. He did his PhD at Copenhagen in 1911. He joined the research group under J.J. Thomson at Cambridge but only for eight months after which Niels moved to Manchester and worked for 4 years under Rutherford. In 1913, he published his serial paper in the philosophic magazine. He went back to Copenhagen in 1916 and in 1918 became the first director of the Institute of Theoretical Physics there. A third nodal centre under the supervision of Sommerfeld was formed at Munich for the development of quantum physics with Niels Bohr at Theoretical Physics Institute, Copenhagen. He got Nobel prize for physics in the year 1922. He was loved by most of the contemporary physicists for his pleasing personality and novel ideas. In 1939 he and John Archibald Wheeler put forward their liquid drop model of nuclear fusion. It was Niels Bohr who was the originator of the famous principle known as the principle of correspondence which says that any modification coverage over an existing classical theory will be accepted if it leads to the classical form as in the approximation limit.

9. **Broglie, Louis de (1892-1987):** Originator of a revolutionary idea about the wave aspect of particle. It was Planckian concept of blackbody radiation to consist of light quanta of energy $h\nu$ and momenta $h\nu/c$ reflecting the particle aspect of a wave. Louis de Broglie was thinking of the reversal. If a wave has a particle character, cannot a particle exhibit wave character? This revolutionary idea and the formula $\lambda = \frac{h}{mv}$ appeared in his doctoral thesis submitted to the university at Sorbonne in 1924. It was his thesis paper of about 100 pages which spoke about wave particle dualism and appeared in Annales de Physique in 1925. For this thesis paper propagating a doctrine of philosophy about the duality of matter-wave, he was awarded Nobel Prize of Physics in 1929.

10. **Carl Friedrich Gauss (see Gauss, Johann Friedrich Carl).**

11. **Carnot, Nicolas-Leonard-Sadi (1796-1832):** French Engineer cum Physicist, done a lot of work on the efficiency of heat engines. His work was published in a monograph entitled 'Rèflexions sur la puissance motrice du fue et sur les machines propres à

développer cette puissance.'—reflection on the motive power of fire. He put forward a theorem on the maximum efficiency attainable by a heat engine. Like his father he was also an engineer who was educated by his father at the beginning and them at the École polytechnique. He served as an Engineer in the army. He at a young age died of plague epidemic.

12. **Chakrabarti, P.K. (1941):** When a child he had to migrate from Rangpur, now in Bangladesh to Kolkata with his family. After facing much of hardship, he graduated from the Presidency College (now Presidency University), Kolkata with honours in Physics and subsequently did his postgraduation from the same University. He then joined St. Xaviers' College in the year 1965 as a faculty member in the department of Physics and still teaching in the postgraduate department of Computer Science.

    He was awarded the PhD from Calcutta University in the year 1982 for his theoretical investigation of the transport properties in liquids and molten metals. He published about a dozen of papers both in national and international science journals.

    He has become very popular as an author of a good number of textbooks at the undergraduation level.

13. **Charles-Augustinde Coulomb: (1736-1806):** Coulomb discovered the inverse-square law of electric interaction by using his very sensitive torsion balance.

14. **Clausius, Rudolf Julius Emanuel (1822-1828):** A German physicist, Clausius enunciated the second law of Thermodynamics. He introduced the concept of entropy. He made many modifications in kinetic theory, Thermodynamics. He was first to introduce the idea of mean free path, etc.

15. **Clerk Maxwell, James (1831-79):** In his school days Maxwell was a shy type of boy by his home-designer clothes and lifestyle. But he was quite happy when he joined Edinburgh University.

    As early as 1831, he declared that Red, Green and Blue are the fundamental colours from which other colours could be derived. The idea of colour photography could be traced to his credit.

    Maxwell was a Scottish physicist whose contribution was in the form of unification of electromagnetism. In the year 1850 he joined Trinity College, Cambridge. He became a professor at Aberdeen and then moved to Trinity College, London. He went back to Scotland after his father's death. He developed the Kinetic Theory of gases. He had experimental interest also and he did extensive experiment on the viscosity of gases. Because of his experimental interest and skill, he was almost forced to accept the post of Cavendish Professor of Experimental Physics at Cambridge. He developed the theory of displacement current and also developed electromagnetic field equations to show that light is a kind of electromagnetic wave.

16. **Dalton, John (1766-1844):** A British chemist and physicist, John Dalton was born of a poor weaver family, was the proposer of the first quantitative atomic theory; he had deepen interest in Meteorology and he used to keep daily records of weather conditions. From his early age, he was very much interested in teaching and was appointed teacher

of Mathematics and Natural Philosophy at the New College, Manchester, in 1793. Study of atmosphere led him to investigate the gases in general and in 1802 he presented his law of partial pressure for a gas mixture and in 1823 he presented the law of multiple proportions of chemical combination of elements forming compounds. He published in three parts 'New System of Chemical Philosophy'.

17. **Davisson, Clinton Joseph (1881-1958):** Davisson together with Lester Germer discovered the diffraction of electrons by crystals confirming the Broglie's hypothesis of the wave nature of matter. G.P. Thomson in a different way made an experiment to confirm the wave nature of matter by producing the diffraction effect due to electron beam diffracted by Nickel crystal. For this revolutionary experiment, Clinton Davisson, Lester Germer and G.P. Thomson shared the Nobel Prize of 1937.

18. **Debye, Peter Joseph William (1884-1966):** Dutch-American chemical physicist, made enormous contribution to the theory of electrolytic solution; contributed to the theory of specific heat of solid. He contributed to the experiment of crystal diffraction of X-ray.

19. **Dirac, P.A.M. (1902-84):** Electrical Engineer cum mathematician; studied Engineering from Bristol and Mathematics from Cambridge. P.A.M. Dirac laid down the general principles of quantum mechanics. He proposed the relativistic quantum theory of electrons which brought in its train the idea of antiparticles and positron.

20. **Einstein, Albert (See Albert Einstein).**

21. **Ernest Rutherford (1871-1937):** British physicist born at New Zealand in 1871 as a son of a wheelwright. In 1895, he obtained a scholarship to work under J.J. Thomson at Cambridge. His work was mostly related to radioactivity. He observed that α-particles are doubly charged He-nucleus, suffer large angle scattering from which he concluded that atoms possess a very small but massive nucleus holding all positive charge. He along with Geiger made a very sensitive counter for detection of charged particle. He made noteworthy discoveries in nuclear physics and in 1919 became the successor of J.J. Thomson as the Cavendish Professor at Cambridge. He received the Nobel Prize in 1908.

22. **Ernst P. Jordan (1902-80):** Theoretical and mathematical physicist of German origin; had his degree at Hanover and PhD at Göttingen, Germany. With Max Born in 1926 he laid the foundation of quantum mechanics.

23. **Erwin Rudolf Josef Alexander Schrödinger (1887-1961):** Austrian physicist, studied (between 1906-10) at the University of Vienna. Schrödinger was an artillery officer during the First World War. In 1920, he was an assistant to Wilhelm Wien. Then he moved to the University of Zurich replacing von Laue. He was there for six years.

    In 1924, the thesis of de Broglie heralding a ravishingly beautiful idea appeared and Schrödinger was simply amazed and he in 1925 wrote to Einstein, 'A few days ago I read with the greatest interest the ingenious thesis of Louis de Broglie which I finally got hold of ...' On this thesis Schrödinger gave a seminar where a student of Sommerfeld asked whether there were equations for the de Broglie waves. In a few weeks time Schrödinger formulated his famous wave equation and opened an approach road to study Quantum Mechanics. In 1933, he won Nobel Prize jointly with de Broglie.

24. **Felix Klein, Christian (1849-1925):** German mathematician, worked on transformation groups; became professor of mathematics at Erlangen, Klein tried to unify mathematics of geometry and theory of invariants.

25. **Fourier, Baron Jean-Baptiste-Joseph (1768-1830):** French mathematician whose early life is obscure, became an orphan at the age of 8 years. Fourier attended military school when he was in an abbey and he studied at the Ecote Polytechnique in Paris. He was very much interested in heat conduction problems and established linear partial differential equations and presented an analysis in the form of a series, called Fourier Series.

26. **Gauss, Johann Friedrich Carl (1777-1855):** Mathematician of German origin, son of a gardener. At the age of three he was told in school to add all the numbers from zero to 100. His prompt answer was 5050 from his knowledge of arithmetic progression, developed by himself. Using a straight edge and compass Gauss could draw at the age of 18 a regular heptadecagon (equilateral figure), thus improving plane geometry. This was the best achievement of Gauss according to him. He discovered the method of least square fitting. He explored conformal mapping, theory of surfaces and electromagnetism. He transformed almost all branches of mathematics. The concept of Gaussian curve is due to him. His contribution to topology and geometry is really epoch making. Gauss thought that knowledge is endless as it is expressed when he wrote to this geometrician friend Bolyai, "It is not knowledge but the act of learning, not possession but the act of getting there, which grants the greatest enjoyment. When I have clarified and exhausted a subject, then I turn away from it, in order to go into darkness again; ...".

27. **Gibbs Josiah Willard (1839-1903):** Father of chemical thermodynamics. American Scientist, a PhD from Yale where he remained throughout his life. He could not come to limelight until he was discovered by Planck, Einstein and others. His works worthy of mention include phase rule, Gibbs' chemical potential, etc.

28. **Heinrich Rudolf Hertz (1857-94):** A German physicist and an assistant under Helmholtz; Hertz was inspired by Helmholtz to work on electromagnetic wave and was the first to show that electromagnetic waves may be generated by electric sparks. The SI unit of frequency is given after the name of Hertz to pay respect to him.

29. **Heisenberg, Karl Werner (1901-76):** German Physicist, his father was a professor of Greek. His influence on Heisenberg's early life was very much discernible. Later Heisenberg felt inclined to number theory. At Göttingen he worked with Born and at Copenhagen with Bohr. He himself wrote, 'I learnt optimism from Sommerfeld, mathematics at Göttingen from Born and Physics from Bohr'. He was central to development of quantum mechanics.

In 1932 at the age of 31 only for his work in formulating a new mechanics for micro-systems in the name of Matrix Mechanics and alternative approach to study quantum mechanics, Heisenberg was awarded Nobel Prize. He was the formulator of the famous uncertainty principle. In 1932, after Chadwick discovered neutrons, Heisenberg proposed the isospin symmetry of protons and neutrons. He was of the opinion that there exists a matrix to each and every variable. This fact was stated by Schrödinger in a slightly different way saying that to each and every state of a microsystem there corresponds a suitable wave.

In 1941, he became the Director of Kaiser Wilhelm Institute of Berlin; later he became the Director of Max Planck Institute which Heisenberg himself helped to build.

30. **Hermann Ludwig Ferdinand Helmholtz (1821-1894):** German physicist and physiologist, discoverer of conservation of energy. He worked deeply into electricity and magnetism and showed much of interest in building up the theory of hearing and vision. Hertz was his student who was inspired to work on Maxwell's electromagnetic radiation.

31. **Isaac Newton (1642-1727):** A versatile genius was born in a small village of Woolsthorpe in England. What was he? A physicist? A chemist? A mathematician? A discoverer? An inventor? And what not? He was a combination of everything, a rare genius for all times. After grammar school educations Newton went to Trinity College for higher studies in 1661. In 1665, he went back to his own village, because in the city Great Plague Epidemic broke out. It was a blessing in disguise, because remaining in isolation in the next two years, he did all the miracles; he invented and discovered the Differential Calculus, Dispersion of light, Integral Calculus, Planetary orbits, Gravitation and Gravity.

    In 1684 he was requested by Halley to write a book on Dynamics and Newton took about twelve months of time to complete the work. It came out as 'Philosophial Naturalis Principia Mathematica'.

32. **J.J. Thomson (1856-1940):** British physicist who discovered electron, by his experimental arrangement could find the ratio (e/m). Received Nobel Prize in 1906. His credit was to build up the Cavendish Laboratory at Cambridge with seven research assistants, all of them received Nobel Prize in the subsequent years.

33. **James Hopwood Jeans (1877-1946):** He was a British mathematician, physicist and astronomer, professor of astronomy at the Royal Society during 1912-16. He did pioneering work on Heat, Thermodynamics and Cosmology. Well-known writer of books on astronomy. His book entitled 'Dynamical theory of heat' is still unsurpassed in popularity and contents. He was the first to propose that matter is continuously created throughout the universe.

34. **John Williams Strutt (Lord Rayleigh) (1842-1919):** Known for his brilliant work on acoustics, a book which he wrote on a house-boat on the River Nile, an authentic book on sound. While writing a book on the house-boat on Nile, Rayleigh thought of why the sky was blue. He gave the theory of scattering called Rayleigh scattering which could explain the blue of the sky. Blackbody radiation was at that time a favourite topic and Rayleigh formulated a law of radiation which could explain the energy distribution for the longer wavelength side of the spectrum but failed at shorter

wavelength side. Rayleigh and Ramsay discovered Argon from their observations on the discrepancies of sound velocity in nitrogen. He got Nobel Prize for Physics in 1904.

35. **Kelvin, William Thomson (1824-1907):** From Glasgow Kelvin went to Paris to work on heat with Renault; second law of Thermodynamics was formulated independently by Kelvin. He worked with Joule on the relationship between heat and work. He proposed the absolute scale of temperature known as Kelvin's scale.

36. **Ludwig Eduard Boltzmann (1844-1906):** An Austrian mathematician cum physicist, established relation between statistical mechanics and kinetic theory of gases. He is very much known for his formula $s = k \ln W$.

37. **Max Karl Ernst Ludwig Planck (1858-1947):** Professor in Kiel, submitted thesis on Thermodynamics but in 1894 he devoted himself to study thermal radiation. He correctly and very wisely formulated the blackbody radiation formula which contained the seed of the revolutionary idea of energy quantisation. Using the quantum idea in the energy distribution of blackbody radiation, Planck was able to put forward finally the equation for energy distribution in the blackbody radiation spectrum. He discovered a new universal constant in the name of Planck's constant. It was Planck who made 1900 the year of transition from classical to quantum physics. He became the president of the Kaiser Wilhelm Society which he resigned as a protest against the atrocities of the Nazi German and attitude towards the Jewish scientists. In spite of the great tragedy in his life, losing his two sons, the first dying in the first world war and his second son was executed for his participation in the revolt against Adolf Hitler in 1944, this spirit of search did not stop. After the second world war, Kaiser was moved to Göttingen and renamed as Max Planck Society in 1948 in honour of its former president.

38. **Max Theodor Felix von Laue (1879-1960):** Max von Laue first released that X-ray diffraction can be possible only by a crystal where the inter-particle distance is of the order of X-ray wavelength. First he passed a beam of polychromatic X-ray through a copper sulphate crystal, then through alkali halide crystal and zinc sulphide crystal; in each case he obtained beautiful diffraction pattern showing that X-rays are waves. He was awarded Nobel Prize in Physics in 1914.

39. **Michael Faraday (1791-1867):** British physicist and chemist; he was the first to liquefy chlorine, discovered benzene.

After 1830 Faraday's interest shifted from chemistry to physics when he studied electricity, formulates the laws of electrolysis and framed the laws of electromagnetic induction. Son of an ailing blacksmith, Faraday at the age of thirteen was an apprentice in a book-binding shop. He used to read the science books, particularly of physics and chemistry with apt attention and keen interest. It so happened that he got a ticket to attend last course of lectures delivered by Prof. H. Davy at the Royal Society. He was so impressed and inspired that he sent the entire note of the lecture with his comments to Prof. Davy. Davy was so pleased that he offered Faraday the post of an assistant in his laboratory. That was the beginning of an eventful life of Michael Faraday. It is surprising that without any formal educational background, a man could reach such a great height.

40. **Nernst, Walther Hermann (1864-1941):** Physical chemist of German origin, studied and worked in different German Universities, finally in 1905 he was appointed Professor in Berlin University. Nernst enunciated third law of thermodynamics; he did extensive work on finding specific heat of solid in Nernst vacuum calorimeter at low temperature.

41. **Niels Bohr: See Bohr, Niels.**

42. **Pieter Zeeman (1865-1943):** A Dutch physicist, famous for discovery of splitting of spectral lines in magnetic fields known as Zeeman effect.

43. **W. Lawrence Bragg (1890-1971):** Son of William Brag and was born at Adelaide where his father was a physics teacher. After obtaining degree from the University there, he moved to England with his father and entered Trinity College, Cambridge, where in 1912 he got Tripos in natural science. Between 1912-14 L. Bragg worked extensively with his father on X-ray crystallography. In 1915, he and his father jointly received the Nobel Prize for physics.

44. **Walther Nernst: See Nernst Hermann Walther.**

45. **William Henry Bragg (1862-1942):** Obtained degree in mathematics from Cambridge; he accepted a position to teach physics at Adelaide in Australia. He changed his interest from radio activity and ionising radiation to X-ray crystallography. He shared Nobel Prize in Physics with his son in 1915.

46. **Zeeman Pieter (See Pieter Zeeman).**